Advances in Intelligent Systems and Computing

Volume 214

Series Editor

J. Kacprzyk, Warsaw, Poland

For further volumes:
http://www.springer.com/series/11156

Fuchun Sun · Tianrui Li
Hongbo Li
Editors

Knowledge Engineering and Management

Proceedings of the Seventh International Conference on Intelligent Systems and Knowledge Engineering, Beijing, China, Dec 2012 (ISKE 2012)

Editors
Fuchun Sun
Hongbo Li
Department of Computer Science
and Technology
Tsinghua University
Beijing
People's Republic of China

Tianrui Li
School of Information Science
Southwest Jiaotong University
Chengdu
People's Republic of China

ISSN 2194-5357 ISSN 2194-5365 (electronic)
ISBN 978-3-642-37831-7 ISBN 978-3-642-37832-4 (eBook)
DOI 10.1007/978-3-642-37832-4
Springer Heidelberg New York Dordrecht London

Library of Congress Control Number: 2013942994

Printed on acid-free paper

Springer is part of Springer Science+Business Media (www.springer.com)

Preface

This book is part of the Proceedings of the 7th International Conference on Intelligent Systems and Knowledge Engineering (ISKE2012) and the 1st International Conference on Cognitive Systems and Information Processing (CSIP2012) held in Beijing, China, during December 15–17, 2012. ISKE is a prestigious annual conference on Intelligent Systems and Knowledge Engineering with the past events held in Shanghai (2006, 2011), Chengdu (2007), Xiamen (2008), Hasselt, Belgium (2009), and Hangzhou (2010). Over the past few years, ISKE has matured into a well-established series of international conferences on Intelligent Systems and Knowledge Engineering and related fields over the world. CSIP 2012 is the first conference sponsored by Tsinghua University and Science in China Press, and technically sponsored by IEEE Computational Intelligence Society, Chinese Association for Artificial Intelligence. The aim of this conference is to bring together experts from different expertise areas to discuss the state of the art in cognitive systems and advanced information processing, and to present new research results and perspectives on future development. Both ISKE 2012 and CSIP 2012 provide academic forums for the participants to disseminate their new research findings and discuss emerging areas of research. It also creates a stimulating environment for the participants to interact and exchange information on future challenges and opportunities of intelligent and cognitive science research and applications.

ISKE 2012 and CSIP received 406 submissions in total from about 1020 authors in 20 countries (United States of American, Singapore, Russian Federation, Saudi Arabia, Spain, Sudan, Sweden, Tunisia, United Kingdom, Portugal, Norway, Korea, Japan, Germany, Finland, France, China, Argentina, Australia, and Belgium). Based on rigorous reviews by the Program Committee members and reviewers, among 220 papers contributed to ISKE 2012, high-quality papers were selected for publication in the proceedings with the acceptance rate of 58.4 %. The papers were organized in 25 cohesive sections covering all major topics of intelligent and cognitive science and applications. In addition to the contributed papers, the technical program includes four plenary speeches by Jennie Si (Arizona State University, USA), Wei Li (California State University, USA), Chin-Teng Lin (National Chiao Tung University, Taiwan, China), and Guoqing Chen (Tsinghua University, China).

As organizers of both conferences, we are grateful to Tsinghua University, Science in China Press, Chinese Academy of Sciences for their sponsorship, grateful to IEEE Computational Intelligence Society, Chinese Association for Artificial Intelligence, State Key Laboratory on Complex Electronic System Simulation, Science and Technology on Integrated Information System Laboratory, Southwest Jiaotong University, University of Technology, Sydney for their technical co-sponsorship.

We would also like to thank the members of the Advisory Committee for their guidance, the members of the International Program Committee and additional reviewers for reviewing the papers, and members of the Publications Committee for checking the accepted papers in a short period of time. Particularly, we are grateful to the publisher, Springer, for publishing the proceedings in the prestigious series of Advances in Intelligent Systems and Computing. Meanwhile, we wish to express our heartfelt appreciation to the plenary speakers, special session organizers, session chairs, and student helpers. In addition, there are still many colleagues, associates, and friends who helped us in immeasurable ways. We are also grateful to them all. Last but not the least, we are thankful to all authors and participants for their great contributions that made ISKE 2012 and CSIP 2012 successful.

December 2012

Fuchun Sun
Tianrui Li
Hongbo Li

Organizing Committee

General Chair:	Jie Lu (Australia)
	Fuchun Sun (China)
General Co-Chairs:	Yang Xu (China)
Honorary Chairs:	L. A. Zadeh (USA)
	Bo Zhang (China)
Steering Committee Chairs:	Etienne Kerre (Belgium)
	Zengqi Sun (China)
Organizing Chairs:	Xiaohui Hu (China)
	Huaping Liu (China)
Program Chairs:	Tianrui Li (China)
	Javier Montero (Spain)
Program Co-Chairs:	Changwen Zheng (China)
	Luis Martínez López (Spain)
Sessions Chairs:	Fei Song (China)
	Victoria Lopez (Spain)
Publications Chairs:	Yuanqing Xia (China)
	Hongming Cai (China)
Publicity Chairs:	Jiacun Wang (USA)
	Zheying Zhang (Finland)
	Michael Sheng (Australia)
	Dacheng Tao (Australia)
Poster Chairs:	Guangquan Zhang (Australia)
	Hongbo Li (China)
Program Chairs:	Tianrui Li (China)
	Javier Montero (Spain)

Members

Abdullah Al-Zoubi (Jordan)
Andrzej Skowron (Poland)
Athena Tocatlidou (Greece)
B. Bouchon-Meunier (France)
Michael Sheng (Australia)
Mihir K. Chakraborty (India)
Mike Nachtegael (Belgium)
Mikhail Moshkov (Russia)

Benedetto Matarazzo (Italy)
Bo Yuan (USA)
Bo Zhang (China)
Cengiz Kahraman (Turkey)
Changwen Zheng
Chien-Chung Chan (USA)
Cornelis Chris (Belgium)
Dacheng Tao (Australia)
Davide Ciucci (Italy)
Davide Roverso (Norway)
Du Zhang (USA)
Enrico Zio (Italy)
Enrique Herrera-Viedma (Spain)
Erik Laes (Belgium)
Etienne E. Kerre (Belgium)
Francisco Chiclana (UK)
Francisco Herrera (Spain)
Fuchun Sun (China)
Gabriella Pasi (Italy)
Georg Peters (Germany)
Germano Resconi (Italy)
Guangquan Zhang (Australia)
Guangtao Xue (China)
Gulcin Buyukozkan (Turkey)
Guolong Chen (China)
Guoyin Wang (China)
H.-J. Zimmermann (Germany)
Huaping Liu (China)
Hongbo Li (China)
Hongjun Wang (China)
Hongming Cai (China)
Hongtao Lu (China)
I. Burhan Turksen (Canada)
Irina Perfilieva (Czech Republic)
Jan Komorowski (Sweden)
Janusz Kacprzyk (Poland)
Javier Montero (Spain)
Jer-Guang Hsieh (Taiwan, China)
Jesús Vega (Spain)
Jiacun Wang (USA)
Jianbo Yang (UK)
Jie Lu (Australia)
Jingcheng Wang (China)
Min Liu (China)
Peijun Guo (Japan)
Pierre Kunsch (Belgium)
Qi Wang (China)
Qingsheng Ren (China)
Rafael Bello (Cuba)
Richard Jensen (UK)
Ronald R. Yager (USA)
Ronei Marcos de Moraes (Brasil)
Ryszard Janicki (Canada)
S. K. Michael Wong (Canada)
Shaojie Qiao (China)
Shaozi Li (China)
Sheela Ramanna (Canada)
Su-Cheng Haw (Malaysia)
Suman Rao (India)
Sushmita Mitra (India)
Takehisa Onisawa (Japan)
Tetsuya Murai (Japan)
Tianrui Li (China)
Tzung-Pei Hong (Taiwan, China)
Ufuk Cebeci (Turkey)
Victoria Lopez (Spain)
Vilem Novak (Czech Republic)
Weiming Shen (Canada)
Weixing Zhu (China)
Wensheng Zhang (China)
Witold Pedrycz (Canada)
Wujun Li (China)
Xiaohui Hu (China)
Xianyi Zeng (France)
Xiaogang Jin (China)
Xiaoqiang Lu (China)
Xiaoyan Zhu (China)
Xiao-Zhi Gao (Finland)
Xuelong Li (China)
Xun Gong (China)
Yan Yang (China)
Yangguang Liu (China)
Yanmin Zhu (China)
Yaochu Jin (Germany)
Yasuo Kudo (Japan)
Yi Tang (China)

Jitender S. Deogun (USA)
Jouni Jarvinen (Finland)
Juan-Carlos Cubero (Spain)
Jun Liu (UK)
Jyrki Nummenmaa (Finland)
Koen Vanhoof (Belgium)
Krassimir Markov (Bulgaria)
Liliane Santos Machado (Brasil)
Lisheng Hu (China)
Luis Magdalena (Spain)
Luis Martinez López (Spain)
Lusine Mkrtchyan (Italy)
Madan M. Gupta (Canada)
Martine De Cock (Belgium)
Masoud Nikravesh (USA)
Yinglin Wang (China)
Yiyu Yao (Canada)
Yongjun Shen (Belgium)
Yuancheng Huang (China)
Yuanqing Xia (China)
Zbigniew Suraj (Poland)
Zbigniew W. Ras (USA)
Zengqi Sun (China)
Zheying Zhang (Finland)
Zhong Li (Germany)
Zhongjun He (China)
Zhongzhi Shi (China)
Bo Peng (China)
Fei Teng (China)

Sponsors Logo

Tsinghua University

Science in China Press

Chinese Academy of Sciences

Institute of Electrical and Electronics Engineers

Contents

Function Set-Valued Information Systems

Hongmei Chen, Tianrui Li, Chuan Luo, Junbo Zhang and Xinjiang Li

Abstract Set-valued Information System (SIS) aims at dealing with attributes' multi-values. It is an extension of the single-valued information system. In traditional SIS, the tolerance and dominance relations were introduced to deal with the relationship between objects. But the partial orders between the attributes values have not been taken into consideration. In this paper, a function set-valued equivalence relation is proposed first considering the applications of SIS. Then the properties of the function set-valued equivalence classes are analyzed. Furthermore, the definitions of approximations under the function SIS are presented and the properties w.r.t. approximations under the tolerance and dominance relations are investigated.

This work is supported by the National Science Foundation of China (Nos. 61175047, 61100117) and NSAF (No. U1230117), the Youth Social Science Foundation of the Chinese Education Commission (No. 11YJC630127), and the Fundamental Research Funds for the Central Universities (SWJTU11ZT08, SWJTU12CX091).

H. Chen (✉) · T. Li · C. Luo · J. Zhang · X. Li
School of Information Science and Technology, Southwest Jiaotong University, 610031 Chengdu, China
e-mail: hmchen@swjtu.edu.cn

T. Li
e-mail: trli@swjtu.edu.cn

C. Luo
e-mail: luochuan@my.swjtu.edu.cn

J. Zhang
e-mail: JunboZhang86@163.com

F. Sun et al. (eds.), *Knowledge Engineering and Management*,
Advances in Intelligent Systems and Computing 214,
DOI: 10.1007/978-3-642-37832-4_1,

1 Introduction

Granular Computing (GrC) proposed by Zadeh [1, 2] in 1989 has been successfully used in many areas, e.g., image processing, pattern recognition, and data mining. A granule is a chunk of knowledge made of different objects "drawn together by indistinguishability, similarity, proximity or functionality" [2]. Different levels of concepts or rules are induced in GrC. Rough set theory proposed by Pawlak in 1982 is considered as an important branch of GrC, which has been used to process inconsistent information [3].

In Traditional Rough Set theory (TRS), an attribute of one object in the information system can only have one value. But in real-life applications, attributes may have multi-values. For example, the degree of a person may be a combination of bachelor, master, and doctor. Set-valued systems are used to handle the case when objects' attribute have multi-values. In addition, there exists a lot of data in the information system which have uncertainty and incompleteness. Set-valued Information System (SIS) can also be used to process incomplete information. In this case, the missing value can be only one of the values in the value domain [4]. In [4], Guan defined the tolerance relation and the maximum tolerance block as well as the relative reduct based maximum tolerance block in SIS. Qian and Liang defined disjunctive and conjunctive SIS in [5] in view of different meaning of attributes' values and they studied the approach of attribute reduction and rule extraction in these two different SISs. Song and Zhang et al. defined the partly accordant reduction and the assignment reduction in an inconsistent set-valued decision information system [6, 7]. Huang et al. introduced a degree dominance relation to dominance set-valued intuitionist fuzzy decision tables and a degree dominance set-valued rough set model [8]. In [9, 10], Chen et al. introduced the probability rough sets into SIS and they studied the method for updating approximations incrementally under the variable precision set-valued ordered information while attributes' values coarsening and refining.

The equivalence relation is employed in TRS to deal with the relationship among objects and then equivalence classes induced by different equivalence relations form the elementary knowledge of the universe. Any set in the universe is described approximately by equivalence classes. Hence, a pair of certain sets, i.e., upper and lower approximations are used to describe the set approximately. The approximations partition the universe into three different regions, namely, positive, negative, and boundary regions. Then we can induce certain rules from the positive and negative regions and uncertain rules from the boundary region, i.e., a tree-way decision [11]. The SIS is a general model of single-valued information system. The equivalence relation is substituted by the tolerance relation and the dominance relation. Because the value of the missing data may be any value in the values' domain, the SIS has been used to process data in the incomplete information system. The tolerance and dominance relations in SIS consider multi-values of attributes. But the partial order among the attributes' values is not taken into consideration. In this paper, by analyzing the meaning of the tolerance and dominance relations, a new relationship in SIS is proposed, i.e., a function set-valued equivalence relation. Then the properties of the function set-valued equivalence relation are discussed. Furthermore, a more general relationship, a general function relation, is investigated and its properties are analyzed.

The paper is organized as follows. In Sect. 2, basic concepts in SIS are reviewed. In Sect. 3, a function SIS is proposed. A function equivalence relation and its properties are discussed. Then the general function relation is investigated. The paper ends with conclusions and further directions in Sect. 4.

2 Preliminary

In this section, Some basic concepts in SIS are briefly reviewed [4–6].

Definition 21 Let $S = (U, A, V, f)$ be a SIS, where $U = \{x_1, x_2, \ldots, x_n\}$ is a non-empty finite set of objects, called the universe. $A = \{a_1, a_2, \ldots, a_l\}$ is a non-empty finite set of attributes. The element in A is called an attribute. $A = C \cup D$, $C \cap D = \emptyset$, C is the set of condition attributes and D is the set of decision attributes. $V = \{V_{a_1}, V_{a_2}, \ldots, V_{a_l}, a_i \in A\}$, $V_{a_i}(i = 1, 2, \ldots, l)$ is the domain of attribute $a_i(a_i \in A)$. V_C is the domain of condition attributes. V_D is the domain of decision attributes. $V = V_C \cup V_D$ is the domain of all attributes. $f : U \times C \to 2^{V_C}$ is a set-valued mapping.$f : U \times D \to V_D$ is a single-valued mapping. $f(x_i, a_l)$ is the value of x_i on attribute a_l. An example of SIS is shown in Table 1.

The tolerance and dominance relation have been proposed to deal with the different applications in SIS. For example, if two persons have the tolerance relation on the attribute "language", it means they can communicate through the common language. The tolerance relation is defined as follows.

Definition 22 Let $S = (U, A, V, f)$ be an SIS, for $B \subseteq C$, the tolerance relation in the SIS is defined as follows:

$$R_B^{\cap} = \{(y, x) \in U \times U | f(y, a) \cap f(x, a) \neq \emptyset (\forall a \in B)\} \tag{1}$$

Table 1 A set-valued information system

U	Price	Mileage	Size	Max-speed
1	{High, Medium}	High	Full	{High, Medium}
2	Low	{High, Medium}	Full	Medium
3	Medium	Low	Full	Medium
4	{Low, Medium}	{Low, Medium}	Compact	Low
5	High	{High, Medium}	Compact	{Low, Medium}
6	Medium	{Low, Medium}	Full	High
7	Low	High	Compact	High
8	High	Low	Compact	{Low, Medium}

The tolerance relation is reflexive and symmetric, but not transitive. Let $[x]_B^{\cap}$ denote the tolerance class of object x, where $[x]_B^{\cap} = \{y \in U | (y,x) \in R_B^{\cap}\} = \{y \in U | f(x,a_i) \cap f(y,a_i) \neq \emptyset, a_i \in B\}$.

Definition 23 Let $S = (U,A,V,f)$ be an SIS, $\forall X \subseteq U$, the approximations of X under the tolerance relation $R_C^{\cap}$ are defined as follows, respectively:

$$\underline{apr_B^T}(X) = \{x \in U | [x_i]_C^{\cap} \subseteq X\} \tag{2}$$

$$\overline{apr_B^T}(X) = \{x \in U | [x_i]_C^{\cap} \cap X \neq \emptyset\} \tag{3}$$

Definition 24 Given an SIS (U,A,V,f), for $B \supseteq C$, a dominance relation in the SIS is defined as follows:

$$\begin{aligned} R_B^{\subseteq} &= \{(y,x) \in U \times U | f(y,a) \supseteq f(x,a)(\forall a \in B)\} \\ &= \{(y,x) \in U \times U | y \succeq x\} \end{aligned} \tag{4}$$

$R_B^{\subseteq}$ is reflexive, dissymmetric, and transitive.

We denote $[x]_B^{\subseteq} = \{y \in U | (y,x) \in R_B^{\supseteq}\}$, $[x]_B^{\subseteq} = \{y \in U | (x,y) \in R_B^{\subseteq}\}$. $[x]_B^{\subseteq}$ $(x \in U)$ are granules of knowledge induced by the dominance relation, which are the set of objects dominating x. The dominance relation in SIS means in some degree, the more values in an attribute of one object the more the probability that it dominates the other objects. It considers the cardinality of the attributes' values.

Definition 25 Let $S = (U,A,V,f)$ be an SIS, $\forall X \subseteq U$, the approximations of X under the dominance relation $R_C^{\supseteq}$ are defined as follows, respectively:

$$\underline{apr_B^D}(X) = \{x \in U | [x_i]_C^{\supseteq} \subseteq X\} \tag{5}$$

$$\overline{apr_B^D}(X) = \{x \in U | [x_i]_C^{\supseteq} \cap X \neq \emptyset\} \tag{6}$$

3 Function Set-Valued Information Systems

Multi-values of attributes in SIS give more information than single-valued information system. In real-life applications, a partial order may exist in the attribute values, i.e., for $V_{a_i} = \{v_{a_1}^1, v_{a_1}^2, \cdots, v_{a_1}^k\}$ $(k = |V_{a_i}|)$, $v_{a_1}^1 \preceq v_{a_1}^2 \preceq \cdots \preceq v_{a_1}^k$ is the partial order among the attribute values on the attribute a_i. In Definitions 22 and 24, the multi-value property of attributes is taken into consideration but the partial order in the attribute values is neglected. Then, we present a new definition, namely, a function set-valued equivalence relation.

Definition 31 Let $S=(U,A,V,f)$ be a SIS, $V_{a_i}=\{v_{a_1}^1,v_{a_1}^2,\ldots,v_{a_1}^k\}(k=|V_{a_i}|)$ is the attribute value domain of attribute a_i, where $v_{a_1}^i(1\leq i\leq k)$ *s.t.* the $v_{a_1}^1\preceq v_{a_1}^2\preceq\ldots\preceq v_{a_1}^k$. Then an attribute function F_{a_i} is defined as $F_{a_i}:V_{a_i}\to P(V_{a_i})$, which is a mapping from V_{a_i} to the power set of V_{a_i}.

The definition of F_{a_i} is given by the requirement of real-life applications. Usually, F_{a_i} may be maximum, minimum, average of the set-valued. For example, the degree a person awarded may be bachelor, master, doctor; the employer in universities may be more interesting in the person with the highest degree. Let $F(f(x,a_i))$ denote the value of the function F_{a_i} on the attribute value $f(x_i,a_l)$. Then the function equivalence relation is defined as follows:

Definition 32 Let $S=(U,A,V,f)$ be an SIS, the function equivalence relation is defined as follows:

$$R_B^F=\{(y,x)\in U\times U|F(f(x,a_i))=F(f(y,a_i)),a_i\in B\} \tag{7}$$

Let $[x]_B^F$ denote the function equivalence class of object x, where $[x]_B^F=\{y\in U|(y,x)\in R_B^F\}$.

Property 31 For R_B^F, the following hold:

1. Reflexive: $\forall x\in U$, xR_B^Fx;
2. Symmetric: $\forall x,y\in U$, if xR_B^Fy, then yR_B^Fx;
3. Transitive: $\forall x,y,z\in U$, if xR_B^Fy and yR_B^Fz, then xR_B^Fz;

$\bigcup_{i=1}^{|U|}[x_i]_B^F=U$ and $[x_i]_B^F\cap[x_j]_B^F=\emptyset$, i.e., U/R_B^F forms a partition of the university.

For any $B\subseteq C$, $[x_i]_C^F\subseteq[x_i]_B^F$. The tolerance, dominance, and function equivalence relation meet the different requirements of real-life applications. In the following, we compare these three different relations defined in SIS. Let $[x_i]_B^*$ denote the classes induced by different relations in a SIS and $\widetilde{I^*}=\{[x_i]_B^*|x_i\in U\}$ denote a classes cluster set, where the superscript * may be replaced by $\cap$, $\supseteq$, F and $\widetilde{I^\cap}$, $\widetilde{I^\supseteq}$, $\widetilde{I^F}$ denote tolerance, dominance, and function equivalence classes, respectively. Liang et al. defined the granularity of knowledge under the tolerance relation in incomplete information system [12]. Similarly, the granularity and the rough entropy of B in an SIS is defined as follows:

Definition 33 Let $S=(U,A,V,f)$ be a SIS, $[x_i]_B^*(x_i\in U,i=1,2,\ldots,|U|)$ is the granule induced by different relation in a set-valued information system, $\forall B\subseteq C\subseteq A$, where the superscript $*$ may be substituted by $\cap$, $\supseteq$, F and the granularity of B is defined as follows:

$$GK^*(B) = -\sum_{i=1}^{|U|} \frac{1}{|U|} \log_2 \frac{|[x_i]_B^*|}{|U|} \quad (8)$$

Definition 34 Let $S = (U, A, V, f)$ be an SIS, $[x_i]_B^*(x_i \in U, i = 1, 2, \ldots, |U|)$ is granule induced by different relation in a set-valued information system, $\forall P \subseteq C \subseteq A$. Then the rough entropy of P is defined as follows:

$$E_r^*(P) = -\sum_{i=1}^{|U|} \frac{1}{|U|} \log_2 \frac{1}{|[x_i]_B^*|} \quad (9)$$

For convenience, we define a partial order binary relation $\preceq$ first.

$I^* \preceq I^\Delta \Leftrightarrow$ *if* $\forall x_i \in U, \exists [x_i]_B^* \subseteq [x_i]_B^\Delta$, where $[x_i]_B^* \in \widetilde{I^*}$ and $[x_i]_B^\Delta \in \widetilde{I^\Delta}$.

Note: The superscript Δ may also be replaced by $\cap$, $\supseteq$, F. Then the following properties hold:

Property 32 For $[x_i]_B^*$ and $\widetilde{I^*}$, the following properties hold:

1. For any $B_1 \subset B_2 \subseteq C$, $[x_i]_{B_2}^F \subseteq [x_i]_{B_1}^F \subseteq [x_i]_C^F$;
2. For any $B \subseteq C$, $[x_i]_B^F \subseteq [x_i]_B^\cap$;
3. For any $B \subseteq C$, $I^F \preceq \widetilde{I^\cap}$;
4. For any $B \subseteq C$, $GK^{\supseteq}(B) \le GK^\cap(B)$;
5. For any $B \subseteq C$, $E_r^\cap(P) \le E_r^{\supseteq}(P)$.

Proof Proof 1. $\forall x_j \in [x_i]_{B_2}^F$, $x_j R_{B_2}^F x_i$, i.e., $F(f(x_i, a_i)) = F(f(x_j, a_i)), \forall a_i \in B_2$. Since $B_1 \subset B_2$, then $x_j \in [x_i]_{B_1}^F$. On the other hand, if $\forall x_j \in [x_i]_{B_1}^F$, $x_j R_{B_1}^F x_i$, i.e., $F(f(x_i, a_i)) = F(f(x_j, a_i))$, $\forall a_i \in B_1$. Because $B_1 \subset B_2$, then if $\exists F(f(x_i, a_i)) \neq F(f(x_j, a_i))$, $\forall a_i \in B_2 - B_1$. Then $x_j \not R_{B_2}^F x_i$, i.e., $x_j \notin [x_i]_{B_1}^F$. Therefore, we have $[x_i]_{B_2}^F \subseteq [x_i]_{B_1}^F$. Consequently, for any $B_1 \subset B_2 \subseteq C$, $[x_i]_C^F \subseteq [x_i]_{B_2}^F \subseteq [x_i]_{B_1}^F$.
2. $\forall x_j \in [x_i]_C^F$, $F(f(x_i, a_i)) = F(f(x_j, a_i)), a_i \in B$. Because $F_{a_i} : V_{a_i} \to P(V_{a_i})$, then $f(x_i, a_i) \cap f(x_j, a_i) \neq \emptyset$, i.e., $\forall x_j \in [x_i]_C^\cap$. $\forall x_j \in [x_i]_C^\cap$, we have $f(x_i, a_i) \cap f(x_j, a_i) \neq \emptyset$, if $\exists F(f(x_i, a_i)) \neq F(f(x_j, a_i)), a_i \in B$, $x_j \notin [x_i]_C^F$. Therefore, $[x_i]_B^F \subseteq [x_i]_B^\cap$. Consequently, for any $B \subseteq C$, $[x_i]_B^F \subseteq [x_i]_B^\cap$. Since 2 is always true, then 3, 4, and 5 hold.

The tolerance, dominance, and function equivalence relations are used to deal with different cases in a SIS. In most cases, three kinds of relations may coexist on different attributes. In the following, a general relation is given.

Definition 35 Let $S = (U, A, V, f)$ be a SIS, where $B_T \subseteq C$, $B_D \subseteq C$, $B_F \subseteq C$, $B_T \cap B_D \cap B_F = \emptyset$, $B_T \cup B_D \cup B_F = B \subseteq C$, the general function relation is defined as follows:

$$\begin{aligned} R_B^G = \{(y,x) \in U \times U | f(x,a_t) \cap f(y,a_t) \neq \emptyset \wedge f(x,a_d) \subseteq f(y,a_d) \\ \wedge F(f(x,a_f)) = F(f(y,a_f)), a_t \in B_T, a_d \in B_D, a_f \in B_F\} \end{aligned} \quad (10)$$

Let $[x]_C^G$ denote the function equivalence class of object x, where $[x]_B^G = \{y \in U | (y,x) \in R_B^G\}$.

Property 33 For R_B^G,

1. Reflexive: $\forall x \in U$, $xR_B^G x$;
2. Dissymmetric: $\forall x, y \in U$, if $xR_B^F y$, then $\overline{yR_B^G x}$;
3. Non-transitive.

$\bigcup_{i=1}^{|U|} [x_i]_B^G = U$and $[x_i]_B^G \cap [x_j]_B^G = \emptyset$, i.e., U/R_B^G forms a covering of the university.

Then, the approximations based on the general function relation are defined as follows:

Definition 36 Let $S = (U, A, V, f)$ be a SIS, $\forall X \subseteq U$, the lower and the upper approximations of X under the general function relation R_C^G are defined as follows, respectively:

$$\underline{apr_B^G}(X) = \{x \in U | [x_i]_C^G \subseteq X\} \quad (11)$$

$$\overline{apr_B^G}(X) = \{x \in U | [x_i]_C^G \cap X \neq \emptyset\} \quad (12)$$

Based on approximations of X, one can partition the universe U into three disjoint regions, i.e., the *positive region* $POS_B^G(X)$, the *boundary region* $BNR_B^G(X)$, and the *negative region* $NEG_B^G(X)$:

$$POS^G(X) = \underline{apr_B^G}(X); \quad (13)$$

$$BNR^G(X) = \overline{apr_B^G}(X) - \underline{apr_B^G}(X); \quad (14)$$

$$NEG^G(X) = U - \overline{apr_B^G}(X). \quad (15)$$

We can extract certain rules from the position and negative regions and possible rules from the boundary region too.

Property 34 For $\underline{apr}_B^G(X)$ and $\overline{apr}_B^G(X)$, the following hold:

$$\underline{apr}_B^T(X) \subseteq \underline{apr}_B^G(X);$$

$$\overline{apr}_B^D(X) \subseteq \overline{apr}_B^G(X) \subseteq \overline{apr}_B^T(X);$$

$$If\ A \subseteq B \subseteq C, then\ \underline{apr}_A^G(X) \subseteq \underline{apr}_B^G(X);$$

$$If\ A \subseteq B \subseteq C, then\ \overline{apr}_B^G(X \subseteq \overline{apr}_A^G(X);$$

Proof (1) From Property 32, for any $B \subseteq C$, $[x_i]_B^F \subseteq [x_i]_B^{\cap}$, then $[x_i]_B^G \subseteq [x_i]_B^{\cap}$. Therefore, we have $\underline{apr}_B^T(X) \subseteq \underline{apr}_B^G(X)$.

4 Conclusions

In this paper, the partial order among the multi-values in an SIS is taken into consideration. The function equivalence relation is proposed. Then the properties of function equivalence classes are analyzed. Furthermore, a general function relation is defined in the SIS. Then, approximations and its properties under the general function relation are discussed. In the future work, we will study the attribute reduction under the general function relation in the SIS.

References

1. Zadeh LA (1997) Towards a theory of fuzzy information granulation and its centrality in human reasoning and fuzzy logic. Fuzzy Sets Syst 19(1):111–127
2. Zadeh LA (2008) Is there a need for fuzzy logic? Inf Sci 178:2751–2779
3. Pawlak Z, Skowron A (2007) Rudiments of rough sets. Inf Sci 177:3–27
4. Guan YY, Wang HK (2006) Set-valued information systems. Inf Sci 176(17):2507–2525
5. Qian YH, Dang CY, Liang JY, Tang DW (2009) Set-valued ordered information systems. Inf Sci 179(16):2809–2832
6. Song XX, Zhang WX (2009) Knowledge reduction in inconsistent set-valued decision information system. Comput Eng Appl 45(1):33–35 (In chinese)
7. Song XX, Zhang WX (2006) Knowledge reduction in set-valued decision information system. In: Proceedings of 5th international conference on rough sets and current trends in computing (RSCTC 2006), 4259 LNAI, pp 348–357
8. Huang B, Li HX, Wei DK (2011) Degree dominance set-valued relation-based RSM in intuitionistic fuzzy decision tables. In: Proceedings of the international conference on internet computing and information services, pp 37–40
9. Chen HM, Li TR, Zhang JB (2010) A method for incremental updating approximations based on variable precision set-valued ordered information systems. In: Proceedings of the 2010 IEEE international conference on granular computing (GrC2010), pp 96–101

10. Zou WL, Li TR, Chen HM, Ji XL (2009) Approaches for incrementally updating approximations based on set-valued information systems whileattribute values' coarsening and refining. In: Proceedings of the 2009 IEEE international conference on granular computing (GrC2009), pp 824–829
11. Yao YY (2010) Three-way decisions with probabilistic rough sets. Inf Sci 180(7):341–353
12. Liang JY, Shi ZZ (2004) The information entropy, rough entropy and knowledge granulation in rough set theory. Int J Uncertainty Fuzziness Knowl Based Syst 12(1):37–46

Neutrality in Bipolar Structures

Javier Montero, J. Tinguaro Rodriguez, Camilo Franco, Humberto Bustince, Edurne Barrenechea and Daniel Gómez

Abstract In this paper, we want to stress that bipolar knowledge representation naturally allows a family of middle states which define as a consequence different kinds of bipolar structures. These bipolar structures are deeply related to the three types of bipolarity introduced by Dubois and Prade, but our approach offers a systematic explanation of how such bipolar structures appear and can be identified.

Keywords Bipolarity · Compatibility · Conflict · Ignorance · Imprecision · Indeterminacy · Neutrality · Symmetry

J. Montero (✉) · J. T. Rodriguez
Faculty of Mathematics, Complutense University, Madrid, Spain
e-mail: monty@mat.ucm.es

J. T. Rodriguez
e-mail: jtrodrig@mat.ucm.es

C. Franco
Centre for Research and Technology of Agro-Environmental and Biological Sciences, Tras-os-Montes e Alto Douro University, Vila Real, Portugal
e-mail: camilo@utad.pt

H. Bustince · E. Barrenechea
Departamento de Automática y Computación, Universidad Pública de Navarra, Pamplona, Spain
e-mail: bustince@unavarra.es

E. Barrenechea
e-mail: edurne.barrenechea@unavarra.es

D. Gómez
School of Statistics, Complutense University, Madrid, Spain
e-mail: dagomez@estad.ucm.es

F. Sun et al. (eds.), *Knowledge Engineering and Management*,
Advances in Intelligent Systems and Computing 214,
DOI: 10.1007/978-3-642-37832-4_2,

1 Introduction

Dubois and Prade [8–10] distinguished between three types of bipolarity, named Type I, Type II, and Type III bipolarity, but in our opinion a unified approach for such a classification is missing.

Our approach in this paper focuses on how such bipolar models are being built in knowledge representation. Our main point is that bipolarity appears whenever two opposite arguments are taken into account, being the key issue how intermediate *neutral* stages are being generated from such an opposition. Such intermediate stages can be introduced from different circumstances and generate a different relationship with the basic two opposite arguments.

As pointed out in [17], different roles should be associated to different structures, and different structures justify different concepts. In particular, we claim that a semantic approach to bipolarity will allow different *neutral* stages between the two extremes under consideration, depending on the nature of these two extreme values. In this way, concepts such as *imprecision*, *indeterminacy*, *compatibility*, and *conflict* between poles will be distinguished. These *neutral* stages are somehow frequently understood as different forms of a heterogeneous *ignorance*. For example, ignorance—lack of information—is sometimes confused with symmetry in decision making—difficulties to choose between two poles despite clear available information. The lack of information can be based on simple imprecision or a deeper conceptual problem, whenever the two considered poles are not enough to fully explain reality. Similarly, a decision maker cannot be able to choose a unique pole when both poles simultaneously hold or when a conflict is being detected (random decision can be acceptable in the first case but not in the second place, where each pole can be rejected because of different arguments).

We should remind that Dubois and Prade [8] classify bipolarity in terms of the nature of the scales that are used and the relation between positive and negative information, differentiating as a consequence two types of bipolar scales: *univariate bipolar* and *bivariate unipolar* scales. Univariate bipolarity was associated in [8] and [14] to a linearly ordered set L in which the two ends are occupied by the poles $+$ and $-$, and certain *middle* value 0 might separate positive evaluations from the negative evaluations. An object is evaluated by means of a single value on L. On the other hand, bivariate unipolar scales admit in [8] positive and negative information to be measured separately by means of two unipolar scales, each one being occupied by a pole and a *neutral* state 0 that somehow appears in between both scales, but not as a middle value as in type I bipolarity. An object can receive two evaluations, which can be perceived as neither positive nor negative, as well as both positive and negative. This is the way type II and type III bipolarities are introduced in [8].

In the following sections, we shall offer an explanation to all those neutral stages between poles which will allow a systematic characterization of each different kind of bipolarity. In fact, we will see that the nature of the middle stage can be used to identify which bipolarity we are dealing with, although it should be acknowledged that complex problems cannot be explained by means of a unique bipolar structure: different bipolarities will simultaneously appear in practice, showing that several semantics might coexist. Our approach will lead to the same differentiation proposed by Dubois and Prade, but giving a key role to the underlying structure-building process generated from the opposite poles and their associated semantic, that will produce in each circumstance a specific neutral middle stage.

2 Building Type I Bipolar Fuzzy Sets

Type I bipolarity assumes a scale that shows tension between a pole and its own negation. It suggests gradualness within a linear scale, and the middle intermediate value simply represents a scale *point of symmetry* (not that both poles hold). This linear structure is the characteristic property of type I bipolarity, and a single value in this scale simultaneously gives the distance to each pole. But such a symmetry value cannot be confused with compatibility or ignorance; it does not properly represent a new concept.

For example, meanwhile "tall" and "short" are viewed as two opposite grades of a unique "tallness" concept, such a "tallness" is being modeled as a type I bipolarity. In this case, we neither estimate "tallness" or "shortness", but "height".

Neutrality within two type I bipolar poles can more properly be associated to the unavoidable imprecision problem (uncertainty about the right value). *Imprecision* represents a very particular kind of *ignorance*, indeed an epistemic state different than both poles. When a decision maker has no information about the exact value, imprecision is maximum and the more information we get the more accurate we are.

Imprecision is the characteristic neutral state in type I bipolarity.

3 Building Type II Bipolar Fuzzy Sets

On the contrary, type II bipolarity requires the existence of two dual (perhaps antagonistic) concepts, related because they refer to a common concept but containing different information. Distance to one pole cannot be deduced from the distance to the other pole, so two separate evaluations are needed for each pole, although they share a common nature. Poles are not viewed as the extreme values of a unique gradualness scale like in type I bipolarity. Depending on the nature of such a duality, we may find different neutral intermediate states with differentiated meaning or semantics.

For example, two opposite concepts not necessarily cover the whole universe of discourse (as shown in [2] within a classification framework when compared to [19]). An object can neither fulfill a concept nor its opposite, a situation that by definition cannot happen within type I bipolarity, where negation connects both poles. If classification poles are too distant, this situation suggests the need of some additional classification concept [1], and meanwhile we do not find such an additional concept, we have to declare an *indeterminacy* that is another kind of neutral *ignorance*, different in nature to previous *imprecision* (see [7]). This is the case when the dual of the concept is strictly contained in its negation (for example, "very short" is strictly contained in "not tall"). Poles do not cover the whole space of possibilities. There exists a region where none of both poles hold.

Alternatively, negation can be implied by its dual concept, in such a way that both poles overlap (as "not very tall" overlaps "not very short", see [6]). In this case, type II bipolarity generates a different specific intermediate concept that cannot be associated to *indeterminacy*, but to a different kind of *neutrality*. We are talking about a simultaneous verification of both poles which is not a point of symmetry. In type I bipolarity, symmetry refers to a situation *in between* poles; meanwhile, in this type II bipolarity, *compatibility* means that both poles simultaneously hold.

In this way, we find that type II bipolarity may generate two different neutral intermediate concepts, depending on the semantic relation between poles. While in type I bipolarity one of the poles is precisely the negation of the other, in type II bipolarity, two alternative type II bipolarities may appear (but they will not simultaneously appear).

4 Building Type III Bipolar Fuzzy Sets

Type III bipolarity refers to negative and positive pieces of information (see [11]), implying the existence of two families of arguments, that should be somehow aggregated. Poles here represent like bags of arguments. Poles are not direct arguments like in type I and type II bipolarity.

Type III bipolarity suggests a different construction than the one used for type I or type II bipolarity. In such type I or type II bipolarity, we start from the opposition between two extreme values within a single characteristic or between two dual poles, but in both cases assuming one single common concept. A concept generates its negation or some kind of dual concept, and their description may need one single value or two values, which can be directly estimated.

Type III bipolarity appears like a second degree bipolar type II structure, being both poles complex concepts like *positive–negative* or *good–bad*, for example, each pole needing a specific description or decomposition to be understood.

Type III bipolarity is essentially more complex than type II bipolarity, for example when we are asked to list on one side "positive" arguments and "negative" arguments on the other side.

This is the standard situation in multi-criteria decision making. It is in this context where *conflict* can naturally appear as another neutrality stage, besides *imprecision*, *compatibility*, and *indeterminacy*. In a complex problem, of course we can find at the same time strong arguments supporting both poles (see, e.g., [18]).

Such a *conflict* stage is natural within type III bipolarity, but of course it may happen that those arguments as a whole do not suggest a complete description of reality, suggesting *indeterminacy*, similarly to type II *indeterminacy*. Notice that in type II bipolarity *compatibility* can appear, but *conflict* should not be expected when dealing with a pair of extreme values coming from simple (1-dimensional) argument. Such a *conflictive* argument belongs to the type III bipolarity context.

It is worth noticing that meanwhile type I and type II bipolarity are built up from symmetrical poles, type III bipolarity is built from asymmetrical poles (see [8, 9]).

The key issue is the different semantic relation between poles. A different semantic relation produces a different intermediate *neutral* state.

Finally, it must be pointed out that in this third more complex problem all previous situations might be implied. Underlying criteria can define a conflict, but they can also define *indeterminacy* situation, resembling the type II *indeterminacy* as already pointed out (too poor descriptions suggesting *ignorance*), and in addition, each underlying criteria is subject to a symmetrical bipolarity framework, allowing *imprecision* or *compatibility*.(besides type II *indeterminacy*). For example, the semantic relation between "very good" and "very bad" is the same relation as "very tall" and "very short", allowing when something is neither "very good" or "very bad" a similar *indeterminacy* to the one that appears when something is neither "very tall" or "very short".

But *conflict* is essentially a type III bipolarity performance.

5 Building General Bipolarities

From the above comments it is clear that some bipolar problems require a quite complex structure.

Type I bipolarity use to imply the existence of an *imprecision* state besides a possible *point of symmetry*.

Type II bipolar implies two potential different intermediate states (*indeterminacy* and *compatibility*).

Type III bipolarity implies the possibility of a *conflict* stage, but previous intermediate stages can also appear associated to each underlying criteria, if not directly.

A general bipolarity model should allow all those four semantic neutral states between poles beside the non semantic *points of symmetry* that may appear within type I bipolarity. In principle, a general bipolar representation should be prepared to simultaneously deal with and evaluate all these states.

The semantic argument is anyway needed to distinguish between different bipolarities, and to produce structured type-2 fuzzy sets [16], as proposed in [17].

It is also interesting to realize that this semantic approach allows an alternative explanation to Atanassov's intuitionistic fuzzy sets (see [3–5]), realizing how different examples given by this author can be explained by means of different semantics and therefore different bipolarities.

6 Conclusion

When poles are defined within a binary context, in terms of a single concept and its *negation* (e.g., "tall" and "not tall"), a linear order of gradation states is allowed around the *point of symmetry*, simply meaning equal distance to both concepts within a linear order, like and in Probability Theory [15] or Fuzzy Sets [20] (see also [12]). An object is associated *to some extent* to both poles by means of a unique family of intermediate *gradation* states. This is the framework for type I bipolarity.

In type II bipolarity, we have dual concepts as poles. These two dual poles can overlap (like "more or less tall" and "more or less short") or they can create a region that is far away from poles (like "very tall" and "very short"). Anyway, opposite poles are not complementary, so two different situations can be generated. If a pole is much smaller than the negation of the opposite pole, *indeterminacy* will be natural. If a pole is much bigger than the negation of the opposite pole, *compatibility* will naturally appear.

Type III bipolarity is the natural framework for multicriteria decision making (see [13]). Poles are complex concepts and need a multicriteria description, and of course different criteria can produce a *conflict*. But also each one of the other situations can appear within each simple criterion. In addition, such a multicriteria description can be so poor that a strict *ignorance* can appear.

Acknowledgments This research has been partially supported by the Government of Spain, grants TIN2009-07901, TIN2012-32482 and TIN2010-15055.

References

1. Amo A, Gomez D, Montero J, Biging G (2001) Relevance and redundancy in fuzzy classification systems. Mathware and Soft Computing 8:203–216
2. Amo A, Montero J, Biging G, Cutello V (2004) Fuzzy classification systems. Eur J Oper Res 156:495–507
3. Atanassov KT (1983) Intuitionistic fuzzy sets. VII ITKR's Session, Sofia, deposed in Central Science-Technical Library of Bulgarian academy of Science, 1697/84 (in Bulgarian)
4. Atanassov KT (1986) Intuitionistic fuzzy sets. Fuzzy Sets Syst 20:87–96
5. Atanassov KT (2005) Answer to D. Dubois, S. Gottwald, P. Hajek, J. Kacprzyk and H. Prade's paper "Terminological difficulties in fuzzy set theory—the case of "Intuitionistic fuzzy sets". Fuzzy Sets Syst 156:496–499
6. Bustince H, Fernandez J, Mesiar R, Montero J, Orduna R (2010) Overlap functions. Nonlinear Analysis 72:1488–1499
7. Bustince H, Pagola M, Barrenechea E, Fernandez J, Melo-Pinto P, Couto P, Tizhoosh HR, Montero J (2010) Ignorance functions. An application to the calculation of the threshold in prostate ultrasound images, Fuzzy sets and Systems 161:20–36
8. Dubois D, Prade H (2008) An introduction to bipolar representations of information and preference. Int J Intell Syst 23:866–877
9. Dubois D, Prade H (2009) An overview of the asymmetric bipolar representation of positive and negative information in possibility theory. Fuzzy Sets Syst 160:1355–1366

10. Dubois D, Prade H Gradualness uncertainty and bipolarity: making sense of fuzzy sets. In: Fuzzy sets and Systems (in press)
11. Franco C, Montero J, Rodriguez JT (2011) On partial comparability and preference-aversion models. In: Proceedings ISKE 2011, Shangai, China, 15–17 Dicimbre 2011
12. Goguen J (1967) L-fuzzy sets. Journal of Mathematical Analysis and Applications 18:145–174
13. Gonzalez-Pachón J, Gómez D, Montero J, Yáñez J (2003) Soft dimension theory. Fuzzy Sets Syst 137:137–149
14. Grabisch M, Greco S, Pirlot M (2008) Bipolar and bivariate models in multi-criteria decision analysis: descriptive and constructive approaches. Int J Intell Syst 23:930–969
15. Kolmogorov AN (1956) Foundations of the theory of probability. Chelsea Publishing Co., New York
16. Mendel JM (2007) Advances in type-2 fuzzy sets and systems. Inf Sci 177:84–110
17. Montero J, Gomez D, Bustince H (2007) On the relevance of some families of fuzzy sets. Fuzzy Sets Syst 158:2429–2442
18. Roy B (1991) The outranking approach and the foundations of ELECTRE methods. Springer, Berlin
19. Ruspini EH (1969) A new approach to clustering. Inf Control 15:22–32
20. Zadeh LA (1965) Fuzzy sets. Inf Control 8:338–353

Multirate Multisensor Data Fusion Algorithm for State Estimation with Cross-Correlated Noises

Yulei Liu, Liping Yan, Bo Xiao, Yuanqing Xia and Mengyin Fu

Abstract This paper is concerned with the optimal state estimation problem under linear dynamic systems when the sampling rates of different sensors are different. The noises of different sensors are cross-correlated and coupled with the system noise of the previous step. By use of the projection theory and induction hypothesis repeatedly, a sequential fusion estimation algorithm is derived. The algorithm is proven to be optimal in the sense of Linear Minimum Mean Square Error(LMMSE). Finally, a numerical example is presented to illustrate the effectiveness of the proposed algorithm.

Keywords State estimation · Data fusion · Cross-correlated noises · Asynchronous multirate multisensor

1 Introduction

Estimation fusion, or data fusion for estimation, is the problem of how to best utilize useful information contained in multiple sets of data for the purpose of estimating a quantity, e.g., a parameter or process [1]. It originated in the military field, and is now widely used in military and civilian fields, e.g., target tracking and localization, guidance and navigation, surveillance and monitoring, etc., due to its improved estimation accuracy, enhanced reliability, and survivability, etc.

Y. Liu (✉) · L. Yan · B. Xiao · Y. Xia · M. Fu
Key Laboratory of Intelligent Control and Decision of Complex Systems,
School of Automation, Beijing Institute of Technology,
100081 Beijing, China
e-mail: lyulei0929@gmail.com

L. Yan
e-mail: liping.yan@gmail.com

F. Sun et al. (eds.), *Knowledge Engineering and Management*,
Advances in Intelligent Systems and Computing 214,
DOI: 10.1007/978-3-642-37832-4_3, © Springer-Verlag Berlin Heidelberg 2014

Most of the earlier works were based on the assumption of cross-independent sensor noises. Bar-Shalom [2], Chong et al. [3], Hashmipour et al. [4] proposed several optimal state estimation algorithms based on Kalman filtering, respectively, in which all sensor measurements are fused by use of centralized fusion structure. In the practical applications, most of multisensor systems often have the correlated noises when the dynamic process is observed in a common noisy environment [5]. Moreover, because most of the real systems are described in continuous forms, discretization is necessary when to get the state estimation on line, and in the process, the system noise and the measurement noises are shown to be coupled. Of course, the centralized filter can still be used for the systems with correlated noises as it is still optimal in the sense of LMMSE. However, the computation and power requirements are too huge to be practical.

Hence, a few pieces of work deal with coupled sensor noises. Duan proposed one systematic way to handle the distributed fusion problem based on a unified data model in which the measurement noises across sensors at the same time may be correlated [6]. Song also dealt with the state estimation problem with cross-correlated sensor noises, and proved that under a mild condition it is optimal [5]. A small amount of papers consider the coupled sensor noises and the correlation between sensor noises and system noises. Xiao et al. [7] considered the two kinds of correlations by augmentation and the computation is complex.

In all the papers mentioned above, the sampling rates of different sensors are the same. Based on the multi-sensor dynamic system in which different sensors observe the same target state with different sampling rates, Yan et al. [8] put forward a kind of optimal state estimation algorithm. The algorithm has stronger feasibility and practicality than the traditional state fusion algorithm, but it does not take the correlations of noises into account. Shi et al. [9] discussed the estimation when multisensors have multirate asynchronous sampling rates. However, it does not consider the sensor-correlations either.

In this paper, when the noises of different sensors are cross-correlated and when they are also coupled with the system noise of the previous step, also, when the sampling rates of different sensors are different, by use of the projection theory, a sequential algorithm is formulated. We analyzed the performance of the algorithm, and it is shown to be optimal in the sense of LMMSE.

The paper is organized as follows. In Sect. 2, the problem formulation is presented. Section 3 describes the optimal state estimation algorithm. Section 4 is the simulation results and Sect. 5 draws the conclusion.

2 Problem Formulation

Consider the following generic linear dynamic system,

$$x(k+1) = A(k)x(k) + w(k), k = 0, 1, \ldots \tag{1}$$

$$z_i(k_i) = C_i(k_i)x(k_i) + v_i(k_i), i = 1, 2, \ldots, N \quad (2)$$

where, $x(k) \in R^n$ is the system state, $A(k) \in R^{n \times n}$ is the state transition matrix, and $w(k)$ is the system noise and is assumed to be Gaussian distributed with zero mean and variance being $Q(k)$, where $Q(k) \geq 0$. $z_i(k_i) \in R^{m_i}$ is the measurement of sensor i at time k_i. Assume the sampling rate of sensor i is S_i, and $S_i = S_1/n_i$, where n_i is known positive integers. Without loss of generality, let the sampling period of sensor 1 be the unit time, that is, $k_1 = k$. Then, sensor i has measurement at sampling point $n_i k$, in other words, $k_i = n_i k$. $C_i(k_i) \in R^{m_i \times n}$ is the measurement matrix. Measurement noise $v_i(k_i)$ is zero-mean and is Gaussian distributed with variance being $R(k)$, and

$$E\{w(k_i - 1)v_i^T(k_i)\} = S_i(k_i) \quad (3)$$

From the above equation, we can see that the measurement noises are coupled with the previous step system noise. Namely, $v_i(k_i)$ is correlated with $w(k_i - 1)$ at time $k = 0, 1, \ldots, i = 1, 2, \ldots, N$. If different sensors have measurements at the same time, their measurement noises are cross-correlated, i.e., $v_i(k_i)$ and $v_j(l_j)$ are coupled when $k_i = l_j$. That is, $E\{v_i(k_i)v_j^T(k_i)\} = R_{ij}(k_i) \neq 0$ where $i, j = 1, 2, \ldots, N$. For simplicity, denote $R_i(k_i) \stackrel{\Delta}{=} R_{ii}(k_i) > 0, i = 1, 2, \ldots, N$.

The initial state $x(0)$ is independent of $w(k)$ and $v_i(k_i)$, where $k = 1, 2, \ldots$, $i = 1, 2, \ldots, N$, and is assumed to be Gaussian distributed with

$$\begin{cases} E\{x(0)\} = x_0 \\ cov\{x(0)\} = E\{[x(0) - x_0][x(0) - x_0]^T\} = P_0 \end{cases} \quad (4)$$

where $cov\{x(0)\}$ means the covariance of $x(0)$.

It can be seen from the above description that sensor i will participate in the fusion process at time $n_i k$. Generally speaking, assume there are p sensors that have measurements at time k with measurements $z_{i_1}(k), z_{i_2}(k), \ldots, z_{i_p}(k)$. Then, to generate the optimal state estimate of x(k) in the measurement update step, the above p sensors shall be fused. The estimation of $x(k)$ is the information fusion of the above p sensors.

3 Optimal State Estimation Algorithm

Theorem 1 Based on the descriptions in Sect. 2, suppose we have known the optimal fusion estimation $\hat{x}(k-1|k-1)$ and its estimation error covariance $P(k-1|k-1)$ at time $k-1$, then the optimal state estimation of $x(k)$ at time k could be computed as follows,

$$\hat{x}_{i_j}(k|k) = \hat{x}_{i_{j-1}}(k|k) + K_{i_j}(k)[z_{i_j}(k) - C_{i_j}(k)\hat{x}_{i_{j-1}}(k|k)] \quad (5)$$

$$P_{i_j}(k|k) = P_{i_{j-1}}(k|k) - K_{i_j}(k)[C_{i_j}(k)P_{i_{j-1}}(k|k) + \Delta^T_{i_{j-1}}(k)] \tag{6}$$

$$\begin{aligned} K_{i_j}(k) = {} & [P_{i_{j-1}}(k|k)C^T_{i_j}(k) + \Delta_{i_{j-1}}(k)][C_{i_j}(k)P_{i_{j-1}}(k|k)C^T_{i_j}(k) \\ & + C_{i_j}(k)\Delta_{i_{j-1}}(k) + \Delta^T_{i_{j-1}}(k)C^T_{i_j}(k)]^{-1} + R_{i_j}(k) \end{aligned} \tag{7}$$

$$\begin{aligned} \Delta_{i_j}(k) = {} & \prod_{u=j}^{1}[I - K_{i_u}(k)C_{i_u}(k_{i_u})]S_{i_{j+1}}(k) - K_{i_j}(k)R_{i_j,i_{j+1}}(k) \\ & - \sum_{q=2}^{j}\prod_{u=j}^{q}[I - K_{i_u}(k)C_{i_u}(k_{i_u})]K_{i_{q-1}}(k)R_{i_{q-1},i_{j+1}}(k) \end{aligned} \tag{8}$$

where $j = 1, 2, \ldots, p$. For $j = 0$,

$$\hat{x}_{i_0}(k|k) = \hat{x}(k|k-1) = A(k-1)\hat{x}(k-1|k-1) \tag{9}$$

$$\begin{aligned} P_{i_0}(k|k) & = P(k|k-1) \\ & = A(k-1)P(k-1|k-1)A^T(k-1) + Q(k-1) \end{aligned} \tag{10}$$

$$\Delta_{i_0}(k) = S_{i_1}(k) \tag{11}$$

The above $\hat{x}_{i_j}(k|k)$ and $P_{i_j}(k|k)$ denote the state estimation of $x(k)$ and the corresponding estimation error covariance based on observations of sensors $i_1, i_2, \ldots, i_j$ respectively. When $j = p$, we have $\hat{x}_s(k|k) = \hat{x}_{i_p}(k|k)$ and $P_s(k|k) = P_{i_p}(k|k)$, which are the optimal state fusion estimation and the corresponding estimation error covariance, where subscript 's' means the sequential fusion.

In addition, from (5), we can see that the sensor with the highest sampling rate is the first sensor whose sampling period is assumed to be the unit time, so sensor 1 is sensor i_1. That is, $\hat{x}_{i_1}(k|k) = \hat{x}_1(k|k)$, $P_{i_1}(k|k) = P_1(k|k)$.

Proof The theorem derives from gradually use of the projection theorem. For $i = 1, 2, \ldots, N$, denote

$$Z_i(k) = \{z_i(1), z_i(2), \ldots, z_i(k)\} \tag{12}$$

$$Z^i_1(k) = \{z_1(k), z_2(k), \ldots, z_i(k)\} \tag{13}$$

$$\bar{Z}^i_1(k) = \{Z^i_1(l)\}^k_{l=1} \tag{14}$$

where $Z_i(k)$ is the measurements of sensor i up to time k. If sensor i has no measurement at time l, we denote $z_i(l) = 0$, therefore, the above descriptions are meaningful. $Z^i_1(k)$ is the measurement of sensors $1, 2, \ldots, i$ at time k. $\bar{Z}^i_1(k)$ is the measurements of all sensors at time k and before.

In the sequel, we will prove Theorem 1 deductively by applying the projection theorem. Suppose, we have obtained $\hat{x}_{i_{j-1}}(k|k)$ and the corresponding estimation error covariance $P_{i_{j-1}}(k|k)$, next we will show how to get $\hat{x}_{i_j}(k|k)$ and $P_{i_j}(k|k)$.

Applying the projection theorem, we have

$$\begin{aligned}\hat{x}_{i_j}(k|k) &= E\{x(k)|\bar{Z}_1^N(k-1), Z_1^{i_j}(k)\}\\ &= E\{x(k)|\bar{Z}_1^N(k-1), Z_1^{i_j-1}(k), z_{i_j}(k)\}\\ &= \hat{x}_{i_{j-1}}(k|k) + cov\{\tilde{x}_{i_{j-1}}(k|k), \tilde{z}_{i_{j-1}}(k)\}\cdot var\{\tilde{z}_{i_j}(k)\}^{-1}\tilde{z}_{i_j}(k)\end{aligned} \tag{15}$$

where $\tilde{z}_{i_j}(k) = z_{i_j}(k) - \hat{z}_{i_j}(k)$, and

$$\begin{aligned}\hat{z}_{i_j}(k) &= E\{z_{i_j}(k)|\bar{Z}_1^N(k-1), Z_1^{i_j-1}(k)\}\\ &= E\{C_{i_j}(k)x(k) + v_{i_j}(k)|\bar{Z}_1^N(k-1), Z_1^{i_j-1}(k)\}\\ &= C_{i_j}(k)\hat{x}_{i_{j-1}}(k|k)\end{aligned} \tag{16}$$

So

$$\begin{aligned}\tilde{z}_{i_j}(k) &= z_{i_j}(k) - \hat{z}_{i_j}(k)\\ &= z_{i_j}(k) - C_{i_j}(k)\hat{x}_{i_{j-1}}(k|k)\\ &= C_{i_j}(k)x_{i_j}(k) + v_{i_j}(k) - C_{i_j}(k)\hat{x}_{i_{j-1}}(k|k)\\ &= C_{i_j}(k)\tilde{x}_{i_{j-1}}(k|k) + v_{i_j}(k)\end{aligned} \tag{17}$$

Therefore

$$\begin{aligned}cov\{\tilde{x}_{i_{j-1}}(k|k), \tilde{z}_{i_j}(k)\} &= E\{\tilde{x}_{i_{j-1}}(k|k)\tilde{z}_{i_j}^T(k)\}\\ &= E\{\tilde{x}_{i_{j-1}}(k|k)[C_{i_j}(k)\tilde{x}_{i_{j-1}}(k|k) + v_{i_j}(k)]^T\}\\ &= P_{i_{j-1}}(k|k)C_{i_j}^T(k) + \varDelta_{i_{j-1}}(k)\end{aligned} \tag{18}$$

and

$$\begin{aligned}var\{\tilde{z}_{i_j}(k)\} &= E\{\tilde{z}_{i_j}(k)\tilde{z}_{i_j}^T(k)\}\\ &= E\{[C_{i_j}(k)\tilde{x}_{i_{j-1}}(k|k) + v_{i_j}(k)]\cdot[C_{i_j}(k)\tilde{x}_{i_{j-1}}(k|k) + v_{i_j}(k)]^T\}\\ &= C_{i_j}(k)P_{i_{j-1}}(k|k)C_{i_j}^T(k) + R_{i_j}(k) + C_{i_j}(k)\varDelta_{i_{j-1}}(k) + \varDelta_{i_{j-1}}^T(k)C_{i_j}^T(k)\end{aligned} \tag{19}$$

where

$$\varDelta_{i_{j-1}}(k) = E\{\tilde{x}_{i_{j-1}}(k|k)v_{i_j}^T(k)\} \tag{20}$$

By use of the inductive assumption, we have

$$\begin{aligned}\tilde{x}_{i_{j-1}}(k|k) &= x(k) - \hat{x}_{i_{j-1}}(k|k)\\ &= x(k) - \hat{x}_{i_{j-2}}(k|k) - K_{i_{j-1}}(k)[z_{i_{j-1}}(k) - C_{i_{j-1}}(k)\hat{x}_{i_{j-2}}(k|k)]\\ &= \tilde{x}_{i_{j-2}}(k|k) - K_{i_{j-1}}(k)[C_{i_{j-1}}(k)x(k) + v_{i_{j-1}}(k) - C_{i_{j-1}}(k)\hat{x}_{i_{j-2}}(k|k)]\\ &= [I - K_{i_{j-1}}(k)C_{i_{j-1}}(k)]\tilde{x}_{i_{j-2}}(k|k) - K_{i_{j-1}}(k)v_{i_{j-1}}(k)\end{aligned} \tag{21}$$

Substitute (21) into (20), and by use of the inductive hypothesis, we have

$$\begin{aligned}\Delta_{i_{j-1}}(k) &= E\{\tilde{x}_{i_{j-1}}(k|k)v_{i_j}^T(k)\} \\ &= [I - K_{i_{j-1}}(k)C_{i_{j-1}}(k)]E\{\tilde{x}_{i_{j-2}}(k|k)v_{i_j}^T(k)\} - K_{i_{j-1}}(k)E\{v_{i_{j-1}}(k)v_{i_j}^T(k)\} \\ &= \prod_{u=j-1}^{1}[I - K_{i_u}(k)C_{i_u}(k)]S_{i_j}(k) - K_{i_{j-1}}(k)R_{i_{j-1},i_j}(k) \\ &\quad - \sum_{q=2}^{j-1}\prod_{u=j-1}^{q}[I - K_{i_u}(k)C_{i_u}(k)]K_{i_{q-1}}(k)R_{i_{q-1},i_j}(k)\end{aligned} \tag{22}$$

Substitute (18), (19) and the second equation of (17) into (15), we have

$$\hat{x}_{i_j}(k|k) = \hat{x}_{i_{j-1}}(k|k) + K_{i_j}(k)[z_{i_j}(k) - C_{i_j}(k)\hat{x}_{i_{j-1}}(k|k)] \tag{23}$$

where

$$\begin{aligned}K_{i_j}(k) &= cov\{\tilde{x}_{i_{j-1}}(k|k), \tilde{z}_{i_{j-1}}(k)\}var\{\tilde{z}_{i_{j-1}}(k)\} \\ &= [P_{i_{j-1}}(k|k)C_{i_j}^T(k)+\Delta_{i_{j-1}}(k)]\cdot[C_{i_j}(k)P_{i_{j-1}}(k|k)C_{i_j}^T(k)+R_{i_j}(k) \\ &\quad + C_{i_j}(k)\Delta_{i_{j-1}}(k) + \Delta_{i_{j-1}}^T(k)C_{i_j}^T(k)]^{-1}\end{aligned} \tag{24}$$

The estimation error covariance should be computed by

$$\begin{aligned}P_{i_j}(k|k) &= E\{\tilde{x}_{i_j}(k|k)\tilde{x}_{i_j}^T(k|k)\} \\ &= E\{[x(k) - \hat{x}_{i_j}(k|k)][x(k) - \hat{x}_{i_j}(k|k)]^T\} \\ &= E\{[(I - K_{i_j}(k)C_{i_j}(k))\tilde{x}_{i_{j-1}}(k|k) - K_{i_j}(k)v_{i_j}(k)] \\ &\qquad \cdot[(I - K_{i_j}(k)C_{i_j}(k))\tilde{x}_{i_{j-1}}(k|k) - K_{i_j}(k)v_{i_j}(k)]^T\} \\ &= [I - K_{i_j}(k)C_{i_j}(k)]P_{i_{j-1}}(k|k)[I - K_{i_j}(k)C_{i_j}(k)]^T \\ &\quad + K_{i_j}(k)R_{i_j}(k)K_{i_j}^T(k) - [I - K_{i_j}(k)C_{i_j}(k)]\Delta_{i_{j-1}}(k)K_{i_j}^T(k) \\ &\quad - K_{i_j}(k)\Delta_{i_{j-1}}^T(k)[I - K_{i_j}(k)C_{i_j}(k)]^T \\ &= P_{i_{j-1}}(k|k) - K_{i_j}(k)[C_{i_j}(k)P_{i_{j-1}}(k|k) + \Delta_{i_{j-1}}^T(k)\end{aligned} \tag{25}$$

where Eq. (24) is used.

Combine (22), (23), (24), and (25), we obtain

$$\hat{x}_{i_j}(k|k) = \hat{x}_{i_{j-1}}(k|k) + K_{i_j}(k)[z_{i_j}(k) - C_{i_j}(k)\hat{x}_{i_{j-1}}(k|k)] \tag{26}$$

$$P_{i_j}(k|k) = P_{i_{j-1}}(k|k) - K_{i_j}(k)\cdot[C_{i_j}(k)P_{i_{j-1}}(k|k) + \Delta_{i_{j-1}}^T(k)] \tag{27}$$

$$K_{i_j}(k) = [P_{i_{j-1}}(k|k)C_{i_j}^T(k) + \Delta_{i_{j-1}}(k)] \cdot [C_{i_j}(k)P_{i_{j-1}}(k|k)C_{i_j}^T(k) + R_{i_j}(k) + C_{i_j}(k)\Delta_{i_{j-1}}(k) + \Delta_{i_{j-1}}^T(k)C_{i_j}^T(k)]^{-1} \quad (28)$$

$$\Delta_{i_j}(k) = \prod_{u=j}^{1}[I - K_{i_u}(k)C_{i_u}(k_{i_u})]S_{i_{j+1}}(k) - K_{i_j}(k)R_{i_j,i_{j+1}}(k) - \sum_{q=2}^{j}\prod_{u=j}^{q}[I - K_{i_u}(k)C_{i_u}(k_{i_u})]K_{i_{q-1}}(k)R_{i_{q-1},i_{j+1}}(k) \quad (29)$$

Let $\hat{x}_s(k|k) = \hat{x}_{i_p}(k|k)$ and $P_s(k|k) = P_{i_p}(k|k)$, then we obtain the state estimation of sequential fusion $\hat{x}_s(k|k)$ and $P_s(k|k)$, and the proof is completed.

4 Simulation

To illustrate the effectiveness of the proposed algorithm, a numerical example is provided in this section.

A target is observed by three sensors, which could be described by Eqs. (1) and (2). Sensor 1 has the highest sampling rate S_1, and the sampling rates of sensor 2 and sensor 3 are S_2 and S_3, respectively, which meet $S_1 = 2S_2 = 3S_3$. And

$$A = \begin{bmatrix} 1 & 1 \\ 0 & 1 \end{bmatrix}, C_1 = [1 \quad 0], C_2 = [1 \quad 0], C_3 = [0 \quad 1]$$

Sensor 1 and sensor 2 observe the position, and sensor 3 observes the velocity.

At time k, the correlations of measurement noises covariance are given by

$$R_1(k) = cov(v_1(k)) = 0.048, R_2(k) = cov(v_2(k)) = 0.064$$
$$R_3(k) = cov(v_3(k)) = 0.064, R_{12}(k) = E[v_1(k)v_2^T(k)] = 0.032$$
$$R_{13}(k) = E[v_1(k)v_3^T(k)] = 0.016, R_{23}(k) = E[v_2(k)v_3^T(k)] = 0.016$$

So, the measurement noises covariance is

$$R(k) = \begin{bmatrix} 0.048 & 0.032 & 0.016 \\ 0.032 & 0.064 & 0.016 \\ 0.016 & 0.016 & 0.064 \end{bmatrix}.$$

and $Q(k) = cov(w_k) = \begin{bmatrix} 0.02 & 0.01 \\ 0.01 & 0.04 \end{bmatrix}$. The covariances between the system noise and the measurement noises are given by

$$S_1(k) = E\{w(k-1)v_1^T(k)\} = \begin{bmatrix} 0.0050 \\ 0.0025 \end{bmatrix}$$

$$S_2(k) = E\{w(k-1)v_2^T(k)\} = \begin{bmatrix} 0.0050 \\ 0.0025 \end{bmatrix}$$

$$S_3(k) = E\{w(k-1)v_3^T(k)\} = \begin{bmatrix} 0.0025 \\ 0.0100 \end{bmatrix}$$

To derive $\hat{x}_s(k|k)$, in the measurement update step, if k could be divided by 2 but could not be divided by 3, then we will use the observations of sensor 1 and sensor 2. Similarly, if k could be divided by 3 but could not be divided by 2, then the observations of sensor 1 and sensor 3 should be fused to generate the estimate of $x(k)$. However, if k could be divided by both 2 and 3, that is, k is a multiple of 6, the observations of sensor 1, sensor 2, and sensor 3 should be used. Otherwise, we only use the observations of sensor 1.

The initial conditions are $x_0 = \begin{bmatrix} 10 \\ 0.1 \end{bmatrix}$, $P_0 = 3 \cdot \begin{bmatrix} 1 & 0 \\ 0 & 1 \end{bmatrix}$.

The Monte Carlo simulation results are shown in Figs. 1, 2, 3, and 4. Using only the measurements of sensor 1, "KF" denotes the Kalman filtering when the sensor noises are cross-correlated and are coupled with the previous step system noise, and "NKF" denotes Kalman filtering when the noises are all independent of each other. Using the measurements of all three sensors, "SFKF" denotes the algorithm given in Theorem 1, and "NSFKF" denotes the sequential fusion algorithm when the noises are treated as independent.

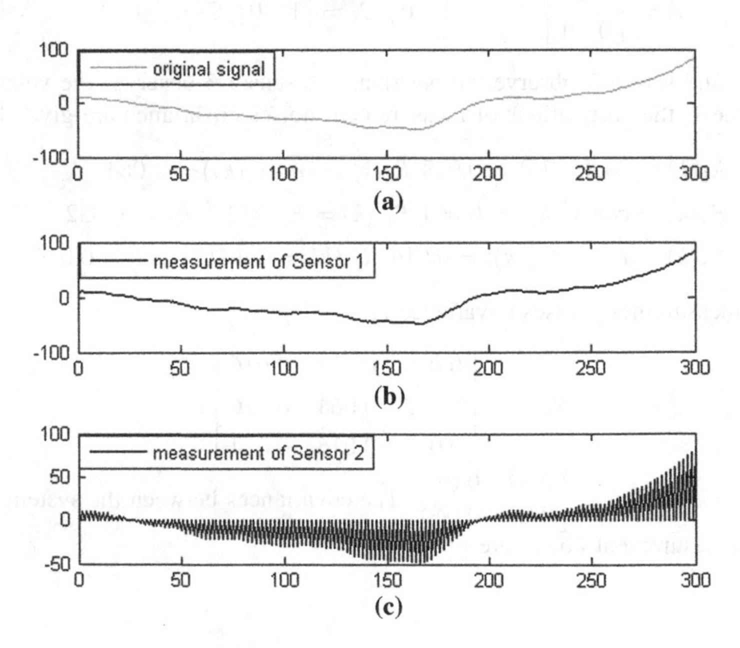

Fig. 1 Position and measurements of sensor 1 and 2

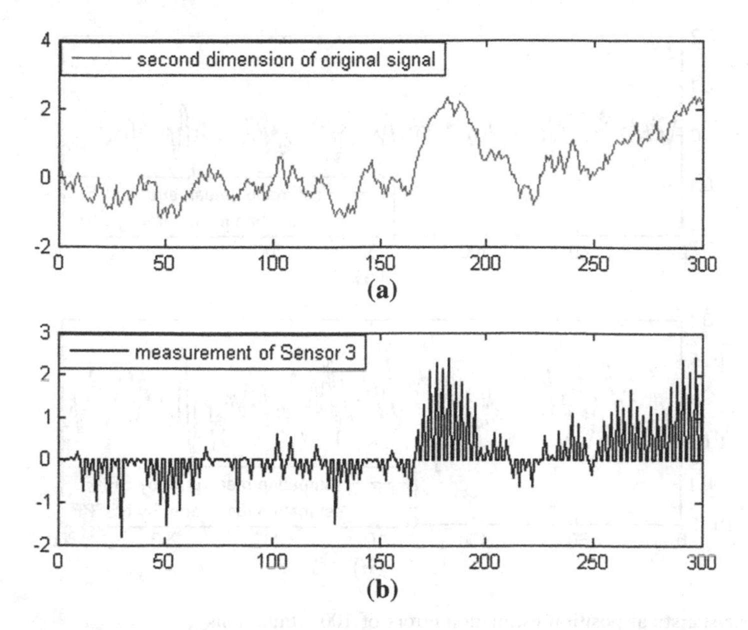

Fig. 2 Velocity and measurements of sensor 3

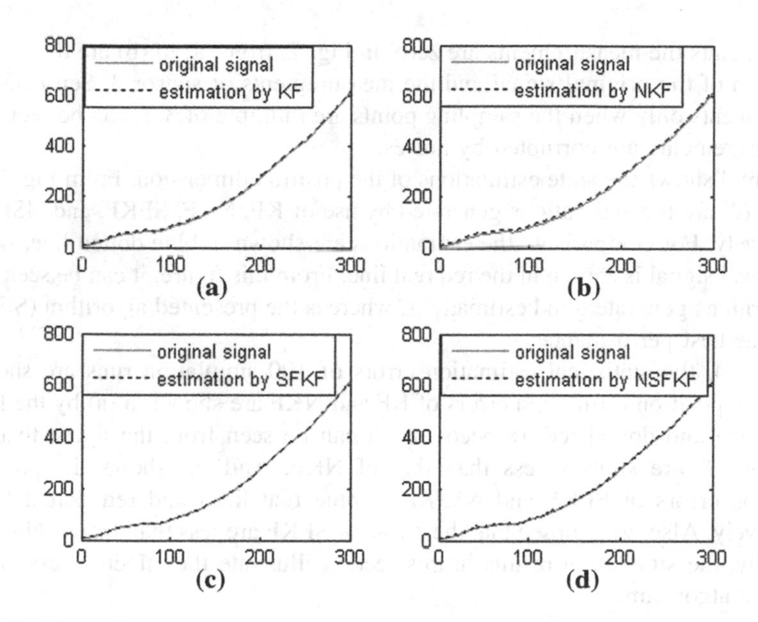

Fig. 3 Position estimations

In Fig. 1, from (a) to (c) are the first dimension of the original signal, the measurement of sensor 1 and the measurements of sensor 2. From the figure, we can see that sensor 2 only has measurements in the even number points and in odd

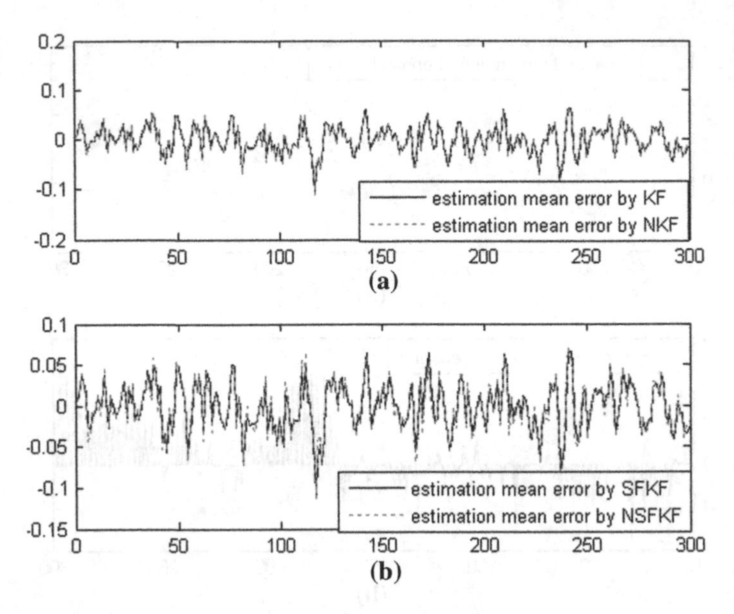

Fig. 4 The statistical position estimation errors of 100 simulations

number points the measurements are zero. In Fig. 2, from (a) to (b) are the second dimension of the original signal and the measurements of sensor 3. Sensor 3 has measurements only when the sampling points are multiple of 3. It can be seen that the measurements are corrupted by noises.

Figure 3 shows the state estimations of the position dimension. From Fig. 3 (a) through (d) are the estimations generated by use of KF, NKF, SFKF, and NSFKF, respectively. For comparison, the estimations are shown in blue dotted line, while the original signal is shown in the red real line. From this figure, it can be seen that all algorithms generate good estimations, whereas the presented algorithm (SFKF) shows the best performance.

In Fig. 4, the statistical estimation errors of 100 simulation runs are shown, where, the position estimation errors of KF and NKF are shown in (a) by the lines of real-blue and dotted-red, respectively. It can be seen from the figure that the errors of KF are slightly less than that of NKF. And (b) shows the position estimation errors of SFKF and NSFKF in blue real lines and red dotted lines, respectively. Also, we can see that the errors of SFKF are less than that of NSFKF.

Briefly, the simulation results in this section illustrate the effectiveness of the presented algorithm.

5 Conclusion

When the sampling rates of different sensors are different and when the measurement noises are cross-correlated and are also coupled with the system noise of the previous step, by use of the projection theory and induction hypothesis repeatedly, a sequential fusion algorithm is generated. The algorithm is proven to be optimal in the sense of Linear Minimum Mean Square Error (LMMSE) mathematically and is applicable to more general cases compared to the existed algorithms.

Acknowledgments The corresponding author of this article is Liping Yan, whose work was supported by the NSFC under grants 61004139 and 91120003, the Scientific research base support, and the outstanding youth foundation of Beijing Institute of Technology. The work of Yuanqing Xia and Mengyin Fu was supported by the NSFC under grants 60974011 and 60904086, respectively. The work of Bo Xiao was supported by Beijing Natural Science Foundation under Grant 4123102, and the innovation youth foundation of Beijing University of Posts and Telecommunications.

References

1. Li XR, Zhu YM, Wang J, Han C (2003) Optimal linear estimation fusion-Part 1:unified fusion rules. IEEE Trans Inf Theory 49:2192–2208
2. Bar-Shalom Y (1990) Multitarget-multisensor tracking: advanced applications, vol 1. Artech House, Norwood
3. Chong CY, Chang KC, Mori S (1986) Distributed tracking in distributed sensor networks. In: 1986 American control conference, Seattle, WA, pp 1863–1868
4. Hashemipour HR, Roy S, Laub AJ (1988) Decentralized structures for parallel Kalman filtering. IEEE Trans Autom Control 3(1):88–93
5. Song E, Zhu Y, Zhou J, You Z (2007) Optimal Kalman filtering fusion with cross-correlated sensor noises. Automatica 43:1450–1456
6. Duan Z, Li XR (2008) The optimality of a class of distributed estimation fusion algorithm. IEEE Inf Fusion 16:1–6
7. Xiao CY, Ma J, Sun SL (2011) Design of information fusion filter for a class of multi-sensor asynchronous sampling systems. In: Control and decision conference, pp 1081–1084
8. Yan LP, Zhou DH, Fu MY, Xia YQ (2010) State estimation for asynchronous multirate multisensor dynamic systems with missing measurements. IET Signal Process 4(6):728–739
9. Shi H, Yan L, Liu B, Zhu J (2008) A sequential asynchronous multirate multisensor data fusion algorithm for state estimation. Chin J Electron 17:630–632

Task Based System Load Balancing Approach in Cloud Environments

Fahimeh Ramezani, Jie Lu and Farookh Hussain

Abstract Live virtual machine (VM) migration is a technique for transferring an active VM from one physical host to another without disrupting the VM. This technique has been proposed to reduce the downtime for migrated overload VMs. As VMs migration takes much more times and cost in comparison with tasks migration, this study develops a novel approach to confront with the problem of overload VM and achieving system load balancing, by assigning the arrival task to another similar VM in a cloud environment. In addition, we propose a multi-objective optimization model to migrate these tasks to a new VM host applying multi-objective genetic algorithm (MOGA). In the proposed approach, there is no need to pause VM during migration time. In addition, as contrast to tasks migration, VM live migration takes longer to complete and needs more idle capacity in host physical machine (PM), the proposed approach will significantly reduce time, downtime memory, and cost consumption.

Keywords Cloud computing · Multi-objective genetic algorithm · Virtual machine migration · Task based system load balancing algorithm

F. Ramezani (✉) · J. Lu · F. Hussain
Decision Systems and e-Service Intelligence Laboratory, Faculty of Engineering and IT, School of Software, Centre for QCIS, University of Technology, Sydney, Australia
e-mail: Fahimeh.Ramezani@students.uts.edu.au

J. Lu
e-mail: Jie.Lu@uts.edu.au

F. Hussain
e-mail: Farookh.Hussain@uts.edu.au

F. Sun et al. (eds.), *Knowledge Engineering and Management*, Advances in Intelligent Systems and Computing 214,
DOI: 10.1007/978-3-642-37832-4_4,

1 Introduction

Cloud computing provides new business opportunities for both service providers and requestors clients (e.g., organizations, enterprises, and end users), by means of a platform and delivery model for delivering Infrastructure as a Service (IaaS), Platform as a Service (PaaS), and Software as a Service (SaaS). A cloud encloses the IaaS, PaaS, and/or SaaS inside its own virtualized infrastructure, in order to carry out an abstraction from its underlying physical assets. Typically, the virtualization of a service implies the aggregation of several proprietary processes collected in a virtual environment, called Virtual Machine (VM) [1, 2].

Often, clouds are also spread over distributed virtualization infrastructure covering larger geographical areas (example, let us think about Amazon ('Amazon Elastic Compute Cloud [24]'), Azure [25], and RESERVOIR (an European project facing the cloud computing IaaS topic [3]). In addition, the perspective of cloud federation [4, 5], where cloud providers use virtualized infrastructures of other federated clouds, opens toward new scenarios in which more and more types of new services can be supplied. In fact, clouds exploiting distributed virtualization infrastructures are able to provide new types of "Distributed IaaS, PaaS, and SaaS" [2].

Cloud computing platform using virtualization technology for resource management, achieves dynamic balance between the servers. Using online VM migration technology [6] can online achieve the remapping of VMs and physical resources, and dynamic achieve the whole system load balancing [7]. In modern data center (DC) or cloud environment, virtualization is a critical element since using virtualization the resources can be easily consolidated, partitioned, and isolated. In particular, VM migration has been applied for flexible resource allocation or reallocation, by moving VM from one physical machine to another for stronger computation power, larger memory, fast communication capability, or energy savings [8].

Although a significant amount of research has been done to achieve the whole system load balancing ([6–9], etc.), more improvement is still needed as most of these approaches tried to migrate VMs, when they became overloaded. As VMs migration takes much more times and cost in comparison with tasks migration, we believe migrating tasks from overloaded VMs instead of migrating overloaded VMs, will significantly reduce transfer time and total cost. In addition, to migrate VMs, we have to find a new PM which can accommodate the VM being migrated, and we can rarely avoid choosing an idle PM to optimize power consumption. But in task migration, we just need to find another VM which is located on an active PM and has the same features and number of CPU and more capacity just for executing a task.

Considering these facts and lack of resources in this area, we developed a novel Task Based System Load Balancing (TBSLB) approach to achieve system load balancing and confront with the lack of capacity for executing new task in one VM, by assigning the task to another homogeneous VM in cloud environment. In addition, we proposed an algorithm to solve the problem of migrating these tasks to new VM host which is a multi-objective problem subject to minimizing cost, minimizing execution time, and transferring time. To solve this problem, we applied multi-objective genetic algorithm (MOGA).

The rest of this paper is organized as follows. In Sect. 2 related works about VM migration are described. In Sect. 3 we propose a conceptual model and the algorithm of TBSLB approach for solving the problem of overloaded VMS by optimal tasks migration from overloaded VMs. The MOGA algorithm is described in Sect. 4. Our developed algorithm for solving multi-objective tasks scheduling problem and completing TBSLB algorithm, is described in Sect. 5. The proposed approach is evaluated in Sect. 6. Finally, we present our conclusion and future work in Sect. 7.

2 Related Works for VM Migration

Virtualization has delivered significant benefits for cloud computing by enabling VM migration to improve utilization, balance load, and alleviate hotspots [10]. Several mechanisms have been proposed to migrate a running instance of a VM (a guest operating system) from one physical host to another to optimize cloud utilization.

VM migration is a hot topic of computing system virtualization. Primary migration relies on process suspend and resume. Many systems [11–13] just pause the VM and copy the state data, then resume the VM on the destination host. This forces the migrated application to stop until all the memory states have been transferred to the migration destination where it is resumed. These methods cause the application to become unavailable during the migration process. ZAP [14] could achieve lower downtime of the service by just transferring a process group, but it still uses stop-and-copy strategy. To reduce the migration downtime and move the VM between hosts in local area network without disrupting it, VMotion [15] and Xen [6] utilize precopy migration technique to perform live migration and support seamless process transfer. Based on their works, [16] tried migrating running VM on a wide area network [8].

In precopy migration technique, VMs migrate by precopying the generated run-time memory state files from the original host to the migration destination host. If the rate for such a dirty memory generation is high, it may take a long time to accomplish live migration because a large amount of data needs to be transferred. In extreme cases, when dirty memory generation rate is faster than precopy speed, live migration will fail. Considering this fact, [8] presented the basic precopy model of VM live migration and proposed an optimized algorithm to improve the performance of live migration by limiting the speed of changing memory through controlling the CPU scheduler of the VM monitor [8].

[7] designed an IPv6 live migration framework for VM based on IPv6 network environment [7]. The framework has been used IPv6 VM live migration in

different IPv6 network. They designed a global control engine as a core complete IPv6 live migration for VM, and provided IPv6 cloud computing service for IPv4/ IPv6 client. In their approach, during VM migration process, the source VM would continue to offer services, and the source VM is still not stopped, but no longer provides new services until the old service completion then stop the source VM.

Since power is one of the major limiting factors for a DC or for large cluster growth, [9] proposed a runtime VM mapping framework in a cluster or DC to save energy [9]. Their placement module focused on reducing the power consumption. The main point of their approach is how to map VMs onto a small set of PMs without significant system performance degradation. Actually, they tried to turn off the redundant nodes or PMs to save energy, while the remaining active nodes guarantee the system performance. In their GreenMap framework, one probabilistic, heuristic algorithm is designed employing the idea from simulated annealing (SA) optimization, for the optimization problem: mapping VMs onto a set of PMs under the constraint of multi-dimensional resource consumptions.

[17] believe that most of the proposed methods for on-demand resource provisioning and allocation, focused on the optimization of allocating physical resources to their associated virtual resources, and migrating VMs to achieve load balance and increase resource utilization. Unfortunately, these methods require the suspension of the executing cloud computing applications due to the mandatory shutdown of the associated VMs [17]. To overcome this drawback, they proposed a threshold-based dynamic resource allocation scheme for cloud computing that dynamically allocates the VMs among the cloud computing applications based on their load changes. In their proposed method, they determined when migration should be done but they did not specify the details of how the reallocation will occur.

A fundamental shortcoming of the most existing research is that they consider complete VM migration to overcome overload VM and achieve system load balance. To improve previous approaches and reduce time and cost consumption in such situation, we proposed a new TBSLB approach which migrate tasks from overloads VMs instead of whole VM migration, and to decrease power consumption, a set of VMs on active PMs will be chosen as a new tasks' host. In addition, our approach not only eliminates the process suspend and resume which will happen in VM migration, but also omits precopy mechanism and producing dirty memory in live VM migration.

3 A Conceptual Model and Main Algorithm for Task Based System Load Balancing

In this section, we describe proposed TBSLB approach. This approach contains a conceptual model and TBSLB algorithm which are designed to achieve whole system load balancing by migrating tasks from overloaded VMs. In this approach to decrease energy consumption and costs, we avoid choosing idle PMs or Computer Nodes (CNs) as a new PM host, because if we transfer tasks to an idle PM, we have to turn it on and this action will increase energy consumption and costs [9].

As cloud computing has the advantages of delivering a flexible, high-performance, pay-as-you-go, on-demand offering service over the Internet, common users and scientists can use cloud computing to solve computationally complex problems (complex applications). The complex applications can be divided into two classes. The one is computing intensive, the other is data intensive. For transferring data intensive applications, the scheduling strategy should decrease the data movement which means decreases the transferring time; but for transferring computing intensive tasks, the scheduling strategy should schedule the data to the high-performance computer [18]. In this paper, we consider bandwidth as a variable to minimize the tasks transferring time for data intensive applications. In addition to enhance performance utilization for computing intensive applications, we consider new host PM's properties (memory, hard disk, etc.).

In cloud environment, there are some tasks schedulers that consider task types, priorities, and their dependencies to schedule tasks in optimal way considering their specific VM's resources. In our proposed system, we design a schedulers' blackboard, where all cloud schedulers (which manage VMs on clouds (see Fig. 1)) share their information about VMs, their features, and their tasks. We apply the information of this blackboard to find an appropriate host VM for the task. Furthermore, the criteria of QoS as SLA information are mentioned in this blackboard.

Every VM has already some tasks to execute and they have limited workload. To determine the time of tasks migration from an overloaded VM, we have to determine online remained workload capacity of a VM (VMs workload information), we defined it as:

$$\mathrm{VM}_{\mathrm{rw}} = \mathrm{VM}_{w} - \mathrm{VM}_{\mathrm{et}} \tag{1}$$

where VM_w is VM workload, and $\mathrm{VM}_{\mathrm{et}}$ is the number of executing tasks in VM. The VM will be overload and arrival tasks should be migrated to another similar VM to execute, when:

$$\mathrm{VM}_{\mathrm{rw}} \leq 1 \tag{2}$$

Considering all these facts, we propose a novel TBSLB algorithm which prepares another scheduler to transfer tasks from an overhead VM to a new similar and appropriate VM according to following steps:

Step 1 Gathering data and information about VMMs, VMs, PMs, and SLA information, in the global blackboard as ***inputs*** of TBSLB algorithm as follow:

1. VMs tasks information:

    1.1 The number of executing tasks
    1.2 Tasks' execution time
    1.3 Tasks' performance model

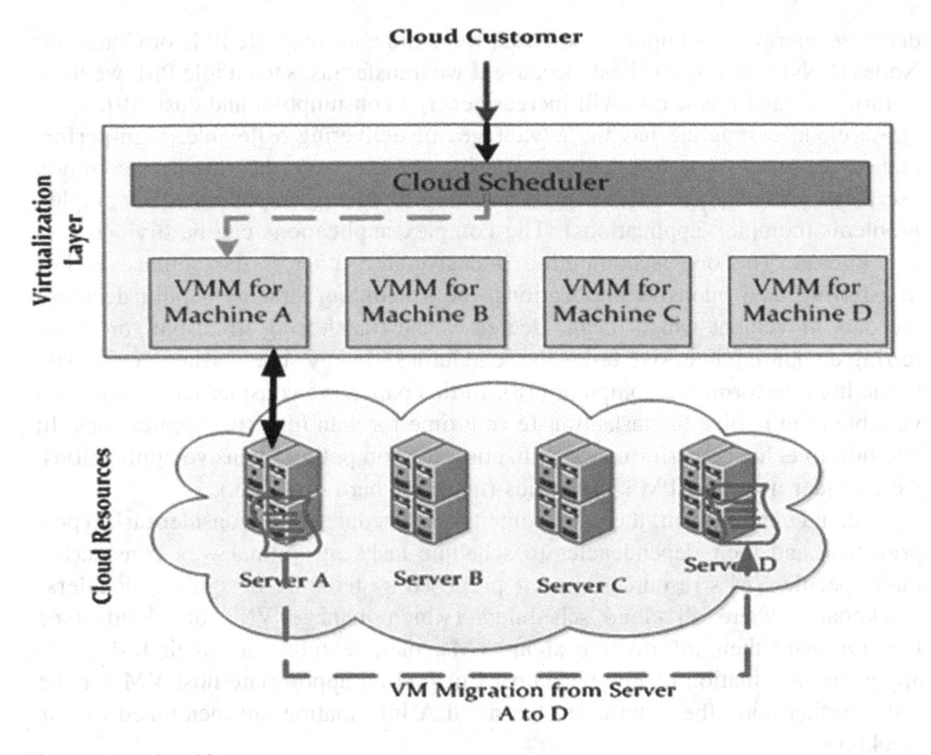

Fig. 1 Cloud architecture

1.4 Tasks' locations
1.5 Tasks' required resources (number of required processors)

2. PMs' Criteria (total/current)

 2.1 CPU (number and speed of the processors)
 2.2 Free Memory and Hard disk
 2.3 Bandwidth
 2.4 Idle or active
 2.5 Its host VMM

3. SLA information
4. The objectives of the tasks migration optimization model and their information:

 4.1 Minimizing cost

  4.1.1 Cost information

 4.2 Minimizing execution time and transferring time

  4.2.1 Execution information
  4.2.2 Bandwidth information

Step 2 Monitoring data and information to determine VMs' workflow situation and determining:

1 VMs workload information
2 Overloaded VMs
3 The tasks which should be migrated from overloaded VM's
4 Migration time

Step 3 Finding optimal homogeneous VMs as a new host for executing the tasks of the overloaded VMs, which is a multi-objective task migration problem, applying MOGA (this step will be described in Sect. 5).

Step 4 Considering obtained optimal tasks migration schema, determining following information as the *outputs* of TBSLB algorithm:

1 New optimal cost
2 New optimal execution time
3 Current VMs properties (Executing tasks, CPU, etc.)

Step 5 Transferring tasks and their corresponding data to the optimal host VMs

Step 6 Updating blackboards and schedulers' information according to the outputs of Step 4.

Step 7 End.

The conceptual model of the proposed approach is summarized in Fig. 2.

4 A Multi Objective Genetic Algorithm

In Step 3 of the algorithm, a multi-objective problem which is described in Sect. 5, should be solved to optimize tasks migration from overload VM and find the best VMs as new tasks hosts. In multi-objective optimization problems, each objective function interacts on each other; they almost cannot be optimal at the same time. In other words, one objective function optimization often means a bad developing direction of other objective functions. Therefore, a compromise strategy can be used among the objective functions so as to make them reach optimization at the same time. Now the most popular MOGAs abroad are Corne's PESA2 [19] and PAES [20], SPEA2 [21] proposed by Ziltler, Deb's NSGAII, and so on. Among them, Deb's NSGAII not only has good convergence and distribution but also has higher convergence speed, and solve the shortcomings that shared parameters are difficult to determine [22].

According to MOGA, first initial population whose scale is N is generated randomly. The first generation child population is gained through non-dominated sorting [23] and basic operations such as selection, crossover and mutation. Then, from the second generation on, the parent population and the child population will be merged and sort them based on fast non-dominated. Calculate crowding distance among individuals on each non-dominated layer. According to non-dominant relationship and crowding distance among individuals, select the appropriate

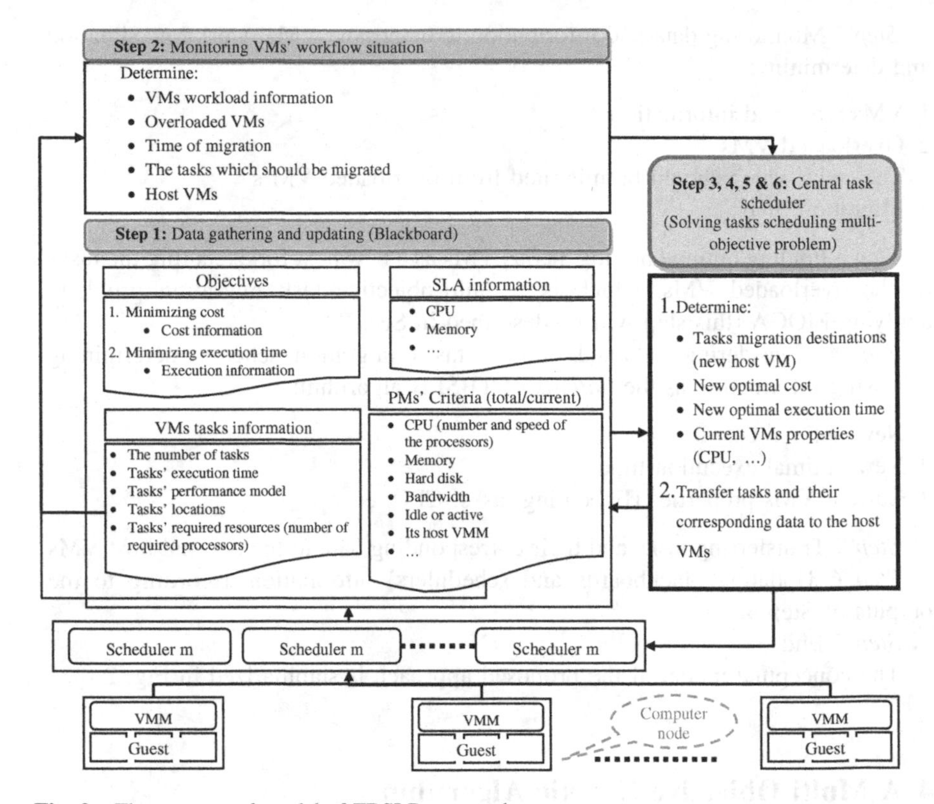

Fig. 2 The conceptual model of TBSLB approach

individuals to form a new parent population. Finally, new child population is generated through basic operations of genetic algorithm. And so on, until the conditions of the process end can be met [22].

5 An Algorithm for Solving Multi-Objective Tasks Migration Problem Using MOGA

In this section, we will describe the sub-TBSLB algorithm which is developed to complete the Step 3 of TBSLB algorithm and solve Multi-objective tasks migration problem. This sub-algorithm will determine the most appropriate VMs to assign the tasks of the overloaded VMs and find optimal tasks scheduling model applying MOGA.

This sub-algorithm applies data and information which are determined in Step 1 and Step 2 of the TBSLB algorithm as its ***inputs***. In this algorithm, we first eliminate those VMs which do not satisfy all constraints to reduce the population of the candidate solution set. Then we apply MOGA method to find the optimal solution.

According to sub-TBSLB algorithm, to find optimal host VMs to assign new overloaded VMs' tasks, following steps should be conducted:

Step 3.1 Determining candidate host VMs set by choosing the set of VMs which satisfy the constraints about host VMs' properties as $VM_{set} = \{vm_1, \ldots, vm_m\}$

Step 3.2 Determining overloaded VMs applying Eq. 2, and eliminating them from candidate host VMs set.

Step 3.3 Determining the set of tasks which should migrate from overloaded VMs as immigrating tasks set: $T_{set} = \{t_1, \ldots, t_n\}$

Step 3.4 Applying MOGA to solve multi-objective problem and assign the immigrating tasks to the optimal host VMs minimizing execution time, transferring time and processing cost. To achieve this goal, following steps should be conducted:

Step 3.4.1 Initializing population P_0 which is generated randomly

Step 3.4.2 Assigning rank to each individual based on non-dominated sort

Step 3.4.3 Implementing binary tournament selection, crossover and mutation on the initial population and creating a new population Q_0 and set $t = 0$

Step 3.4.4 Merging the parent P_t and the child Q_t to form a new population $R_t = P_t \cup Q_t$

Step 3.4.5 Adopting non-dominated relationship to sort population and calculate the crowding distance among population on each layer

Step 3.4.6 Selecting the former N individuals as the parent population, namely $P_{t+1} = P_{t+1}$ [1: N] (Elite strategy)

Step 3.4.7 Implementing reproduction, crossover and mutation on population P_{t+1} to form population Q_{t+1}

Step 3.4.8 If the termination conditions are met, *output* results as optimal tasks migration schema; otherwise, update the evolutionary algebra counter $t = t{+}1$ and go to step 3.4.4

Step 3.5 End.

6 Evaluation

We determine two parameters to evaluate our proposed TBSLB approach and compare it with traditional whole VM migration methods. The first parameter is related to "power consumption". As, the less number of active PM means the less power consumption [9], we applied following ratio to compare power consumption after load balancing:

$$R_{pc} = \frac{\text{Number of active PM}}{\text{Number of overloaded VMs}} \tag{3}$$

In proposed approach, to execute some tasks of overloaded VM, we need to find a new similar VM on an active PM as a new host and there will be no need to turn a new PM on. In contrast, for whole VM migration, more hardware capacity will be needed and it is impossible for every case to avoid choosing idle PM. So, in our approach, R_{pc} should be less than whole VM migration attitude for load balancing. Therefore, we will have less "power consumption" after load balancing and:

$$R_{\mathrm{pc}_{\text{Offline VM}}} \geq R_{\mathrm{pc}_{\text{Online VM}}} > R_{\mathrm{pc}_{\text{NewApproach}}}$$

To compare the efficiency of TBSLB approach, we applied "downtime VM pause time)" as second parameter and estimate the amount of "idle memory" which is prepared during the time of solving the problem of overloaded VM as:

$$M_{\mathrm{im}}(t) = \mathrm{OriginalVM_m}(t) + \mathrm{HostVM_m}(t) \tag{4}$$

where $OriginalVM_m$ and $HostVM_m$ are the amount of original VM memory and host VM, respectively.

In offline VMs migration method, during VM migration time, the original VM should be suspend and its memory and the amount of memory in new host PM which is determined for host VM will be idle. In online VMs migration method, although VM will not be suspended during migration process, the amount of memory in new host PM will be idle in this time. Meanwhile, in TBSLB approach, we eliminate process of suspend and resume in primary VM migration which is mentioned in [8] and there will be no downtime for VMs and no idle memory. As the results:

$$M_{\mathrm{im}}(t)_{\text{Offline VM}} > M_{\mathrm{im}}(t)_{\text{Online VM}} > M_{\mathrm{im}}(t)_{\text{NewApproach}}$$

7 Conclusion and Future Work

VM migration has been applied for flexible resource allocation or reallocation, by moving overload VM from one PM to another to achieve stronger computation power, larger memory, fast communication capability, or energy savings.

This paper proposed a new TBSLB approach to confront with the problem of overload VM by migrating arrival tasks to another homogeneous VM. This algorithm contains multi-objective tasks migration model subject to minimizing cost, execution time, and transferring time. In proposed approach, there is no need to pause VM during migration time. In addition, as contrast to tasks migration, VM live migration takes longer to complete and needs more idle capacity in host PM, the proposed approach will significantly reduce time, downtime memory, and cost consumption. Furthermore, proposed approach will decrease energy consumption by avoiding choosing idle PMs or CNs as a new host PM. In our future work, we will propose a method to predict the time of task migration from an overload VM to accelerate load balancing process in our proposed approach.

References

1. Buyya R, Broberg J, Goscinski A (eds) (2011) Cloud computing: principles and paradigms
2. Celesti A, Fazio M, Villari M, Puliafito A (2012) (VM) provisioning through satellite communications in federated cloud environments. Future Gener Comput Syst 28(1):85–93
3. Rochwerger B, Breitgand D, Epstein A, Hadas D, Loy I, Nagin K, Tordsson J, Ragusa C, Villari M, Clayman S (2011) Reservoir-when one cloud is not enough. Comput 44(3):44–51
4. Goiri I, Guitart J, Torres J (2010) Characterizing cloud federation for enhancing providers' profit. In: IEEE 3rd international conference on cloud computing (CLOUD), pp 123–130
5. Ranjan R, Buyya R (2008): Decentralized overlay for federation of enterprise clouds, Arxiv preprint arXiv:0811.2563
6. Clark C, Fraser K, Hand S, Jacob GH (2005) Live migration of (VM)s. In: Proceedings of 2nd ACM/USENIX symposium on network systems, design and implementation (NSDI)
7. Jun C, xiaowei C(2011): IPv6 (VM) live migration framework for cloud computing, Energy procedia, vol 13(0):5753–5757
8. Jin H, Gao W, Wu S, Shi X, Wu X, Zhou F (2011) Optimizing the live migration of (VM) by CPU scheduling'. J of Netw and Comput Appl 34(4):1088–1096
9. Liao X, Jin H, Liu H (2012) Towards a green cluster through dynamic remapping of (VM)s. Future Gener Comput Syst 28(2):469–477
10. Jain N, Menache I, Naor J,.Shepherd F(2012) Topology-aware VM migration in bandwidth oversubscribed datacenter networks, automata, languages, and programming, pp 586–597
11. Kozuch M, Satyanarayanan M (2002) Internet suspend/resume, Mobile computing systems and applications. Proceedings fourth IEEE workshop on, pp 40–46
12. Sapuntzakis CP, Chandra R, Pfaff B, Chow J, Lam MS, Rosenblum M (2002) Optimizing the migration of virtual computers. ACM SIGOPS operating systems review, vol 36, no. SI, pp 377–390
13. Whitaker A, Cox RS, Shaw M, Gribble SD(2004) Constructing services with interposable virtual hardware. In: Proceedings of the 1st symposium on networked systems design and implementation (NSDI), pp 169–182
14. Osman S, Subhraveti D, Su G, Nieh J (2002): The design and implementation of Zap: A system for migrating computing environments, ACM SIGOPS Operating systems review, vol 36, no. SI, pp 361–376
15. Nelson M, Lim BH, Hutchins G (2005) Fast transparent migration for (VM)s, pp 25–25
16. Travostino F, Daspit P, Gommans L, Jog C, De Laat C, Mambretti J, Monga I, Van Oudenaarde B, Raghunath S, Yonghui Wang P (2006) Seamless live migration of (VM)s over the MAN/WAN. Future Gener Comput Syst 22(8):901–907
17. Lin W, Wang JZ, Liang C, Qi D (2011) A threshold-based dynamic resource allocation scheme for cloud computing. Proced Eng 23:695–703
18. Guo L, Zhao S, Shen S, Jiang C (2012) Task scheduling optimization in cloud computing based on heuristic algorithm. J Netw 7(3):547–553
19. Corne DW, Jerram NR, Knowles JD, Oates MJ (2001) PESA-II: Region-based selection in evolutionary multiobjective optimization, Citeseer
20. Knowles JD, Corne DW (2000) Approximating the nondominated front using the Pareto archived evolution strategy. Evolut compu 8(2):149–172
21. Zitzler E, Laumanns M, Thiele L(2001) SPEA2: Improving the strength pareto evolutionary algorithm

22. Zhang Y, Lu C, Zhang H, Han J(2011) Active vibration isolation system integrated optimization based on multi-objective genetic algorithm, computing, control and industrial engineering (CCIE), IEEE 2nd international conference on, vol 1, pp. 258–261
23. Srinivas N, Deb K (1994) Muiltiobjective optimization using nondominated sorting in genetic algorithms. Evol Comput 2(3):221–248
24. Amazon Elastic Compute Cloud (Amazon EC2), http://aws.amazon.com/ec2/
25. Azure: Microsoft's service Cloud platform, http://www.microsoft.com/windowsazure

Accurate Computation of Fingerprint Intrinsic Images with PDE-Based Regularization

Mingyan Li and Xiaoguang Chen

Abstract The intrinsic images of fingerprint, such as orientation field and frequency map, represent the particular and basic characteristics of fingerprint ridge/valley patterns, and play a key role in feature extraction and matching of the fingerprint recognition system. In this paper, a novel algorithm is presented for accurate computation of fingerprint intrinsic images with PDE-based regularization technique. First, the coarse orientation field is estimated using general gradient-based method, and the frequency map is estimated by Fourier spectrum analysis of the projected curve of local window. Then, to measure the reliability of the intrinsic images, the quality map is computed based on the gray-scale intensity and local structural information. Finally, accurate orientation field and frequency map are reconstructed with PDE-based regularization of nonlinear diffusion filtering which is controlled by the quality-based diffusivity. Experimental results illustrated the efficiency of the proposed approach.

Keywords Fingerprint recognition · Orientation field · Frequency map · Image quality measurement · Nonlinear diffusion filtering

1 Introduction

In fingerprint recognition systems, efficient algorithms of fingerprint feature extraction and matching play most important roles [1, 2]. Due to the various reasons, the captured fingerprint images are often degraded and have blurred and

M. Li (✉)
College of Sciences, Beijing Forestry University, 100083 Beijing, China
e-mail: limingyan55@sina.com

X. Chen
Institute of Image Processing and Pattern Recognition, College of Sciences,
North China University of Technology, 100144 Beijing, China
e-mail: chenxg1018@sina.com

F. Sun et al. (eds.), *Knowledge Engineering and Management*,
Advances in Intelligent Systems and Computing 214,
DOI: 10.1007/978-3-642-37832-4_5, © Springer-Verlag Berlin Heidelberg 2014

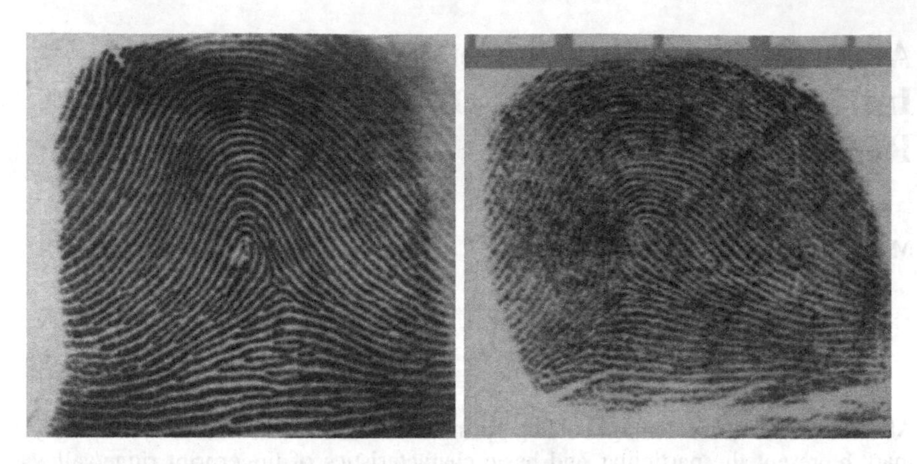

Fig. 1 Two sample fingerprint images with fair (*left*) and bad (*right*) quality. The two images are "F0009-08" and "F0022-05" in NIST SD4 [4] respectively

broken ridge structures which decrease the system performance dramatically [1]. To extract reliable features such as minutiae, enhancement algorithms usually are performed to improve the image quality. Most of the existing enhancement algorithms employ some intrinsic characteristics of fingerprint such as ridge orientation and frequency which help to control or adjust the behavior of the designed filters [3].

The most evident structural characteristic of a fingerprint is a pattern of interleaved ridges and valleys, as shown in Fig. 1. The orientation of the ridge pattern flow and the period width of the ridge/valley structure are the critical intrinsic characteristics. For fingerprint images with fair quality, such as the left image in Fig. 1, the intrinsic characteristics can be accurately computed as for most algorithms. However, it remains difficult for low-quality fingerprint images, such as the right image in Fig. 1, because of the heavy noise and serious blurring.

In this paper, we propose a novel approach for computing accurate orientation field and frequency map by utilizing a PDE-based regularization. First, coarse orientation filed is computed by estimating the dominant direction in each local window, and ridge frequency map is computed in Fourier domain by estimating the dominant frequency of the projected curve along the perpendicular direction of the ridge. Then, local quality map of the fingerprint is computed, and it represents the reliability of the computed intrinsic characteristics of the associated local ridge structure. To achieve more accurate estimation, nonlinear diffusion filtering is employed to regularize the coarse orientation and frequency map, and the diffusion process is controlled by the diffusivity which is decided by the local quality score of the fingerprint image. Thus, the accuracy of the intrinsic images is greatly improved after the quality-controlled diffusion process.

The rest of this paper is organized as follows. Section 2 describes the algorithms for computing coarse intrinsic images including orientation field, frequency map, quality map, and segmentation map. In Sect. 3, a PDE-based regularization algorithm is presented to compute accurate orientation field and frequency map by nonlinear diffusion. Experimental results are presented in Sect. 4. Finally, the paper is concluded with a summary in Sect. 5.

2 Coarse Intrinsic Images Computation

Fingerprint intrinsic images, such as orientation filed and frequency map, represent the essential characteristic of digital image and reflect the main feature of the ridge patterns. In this work, the quality map and segmentation map are also regarded as intrinsic images, since the former indicates the clarity of the ridge patterns and the later indicates the valid region indeed. Therefore, the intrinsic images computed in this paper consist of orientation filed, frequency map, quality map, and segmentation map. It is noted that the mean and variance of the gray-scale intensity of the input fingerprint image are firstly normalized by using the method in [4] before the computation of intrinsic images.

In proposed approach, the input fingerprint image I is divided into nonoverlapped blocks with size $b \times b$. Then the intrinsic characteristics, such as orientation, frequency, quality score, are computed for each block, and the value is assigned to each pixel in the corresponding block. To trade off between computational accuracy and complexity, the block size b is chosen empirically according to the image size. If the image size is small, the value of b even can be set to 1 to perform the computation pixel by pixel.

2.1 Orientation Field

The ridge orientation plays an important role in fingerprint image processing and matching. In the literatures, many approaches have been proposed for estimating the orientation field. In our approach, we adopt the Least-Mean-Square method [3] to coarsely estimate the ridge orientation at each pixel.

For each block centered at pixel (x, y), the orientation of the block is defined as the local dominant ridge orientation of the local window $\mathcal{N}_w(x, y)$ centered at pixel (x, y) with size $w \times w$. The orientation $O(x, y)$ is computed as follows,

$$O(x,y) = \frac{\pi}{2} + \frac{1}{2}\angle\big(S_y(x,y), S_x(x,y)\big), \tag{1}$$

$$S_x(x,y) = \sum_{(i,j)\in\mathcal{N}_w(x,y)} 2I_x(i,j)I_y(i,j), \tag{2}$$

$$S_y(x,y) = \sum_{(i,j)\in\mathcal{N}_w(x,y)} \left(I_x^2(i,j) - I_y^2(i,j)\right), \tag{3}$$

where I_x and I_y are the x- and y-component of the image gradient computed by Sobel operators. Here, the size of neighborhood region w is chosen to satisfy $w > b$ which allows a continuous omputation among the adjacent overlapped blocks.

2.2 Frequency Map

The ridge frequency (or period, or width) also is one of the intrinsic characteristics of fingerprint. At the region without minutiae or singular points, the gray intensities along the direction at the alternating ridge and valley forms a curve like sinusoidal wave. Based on this observation, we compute the ridge frequency based on the Fourier spectrum of the projected curve of local oriented window as shown in Fig. 2.

For the block centered at pixel (x, y), a local oriented window with size $L \times w$ centered at pixel (x, y) is defined according to the local ridge orientation as shown in Fig. 2. The pixel intensities in the local window are projected along the ridge orientation to to form a curve. Unlike the method in [4] which is based on the analysis of the peaks of the curve, our approach estimates the ridge frequency based on the Fourier spectrum analysis of the curve. The dominant frequency $F(x, y)$ is calculated by the following formula,

$$F(x,y) = \sum_{k=1}^{L/2} \left(S[k] \frac{k}{L} \right) / \sum_{k=1}^{L/2} S[k], \tag{4}$$

where $S[k]$ is the Fourier spectrum. It is known that the period T of the ridge/valley structure is given by $1/F(x, y)$.

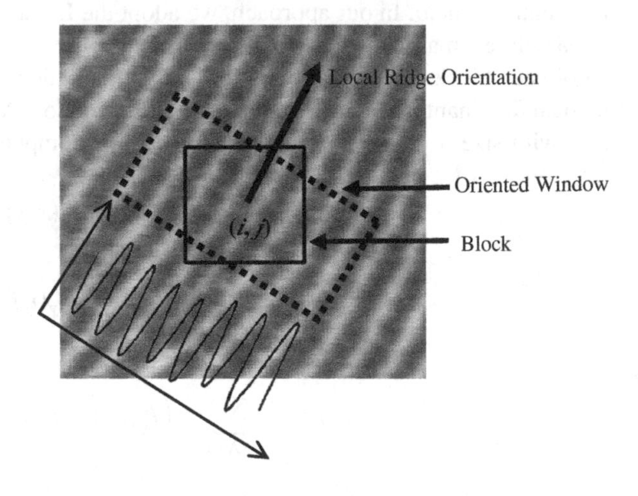

Fig. 2 Ridge frequency estimation based on Fourier spectrum analysis of the projected curve of local oriented window

2.3 Quality Map and Segmentation Map

To measure the accuracy of the orientation and frequency computed in the previous subsections, a quality score for each block is computed based on special features of the local window. Four features are used and there are the range, variance, uniformity of the pixel intensities, and the Fourier spectrum energy.

Range and Variance of Pixel Intensity. The range reflects the difference of the pixel intensity in the window. For the window containing clear ridge/valley pattern, the difference should be large; however, for the noisy or blurred regions, the difference will have smaller range of the pixel intensity. The range is computed as follows,

$$range(x,y) = \overline{max}(I,x,y,w) - \overline{min}(I,x,y,w), \tag{5}$$

where $\overline{max}$ and $\overline{min}$ are the mean value of the αw^2 largest and the αw^2 smallest intensities in the local window $\mathcal{N}_w(x,y)$ respectively. In our experiments, the parameter α is 0.15.

The variance is computed as follows,

$$var(x,y) = \frac{1}{w^2} \sum_{(i,j)\in\mathcal{N}_w(x,y)} (I(i,j) - mean(x,y))^2, \tag{6}$$

where $mean(x,y)$ is the mean value of the pixel intensities in the local window.

Uniformity of Pixel Intensity. The uniformity reflects the distribution on each gray-scale intervals of the pixel intensity . If the block is a normal region containing clear ridge/valley patterns, the pixel intensity should have a uniform distribution. The general gray-scale range [0, 255] is quantized into 32 intervals, and then the statistical histogram is calculated for the window. The uniformity is computed as follows,

$$unif(x,y) = \frac{1}{w^2} \sum_{i=1}^{32} (h[k])^2, \tag{7}$$

where $h[k]$ is the value at the kth bin of the histogram.

Fourier Spectrum Energy. The Fourier spectrum energy reflects the existence of ridge/valley pattern in certain degree, and this feature is computed together with the frequency estimation as described in previous subsection. The computation formula is given by

$$f_{se}(x,y) = \sum_{k=1}^{L/2} c(\|k - k_F\|)s[k], \tag{8}$$

where the coefficient $c(\cdot)$ is Gaussian-like weight function, and $k_F = [F(x,y)/L]$ is the array index corresponding to the dominant frequency.

After the four features are computed as above mentioned, the quality score of the block is computed as follows:

$$Q(x,y) = Q_r(x,y) \times Q_v(x,y) \times Q_u(x,y) \times Q_f(x,y), \tag{9}$$

where Q_r, Q_v, Q_u, Q_f are the quality factors corresponding to the four features respectively, and they are defined as follows,

$$\begin{aligned} Q_r &= 0.5 + 0.5\tanh((range - r_1)/r_2), \\ Q_v &= 0.5 + 0.5\tanh((var - v_1)/v_2), \\ Q_u &= \max\{0, 1 - unif\}, \\ Q_f &= 0.5 + 0.5\tanh((f_{se} - f_1)/f_2), \end{aligned}$$

where parameters $r_1, r_2, v_1, v_2, f_1, f_2$ are chosen empirically.

Based on the quality map Q, the segmentation map R is easily determined by a threshold as follows:

$$R(x,y) = \begin{cases} 1(\text{foreground}) & \text{if } Q(x,y) > T_Q \\ 0(\text{background}) & \text{otherwise,} \end{cases} \tag{10}$$

where T_Q is the threshold. In our experiment, $T_Q = 0.15$. Post-processing with as morphological operations *close* and *open* are performed to obtain a compact segmentation result.

3 PDE-based Regularization

Due to the continuity of ridge flow, the ridge orientation and frequency also vary continuously and smoothly. However, because of the low quality of the fingerprint image, the computed orientation field and frequency map are not accuracy, and thus the continuity and smoothness are often destroyed. To repair the intrinsic images, post-processing is usually carried out to achieve more accurate ones, and many approaches are proposed in the literatures for this purpose. In [3], low-pass filtering is performed among the adjacent blocks to reduce the noise. In [5], specific filters are designed to estimate the orientation and frequency parameters. Besides, some complicated orientation field models are developed to amend the orientation field based on the singular points, such as in the literatures [6–11]. Considering the effect, the former often fail in the low-quality region, and the latter endures the algorithm complexity and computational burden.

To reconstruct more accurate orientation field and frequency map, a PDE-based regularization approach is proposed which is based on the quality map and non-linear diffusion filtering. In our approach, the local quality score is regarded as the reliability of the computed ridge orientation and frequency, and it determines the diffusivity of the diffusion processing.

Since the orientation value is in $[0, \pi)$, there is ambiguity at the direction 0 and π. To eliminate the ambiguity, the original orientation field O is mapped to a two-dimensional vector field $\Phi = [\Phi_x, \Phi_y]$ by Eq. (11) before regularization, and then regularized Φ will be mapped back to the orientation field O by Eq. (12). Here, Φ is called *Squared Orientation Field (SOF)*.

$$O \mapsto \Phi: \qquad \Phi(x,y) = [\cos(2O(x,y)), \sin(2O(x,y))], \tag{11}$$

$$\Phi \mapsto O: \qquad O(x,y) = \frac{1}{2}\arctan\left(\Phi_y(x,y)/\Phi_x(x,y)\right). \tag{12}$$

For the SOF Φ and frequency map F, we propose to regularize them by the following nonlinear isotropic diffusion,

$$\frac{\partial \Phi(x,y)}{\partial t} = div(C(x,y)\nabla\Phi(x,y)), \tag{13}$$

$$\frac{\partial T(x,y)}{\partial t} = div(C(x,y)\nabla T(x,y)), \; T = 1/F, \tag{14}$$

where C is the diffusivity and it control the diffusion strength adaptively.

The local quality score Q indicates the reliability of the ridge information in the region. For low-quality region, stronger diffusion filtering is needed since the information is not reliable; otherwise not much regularization is needed. Therefore, we propose the following formula to map the quality score Q to diffusivity C,

$$C = C_{min} + (1 - C_{min})\left(1 - \tanh\left(\frac{Q - 0.5}{0.15}\right)\right), \tag{15}$$

where C_{min} is the minimum diffusivity to ensure that diffusion filtering will be performed in all valid region. Figure 3 shows the Q–C map function.

As to the numerical implementation of the two PDEs in Eqs. (13) and (14), the semi-implicit scheme with additive operator splitting (AOS) proposed by Weickert et al. [12] is employed in our approach. By using AOS, the computational complexity of solving the PDEs is reduced greatly, and reliable solution can be obtained.

4 Experimental Results

To evaluate the performance, the comparative experiment between proposed algorithm and the algorithm presented in [3] is carried out. A collection of typical fingerprint images from NIST SD4 [4], which covers a variety of different quality levels, is used as benchmark dataset. Two example images are illustrated in Fig. 1.

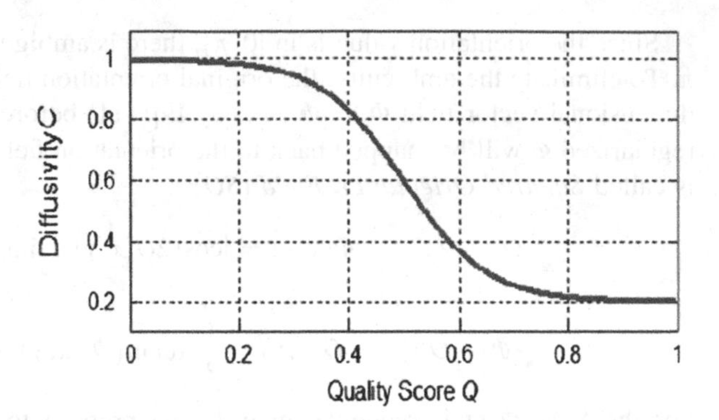

Fig. 3 The map function between quality score Q and diffusivity C. The minimum diffusivity $C_{min} = 0.2$

The results of intrinsic images computation for fingerprint image "F0009-08" in Fig. 1 are depicted in Fig. 4 visually. As seen from the mean/variance-normalized image (Fig. 4a) and estimated quality map (Fig. 4b), the central part has good quality with clear ridge patterns; on the contrary, the upper and lower parts are noisy. What is more, the ridge period in the image is not distributed uniformly, and has a large variety from the lower to the upper parts of the image. As a result, the coarse orientation field (Fig. 4e) and the coarse frequency map (Fig. 4g) do not have sufficient smoothness among different parts. After regularized by diffusion filtering with local quality-based diffusivity (Fig. 4d), continuous and smooth orientation field (Fig. 4f), and frequency map (Fig. 4h) are obtained, which lead to more accurate and robust enhancement processing for feature extraction.

The effectiveness of proposed algorithm is also verified quantitatively by the experiment of minutiae extraction. In this experiment, the intrinsic images of each fingerprint are firstly computed by proposed algorithm and the methods in reference [3], respectively. Then, the Gabor filter-based enhancement algorithm in [3] with different intrinsic images as algorithm parameters is used to improve the quality of fingerprints. Finally, minutiae are extracted from these enhanced fingerprint images. Here, three indicators are computed to describe the performance of minutiae extraction as follows:

$$r_{false} = (n_E - n_M)/n_E, \tag{16}$$

$$r_{miss} = (n_M - n_E)/n_M, \tag{17}$$

$$r_{true} = n_{M \cap E}/n_E, \tag{18}$$

where M and E represent the minutiae sets extracted by human experts and automatic algorithm respectively. n_M, n_E and $n_{M \cap E}$ represent the number of the minutiae in the corresponding sets respectively. By the definitions, r_{false} represents the ratio of spurious minutiae to the extracted minutiae. r_{miss} represents the ratio of missing minutiae to the genuine minutiae. r_{true} represents the ratio of correct

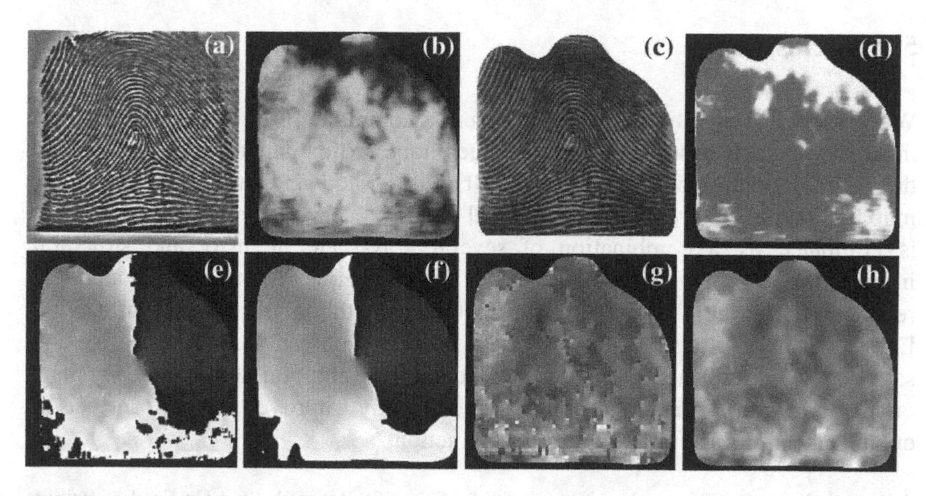

Fig. 4 Intrinsic images computation of the *left* image in Fig. 1. **a** Mean/variance-normalized. **b** Quality map. **c** Segmented fingerprint. **d** Diffusivity. **e** Coarse orientation field. **f** Regularized orientation field. **g** Coarse frequency map. **h** Regularized frequency map. All values displayed in *gray scale*, and frequency maps depict the inverse of the frequency

Table 1 Performance comparison of minutiae extraction

Image	Algorithm in [3]			Proposed algorithm		
	r_{false}	r_{missed}	r_{true}	r_{false}	r_{missed}	r_{true}
F0001_01	0.429	0.075	0.571	0.316	0.075	0.684
F0002_05	0.469	0.254	0.531	0.383	0.186	0.617
F0003_10	0.489	0.274	0.511	0.396	0.294	0.604
F0004_05	0.338	0.122	0.662	0.294	0.102	0.706
F0005_03	0.148	0.137	0.852	0.102	0.041	0.898
F0006_09	0.232	0.059	0.768	0.216	0.029	0.784
F0007_09	0.389	0.121	0.611	0.287	0.060	0.713
F0008_10	0.411	0.166	0.589	0.341	0.083	0.659
F0009_08	0.123	0.040	0.877	0.087	0.013	0.913
F0010_01	0.141	0.106	0.859	0.135	0.146	0.865
Average	0.317	0.135	0.683	**0.256**	**0.103**	**0.744**

minutiae to the extracted minutiae. The larger r_{true} and the smaller r_{false} and r_{miss}, the better the minutiae extraction results. Since the same enhancement algorithm and minutiae extraction algorithm are used, the better performance of minutiae extraction implies the better computation approach of intrinsic images. The performance comparison of minutiae extraction is given in Table 1. From the table, we can observe that the performance of minutiae extraction is substantially improved by using proposed approach.

5 Conclusion

We presented a novel algorithm for accurate computation of fingerprint intrinsic images with PDE-based regularization technique in this paper. First, we estimate the coarse orientation field by gradient-based method and the coarse frequency map by Fourier spectrum analysis of local projected curve. Then, local quality map is computed by a combination of several gray-scale intensity and structural information. Finally, the orientation field and frequency map are adaptively regularized by nonlinear diffusion filtering with quality-based diffusivity. Experimental results show that our proposed approach can compute accurate and satisfactory fingerprint intrinsic images including orientation field, frequency map, quality map and segmentation map, and efficiently improves the performance of enhancement algorithm and minutiae extraction.

Acknowledgments This work was supported by Scientific Research Start-Up Fund of Beijing Forestry University (Grant No. 2010BLX12), Beijing Outstanding Talents Project (No.2011D005002000003), and Beijing Municipal Education Committee Surface Technology Development Plan (No.KM201210009012).

References

1. Maltoni D, Maio D, Jain AK, Prabhakar S (2003) Handbook of fingerprint recognition. Springer, New York
2. Luo X, Tian J, Wu Y (2000) A minutia matching algorithm in fingerprint verification. In: Proceedings of 15th international conference on pattern recognition (ICPR'00), vol 4, pp 833–836
3. Hong L, Wan Y, Jain AK (1998) Fingerprint image enhancement: algorithm and performance evaluation. IEEE Trans Pattern Anal Mach Intell 20(8):777–789
4. NIST fingerprint databases. http://www.nist.gov/srd/biomet.htm
5. Gottschlich C (2012) Curved-region-based ridge frequency estimation and curved Gabor filters for fingerprint image enhancement. IEEE Trans Image Process 21(4):2220–2227
6. Sherlock BG, Monro D (1993) A model for interpreting fingerprint topology. Pattern Recogn 26(7):1047–1055
7. Vizcaya P, Gerhardt L (1996) A nonlinear orientation model for global description of fingerprints. Pattern Recogn 29(7):1221–1231
8. Gu J, Zhou J, Zhang D (2004) A combination model for orientation field of fingerprints. Pattern Recogn 37:543–553
9. Tao X, Yang X, Cao K, Wang R, Li P, Tian J (2010) Estimation of fingerprint orientation field by weighted 2D Fourier expansion model. In: Proceedings of 20th international conference on pattern recognition (ICPR'10), pp 1253–1256
10. Hou Z, Yau W-Y (2010) A variational formulation for fingerprint orientation modeling. In: Proceedings of 20th international conference on pattern recognition (ICPR'10), pp 1626–1629
11. Wang Y, Hu J (2011) Global ridge orientation modeling for partial fingerprint identification. IEEE Trans Pattern Anal Mach Intell 33(1):72–87
12. Weickert J, Romeny B, Viergever M (1998) Effcient and reliable schemes for nonlinear diffusion filtering. IEEE Trans Image Process 7(3):398–411

A Context Ontology Modeling and Uncertain Reasoning Approach Based on Certainty Factor for Context-Aware Computing

Ping Zhang and Hong Huang

Abstract Context-aware computing is an emerging intelligent computing model. Its core idea is to have computing devices understand the real world and automatically provide appropriate services without much concern from users. The research on context modeling and reasoning approaches is the important content of realizing context-aware computing. In this paper, an ontology modeling approach based on certainty factor is proposed in order to not only share and structurally represent context information but also to support the uncertain reasoning by the uncertainty representation for the domain knowledge. Then a certainty factor reasoning approach based on weights is presented to deal with the uncertain context and at last we give a case on the home health telemonitoring application to show the feasibility and validity of the proposed approach.

Keywords Context information · Context modeling · Ontology · Certainty factor · Uncertain reasoning

1 Introduction

With the rapid developments of sensors, computer science, and communication technologies, the intelligence degree of calculation also becomes higher and higher. As a kind of intelligent computing mode, context-aware computing has begun to be recognized by the industry and attracted widespread concerns. It

P. Zhang (✉) · H. Huang
School of Automation, Beijing Institute of Technology, Beijing 100081, People's Republic of China
e-mail: zhangping10003@163.com

H. Huang
e-mail: honghuang@bit.edu.cn

F. Sun et al. (eds.), *Knowledge Engineering and Management*, Advances in Intelligent Systems and Computing 214,
DOI: 10.1007/978-3-642-37832-4_6,

requires that the system should automatically detect the context information (such as users' location, time, environmental parameters, users' activities, and so on), and use these information effectively to compute and provide appropriate services without much concern from users [1]. However, on one hand, various kinds of context information and different ways of perception in the real world make the expression of knowledge and realization of knowledge sharing and reuse become more and more difficult, which will make it hard to implement the knowledge-based reasoning; on the other hand, due to the limitations of sensing devices and network transmission, the context information acquired may be incomplete and imprecise [2]. Meanwhile, because of the inappropriate reasoning method, the high-level context information which is inferred from raw sensing data also has a certain degree of uncertainty [3]. Therefore, for the context-aware computing, how to explicitly describe the context information, realize knowledge sharing and reuse, and at the same time have the capability of modeling and reasoning about the uncertain context are the problems which need to be urgently solved.

The emergence of the ontology model and certainty factor model theory provides a good chance to resolve the problems above. Ontology-based models are more expressive for complex context information than other modeling methods. It becomes possible to share context by providing the formal semantics for context information [4]. But ontology-based models are difficult to support the uncertain reasoning. To overcome the difficulty, we introduce the certainty factor model, which is useful and successful for the uncertain reasoning. Its representative application is MYCIN system [5]. But the certainty factor model is weak in supports for semantic information and it is difficult to achieve the full expression of context information. So in this paper, we put forward an ontology modeling approach based on certainty factor in order to make full use of the complementary advantages of the ontology model and certainty factor model theory. Moreover, the weighting factors are introduced to the model to represent the importance and independence of evidences. On this basis, a certainty factor reasoning approach based on weights is presented to deal with the uncertain context.

This paper is structured as follows. Section 2 outlines the context-aware system. Section 3 gives the research on the ontology model based on certainty factor. Section 4 introduces the certainty factor reasoning approach based on weights. In Sect. 5, we show a simple application case. And Sect. 6 concludes the paper.

2 Overview of the Context-Aware System

Context is any information that can be used to characterize the situation of entities (e.g., a person, place, or object) which are considered relevant to the interaction between a user and an application, including the user and the application themselves [6]. Depending on different properties and purposes, context information can be divided into the following categories:

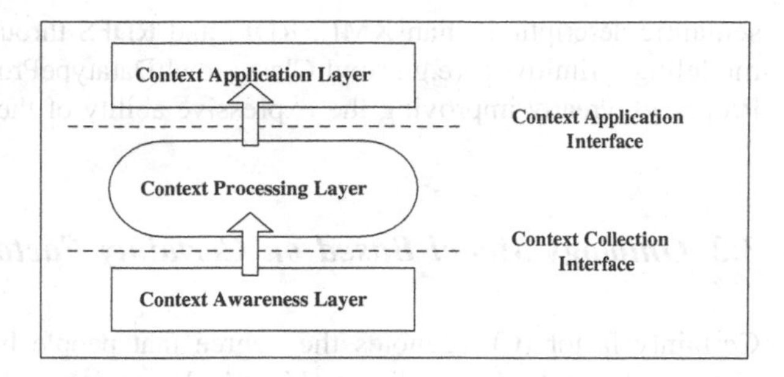

Fig. 1 Hierarchical model of the context-aware system

- User context: user location, personal preferences, social relations.
- Computing context: computing power, network bandwidth, and costs of communication.
- Physical context: temperature, light intensity, noise level.

A system is context-aware if it can acquire, interpret, use context information, and adapt its behavior to the current context of use. As shown in Fig. 1, the context-aware system is logically divided into three layers:

- Context awareness layer: It is mainly used to obtain context information through user inputs or sensors.
- Context processing layer: This layer, which reflects the intelligence of the context-aware system, uses the acquired context information to reason and compute, and get the appropriate processing results.
- Context application layer: According to the context processing results, it will dynamically call the corresponding services, and adjust the device behaviors.

3 Context Model

3.1 Ontology-Based Model

An ontology may be defined as a formal, explicit specification of a shared conceptualization [7]. It helps modeling a world phenomenon by strictly defining its relevant concepts and relationships between the concepts. The ontology-based model has its advantages in: (1) facilitating knowledge sharing by providing a formal specification of the semantics for context information; (2) supporting for logic reasoning, referring to the capability of inferring new context information based on the defined classes and properties; (3) enabling knowledge reuse by use of existing and mature ontology libraries without starting from scratch; (4) having the stronger ability for expressing complex context information.

The Web ontology language (OWL) [8] has become the widely used ontology language for its benefits of the well-defined syntax, efficient reasoning support, formal semantics and the full expression ability, and so on. It can provide more semantic descriptions than XML, RDF, and RDFS through the increase of some modeling primitives (e.g., owl:Class, owl:DatatypeProperty, and owl:ObjectProperty), greatly improving the expressive ability of the model.

3.2 Ontology Model Based on Certainty Factor

Certainty factor (CF) denotes the degree that people believe that a thing or a phenomenon is true according to historical experiences [9]. It is a subjective and empirical concept and therefore it is difficult to grasp its accuracy. However, for a specific domain, experts have abundant professional knowledge and practical experiences, so it is not difficult to give the certainty factor of domain knowledge. Its value varies from -1 to 1.

The uncertainties of evidences and knowledge are both measured by certainty factor. CF(E) shows the credibility of the evidence E. CF(E) > 0 represents a degree of belief for the evidence E, and CF(E) < 0 represents a degree of disbelief for it. CF(H, E) means the support degree of the evidence E for the conclusion H. Its definition is given as follows [10]:

$$\mathrm{CF}(H,E)=\begin{cases}\frac{P(H/E)-P(H)}{1-P(H)} & P(H/E)\geq P(H)\\ -\frac{P(H)-P(H/E)}{P(H)} & P(H/E)<P(H)\end{cases}\tag{1}$$

According to (1), we can draw that when CF(H, E) > 0, there is $P(H/E)-P(H) > 0$, which means the probability that the conclusion H is true increases because of the presence of the evidence E, and when CF(H, E) $= 0$, there is $P(H/E) = P(H)$, which shows the evidence E has no effects on the conclusion H, and when CF(H, E) < 0, there is $P(H/E)-P(H) < 0$, which indicates the probability that the conclusion H is false increases due to the impacts of the evidence E.

Considering advantages of the ontology in context modeling and supports for uncertain reasoning, this paper proposes an ontology conceptual model based on certainty factor, as is shown in Fig. 2. The context model adopts the hierarchical structure that defines the upper general ontology and the domain-specific ontology. The upper general ontology is composed of abstract concepts of various entities related to context information, in order to achieve ontology sharing in different application environments. The domain-specific ontology defines concept classes closely associated with the specific environments and their concrete subclasses in each sub-domain. The two-tier structure ontology model establishes a loosely coupled relationship between sharing knowledge and specific knowledge, and thus has a certain degree of commonality and expansibility.

In Fig. 2, we have introduced Evidences class and Conclusions class in the upper general ontology, which are used to respectively express the underlying context information obtained from sensors and high-level context information derived by inference. Meanwhile, Certainty Factor class is added to describe the

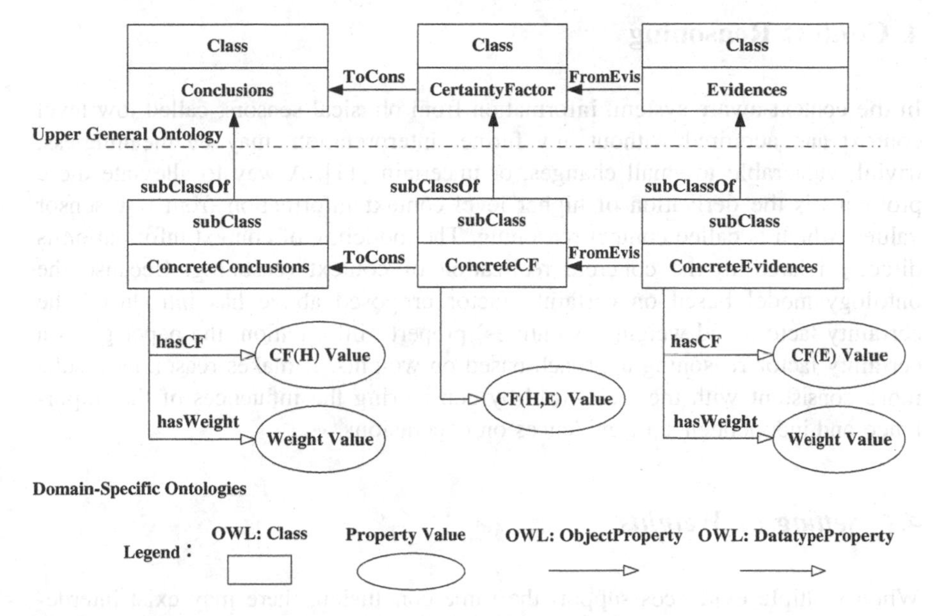

Fig. 2 Ontology conceptual model based on certainty factor

support degree of the evidence for the conclusion. The object properties, FromEvis and ToCons, show relations between the classes, i.e., the interdependencies between conclusions and evidences. In the domain-specific ontology, ConcreteEvidences class, ConcreteConclusions class, and ConcreteCF class, which are subclasses derived through inheritance, are used to denote all possible values involved in a specific domain. The new data type properties, hasCF and hasWeight, are added to these subclasses to express the credibilities and weights of evidences or conclusions, so that the model can support the dynamic nature and uncertainty of context information, and different importance of evidences on conclusions in the context-aware environments. Among them, the weights of evidences are directly given by domain experts. If the conclusion is the final result of reasoning, its weight is set to 1; if the conclusion is the intermediate result of reasoning, it needs to be assigned to the corresponding weight by domain experts in order to continue to serve as the evidence for the upper reasoning.

As the reasoning model of uncertain information, the model not only realizes the precise expression of context information and knowledge sharing, but also supports the uncertain reasoning by the uncertainty representation for the domain knowledge.

4 Context Reasoning

In the context-aware system, information from physical sensors, called low-level context and acquired without any further interpretation, may be meaningless, trivial, vulnerable to small changes, or uncertain [11]. A way to alleviate these problems is the derivation of higher level context information from raw sensor values, which is called context reasoning. The modeling of context information is directly related to the concrete realization of context reasoning. Because the ontology model based on certainty factor proposed above has introduced the certainty factors and weights as entities' property information, the paper gives a certainty factor reasoning approach based on weights. It makes reasoning results more consistent with the real world by considering the influences of the importance and independence of evidences on conclusions.

4.1 Setting of Weights

When multiple evidences support the same conclusion, there may exist interdependencies between them, i.e., they are not mutually independent. What's more, each evidence may contain a different amount of information and thus has different degrees of importance on the corresponding conclusion. Therefore, the weighting factors are introduced to represent the importance and independence of evidences. Usually, the weighting factors of evidences are directly given by domain experts based on the following two factors:

- Importance of evidences on conclusions: the more important the evidence is, the greater the corresponding weight is.
- Independence of evidences: the more reliance the evidence has on other evidences, the smaller weight it has.

4.2 Uncertain Reasoning Process Description

Let's introduce a definition first: the certainty factor of the weighted evidence is called the absolute certainty factor of the evidence. In the uncertain reasoning process, the absolute certainty factor of the evidence is made use of to participate in the synthesis of conclusion uncertainty. In the following, we have three steps to show the process of uncertain reasoning.

Step 1 We use (2) to get the absolute certainty factor of the evidence:

$$\mathrm{CF}'(E_i) = \frac{w_i}{\operatorname*{Max}_i(w_i)} \cdot \mathrm{CF}(E_i) \tag{2}$$

For this, $\mathrm{CF}(E_i)$ is the certainty factor of the evidence E_i; w_i is the weight factor of the evidence E_i; $\underset{i}{\mathrm{Max}}(w_i)$ is the maximum value of w_i, $i = 1, 2, \ldots, n$.

Step 2 The certainty factor of the conclusion for each evidence is obtained as follows:

$$\mathrm{CF}_i(H) = \mathrm{CF}'(E_i) \cdot \mathrm{CF}(H, E_i) = \frac{w_i}{\underset{i}{\mathrm{Max}}(w_i)} \cdot \mathrm{CF}(E_i) \cdot \mathrm{CF}(H, E_i) \tag{3}$$

where $\mathrm{CF}(H, E_i)$ is the certainty factor of knowledge corresponding to each evidence and $\mathrm{CF}'(E_i)$ is the absolute certainty factor of the evidence E_i computed above.

Step 3 The synthesis of $\mathrm{CF}_i(H)$ is calculated as follows:

Suppose there are two evidences E_1 and E_2 reasoning out the same conclusion. The synthesis formula for the conclusion is shown by

$$\mathrm{CF}(H) = \begin{cases} \mathrm{CF}_1(H) + \mathrm{CF}_2(H) - \mathrm{CF}_1(H) \cdot \mathrm{CF}_2(H) & \mathrm{CF}_1(H),\ \mathrm{CF}_2(H) \geq 0 \\ \mathrm{CF}_1(H) + \mathrm{CF}_2(H) + \mathrm{CF}_1(H) \cdot \mathrm{CF}_2(H) & \mathrm{CF}_1(H),\ \mathrm{CF}_2(H) < 0 \\ \dfrac{\mathrm{CF}_1(H) + \mathrm{CF}_2(H)}{1 - \min\{|\mathrm{CF}_1(H)|, |\mathrm{CF}_2(H)|\}} & \mathrm{CF}_1(H) \cdot \mathrm{CF}_2(H) < 0 \end{cases} \tag{4}$$

If there are more evidences E_3, E_4, and so on, we will still follow the steps above. First, $\mathrm{CF}_i(H)$ of each evidence is acquired according to (3), and then $\mathrm{CF}_{1,2}(H)$ is calculated for the first two evidences by using (4). Next, we still use (4) for the synthesis of $\mathrm{CF}_3(H)$ and $\mathrm{CF}_{1,2}(H)$, then the fourth, and so on. Finally, the synthetic certainty factor $\mathrm{CF}(H)$ is obtained after the combination of all the evidences.

5 Case Study

As an important application field of context-aware computing, the home health telemonitoring system has achieved rapid developments and integrated into our lives. In this application scenario, we assume that the biomedical parameters (e.g., blood pressure, body temperature, and heart rate) of the elderly are monitored via a variety of sensors (e.g., embedded sensors and wearable sensors) deployed in the room, and then the sensed context information is automatically analyzed and processed. Eventually, the conclusion related to the physical condition of the elderly is drawn, and moreover, the monitoring system will call the executive equipment to provide appropriate services (for instance, alarm services, E-mail, and SMS) for critical situations. According to this application case, we show how to construct the ontology model based on certainty factor and apply the uncertain reasoning approach.

5.1 Reasoning Model of Uncertain Context

We use the ontology editor tool Protégé 4.1 to construct the reasoning model of the home health telemonitoring process described above, i.e., the ontology model based on certainty factor. Its class diagram appears in Fig. 3.

In this model, it is assumed that there are three evidences: blood pressure (E_1), body temperature (E_2), and heart rate (E_3), which are respectively expressed as the three subclasses of Evidences class. Abnormal class, the subclass of Conclusions class, is added to represent the reasoning conclusion, i.e., it means that the old people are critically ill. The three subclasses, CF_1, CF_2, and CF_3, are also added to describe the support degree of each evidence for the conclusion. The object properties, FromEvis and ToCons, are used to show the interdependencies between each evidence and the conclusion. And the certainty factors and weights of evidences or conclusions are realized by data type properties hasCF and hasWeight. The partial OWL descriptions for the model are illustrated in Fig. 4.

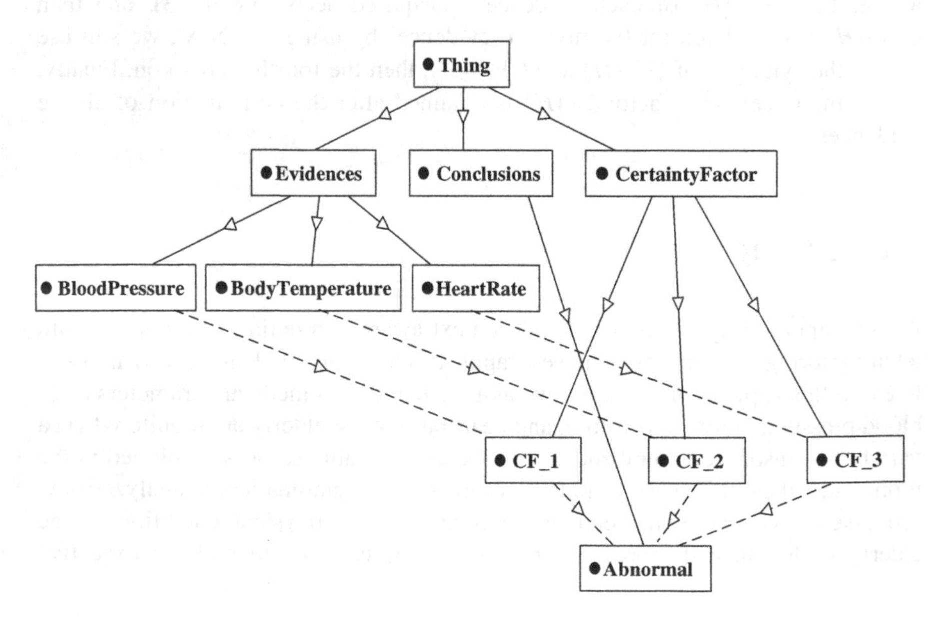

Fig. 3 Class diagram of the ontology model

Fig. 4 The partial OWL descriptions for the model

```
<owl:Class rdf:ID="Abnormal">
    <rdfs:subClassOf rdf:resource="#Conclusions"/>
</owl:Class>
<owl:Class rdf:ID="BloodPressure">
    <rdfs:subClassOf rdf:resource="#Evidences"/>
</owl:Class>
<owl:Class rdf:ID="CF_1">
    <rdfs:subClassOf rdf:resource="#CertaintyFactor"/>
</owl:Class>
<owl:ObjectProperty rdf:ID="FromEvis">
    <rdfs:domain rdf:resource="#BloodPressure"/>
    <rdfs:range rdf:resource="#CF_1"/>
</owl:ObjectProperty>
<owl:ObjectProperty rdf:ID="ToCons">
    <rdfs:domain rdf:resource="#CF_1"/>
    <rdfs:range rdf:resource="#Abnormal"/>
</owl:ObjectProperty>
<owl:DatatypeProperty rdf:ID="hasCertaintyFactor">
    <rdfs:domain rdf:resource="#CF_1"/>
    <rdfs:range rdf:resource="&xsd;float"/>
</owl:DatatypeProperty>
<owl:DatatypeProperty rdf:ID="hasCF">
    <rdfs:domain rdf:resource="#BloodPressure"/>
    <rdfs:range rdf:resource="&xsd;float"/>
</owl:DatatypeProperty>
<owl:DatatypeProperty rdf:ID="hasWeight">
    <rdfs:domain rdf:resource="#BloodPressure"/>
    <rdfs:range rdf:resource="&xsd;float"/>
</owl:DatatypeProperty>
```

We can read the OWL ontology via Jena [12] and extract the reasoning-related information from the above OWL ontology model, so as to get prepared for the implementation of the uncertain reasoning.

5.2 Application of Uncertain Reasoning

According to expert experiences, we can always get the mapping function from the raw context data to CF (H, E). Based on the mapping function curves, we acquire the CF (H, E) corresponding to each biomedical parameter. As the information obtained from sensors is certainty information, the certainty factors of evidences are all endowed with 1, i.e., $\mathrm{CF}(E) = 1$. Table 1 gives weights of evidences set by experts and the certainty factors of knowledge at some point.

The uncertain reasoning process is as follows:

Table 1 Weights and certainty factors of knowledge

Evidences	Values	w_i	CF(*H*, *E*$_i$)
E$_1$: BloodPressure	161/78 mmHg	0.3	0.8
E$_2$: BodyTemperture	38.5 °C	0.4	0.9
E$_3$: HeartRate	75 bpm	0.3	−0.65

$$\begin{aligned} \mathrm{CF}_1(H) &= \mathrm{CF}'(E_1) \cdot \mathrm{CF}(H, E_1) \\ &= \frac{w_1}{\underset{i}{\mathrm{Max}}(w_i)} \cdot \mathrm{CF}(E_1) \cdot \mathrm{CF}(H, E_1) \\ &= 0.6 \end{aligned}$$

Similarly, we get $\mathrm{CF}_2(H) = 0.9$ and $\mathrm{CF}_3(H) = -0.49$. And thus the synthesis results are:

$$\begin{aligned} \mathrm{CF}_{1,2}(H) &= \mathrm{CF}_1(H) + \mathrm{CF}_2(H) - \mathrm{CF}_1(H) \cdot \mathrm{CF}_2(H) \\ &= 0.6 + 0.9 - 0.6 \times 0.9 \\ &= 0.96 \end{aligned}$$

$$\begin{aligned} \mathrm{CF}_{1,2,3}(H) &= \frac{\mathrm{CF}_{1,2}(H) + \mathrm{CF}_3(H)}{1 - \min\{|\mathrm{CF}_{1,2}(H)|, |\mathrm{CF}_3(H)|\}} \\ &= \frac{0.96 - 0.49}{1 - \min\{0.96, 0.49\}} \\ &= 0.92 \end{aligned}$$

The system credibility threshold λ is introduced to determine whether the conclusion will happen. If the synthetic certainty factor of the conclusion is greater than it, then the conclusion is true. For the case, we set $\lambda = 0.9$. As $\mathrm{CF}_{1,2,3}(H) = 0.92 > \lambda$, it is believed that the conclusion is true, i.e., the old people are in a critical condition. The monitoring system needs to make the real-time response to the critical situation, such as triggering an alarm, and notifying family members or the emergency center, etc.

6 Conclusion

Context modeling and reasoning are the key problems of realizing context-aware computing. In this paper, we put forward the ontology model based on certainty factor. On this basis, a certainty factor reasoning approach based on weights is presented to deal with the uncertain context. The advantages of the proposed approach lie in: (1) realizing the precise expression of context information and knowledge sharing; (2) extending the representation of ontology for the uncertain context and thus supporting the uncertain reasoning; (3) making reasoning results more consistent with the real world by taking into consideration the influences of the importance and independence of evidence sources on conclusions. Finally the feasibility and validity of the proposed approach are illustrated through the case study and analysis.

References

1. Anind KD, Gregory DA (2000) Towards a better understanding of context and context-awareness. In: Proceedings of the CHI 2000 workshop on "The What, Who, Where, When, and How of Context-Awareness", Hague
2. Truong BA, Lee YK, Lee SY (2005) Modeling uncertainty in context-aware computing. In: Proceedings of fourth annual ACIS international conference on computer and information science, pp 676–681
3. Hong X, Nugent C, Mulvenna M, McClean S (2009) Evidential fusion of sensor data for activity recognition in smart homes. Pervasive Mobile Comput 5(3):236–252
4. Paganelli F, Giuli D (2011) An ontology-based system for context-aware and configurable services to support home-based continuous care. IEEE Trans Inf Technol Biomed 15(2):324–333
5. Buchanon BG, Shortliffe EH (1984) Uncertainty and evidential support. Rule-based expert systems. Addison-Wesley, Reading, pp 209–232
6. Dey AK, Abowd GD (2001) A conceptual framework and a toolkit for supporting the rapid prototyping of context-aware applications. Hum Comput Interact 16:97–166
7. Studer R, Benjamins VR, Fensel D (1998) Knowledge engineering: principles and methods. Data Knowl Eng 25(1):161–197
8. Web Ontology Language-W3C Recommendation (2004). http://www.w3.org/TR/owl-guide/
9. Hao Q, Lu T (2009) Context modeling and reasoning based on certainty factor. In: Proceedings of 2009 second Asia-Pacific conference on computational intelligence and industrial applications, pp 38–41
10. Wang WS (2007) Principles and applications of artificial intelligence. Publishing House of Electronics Industry, Beijing, pp 173–178
11. Ranganathan A, Al-Muhtadi J, Campbell RH (2004) Reasoning about uncertain contexts in pervasive computing environments. IEEE Pervasive Comput 3(2):62–70
12. Jena Semantic Web Framework (2009) http://jena.sourceforge.net/

Research on Mental Coefficient of a Multi-Agent Negotiation Model

Yu-Mei Jian and Ming Chen

Abstract In Multi-Agent system, negotiations model was commonly used by contract net. A multi-agent negotiation model based on the acquaintance coalition and the mental coefficient contract net protocol are presented to improve the efficiency of negotiation. The structure of the multi-agent negotiation model is given to support the acquaintance coalition contract net protocol. The trust degree parameter, familiar degree, reliability degree, busy degree of mental state, and the update rules are introduced. Finally, through an example and analysis of a Robot soccer system which uses the model, improvements of the negotiation efficiency and negotiation communication traffic is proven.

Keywords Contract net · Acquaintance coalition · Mental coefficient · Multi-agent

1 Introduction

In Multi-Agent system, negotiation plays a very important role in solving the conflict of goals and resources. Many thinning and complement were made in the actual apply process. Sand Holm drew boundary cost count into the process of bidding and awarding of contract net [1, 2], Chen used bid threshold value [3] to restrict calculate and communication cost during the negotiation; Fischer etc. [4] optimized task allocation by drawing provisional trust refuse, and mocked trading into contract net; Collins [5] mixed arbitration mechanism into the negotiation process to prevent fraudulent conduct in tendering process; Lee [6] drew acceptance term into contract net to adapt task environmental changes; Conroy [7]

Y.-M. Jian (✉) · M. Chen
College of Information Technology, Shanghai Ocean University, Shanghai 201306, China
e-mail: jianyumei0628@126.com

F. Sun et al. (eds.), *Knowledge Engineering and Management*,
Advances in Intelligent Systems and Computing 214,
DOI: 10.1007/978-3-642-37832-4_7,

introduced the multi-step negotiation into contract net, repeatedly negotiations were allowed during bidding or to win the bidding; the literature [8] mentions a Multi-Agent Cooperation Model based on Improved Dynamic Contract Net Protocol (MACMIDCNP). To bid by making the alliance as a unit in MACMIDCNP, to calculate the optimal alliance which can finish the task by using mingle inheritance ant algorithm, and to choose the alliance as contractor directly base on reliability. Thus it could greatly reduce communication cost and save systematic run time by lessening the communication boundary to alliance itself inside, and also it enhances the whole performance of the system. The literature [9] mentions dynamic contract net with fault-tolerant ability, the contractor had the strategy of task option, the recognition of fault contractor, the task recovery and secondary scheduling and the rules for reliability updating, draw trust, steady, cooperation frequent, and positive level in the process of bidding and awarding, at the same time, the literature [9] proposes a algorithm-multi-rank top-n random select algorithm, which considered cooperator to complete the task's history situation and abilities' change. The literature [10] adds mental coefficient based on acquaintance alliance, it packed the task as a bid book, and published bid book to commonality blackboard, when there were no bid agents for a given task.

The multi-agent cooperation model which based mental state mentioned in this paper, introducing the strategy of acquaintance alliance and mental state into typical contract net comparing with DCNP, MACMMS newly increased reliability, busyness, intimate, satisfaction, initiative, etc. This paper presents the classification function and the rules for updating reliability, busyness. On the basis of ensuring the quality of negotiation quality, MACMMS effectively advance the efficiency of negotiation and prove the effectiveness by researching the test of robot football match [11] and its result data.

2 Multi-Agent Cooperation Model

2.1 Negotiation Process

In MACMMS, other agents bid to become a negotiation agent after management agent sends out a task. Negotiation process defines as a five-element group: <G, S, A, Time, Protocol>.

- G: All agents who want negotiation are divided into three types: management agent, negotiation agent, consultant agent.
- S: All tasks.
- A: Negotiation process value. Acceptance means consultant agent accepts to execute the task, refusal means refuse negotiation, rejection means the consultant agent rejects to execute consultant task. Both rejection and refusal may lead to the restart of negotiation.

- Time: System clock arranges by order natural world. It limits negotiation time and must finish if negotiation time exceeds limit time.
- Protocol: Negotiation treaty (Fig. 1)

Step 1 Before negotiation, management agent receives a task T_j, resolves the task to $T_j = \{t_1, t_2, \ldots t_j\}$ and classifies the acquaintance to acquaintances gather, general acquaintances gather, and strangeness gather by mental coefficient of each task.

Step 2 Management agent sends bid information to every task's acquaintances gather and general acquaintances gather.

Step 3 Negotiation agents solve bid wish value Price (t_j) of each task by bid information and negotiation algorithm.

Step 4 After limit time, management agent decides the last consultant agent among negotiation agents' bid condition by win the bidding decision function.

Step 5 Negotiation agents change their initial offer. If the winner (consultant agent) wills to execute task, consultant agent sends affirm to management agent. Management agent turns to step 9 if it did not receive confirm information.

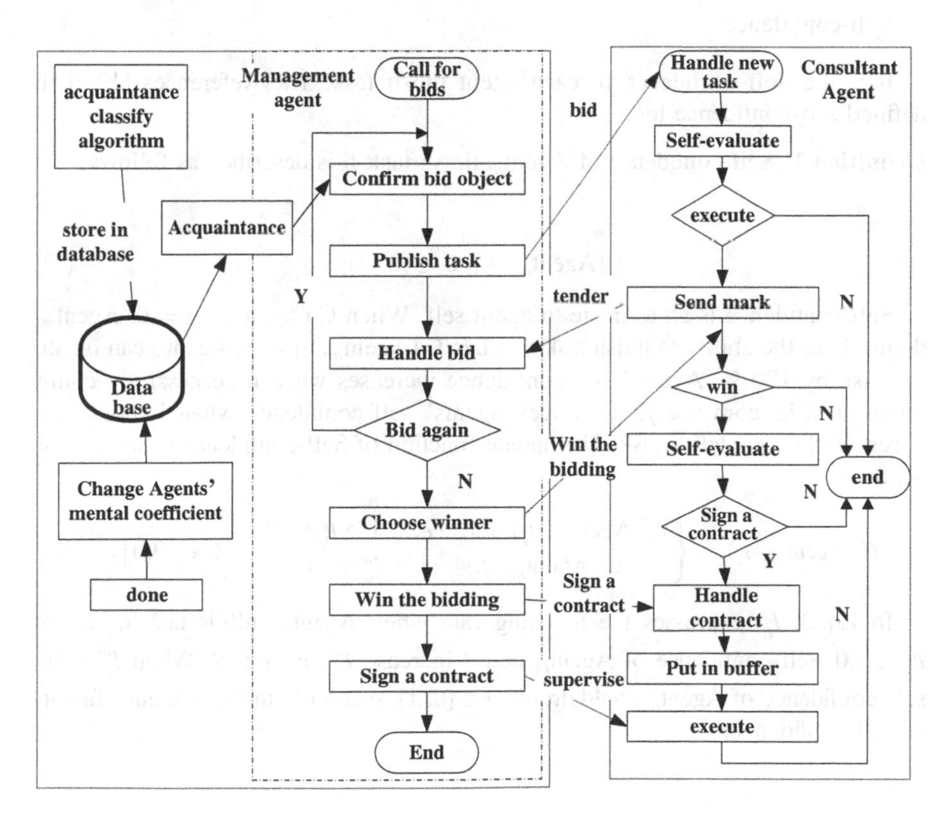

Fig. 1 Negotiation processes of a mental coefficient contract net

Step 6 If management agent receives a confirm information in the limit time, it sends task to consultant agent and supervises consultant agent finish task, or if consultant agent rejects or refuses, algorithm should turn to Step 9 and renegotiate.
Step 7 Consultant agents execute task after it receives it and updates its own self-confidence, reliability, and busy factor.
Step 8 Send information to management agent when task does not finish on time, and updates consultant agent's self-confidence, reliability, and busy factor. Then turn to step 9.
Step 9 One wheel negotiation is over.
Step 10 Management agent decides to run a new bid, turn to step 2.
Step 11 Task negotiation is over.

2.2 Definition and Update Agents' Mental Coefficient

Mental coefficient is the standard of each agent action. The following mental coefficient was defined in the improved contract net model:

1. Self-confidence

It is the Self-confidence of each agent finish task. The References [12, 13] defined self-confidence too.

Definition 1 Self-confidence of Agent_m finish task t_j is described as follows:

$$C(\text{Agent}_m,\ t_j), C \in [0,\ 1]. \tag{1}$$

Self-confidence is an estimate to agent self. When C $(\text{Agent}_m,\ t_j) = 0$, Agent_m do not have the ability to finish task t_j; when C $(\text{Agent}_m,\ t_j) = 1$, Agent_m can finish the task by 100 %. Agents' self-confidence increases when it successfully completes a task; conversely, it reduces agents' self-confidence when it does not successfully complete a task. The update function of Self-confidence is defined as follows:

$$C(\text{Agent}_m,\ t_j) = \begin{cases} C'(\text{Agent}_m,\ t_j) + I_{t_j}^m \cdot e, & 1 \geq I_{t_j}^m > 0 \\ C'(\text{Agent}_m,\ t_j) - r, & I_{t_j}^m = 0 \end{cases}, \quad C \in [0, 1] \tag{2}$$

In Eq. 2, $I_{t_j}^m$ expresses the finishing rate when Agent_m fulfills task t_j. when $I_{t_j}^m > 0$, self-confidence of Agent_m could increase $I_{tj}^m \cdot e \quad e \in R$. When $I_{t_j}^m = 0$, self-confidence of Agent_m could drop r, $r \in [0,\ 1]$, e and r both the influence factor of self-confidence.

2 Busy

Definition 2 busy after Agent_m accept task t_j is described as follows:

$$B(\text{Agent}_m) = \frac{n_{t_j}}{n_{\max}}, \quad B \in [0, 1] \tag{3}$$

Busy is used to present agents' busy state. In the definition, n_{tj} means the number of Agent_m has now (include task t_j); $n_{\max}$ means can handle the maximum of Agent_m in the same time. When B (Agent_m) = 1, it means Agent_m has already achieved task saturation state, Agent_mcannot accept any task. If B (Agent_m) $\neq$ 1, n_{t_j} needs to add 1 and update B (Agent_m) to adjust consultant agents' busy when consultant agent adds one cooperation task every time.

3 Reliability

Definition 3 Agent_m considers the self-confidence of Agent_m successfully fulfilling task t_j is described as reliability, the definition is as follows:

$$T_{\text{Agent}_i}(\text{Agent}_m, t_j) = \frac{\sum_{w=1}^{N_{t_j}{}^m} I_{t_j}{}^m(w)}{N_{t_j}^m}, \quad T \in [0, 1] \tag{4}$$

Reliability means trust degree between Agent_i to another agent, which is reflected in the trust relationship between agents, $I_{t_j}^m(w)$ presents finishing rate of Agent_m successfully finishes Agent_i's wth task t_j; N_{tj}^m is the number of Agent_i relegate tasks t_j to Agent_m. When $T = 1$, Agent_i consider Agent_m can finish task t_j by 100 %, otherwise, Agent_i think Agent_m do not have the ability to finish task t_j.

4 Intimate

Intimate is used to measure the familiar degree of cooperative relationship between agents. In the reference [13] appeared the definition of intimate.

Definition 4 Intimate is considered as the frequent degree of Agent_i and Agent_m working together to deal with task; the definition is as follows:

$$R(\text{Agent}_i, t_j, \text{Agent}_m) = \frac{N_{tj}^m}{N_{tj}^i}, \quad R \in [0, 1] \tag{5}$$

In Eq. 5, $N_{t_j}^m$ is number of Agent_i relegating task t_j to Agent_m, $N_{t_j}^i$ means the total number of Agent_i relegate task t_j. Compared with the literature [14], this paper expansion intimate's boundary, contribute self-confidence and reliability's update.

5 Satisfaction

Satisfaction is the evaluate result of every cooperative, including the task completion quality of the solution and the time of evaluation.

Definition 5 Satisfaction is defined as Agent_m finishing the wth task t_j of Agent_i:

$$S_{\text{Agent}_i}^w(\text{Agent}_m, t_j) = kq \cdot q(\text{Agent}_m, t_j) + kt \cdot t(\text{Agent}_m, t_j) \quad (6)$$

In Eq. 6, $q(\text{Agent}_m, t_j)$ is the quality evaluation of Agent_m finish task t_j; $\text{t}(\text{Agent}_m, t_j)$ is the time evaluation value; K_q, K_t respectively, mean weight of quality evaluation, time evaluation. The total satisfactions of Agent_i consider Agent_m finish task t_j was described as follows:

$$S_{\text{Agenti}}(\text{Agent}_m,\ t_j) = \sum_{w=1}^{N_{t_j}^m} kw \cdot S_{\text{Agenti}}^w(\text{Agent}_m,\ t_j) \quad (7)$$

In Eq. 7, $N_{t_j}^m$ is number of Agent_i relegating task t_j to Agent_m, K_w means the weight of every cooperation satisfaction.

6 Initiative

Definition 6 The initiative was described as follows:

$$A(\text{Agent}_i,\ t_j,\ \text{Agent}m) = \frac{N_{t_j}^{\prime m}}{N_{t_j}^i}, \quad A \in [0, 1]. \quad (8)$$

Initiative is the positive degree of negotiation agent, it has nothing to do with winning the bid, $N_{t_j}^{\prime m}$ is number of Agent_m relegating task t_j to Agent_i, $N_{t_j}^i$ is the total number of Agent_i entrust task t_j.

2.3 Definition of Acquaintance Model and its Sort Management

Definition 7 The acquaintance of Agent_a is defined as some agents which successfully cooperate with task a_i more than a certain frequency. In the negotiating process, acquaintances set and the formation of negotiation model is called acquaintance model.

Definition 8 $E_{\text{Agent}_m} = \left(T_{\text{Agent}_m}, R_{\text{Agent}_m}, S_{\text{Agent}_m}, A_{\text{Agent}_m}\right)$ which means the overall merit of Agent history recording, embodies the task for degree, called a familiar degree.

2.3.1 The Concept of Acquaintance Coalition

The literature [15] introduces alliance, this paper defines acquaintance model as participating in cooperation all the resources of the mental state information Agent of the abstract description.

Definition 9 $F_{\text{Agent}_m} = <C, E(T, R, S, A), B>$, which means the acquaintance model is based on the mental coefficient. C is the self-confidence of Agent_m; E is Intimate; B is the busy of Agent_m.

2.3.2 Classify Acquaintance Coalition

Every agent can classify as acquaintance, general acquaintance, and strangeness.

Agent $=< \text{Agent}_f >+< \text{Agent}_y >+< \text{Agent}_p >$, Agent_f means acquaintance, Agent_y means general acquaintance, Agent_p means strangeness.

Definition 10 The definition formula of Agent_f is described as follows:

$$\text{Agent}_f = \{ \text{Agent}_f | C \geq C_{tj} \cap E \geq \text{eval}_{tj} \cap L(E) < \text{l}_{t_j} \}$$

Definition 11 The definition formula of Agent_y is described as follows:

$$\text{Agent}_y = \{ \text{Agent}_y | C \geq C_{tj} \cap E < \text{eval}_{tj}) \cup (C \geq C_{tj} \cap \text{E} \geq \text{eval}_{tj} \cap L(E) > \text{l}_{tj} \}$$

Definition 12 The definition formula of Agent_p is described as follows: $\text{Agent}_p = \{ \text{Agent}_p | C < C_{tj} \}$

In Definition 10–12, C_{t_j} is the self-confidence lower limit, $eval_{tj}$ is the task history comprehensive evaluation lower limit, L(E) is a sort function, which express the ranking in the similar agent of E, l_{t_j} is the comprehensive evaluation ranking.

2.3.3 The Management of Acquaintance Coalition

In the initial state, acquaintance model, $Agent_f$, $Agent_y$ and $Agent_p$ both initialized to empty. Management agent chooses an agent as cooperation Agent whose self-confidence was the biggest, then updates the cooperation Agent's mental coefficient.

Definition 13 The communication record is described as Contact - F($Agent_i$)= $< L_1, L_2, \ldots L_m >$. Lm is the communication information of acquaintance $Agent_m$, and $L_m = <$ ID, Address , C, L(E), $B >$, ID is the identifier of $Agent_m$, Address is the communication address, Contact - F($Agent_i$) is acquaintance communication record, Contact - Y($Agent_i$) is general acquaintance communication record.

2.4 Bidding and Decision Function

2.4.1 Bidding Document

Definition 14 Bidding was defined as Announce (t_j) = { Des, Ability (t_j), Bid $-deadline, Price(t_j), \ldots\}$. Des describes the particular task. Ability is the capacity gather to solve task t_j, which express as Ability(t_j) = { Ab_1, Ab_2, …, Ab_n} , Price (t_j) is the bidding price (Eq. 9); Bid-deadline is the blocking time.

$$\text{Price}(t_j) = \text{COST\_}T_j * \text{k\_}t_j * \text{s*k}, \quad \text{s, k} \in (0,\ 1] \tag{9}$$

In Eq. 9, $COST\_T_j$ is the planning overhead of task T_j; k_t_j is the weight of subtask t_j; s is the load coefficient, when $s = 1$, the system's load is empty.

2.4.2 Bidding Decision Function

Negotiation agent first checks their own self-confidence C, synthetically considering the busy degree B, the blocking time Bid-deadline of task and the actual expenses(COST) when it received task information in effective time, if $COST\_T_j$>Price(t_j), negotiation agent do not have eligibility to bid.

The evaluation formula of Agent_m to bidding task t_j is described as Eq. 10:

$$\mathrm{J}_{\mathrm{Agent}_m}(t_j){=}k_c * C(\mathrm{Agent}_m,\ t_j) + k_{\mathrm{em}} * \mathrm{EM(Bid\ -\ deadline)} \tag{10}$$

In Eq. 10, $J_{\mathrm{Agent}_m}(t_j)$ means the evaluation value of Agent_m to bidding task t_j; EM (Bid-deadline) is task emergency degree function; K_c is the weight of self-confidence and K_{em} is the weight of emergency degree.

The bidding document of consultant agent sends bidding document to management agent described as Eq. 11:

$$\mathrm{Bid}(\mathrm{Agent}_m,\ t_j){=}\left\{\mathrm{ID},\ \mathrm{Address},\ \mathrm{Ability}_{\mathrm{Agent}_m},\ B\mathrm{Price}_{\mathrm{Agent}_m}(t_j),\ \ldots\right\} \tag{11}$$

In Eq. 11, $B\mathrm{Price}_{\mathrm{Agent}_m}(t_j)$ is bidding price, the computational formula as Eq. 12:

$$B\mathrm{Price}_{Agent_m}(t_j) = \mathrm{COS}T\_T_j * (1 + B) \tag{12}$$

2.4.3 Win the Bidding Decision Function

When the task to the deadline of the bid, according to the received tender, management agent synthetically considering the bid price ($B\mathrm{Price}_{\mathrm{Agent}_m}(t_j)$) and historical records comprehensive evaluation E, select the maximum definition agent as consultant agent according to the decision function (Eq. 13).

$$\mathrm{Sel}(\mathrm{Agent}_m,\ t_j) = K_{BP} * B\mathrm{Price}_{\mathrm{Agent}_m}(t_j){+}K_E * E \tag{13}$$

In Eq. 13, $\mathrm{Sel}(\mathrm{Agent}_m,\ t_j)$means the evaluation value of Agent_m finishing task t_j, K_{bp} is the weight of bid price and K_E is the weight of comprehensive evaluation.

3 MACMMS Algorithm

Contract net algorithm based on mental factor is shown in Fig. 2:

MACMMS Algorithm: Negotiation algorithm of contract net based on mental coefficient.

Input:

subject A = {Agent1, Agent2,.......Agent n}

Tasks T = {T1, T2,......Tn}

Every agent has its initial self-confidence, reliability, busy, intimate, satisfaction, and other related coefficient, each task also contains the minimum confidence and cost of coefficient which were required to complete task.

Output: Task record

Steps:

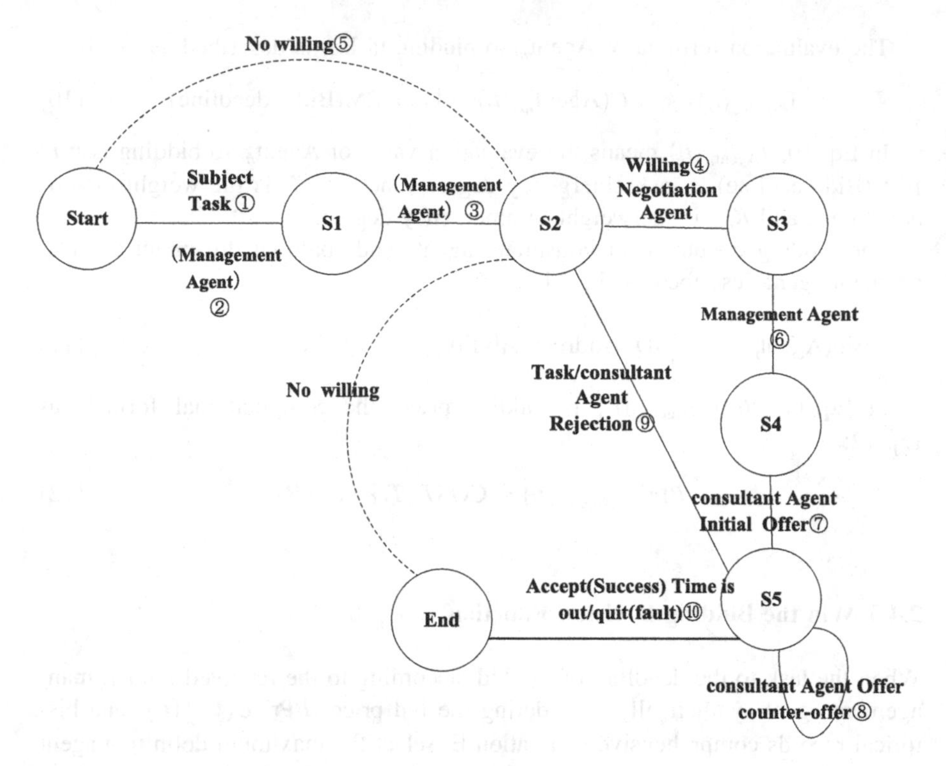

Fig. 2 Negotiation algorithm

Step 1 Find out the comprehensive evaluation value of each Agent by Definition 8, and sort the data in database.

Step 2 Management agent classifies every task into acquaintances gather; general acquaintances gather and strangeness gather according Definitions 10, 11, 12.

Step 3 For each data transmission, negotiation agent tests its bidding qualification by Eq. 9.
If negotiation agent has tender eligible, calculate out its evaluation value by Eq. 10, send bid to management agent in the form of Eq. 11, and after bidding calculate the bid price by Eq. 12.

Step 4 For data transmission task, in the bidding cases, based on the bid decision function, management agent select each task to Sel (Agent$_m$, t_j)'s largest negotiation agent to become consultant agent. According to Task-Number which was the number of tenders for each task:

If ① Task-Number = 0, no tender, renegotiate; Otherwise, if ② Task-Number = 1, choose this tender agent to become consultant agent; Otherwise ③, Task-Number >1, select the largest evaluation value of bidding Agent to complete the Task's as the final consultant agent by Eqs. 12 and 13.

If management agent has not received a consultation intend in the limited time, it could send messages again or end the negotiation (Fig. 1 with dotted lines).

Step 5 It is not until each mission does not cooperate with agent that the calculation ends.

4 Application and Experiment

This paper is against the background of Robocop robot soccer, where the entire team is a typical MAS that considers robots as agents. Cooperation exists between adjacent positions of the players in the soccer process, such as passing, cooperation, etc. Coach (management agent) considers the performance of the players in the past (mental coefficient) when deciding which player to participate in the specific game. Then management agent selects the first 11 players who are most likely to win this game to participate in the game by evaluating all players' parameters.

The following test via relatively using mental coefficient Network and non-mental coefficient choose negotiation cost and the cost of task solving for the player to explain the effect of the mental coefficient's consulting.

Initial data set, when player successfully complete the requirement function, set $I_{t_j}^{m} = 1$, $I_{t_j}^{m}(w) = 1$, otherwise, $I_{t_j}^{m} = 0$, $I_{t_j}^{m}(w) = 0$. Self-confidence change factor $e = 0.1$, $r = 0.1$, initial busy is 0, reliability according to usual training results decision. When player successfully goals or passes, $q(\text{Agent}_m, t_j) = 1$, else, $q(\text{Agent}_m, t_j) = 0$. When player scores in the playing time, $t(\text{Agent}_m, t_j) = 1$, else$t(\text{Agent}_m, t_j) = 0.K_q = 0.8$, $K_t = 0.2$, K_w is a random number.

Assume that there are 20 players, coach (management agent) needs to choose 11 players to become formal players, the rest of the nine for bench players. The number of games is from 5 to 30. The Example of negotiation cost of football match are shown in Fig. 3, and the example of cost of task solving in football match are shown in Fig. 4.

The horizontal axis represents the number of games; the ordinate axis cost represents the negotiation cost in Fig. 3; the ordinate axis cost represents the cost of task solving in Fig. 4. We can see from the above test results that the negotiation cost of classic contract net consultative is far higher than contract net protocol based on mental coefficient. This is mainly due to classified each agent of mental contract net, and only sends a message to acquaintance and general acquaintance which improves the traditional way of broadcasting. However, it is basically flat about the cost of task solving. We can see from Fig. 4 that with the growth of task, the advantage of contract net is based on the mind, so mental coefficient contract net is more suitable for multi-agent.

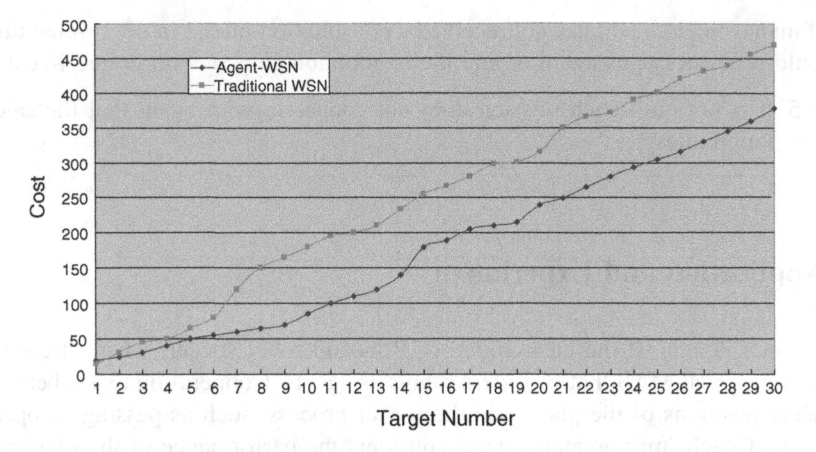

Fig. 3 Example of negotiation cost of football match

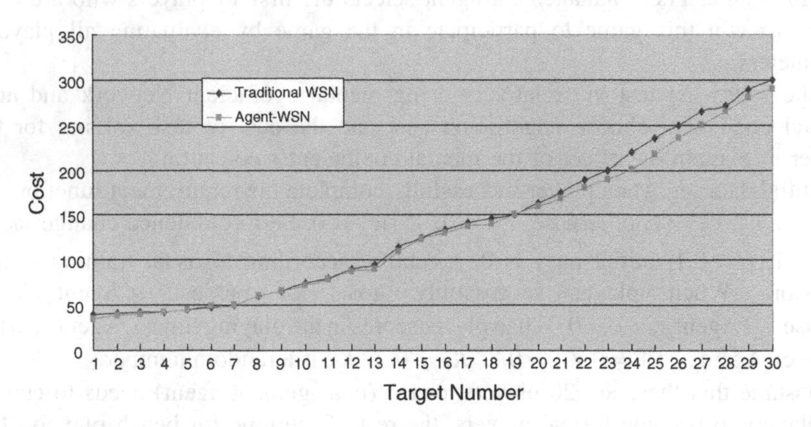

Fig. 4 Example of cost of task solving in football match

5 Conclusion

This paper aims at the large scale of MAS system, which pays the expense of the classic contract net which is large and its spread is slow. Having imported the acquaintance model and the mental coefficient into the classic contract net, it constructs a network based on the mental state contract negotiation model architecture. The confidence of mental state, familiarity, reliability, busy condition described agent's social attribute, and noted down the historical cooperation with the agents.

It has increased the efficiency of the negotiation based on the quality of negotiation. But it is simple being the update role of confidence between mental coefficients just according to the related proportion factor. So it is a possibility that the self-confidence and comprehensive evaluation are maybe generally alike, which will lead to the weakness of the mental coefficients during the mission distributing. In the meantime, it will unexpectedly increase the cost of calculate, which is necessary to how to update every factor of mental coefficients in the subsequent research. This is an important subsequent research.

Acknowledgements Supported: the national high technology research and development Plan of China (863): 2012AA101905.

References

1. Holm TS, Lesser VR (1997) Coalition among computationally bounded agents. Artif Intell 94:99–137
2. Holm TS, Lesser VR (1996) Advantages of a leveled commitment contracting protocol. In: Proceedings 13th national conference on artificial intelligence. AAAI Press, Menlo Park, CA, pp 126–133
3. Chen X-G (2004) Further extensions of FIPA contract net protocol: threshold plus doe. In: Proceedings of 2004 ACM symposium on applied computing. ACM Press, New York, pp 135–141
4. Fischer K, Muller JP (1995) A model for cooperative transportation scheduling. In: Proceedings of the 1st international conference on multi-agent systems. MIT Press, Cambridge, pp 169–175
5. Collins J, Dahl BY et al (1998) A market architecture for multi-agent contracting. In: Proceedings of 2nd international conference on autonomous agents. ACM Press, New York, pp 256–278
6. Jun LK, Ski CY, Kym LJ (2000) Time-bound negotiation framework for electronic commerce agents. Decis Support Syst 26(2):319–331
7. Corny SE, Kawabata, Lesser VR, Meyer RA (1991) Multistage negotiation for distributed satisfaction. IEEE Trans Syst Man Cinematic 21(6):1462–1465
8. Wei Z-W, Qu Y-P Yan J-Y (2007) An improved dynamic contract nets. Comput Eng Appl 43(36):208–210 (in Chinese with English abstract) Beijing
9. Lan S-H (2002) The research of multi-agent technology and its application. Ph.D. thesis, Nanjing University of science and technology, Nanjing (in Chinese with English abstract)
10. Lin L, Feng L (2010) A multi-agent cooperation model based on improved contract net protocol. Comput Technol Dev 20(3):71–75
11. Wen HU (2010) Multi-robot system and key technology research based on MAS. Xihua University, Sichuan
12. Wang L-C Chen S-F (2002) A multi-agent and multi-issue negotiation model. J Softw 13(8):1637–1643 (in Chinese)
13. Zhang X-L, Shi C-Y (2007) A dynamic formation algorithm of multi-agent coalition structure 18(3):574–581 (in Chinese with English abstract)
14. Tao H-J, Wang Y-D, Guo M-Z, Wang H-L (2006) A multi-agent negotiation model bases on acquaintance coalition and extended contract net protocol. J Comput Res Dev 43(7):1155–1160 (in Chinese with English abstract), ISSN, Beijing
15. Luo Y, Shi C-Y (1995) A behavior strategy of agent cooperation alliance. J Comput 40(6):961–965 (in Chinese)

A Novel Radar Detection Approach Based on Hybrid Time-Frequency Analysis and Adaptive Threshold Selection

Zhilu Wu, Zhutian Yang, Zhendong Yin and Taifan Quan

Abstract Due to the increasing complexity of electromagnetic signals, there exists a significant challenge for radar signal detection. In this paper, a novel radar detection approach based on time-frequency distribution (TFD) is proposed. Exploiting the complementation of linear TFD and bilinear TFD approaches, the cross terms of Wigner-Ville distribution (WVD) are suppressed. By using an optimal threshold selected adaptively, the regions of WVD auto terms are determined exactly. And the multicomponent radar signals can be detected efficiently. Simulation results show that this approach can efficiently detect not only linear frequency modulation (LFM) signals, but also normal signals and phase modulation (BPSK and QPSK) signals.

Keywords WVD · Spectrogram · Adaptive threshold selection · Auto term

1 Introduction

Radar signal detection is a subject of wide interest in both civil and military applications [1]. In battlefield surveillance applications, radar signal detection is a critical function in radar reconnaissance system [2]. Concretely, radar signal detection provides an important means to detect targets employing radars, especially those from hostile forces. In civilian applications, the technology can be used to detect and identify navigation radars deployed on ships and cars used for criminal activities [1]. The recent proliferation and complexity of electromagnetic signals encountered in modern environments is greatly complicating the detection of radar signals [2]. Traditional recognition methods are becoming inefficient

Z. Wu · Z. Yang · Z. Yin (✉) · T. Quan
School of Electronics and Information Technology, Harbin Institute of Technology, Harbin, Heilongjiang, China
e-mail: 86166661@qq.com

F. Sun et al. (eds.), *Knowledge Engineering and Management*,
Advances in Intelligent Systems and Computing 214,
DOI: 10.1007/978-3-642-37832-4_8,

against this emerging issue [3]. Many new radar signal detection methods were proposed, e.g., energy detection, time-frequency analysis [4–6] and adaptive detection [7].

Thereinto, time-frequency representation (TFR) has been frequently used for the detection, the analysis and the classification, of nonstationary signals [8–10]. In particular, many studies are done using the WVD to detect the LFM signals [11]. Unfortunately, these approaches are inefficient against other modulation signals, e.g. nonlinear modulation and phase-modulated (BPSK and QPSK) signals. And most of these approaches focus on the case of simple component. About multicomponent radar signal detection, Rong et al. builded the model of multicomponent radar signals and detect radar signals with SVM [11, 12]. Sun et al. proposed a multicomponent radar signal detection approach using WVD and multiple spectrograms, which can detect multicomponent radar signal efficiently and suppress the WVD cross terms. This approach exploits the linear spectrogram to remedy the bilinear WVD's and suppresses the WVD cross terms by using a threshold, which proposes a potential research direction. But this approach is only used to detect LFM signals and the process of determining the threshold is not introduced.

In this paper, a novel radar signal detection approach is proposed. To deal with the drawback of approaches on WVD, multicomponent radar signals are detected via adaptively thresholding superimposition of multiple spectrograms. The WVD auto terms are determined by signal sorting. The simulation results show that the proposed approach can detect LFM and PM radar signals efficiently.

The rest of the paper is organized as follows. In Sect. 2, the model of the radar detection system is introduced. In Sect. 3, theory of adaptive radar signal detection based on hybrid TFD. The performance of the proposed approach is analyzed in Sect. 4, and conclusions are given in Sect. 5.

2 Model of the Radar Detection System

In this paper, a radar detection approach based on multi spectrogram and an adaptive threshold are proposed to deal with the drawback of the recognition approaches using WVD. The process of the radar emitter recognition approach is shown in Fig. 1.

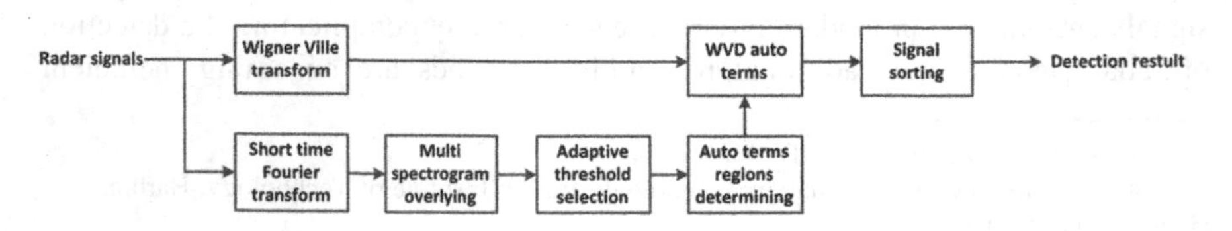

Fig. 1 Model of radar detection system

This recognition approach consists of three distinct functional modules:

1. time-frequency analysis,
2. adaptive threshold selection,
3. signal sorting.

Time-frequency analysis is a well-developed technique [13]. In this paper, the time-frequency analysis consists of two parts, namely, Wigner-Ville transformation and short time Fourier transformation. Especially, the Wigner-Ville distribution (WVD) has advantage to analyze nonstationary signals. However the cross WVD is a serious interference. So the cross WVD is suppressed by determining the regions of WVD auto terms. And these regions are obtained by adaptively thresholding the superimposition of multiple spectrograms method. After that, multicomponent radar signals are sorted based on their WVD auto terms.

3 Adaptive Radar Signal Detection Based on WVD and Spectrogram

3.1 Time-Frequency Analysis Methods

WVD Time-frequency distributions consist of linear and nonlinear methods. The bilinear t-f representation (BTFR) is the latter, which is developed into a general framework by Coben [14]. The WVD is a regular BTFR, which has been widely used for feature extraction and signal detection. Given a signal $x(t)$, the WVD of $x(t)$ is defined by

$$W_x(t,f) = \int_{-\infty}^{+\infty} R_x(t,\tau)\exp(-2\pi jf\tau)d\tau \tag{1}$$

$$R_x(t,\tau) = x\left(t+\frac{\tau}{2}\right)x^*\left(t-\frac{\tau}{2}\right) \tag{2}$$

where $*$, τ and $W_x(t,f)$ denote the complex conjugate, the dummy variable and the energy distribution of $x(t)$ in frequency f at time t. The region of WVD is obtained by applying a window on the signal WVD, described by the shape Ω in the time-frequency plane. The E_Ω denotes the energy of the WVD restricted to the region Ω. E_Ω is defined by

$$E_\pi(t,f) = \int_{-\infty}^{+\infty}\int_{-\infty}^{+\infty} R_x(t,\tau)W_x(t',f')dt'df' \tag{3}$$

It is important to note that any other BTFR derived from WVD by smoothing in *t-f* plane is singular and the smoothing operation will make the information lost [13]. The standard WVD has an optimum information-preserving property. From a signal detection viewpoint, the WVD is real valued for any complex-valued input signal, and it exhibits the least amount of spread in the *t-f* plane. So we choose WVD as the tool for performing the *t-f* analysis that represents the first step in our approach.

A criticism leveled against WVD is the cross terms, due to the presence of two (or more) signal components. In this paper, a method to suppress the cross-terms and detect the auto-terms of WVD, by using superimposition of multiple spectrograms, which is proposed in reference [15].

Spectrogram The spectrogram of input signal $x(t)$ is defined by

$$P_f(t,\omega) = \left|S_f(t,\omega)\right|^2 \tag{4}$$

where the $S_f(t,\omega)$ is the short time Fourier transformation of the input signal $x(t)$, which is defined by

$$S_f(t,\omega) = \int f(u)g^*(t-u)e^{-i\omega u}du \tag{5}$$

The time-frequency resolution of short time Fourier transformation is determined by the time-bandwidth product, namely, $\sigma_t\sigma_\omega$. Due to the uncertainty principle [16], high frequency resolution and high time resolution can not be received at the same time. The WVD of a LFM signal is a linear pulse and the spectrogram is in the band centered with the linear pulse. The sharp of the band is elliptical, and the width of the band can be regarded as the time-frequency resolution. To express the multicomponent signals, the Chirplet transformation is proposed [17]. The Chirplet transformation is given by

$$g_{\sigma_t,\beta}(t) = \frac{1}{\sqrt[4]{\pi\sigma_t^2}}\exp\left(-\frac{t^2}{2\sigma_t^2}+j\frac{\beta}{2}t^2\right) \tag{6}$$

$$W_{g_{\sigma_t,\beta}}(t,\omega) = \frac{1}{\pi}\exp\left[-\frac{t^2}{\sigma_t^2}-\sigma_t^2(\omega-\beta t)^2\right] \tag{7}$$

where β denotes the time-frequency rate of Chirplet transformation. When the parameter β is closer to the time-frequency rate of a LFM signal, the width of the spectrogram is narrower and the energy is more intensive around the instantaneous frequency.

Suppression of WVD cross terms Because short time Fourier transformation is linear, in analysis of multicomponent signals, it can meet the superposition principle and not produce cross terms. So we can determine the regions of WVD auto terms based on spectrograms.

In the case of the discrete signals, let $f(t)$ be the the sampling value of the input signal, i.e., $f(t) = x(n\Delta t)$, where $0 \leq n < N$. And the corresponding discrete short time Fourier transformation is as follow:

$$S_{f,L_t,\beta}(n,k) = \sum_{m=0}^{N-1} f(m) g^*{}_{L_t,\beta}(m-n) \exp\left(-\frac{2\pi mk}{N}\right) \tag{8}$$

$$g_{L_t,\beta}(n) = \frac{1}{\sqrt[4]{\pi L_t^2}} \exp\left(-\frac{n^2}{2L_t{}^2} + j\frac{\beta}{2}n^2\right) \tag{9}$$

where $g_{L_i,\beta}(n)$ is a discrete window function, $L_t = \sigma_t/\Delta t$, $\beta = \eta \Delta t^2$.

Suppose the modulating frequency range of signal components $\delta \in [-\alpha, \alpha]$, the modulating frequencies of the window functions is given by

$$\beta_i = -\alpha + i\Delta\beta, \; \Delta\beta = 2\alpha/I, \; i = 0, 1, \ldots, I \tag{10}$$

The spectrogram of the discrete signal is given by

$$D_f(n,k) = \frac{1}{I+1} \sum_{i=0}^{I} \left|S_{f,L_i,\beta}(n,k)\right|^2 \tag{11}$$

The regions of auto-terms of spectrogram are determined by thresholding $D_f(n,k)$, i.e.,

$$\Omega = \{(n,k) : D_f(n,k) \geq T\} \tag{12}$$

where T denotes the threshold, which has to do with signal noise ratio (SNR). T is determined by adaptive threshold selection, which will be introduced blow.

So the regions of auto-terms of WVD are given by

$$W_{\text{auto}}(n,k) = S_\Omega(n,k) W_f(n,k) \tag{13}$$

$$S_\Omega(n,k) = \begin{Bmatrix} 1, & (n,k) \in \Omega \\ 0, & (n,k) \notin \Omega \end{Bmatrix} \tag{14}$$

3.2 Adaptive Threshold Selection

Adaptive preferences is a popular topic in machine learning. In signal detection, adaptive threshold selection is an important topic. In [18], OTSU proposed an appropriate criterion for evaluating the "goodness" of threshold from a more general standpoint, which is regarded as the most commonly used standard. The standard is confined to the elementary case of threshold selection where only the histogram suffices without other a priori knowledge. It approaches the feasibility of thresholds and selects an optimal threshold automatically [18]. In this paper, the threshold is automatically selected based on this standard.

Let the energy of a given radar signal sampled be represented in L levels $[1,2,\ldots,L]$. The number of sample at level i is denoted by n_i and the total number of samples by $N = n1_1 + n_2 + \cdots + n_L$. To simplify the discussion, the energy-level histogram is normalized and regarded as a probability distribution:

$$p_i = \frac{n_i}{N}, p_i \geq 0. \tag{15}$$

It is obvious that

$$\sum_{i=1}^{L} p_i = 1. \tag{16}$$

We classify the samples into two classes C_0 and C_1 using a threshold at level k. Suppose that C_0 denotes samples in levels $[1,2,\ldots,k]$ and C_1 the samples in levels $[k+1,\ldots,L]$. Then the probability of each class occurrence is given by

$$\omega_0 = \Pr(C_0) = \sum_{i=1}^{k} p_i = \omega(k) \tag{17}$$

$$\omega_1 = \Pr(C_1) = \sum_{i=k+1}^{L} p_i = 1 - \omega(k) \tag{18}$$

The mean level of each class is give by

$$\mu_0 = \sum_{i=1}^{k} i \Pr(i|C_0) = \sum_{i=1}^{k} ip_i/\omega_0 = \mu(k)/\omega(k) \tag{19}$$

$$\mu_1 = \sum_{i=k+1}^{L} i \Pr(i|C_1) = \sum_{i=k+1}^{k} ip_i/\omega_1 = \frac{\mu_T - \mu(k)}{1 - \omega(k)} \tag{20}$$

respectively, where

$$\omega(k) = \sum_{i=1}^{k} p_i \tag{21}$$

and

$$\mu(k) = \sum_{i=1}^{k} ip_i \tag{22}$$

denote the zeroth order and the first order cumulative moments of the histogram up to the kth level, respectively, and

$$\mu_T = \mu(L) = \sum_{i=1}^{L} ip_i \tag{23}$$

denotes the total mean level of the energy of all samples. For any k,

$$\omega_0\mu_0 + \omega_1\mu_1 = \mu_T \tag{24}$$

Class variances are given by

$$\sigma_0^2 = \sum_{i=1}^{k} (i - \mu_0)^2 \Pr(i|C_0) = \sum_{i=1}^{k} (i - \mu_0)^2 p_i/\omega_0 \tag{25}$$

$$\sigma_1^2 = \sum_{i=k+1}^{L} (i - \mu_1)^2 \Pr(i|C_1) = \sum_{i=k+1}^{L} (i - \mu_1)^2 p_i/\omega_1 \tag{26}$$

In order to evaluate the performance of the threshold, we use the measures of class separability, as follow:

$$\eta = \sigma_B^2/\sigma_T^2 \tag{27}$$

where

$$\sigma_B^2 = \omega_0\omega_1(\mu_1 - \mu_0)^2 \tag{28}$$

and

$$\sigma_T^2 = \sum_{i=1}^{L} (i - \mu_T)^2 p_i \tag{29}$$

are the between-class variance, and the total variance of levels, respectively. It is noticed that σ_B^2 is the function of the threshold level k, but σ_W^2 is independent of k.

So the problem is reduced to an optimization problem to search for the threshold k^* that maximizes μ, which is equivalently to maximizes σ_B^2, viz.

$$\sigma_B^2(k^*) = \max_{1 \le k \le L} \sigma_B^2(k) \tag{30}$$

The optimal threshold k^* that maximizes η, is selected in the following sequential search.

3.3 Radar Signal Sorting

Multicomponent radar signal may include LFM signals with different time-frequency rates and other modulation signals. Many nonlinear modulation signals can be approximated by a combination of LFM signals.

After radar signal detection introduced above, WVD cross terms and noise are wiped off. Regions of WVD auto terms are obtained. For each WVD auto term, let β^* is the maximum of the β_i, viz.

$$\beta^* = \max(\beta_i), \ (i = 0, 1, \ldots, I) \tag{31}$$

where β^* can be regarded as the approximate time-frequency rate of the signal according to the WVD auto term. For linear frequency modulation signals, $\beta^* \neq 0$. The region of a LFM signal auto term can be determined based on β^*.

Due to the threshold decision based on spectrograms, phase modulation signals, e.g., QPSK, are presented as a group of LFM auto terms with $\beta^* = 0$, which are narrow in time domain. So the PM signals can be determined by it's group of auto terms, namely, by the position of marginal auto terms.

4 Simulation Results and Analysis

The validity and efficiency of the proposed approach is proved by simulations. In this simulation, a multicomponent radar signal made up of three types radar signals is exploited, which consist of a signal of LFM, NS, BPSK and QPSK, respectively. The multicomponent radar signal is given by

$$s(t) = s_1(t) + s_2(t) + s_3(t) + s_4(t) + n(t) \tag{32}$$

where

$$s_1(t) = \exp\left[j(5 \times 10^7 \pi t) + 2.5 \times 10^{13} \pi t^2\right] \tag{33}$$

$$s_2(t) = \exp\left[j(6 \times 10^7 \pi t)\right] \tag{34}$$

$$s_3(t) = \exp\left[j(5 \times 10^7 \pi t) + \pi d_1(t)\right] \tag{35}$$

$$s_4(t) = \exp\left[j(6 \times 10^7 \pi t) + \pi d_2(t)\right] \tag{36}$$

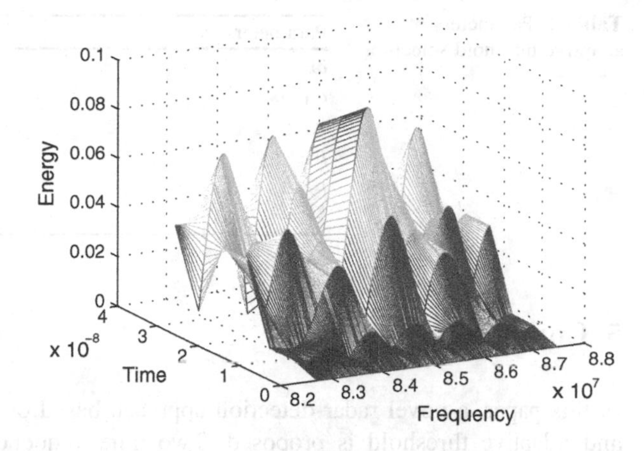

Fig. 2 Model of radar detection system

where $d_1(t)$ is the phase function, which is binary-phase coded, using the 7 Barker codes in 10 μs. $d_2(t)$ is the phase function, which is four-phase coded, using the 16 Frank codes in 20 μs. $n(t)$ is additive white gaussian noise and the $SNR = 10\ dB$.

The WVD of the multicomponent radar signal is shown as Fig. 2.

Spectrograms with different time-frequency resolutions of the multicomponent radar signal is thresholded adaptively. And the parameter of adaptive threshold selection is shown in Table 1.

Using the threshold selected, the regions of spectrogram auto terms is determined, which are the auto term support regions of WVD as well. The auto WVD of each radar component is extracted, which is shown in Fig. 3.

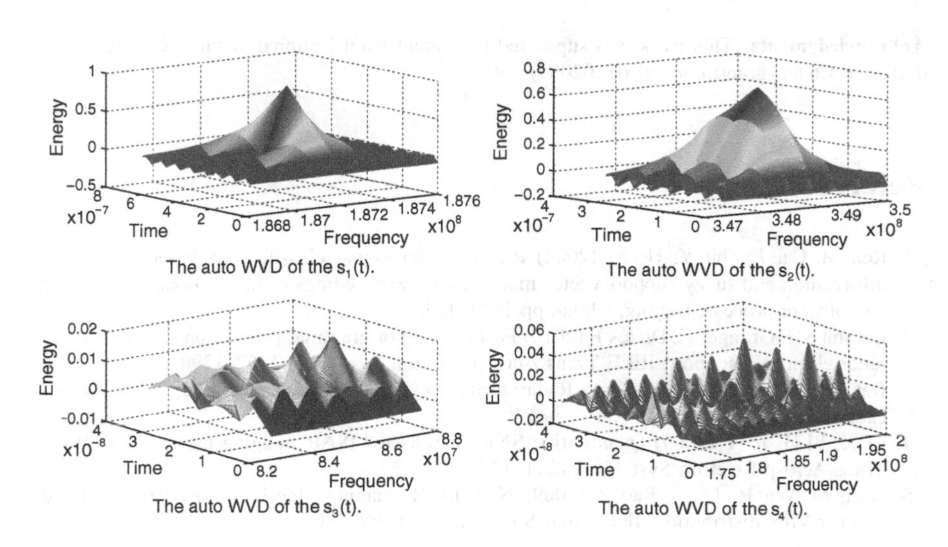

Fig. 3 Model of radar detection system

Table 1 Parameters of adaptive threshold selection

Parameter	Value
ω_0	0.832
ω_1	0.168
μ_0	3.1
μ_1	9.7
μ_T	4.2
k^*	7

5 Conclusion

In this paper, a novel radar detection approach based on time-frequency analysis and adaptive threshold is proposed. Two time-frequency analysis methods are exploited, viz. WVD and spectrogram. Spectrograms with different time-frequency resolutions are superimposed and then is thresholded by the adaptive threshold, to localize the auto-term support region of the WVD.

The additivity of spectrogram is exploited to suppress the bilinearity of WVD. By determining the auto term of the spectrogram, the regions of auto WVD's are obtained and the cross WVD is abandoned. Unlike the traditional kernel function methods to suppress cross terms, this approach not only reduces the interfering cross terms, but also preserves the superb time-frequency concentration of the WVD. In addition, the adaptive threshold selection is exploited to get an optimal threshold for superimposition of spectrograms with different time-frequency resolutions. By using this optimal threshold, the spectrograms can be distinguished from the noise. The regions of spectrogram auto terms are determined and the each radar signal component is detected exactly.

Acknowledgments This work was supported by a grant from National Natural Science Foundation of China (grant number: 61102084).

References

1. Ren M, Cai J, Zhu Y, He M (2008) Radar emitter signal classification based on mutual information and fuzzy support vector machines. In: Proceedings of international conference on software process, Beijing, China, pp 1641–1646
2. Latombe G, Granger E, Dilkes F (2010) Fast learning of grammar production probabilities in radar electronic support. IEEE Trans Aerosp Electron Syst 46(3):1262–1290
3. Bezousek P, Schejbal V (2004) Radar technology in the Czech Republic. IEEE Aerosp Electron Syst Mag 19(8):27–34
4. Li X, Bi G, Ju Y (2009) Quantitative SNR analysis for ISAR imaging using LPFT. IEEE Trans Aerosp Electron Syst 45(3):1241–1248
5. Xing M, Wu R, Li Y, Bao Z (2009) New ISAR imaging algorithm based on modified Wigner-Ville distribution. IET Radar Sonar Navig 3(1):70–80

6. Jianxun L, Qiang LV, Jianming G, Chunhua, X (2007) An intelligent signal processing method of radar anti deceptive jamming. In: Wen TD (ed) ISTM/2007: 7th international symposium on test and measurement, vols 1–7, conference proceedings, Chinese Soc Modern Tech Equipment; Chinese Assoc Higher Educ, (2007) 1057–1060 7th international symposium on test measurement, Beijing, Peoples Republic China, August 05–08
7. Svensson A, Jakobsson A (2001) Adaptive detection of a partly known signal corrupted by strong interference. IEEE Signal Process Lett 18(12):729–732
8. Bin GF, Liao CJ, Li, XJ (2011) The method of fault feature extraction from acoustic emission signals using Wigner-Ville distribution. In: Yang YH, Qu XL, Luo YP, Yang A (eds) Optical, electronic materials and applications, PTS 1-2, volume 216 of advanced materials research, Hunan institute engineering, Hunan university science and technology, Shanghai Jiaotong university, Chongqing normal university, Wuhan university technology, Nanyang technology university, Hebei polytechnic university, international conference on optical, electronic materials and applications, Chongqing, Peoples Republic China, pp 732–737, March 04–06
9. Feng L, Peng SD, Ran T, Yue W (2008) Multi-component LFM signal feature extraction based on improved wigner-hough transform. In: 2008 4th international conference on wireless communications, networking and mobile computing, vols 1–31, international conference on wireless communications, networking and mobile computing, IEEE communication society, IEEE Antennas and propagation society, Dalian university technology, Wuhan university, Scientific research publishing, (2008) 1938–1941 4th international conference on wireless communications, networking and mobile computing, Dalian, Peoples Republic China, October 12–17
10. Li L, Ji HB, Jiang L (2011) Quadratic time-frequency analysis and sequential recognition for specific emitter identification. IET Signal Proc 5(6):568–574
11. Zhang G, Rong H, Jin W (2006) Pulse modulation recognition of unknown radar emitter signals using support vector clustering. Lecture Notes in Artificial Intelligence (FSKDZO06), vol 4223. Springer, Heidelberg, pp 120–429
12. Rong H, Zhang G, Jin W (2006) Application of S-method to multi-component emitter signals. In: Lecture Notes in Artificial Intelligence (FSKDZO06), vol 4223, Springer, pp 420–429
13. Haykin S, Bhattacharya T (1997) Modular learning strategy for signal detection in a nonstationary environment. IEEE Trans Signal Process 45(6):1619–1637
14. Cohen L (1989) Time-frequency distributions-a review. Proc IEEE 77(7):941–981
15. Li Q, Shui P, Lin Y (2006) A new method to suppress cross terms of WVD via thresholding superimposition of multiple spectrograms. J Electron Inf Technol 28(8):1435–1438
16. Mann S, Haykin S (1995) The chirplet transform: physical considerations. IEEE Trans Signal Process 43(11):2745–2761
17. Bultan A (1999) A four-parameter atomic decomposition of chirplets. IEEE Trans Signal Process 47(3):731–745
18. Otsu N (1979) A threshold selection method from gray-level histograms. IEEE Trans Syst Man Cybern 9(1):62–66

A Signed Trust-Based Recommender Approach for Personalized Government-to-Business e-Services

Mingsong Mao, Guangquan Zhang, Jie Lu and Jinlong Zhang

Abstract Recently recommender systems are introduced into the web-based government applications which expect to provide personalized Government-to-Business (G2B) e-Services. For more personalization, we illustrate a subjective signed trust relationship between users, and based on such trust we proposed a recommendation framework for G2B e-services. A case study is conducted as an example of implementing our approach in e-government applications. Empirical analysis is also conducted to compare our approach with other models, which shows that our approach is of the highest. In conclusion, the signed trust relationship can reflect the real preferences of users, and the proposed recommendation framework is believed to be reliable and applicable.

Keywords Recommender system · Signed trust · Distrust · G2B e-services

---

M. Mao (✉) · G. Zhang · J. Lu
Decision Systems and e-Service Intelligence (DeSI) Lab,
Center for Quantum Computation and Intelligent Systems,
Faculty of Engineering and Information Technology,
University of Technology, Sydney, Sydney, Australia
e-mail: Mingsong.mao@student.uts.edu.au

G. Zhang
e-mail: Zhangg@it.uts.edu.au

J. Lu
e-mail: Jielu@it.uts.edu.au

M. Mao · J. Zhang
Institute of Management Information, Department of Management
Science and Information Management, School of Management
, Huazhong University of Science and Technology, Wuhan, China
e-mail: jlzhang@mail.hust.edu.cn

F. Sun et al. (eds.), *Knowledge Engineering and Management*,
Advances in Intelligent Systems and Computing 214,
DOI: 10.1007/978-3-642-37832-4_9,

1 Introduction

The rapid growth of government web-based applications brings various Government-to-Business (G2B) e-Services by which the businesses enable to browse and identify partners on the government websites. However, how to provide personalized e-services is getting more and more emphases in this stage because the information overloading leads to that the online agents (businesses or citizens) are hard to effectively choose the information they expect. A notable technique to provide personalized services is recommender system. Recommender systems can generate recommended products or service to customers using justifications and to ensure the customers like them. These justifications can be obtained either from preferences directly expressed by customers, or induced, using data representing the customer experience. Varied by the source of data, recommendation systems are commonly categorized into three types [1]: (a) content-based (CB) systems, (b) collaborative filtering (CF) systems, and (c) hybrid systems that make use of a combination of the former two approaches and/or knowledge-based methods. In the CB systems, the most interest items are recommended to the target user according to his/her personalized preference abstracted by his/her profiles or previous actions. The CF approaches [2] identify the similar users (user-based) or similar items (item-based) only relies on history ratings. They have been widely employed since they do not request the explicit descriptions of users/items. Commonly, the CF-based recommender systems are believed having advantages over CB-based systems in terms of time and space complexity in large-scale data sets. However, CF approaches suffer from problem known as "cold start problem," which points their inability of making recommendations for the users who rated very few items (new-user problem), or for the items which have not been rated by anyone (new-item problem). To overcome these drawbacks, hybrid approaches combined by both CB and CF techniques are commonly implemented in real systems [3].

Recently, trust-based filtering technique is becoming another improving facet in recommender systems. The most discussed two questions of trust studies in the field of recommender system are (1) the measure of trust and (2) the propagation mechanism of trust (also known as the calculation of inferred trust). Massa et.al constructed a baseline deviation and propagation mechanism of "the web of trust." For example, in the trust model of [4], the indirect trust is estimated based on a simple assumption: "users who are closer to the source user in the trust network have higher trust values.". A more systematic algorithm was proposed by Golbeck in [5]: TidalTrust, a trust inferring approach based on breath-first search algorithm. TidalTrust has been thought to be effective and applicable in several systems [6, 7]. Typically, trust-based approaches are believed enable to improve recommendation coverage while maintaining the level of accuracy contrasted with CF approaches. In most of these studies, the term "trust" only represents a positive relationship. Rare of them discuss the "distrust" relationship (negative relationship). Then in the recommendation making process, only the suggestions from the ones who are strongly trusted are considered, but the information of those who are strongly distrusted is ignored. In order to consider both trust and distrust relationships between users, this paper aims to propose a signed trust-based recommendation framework for personalized G2B e-services. The subsequent sections are organized as follows: Sect. 2 gives a definition of signed trust in recommender systems, and discusses the derivation method of such trust relationship. In Sect. 3, a case study is conducted in order to give a guideline of how to implement our approach in real G2B systems. Then empirical experiments are processed in large-scale data to evaluate the recommendation accuracy of our approach, in Sect. 4. Finally in Sect. 5, the conclusion and future work are presented.

2 Building Trust and Distrust Network

2.1 Definition of Trust and Distrust

The term "trust" in recommender systems is commonly defined as *"the probability that Alice will accept Bob's suggestion base on a belief that Bob's future actions will lead to a good outcome."*[5]. Obviously it represents a positive relationship, because the worst situation is that *"Alice will entirely ignore Bob's suggestion,"* i.e., the accept probability is zero. However, this definition does not consider the negative trust relationships, which reflect the opposite preferences of people. Therefore, we introduce the term "distrust" in our research to define *"the probability that Alice will reject Bob's suggestion base on a belief that Bob's future actions will lead to a bad outcome."* Formally, we use a signed value $t_{uv} \in [-1, +1]$ representing the trust relationship of user u and user v, where the sign "+" expresses "trust" relationship, the sign "−" expresses "distrust" relationship, and the absolute value expresses the strength of the trustworthiness.

2.2 Trust Derivation

In recommender systems, the trust relationships can be expressed by rating history, based on an assumption that other users' ratings can be seemed as recommendations to a certain user [4, 8]. For example, if a user v has delivered highly similar/opposite recommendations to user u in the past, then user v should acquire a high trust/distrust value from user u [9]. For this purpose, we propose a trust inference method to obtain the trust relationship of any two users based on comparing their rating history. Formally, let $U = \{u_k\}$, $k = 1,2,\ldots,m$, be the set of users in a recommender system, $I = \{i_s\}$, $s = 1,2,\ldots,n$, be the set of items, and $R = \{r_{ui}\}$, $u \in \mathrm{U}$, $i \in \mathrm{I}$, be the set of user-item ratings. Conveniently, let I_u be the subset of items that has been rated by a certain user u, let I_{uv} simplify the expression of $I_u \cap I_v$.

Then the average rating $\overline{R_u} = \sum_{i \in I_u} r_{ui} \big/ n(I_u)$ represents the mean level of user u's ratings to the items. Then a user's preferences on rated items can be depicted by a set $P_u = \{p_{ui} | i \in I_u\}$, where the preference p_{ui} is expressed by the deviation of r_{ui} and the average rating:

$$p_{ui} = r_{ui} - \overline{R_u}. \tag{1}$$

More concretely, the preference $p_{ui} > 0$ means that the user u interest item i more than the average level, while $p_{ui} < 0$ means that the user u does not prefer item i. Then the direction (prefer or not) of the preference p_{ui} expresses the user's overall opinion on such item. Furthermore, if we compare two users preference, p_{ui}, and p_{vi} on a same item, then we can illustrate the "one-item based" trust, denoted as t_{uv}^{i} (the inferred trust from u to v on a particular item i), by identifying whether they have the same or opposite preferences on such item. That is, if both of them think that item i is of greater or lower interest than their own average level, they see the ratings of each other as a reliable recommendation, i.e., they trust each other on such item, otherwise, they distrust. Accordingly, the sign of t_{uv}^{i} can be identified by (2).

$$t_{uv}^{i} = \begin{cases} \geq 0, & \text{if } p_{ui} \cdot p_{vi} \geq 0 \\ < 0, & \text{if } p_{ui} \cdot p_{vi} < 0 \end{cases}. \tag{2}$$

The only information for calculating the absolute value of t_{uv}^{i} is the deviation of r_{ui} and r_{vi}, which we denote as $\Delta_{uv}^{i} = |r_{ui} - r_{vi}|$. However Δ_{uv}^{i} makes different contribution for "trust" relationships and "distrust" relationships. For example, if user u and user v trust each other, the higher deviation they have, the lower the absolute value of such "trust" relationship is. On the other hand, if user u and user v distrust each other, then a higher deviation will lead to higher "distrust" strength. To compute the strength of such trust/distrust relationship, the deviation Δ_{uv}^{i} is first normalized by being divided by the span of ratings. Denote $R_{\max}$ and $R_{\min}$ the maximum and minimum value of the rating scale, then the normalized deviation $\tilde{\Delta}_{uv}^{i} = \Delta_{uv}^{i} / (R_{\max} - R_{\min})$. Accordingly, t_{uv}^{i} is calculated by (3).

$$t_{uv}^{i} = \begin{cases} 1 - \tilde{\Delta}_{uv}^{i}, & \text{if } t_{uv}^{i} \geq 0 \\ -\tilde{\Delta}_{uv}^{i}, & \text{if } t_{uv}^{i} < 0 \end{cases} \tag{3}$$

Then the overall trust t_{uv} is combined by weighted averaging the trust values of user u and v on each of the items rated by both of them. The weight we set here is the absolute preference set of the active user, for example, in the calculation of t_{uv}, the preference set of user u, $\{|P_{ui}|\}$ is set as the weight set. That is, we assume that users most concern the items which acquire extreme preferences. Finally, the trust inferring formula is:

$$t_{uv} = \frac{\sum_{i \in I_{uv}} |p_{ui}| \cdot t_{uv}^{i}}{\sum_{i \in I_{uv}} |p_{ui}|}. \tag{4}$$

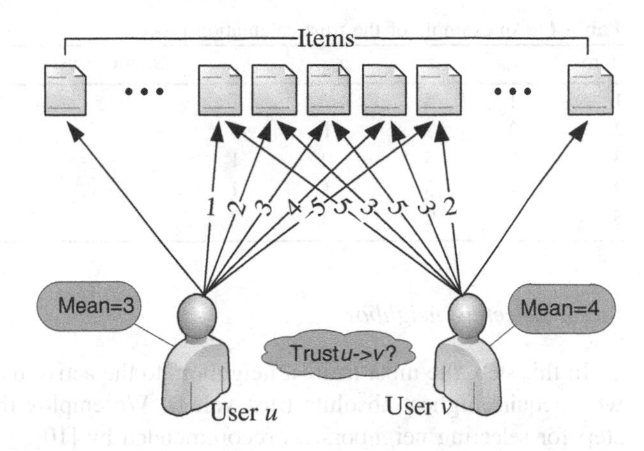

Fig. 1 An example of trust calculation based on history ratings

A toy example is given in Fig. 1, in a given system, user u and user v have rated some items. The co-rated items of them are item 1–5. The average ratings of user u and v are 3, and 4, respectively. Then the middle processes of trust calculating t_{uv} and t_{vu} is detailed in Table 1.

First, the preference sets are calculated to identify the sign of the trust on each item. The results of Table 1 show that they trust each other on item 2 and 3, but distrust on item 1, 4, and 5. On the other hand, the deviations of each pair of ratings are computed and normalized for calculating the absolute value of trust values with regards to the signs. Accordingly, the trust set is $T = [-1, 0.75, 0.5, -0.25, -0.75]^T$, and the weight set for user u is $w_u = [2, 1, 0, 1, 2]$, the weight set for user v is $w_v = [1, 1, 1, 1, 2]$. Then in such example, t_{uv} and t_{vu} are calculated by:

$$\begin{aligned} t_{uv} &= w_u \cdot T \Big/ \sum w_u = -3/6 = -0.5 \\ t_{vu} &= w_v \cdot T \Big/ \sum w_v = -1.5/6 = -0.25. \end{aligned} \tag{5}$$

For the active user, once the prediction process is completed, a certain number of unrated items with the highest predicted ratings will be recommended to the user at the final step.

2.3 Recommendation Framework

To generate recommendations, four main steps will be processed in a trust-based recommender system which considers the distrust information:

Step1 *Calculating the trust relationships between users*

The user–user trust/distrust relationships are inferred by the trust inferring formula (4) at the beginning when a request is received.

Table 1 An example of the trust calculation process

Item	r_{ui}	r_{vi}	p_{ui}	p_{vi}	Sign of trust	Δ_{uv}^{i}	$\tilde{\Delta}_{uv}^{i}$	t_{uv}^{i}
1	1	5	−2	1	−	4	1	−1
2	2	3	−1	−1	+	1	0.25	+0.75
3	3	5	0	1	+	2	0.5	+0.5
4	4	3	1	−1	−	1	0.25	−0.25
5	5	2	2	−2	−	3	0.75	−0.75

Step2 *Selecting neighbors*

In this step, the most nearest neighbors to the active user are chosen from those who acquire highest absolute trust values. We employ the *Top-K* method in this step for selecting neighbors, as recommended by [10].

Step3 *Rating prediction*

Once we identify the set of most trusted or distrusted neighbors, *NB*, generating predictions for the unrated items is the subsequent important step. In our approach, we use the Resinick's formula for calculating predictions [11], however, the similarity metric used in traditional CF approaches is replaced with the signed trust value. The predicted rating from a user u to item i, denoted by $\hat{r}_{ui}$, is hence calculated by (6).

$$\hat{r}_{ui} = \overline{R_u} + \frac{\sum_{v \in NB} \left(r_{vi} - \overline{R_v}\right) \cdot t_{uv}}{\sum_{v \in NB} |t_{uv}|}. \tag{6}$$

Step4 *Recommendation making*

For the active user, once the prediction process of ratings is completed, a certain number of unrated items (*Top-N*) with the highest predicted ratings will be recommended to the user at the final step.

3 A Case Study of G2B Recommendation Application

The Australian Trade Commission (Austrade) (http://www.austrade.gov.au) is the Australian Government's trade and investment development agency. One of the purposes they dedicate to is helping overseas companies to locate and identify the right Australian supplier through the Australian Suppliers Directory (ASD; http://www.austrade.gov.au/ASD/). The ASD promotes Australian goods and services to overseas buyers, as well as assists overseas buyers to search suppliers all over Australia. In this section, we introduce an example of how a recommender system may be deployed by the ASD, as a G2B e-service application, to support overseas buyers throughout the process of searching and selecting Australian's suppliers that match their preferences. For example, assume that there are in total seven Australian supplier businesses (S_1–S_7) in a particular industry listed in the directory. Also, suppose that there are now five overseas buyers (B_1–B_5) who have undertaken business with some of the listed suppliers and have rated them on a numeric 5-point scale from 1 (poor) to 5 (excellent). Accordingly, a raw buyer–supplier rating matrix can be constructed as depicted in Table 2.

Step1 *Calculating the signed trust between buyers*

Employing our approach, the trust values between each pair of the five buyers are calculated first by the trust inferring formula (4), as shown in Table 3, where we can find that both trust and distrust relationships exist.

Step2 *Neighbors selection*

Let the number of nearest neighbors *Top-K* equal three, then based on Table 3, take buyer B_5 for instance, the selected neighbors are $B_1(t_{51} = 0.875)$, $B_4(t_{54} = 0.750)$, and $B_3(t_{53} = -0.625)$.

Step3 *Calculating prediction values*

By the prediction formula (6), and the selected neighbors in last step, the absent ratings are then predicted, as shown in Table 4.

Table 2 Raw buyer–supplier rating matrix. The values in the cells with "?" are expected to be predicted

Suppliers	Buyers				
	B_1	B_2	B_3	B_4	B_5
S_1	1	5	5	3	3
S_2	2	4	4	Null	2
S_3	3	5	1	5	4
S_4	Null	2	Null	3	?
S_5	4	3	5	4	?
S_6	5	4	2	1	?
S_7	Null	Null	3	1	?
Average	3	3.83	3.33	2.83	3

Table 3 Buyer–buyer trust matrix

	B_1	B_2	B_3	B_4	B_5
B_1		−0.208	−0.542	−0.400	+0.667
B_2	−0.214		−0.095	+0.185	+0.550
B_3	−0.076	−0.196		+0.125	−0.268
B_4	+0.062	+0.098	+0.151		+0.768
B_5	+0.875	+0.125	−0.625	+0.750	

Table 4 Predicted buyer–supplier rating matrix

	B_1	B_2	B_3	B_4	B_5
S_1					
S_2				2.194	
S_3					
S_4	3.518		4.517		2.881
S_5					3.315
S_6					4.208
S_7	3.971	2.735			1.152

Step4 *Recommendation making*

For the buyer B_5, let *Top-N* = 3, hence, the most interested three suppliers for B_5 are recommended finally. That is, ASD will recommend the supplier $S_6(\hat{r}_{56} = 4.208)$, $S_5(\hat{r}_{55} = 3.315)$, and $S_4(\hat{r}_{54} = 2.881)$ to buyer B_5.

4 Empirical Analysis and Results

In this section, we implement our approach with a public dataset, MovieLens dataset, in order to compare our approach with other models. The dataset is chosen because it is publicly available online and has been widely used for evaluating recommender systems [12, 13]. The 100 K ratings dataset is selected in our experiments. This dataset consists of 100,000 ratings assigned to 1,682 movies by a total of 943 users. All users have guaranteed rated at least 20 movies on a scale of 1–5, where 1 = "Awful," 2 = "Fairly bad," 3 = "It is OK," 4 = "Will enjoy," and 5 = "Must see." The sparsity of the rating matrix is 93.70 %. Two metrics are commonly selected for evaluating recommender systems: recommendation accuracy and recommendation coverage. However, the relatively low sparsity of Movielens dataset leads to high recommendation coverage (almost 100 %). Therefore, we only evaluate the statistical accuracy metric. This measure evaluates the accuracy of a system by comparing the numerical prediction values against the actual user ratings for the user-item pair in the test data sates. The mean absolute error (MAE) is the most widely used metric in recommendation research [11]. MAE is calculated by the deviation of prediction from the actual rating. In particular, given the set of the actual/predicted rating pair $<r_{ui}, \hat{r}_{ui}>$ for all the n items in the test set, the MAE is computed by (7). Note that a lower MAE value represents higher recommendation accuracy.

$$\mathrm{MAE} = \frac{\sum_{u,i} |r_{ui} - \hat{r}_{ui}|}{n}. \tag{7}$$

Table 5 MAE of each test set under different level of k. At each round of test, the minimum errors are reached by K=50, therefore, K=50 is selected as the optimal neighbourhood size

K	u1.test	u2.test	u3.test	u4.test	u5.test	Average MAE
k = 5	0.7926	0.7849	0.7840	0.7809	0.7897	0.7864
k = 10	0.7668	0.7558	0.7547	0.7519	0.7606	0.7580
k = 20	0.7537	0.7423	0.7402	0.7379	0.7462	0.7441
k = 30	0.7504	0.7387	0.7361	0.7336	0.7418	0.7401
k = 50*	0.7492	0.7372	0.7348	0.7328	0.7399	0.7388
k = 80	0.7496	0.7380	0.7357	0.7337	0.7400	0.7394
k = 120	0.7507	0.7398	0.7368	0.7350	0.7409	0.7406
k = ALL	0.7529	0.7422	0.7388	0.7369	0.7427	0.7427

4.1 Optimal neighborhood size

The experiments are conducted using five-fold cross validation technique. In this dataset, the five training sets (u1.train-u5.train) and five testing sets (u1.test–u5.test) have already been split by the provider for convenient comparisons with other models. To test the effect, the number of selected neighbors (*Top-K*) has on the accuracy of our system, we perform experiments by varying the level of K and compare the errors (MAE). Several levels of the neighborhood size are considered in our experiments: $K = 5$, $K = 10$, $K = 20$, $K = 30$, $K = 50$, $K = 80$, $K = 120$, and $K =$ ALL. The results are shown in Table 5. From Table 5 we find that for each test set, accompany with the increasing of K, first a considerable drop of MAE happens at the beginning until $K = 50$, at which the MAE reaches the minimal level; and then the MAE slightly increase until the end. The algorithm hence achieves the maximum accuracy performance (i.e., minimal level of error) at $K = 50$. Therefore, the neighborhood size 50 is selected as the optimal value for producing the best performance of our approach.

4.2 Comparing with Other Approaches

We prefer to use the average but not the best value of MAE of the five test sets under the optimal neighborhood size as the final evaluation result of our approach. In order to compare our results with those of other models, accuracy reports of several different models found in [12] are imported. We use their reports because all of these models show experiment results using the same metric (MAE) and cross-validation process over the same data sets, also the same five-fold cross validation sets.

In [7], Porcel, et al. presented a hybrid recommender system integrating the 2-tuple fuzzy linguistic approach. They also implemented four other content-based CF-based models for comparison analysis: (1) A pure content-based approach (titled as CB) [14] in which the similarity between two items is calculated using the cosine similarity. (2) A user-based collaborative approach (titled as UBC) [15]

Table 6 MAE values to compare with other models

	Our approach	IBC-C	IBC-P	UBC	CB
Average MAE	0.7388	0.7705	0.7716	0.7848	0.9187

in which predictions are made referring similar users ratings; the similarity between users is computed using Pearson's correlation coefficient. (3) An item-based collaborative approach (titled as IBC-C) [14] in which the predictions are averaged by the active users' ratings to similar items, which are identified using cosine measure. (4) Another item-based collaborative approach (titled as IBC-P) [14] where the similarity between two items is measured by Pearson's correlation coefficient.

Table 6 presents the MAE results obtained by each model, where we find our approach is of the lowest error (0.7388) contrasted with others. CB (0.9187) works the worst in terms of accuracy, followed by UBC (0.7848). The item-based CF approaches, IBC-C (0.7705) and IBC-P (0.7716), seem to be more accurate than UBC. It is believed that our approach achieves good accuracy performance on Movielens dataset. It also indicates that the signed trust model is effective on improving the recommendation quality in terms of accuracy.

5 Conclusions

This paper focus on illustrating trust and distrust relationships between users from their rating history, then based on obtained such signed trust value, the rating prediction is conducted using the similar filtering method of CF approaches. Comparing with the rating similarity metric (Cosine similarity or Pearson similarity for example) employed in traditional CF approaches, the signed trust used in our approach reflects a more personalized user–user relationship, as it is a "subjective" similarity. The good performance we achieved in the large-scale experiments proves that the inferred signed trust is referable for personalized G2B recommendation making. In the next stage, we expect to propose a general method of introducing trust network or social network into business recommender systems.

Acknowledgments The work presented in this paper was partially supported by the Australian Research Council (ARC) under discovery grant DP110103733.

References

1. Balabanovic M, Shoham Y (1997) Fab: content-based, collaborative recommendation. Commun ACM 40:66–72
2. Goldberg D et al (1992) Using collaborative filtering to weave an information tapestry. Commun ACM 35:61–70

3. Shambour Q, Lu J (2011) A hybrid trust-enhanced collaborative filtering recommendation approach for personalized government-to-business e-services. Int J Intell Syst 26:814–843
4. Massa P, Avesani P (2004) Trust-aware collaborative filtering for recommender systems. In: OntheMove confederated international workshop and conference. Springer-Verlag, Berlin, pp 492–508
5. Golbeck JA (2005) Computing and applying trust in web-based social networks. Ph D, University of Maryland
6. Golbeck J (2006) Combining provenance with trust in social networks for semantic web content filtering. In: International provenance and annotation workshop (IPAW 2006). Springer-Verlag, Berlin, pp 101–108
7. Golbeck J, Hendler J (2006) Filmtrust: movie recommendations using trust in web-based social networks. In: 3rd IEEE consumer communications and networking conference. Citeseer, Las Vegas, pp 282–286
8. O'Donovan J, Smyth B (2005) Trust in recommender systems. In: 10th international conference on intelligent user interfaces. ACM, San Diego, pp 167–174
9. Hwang CS, Chen YP (2007) Using trust in collaborative filtering recommendation. In: 20th international conference on industrial, engineering and other applications of applied intelligent systems. Springer-Verlag, Berlin, pp 1052–1060
10. Herlocker J et al (2002) An empirical analysis of design choices in neighborhood-based collaborative filtering algorithms. Inf Retrieval 5:287–310
11. Resnick P et al (1994) GroupLens: an open architecture for collaborative filtering of netnews. In: 1994 ACM conference on computer supported cooperative work. ACM, New York, pp 175–186
12. Porcel C et al (2012) A hybrid recommender system for the selective dissemination of research resources in a technology transfer office. Inf Sci 184:1–19
13. Vozalis MG, Margaritis KG (2007) Using SVD and demographic data for the enhancement of generalized collaborative filtering. Inf Sci 177:3017–3037
14. Barragans-Martinez AB et al (2010) A hybrid content-based and item-based collaborative filtering approach to recommend TV programs enhanced with singular value decomposition. Inf Sci 180:4290–4311
15. Sarwar B et al (2000) Analysis of recommendation algorithms for e-commerce. In: ACM E-Commerce 2000 conference. ACM, New York, pp 158–167

Aspects of Coordinating the Bidding Strategy in Concurrent One-to-Many Negotiation

Khalid Mansour, Ryszard Kowalczyk and Michal Wosko

Abstract Automated negotiation is an important mechanism of interaction between software agents and has been an active research area for more than a decade. When the automated negotiation process involves multiple agents, the problem of interdependency between the actions of agents during negotiation arises and consequently, a coordination mechanism becomes an essential part of the negotiation process. One of the important characteristics of a negotiating agent is its bidding strategy. This work addresses the problem of coordinating the bidding strategy of an agent negotiating concurrently with multiple agents (i.e., one-to-many negotiation) and discusses different interdependency factors affecting it.

Keywords Multiagent systems · Bidding strategy · Negotiation · Coordination

1 Introduction

Negotiation is a method of interaction between different parties that can effectively resolve conflicts [1]. This paper discusses various aspects of coordinating or managing related automated negotiations for a software agent negotiating concurrently with other software agents (i.e., one-to-many negotiation) in terms of deciding on the bidding strategy for each negotiation instance in each negotiation round.

K. Mansour (✉) · R. Kowalczyk · M. Wosko
Swinburne University of Technology, Melbourne, Australia
e-mail: mwmansour@swin.edu.au

R. Kowalczyk
e-mail: rkowalczyk@swin.edu.au

M. Wosko
e-mail: mwosko@swin.edu.au

F. Sun et al. (eds.), *Knowledge Engineering and Management*,
Advances in Intelligent Systems and Computing 214,
DOI: 10.1007/978-3-642-37832-4_10, © Springer-Verlag Berlin Heidelberg 2014

The key point in the coordination theory is managing interdependencies amongst related activities to achieve a common goal [2]. In other words, when the interdependencies between different related activities arise, the need for a coordination mechanism becomes essential. The related activities in the context of automated negotiation are the instances of interactions between autonomous agents. Given that an agent initiates multiple negotiation instances, there is a need to manage the bidding strategy for each instance given the behaviors of the opponents in each negotiation instance. Managing the bidding strategy involves managing the negotiation variables, e.g., a concession parameter.

The coordination as a process can be classified into two types: *centralized coordination* and *distributed coordination*. In this work, we mainly address the centralized approaches for coordinating multiple negotiations conducted by a buyer agent.

In our work, we consider the automated one-to-many multiagent systems and assume the following:

- agents are rational,
- agents are self-interested and aim to maximize their utility,
- agents do not disclose their private information such as their utility structures or deadlines, and
- agents use the alternating offers protocol [3].

In general, automation of any process reduces the time needed to do the job and produces more efficient/effective results. Automation of negotiation has similar objectives where software agents can work on behalf of the users to negotiate with each other for buying, selling, task assignment, resource allocation, etc.

Many negotiation frameworks are proposed in the literature to describe and automate the process of generation offers and counteroffers. The process of offers/counteroffers generation (i.e., bidding strategy) depends on some criteria that control the process. The most used criteria are the agent's internal resources such as time and the behavior of the opponents, e.g., [4].

Many published works address the problem of coordinating multiple related negotiations, e.g., [5–9]. Most of the previous works consider negotiation over one issue (e.g., price) for the purpose of reaching one agreement. When both the number of issues and the number of agreements increase, the coordination process becomes more complicated.

The contribution of this paper can be summarized as follows:

- investigate possible sources of interdependencies between related negotiations,
- discuss possible coordination approaches for different negotiation scenarios, and
- show a sample of empirical results which demonstrate the importance of coordination.

The rest of the paper is organized as follows: Sect. 2 discusses the negotiation model that captures the negotiation settings of our work while Sect. 3 discusses possible sources of interdependency in negotiation domain. Section 4 presents the related work. Section 5 presents the coordination approach while Sect. 6 shows an illustrative example of some empirical results and finally Sect. 7 concludes the paper.

2 Negotiation Model

We consider a buyer agent negotiating with a set of seller agents $\mathbf{S} = \{s_1, s_2, \ldots, s_n\}$ concurrently, see Fig. 1. We assume that the seller agents are independent in their actions, i.e., they do not exchange information. The buyer agent has a set of delegate negotiators $\mathbf{D} = \{d_1, d_2, \ldots, d_n\}$. It creates and destroys delegate negotiators during negotiation in response to the number of the seller agents who enter or leave negotiation. Each delegate d_i negotiates with a seller s_i. The possible negotiation issues over which **D** and **S** negotiate are included in the set $\mathbf{J} = \{j_1, j_2, ..., j_g\}$ and each issue $j_i \in \mathbf{J}$ must be an issue of negotiation by at least one negotiation pair, i.e., (d_i, s_i).

To make our negotiation framework more comprehensive, we introduce the notion of negotiation objects set (**O**) notion. The negotiation object is any item in which agents have interest to negotiate over. A negotiation object represents either a physical item (e.g., a printed book) or a non-physical item, e.g., a web service. Let $\mathbf{O} = \{o_1, o_2, \ldots, o_m\}$. Each $o_i \in \mathbf{O}$ represents an object of negotiation. The illustration of the idea is shown in Fig. 1.

We assume that each negotiation delegate is responsible to negotiate over one object, and many delegates can negotiate over the same object, but a delegate cannot negotiate over more than one object concurrently, see function f_d in Eq. 1.

In our model, each negotiation delegate is mapped onto an object, a deadline $t_{\max} \in \mathbb{N}^*$ and an offer generation tactic $\theta \in \Theta$. Each object is mapped onto a negotiation issue set ($J_l \in 2^{\mathbf{J}}$). Finally, each issue is mapped onto a set of constraints, e.g., the reservation values ([min,max]). The number and types of constraints vary. Equation 1 shows the formal representation of the three functions (i.e., f_d,f_o,f_j).

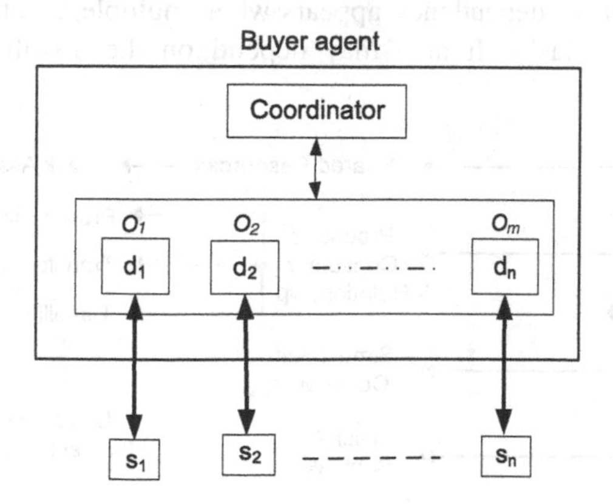

Fig. 1 One-to-many negotiation

$$f_d : \mathbf{D} \xrightarrow[(]{1-1} \mathbf{O} \times \mathbb{N}^* \times \Theta)$$

$$f_o : \mathbf{O} \xrightarrow[2]{1-1} \mathbf{J} \tag{1}$$

$$f_j : \mathbf{J} \xrightarrow[(]{1-1} [\text{min,max}] \times ...)$$

In each negotiation round, the buyer agent may need to execute one or more of the functions in Eq. 1 to reflect some changes in the environment. At the start of a negotiation process, all the functions in Eq. 1 are executed. For example, using f_d, a delegate d_i can be assigned a currency converter web service as a negotiation object, 30 negotiation rounds as $t_{\max}$ and a linear time-dependent counteroffer generation tactic. For the currency converter web service object, the price and response time can be assigned as negotiation issues using f_o. Finally, for the price and response time issues, reservation values are assigned using f_j. Similar assignments can be done to the rest of delegates, objects, and issues.

3 Sources of Interdependency in Negotiation Domain

Interdependency between related activities is the driving force behind the need for coordination. Figure 2 shows a categorization of possible dependencies between related activities that can apply in different application domains. As in Fig. 2, we identify several types of dependencies in the automated negotiation domain.

Objects under negotiation may *share resources*, for example, an agent negotiating for buying a laptop and a camera needs to allocate a certain amount of the available budget (a resource) for buying the laptop and another amount for buying the camera. In that sense, the laptop and the camera share a resource. The agent needs to distribute/redistribute the resource in a way to achieve the negotiation goal, i.e., reach a valuable agreement.

The *task assignment* dependency appears when multiple agents negotiate over performing different tasks. It also may depend on the distribution of certain

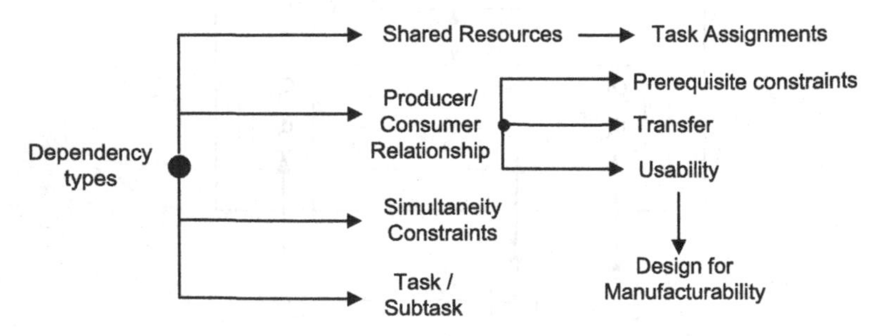

Fig. 2 Common dependency types [2]

resources, i.e., the assignment of tasks to agents is related to the amount of resources allocated to each agent.

The *prerequisite constraints dependency* exists when one activity must be completed before another activity can start or finish [2]. When an agent seeks to buy a hardware and a software given that both the hardware and software must be compatible, it can select to buy the hardware first and then buy a compatible software or vice versa since buying both simultaneously may result in buying incompatible products.

The *simultaneity constraints* dependency determines which negotiations can run concurrently and which negotiations should not run concurrently. In other words, the process of running different negotiations or taking certain actions during negotiation such as quitting a certain instance of negotiation needs to be synchronized.

The *task/subtask dependency* in a negotiation process can be illustrated when one negotiation depends on some other negotiations, i.e., the negotiations are multi-linked [10]. For example, if there are three negotiations a, b, c, but negotiation a depends on negotiations b and c (i.e., negotiation a is successful iff negotiation a is successful and both negotiations b and c are also successful), then we consider that negotiation a has two subtasks (i.e., b and c).

One of the most complicated activities during negotiation is deciding on the value of an offer/counteroffer (i.e., choosing a certain bidding strategy to generate an offer/counteroffer) in each negotiation round. Calculating the value of an offer/counteroffer in each negotiation round is a non-trivial task due to the following:

- the process can be affected by the actions of the outside options,
- the interdependency between the issues of the same object,
- the interdependency between the issues of different objects, and
- the interdependency between different negotiation objects.

In the next few subsections, we elaborate more on the interdependency in one-to-many negotiation looking at the interdependency from the point of view of a buyer agent who negotiates concurrently with multiple seller agents.

3.1 Interdependency Amongst Objects

Accepting a number of agreements by an agent in a certain order while negotiating concurrently with multiple opponents is equivalent to procuring the same number of objects in the same order while negotiating with opponents sequentially. Procuring a certain number of objects sequentially is the easiest solution to solve the problem of procuring a certain number of objects in a certain predefined order. However, using the sequential approach has a few drawbacks. First, because negotiations are conducted once at a time, it is difficult to predict the results of the future negotiations in terms of (1) whether a certain negotiation instance will be able to reach an agreement (2) the expected utility of the agreements. Second, it is difficult to allocate resources for each negotiation instance because we have no knowledge about the demand behavior of the opponents of the future negotiations. For example, we might allocate more resources for the first few negotiation instances to guarantee reaching agreements but the resources for the next negotiations may not be enough to guarantee reaching agreements over the rest of the objects. Finally, the sequential approach takes more time.

The alternative solution to the sequential negotiation approach is adopting the concurrent negotiation approach where a negotiating agent receives feedback during negotiation in terms of the opponents' offers and can act accordingly to fine-tune its strategy and resource allocation pattern. The drawback of the concurrent approach is the need for coordination whenever any type of interdependency exists between objects or between objects' issues, etc. For example, different objects may have interdependency between their attributes such as interface compatibility between two different softwares. Sometimes buying object an (o_1) before buying object an (o_2) causes a loss in utility such as confirming a hotel reservation before confirming a flight. If the flight is canceled for any reason then the buyer is obliged to fulfill his/her obligations towards the hotel reservation.

In some cases, the order of procurement is not defined before the start of negotiation, it could be determined dynamically during negotiation. For example, a person needs to book a flight and an accommodation before starting his/her vacation and at the same time, he/she does not know which one is more difficult to find. During negotiation, the agent working on behalf of that person can detect which one is more difficult to attain and decide on the order of agreements and resource allocation dynamically. The agent may find that booking a flight is way more difficult than booking an accommodation, then it decides to secure an agreement for the flight before securing an agreement for the accommodation.

3.2 Interdependency Amongst Issues

Each object under negotiation is characterized by one or more negotiation issues. Different issues can be interdependent in terms of their acceptable values. For example, an agent may accept to pay high price for a high quality product. When the negotiation issues are interdependent then the utility function is nonlinear, otherwise the utility function can be a weighted sum of the utility of each issue. Apart from dealing with the problem of searching for the best offer/counteroffer that can achieve the highest possible utility in case the utility function is nonlinear, we focus on the problem of allocating shared resources amongst different issues, e.g., price. In our work, we call the issues of different objects that share resources *common issues*.In other words, a *common issue* is an issue that is common amongst multiple objects. For example, multiple services can have the price issue as a *common issue*.

Definition 1 A common negotiation issue is an issue $j_i \in \mathbf{J}$ s.t. at least two subsets $J_k, J_l \in 2^{\mathbf{J}}$ exist where $j_i \in J_k \cap J_l$.

To this end, we propose managing resources shared amongst *common issues* as an approach for coordinating the bidding strategy which takes into consideration the behaviors of the opponents over the *common issues*. Managing the distribution of the available resources (which is part of the bidding strategy) is one solution for managing the interdependency problem.

4 Related Work Review and Analysis

This section addresses the coordination of automated negotiation in literature and analyzes possible negotiation scenarios where the need for some coordination mechanism is essential.

To help presenting and analyzing the related work in a systematic way, Fig. 3 shows possible negotiation scenarios taking into consideration the three main criteria that can determine a particular negotiation scenario, i.e., the *issues of negotiation*, the *number of opponents* and the *number of required agreements*. We consider that each *negotiation object* requires one *object agreement* and each issue requires one *issue agreement*. Formally, if an object o_i has k issues, then we need k *issue agreements* to make an *object agreement* for the object o_i. For the rest of this document, we call *an object agreement* an *agreement*.

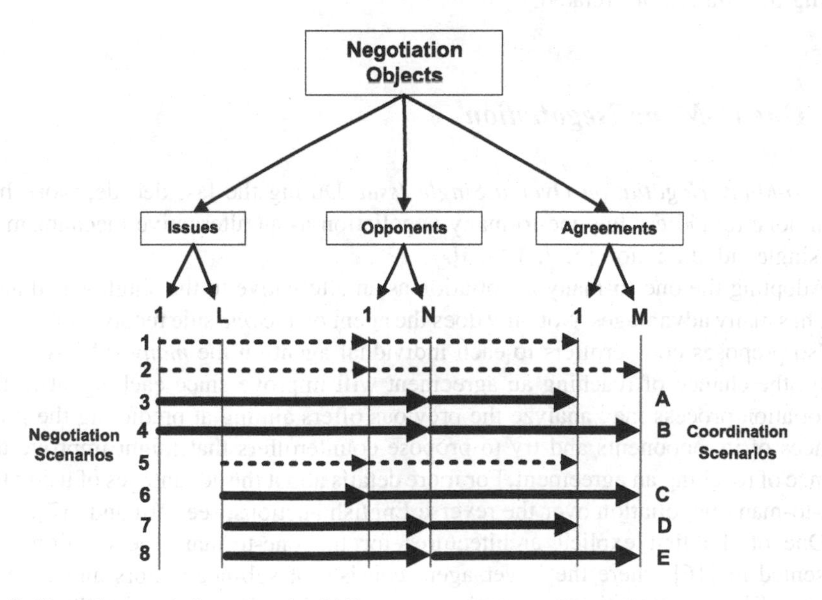

Fig. 3 Possible negotiation scenarios [11]

The *agreements node* in Fig. 3 refers to agreements over objects given that each negotiation object requires one agreement while that agreement requires multiple issue agreements in case the object has multiple issues.

For an agent interacting with other agents through negotiation, we can describe eight possible scenarios of interaction taking into consideration the main criteria of negotiation objects (i.e., the number of issues and the number of opponents) as shown in Fig. 3. The number of agreements indicates the number of negotiation objects under negotiation and vice versa.

Scenario 1 in Fig. 3 shows that an agent has one object characterized by one issue and negotiates with one opponent for the purpose of securing one agreement. As mentioned earlier, we assume that the number of objects is equal to the number of agreements, accordingly in Fig. 3, we can decide whether an agent negotiates over one object or more by looking at the arrow targeting the agreements node, if the arrow ends at 1, then the number of objects is 1, otherwise the number of objects is more than 1.

We consider Fig. 3 as our base for reviewing and analyzing the related work. Scenarios 1, 2, and 5 are basically bilateral negotiations where the number of objects/issues vary. For example scenario 1 represents the bilateral negotiation where an agent interacts with another agent over one issue while scenario 2 is also a bilateral negotiation where two agents negotiate over multiple objects given that each object has one issue.

This paper does not intend to investigate the bilateral negotiation, it rather focuses on the *one-to-many* negotiation where the coordination process is explicitly needed since there are multiple and related negotiation instances that need to be managed. In our work, we assume that the different actions of an agent during negotiation are related.

4.1 One-to-Many Negotiation

One-to-Many Negotiation Over a Single Issue During the last decade, work has been done to address the one-to-many negotiation as an alternative mechanism to the single-sided auction [5, 7, 12–15].

Adopting the one-to-many negotiation as an alternative to the single-sided auction has many advantages. Not only does the agent on the *one* side receive offers, but it also proposes counteroffers to each individual agent on the *many* side. Accordingly, the chance of reaching an agreement will improve since each agent in the negotiation process may analyze the previous offers aiming at predicting the preferences of its opponents and try to propose counteroffers that might improve the chance of reaching an agreement. For more details about the advantages of using the one-to-many negotiation over the reverse English auctions, see [16] and [17].

One of the first explicit architectures for the one-to-many negotiation was presented in [16] where the buyer agent consists of sub-negotiators and a coordinator. The sub-negotiators negotiate concurrently with a set of seller agents given that each sub-negotiator negotiates with one seller. That paper discusses four different coordination strategies: the desperate strategy in which the buyer agent accepts the first agreement and quits negotiations with all other sellers; the patient strategy where the buyer agent makes temporary agreements with some or all sellers during negotiation and holds on to these agreements until all the remaining instances of negotiations are finished, then the buyer agent selects the agreement with the highest utility; the optimized patient strategy is similar to the patient strategy except that it does not accept a new agreement with less utility than the highest accepted one; and finally the manipulation strategies in which the coordinator changes the negotiation strategies of its sub-negotiators during negotiation which were left for future work.

Other existing work [5, 17] develops coordination methods that change the negotiation strategy during negotiation. For example, the decision-making technique in changing the negotiation strategies [5] during negotiation depends on historic information about previous negotiations in terms of agreement rate and utility rate.

While the work of [7, 15] considers a decommitment penalty during negotiation, [9] assumes that the buyer agent incurs no penalty for exercising decommitment during negotiation and proposes a coordination mechanism to change the negotiation strategy during negotiation using only the current information during negotiation, i.e., the sellers's offers during negotiation. [9] argues that granting the buyer agent the privilege of reneging from an agreement without a penalty, while forcing the seller agents to honor their agreements can be a realistic scenario in situations where the number of seller agents is large and/or the seller agents are offering infinite supply, e.g., information. In such cases, a seller agent might be satisfied to make deals with many potential buyers in a hope that some of these buyers will confirm their deals later.

Some heuristic methods were proposed to estimate the expected utility in both synchronized multi-threaded negotiations and dynamic multi-threaded negotiations [18]. The synchronized multi-threaded negotiation model considers the existing outside options for each single negotiation instance, while the dynamic multi-threaded negotiation model considers also the uncertain outside options that might come in the future. In both cases, the methods assume a knowledge of the probability distribution of the reservation prices of the opponents. In many cases, this kind of information is not available.

While [14] proposes a decision-making strategy using Markov chains to decide whether to accept the best available offer or to proceed in negotiation with a hope to achieve a better deal, their work assumes that the buyer cannot make temporary deals with his opponents.

One-to-Many Negotiation Over Multiple Issues In real life, most negotiations involve more than one issue. For example, buying a laptop may involve negotiating the price of the laptop and both, the memory size and processor speed. If the agents participating in negotiation are competitive and self-interested, then the objective of each agent is to reach an agreement with the highest possible utility regardless of the opponents' needs or preferences. However, when negotiation involves multiple issues, agents usually have divergent preferences over different issues which allows reaching an efficient agreement for both parties, i.e., achieving a win–win outcome.

Our previous work investigates negotiation scenario **D** in Fig. 3 where a buyer agent negotiates with multiple seller agents over one object characterized by several issues [19]. For that scenario, we use a meta-strategy that uses two different offer generation tactics: the time-dependent tactics and the trade-off tactic. In each negotiation round, the buyer agent needs to decide on using a time-dependent tactic or the trade-off tactic depending on the behaviors of the opponents. During negotiation, the buyer agent assigns each seller agent to either a favorable group or unfavorable group. The favorable group offers more concessions than the concessions offered by the corresponding buyer agent's delegates. The meta-strategy is applied amongst the favorable group of seller agents.

One-to-Many Negotiation Over Multiple Objects To the best of our knowledge, little work has been done on that scenario. The work in [20] investigates scenario **C** in Fig. 3 where a buyer agent seeks agreements over multiple objects given that each object has several issues and a single provider. The work in [20] investigates the process of adapting the local reservation values during negotiation subject to the behaviors of the existing opponents, while the work in [21] involves adaptation of both the initially generated counteroffer values and the weight matrix of the counteroffers' issues during negotiation.

5 The Coordination Approach

During multi-bilateral concurrent negotiation, the buyer agent needs to coordinate its actions against its opponents in each negotiation round. One of the important actions is deciding on the bidding strategy that can be used to generate the next counteroffer. Part of that process is to distribute/redistribute the available resources amongst the buyer's delegates in a way to achieve the goal of the negotiation process in terms of reaching valuable agreements. Coordinating the buyer's actions in that context means managing the buyer's negotiation strategy during negotiation.

Formally, let Ω^a be the negotiation strategy of an agent a, then $\Omega^a = \langle IV^a, RV^a, T^a, \Theta^a \rangle$, where $IV^a, RV^a, T^a, \Theta^a$ stands for the initial offer value(s), the reservation value(s), the deadline(s), and the set of offer generation strategies of an agent a respectively.

Our representation of an agent's strategy Ω^a is similar to its representation in [21], the difference is that the fourth part of the strategy in [21] represents the β parameter in the time-dependent tactics [4] while the fourth part in our model (Θ^a) has a more general representation which indicates any possible offer/counteroffer generation method, e.g., trade-off, time-dependent, behavior dependent, etc., and their associated parameters. Any change to the components of Ω^a during negotiation means a change in agent a's negotiation strategy.

6 An Illustrative Example

This section shows some empirical results that compare between using a dynamic and a static strategy in coordinating the bidding strategy of different buyer's delegates. It is not in the scope of this paper to explain in detail about the specific dynamic strategy and experimental settings that produced the results shown in Fig. 4. The purpose of displaying Fig. 4 is to demonstrate the difference in performance between using a dynamic strategy that changes some of the negotiation strategy components (i.e., components of Ω) and a static strategy that initializes the strategy components and does not change them during negotiation. The results show that the buyer agent $T2$ who uses the dynamic strategy outperforms the buyer agent $T1$ who uses the static strategy in both utility rate and agreement rate.

For testing the agreement rate, we ran an experiment 500 times for each different number of seller agents, then the results were averaged. For example, when the number of seller agents per object is two, we repeated the experiment 500 times and then the results were averaged. We did the same for testing the utility rate. The dynamic strategy in this case involves assigning a new local reservation value for each issue of each object in each negotiation round depending on the behaviors of the current opponents in terms of their concessions. The experimental results shown in this section are related to scenario **B** in Fig. 3 where a buyer agent seeks to procure multiple objects given that each object has a single issue and multiple providers. The numbers of seller agents are shown on the top of Fig. 4a, b. Figure 4 shows that when the number of opponents increases, it is a favorable situation for both types of buyer agents since more seller agents means better opportunity for reaching agreements and getting better utility.

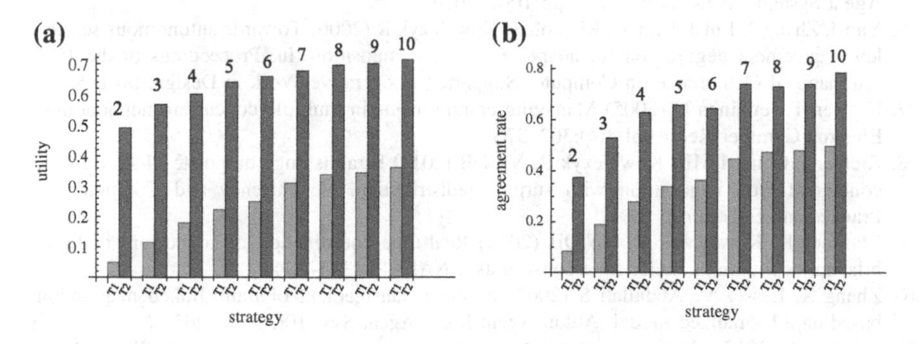

Fig. 4 Comparing between dynamic and static strategies. **a** Average utility. **b** Agreement rate

7 Conclusion

This paper addresses some aspects of coordination in concurrent one-to-many negotiation considering different sources of interdependency between the instances of multiple negotiations conducted concurrently by an agent. In our work, we consider a buyer agent negotiating concurrently with multiple seller agents for the purpose of procuring one or more objects. We propose adapting the negotiation bidding strategy in terms of distributing/redistributing of resources during concurrent negotiations subject to the behaviors of the current opponents. We further need to investigate adapting the bidding strategy of a buyer agent to generate counteroffers that improve the probability of reaching an agreement with the highest possible utility in different coordination scenarios, see Fig. 3. In other words, we need to investigate what negotiation strategy components or parameters to select for adaptation in each scenario to design an effective and robust dynamic negotiation strategy.

Acknowledgments This work was partially supported by Smart Services CRC.

References

1. Raiffa H (1982) The Art and Science of Negotiation. Harvard Univiversity Press, Cambridge
2. Malone TW, Crowston K (1994) The interdisciplinary study of coordination. ACM Comput Surv 26(1):87–119
3. Rubinstein A (1982) Perfect Equilibrium in a Bargaining Model. Econometrica 50(1): 97–109
4. Faratin P (2000) Automated service negotiation between autonomous computational agents. Ph.D thesis, University of London
5. Nguyen TD, Jennings NR (2004) Coordinating multiple concurrent negotiations. In: Proceedings of the Third International Joint Conference on Autonomous Agents and Multi Agent Systems, New York, USA. pp 1062–1069
6. Yan J, Zhang J, Lin J, Chhetri M, Goh S, Kowalczyk R (2006) Towards autonomous service level agreement negotiation for adaptive service composition. In: Proceedings of the 10th International Conference on Computer Supported Cooperative Work in Design, pp 1–6
7. Nguyen T, Jennings N (2005) Managing commitments in multiple concurrent negotiations. Electron Commer Res Appl 4(4):362–376
8. Richter J, Chhetri MB, Kowalczyk R, Vo QB (2011) Establishing composite SLAs through concurrent QoS negotiation with surplus redistribution. Concurrency and Computation: Practice and Experience
9. Mansour K, Kowalczyk R, Vo QB (2010) Real-time coordination of concurrent multiple bilateral negotiations under time constraints. LNAI 6464:385–394
10. Zhang X, Lesser V, Abdallah S (2005) Efficient management of multi-linked negotiation based on a formalized model. Auton Agent Multi-Agent Syst 10(2):165–205
11. Mansour K (2012) Managing concurrent negotiations in multi-agent systems. In: Proceedings of the Canadian Conference on AI, pp 380–383
12. Gerding EH, Somefun DJa, La Poutré JA (2004) Automated bilateral bargaining about multiple attributes in a one-to-many setting. In: Proceedings of the 6th international conference on Electronic commerce-ICEC '04, p 105

13. Yan J, Zhang J, Lin J, Chhetri M, Goh S, Kowalczyk R (2007) Autonomous service level agreement negotiation for service composition provision. Future Gener Comput Syst 23(6):748–759
14. An B, Sim KM, Miao CY, Shen ZQ (2008) Decision making of negotiation agents using markov chains. Multiagent Grid Syst 4:5–23
15. An B, Lesser V, Irwin D, Zink M (2010) Automated negotiation with decommitment for dynamic resource allocation in cloud computing. In: Proceedings of the 9th International Conference on Autonomous Agents and Multiagent Systems (AAMAS 2010), Toronto, pp 981–988
16. Rahwan I, Kowalczyk R, Pham HH (2002) Intelligent agents for automated one-to-many e-commerce negotiation. In: Proceedings of the Twenty-Fifth Australian Computer Science Conference, Melbourne, Australia, pp 197–204
17. Nguyen TD, Jennings NR (2003) Concurrent bi-lateral negotiation in agent systems. In: Proceedings of the Fourth DEXA Workshop on E-Negotiations
18. Cuihong L, Giampapa J, Sycara K (2006) Bilateral negotiation decisions with uncertain dynamic outside options. IEEE Transactions on Systems, Man and Cybernetics, Part C (Applications and Reviews), 36(1):31–44
19. Mansour K, Kowalczyk R (2012) A meta-strategy for coordinating of one-to-many negotiation over multiple issues. In: Wang Y, Li T (eds) Foundations of Intelligent Systems, Shanghai, Springer, Heidelberg, pp 343–353
20. Mansour K, Kowalczyk R, Chhetri MB (2012) On effective quality of service negotiation. In: Proceedings of the 12th IEEE/ACM International Symposium on Cluster, Cloud and Grid Computing, pp 684–685
21. Fatima S (2004) An agenda-based framework for multi-issue negotiation. Artificial Intelligence 152(1):1–45

Relevance in Preference Structures

Camilo Franco, Javier Montero and J. Tinguaro Rodriguez

Abstract Fuzzy preference and aversion relations allow measuring in a gradual manner the attitude of the individual regarding some pair of alternatives. Following the Preference-Aversion (P-A) model, previously introduced for identifying the subjective cognitive state for some decision situation; here, we explore a methodology for learning relevance degrees over the complete system of alternatives. In this way it is possible to identify in a quick way, the pieces of information that are more important for solving a given decision problem.

Keywords Preference structures · Preference-aversion · Relevance

1 Introduction

Preference relations express information obtained under natural conditions of subjectivity and uncertainty characteristic of human intelligence and rationality (see, e.g., [1–3]). Therefore, the decision process of an individual can be better described if the representation possibilities of the preference model take into consideration the rational capabilities of the individual.

C. Franco (✉)
Centre for Research and Technology of Agro-Environmental and Biological Sciences, Tras-os-Montes E Alto Douro University, Vila Real, Portugal
e-mail: camilo@utad.pt

J. Montero · J. T. Rodriguez
Complutense University, Madrid, Spain
e-mail: monty@mat.ucm.es

J. T. Rodriguez
e-mail: jtrodrig@mat.ucm.es

F. Sun et al. (eds.), *Knowledge Engineering and Management*,
Advances in Intelligent Systems and Computing 214,
DOI: 10.1007/978-3-642-37832-4_11,

In this sense, the epistemic states of a decision problem can be explored and understood using the basic and general attributes of human rationality. Here, such rationality refers to the brain's capability of distinguishing between perceptions with positive (gains) or negative (losses) character (see e.g., [2, 4–6]). We do not refer only to monetary gains or losses, but in a more general way, to the positive and negative sides for each alternative when being compared with another one. Then, two *independent* evaluations for the positive and the negative perceptions are needed (as in [7]), following the general approach of the Preference-Aversion (P-A) model formally introduced in [8, 9].

In this paper, we focus on how the P-A model allows understanding the decision problem, where the general *attitude* (in the sense of [10, 11]) of the individual can be described considering *bipolar* (see [12–14], but also [15]) knowledge representation, and at the same time, allows identifying the *relevance* of each alternative, which can be measured according to the whole set of preference and aversion relations, i.e., the system of alternatives. Our main point is that relevance refers not only to alternatives that are strictly preferred, but also to alternatives that are strictly rejected, offering complementary information over the decision process of the individual and the complete system of alternatives.

2 Modeling Fuzzy Preference-Aversion Relations

Standard fuzzy preference models (see e.g. [16–19]) examine the subjective decision-making process using binary preference relations over a finite set of alternatives A. Such models explore preference relations as gradual predicates (in the sense of [20]), where some basic properties can be verified up to a certain degree.

From this standing point, the characterization of a binary fuzzy preference relation is given by (see e.g., [20, 21])

$$R(a,b) = \{\langle a,b,\mu_R(a,b)\rangle | a,b \in A\}, \tag{1}$$

where $\mu_R : A \times A \rightarrow [0,1]$ is the membership function for the fuzzy relation R, such that $\mu_R \in [0,1]$ is the membership intensity for every $(a,b) \in A \times A$, according to the verification of the property of "being at least as desired as."

Introducing an independently opposite counterpart for this preference predicate, the preference relation $R(a, b)$ can now be characterized in the following way (see e.g., [7, 15, 22, 23]),

$$R(a,b) = \{\langle a,b,\mu_R(a,b),\nu_R(a,b)\rangle | a,b \in A\}, \tag{2}$$

where $\mu_R, \nu_R : A \times A \rightarrow [0,1]$ are the membership and non-membership functions, respectively, such that $\mu_R, \nu_R \in [0,1]$ are the corresponding membership and non-membership intensities for every $(a,b) \in A \times A$, according to the verification of the opposite properties of being "at least as desired as" and "at least as rejected as," respectively.

3 The Fuzzy Preference-Aversion Model

The preference-aversion relation R is now composed by two input values or data parts, its positive part, the weak preference intensity $\mu_R(a,b)$, and its negative part, the weak aversion intensity $\nu_R(a,b)$. The inclusion of these two weak relations in the preference structure allows defining the basic P-A structure [8, 9],

$$R = \langle (P,I,J),(Z,G,H) \rangle \tag{3}$$

composed by six relations which are, strict preference P, indifference I, incomparability J, and strict aversion Z, negative indifference G, and incomparability on weak aversion H.

In this way, there exist six functions [8, 9],

$$p,i,j,z,g,h : [0,1]^2 \rightarrow [0,1], \tag{4}$$

such that

$$P(a,b) = p(\mu_R(a,b),\ \mu_R(b,a)), \tag{5}$$

$$I(a,b) = i(\mu_R(a,b),\ \mu_R(b,a)), \tag{6}$$

$$J(a,b) = j(\mu_R(a,b),\ \mu_R(b,a)), \tag{7}$$

$$Z(a,b) = z(\nu_R(a,b),\ \nu_R(b,a)), \tag{8}$$

$$G(a,b) = g(\nu_R(a,b),\ \nu_R(b,a)), \tag{9}$$

$$H(a,b) = h(\nu_R(a,b),\ \nu_R(b,a)). \tag{10}$$

Using standard fuzzy logic operators [24], where the valued union or disjunction can be represented by a continuous t-norm S, the valued intersection or conjunction by a continuous t-norm T, and n is a strict negation, the following system of equations holds (following [9, 16]),

$$S(p(\mu_R,\mu_{R^{-1}}),\ i(\mu_R,\mu_{R^{-1}})) = \mu_R, \tag{11}$$

$$S(p(\mu_R,\mu_{R^{-1}}),\ i(\mu_R,\mu_{R^{-1}}),\ p(\mu_{R^{-1}},\mu_R)) = S(\mu_R,\mu_{R^{-1}}), \tag{12}$$

$$S(p(\mu_{R^{-1}},\mu_R),\ j(\mu_R,\mu_{R^{-1}})) = n(\mu_R), \tag{13}$$

$$S(z(\nu_R,\nu_{R^{-1}}),\ g(\nu_R,\nu_{R^{-1}})) = \nu_R, \tag{14}$$

$$S(z(\nu_R,\nu_{R^{-1}}),\ g(\nu_R,\nu_{R^{-1}}),\ z(\nu_{R^{-1}},\nu_R,)) = S(\nu_R,\nu_{R^{-1}}), \tag{15}$$

$$S(z(\nu_{R^{-1}},\nu_R),\ h(\nu_R,\nu_{R^{-1}})) = n(\nu_R). \tag{16}$$

Therefore, the P-A model takes into account a type of rationality that frames alternatives in terms of gains and losses, assigning two different and separate values for expressing weak preference and weak aversion. In this way, different

Table 1 The complete P-A cognitive space

R_{PA}	z	z^{-1}	g	h
P	pz	pa	pg	ph
P^{-1}	pa^{-1}	pz^{-1}	pg^{-1}	ph^{-1}
I	iz	iz^{-1}	ig	ih
J	jz	jz^{-1}	jg	jh

pieces of information can be ordered according to the strength of the positive and negative sides of things.

The conjunctive meaning of the basic P-A structure can now be examined, as there are ten different situations for describing the cognitive state of the individual in the face of decision. Hence, the complete P-A structure, defined by [8, 9],

$$R_{PA} = \langle \mu_R, \nu_R \rangle = T(\langle p, i, j \rangle, \langle z, g, h \rangle), \tag{17}$$

represents the system of gradual situations that arise from the independent reasoning over gains and losses. Such states constitute a complete semantic space for characterizing the subjective perceptions of a *gains-losses rational* individual.

Such valuation space is represented in Table 1, and it is composed by [8, 9],

- Ambivalence: $pz = T(p, z)$,
- Strong preference: $pa = T(p, z^{-1})$,
- Pseudo-preference: $pg = T(p, g)$,
- Semi-strong preference: $ph = T(p, h)$,
- Pseudo-aversion: $iz = T(i, z)$,
- Strong indifference: $ig = T(i, g)$,
- Positive indifference: $ih = T(i, h)$,
- Semi-strong aversion: $jz = T(j, z)$,
- Negative indifference: $jg = T(j, g)$,
- Strong incomparability: $jh = T(j, h)$.

Notice the different characteristics of the epistemic states represented by each one of these relations. For example, it can be the case that a decision is seen clearer by valuing the negative aspects of alternatives, as in *strong preference* (pa), where a is better than b and b is worse than a. Or it can be the case that there exists certain uneasiness over the available options in the face of decision, like in *pseudo-preference* (pg), where a is better than b but the two alternatives are considered equally bad. But it can also be the case of *negative indifference* (jg), where the two possibilities are just as bad, i.e., full discomfort on the options is expressed in this way. Or it can also be the case of *ambivalence* (pz), where one alternative is considered at the same time better and worse, a major conflict typical of real-life decision problems.

The relation of conflict toward a decision can then be carefully examined by the different aspects of ambivalence, which rises from the joint consideration of aversion and preference. In this sense, the P-A model offers a formal methodology for representing ambivalence in a non-symmetric and expressive way (for more details see [8, 9]).

4 Relevance Degrees for Fuzzy Preference-Aversion Relations

The determination of which part of the available information is most relevant for solving some decision problem relies on our objective and knowledge of the problem at hand (following the general insights presented in [25–27]). In this context, it is necessary to take into consideration the concept of relevance for the representation of the individual's subjective perceptions and judgments. Here we propose relevance degrees over the individual's preference-aversion relations, making reference to the relative importance of valued arguments. Our objective is to obtain a balanced evaluation between the amount of available information and the existing knowledge and ignorance over the complete system of alternatives.

Therefore, relevance degrees have to take into consideration the relative importance of a given alternative. Here we make a proposal for obtaining such degrees using the independent methodology of the P-A model, where a positive order, O^+, and a negative one, O^-, are separately induced over the alternatives in A. Following [28] where the concept of dimension for a simple order is examined (not to be confused with the dimension approach used in [29]), the dimension $d[a]$ of an element $a \in A$ can be understood as the maximum length d of chains $c \prec b \prec \ldots \prec a$ in O^+ having a for greatest element, where $b \prec a$ holds if $p(\mu_R, \mu_{R^{-1}}) > \varepsilon$ holds for certain threshold $\varepsilon > 0$.

Depending on the objective of the decision maker, it may be of interest to identify not only the strict preference chain, but also the indifference or the incomparability ones. Then, we say that $d[a]$, as it has been just defined, makes reference to the preference chain $d_p[a]$, and that the dimension $d_i[a]$ or $d_j[a]$ of an element $a \in A$ can also be understood as the maximum length of chains $c \sim b \sim \ldots \sim a$ or $c \nsim b \nsim \ldots \nsim a$ in O^+, respectively, having a for greatest element, where $b \sim a$ holds if $i(\mu_R, \mu_{R^{-1}}) > \varepsilon$ holds for certain threshold $\varepsilon > 0$ and $b \nsim a$ holds if $j(\mu_R, \mu_{R^{-1}}) > \varepsilon$ holds for certain threshold $\varepsilon > 0$.

Similarly, by the aversion dimension $d_{z,g,h}[a]$ of an element $a \in A$ it can be understood the maximum length d of chains $c \lhd b \lhd \ldots \lhd a$, $c \approx b \approx \ldots \approx a$, or $c \not\approx b \not\approx \ldots \not\approx a$ in O^- having a for greatest element, where $b \lhd a$ holds if $z(\nu_R, \nu_{R^{-1}}) > \delta$ holds for certain threshold, $\delta > 0$, $b \approx a$ holds if $g(\nu_R, \nu_{R^{-1}}) > \delta$ holds for certain threshold $\delta > 0$, and $b \not\approx a$ holds if $h(\nu_R, \nu_{R^{-1}}) > \delta$ holds for certain threshold $\delta > 0$. Finally, by the dimension of O^+, $d_{p,i,j}[O^+]$, and the dimension of O^-, $d_{z,g,h}[O^-]$, it is meant the maximum length of a chain in O^+ and O^-, respectively.

Definition 1 For every alternative $a \in A$, the relevance degree of a regarding the order given by $\langle p, i, j \rangle$ is defined as,

$$\lambda_a^{p,i,j} = \frac{d_{p,i,j}[a]}{d_{p,i,j}[O^+]}, \tag{18}$$

and the relevance degree of a regarding the order given by $\langle z, g, h \rangle$ is defined as,

$$\lambda_a^{z,g,h} = \frac{d_{z,g,h}[a]}{d_{z,g,h}[O^-]}. \tag{19}$$

In this way, for a given alternative a and for every $a, b \in A \times A$, the positive relevance degree of a, $\lambda_a^{p,i,j}$, can be built by counting over how many alternatives a is, up to a certain degree greater than ε, strictly preferred, indifferent, or incomparable, while the negative one, $\lambda_a^{z,g,h}$, can be built by counting over how many alternatives a is, up to a certain degree greater than δ, strictly worse, just as bad, or incomparable on aversion attributes. Hence, these relevance degrees make use of all of the available information about some alternative a regarding all of the other alternatives in A. In this sense, such relevance degrees measure the relative importance of one alternative with respect to the complete system of alternatives.

5 Construction of Relevance Degrees Under the Preference-Aversion Model: An Example

As we have seen, relevance is a concept that deals jointly with the importance and the meaning of things. In this approach, we make a first attempt to treat the concept of relevance over a given set of alternatives, focusing on some degree for its quantification. Then, we assume that relevance refers to the pieces of information that attracts with more intensity our attention.

Taking these observations into consideration, we illustrate now the use of relevance degrees over the P-A model. For example, consider a set of alternatives $A = \{a, b, c, d\}$ and an individual with the preference intensities $(\mu_R, \mu_{R^{-1}})$ of Table 2, where, e.g., $\mu_R(a, b) = 0.70$, and the aversion intensities $(\nu_R, \nu_{R^{-1}})$ given in Table 3.

So, we have to find an overall preference-aversion order reflecting the individual's attitude toward the available alternatives in A. For all (a, b) in $A \times A$, a person is supposed to perceive preference of a over b if a is strictly preferred to b, i.e., $p(\mu_R, \mu_{R^{-1}}) > \varepsilon$ holds, and similarly to support the aversion for a over b if a is strictly worse than b, i.e., $z(\nu_R, \nu_{R^{-1}}) > \delta$ holds.

Table 2 Preference intensities

$\mu_R, \mu_{R^{-1}}$	a	b	c	d
a	1.00	0.70	0.40	0.30
b	0.30	1.00	0.60	0.70
c	0.70	0.90	1.00	0.50
d	0.70	0.90	0.50	1.00

Table 3 Aversion intensities

$\nu_R, \nu_{R^{-1}}$	a	b	c	d
a	1.00	0.80	0.60	0.40
b	0.80	1.00	0.90	0.90
c	0.40	0.10	1.00	0.00
d	0.10	0.80	0.20	1.00

Table 4 Strict preference intensities

P	a	b	c	d
a	0.00	0.40	0.00	0.00
b	0.00	0.00	0.00	0.00
c	0.30	0.30	0.00	0.00
d	0.40	0.20	0.00	0.00

Then (see e.g., [8, 9, 16]), knowing that p has a lower bound in

$$p(\mu_R, \mu_{R^{-1}}) = T^L(\mu_R, n(\mu_{R^{-1}})), \tag{20}$$

where T^L is the Lukasiewicz t-norm, and in an analogous way, z has an upper bound in

$$z(\nu_R, \nu_{R^{-1}}) = T^m(\nu_R, n(\nu_{R^{-1}})), \tag{21}$$

where T^m is the t-norm of the minimum, we obtain the strict preference intensities shown in Table 4 and the strict aversion intensities of Table 5.

Notice that we have followed the basic intuition of Cummulative Prospect Theory [4, 30], where it is argued that losses loom larger than gains in decision under uncertainty, so aversion is valued by a greater t-norm than preference.

So, taking $\varepsilon = \delta = 0.01$, we identify the maximum length of a chain in P, such that $d_p[O^+] = 2$. Hence, $\lambda_a^p = 1/2$, $\lambda_b^p = 0$, $\lambda_c^p = 1$, and $\lambda_d^p = 1$. In the same way, we can see that $d_z[O^-] = 3$, and $\lambda_a^z = 1$, $\lambda_b^z = 1$, $\lambda_c^z = 2/3$, and $\lambda_d^z = 1$.

Therefore, we can see that alternatives c and d are the most relevant ones weighing only the strict preference intensities, while c is the less relevant one regarding the strict aversion ones. Following a basic decision rule where negative or harmful aspects have to be avoided in order to reach a satisfactory outcome (see [9]), alternative c stands out because of its positive and non-negative relevance.

Applying the P-A model over the two most relevant alternatives $\{c, d\}$, using the t-norm T^m of the minimum for the construction of R_{PA}, we find that the decision situation between c and d can be described by positive indifference and

Table 5 Strict preference intensities

Z	a	b	c	d
a	0.00	0.20	0.60	0.40
b	0.20	0.00	0.90	0.20
c	0.40	0.10	0.00	0.00
d	0.10	0.10	0.20	0.00

Table 6 The P-A description of the decision problem between alternatives $\{c, d\}$

$R(c,d)$	z	z^{-1}	g	h
p	0.00	0.00	0.00	0.00
p^{-1}	0.00	0.00	0.00	0.00
i	0.00	0.20	0.00	0.50
j	0.00	0.20	0.00	0.50

strong incomparability, each one with intensities 0.5, and by pseudo-aversion and semi-strong aversion of *d* over *c*, with intensities of 0.2 (see Table 6). These results help us explain that there is a strong conflict between both alternatives (existence of strong incomparability).

Combining the Preference-Aversion model with relevance degrees, we obtain a decision support system that recommends choosing alternative *c* over *d*, given that its positive aspects are always stronger than the negative ones. In this way, by obtaining the relevance degrees and organizing the information according to the P-A model, the final results can be understood and explained, arriving to a descriptively satisfactory answer.

6 Conclusion

The P-A model is a natural framework for understanding the cognitive process behind decision making, where relevance degrees can be introduced for identifying the pieces of information that are most important in a quick and direct way. Relevance is a complex concept that needs to be addressed with more detail, following the general insights of [26, 27], under conditions of abundance of information and decision making.

Acknowledgments This research has been partially supported by the Government of Spain, grant TIN2009-07901.

References

1. Amo A, Montero J, Biging G, Cutello V (2004) Fuzzy classification systems. Eur J Oper Res 156:495–507
2. Cacioppo J, Gardner W, Berntson G (1997) Beyond bipolar conceptualizations and measures: the case of attitudes and evaluative space. Pers Soc Psychol Rev 1:3–25
3. Montero J, Gómez D, Bustince H (2007) On the relevance of some families of fuzzy sets. Fuzzy Sets Syst 158:2429–2442
4. Kahneman D, Tversky A (1979) Prospect theory: an analysis of decision under risk. Econometrica 47:263–291
5. Keil A, Stolarova M, Moratti S, Ray W (2007) Adaptation in human visual cortex as a mechanism for rapid discrimination of aversive stimuli. NeuroImage 36:472–479

6. O'Doherty J, Kringelback M, Rolls E, Hornak J, Andrews C (2001) Abstract reward and punishment representations in the human orbitofrontal cortex. Nat Neurosci 4:95–102
7. Atanassov K (1986) Intuitionistic fuzzy sets. Fuzzy Sets Syst 20:87–96
8. Franco C, Montero J, Rodríguez JT (2011) On partial comparability and fuzzy preference-aversion models. In: Proceedings of the ISKE conference, Shangai, paper 1448 15–17 December
9. Franco C, Montero J, Rodríguez JT (2012) A fuzzy and bipolar approach to preference modeling with application to need and desire. Fuzzy Sets Syst 214:20–34. doi:10.1016/j.fss.2012.06.006
10. Kaplan K (1972) On the ambivalence-indifference problem in attitude theory and measurement: a suggested modification of the semantic differential technique. Psychol Bull 77:361–372
11. Osgood Ch, Suci G, Tannenbaum P (1958) The measurement of meaning. University of Illinois Press, Urbana
12. Benferhat S, Dubois D, Kaci S, Prade H (2006) Bipolar possibility theory in preference modeling: representation, fusion and optimal solutions. Inform Fus 7:135–150
13. Bonnefon JF, Dubois D, Fargier H, Leblois S (2008) Qualitative heuristics for balancing the pros and the cons. Theor Decis 65:71–95
14. Dubois D, Prade H (2008) An introduction to bipolar representations of information and preference. Int J Intell Syst 23:866–877
15. Rodríguez JT, Franco C, Montero J (2011) On the relationship between bipolarity and fuzziness. In: Proceedings of the EUROFUSE conference, Régua, 193–205, September 21–23
16. Fodor J, Roubens M (1994) Fuzzy preference modelling and multicriteria decision support. Kluwer Academic, Dordrecht
17. Montero J, Tejada J, Cutello C (1997) A general model for deriving preference structures from data. Eur J Oper Res 98:98–110
18. Roubens M, Vincke Ph (1985) Preference modeling. Springer, Berlin
19. Van der Walle B, de Baets B, Kerre E (1998) Characterizable fuzzy preference structures. Ann Oper Res 80:105–136
20. Goguen J (1969) The logic of inexact concepts. Synthese 19:325–373
21. Zadeh L (1975) The concept of a linguistic variable and its application to approximate reasoning-I. Inf Sci 8:199–249
22. Franco C, Montero J (2010) Organizing information by fuzzy preference structures—Fuzzy preference semantics. In: Proceedings of the ISKE conference, Hangzhou, 135–140, 15–16 November
23. Franco C, Rodríguez JT, Montero J (2010) Information measures over intuitionistic four valued fuzzy preferences. In: Proceedings of the IEEE-WCCI, Barcelona, 18–23 July, 1971–1978
24. Scheweizer B, Sklar A (1983) Probabilistic metric spaces. North-Holland, Amsterdam
25. Keynes J (1963) A treatise on probability. MacMillan, London
26. Yager R (2007) Relevance in systems having a fuzzy-set-based semantics. Int J Intell Syst 22:385–396
27. Yager R, Petry F (2005) A framework for linguistic relevance feedback in content-based image retrieval using fuzzy logic. Inf Sci 173:337–352
28. Birkhoff G (1964) Lattice theory. American Mathematical Society, Providence
29. Gonzalez-Pachón J, Gómez D, Montero J, Yáñez J (2003) Soft dimension theory. Fuzzy Sets Syst 137:137–149
30. Tversky A, Kahneman D (1992) Advances in prospect theory: cumulative representation of uncertainty. J Risk Uncertain 5:297–323

Segregation and the Evolution of Cooperation

Noureddine Bouhmala and Jon Reiersen

Abstract Thirty years have passed since Robert Axelrod and William Hamilton published their influential contribution to the problem of cooperation. They showed, with the help of both an experiment and analytical techniques, that cooperation is the most likely evolutionary outcome of a Prisoner's Dilemma game when individuals interact repeatedly. Building on Hamilton's earlier work they also demonstrated that, when pairing of individual is not completely random, cooperating behavior can evolve in a world initially dominated by defectors. In this paper, Computer simulations are used to study the relation between non-random pairing and the maintenance of cooperative behavior under evolutionary dynamics. We conclude that cooperation can survive also when the possibility of repeated interaction and reciprocity is ruled out.

Keywords Segregated interaction · Cooperation · Evolution

1 Introduction

The paper by Axelrod and Hamilton [1] has inspired much theoretical and empirical work on the problem of cooperation. Their famous model of cooperation is based on letting individuals interact repeatedly over time and that each member of a pair has the opportunity to provide a benefit to the other at a cost to himself by cooperating. Now consider a population of Tit-for-Tatters which cooperates on the first interaction and keeps on cooperating only as long as their partner cooperates.

N. Bouhmala (✉)
Department of Computer Science, Vestfold University College,
Raveien 215, 3184 Borre, Norway
e-mail: noureddine.bouhmala@hive.no

J. Reiersen
Department of Business and Management, Vestfold University College,
Raveien 215, 3184 Borre, Norway

F. Sun et al. (eds.), *Knowledge Engineering and Management*,
Advances in Intelligent Systems and Computing 214,
DOI: 10.1007/978-3-642-37832-4_12,

Axelrod and Hamilton [1] showed that Tit-for-Tatters can resist invasion by defectors who never cooperate as long as the long-run benefit of mutual cooperation is greater than the short-run benefit that a defector gets by exploiting a cooperator. However, as shown by Axelrod and Hamilton [1], a population of Tit-for-Tatters is not the only one that is evolutionary stable. In fact, a population where all are defectors is also evolutionary stable. If (almost) all players in a population are defectors, a cooperator will have no one to cooperate with. Therefore, a player cannot do any better than playing defect. The long-run benefit associated with sustained cooperation becomes irrelevant. This raises the problem concerning initiation of cooperation from a previous asocial state. How could an evolutionary trend towards cooperative behavior have started in the first place?

To study this question more closely Axelrod and Hamilton introduce the concept of *segregation*. Segregated interaction means that the probability for a Tit-for-Tatter to meet another Tit-for-Tatter is higher than the proportion of Tit-for-Tatters in the population. Axelrod and Hamilton then show that if there are few Tit-for-Tatters in the population, and if the long-run benefit of cooperation is big, only a small amount of segregation is needed in order to secure Tit-for-Tatters a higher expected payoff than defectors. An evolutionary trend towards universal cooperation can then start. The results established by Axelrod and Hamilton are generated within a setup where pairs of individuals interact *repeatedly* over time, and where everybody is able to remember the action taken by each member of the population in previous interactions. However, in many human social environments, Axelrod and Hamilton's conditions favoring cooperation can be questioned. Individuals do not always interact repeatedly over long periods of time, and in large groups it can be difficult to remember the action taken by a potential exchange partner in previously interactions. This leads us to the main question of this paper: Since segregation is a powerful mechanism for the promotion of cooperation in a repeated Prisoner's Dilemma game, can segregation also promote the evolution of cooperation in a non-repeated version of the game? If so, how much segregation is needed, and how does cooperative behavior evolve over time depending on the degree of segregation?

2 The Problem of Cooperation

Consider a large population of players who interact in pairs with available actions and payoffs describing a Prisoner's Dilemma game. We have the following payoff matrix, where $a > b > c > d$.

	Cooperate	Defect
Cooperate	b , b	d , a
Defect	a , d	c , c

If both players cooperate, they both receive a payoff of b. If both defect, they both receive payoffs of c. If one cooperates and the other defects, the cooperator receives a payoff of d, while the defector does very well with a payoff of a. Assume further that individuals in the larger population are either (perhaps due to cultural experiences, perhaps due to genes) *cooperators* (C) or *defectors* (D) in a single period Prisoner's Dilemma. Let p denote the proportion of the population that are cooperators and $(1-p)$ the proportion of defectors. If the members of the population are randomly paired, the expected payoffs are given by

$$V(C) = pb + (1-p)d \tag{1}$$

$$V(D) = pa + (1-p)c \tag{2}$$

where $V(C)$ and $V(D)$ are the expected payoff for a cooperator and a defector respectively. Equation (1) says that with probability p a cooperator is paired with another cooperator producing a payoff b, and with probability $(1-p)$ is paired with a defector producing a payoff d. Equation (2) has a similar interpretation: With probability p a defector is paired with a cooperator producing a payoff a, and with probability $(1-p)$ is paired with a another defector producing a payoff c.

Assume now the following simple evolutionary dynamics: At any time, the growth rate of the proportion of cooperators (p) is positive or negative, depending on whether the expected payoff for cooperators is higher or lower than the expected payoff for defectors. The population distribution (p) will be unchanging, producing an equilibrium, if

$$V(C) = V(D) \tag{3}$$

It is easy to see from (1) and (2) that the only evolutionary stable equilibrium in this game is $p = 0$, where all members of the population defects.

This result follows from the fact that $a > b$ and $c > d$, which gives $V(C) < V(D)$ for all $p \in (0,1)$. Cooperators cooperate irrespective of the type of player whom they meet. Defectors take advantage of such indiscriminate cooperative behavior and get a higher expected payoff compared to cooperators. Defectors increase in numbers, and in the long run take over the whole population. This result motivated Axelrod and Hamilton to examine more closely conditions, not captured in the situation just studied, that can lead to the evolution of cooperation when cooperators and defectors meet to play the Prisoner's Dilemma game.

3 Evolution of Cooperative Behavior

3.1 Repeated Interaction

Assume that the Prisoner's Dilemma game introduced above is repeated with an unknown number of rounds. After each round there is a probability β that another round will be played. Hence, the expected number of rounds is $1/(1-\beta)$. Assume also that the population consists of two types of players, unconditional defectors and conditional cooperators. The unconditional defectors always defect, while the conditional cooperators are endowed with the Tit-for-Tat strategy. The Tit-for-Tat strategy dictates cooperators to cooperate on the first round, and on all subsequent rounds do what the partner did on the previous round. The fraction of the population adopting Tit-fot-Tat is p, while the remaining is adopting unconditional Defect. The expected payoff for cooperators adopting Tit-for-Tat and defectors, respectively, are then

$$V(C) = p\left(\frac{b}{1-\beta}\right) + (1-p)\left(d + \frac{c\beta}{1-\beta}\right) \tag{4}$$

$$V(D) = p\left(a + \frac{c\beta}{1-\beta}\right) + (1-p)\left(\frac{c}{1-\beta}\right) \tag{5}$$

Equation (4) says that when two Tit-for-Tatters meet, they will both cooperate on the first interaction and then continue to do so until the interaction terminated, giving a expected payoff of $b/(1-\beta)$. When a Tit-for-Tatter meets a defector, the former gets d on the first interaction while the defector gets a. Then both will defect until the game terminates, the expected number of iterations after the first round being $(1/(1-\beta)) - 1 = \beta/(1-\beta)$. Equation (5) has a similar interpretation.

According to (3) the condition for equilibrium is that the expected payoff for the two types is equal, giving

$$p^* = \frac{c-d}{\frac{b-c\beta}{1-\beta} + c - d - a} \tag{6}$$

Since the nominator is positive, the denominator of (6) must also be positive. In addition, for $p^* \in (0,1)$, the denominator must be greater than the nominator. Both conditions are satisfied if

$$\frac{b-c\beta}{1-\beta} - a > 0 \tag{7}$$

which gives

$$\beta > \frac{a-b}{a-c} \tag{8}$$

When (8) holds, p^* is an interior equilibrium. This situation can be explained as follows: Suppose that the initial frequency of cooperators is lower than p^*, when there are many defectors, rare cooperators are likely to be paired with defectors, producing a low payoff for cooperators. If, however, the initial frequency of cooperators is higher than p^*, then $V(C) > V(D)$. Cooperators often meet other cooperators with whom to associate. Expected payoff for Tit-for-Tatters is higher than that of the defectors, which causes cooperating behavior to spread. Hence, p^* is an interior *unstable* equilibrium (a tipping point) which marks the boundary between the ranges of attraction of the two *stable* equilibrium, $p = 0$ and $p = 1$.

We can then draw the following conclusion from the model: *In a population where defecting behavior is not too common, the cooperating Tit-for-Tat strategy leads to universal cooperation if pairs of individuals are likely to interact many times.* From (6) we get

$$\frac{dp^*}{d\beta} = \frac{p^* \frac{c-b}{(1-\beta)^2}}{\frac{b-c\beta}{1-\beta} + c - d - a} < 0 \tag{9}$$

saying that an increase in the probability for the game to be continued moves p^* to the left. A smaller fraction of Tit-for-Tatters is then needed in order to secure an evolutionary stable survival of cooperative behavior.

However, even if β is high we still need a certain fraction of Tit-for-Tatter in order to start a process where Tit-for-Tatters increase in numbers. This illustrates that the model fails to answer what many consider as the most fundamental problem related to the evolution of cooperation: How could cooperation ever have started from a previous asocial state where (almost) all are defectors? To solve this puzzle Axelrod and Hamilton introduce the concept of segregation (or clustering as they name it). When there is some segregated interaction, Tit-for-Tatters are more likely paired with each other than chance alone would dictate. If the long-run benefit of cooperation is big, even a small amount of segregation can cause the expected payoff of Tit-for-Tatters to exceed the expected payoff of defectors. An evolutionary trend towards universal cooperation can then get started.

3.2 Segregation

A main result in the work by Axelrod and Hamilton is that segregation can be very effective for the evolution of cooperation in a repeated Prisoner's Dilemma game. But what about the no-repeated version of the game ? Can segregation also promote the evolution of cooperation when the players meet to play the one-shot Prisoner's Dilemma game, that is when $\beta = 0$. It is immediately clear that *complete segregation* of cooperators and defectors within a large population secure cooperation. Complete segregation means that cooperators always meet cooperators, and defectors always meet defectors. Cooperators get a payoff of b, while defectors get c. Since $b > c$ cooperating behavior will spread, and in the long run take over the whole population.

The case where cooperators and defectors are only partly segregated can be modeled by using the following formulation, adopted from Boyd and Richerson [3]. Let $r \in (0, 1)$ be a measure of the degree of segregation. When p is the fraction of cooperators in the population, the probability that a cooperator meets another cooperator is no longer p but $r + (1 - r)p$. Correspondingly, the probability that a defector meets another defector is $r + (1 - r)(1 - p)$. If $r = 1$, we have complete segregation, implying that cooperators never interact with defectors. If $r = 0$, we are back to the situation with random matching. Adopting this formulation, the expected payoff for cooperators and defectors, respectively, are

$$V(C) = [r + (1 - r)p]b + [(1 - r)(1 - p)]d \tag{10}$$

$$V(D) = [(1 - r)p]a + [r + (1 - r)(1 - p)]c \tag{11}$$

From (10) and (11) we see that with random matching ($r = 0$), we are back to the situation analyzed in Sect. 2. Defectors do it better than cooperators for every $p \in (0, 1)$, giving $p = 0$ as an evolutionary stable equilibrium. With complete segregation ($r = 1$), we reach the complete opposite conclusion, as noted above. Cooperators do it better than defectors for every $p \in (0, 1)$, giving $p = 1$ as an evolutionary stable equilibrium. In the simulation we are therefore interested in analyzing the situation where the segregation parameter (r) lies between these two extreme cases. In particular we are interesting in finding out how small r can be in order to support an evolutionary stable proportion of cooperators. However, as it has been shown in earlier work, in addition to r, the expected payoffs are also influenced by the proportion cooperators (p) and defectors ($1 - p$) in the population. In the simulation we therefore have to vary both the segregation parameter and the initial proportion of cooperators and defectors in the population. This makes it possible to study how different combinations of r and p affect the evolution of cooperators and defectors.

4 The Simulation

There has been a lot of research on the simulation of the PD [2–4]. As a simulation model, we use an agent-based simulation approach in which an agent represents a player with a predefined strategy. The basic activity of the agent is to play the iterated Prisoner's Dilemma. Each agent is identified using a unique label. The label C is used to identify the agents choosing the cooperative strategy, while the label D is used for those choosing the defective strategy. Each agent's label can be viewed as a mapping from one state of the game to a new state in the next round, and the simulation experiments searches for the ability of an agent to survive the evolution process.

The simulation of the iterated Prisoner's Dilemma is described as follows. Initially, a population of agents is generated. A user-defined parameter will determine the percentage of agents playing the cooperative strategy against those playing the defective strategy. The payoff of all agents is set to 0. The next step of the algorithm proceeds by pairing off agents to play one game of PD. This step can be viewed as a matching process. To begin with, a random number *random* is drawn uniformly on the interval (0,1). Thereafter, an $agent_k$ is drawn randomly from the set of unmatched agents. If $agent_k$ is assigned the label C, then the matching scheme will select a randomly unmatched agent with the label C provided the following inequality ($\mathbf{random} < r + (1-r) * p_c$) holds, otherwise the matching mate of $agent_k$ will be a randomly chosen unmatched agent with the label D. The value of p_c represents the proportion of agents playing the cooperative strategy. On the other hand, if $agent_k$ is assigned the label D, then its matching mate will be chosen with the label D provided the inequality ($\mathbf{random} < r + (1-r) * (1-p_c)$) holds, otherwise the matching mate of $agent_k$ will be selected with the label C. If by the chance, the matching scheme is unable to locate the matching mate with the required label, then $agent_k$ will be left unmatched. At the end of each tournament, the agents of the current population P_t are transformed into a new population P_{t+1} that engages in a new round of PD based on each agent's payoff. In the simulation we use the same payoff parameters as Axelrod and Hamilton [1]. These are shown in the payoff matrix below.

	Cooperate	Defect
Cooperate	3, 3	0, 5
Defect	5, 0	1, 1

The payoff received will determine whether an agent is removed from the game or allowed to continue. It is assumed that the size of the entire population stays fixed during the whole simulation process. All the unmatched agents from P_t will automatically be allowed to be part of the new population P_{t+1}. The agents that were engaged in the one-shot Prisoner's Dilemma game are ranked according to their payoff from best to worse (i.e., sorting agents to decreasing payoff values) and those with the highest payoff will be allowed to proceed to the next round and multiplies by cloning a duplicate agent with similar strategy. Each agent resets its payoff to 0 before starting a new round of PD. The simulation process is assumed to have reached a stabilization of its convergence when all the agents have similar strategy.

5 Experiments

5.1 Experimental Setup

The simulation model has a number of user-defined parameters such as the segregation parameter, and the starting initial conditions (i.e., percentages of cooperators and defectors). We perform several simulations using instances defined by the 4-duple $<n, p_c, p_d, r>$, where n denotes the number of agents, p_c denotes the percentage of cooperators, p_d denotes the percentage of defectors, and r the segregation parameter. We set the number of agents to $1,000$. In order to obtain a more fair understanding of the simulation process, we vary the parameters r and p_c from 0.1 to 0.9 with a step size of 0.1, and p_c from 10 to 90 % with a step size of 10. Thereby, producing 81 different pairs of r and p_c. Because of the stochastic nature of the simulation process, we let each simulation do 100 independent runs, each run with a different random seed. In this way every result we present is averaged over 100 runs. The simulation process ends when the population of agents converges to either 100 % C's or 100 % D's, or a maximum of 10^6 generations have been performed.

5.2 The Benchmark Case

In this section, we conduct an experiment using $p_c = 90$, $p_d = 10$ and setting the segregation parameter r to 0. Figure 1 shows one typical run of the simulation experiment. The course of the percentage function suggests an interesting feature which is the existence of two phases. The first phase starts with a steady decline of agents with the cooperative strategy over the first generations before it flattens off as we mount the plateau, marking the start of the second phase. The plateau spans a region where the percentage of C's and D's fluctuates around 50 %. The plateau is rather short and becomes less pronounced as the number of generation increases. Then the percentage of C's start to decrease before finally it jumps to 0 %. This example illustrates how agents tend to evolve strategies that increasingly defect in the absence of the segregation. The explanation is rather a simple one. The agents evolve in a random environment, and therefore the agents that manage to survive the simulation process are those willing to always defect.

5.3 Phase Transition

Table 1 gives the results of the simulations with different values of r and p_ck. A quick look at this table reveals the existence of three different regions. The first region lies in the upper left corner where the values of p_c and r are low shows that

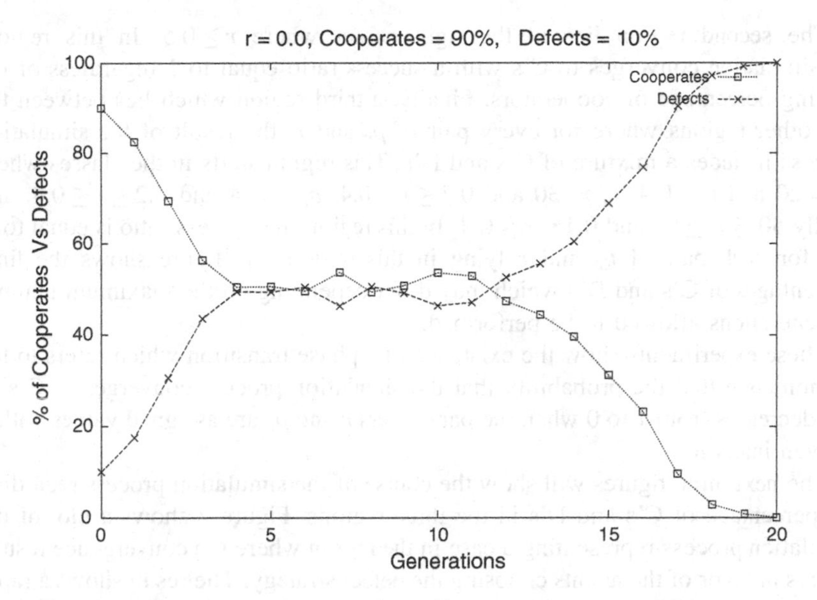

Fig. 1 Evolution process of cooperators and defectors with r = 0

Table 1 Convergence ratios for cooperates and defects

		Segregation : r								
C_{pc}	C,D	0.1	0.2	0.3	0.4	0.5	0.6	0.7	0.8	0.9
10	C	0	0	0	0	1	1	1	1	1
	D	1	1	1	1	0	0	0	0	0
20	C	0	0	0	0.86	1	1	1	1	1
	D	1	1	1	0.14	0	0	0	0	0
30	C	0	0	0.70	0.85	1	1	1	1	1
	D	1	1	0.30	0.15	0	0	0	0	0
40	C	0	0.65	0.72	0.86	1	1	1	1	1
	D	1	0.35	0.28	0.14	0	0	0	0	0
50	C	0.53	0.62	0.70	0.85	1	1	1	1	1
	D	0.47	0.38	0.30	0.15	0	0	0	0	0
60	C	0.56	0.64	0.71	0.84	1	1	1	1	1
	D	0.44	0.36	0.29	0.16	0	0	0	0	0
70	C	0.55	0.61	0.72	0.84	1	1	1	1	1
	D	0.46	0.39	0.28	0.16	0	0	0	0	0
80	C	0.56	0.62	0.69	0.85	1	1	1	1	1
	D	0.44	0.38	0.31	0.15	0	0	0	0	0
90	C	0.53	0.63	0.67	0.82	1	1	1	1	1
	D	0.47	0.64	0.33	0.18	0	0	0	0	0

the result of the simulation converges globally to D's with a success ratio equal to 1. This region starts in the classes where $p_c = 10$ and $r \leq 0.4$, $p_c = 20$ and $r \leq 0.3$, and finally $p_c = 30$ and $r \leq 0.2$.

The second region lies in the right corner where $r \geq 0.5$. In this region, the simulation converges to C's with a success ratio equal to 1 regardless of the starting percentage of cooperators. Finally, a third region which lies between the two other regions where for every pair of p_c and r, the result of the simulation process includes a mixture of C's and D's. This region starts in the classes where $p_c = 20$ and $r = 0.4$, $p_c = 30$ and $0.3 \leq r \leq 0.4$, $p_c = 0.4$ and $0.2 \leq r \leq 0.4$, and finally $50 \leq p_c \geq 90$ and $0.1 \leq r \leq 0.4$. In this region the success ratio is equal to 0, and for each pair of p_c and r lying in this region, the figure shows the final percentages of C's and D's which may differ depending on the maximum number of generations allowed to be performed.

These experiments show the existence of a phase transition which refers to the phenomenon that the probability that the simulation process converges to C's or D's decreases from 1 to 0 when the parameters r and p_c are assigned values within a given interval.

The next three figures will show the course of the simulation process regarding the percentage of C's and D's in the three regions. Figure 2 shows a plot of the simulation process representing a case in the region where the convergence results always in favor of the agents choosing the defect strategy. The result shows a rapid rise of D's before it reaches 100 % at about the sixteenth generation. Choosing the values of r and p_c in this region prevent agents with the cooperative strategy to develop leading to a random working environment where the agents with the defect strategy proliferate. Figure 3 shows a plot of the simulation process representing a case in the phase transition with $r = 0.3$, 50 %, and $p_d = 50$ %, where the convergence results always in a mix population of C's and D's. Notice the rapid increase in the percentage of C's and the rapid decline in the percentage of

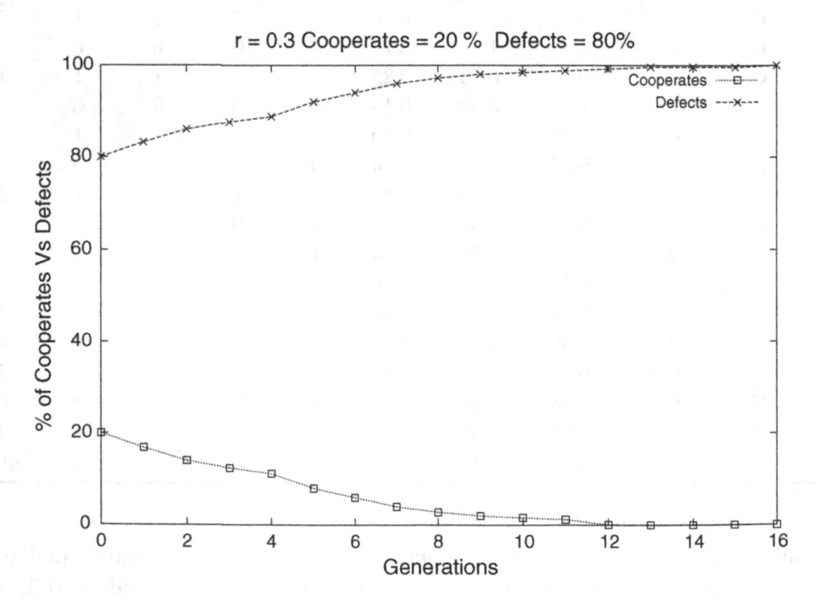

Fig. 2 Evolution process of cooperators and defectors with $r = 0.3$, $p_c = 80\%$, $p_d = 20\%$

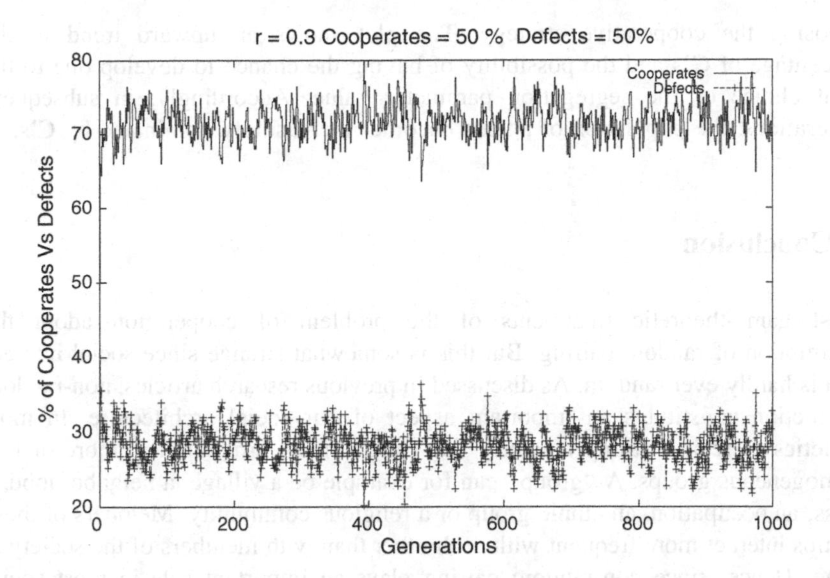

Fig. 3 Evolution process of cooperators and defectors with $r = 0.3$, $p_c = 50\%$, $p_d = 50\%$

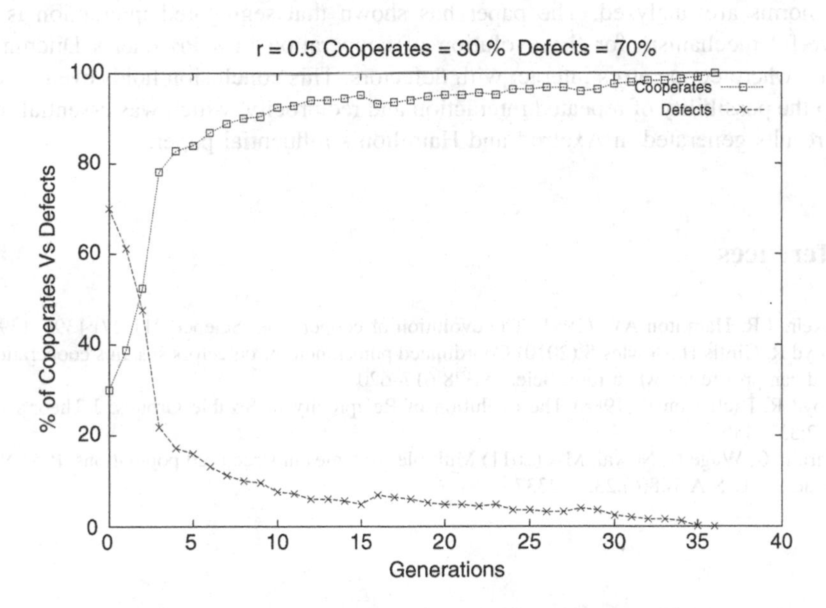

Fig. 4 Evolution process of cooperators and defectors with $r = 0.5$, $p_c = 30\%$, $p_d = 70\%$

D' during the first generations. Both strategies reach a peak value at about 400 generations and periodically fluctuates between a low and a high percentage range and remain there indefinitly. Finally, Fig. 4 shows a plot representing a case in the third region characterized by a convergence resulting always in favor of the agents choosing the cooperative strategy. The plot shows an upward trend in the percentage of C's and the possibility of having the chance to develop due to the right choice of the segregation parameter value. Accordingly, in subsequent generations, the population of agents becomes increasingly dominated by C's.

6 Conclusion

Most game-theoretic treatments of the problem of cooperation adopt the assumption of random pairing. But this is somewhat strange since social interaction is hardly ever random. As discussed in previous research articles, non-random interaction constitutes an important aspect of our social architecture. In most societies there is a strong tendency that members are structured in more or less homogeneous groups. A "group" can for example be a village, a neighborhood, a class, an occupation, an ethnic group or a religious community. Members of these groups interact more frequent with each other than with members of the society at large. Hence, since non-random pairing plays an important role in most social interaction, it should be taken into consideration when the evolution of behavior and norms are analyzed. The paper has shown that segregated interaction is a powerful mechanism for the evolution of cooperation in a Prisoner's Dilemma game, where cooperators interact with defectors. This conclusion holds even if we drop the possibility of repeated interaction and reciprocity, which was essential for the results generated in Axelrod and Hamilton's influential paper.

References

1. Axelrod R, Hamilton AW (1981) The evolution of cooperation. Science 211(27):1390–1396
2. Boyd R, Gintis H, Bowles S (2010) Coordinated punishment of defectors sustains cooperation and can proliferate when rare. Science 328:617–620
3. Boyd R, Richerson P (1988) The evolution of Reciprocity in Sizable Groups. J Theor Biol 132:337–356
4. Tarnita C, Wage N, Nowak MA (2011) Multiple strategies in structured populations. Proc Nat Acad Sci U S A 108(6):2334–2337

Characteristic Analysis and Fast Inversion of Array Lateral-Logging Responses

Yueqin Dun and Yu Kong

Abstract The characteristic between responses of different models and parameters are analyzed, which is crucial to the inversion problem of logging and useful for logging engineers. The inversion of the multi-parameters of the array lateral-logging is a nonlinear and multi-solution problem. To avoid the defects of the traditional inversion algorithms and enhance the inversion speed, the three-layer back propagation (BP) neural network is built to inverse formation parameters. The inputs of the BP neural networks are the logging responses and the target outputs are the formation parameters. The training results of different networks are given and compared, and the inversion results calculated with the trained nets are close to that of company, it only takes 3 s to inverse 100 groups of parameters.

Keywords Fast inversion · Array lateral-logging · BP neural networks

1 Introduction

To judge which layer or what depth of the earth is of petroleum or gas, a measurement sonde is put into a borehole and moved down to measure the physical characteristics of the soil, which is called logging. Extracting the electrical and structure parameters of the formations from the measured data, which is an inversion problem, is the most important objective of logging. Rapid inversion is

Y. Dun (✉)
School of Electrical Engineering, University of Jinan, Jinan 250022 Shandong, People's Republic of China
e-mail: dunyq828@163.com

Y. Kong
Information Center, Shandong Medical College, Jinan 250002 Shandong, People's Republic of China
e-mail: kongy@sdmc.net.cn

F. Sun et al. (eds.), *Knowledge Engineering and Management*,
Advances in Intelligent Systems and Computing 214,
DOI: 10.1007/978-3-642-37832-4_13,

always needed at the logging field to provide real-time and rough formation model parameters. The principal parameters that need to be inversed involve the resistivity of invasion (Rxo) surrounding the borehole, the resistivity of the true formation (Rt), and the diameter of the invasion (Di).

The inversion of parameters is a nonlinear and multi-solution problem. The linearization methods are often used to solve the problems, which gets a convergent solution by an iterative process with suitable initial parameters. The linearization methods may converge to a local optimal solution if the initial parameters deviate far from the real solution; and the forward problem must be calculated many times during the iterative process. To avoid calculating forward responses and enhance the inversion speed, the Look-Up Tables of precalculated forward responses are generated over a set of cylindrical 1D earth model covering a large range of resistivity contrasts and invasion length [1–3]. However, large numbers of Look-Up Tables must be stored in advance and many interpolations need to be calculated.

To avoid these defects in the iterative algorithms, the back propagation (BP) neural network is applied to inverse the formation parameters in this paper. To get better training results, analyzing the characteristic between the responses and the parameters is necessary, which is useful to simplify and reduce the number of training networks.

2 Characteristic Analysis

The response characteristics are different due to the influence of the different parameters. To analyze the characteristic of the logging responses, different models are calculated and analyzed.

2.1 Formation Model

By the array lateral-logging sonde as shown in Fig. 1, a total of six resistivity (Ra) responses are measured, among which the shallowest apparent resistivity (Ra0) is most sensitive to the mud (Rm) in the borehole and can be used to estimate Rm, and the other five responses, Ra1 to Ra5 (the detection depth increase gradually from Ra1 to Ra5), are all sensitive to the formation parameters (Rxo, Di and Rt) as shown in Fig. 1. If the thickness (H) of the formation is quite thick, the effect of shoulder (Rs) to the apparent resistivity Ra1 to Ra5 is very small, and the 2-D formation can be simplified to a 1-D problem, namely non-shoulder. In this paper, we mainly pay attention to Ra1 $\sim$ Ra5, which can provide the principal information of the formation. The logging responses Ra1 to Ra5 are used as the input training samples, and the corresponding formation parameters as the training targets. Then, a supervised BP network can be trained based on the simplified 1-D

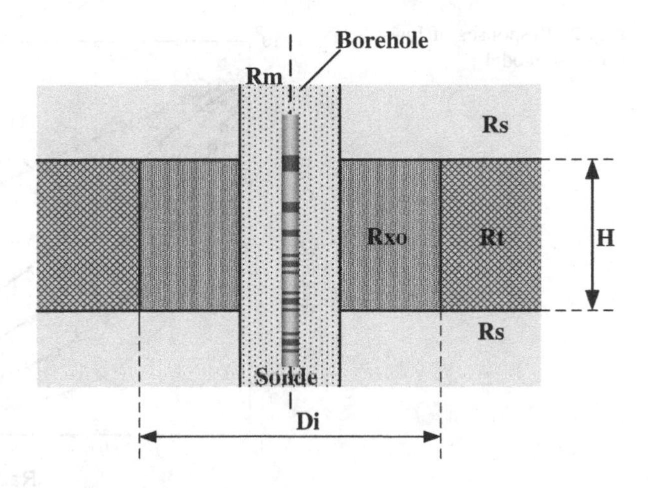

Fig. 1 The parameters of formation model

model, but the characteristic of the responses should be analyzed before training the network.

2.2 Characteristic of Non-shoulder Model

Figure 2 gives the responses of the low-invasion model (Rxo < Rt). The horizontal axis is invasion radius with rectangular coordinate and longitudinal axis is resistivity with logarithmic coordinates. The formation model is: $Rm = 1.0\,\Omega \cdot m$, $Rxo = 10\,\Omega \cdot m$, $Rt = 100\,\Omega \cdot m$, and the borehole diameter is 8 inch. With the increase in the invasion radius, the influence of invasion zone to the resistivity (Ra1 ~ Ra5) enhances. The detection depth is shallow, and the influence is more serious. The resistivity follows a law of increasing by degrees, namely, Ra1 < Ra2 < Ra3 < Ra4 < Ra5.

Figure 3 shows the responses of the high-invasion model (Rxo > Rt). The horizontal axis and the longitudinal axis are the same with the low-invasion model. The formation model is: $Rm = 1.0\,\Omega \cdot m$, $Rxo = 100\,\Omega \cdot m$, $Rt = 10\,\Omega \cdot m$, and the borehole diameter is 8 inch. The influence of invasion zone to the resistivity is similar to the low-invasion model, but the resistivity follows another law of decreasing by degrees, that is, Ra1 > Ra2 > Ra3 > Ra4 > Ra5.

The borehole diameter is another factor to influence the resistivity. Here, only the resistivity of the common 8-inch borehole diameter is given. The law between the different resistivity of low-invasion model will not change and will also increase by degrees from Ra1 to Ra5. However, that of high-invasion model will change with the increase in borehole diameter, and there will appear a cross between the different resistivity. The characteristic of the high-invasion model is more complex than that of the low-invasion.

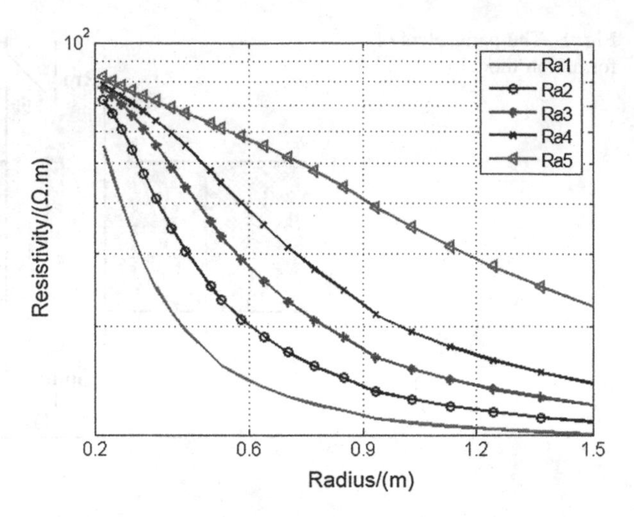

Fig. 2 Responses of low-invasion model

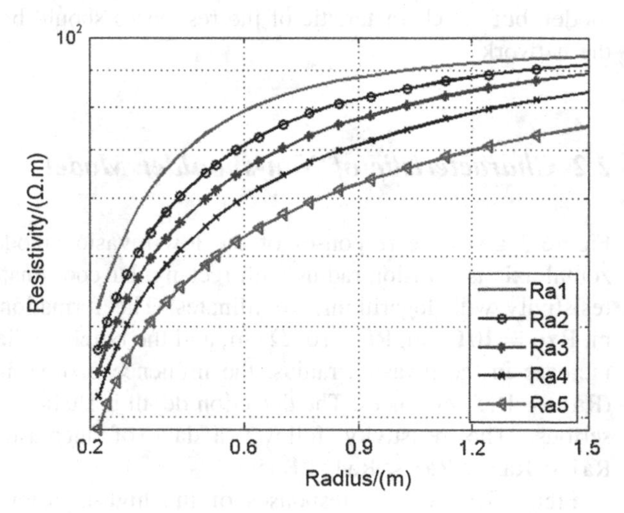

Fig. 3 Responses of high-invasion model

2.3 Characteristic of Models with Shoulder

Figures 4, 5, and 6 give the responses of different layer thicknesses devoted by H under different models. Here, the resistivity of mud Rm is $1\,\Omega \cdot \mathrm{m}$, the diameter of the borehole is 8 inches, and Rt is $100\,\Omega \cdot \mathrm{m}$. Figure 4 has no mud invasion, and Rs is $20\,\Omega \cdot \mathrm{m}$, the responses are very close to Rt. Figure 5 has invasion and Rxo is 40 $40\,\Omega \cdot \mathrm{m}$, which is bigger than Rs ($20\,\Omega \cdot \mathrm{m}$), the responses are smaller than Rt due to the effect of the invasion zone and the shoulder. Figure 6 has invasion, but Rxo ($5\,\Omega \cdot \mathrm{m}$) is smaller than Rs ($20\,\Omega \cdot \mathrm{m}$), and the responses are also smaller than Rt, while in the responses nearby the interfaces, there occurs oscillation.

The main purpose of the inversion is to inverse Rt from the responses; analyzing and finding out the characteristic between the responses and the parameters

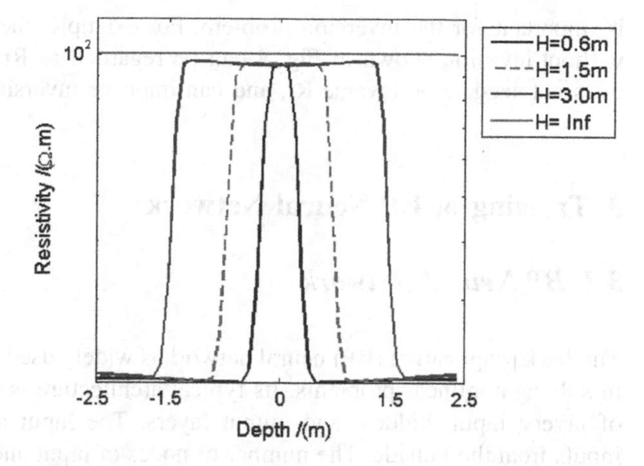

Fig. 4 Responses of non-invasion model with shoulder

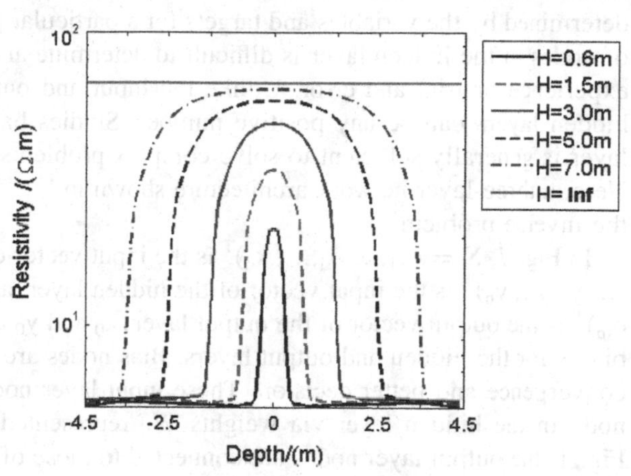

Fig. 5 Responses of high-invasion model with shoulder

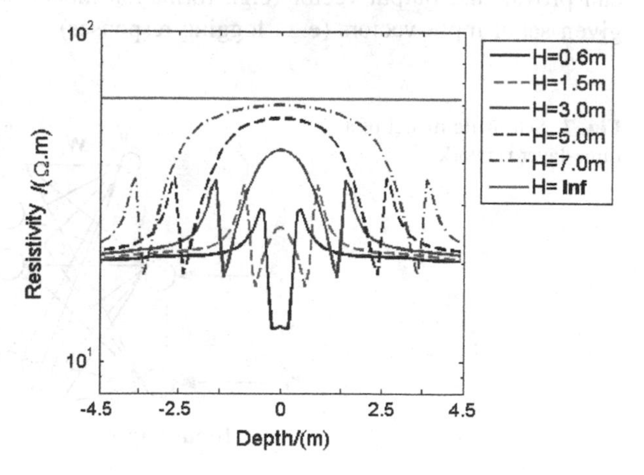

Fig. 6 Responses of low-invasion model with shoulder

is important for the inversion problem. For example, the responses of the model without invasion shown in Fig. 4 can be regarded as Rt approximatively, which does not need to be inverse Rt, and can improve inversion speed.

3 Training of BP Neural Network

3.1 BP Neural Network

The back propagation (BP) neural network is widely used as supervised neural nets in solving nonlinear problems. Its typical architecture is composed of three kinds of layers: input, hidden, and output layers. The input layer nodes only receive inputs from the outside. The number of nodes of input and output layers is usually determined by the variables and targets for a particular problem, while the number of nodes in the hidden layer is difficult to determine and is usually determined by experience or trial and error. Unlike the input and output layers, the number of hidden layers can be any positive number. Studies have shown that one hidden layer is generally sufficient to solve complex problems if enough nodes are used. Here, a three-layer network architecture shown in Fig. 7 has been chosen to solve the inverse problem.

In Fig. 7, $X = (x_1, \ldots, x_i, \ldots, x_n)^T$ is the input vector of the input layer, $Y = (y_1, \ldots, y_j, \ldots, y_h)^T$ is the input vector of the hidden layer, and the $O = (o_1, \ldots, o_k, \ldots, o_m)^T$ is the output vector of the output layer. x_0 and y_0 are both set to be -1 as the biases for the hidden and output layers. Bias nodes are usually included for faster convergence and better decision. These input layer nodes are connected to every node in the hidden layer via weights W_{ij} represented by the lines as shown in Fig. 1, the output layer nodes are connected to those of hidden layer with W_{jk} and can provide the output vector (e.g., formation model parameters) according to a given set of input vectors (e.g., logging responses).

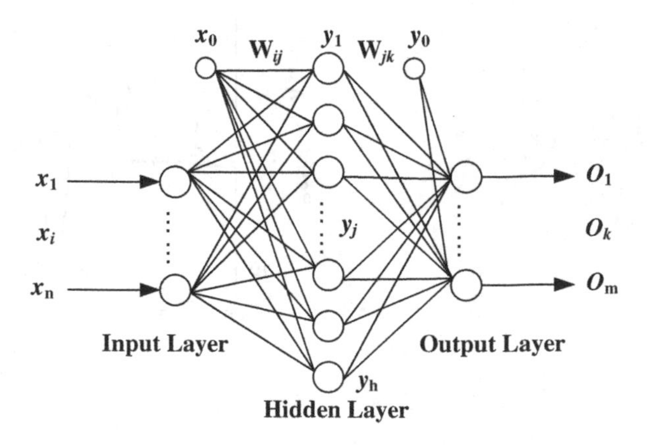

Fig. 7 Schematic model of a three-layer network

The training is an essential part in using neural networks. This process requires training a number of input samples paired with target outputs. If the target vector is $d = (d_1,\ldots, d_k, \ldots, d_m)^T$, the aim of training is to minimize the differences between the target outputs for all the training pairs, which can be defined as the error function

$$E = \frac{1}{2}\sum_{k=1}^{m} (d_k - o_k)^2 \tag{1}$$

Before beginning training, some small random values are generally used to initialize the weights and biases. BP neural network is a forward-propagation step followed by a backward-propagation step. The forward-propagation step begins by sending the inputs through the nodes of each layer. A nonlinear function, called the sigmoid function, is used at each node for the transformation of the inputs to outputs. This process repeats until an output vector is calculated. The backward-propagation step calculates the error vector by comparing the calculated outputs and targets, and the network weights and biases are modified to reduce the error.

The weight adjustment in hidden layer and output layer is:

$$\begin{gathered}\Delta w_{jk} = -\eta \frac{\partial \mathrm{E}}{\partial w_{jk}} = \eta(d_k - o_k)o_k(1 - o_k)y_j = \eta\delta_k^o y_j \\ (j = 0, 1, 2, \ldots, h, \quad k = 1, 2, \ldots, m)\end{gathered} \tag{2}$$

where η is the learning rate, and $\delta_k^o = (d_k - o_k)o_k(1 - o_k)$.

The weight adjustment in input layer and hidden layer is:

$$\begin{gathered}\Delta w_{ij} = -\eta \frac{\partial \mathrm{E}}{\partial w_{ij}} = \eta\left(\sum_{k=1}^{m} \delta_k^o w_{jk}\right)y_j(1 - y_j)x_i = \eta\delta_j^y x_i \\ (i = 0, 1, 2, \ldots, n, \quad j = 1, 2, \ldots, h)\end{gathered} \tag{3}$$

where $\delta_j^y = \left(\sum_{k=1}^{m} \delta_k^o w_{jk}\right)y_j(1 - y_j)$.

By training, a set of weights and biases will be produced finally and can be used for calculating target values (e.g., the formation model parameters) when the actual output is unknown. More details about BP neural network can be found in [4] and [5].

3.2 Training of BP Neural Network

Before training, 3,166 representative input samples paired with targets according to the practical engineering logging are precalculated by the forward problem solution. The input samples (Ra1 to Ra5) should be preprocessed by two steps. The first is borehole correction which can eliminate the effect of different Rm, and only one set of network needs to be trained under some Rm (such as 1.0). The second

Table 1 Training Results of Different Hidden Layer Nodes

Network size	MAE	Network size	MAE
6-8-3	0.006715	6-16-3	0.001998
6-10-3	0.003715	6-18-3	0.001086
6-12-3	0.002722	6-20-3	0.001039
6-14-3	0.002704	6-22-3	0.000928

step is calculating the logarithm (base 10) of the input samples, while the training targets just need to calculate its logarithm (base 10).

A supervised BP networks with three layers is built to inverse the formation parameters. The transfer function between the input layer and the hidden layer is sigmoid function, between the hidden layer and the output layer is linear function, and the Levenberg–Marquardt method is adopted as the training function. The learning rate of the network is set to 0.05, the stop iteration is set to 2,000, and the nodes of hidden layer are 8, 10, 12,…, 22, respectively. The mean absolute error (MAE) threshold is set to 0.001, and the formula of MAE is as follows:

$$MAE = \frac{1}{pm}\sum_{s=1}^{p}\sum_{k=1}^{m}|d_{sk} - o_{sk}| \tag{4}$$

where p is the number of training samples, m is the number of output nodes, d_{sk} and o_{sk} is the target output and net output of sample p at output node k. The training result is shown in Table 1, and 6-8-3 network means 6 inputs, 8 hidden layer nodes, and 3 outputs. The error will be lower than 0.001 when the number of hidden layer nodes is 22, so the following calculation is based on the network with 22 hidden layer nodes.

4 Inversion Results

Figure 8 shows the inverse results of the measured data. The first column is the depth of borehole, from 768 to 783 m. The second column is the borehole diameter, DH is the actual borehole diameter, DI_C is the inverse result of some company, and DI_P is the inverse result of this paper. The third column has nine curves, the dotted lines are the actual measured data Ra1 ~ Ra5, RT_C and RXO_C are the results of company, and RT_P and RXO_P are the results of this paper. The fourth column has 12 curves, the dotted lines are the actual measured data Ra0 ~ Ra5, and the full lines are the results of forward calculation based on the inverse results of this paper, namely Ra0_P ~ Ra5_P.

The average errors of the three parameters (DI, RXO and RT) are 28.3, 34.9 and 8.4 %, separately. RT_C and RT_P agree well with each other by comparing the result between the company and this paper. While DI_C and DI_P have a slight difference, RXO_C and RXO_P have some differences. To ensure the rationality

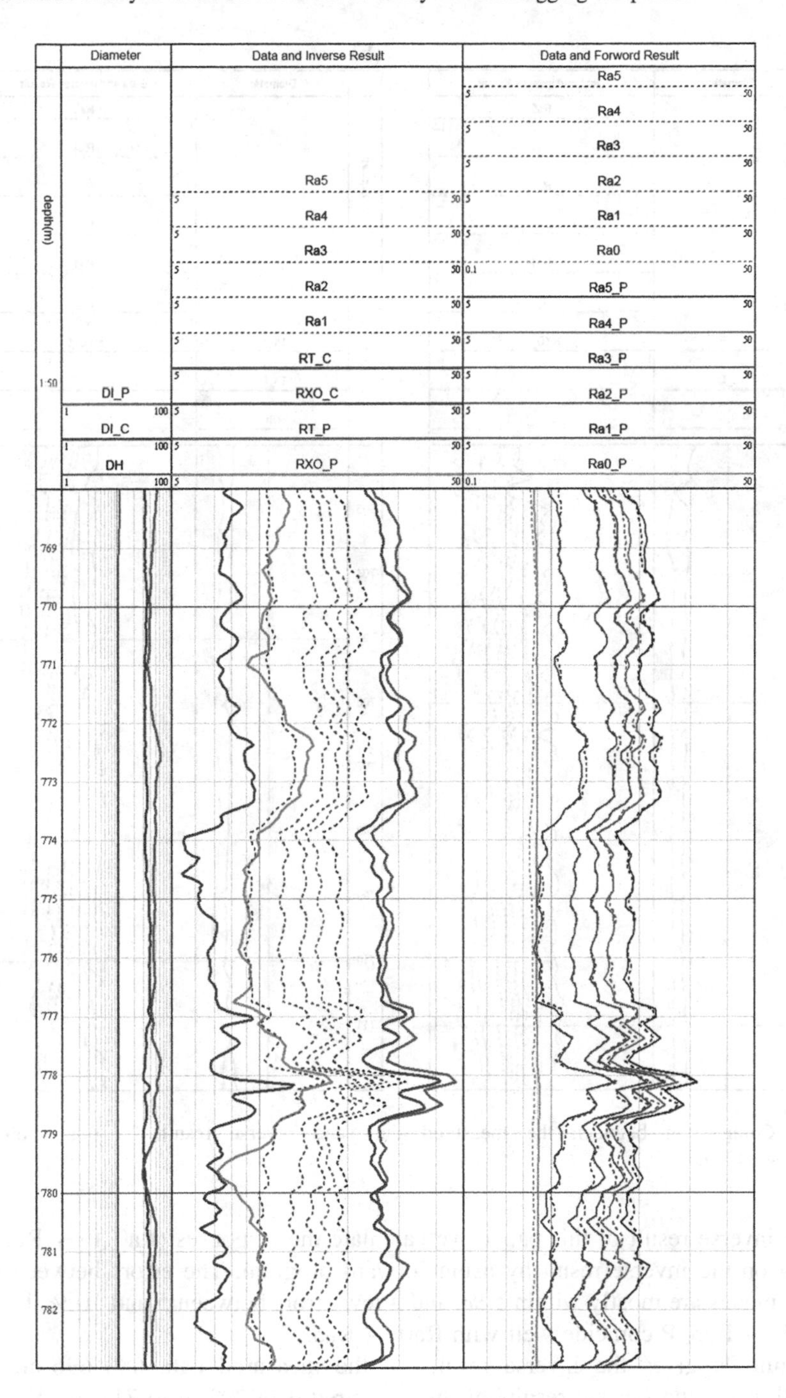

Fig. 8 Comparison between the measured data and inverse results

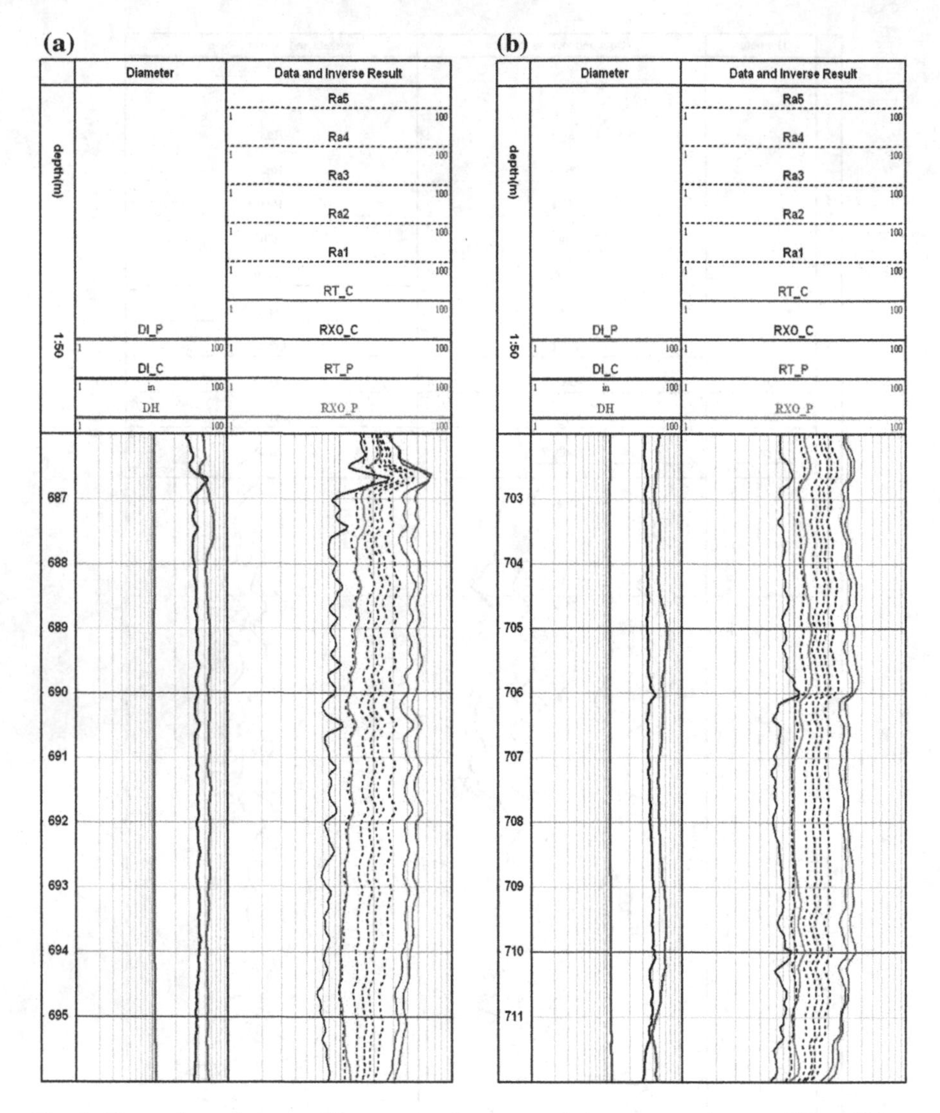

Fig. 9 Comparison between the measured data and inverse results. **a** 686 ~ 696 m, **b** 702 ~ 712 m

of the inverse result of this paper, we calculate the responses (Ra0_P ~ Ra5_P) basing on the inverse result by using forward program. The errors between different modes are mostly within 5 %, and a few errors between 5 and 10 %, hence, Ra0_P ~ Ra5_P coincide well with Ra0 ~ Ra5.

Figure 9a gives the inverse results of the measured data 686–696 m, and Fig. 9b gives the inverse results of the measured data 702 m to 712 m. We also find that errors between RT_C and RT_P are small, and DI_C and DI_P (RXO_C and RXO_P) have largish errors, which are due to the multi-solution characteristic of the nonlinear multi-parameter inversion problem, and also, it is difficult to simultaneously get the exact solution for the multi-parameter. Therefore, it is improper to judge which result is right or not. In the actual logging engineering, the real resistivity RT is particularly important, and the engineer can interpret the logging curves according to the inversion results of RT. Generally, the main objective of inversion is to obtain the accurate RT, DI and RXO can have larger errors. It only takes 3 s to calculate 100 groups of parameters on the computer with 3GH main frequency.

5 Conclusion

In the study of multi-parameter inversion of array lateral-logging, we adopt the three-layer BP neural network to inverse the thick formation parameters. According to the characteristic of the array lateral-logging responses, a great number of preprocessed training samples pared with targets have been trained. The hidden layer nodes have been chosen to be 22 by comparing the training results under different hidden layer nodes. The inversion results under different Rm with the trained net are close to the real values. It shows the validity of borehole correction of the logging responses. Once the BP neural network has been trained successfully, it does not need any initial parameters and can be used without any iterative calculation. Hence the inversion speed will be quite fast, and can be used as the rapid inversion at the logging field.

Acknowledgments The work was supported by the Program for the Natural Science Foundation of Shandong Province (ZR2010EQ012).

References

1. Yin H, Wang H (2002) Method for 2D inversion of dual lateralog measurements. United Stated Patent, Pub. No. US 0040274
2. Frenkel MA, Mezzatesta AG (1999) Well logging data interpretation system and methods. United Stated Patent, Pub. No. US 5889729
3. Frenkel MA (2006) Real-time interpretation technology for new multi-laterolog array logging tool. In: The 2006 SPE Russian oil and gas technical conference and exhibition. SPE 102772, pp 1–11, 2006
4. Ham FM (2000) Principles of neurocomputing for science and engineering. McGraw-Hill, New York
5. Hagan MT, Demuth HB (1996) Neural network design. PWS Publishing Company, Boston

A Model of Selecting Expert in Sensory Evaluation

Xiaohong Liu and Xianyi Zeng

Abstract Sensory evaluation is a subjective method that is mainly dependent on expert knowledge and experience. This method is widely used in industrial products, consumer preference measurement, and management decision and so on. It is an important premise in sensory evaluation of selecting appropriate experts to participate in the test. On discussing the basic assumption and selection criteria, the ideal process of selecting expert in sensory evaluation is put forward in this paper. The method of fuzzy comprehensive evaluation is used in establishing an algorithm of selecting expert in sensory evaluation, and the application of this algorithm is explained by one example.

Keywords Expert · Sensory evaluation · Fuzzy comprehensive evaluation

1 Introduction

The method of sensory evaluation was put forward in 1975 by the United States Association of Food [1]. This method has been extended to other industry fields, service, and management areas from the food evaluation [2]. There are a lot of applications in the different fields of sensory evaluation test, such as consumers' preference evaluation based on their perception in the field of clothing, cosmetics, automotive products and so on. According to the program of working, the step of sensory evaluation is divided into designing, testing, data processing, report

X. Liu (✉)
College of Management, Southwest University for Nationalities, Chengdu, People's Republic of China
e-mail: lxhdoctor@163.com

X. Zeng
The ENSAIT Textile Institute, 9 rue de l'Ermitage F-59100 Roubaix, France
e-mail: Xianyi.zeng@Ensait.fr

F. Sun et al. (eds.), *Knowledge Engineering and Management*, Advances in Intelligent Systems and Computing 214,
DOI: 10.1007/978-3-642-37832-4_14,

writing and so on. Broadly speaking, one new type of sensory evaluation is based on expert knowledge and experience, which is used in making investment decision, analyzing and managing risk and safety, marketing planning, formulating business strategy, handling public emergency and so on. Sensory evaluation is a scientific discipline used to evoke, measure, analyze, and interpret reactions to chose characteristics of products or materials as they are perceived by the senses of sight, smell, taste, touch, and hearing [3]. Scholars studied and improved the test method of sensory evaluation test; especially, in recent years put forward some intelligent techniques and methods of sensory evaluation according to different demands [4, 5]. In sensory evaluation, one of the traditional test methods is to test with a pair for samples, i.e., evaluators determine the quality level of each sample according to their results of comparison between any two samples. It is obvious that the number of tests is very large when the samples are more than 50. The prerequisite of sensory evaluation test is data quality meet user requirements. We should optimize the method of test with a pair in sensory evaluation to decrease the number of test [6]. According to the knowledge of project management, we can consider one test of sensory evaluation as one project. Therefore, there are three key factors in the process of sensory evaluation test, i.e., quality, time, and cost. In fact, these key factors are related to evaluators, we called them as experts. As we know, the quality of sensory evaluation test is mainly dependent on the expert's knowledge and experience, and the cost of sensory evaluation test is mainly to pay the expenses of experts and buy materials. The time of sensory evaluation test is mainly expert working time. Therefore, it is one of the key steps of guaranteeing the success in sensory evaluation test of selecting appropriate experts. Moreover, as for the result of selecting experts, there are two aspects of expectation for experts, i.e., data effectiveness and economic benefits of test, which mainly restrict its application of sensory evaluation test. There are multi factors and links, including design, implementation, result treatment, and application of sensory evaluation, that affect the data's validity [7]. Because the boundary of much content about expert is fuzzy, it is difficult to accurately measure expert's knowledge and experience. Fuzzy comprehensive evaluation method is widely used in treating with these questions with fuzzy multiple factors and their comprehensive evaluation, such as in the fields of engineering, economics, management and so on.

In this paper, we hope to improve the effectiveness of selecting expert in sensory evaluation test. First, according to the actual situation of sensory evaluation test, we discuss the basic assumption and selection criteria related to evaluation experts. Second, in order to improve the contribution of expert group in sensory evaluation test, we put forward an ideal process of selecting experts. Third, we apply the method of fuzzy comprehensive evaluation in establishing an algorithm of selecting experts in sensory evaluation test. Finally, we present one numerical example to explain the application of selecting experts in sensory evaluation test.

2 Basic Assumption and Selection Criteria

2.1 Basic Assumption

Assumption of expert classification: (1) according to the professional field, experts are divided into domain experts and general experts; (2) according to the social influence, experts are divided into common experts and well-known experts. The domain expert is familiar with the background, reason, and destruction and development of object evaluated, as well as effectively participated in evaluating the same or similar question. The general expert mainly studies general question of science and technology, such as natural science, social science, engineering disciplines and so on. However, the general expert had almost no success and fame in the specific field of evaluating, such as earthquake area, industrial disputes, or demolition contradictions and so on.

Assumptions of expert number: (1) the number of domain expert is less than that of general expert, namely the quantity of domain expert is relatively scarce for the desired number by user; (2) the number of general expert is relatively stable in a certain scope and period; however, the number of domain expert is often changing, i.e., when the same type or similar events occurred frequently, the number of experts in this field is increasing; conversely, the number of expert in the field is decreasing continuously.

Assumption of expert cost: (1) the well-known expert's comprehensive cost, which includes travel fee, service fee, receive fees etc., is more than the common expert's relatively; (2) the common expert's free time is more than well-known expert's relatively. In general, the common expert is relatively more likely to meet users' requirements for the view of expert time.

Assumption of expert returns: (1) the domain expert is more familiar with evaluating event than the general expert, therefore, the result made by the domain expert has more practical effect than that of the general expert; (2) the popularity of expert may affect the preference of expert group, and the well-known expert in the field of evaluating event has a greater role in contribution of the group evaluating than the common expert.

2.2 Selection Criteria

We present two aspects of criteria that are related to selecting expert in sensory evaluation test, i.e., necessary conditions and sufficient conditions. The principle of these two conditions in selecting expert in sensory evaluation test is same as that of mathematics.

Necessary conditions: (1) professional demand, which is the expert selected in sensory evaluation test should meet the requirement of knowledge and experience in evaluating field; (2) time demand, which is the expert selected in sensory evaluation test has enough time to participate in test.

Sufficient conditions: (1) group demand, which is the expert selected in sensory evaluation test should support their group and become true group consensus in the process of evaluation; (2) type demand, which is the expert selected in sensory evaluation test is domain expert; (3) return demand, which is the actual contribution of the expert selected in sensory evaluation test is bigger than its actual cost.

3 An Ideal Process of Selecting Experts in Sensory Evaluation

According to the basic assumptions and selection criteria, in order to effectively select expert and improve their contribution of the group in sensory evaluation test, an ideal process of selecting expert in sensory evaluation test is put forward in this paper, and is shown in Fig. 1.

We explain these terms that are related to Fig. 1 as follows:

(1) *Expert database*: it is an information database of experts. Government departments or other agencies should legally collect and preserve experts' information. They are responsible for keeping personal privacy of experts. It is a foundation work in sensory evaluation test of establishment database of experts, which need to be maintained and updated in a long-term.

(2) *New expert*: are those experts found in the process of evaluating one event. These experts include two types of human beings. One type is these peoples who are familiar with background, reason, destruction, and development of the event found by government departments or other agencies first, and the

Fig. 1 An ideal process of selecting expert

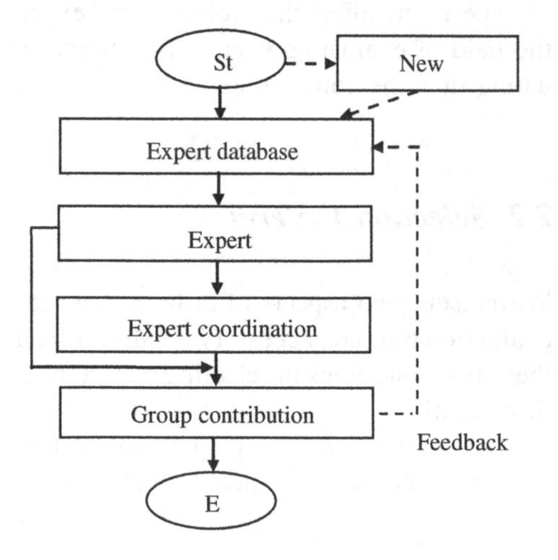

another type is these peoples whose major is related to study and research the new event evaluated. New expert is not only an important resource of supplementing expert database, but may be an effective expert of sensory evaluation test for the event evaluated.

(3) *Expert individualization*: that is expert personalized characteristic information. The effectiveness is one of key problems of expert individualization in sensory evaluation test, which is reflected using different dimensions and indexes. While the knowledge and experience of experts meet special evaluating requirements, these experts are called as effective experts. Expert individualization evaluation is a foundation work for guaranteeing experts' coordination, and improving the contribution of group in sensory evaluation test.

(4) *Expert coordination*: it is the relationship of working among experts. The working state of experts can be reflected using different dimensions and indexes also in sensory evaluation test. The coordination of one group among experts is one of the key links to achieve success in sensory evaluation test.

(5) *Group contribution*: it is the result made by expert group. The result of one group contribution may be forecasted through indexes of improving expert individuation and expert coordination. The main contribution of the group in sensory evaluation test is that the most important suggestion used in supporting decision-making was put forward by the expert.

(6) *Feedback*: it is evaluating the expert's attitude, ability, and performance in the process of sensory evaluation through the event evaluated. It is not only one of main information resources of updating the expert database, but is an effective way of selecting expert for the next sensory evaluation test.

4 An Algorithm of Selecting Expert in Sensory Evaluation

In order to ensure the validity of data obtained by sensory evaluation test, it is necessary to carefully select expert to do test. In fact, there are two factors that affect the result of selecting expert. One is the process of selecting expert, and another is the algorithm of selecting expert.

4.1 A Process of Selecting Expert

In this paper, a process of selecting expert in sensory evaluation test is presented, and shown in Fig. 2, which mainly includes three aspects of content, i.e., expert individualization evaluation, expert coordination evaluation, and group contribution evaluation.

Fig. 2 A process of selecting expert

We explain the mainly content related to Fig. 2 as follows:

(1) *Expert individualization evaluation*: called as S1 evaluation, which aims at evaluating expert personalized characteristic information. According to the model of general evaluation, we can present the target layer, criterion layer, and index layer of this evaluation, respectively. The target of S1 evaluation is to select effective expert. The criterion of this evaluation includes field, specialty, and contribution and so on. The index of this evaluation can be determined according to different criterion layer respectively.
(2) *Expert coordination evaluation:* called as S2 evaluation, which aims at evaluating group coordination based among experts working together. The principle and structure of S2 evaluation is same as S1 evaluation: The difference between S2 evaluation and S1 evaluation is specific content of target layer, criterion layer, and index layer. For example, the target of S2 evaluation is to improve the coordination among experts working together, and its criterion includes complementary, preference, cooperation and so on.
(3) *Group contribution evaluation*: called as S3 evaluation, which aims at forecasting the result of group contribution. The principle of S3 evaluation is applying different mathematics.
(4) *Satisfaction*: that is an evaluation criterion in the process of selecting expert in sensory evaluation test. For each evaluation, the condition of pass is satisfaction. We may use Likert scale with five grades to establish a standard for satisfaction.

4.2 Evaluation Model of Three Stages

According to the process of selecting expert, a three stages evaluation model is presented by this paper, and shown in Fig. 3.

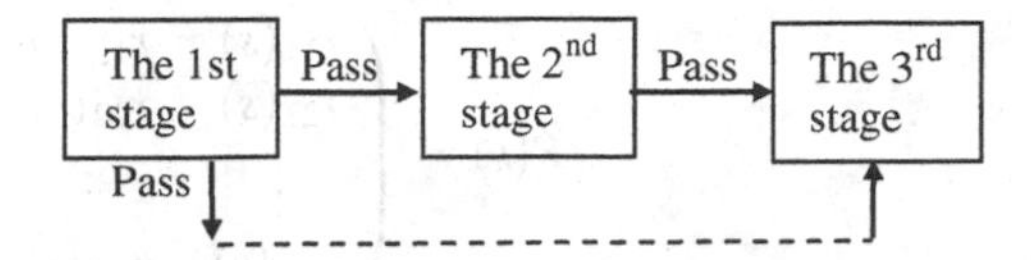

Fig. 3 An evaluation model of three stages

In Fig. 3, the first stage of evaluation is called as a necessary condition evaluation. In the filed of sensory evaluation test, evaluating expert individualization is a necessary condition. In order to insure the quality of data obtained by sensory evaluation test, these experts whose result of individualization evaluated does not meet users' requirements should be given up. Otherwise, they will have a direct negative effect on expert group contribution. Only after passing the result of the first stage of evaluation, the second stage of evaluation will start; moreover, its result will affect the third stage of evaluation in some way.

In Fig. 3, the second stage of evaluation is called as a sufficiency condition evaluation. In the field of sensory evaluation test, evaluating expert coordination is a sufficiency condition. If the result of expert coordination evaluated did not meet users' and expert group's requirements, we should try our best to look for new experts and replace those experts who destroyed the group cooperation. For the new expert, it is necessary to return to the first stage of evaluation. Only after passing the result of the second stage of evaluation, the satisfied result of the third stage of evaluation would be obtained.

In Fig. 3, the third stage of evaluation is a prediction of contribution of expert group. In the field of sensory evaluation test, due to many factors, such as expert motivation and behavior, test environment and others uncertainty be dynamic, it is very difficult to accurately predict the result of contribution of expert group. In any case, the most important work is to obtain en effective prediction through the third stage of evaluation in sensory evaluation. In general, according to results of the first two stages evaluation, we can apply appropriate mathematics tool, such as fuzzy comprehensive evaluation, in predicting the result of the third stage of evaluation.

4.3 Algorithm of Selecting Expert

An algorithm of selecting effective expert in the field of sensory evaluation test is presented in this paper. First, this algorithm is based on one multi stages evaluation model, which shown in Fig. 3. Second, this algorithm is based on fuzzy relation and fuzzy comprehensive evaluation. Third, this algorithm will be applied in selecting effective expert in sensory evaluation test.

Set U be an object sets of multi stages fuzzy evaluation, the number of comment and stage is m and N, respectively. The number of comment index of the $s(s = 1, 2, \ldots, N)$ stage and the value of index is p_s and y_j, respectively. The degree belonging to the comment of $j(j = 1, 2, \ldots, m)$ is $r_{ij}(t_k)$. The fuzzy coefficient matrix of evaluation of the s stage is $\tilde{R}(s)$, and has

$$\tilde{R}(s) = \begin{pmatrix} r_{11}(s) & r_{12}(s) & \cdots & r_{1m}(s) \\ r_{21}(s) & r_{22}(s) & \cdots & r_{2m}(s) \\ \vdots & \vdots & & \vdots \\ r_{p_s1}(s) & r_{p_s2}(s) & \cdots & r_{p_sm}(s) \end{pmatrix}$$

that is

$$\tilde{R}(s) = (r_{ij}(s))_{p_s \times m} \tag{1}$$

In Eq. (1)

$$0 \le r_{ij}(t_k) \le 1, \quad s = 1, 2, \ldots, N$$

Let $B(s)$ be the result of the s stage of evaluation.
and has

$$B(s) = \tilde{R}(s) \circ W_{p_s}$$

that is

$$B(s) = (\, b_{s1} \quad b_{s2} \quad \cdots \quad b_{\mathrm{sm}}\,), j = 1, 2, \ldots, m \tag{2}$$

In Eq. (2):
W_{P_j} is the weight vector set, and $\circ$ is a fuzzy set operator.
On the $B(s)$ normalization,
set

$$C(s) = B(s) \wedge A(s) = \left((bs_j \wedge a_{\mathrm{sj}})\right)_{1\times m} \tag{3}$$

In Eq. (3):
$A(s)$ is called insurance coefficient matrix, and a_{sj} insurance coefficient.

$$A(s) = (\, a_{s1} \quad a_{s2} \quad \cdots \quad a_{\mathrm{sm}}\,) \tag{4}$$

and

$$a_{sj} = \begin{cases} 1 & \text{if} \quad b_{\mathrm{sj}} \ge \alpha \\ 0 & \text{otherwise} \end{cases} \tag{5}$$

Set result of comprehensive evaluation of U is $\Im$, then

$$\Im = C(s) \circ W = \begin{pmatrix} c_{11} & c_{12} & \cdots & c_{1m} \\ c_{21} & c_{22} & \cdots & c_{2m} \\ \vdots & \vdots & & \vdots \\ c_{s1} & c_{s2} & \cdots & c_{\mathrm{sm}} \end{pmatrix} \circ \begin{pmatrix} w_1 \\ w_2 \\ \cdots \\ w_s \end{pmatrix}$$

that is

$$\Im == (\, c_1 \quad c_2 \quad \ldots \quad c_m\,) \tag{6}$$

According to the principle of the maximum degree of membership or weighted average operator, we will get the final result of evaluation.

5 One Numerical Example

Invite some domain experts to evaluate one process of selecting expert in the field of sensory evaluation test.

Set the number of evaluating index of the first stage, the second stage, and the third stage of evaluation are four, and using the Likert scale with five grades as evaluation set. The evaluation comments are *very satisfied, satisfied, general, unsatisfied,* and *very unsatisfied,* respectively.

The results integrated different domain experts' opinion were obtained of three stages evaluation by some way, and their fuzzy matrix are $\tilde{R}(1)$, $\tilde{R}(2)$, and $\tilde{R}(3)$, respectively.

Set the insurance coefficient $\alpha = 0.65$, and the weight vector set of each stage is W_{p_1}, W_{p_2}, and W_{p_3}, respectively, and the weight vector of three stages is W, and has

$$W_{p_1} = W_{p_2} = W_{p_3} = (0.25 \quad 0.25 \quad 0.25 \quad 0.25).$$
$$W = (0.2 \quad 0.5 \quad 0.3)$$
$$\tilde{R}(1) = \begin{pmatrix} 0.6 & 0.4 & 0 & 0 & 0 \\ 0.6 & 0.4 & 0 & 0 & 0 \\ 0.7 & 0.3 & 0 & 0 & 0 \\ 0.7 & 0.3 & 0 & 0 & 0 \end{pmatrix} \quad \tilde{R}(2) = \begin{pmatrix} 0.7 & 0.3 & 0 & 0 & 0 \\ 0.7 & 0.3 & 0 & 0 & 0 \\ 0.6 & 0.4 & 0 & 0 & 0 \\ 0.6 & 0.4 & 0 & 0 & 0 \end{pmatrix}$$
$$\tilde{R}(3) = \begin{pmatrix} 0.65 & 0.35 & 0 & 0 & 0 \\ 0.65 & 0.35 & 0 & 0 & 0 \\ 0.65 & 0.35 & 0 & 0 & 0 \\ 0.65 & 0.35 & 0 & 0 & 0 \end{pmatrix}$$

Select the operator of fuzzy sets is $(\bullet, +)$, according to Eq. (2), and the result of $B(1)$ is

$$B(1) = (0.65 \quad 0.35 \quad 0 \quad 0 \quad 0)$$

Because $b_{11} \geq \alpha = 0.65$, the result evaluated of this expert's individualization pass the first stage of evaluation. The result of $B(2)$ is

$$B(2) = (0.65 \quad 0.35 \quad 0 \quad 0 \quad 0)$$

In the same way, because $b_{21} \geq \alpha = 0.65$, the result evaluated of this expert group's coordination pass the second stage evaluation, and the result of $B(3)$ is

$$B(3) = (0.65 \quad 0.35 \quad 0 \quad 0 \quad 0)$$

According to Eqs. (3), (4), and (5), the result of C is

$$C = \begin{pmatrix} 0.65 & 0 & 0 & 0 & 0 \\ 0.65 & 0 & 0 & 0 & 0 \\ 0.65 & 0 & 0 & 0 & 0 \end{pmatrix}$$

Select the operator of the fuzzy set is $(\bullet, +)$, according to Eq. (6), the result of $\Im$ is

$$\Im = C \circ W = \begin{pmatrix} 0.65 & 0 & 0 & 0 & 0 \\ 0.65 & 0 & 0 & 0 & 0 \\ 0.65 & 0 & 0 & 0 & 0 \end{pmatrix} \circ \begin{pmatrix} 0.2 & 0.5 & 0.3 \end{pmatrix}$$
$$\Im == \begin{pmatrix} 0.65 & 0 & 0 & 0 & 0 \end{pmatrix}$$

According to the principle of maximum degree of membership, the result of evaluation of this expert group contribution is *very satisfied.*

6 Conclusion

Sensory evaluation has been an important way of getting data for a lot fields. There are many factors and links affecting the validity of data in sensory evaluation. The most important step in sensory evaluation is selecting effective expert to do test. In this paper, we proposed an assumption and criterion related to selecting expert, and presented an ideal process and its algorithm based on fuzzy comprehensive evaluation. As we know, it has theoretical and practical significance of effectively selecting expert in sensory evaluation test. Moreover, it is one of the foundations for intelligent sensory evaluation. It should also be pointed out that, only is not enough from a rational selecting expert in sensory evaluation test. The performance of expert and its group in the process of sensory evaluation test is really important; however, some factors affecting expert and its group are randomness and unpredictability. Therefore, we should do further research of how to strengthen management in the course of sensory evaluation test.

Acknowledgments We gratefully acknowledge the support of State Bureau of foreign expert's affairs project of China (grant No. 2012-12), the support by grant 2010GXS5D253 of National Soft Science of China, and thte support by grant 2012XWD-S1202 of Southwest University for Nationalities.

References

1. Dijksterhuis GB (1997) Multivariate data analysis in sensory and consumer science. Food & Nutrition Press, Inc. Trumbull, Connecticut
2. Zeng XY, Ding YS (2004) An introduction to intelligent evaluation. J Donghua Univ 2:1–4

3. Stone H, Sidel JL (2004) Sensory evaluation practices, 3rd edn. Academic Press, Inc., San Diego
4. Ruan D, Zeng XY (2004) Intelligent sensory evaluation—methodologies and applications. Spinger, Berlin
5. Etaio I, Albisu M, Ojeda M, Gil PF, Salmerdn J, Perez EFJ (2010) Sensory quality control for food certification case study on wine method development. Food Control 21:533–541
6. Liu XH, Zeng XY, Xu Y, Koehl L (2006) A method for optimizing dichotomy in sensory evaluation.2006. In: International conference on intelligent systems and knowledge engineering (ISKE2006), Shanghai, China, 6–7 April 2006
7. Deng FM, Liu XH (2006) A model for evaluating the rationality of experiment system of sensory evaluation. J Southwest Univ Natl 32:576–580

Software Development Method Based on Structured Management of Code

Xia Chen and Junsuo Zhao

Abstract Code management based on files has a lot of problems and shortages. This paper puts forward the software development method based on structured management of code. Structured management of code reduces the source code unit from files to minimum unit code, and achieves the fine granular and structuration. The software development based on code structured management has shown obvious advantages in many aspects, such as improving personal and collaborative working efficiency, statistical analysis, roles, and authority management etc.

Keywords Software development · Structured management of code · File-based management of code · Database

1 Introduction

The scale and complexity of software has increased dramatically along with the increasing demand for software [1]. Software development gradually changes from stand-alone to network, centralized to distributed, exclusive to shared, and last but not least, isolation to collaboration. This phenomenon is referred to as Global Software Development, Distributed Software Development, or Multi-Site Software Development [2]. Nowadays, Software development is no longer a short-term

X. Chen (✉) · J. Zhao
Science and Technology on Integrated Information System Laboratory,
Institute of Software, Chinese Academy of Sciences, Beijing, China
e-mail: chen918xia@sina.com

J. Zhao
e-mail: junsuo@iscas.ac.cn

X. Chen
University of Chinese Academy of Sciences, Beijing, China

F. Sun et al. (eds.), *Knowledge Engineering and Management*,
Advances in Intelligent Systems and Computing 214,
DOI: 10.1007/978-3-642-37832-4_15, © Springer-Verlag Berlin Heidelberg 2014

individual behavior, but a long-term group collaboration process. More and more software development activities are facing severe problems including the rapid increase of software development tasks and number of developers in one project, the extension of software development cycle, as well as dramatic decrease in average efficiency. Many factors will lead to the decrease in average efficiency. On one hand, personal ability differs from each other, on the other hand, the cost of communication and coordination between groups cannot be ignored [3, 4]. DeMarco and Jones pointed out that during large-scale system development, about 70–85 % of the development cost is spend on teamwork activities [5, 6].

Until now, there have been many theories and methods focusing on improving software engineering, such as role-based collaboration modeling method [7], component-based software development [8] paradigm of software development for mass customization [9], etc. However, these methods ignore the important roles that code management and software development tools played in improving personal and collaborative work efficiency. For so long software development proceeds based on file-based code management which means the source code space consists of multiple source code files which contain partial codes, respectively. There are many disadvantages for file-based code management such as uneasy to read, unobvious relevance, etc. These shortages will result in the increasing personal and collaborative work costs.

In this paper, we put forward software development method based on structured management of code. We reduce the source code management unit from files to minimum unit code, and utilize database to realize the fine granular and structuration. It should be noted that code management is the foundation of software engineering. Therefore, structured management of code which based on database will change the way of former code management (file-based code management) fundamentally, cause changes in many steps of software engineering, reduce collaboration costs, improve personal efficiency, and bring more convenience to project management.

2 Software Development Based on Files and Documentations

2.1 Environment of Software Development Based on Files and Documentations

Structure of Software Development based on files and documentations is shown in Fig. 1. Source code files and documentations are the core of software product. The activity of software development is proceeded under source code files and related documentations. The essential is the process of increasing, deleting, and updating of source code files and documentations.

Each source code file contains an independent section code under the form of code management based on files. Many source code files consist of the code space

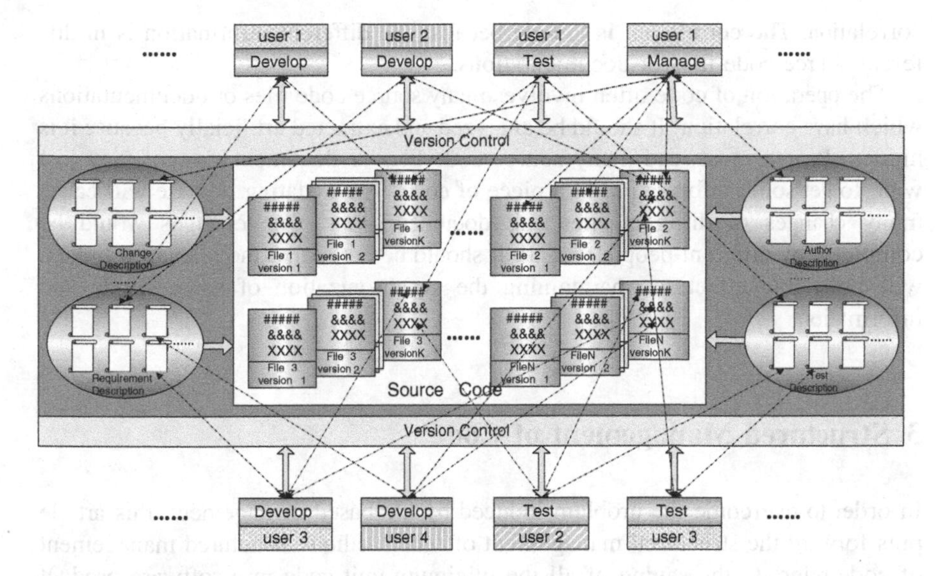

Fig. 1 Environment of software development

of software product, and also as the documentations. In the life cycle of the software product, different roles, such as development people and testing people, operate the source code and related documentations via development environment, testing environment and management environment. Source code files and documentations are evolutionary. As the increasing of software products and functional modules, the number of file and documentation is increasing. The content of each source code file or documentation is becoming perfect along with the advance of development. The increasing of files and documentations in space and time forms the process of software product.

2.2 The Problems of Code Management Based on files

Uneasy to Read. Source code files are transparent before people read them. And people can only capture the functions and interfaces of the code but not the amount of code, content, structure of the code and other details.

Every time the smallest unit people can get is the whole codes of a source code file if they manage codes based on file. But they actually only need to modify a piece of codes (Fig. 1). As the source code files are transparent, only after reading the whole codes, can developers modify relevant part. This kind of repetition is time consuming.

Unobvious Relevance. Source code includes lots of related information, such as compiling, testing, author, functions, version, modules, references etc. Code has strong correlation with its related information logically. Different codes also have correlation. The correlation is hidden because the different information is in different source code files or documentations.

The operation of code often involves many source code files or documentations which have correlation. It should be analyzed and extracted artificially because it is hidden. People must read many source code files or documentations if they just want to get some information of a piece of code. The updating of code will cause many changes of different files and documentations. The changes would be completed by different people related. It should be noted that the relevance hidden will bring bad effect on maintaining the synchronization of source code and information.

3 Structured Management of Code

In order to overcome the problem induced by file-based management, this article puts forward the structured management of code method. Structured management of code refers to the saving of all the minimum unit code in a software product source code space in relevance database, and realizing the easy, accurate retrieval, up-to-date operation of any minimum unit code. The structured management of source code brings about the fine-grained control of source code space. In the mean time, it also makes the combination of source code space and various information described space come true.

3.1 Features of Structured Management of Code

Completeness. Source code and related information are saved in database. The information is complete therefore we can retrieve any source code or related information in a very short time.

Fine Granular. Fine-grained means abandoning the concept of files, and adopting the management method which combine minimum unit code and modules.

Minimum unit code means the minimum code part which has complete and accurate physical meaning. Take C language as an example, function definition, function declaration, type definition, global variable definition, variable declaration, macro definition are all Minimum unit codes.

Fine-grained management of code has two meanings: first, source code space consists of minimum unit code instead of file, and plenty of minimum unit code together forms the source code space; second, minimum unit codes in source codes space are not saved disorderly but managed by detailed multi-level modules. The properties of minimum unit code determine which modules level it should belong to.

Therefore, adding, deleting, updating, and testing minimum unit codes or modules can be operated based on either the modulus or minimum unit code.

Obvious Relevance. Relevance is unobvious in file-based code management method. If adopting the relevance database, the code and its relevant information will be one or several entries of the relevance database table. The relationships between code and its relevant information, code and code could be easily presented through the connection in relevance tables.

Fast and Accurate Retrieval. All the content saved in database is accurate, complete, and clearly correlated because of the completeness, fine granular, and obvious relevance of database management. All of these characters will ensure us to easily, fast, and accurately retrieve the source codes as well as its related information by simply operating on database such as selecting, connecting, and projection.

3.2 *Implementation of Structured Management of Code Based on Database*

Figure 2 illustrates the design of database for the project in C programming language. The project information, module information, and code information related to a project are all stored in database. Each information record is comprised of several relationship tables and each relationship table including several table items saves the source code or related description information. In function information table's case, the main content of table item including function ID, function name, return values, function description, function declaration statement, status(have been written or have been tested), module ID, public or private, programmer's ID, version, whether is the latest, creation time, the latest modification time, lines of source code, compiler-generated file name, etc.

Function is the most frequently minimum unit code appearing in source code. For example, by using function information table and function body table, source code can be obtained. The code information such as type, parameters, return values, local variables, and lines of source code can be obtained from function information table, function parameters table, local variables table, and function return values table without function code analysis, respectively. Functional description, versions (such as the version number and whether it is the latest), status (writing or testing), and history (creation time and the latest modification time) can be obtained from function information table. Through the combination of selecting operation of module information table, function reference table and function information table, the module information which belongs to function, visibility, attribute of public or private, and the relationship of function calls can also be obtained. The author information about the function in different versions can be obtained from function information table and user information table.

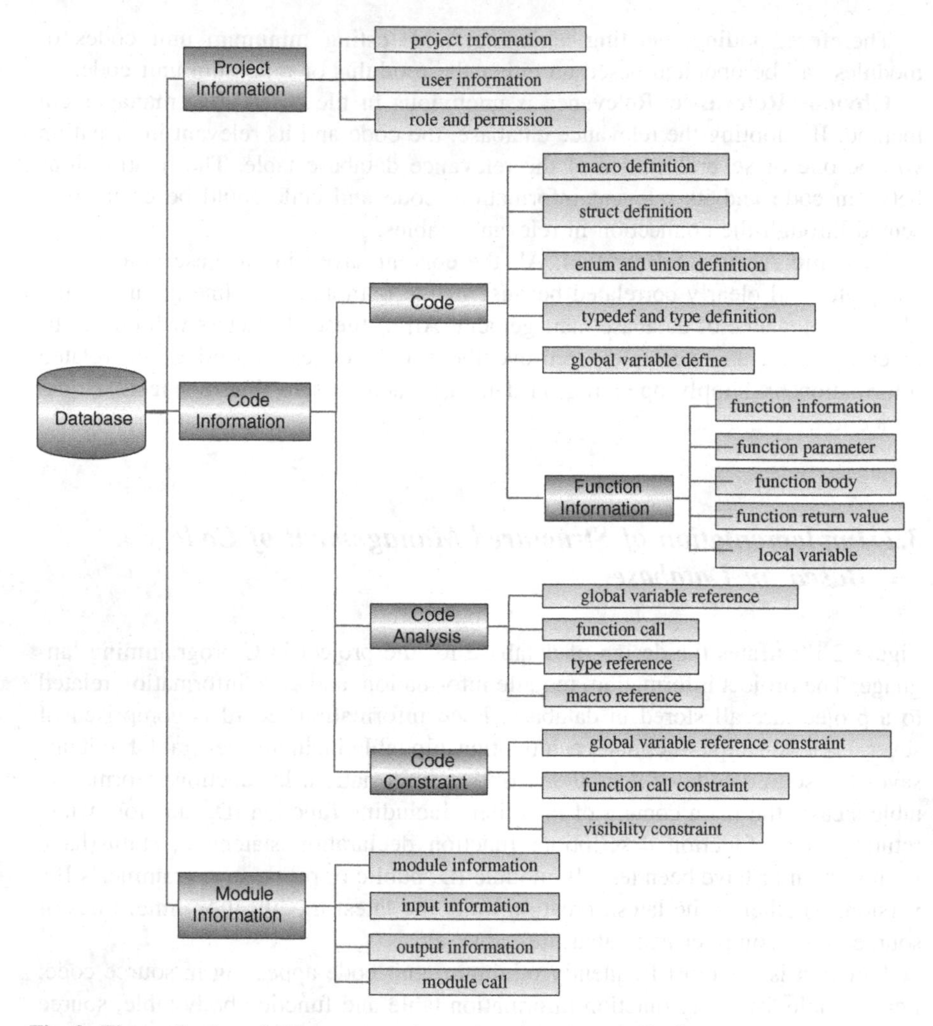

Fig. 2 The realization of C language code structured storage via database

4 Software Development Environment Based on the Structured Management of Code

4.1 System Design

The software development environment and the server architecture based on the code structured management are shown in Figs. 3 and 4, respectively. We use C/S structure to establish the connection between software development client and software development server. The client sends data to the server and requests services. According to the service request, the server combining with user roles

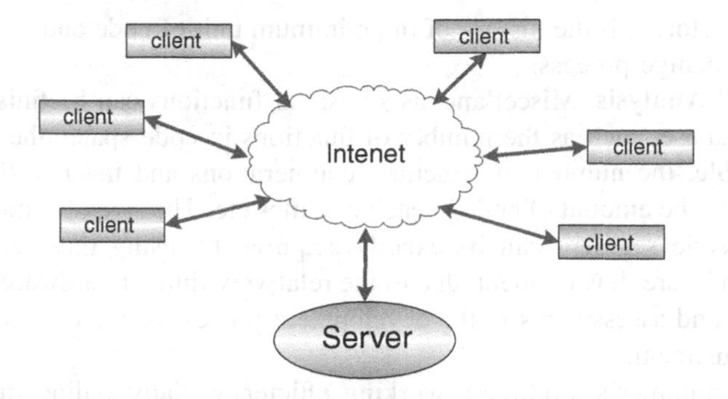

Fig. 3 The software development environment based on internet

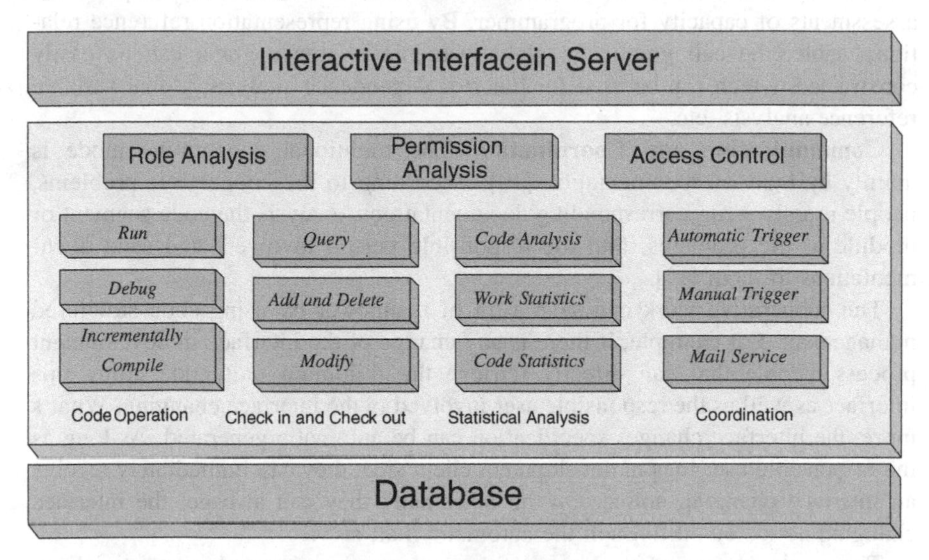

Fig. 4 The architecture diagram of the software development environment on the server side

and access control returns the results to the user through the database and service mechanism. All the source code and related information are stored on the server side in the database. Users log in software development from terminal side and work cooperatively in a unified, comprehensive and up-to-date space.

Server provides services mainly include the following issues.

Incremental Compiling, Debugging, and Running. The server provides code compiling, debugging, and operating environment. When a minimum unit code has been written, it can be submitted to a server to build, run, and view the results in a real time rather than only after writing a lot of code can they compile and error check.

Check in and Check out. Combining with roles and permissions, the user can get the latest and most complete view about code space and related information on their own authority. Writing operation on code and related information within the scope of user's own authority will be saved to the database in the first time.

The database stores all the history of the minimum unit of code and its associated information change process.

Statistical Analysis. Miscellaneous statistical functions can be finished easily by using database, such as the number of functions in code space, the number of global variable, the number of structure, enumerations and macros, the rows of function body, the amount of code in each modules etc. The size of static data area, the size of code segment can be exactly acquired by using these statistics. In embedded software development, due to the relatively limited hardware resources, the statistics and assessments in the development process can avoid the huge cost of late readjustment.

Each programmer's workload, working efficiency, daily online time can be counted by using database and client terminal. It is easier to complete the assessments of capacity for programmer. By using representation reference relational table, the call graph and global variable reference graph can be easily constructed, which can be used for function dependence analysis, global variable reference analysis, etc.

Communication and Coordination. The traditional cooperative mode is mainly by mail or documentation [10]. According to the cooperative problems, people usually write corresponding documentations, analysis the code segment or module of the problems, find the responsible person involved, and send documentations to them [11].

The cooperative work can save a lot of manpower by using code structured management. For example, if there is any change of the interface in development process, people that can directly retrieve the minimum unit code citing this interface as well as the responsible user involved in the interface changing. What's more, the interface changes specification can be automatic generated. As long as the responsible user logs in development client side, they can immediately receive an interface changing notice. On the other hand they can also get the interface changing notice email through the automatic mail service.

The whole process of the cooperative work can be triggered automatically or manually. The completeness of the database information and powerful search function can reduce the manpower cost on communication and cooperation, improve response time, and ensure the efficiency and accuracy of the cooperation.

4.2 Improving the Efficiency of Development

This method improves the efficiency and accuracy of personal development perform in many aspects, such as detecting an error as soon as code incremental compiling, doing named conflict detection by using statistical analysis of the database, preventing renaming error finding at link stage with high cost, and reducing duplication development and improving the code reusability. Integration testing can be on as soon as possible which means that in the file-based code management mode before, waiting for the completing developing all the files involved in integrated test function is needed, now as long as all minimum unit codes involved in integration testing complete developing, process can continue. Minimum compiling can be affected by local changes. By using minimum compiling strategy, minimum unit code as basic element can be used as much as possible. If local changes appear, people only need to recompile the minimum amount of local codes.

5 Conclusion

By analyzing the problems of the file-based code management, this article puts forward the concept of structured management, gives the designs and implements of a software development environment based on this method. Structured management of code minimizes the management unit of source code and provides a detailed, organically organized view of the space of source code. The development of software based on the structured management of code changes the way of code's organization and management fundamentally. It shows advancement in the following four aspects: (1) overcoming problems in the file-based code management, (2) conveniently providing functions such as the control of roles and permission, the statistical analysis, and the latest systematic view etc. (3) implementing the communication and teamwork accurately and efficiently, (4) accelerating individuals' efficiency and accuracy etc.

References

1. Source lines of code. http://en.wikipedia.org/wiki/
2. Ultra-Large-Scale-Systems The software challenge of the future. software engineering institute, Carnegie Mellon (2006)
3. PJim W (2007) Collaboration in software engineering: a roadmap. Future Softw Eng 214–225
4. Kraut RE, Streeter LA (1995) Coordination in software development. communications of the ACM. ACM Press, New York, pp 69–81
5. DeMarco T, Lister T (1987) Peopleware: productive projects and teams. DorsetHouse, New York
6. Jones TC (1986) Programming productivity. McGraw-Hill, New York
7. Ge S, Sun YL, Du ZX (2003) A role based collaboration modeling method. Comput Eng Appl pp 14–18
8. Wei HY, Zhan JF Wang Q (2004) Overview of distributed component technologies. Appl Res Comput 12–15
9. Cao YF, Zhao JW, Han YS, Dai GZ (2002) A paradigm of software development for mass customization. J Comput Res Dev 593–598
10. Jeff K (1994) Distributed software engineering. In: Proceedings of the 16th international conference on software engineering, pp 253–263 1994
11. Maryam P, Martin P, Bastin TRS (2006) Facilitationg collaboration in a distributed software development environment using P2P architecture. The information science discussion paper series

A Collaborative Filtering Recommender Approach by Investigating Interactions of Interest and Trust

Surong Yan

Abstract Collaborative filtering-based recommenders operate on the assumption that similar users share similar tastes; however, due to data sparsity of the input ratings matrix, traditional collaborative filtering methods suffer from low accuracy because of the difficulty in finding similar users and the lack of knowledge about the preference of new users. This paper proposes a recommender system based on interest and trust to provide an enhanced recommendations quality. The proposed method incorporates trust derived from both explicit and implicit feedback data to solve the problem of data sparsity. New users can highly benefit from aggregated trust and interest in the form of reputation and popularity of a user as a recommender. The performance is evaluated using two datasets of different sparsity levels, viz. Jester dataset and MovieLens dataset, and are compared with traditional collaborative filtering-based approaches for generating recommendations.

Keywords Collaborative filtering · Data sparsity · Cold start · Interest and trust

1 Introduction

Recommender systems have been developed as a solution to the well information overload problem [1, 2], by providing users with more proactive and personalized information services. The design of recommender systems that has seen wide use is collaborative filtering (CF). CF is based on the fact that word of mouth opinions

S. Yan (✉)
College of Computer Science and Technology, Zhejiang University,
Hangzhou 310027, China
e-mail: bkspace3000@gmail.compersephone@zufe.edu.cn

S. Yan
College of Information, Zhejiang University of Finance and Economics,
Hangzhou 310018, China

F. Sun et al. (eds.), *Knowledge Engineering and Management*,
Advances in Intelligent Systems and Computing 214,
DOI: 10.1007/978-3-642-37832-4_16,

of other people have considerable influence on the decision making of buyers [3]. If reviewers have similar preferences with the buyer, he/she is much more likely to be affected by their opinions.

In CF-based recommendation schemes, two approaches have mainly been developed: memory-based CF (also known as user-based CF) [1, 4] and item-based CF [5, 6]. User-based CF approaches have seen the widest use in recommendation systems. User-based CF uses a similarity measurement between neighbors and the target users to learn and predict preference towards new items or unrated products by a target user. An item-based CF reviews a set of items the target user has rated (or purchased) and selects the most similar items based on the similarities between items. Item-based CF has been applied to commercial recommender systems such as Amazon.com [7].

CF systems encounter two serious limitations with quality evaluation: the sparsity problem and the cold start problem. The sparsity problem occurs when an available data is insufficient for identifying similar users or items (neighbors) due to an immense amount of users and items [5]. The second problem, the so-called cold start problem, can be divided into cold-start items and cold-start users [8]. New items cannot be recommended until some user rate it, and new users are unlikely to receive good recommendations because of the lack of their rating or purchase history. A cold-start user is the focus of the present research. In this situation, the system is generally unable to make high quality recommendations [9]. A number of studies have attempted to resolve these problems in various applications, such as [10] and [11] tried to address the new user (or cold start) problem, [12] and [13] discussed the new item (or the first rater) problem, and [10], [11], [14], [15], and [16] studied the sparsity problem.

Approaches incorporating trust models into recommender systems are gaining momentum [17]. The emerging channel of SNS not only permits users to express opinions, but also enables users to build various social relationships. For example Epinions.com allows users to review various items (cars, books, music, etc.), and to assign a trust rating on reviewers as well as to build a trust relationship with another by adding him or her to a trust list, or block him or her with a block list. Some researchers [18–20] argued that this trust data can be extracted and used as part of the recommendation process, especially as a means to relieve the sparsity problem. Two computational models of trust namely profile-level trust and item-level trust have been developed and incorporated into standard CF frameworks [17]. The explicit trust-based recommender system models have high rating prediction. New users can also benefit from trust propagation as long as the users provide at least one trusted friend. However, these approaches rely on models of trust that are built from the direct feedback of users. Actually, in reality, the explicit trust statements may not always be available [21].

Also, there are needs for producing valuable recommendations when input data (trust data) is sparse and new user problem is present. These problems are intrinsic in the process of finding similar neighbors. Bedi and Sharma [22] tried to propose trust-based Ant recommender system (TARS) by incorporating a notion of dynamic trust between users to provide better recommendations. New users can benefit from pheromone updating strategy. But TARS still suffers from low accuracy due to the lack of the feedback rating of user.

To address the problems above, this paper proposes a novel approach, combination of interest and trust (CIT), to improve recommendation quality and value. Our wok focuses on sparseness in user similarity due to data sparsity of input matrix and difficulty in solving the new user's due to the lack of knowledge about preference of new user, and differs from the above works in a number of ways [17, 22]. First, the taste similarity measurement considers the number of overlapping items and distance between ratings of overlapping items. Second, as opposite to traditional predication methods, the prediction rating takes into account the difference of rating scale of user. Third, interest and trust score is generated based on user's explicit and implicit feedback behaviors. Fourth, definition of heavy users considers both virtual community-specific properties and trust-specific properties.

The rest of this paper is organized as follows: Section 2 describes the proposed CIT approach and provides a detailed description of how the system uses CIT for recommendation. Section 3 presents the effectiveness of CIT approach through experimental evaluations compared with existing work using the two datasets. Finally, in Sect. 4, the paper concludes with a discussion.

2 Proposed Recommender Approach by Investigating Interactions of Interest and Trust

In this section, the paper first derives ratings of users on items by identifying user's explicit and implicit feedback behavior; then estimates the interest similarity and trust intensity between users; third generates recommendations for target user by selecting small and best neighborhood; finally, discusses a solution to new user (cold start) problem by finding the migratory behavior of heavy users based on the Pareto principle rule.

2.1 Deriving Implicit Ratings of Users on Items

Feedback behaviors of a user not only show her/his personal interest in item recommendation list but also exhibit her/his degree of satisfaction on these items or products.

Deriving implicit item rating based on feedback behaviors: Target user can directly rate on recommended item by voting or clicking like/dislike button, or exhibit the degree of satisfaction on recommended item by indirect feedback behaviors. Such behaviors include buying the product, browsing the information of recommendation list, or adding the product into the favorites list or wish list. Leveraging implicit user feedback [23] has become more popular in CF systems.

In this paper, we leverage implicit user feedback behaviors, which can be viewed as a variant of user feedback. Therefore, rating of target user u on item j can be derived by direct or indirect behaviors of u.

If target user does not give a rating $r_{u,j}$ about recommended item j, recommender system will evaluate the feedback behaviors of target user based on association rule [14] and fuzzy relation [24] to generate the rating $r_{u,j}$ of target user on item j.

Deriving implicit feedback rating based on feedback behaviors: Referral feedback can be derived by directly feedback behavior of target user such as voting or clicking like/dislike button, or by indirectly feedback behavior of target user such as buying the product, browsing the information of recommendation list, or putting it into her/his favorites list or wish list. If target user does not give a vote/feedback rating $f_{u,v}(j)$, recommender system will evaluate the feedback behaviors of target user as described above and then generate the feedback rating.

Definition 1 Feedback rating: let $r_{u,i}$ and $p_{u,i}$ denote the real rating and predicted rating of target user u on item j respectively, the feedback rating of u on the item j recommended by v in time t based on fuzzy relation is

$$f^t_{u,v\in \mathrm{reco}(u,j)}(j) = M(|r_{u,j} - p_{u,j}|) \tag{1}$$

where reco(u, j) is the set of recommendation partners that have recommended target user u with item j in the latest recommendation round and M is the membership function that maps feedback rating of item j in different intervals. Figure 1 shows a membership function for the feedback rating.

2.2 Calculating Interest and Trust Scores

Calculating interest similarity score: There are a number of possible measures for computing the similarity, for example, the distance metric [11, 25], cosine/vector similarity [13, 15] and the Pearson correlation metric [2, 13]. The proposed approach defines a new taste similarity measurement for computing interest similarity between two users based on the overlapping part of their rating profiles, i.e., on the items that are rated by them in common:

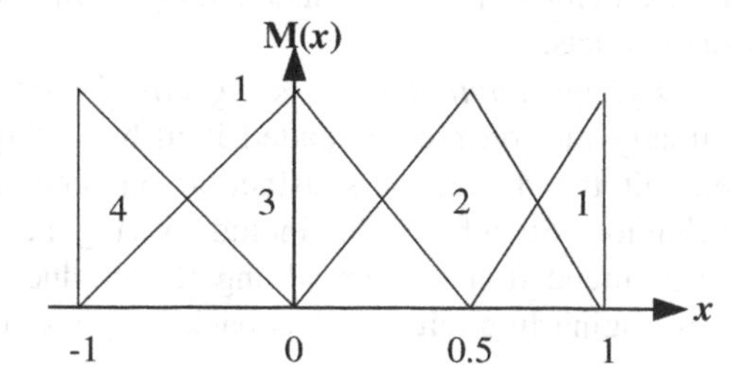

Fig. 1 Membership function for feedback rating on rating scale (1–5)

Definition 2 Similarity rating: the similarity rating models the difference between ratings of two users on given common item. Let $r_{u,\ i}$ and $r_{v,\ i}$ denote the ratings of users u and v for item i respectively, the similarity rating of the item i between u and v based on fuzzy relation is

$$s_{u,v}(i) = M(|r_{u,i} - r_{v,i}|) \tag{2}$$

where M is the membership function that maps the absolute difference of rating between u and v on item i to the similarity rating of u and v on item i at different intervals.

The similarity function measures the taste similarity degree between the two users based on their ratings on common items and characterizes each participant's contextual view of the taste similarity. Let τ be the total number of items that u have rated, com(u, v) be the set of items rated by u, and v in common and q be the size of the set, $p = \sum_{i \in \text{com}(u,v)} s_{u,v}(i)$ be the sum of the similarity rating between u and v on the ratings of the common items, and $\ell = \sqrt{\frac{1}{n}\sum_{i \in \text{com}(u,v)} |r_{u,i} - r_{v,i}|^2}$ be the average of the absolute difference square of common rating between u and v. The taste similarity degree between u and v from the perspective of u is

$$\text{Sim}(v/u) = \begin{cases} \frac{2p*q}{\tau(p+q)} * \exp(-\kappa\ell) & \text{if } p \neq 0 \text{ and } p+q \neq 0 \\ \eta\frac{q}{\tau} & \text{if } q \neq 0 \text{ and } p = 0 \\ 0 & \text{else} \end{cases} \tag{3}$$

where $0 < k, \eta < 1$ are adjust arguments. This taste similarity degree captures the intuition that user v should have rated many of the items that user u has rated and their ratings on these common items are similar. As seen from Eq. (3), sim (v/u) is large if rating overlaps between the two partners u and v is high and less otherwise.

Calculating trust scores: Trust intensity being dynamic information, increases or decreases with the involvement (positive experience or negative experience) or non-involvement (over time without any experience) of user as a recommender. Initially at time $t = 0$, dynamic trust intensity depends on the existed trust or distrust relationships of target user such as friend list or follow list or block list. A default trust value is preset. If there are any trust relationships absent, the paper will discuss this problem later.

At time $t > 0$, dynamic trust intensity is calculated as follows: Trust intensity determines the amount of confidence a user should have in (on) other user and characterizes each participant's contextual view of the trust intensity. Let $f^t_{u,v}(i) \in [-1, +1]$, denote the feedback rating of target user u on item i recommended by v in the time t, n denote the total number of feedback ratings of u on v, $m = \sum_{t=1}^{n} f^t_{u,v}(i)$ denote the sum of feedback ratings, and Γ denote the total number of recommendations that u have received from v. The amount of confidence a user u should have on user v is given as:

$$\text{Trust}(v/u) = \begin{cases} \frac{2m*n}{\Gamma(m+n)} * \exp(-\kappa\ell) & \text{if } m \neq 0 \text{ and } m+n \neq 0 \\ \eta\frac{n}{\Gamma} & \text{if } n \neq 0 \text{ and } m = 0 \\ 0 & \text{else} \end{cases} \quad (4)$$

Combination of taste similarity and trust intensity (CIT): Let $C(v/u)^t$ denote the value of CIT in time t, it is a weighted sum of taste similarity and trust intensity between u and v from the perspective of u:

$$C(v/u)^t = \chi \text{sim}(v/u)^t + (1-\chi)\text{trust}(v/u)^t \quad (5)$$

where $0 \leq \chi \leq 1$ is a small constant value. The value of $C(v/u)^t$ lies in between >−1 to 1 indicating the degree of confidence ranging in between 0 to 1. Since $C(v/u)^t \leq 0$ indicates that u and v are not correlated, therefore zero and negative correlations are not considered and all such outcomes are set to zero.

2.3 Generating Prediction Scores

Scale and translation invariance states that each user may have its own scale of rating items and is a stronger condition than normalization invariance [4]. For example, some users might rate most items as roughly similar while others tend to use more often extreme ratings. For compensating the difference in rating scales between users, the paper adopts the scale and translation invariance to produce predictions, which depends on the following factors:

Definition 3 Complete factor: Complete factor w is a weighted sum of difference between ratings $r_{v,\,j}$ ($v = 1, 2 \ldots$) of selected best neighborhood ner(u) of u on given item j and their average rating $\overline{r_v}$ on all rated items respectively. The complete rating is defined as

$$w_{u,j}^t = \frac{\sum_{v \in \text{ner}(u)} C(v/u)^t (r_{v,j} - \overline{r_v})}{\sum_{v \in \text{ner}(u)} |C(v/u)^t|} \quad (6)$$

where $c(v/u)^t$ denotes the value of CIT that user u holds about user v at time t.

Definition 4 Adjust scale factor: Adjust scale factor $\Im$ is an adjusting about scales differences between ratings of target user and ratings of selected neighborhood on common items.

$$\Im^t = \frac{\sum_{i \in \text{com}(u,v)} (r_{u,i} - \overline{r_u})(w_{u,i} - \overline{w_u})}{\sum_{i \in \text{com}(u,v)} (w_{u,i} - \overline{w_u})(w_{u,i} - \overline{w_u})} \quad (7)$$

where $r_{u,\,i}$ and $w_{u,\,i}$ denote the actual ratings and the predicted complete ratings of user u on item i respectively, $\overline{r_u}$ and $\overline{w_u}$ denote the average actual ratings and the average predicted complete ratings of users u respectively.

Combining the above two weighting factors, the final prediction for the target user u on item j at time t ($t > 0$) is produced as follows:

$$p_{u,j}^t = \overline{r_u} + \Im^t(w_{u,j}^t - \overline{w_u}) \tag{8}$$

where r_u denotes the average ratings of users u on all rated items.

2.4 Addressing to New User (Cold Start) Problem

In a user's interest network, local similarity is aggregated on each edge connecting similar users and the user. Summing up the aggregated local similarity corresponding to all incoming links of the user and number of incoming links to the user, we get the global taste similarity. Global taste similarity acts as a positive indicator reflecting the user's taste that is popular or unpopular, and the popular users are defined as candidate heavy users over a period of time.

In a user's trust network, local trust is aggregated on each edge connecting trustworthy users and the user. Summing up the aggregated local trust corresponding to all incoming links of the user and number of incoming links to the user, we get the global trust. Users benefit from this wisdom of crowds because global trust acts as positive feedback reflecting the user's behavior that is universally considered good or bad, and high reputation users are defined as candidate heavy users over a period of time.

Definition 5 Heavy user: User with a long-term recommender history, high global taste similarity, and high trust intensity is defined as a heavy user.

List of heavy user can act as default friends list at time t and be added as friends of new user at initial time $t = 0$. Combining ratings of the top_n heavy users on item j, the final prediction for the new user u on item j at time t ($t = 0$) is produced as follows:

$$p_{u,j}^0 = \frac{\sum_{v=1}^{\text{top\_}n} \overline{r_v}}{\text{top\_}n} + \frac{\sum_{v=1}^{\text{top\_}n} G(v)(r_{v,j} - \overline{r_v})}{\sum_{v=1}^{\text{top\_}n} G(v)} \tag{9}$$

where $G(v)$ is a weight sum of global taste similarity and global trust intensity of v.

3 Experimental Study

In this section, the paper describes the experimental methodology and metrics that are used to compare different prediction algorithms; and presents the results of the experiments.

3.1 Experimental Configuration

The experimental study was conducted on two different types of datasets. The Jester dataset and MovieLens dataset available online on websites www.ieor.berkeley.edu/~goldberg/jester-data and www.grouplens.org/node/73, respectively. Jester dataset has 73,421 users who entered numeric ratings for 100 jokes ranging on real value scale from −10 to +10. The MovieLens dataset consists of 100,000 ratings ranging in the discrete scale 1 to 5 from 943 users on 1,682 movies.

Benchmark algorithms: The experiments compare the proposed method CIT to some typical alternative prediction algorithms: pure Pearson predictor [1]; Pearson-like predictor (scale and translation invariant Pearson (STIPearson_2.0) [4], Non-personalized (NP) [26], Slope one-like predictor (Bi-Polar Slope One) [27], and compositions of adjust scale [4] + approach (Non personalized/Bi-Polar Slope One). The average predictor simulates the absence of a recommender and uses the target user's average rating as the predicting rating. It is only for providing a baseline for comparison purposes.

Methodology: For the purposes of comparison, a subset of the datasets is adopted.

MovieLens: The full u data set, 100,000 ratings by 943 users on 1,682 items.

MovieLens dataset 1: The data sets u1.base and u1.test through u5.base and u5.test are 80 %/20 % splits of the u data into training and test data. Each of u1, ..., u5 has disjoint test sets. MovieLens dataset 2: The data sets ua.base, ua.test, ub.base, and ub.test split the u data into a training set and a test set with exactly 10 ratings per user in the test set. The sets ua.test and ub.test are disjoint. MovieLens 2 represents the sparse user-item rating matrix.

Jester dataset:

Jester dataset 1: Data from 24,983 users who have rated 36 or more jokes, a matrix with dimensions 24,983 × 101. Jester dataset 2: Data from 24,938 users who have rated between 15 and 35 jokes, a matrix with dimensions 24,938 × 101. Jester dataset 2 represents the sparse user-item rating matrix.

In every test round, for each user in the test sets, ratings for one item were withheld. Predictions will be recomputed for the withheld items using each of the different predictors. The quality of the various prediction predictors is measured by comparing the predicted values for the withheld ratings to the actual ratings.

Metrics: The type of metrics used depend on the type of CF applications. To measure predictive accuracy, we use the All But One MAE (AllBut1MAE) [2, 28] by withholding a single element j from ratings of target user to compute the MAE on the test set for each user and then average over the set of test users.

$$\text{AllBut1MAE} = \frac{1}{n}\sum\nolimits_{u,j} |p_{u,j} - r_{u,j}| \tag{10}$$

where n is the total number of ratings over all users, $p_{u,\,j}$ and $r_{u,\,j}$ are the predicted rating and actual rating of u on withheld item j respectively. The lower the AllBut1MAE, the better the prediction.

To measure decision-support accuracy, the experiments use receiver operating characteristic (ROC) sensitivity and specificity. Sensitivity [true positive rate (TPR)] is defined as the probability that a good item is accepted by the filter; and specificity (SPC) (1 − false positive rate) is defined as the probability that a bad item is rejected by the filter. For the confusion matrix [29], it has

$$\begin{aligned} \mathrm{TPR} &= \frac{\mathrm{TP}}{\mathrm{TP}+\mathrm{FN}} \\ \mathrm{SPC} &= 1-\frac{\mathrm{FP}}{\mathrm{FP}+\mathrm{TN}} = \frac{\mathrm{TN}}{\mathrm{FP}+\mathrm{TN}} \end{aligned} \tag{11}$$

The higher the TPR and the SPC, the better the prediction. By tuning a threshold value, all the items ranked above it are deemed observed by the user, and below unobserved, thus the system will get different prediction values for different threshold values. In Jester dataset, the experiments consider an item good if the user gives it a rating greater than or equal to 4, 6, 8, or above, otherwise they consider the item bad. In MovieLens dataset, the experiments consider an item good if the user gave it a rating greater than or equal to 3, 4, or above, otherwise they consider the item bad. ROC Sensitivity and specificity ranges from 0 to 1.

3.2 Experimental Results for Data Sparsity in User-Item Rating Matrix

The data sparsity in user-item rating matrix results in sparseness of user similarity matrix. CIT and all compared prediction algorithms are implemented in java on Microsoft windows operating system by modifying the Cofi: A Java-Based Collaborative Filtering Library.

The performance of our methods: The test results (Figs. 2, 3) indicate that CIT (our method) performs well when facing two kinds of dataset under three kinds of metrics. The AllBut1MAE of CIT slightly decreases while the TPR (ROC sensitivity) clearly increases when facing the Jester dataset 2 and MovieLens dataset 2 that represent the sparse user-item rating matrix. The metrics scoring of CIT shows

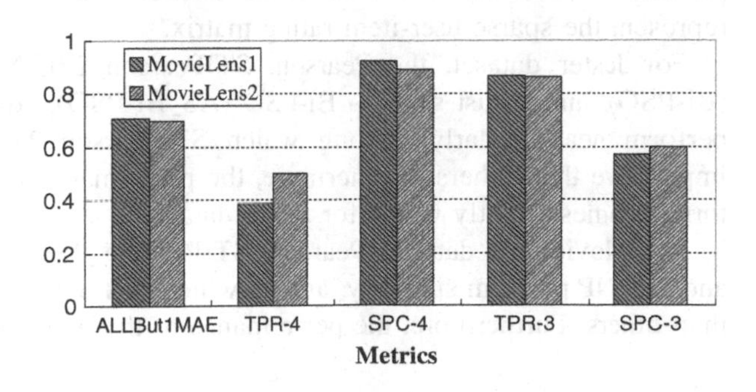

Fig. 2 Metrics scoring of CIT for MovieLens dataset

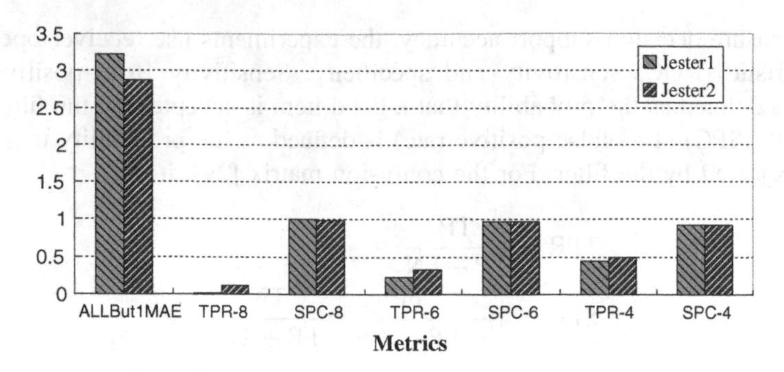

Fig. 3 Metrics scoring of CIT for Jester dataset

Table 1 Metrics scoring of all predictor compared for Jester dataset 1

Predictors	MAE	TPR-8	SPC-8	TPR-6	SPC-6	TPR-4	SPC-4
CIT	3.2177	0.0116	0.9993	0.2263	0.9783	0.4517	0.9251
Pearson	3.4138	0.0077	0.9993	0.1735	0.9800	0.3744	0.9344
STIPearson_2.0	3.3362	0.0886	0.9927	0.2972	0.9650	0.4635	0.9152
STINP_2.0	3.4625	0.0077	0.9989	0.1496	0.9803	0.3608	0.9356
Bi-PSO	3.4481	0.0039	0.9989	0.2196	0.9745	0.4790	0.9015
AS_Bi-PSO	3.4586	0.0039	0.9993	0.1323	0.9838	0.3620	0.9365
NP	3.5306	0.0058	0.9978	0.1275	0.9833	0.3558	0.9298
AS_NP	3.4614	0.0116	0.9987	0.1477	0.9795	0.3608	0.9339
Average	3.7785	0.0000	1.0000	0.0364	0.9935	0.2779	0.9266

that it is insensitive to dataset and is adaptive to sparse user-item rating matrix. The test results also show that TPR increases and SPC (ROC specificity) declines as the threshold of ROC become smaller. Meanwhile, the smaller the threshold of ROC, the bigger the TPR; the smaller the threshold of ROC, the smaller the SPC.

Compare with other methods: Tables 1, 2 and Fig. 4 show experimental results for Jester dataset while Tables 3, 4 and Fig. 5 show experimental results for MovieLens dataset. Overall, any compared predictors with metrics scoring outperform base line average. CIT is very successful and outperforms all other predictors compared. Especially CIT outperforms significantly all other compared predictors when meeting with Jester dataset 2 and MovieLens dataset 2 that represent the sparse user-item rating matrix.

For Jester dataset, the Pearson, STIPearson_2.0, NP, (Bi-Polar Slope One (Bi-PSO), and adjust scale + Bi-PSO (AS_Bi-PSO), adjust scale + NP(AS_NP) perform near-similarly, among which STIPearson_2.0 and AS_NP are more impressive than others. Furthermore, the performance of these compared predictors becomes slightly worse for Jester dataset 2.

For MovieLens dataset, Pearson, STIPearson_2.0, NP, Bi-PSO, AS_Bi-PSO, and AS_NP perform similarly, among which Bi-PSO and NP are more impressive than others. Furthermore, the performance of these compared predictors becomes

Table 2 Metrics scoring of all predictor compared for Jester dataset 2

Predictors	MAE	TPR-8	SPC-8	TPR-6	SPC-6	TPR-4	SPC-4
CIT	2.8674	0.1234	0.9951	0.3329	0.9754	0.5022	0.9311
Pearson	3.5507	0.0127	1.0000	0.1163	0.9878	0.3345	0.9416
STIPearson_2.0	3.4586	0.0823	0.9964	0.2475	0.9744	0.4396	0.9185
STINP_2.0	3.4996	0.0538	0.9959	0.1696	0.9801	0.3612	0.9405
Bi-PSO	3.5271	0.0000	0.9998	0.1003	0.9919	0.3367	0.9463
AS_Bi-PSO	3.5464	0.0506	0.9979	0.1498	0.9830	0.3568	0.9458
NP	3.5652	0.0095	0.9991	0.0891	0.9926	0.2928	0.9475
AS_NP	3.5000	0.0538	0.9955	0.1696	0.9809	0.3648	0.9408
Average	3.9317	0.0000	1.0000	0.0198	0.9981	0.1115	0.9694

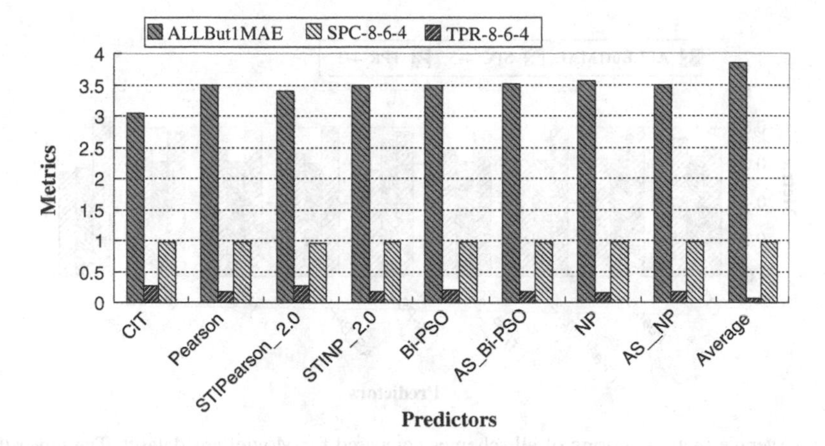

Fig. 4 Average metrics scoring of all schemes compared for Jester dataset: The lower the AllBut1MAE, the better the prediction accuracy is; while the higher the TPR and the SPC, the better the decision-support accuracy is

Table 3 Metrics scoring of all predictor compared for MovieLens dataset 1

Predictors	MAE	TPR-4	SPC-4	TPR-3	SPC-3
CIT	0.7077	0.3867	0.9227	0.8670	0.5722
Pearson	0.7687	0.3218	0.9125	0.8707	0.4783
STIPearson_2.0	0.7448	0.4042	0.8937	0.8517	0.5485
STINP_2.0	0.7418	0.3457	0.9190	0.8722	0.5206
Bi-PSO	0.7490	0.5585	0.8109	0.9241	0.3958
AS_Bi-PSO	0.7363	0.3437	0.9215	0.8780	0.5204
NP	0.7561	0.3580	0.9067	0.8572	0.5219
AS_NP	0.7417	0.3446	0.9175	0.8729	0.5175
Average	0.8409	0.1884	0.9314	0.9196	0.2509

worse obviously for MovieLens dataset 2. In addition, TPR of Bi-PSO is above all other predictors but SPC of Bi-PSO is below all other predictors. All methods except CIT are sensitive to the density of dataset.

Table 4 Metrics scoring of all predictor compared for MovieLens dataset 2

Predictors	MAE	TPR-4	SPC-4	TPR-3	SPC-3
CIT	0.6969	0.5014	0.8913	0.8620	0.5997
Pearson	0.8407	0.3649	0.8398	0.8713	0.3615
STIPearson_2.0	0.8370	0.3911	0.8332	0.8675	0.3810
STINP_2.0	0.8246	0.4136	0.8435	0.8495	0.4454
Bi-PSO	0.8125	0.5742	0.7327	0.9063	0.3510
AS_Bi-PSO	0.8769	0.3647	0.8203	0.8514	0.3651
NP	0.8112	0.4213	0.8421	0.8542	0.4449
AS_NP	0.8215	0.4169	0.8411	0.8517	0.4448
Average	0.8736	0.3326	0.8288	0.9079	0.2504

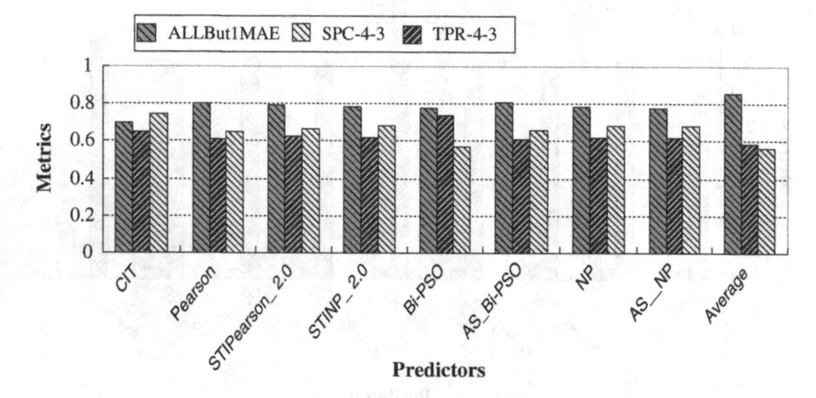

Fig. 5 Average metrics scoring of all schemes compared for MovieLens dataset: The lower the AllBut1MAE, the better the prediction accuracy is; while the higher the TPR and SPC, the better the decision-support accuracy is

As a result, we have evidence that CIT is insensitive to dataset and more adaptive to sparse user-item rating matrix than all other compared predictors. CIT can effectively relieve the data sparsity problem. The reason is that CIT combines the interest and trust to create the user similarity.

3.3 Experimental Results for New User

New user (cold start) problem occurs when recommendations are to be generated for user who has not rated any item or rated very few items since it is difficult for the system to compare her/him with other users and find possible neighborhood. This problem can be solved through "heavy" users. The list of default trusted friends can be captured and automatically added to the new user's friend list at initial stage.

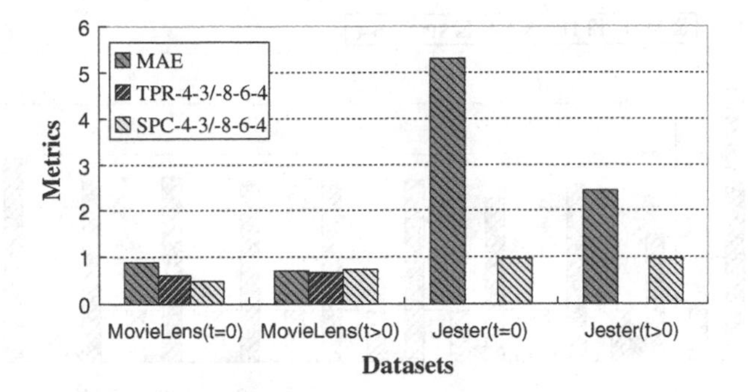

Fig. 6 Average metrics scoring of the CIT approach on MovieLens and Jester datasets

Figure 6 shows that results of CIT approach on MovieLens and Jester Datasets at time $t = 0$ and $t > 0$. Figures 7 and 8 show experimental results of all approaches compared on MovieLens and Jester datasets. It can be observed that the predication on MovieLens dataset is more accurate than on Jester Dataset. This is because the rating scale of MovieLens dataset is smaller than that of Jester datasets.

As a result, we have evidence that CIT is slightly sensitive to dataset for new user. The reason is that the larger the ranging rating scale, the more difficult the prediction. CIT combines the taste similarity and referral trust to compute the user similarity and adopts adjust scale modified Resnicks prediction formula to predict rating. On one hand, CIT incorporates the advantages of STI Pearson and Bi-PSO. On the other hand, feedback-based referral trust considers that target user's feedback is valuable to continually update the similarity between users. Thus, the proposed approach can be used as part of a more broad recommendation explanation which not only improves user satisfaction but also helps users make better decisions.

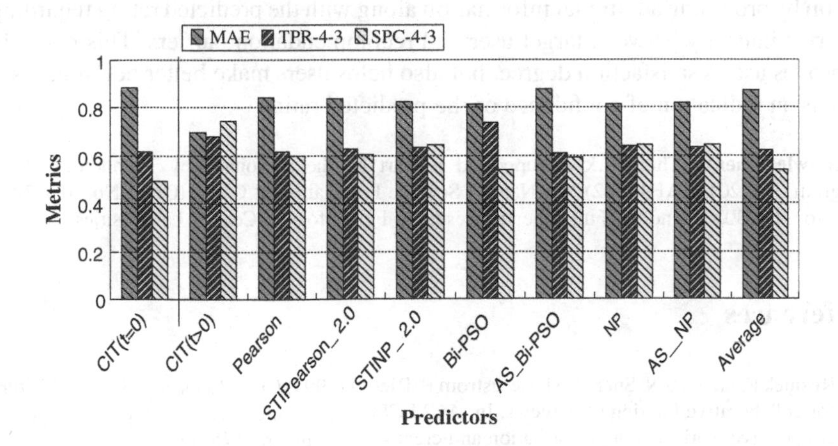

Fig. 7 Average metrics scoring of all schemes compared for MovieLens dataset

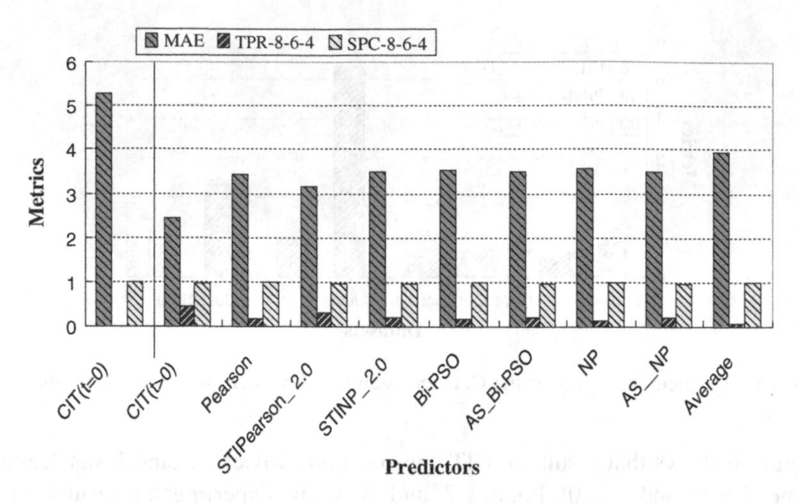

Fig. 8 Average metrics scoring of all schemes compared for Jester dataset

4 Conclusion

Traditionally, CF systems have relied heavily on similarities between the ratings profiles of users as a way to differentially rate the prediction contributions of different profiles. In the parse data environment, where the predicted recommendations often depend on a small portion of the selected neighborhood, ratings similarity on its own may not be sufficient, therefore other factors might also have an important role to play. In the proposed system, sparseness in user similarity due to data sparsity of input matrix is reduced by incorporating trust. Also, new users highly benefit from this wisdom of crowds as positive feedback in the form of aggregated global trust and global interest, and defines heavy user as default recommender over a period of time. The proposed approach is used as an explanation for recommendation by providing additional information along with the predicted rating regarding the trust intensity between target user and recommendation partiers. This not only improves user's satisfaction degree, but also helps users make better decisions as it acts as an indication of confidence of the predicted rating.

Acknowledgments This work is supported in part by the National Key Technology R&D Program (No. 2012BAH16F02), the Natural Science Foundation of China (Grant No. 61003254 and No. 60903038), and the Fundamental Research Funds for the Central Universities.

References

1. Resnick P, Iacovou N, Suchak M, Bergstrom P, Riedl J (1994) GroupLens: an open architecture for collaborative filtering of netnews. In: ACM CSCW'94 conference on computer-supported cooperative work, sharing information and creating meaning, pp 175–186

2. Breese JS, Heckerman D, Kadie C (1998) Empirical analysis of predictive algorithms for collaborative filtering. In: 14th conference on uncertainty in artificial intelligence (UAI'98). Morgan Kaufmann, San Francisco, pp 43–52
3. Shardanand U, Maes P (1995) Social information filtering: algorithms for automating word of mouth. In: Proceedings of the SIGCHI conference on human factors in computing systems, Denver, CO, USA, pp 210–217
4. Lemire D (2005) Scale and translation in variant collaborative filtering systems. Inf Retr 8(1):129–150
5. Sarwar B, Karypis G, Konstan J, Riedl J (2001) Item-based collaborative filtering recommendation algorithms. In: 10th international conference on World Wide Web, pp 285–295
6. Deshpande M, Karypis G (2004) Item-based top-N recommendation algorithms. ACM Trans Inf Syst 22(1):143–177
7. Linden G, Smith B, York J (2003) Amazon.com recommendations: item-to-item collaborative filtering. IEEE Internet Comput 7(1):76–80
8. Schein AI, Popescul A, Ungar LH (2002) Methods and metrics for cold-start recommendations. In: the 25th annual international ACM SIGIR conference on research and development in information retrieval, pp 253–260
9. Massa P, Avesani P (2004) Trust-aware collaborative filtering for recommender systems. In: International conference on cooperative information systems, Agia Napa, Cyprus, pp 492—508
10. Kim HN, Ji AT, Ha I, Jo GS (2010) Collaborative filtering based on collaborative tagging for enhancing the quality of recommendation. Electron Commerce Res Appl 9(1):73–83
11. Park YJ, Chang KN (2009) Individual and group behavior-based customer role model for personalized product recommendation. Expert Syst Appl 36(2):1932–1939
12. Balabanovic M, Shoham Y (1998) Content-based, collaborative recommendation. Commun ACM 40(3):66–72
13. Lee TQ, Park Y, Park YT (2008) A time-based approach to effective recommender systems using implicit feedback. Expert Syst Appl 34(4):3055–3062
14. Huang CL, Huang WL (2009) Handling sequential pattern decay: developing a two stage collaborative recommendation system. Electron Commerce Res Appl 8(3):117–129
15. Jeong B, Lee J, Cho H (2009) An iterative semi-explicit rating method for building collaborative recommender systems. Expert Syst Appl 36(3):6181–6186
16. Lee JS, Olafsson S (2009) Two-way cooperative prediction for collaborative filtering recommendations. Expert Syst Appl 36(3):5353–5361
17. O'Donovan J, Smyth B (2005) Trust in recommender systems. In: IUI'05, SanDiego, CA, USA, pp 167–174
18. Massa P, Bhattacharjee B (2004) Using trust in recommender systems: an experiment analysis. In: 2nd international conference on trust management, Oxford, England, pp 221–235
19. Massa P, Avesani P (2007) Trust-aware recommender systems. In: RecSys. ACM, New York, NY, USA, pp 17–24
20. Massa P, Avesani P (2009) Trust metrics in recommender systems. In: Golbeck J (ed) Computing with social trust. Springer, London, pp 259–285
21. Yuan W, Guan D, Lee YK, Lee S, Hur SJ (2010) Improved trust-aware recommender system using small-worldness of trust networks. Knowl-Based Syst 23(3):232–238
22. Bedi P, Sharma R (2012) Trust based recommender system using ant colony for trust computation. Expert Syst Appl 39(1):1183–1190
23. Claypool M, Le P, Wased M, Brown D (2001) Implicit interest indicators. In: IUI '01, pp 33–40
24. Kuo YL, Yeh CH, Chau R (2003) A validation procedure for fuzzy multi-attribute decision making. In: The 12th IEEE international conference on fuzzy systems, vol 2, pp 1080–1085
25. Kim HK, Kim JK, Ryu YU (2009) Personalized recommendation over a customer network for ubiquitous shopping. IEEE Trans Serv Comput 2(2):140–151

26. Herlocker J, Konstan J, Borchers A, Riedl J (1999) An algorithmic framework for performing collaborative filtering. In: Research and development in information retrieval
27. Lemire D, Maclachlan A (2005) Slope one predictors for online rating-based collaborative filtering. In: SIAM data mining (SDM'05), Newport Beach, CA
28. Canny J (2002) Collaborative filtering with privacy via factor analysis. In: the Special inspector general for Iraq reconstruction
29. Receiver Operating Characteristic (2012). http://en.wikipedia.org/wiki/Receiver_operating_characteristic

Probabilistic Composite Rough Set and Attribute Reduction

Hongmei Chen, Tianrui Li, Junbo Zhang, Chuan Luo and Xinjiang Li

Abstract Composite rough set aims to deal with multiple binary relations simultaneously in an information system. In this paper, probabilistic composite rough set is presented by introducing the probabilistic method to composite rough set. Then, the distribution attribute reduction method under probabilistic composite rough set is investigated. Examples are given to illustrate the method.

1 Introduction

Rough Set Theory (RST) proposed by Pawlak is an important theory to deal with inconsistent and uncertain information [1, 2]. In Traditional Rough Set (TRS), the information system is complete, the data type in the information system is nominal and the relationship between objects is an equivalence relation. However, in real-

This work is supported by the National Science Foundation of China (Nos. 60873108, 61175047, 61100117) and NSAF (No. U1230117), the Youth Social Science Foundation of the Chinese Education Commission (No. 11YJC630127), and the Fundamental Research Funds for the Central Universities (SWJTU11ZT08, SWJTU12CX091).

H. Chen (✉) · T. Li · J. Zhang · C. Luo · X. Li
School of Information Science and Technology, Southwest Jiaotong University, 610031 Chengdu, China
e-mail: trli@swjtu.edu.cn

T. Li
e-mail: hmchen@swjtu.edu.cn

J. Zhang
e-mail: JunboZhang86@163.com

C. Luo
e-mail: luochuan@my.swjtu.edu.cn

F. Sun et al. (eds.), *Knowledge Engineering and Management*,
Advances in Intelligent Systems and Computing 214,
DOI: 10.1007/978-3-642-37832-4_17,

life applications, data missing may exist, data type may be various, and preference order in attributes' values may exist. Then, TRS has been extended to deal with different data types, data missing, and preference ordered data in Extended Rough Set (ERS) by replacing the equivalence relation with other binary relations [3–11]. Different binary relations, e.g., tolerance relations, partial order relations in interval valued information system, set-valued information system, and hybrid data information system have been defined in different conditions. In most ERS, there is a supposition that there exists only one data type and one relationship in an information system.

Some works have been done considering multiple relations or hybrid data types. When considering hybrid data types (nominal data and numerical data) in an information system, Hu et al. proposed Neighborhood Rough Set (NRS) to deal with homogeneous feature selection [12, 13]. Neighborhood relation and nearest neighborhood relation have been defined in NRS. Wei et al. studied the hybrid data in the framework of fuzzy rough set [14]. When considering multiple binary relations may exist in an information system, An and Tong proposed global binary relation and they further defined the approximation on upper and down union [15]. Abu-Donia proposed new types of rough set approximations using multi knowledge base, that is, a family of finite number of (reflexive, tolerance, dominance, equivalence) relations by two ways [16]. Zhang et al. defined composite relation and Composite Rough Set (CRS) model in [17]. Then, they investigated matrix-based rough set approach in CRS. However, the attributes reductions in an information system with multiple relations have not been investigated in the literatures.

Probabilistic rough set is an important extension of TRS by considering conditional probability between equivalence classes in TRS [18]. Yao and Wong proposed a Decision-Theoretic Rough Set model (DTRS) based on well established Bayesian decision procedure [21]. The parameters in DTRS are decided by loss function. Ziarko proposed Variable Precision Rough Set (VPRS) to deal with the errors in data [19, 20]. Slezak and Ziarko investigated Bayesian Rough Set (BRS) when considering the prior probability of an event [22]. Probabilistic rough set have been successfully applied to data mining [23].

In this paper, we investigate the extended model of CRS by introducing probabilistic method. Then, Probabilistic Composite Rough Set (PCRS) is proposed. Considering the arbitrary binary relation existing in PCRS, distribute attribute reduction in PCRS is discussed. Examples are given to illustrate the method presented in the paper.

The paper is organized as follows. In Sect. 1 , we review the basic concepts of CRS. In Sect. 2, PCRS is defined and its properties are discussed. In Sect. 3, we study the distribute reduction method in PCRS. In Sect. 4, we conclude the paper and outline the direction of future work.

Table 1 An Multi-data type decision information system

U	a_1	a_2	a_3	a_4	*d*
x_1	2	0.2	[2.17, 2.86]	{0, 1}	1
x_2	4	0.85	[3.37, 4.75]	{2}	2
x_3	3	0.31	[2.56, 4.10]	{1, 2}	0
x_4	1	0.74	[3.55, 5.45]	{1}	1
x_5	1	0.82	[3.46, 5.35]	{1}	0
x_6	2	0.72	[2.29, 3.43]	{1}	1
x_7	1	0.6	[2.22, 3.07]	{0, 2}	1
x_8	3	0.44	[2.51, 4.04]	{1, 2}	0

2 Composite Rough Set

In this section, we first review some concepts in CRS [17].

The definition of Multi-Data Type Decision Information system (*MDTDIS*) is given as follows:

Definition 21 A quadruple $MDTDIS = (U, A, V, f)$ is *an decision information system,* where U is a non-empty finite set of objects, called the universe. A is a non-empty finite set of attributes, $A = C \cup D, C \cap D = \emptyset$, where C and D denote the sets of condition attributes and decision attributes, respectively. $V = \bigcup_{C_i \in C} V_{C_i}$, V_{C_i} is domain of attributes set C_i, $V_{C_i} \cap V_{C_j} = \emptyset (i \neq j)$, $C = \cup C_i$, $C_i \cap C_j = \emptyset$. The data type of V_{C_i} and V_{C_j} is different. $f : U \times A \rightarrow V$ is an information function, which gives values to every object on each attribute, namely, $\forall a \in A$, $x \in U$, $f(x, a) \in V_a$.

Definition 22 $MDTDIS = (U, A, V, f)$ is an decision information system, $\forall x \in U$, $B = \cup B_k \subseteq C$, $B_k \subseteq C_k$, the composite relation CR_B is defined as

$$CR_B = \{(x, y) \big| (x, y) \in \bigcap_{B_k \in B} R_{B_k}\} \tag{1}$$

where $R_{B_k} \subseteq U \times U$ is an binary relation defined by an attribute set B_k on U. Let $[x]_{CR_B} = \{y | y \in U, \forall B_k \in B, y R_{B_k} x\}$ denotes the composite class of x.

Definition 23 Let $M_{R_{B_i}} = [r_{ij}^{B_i}]_{n \times n} (B_i \subseteq C)$ is the relation matrix of binary relation R_{B_i}, where

$$r_{ij}^{B_i} = \begin{cases} r_{ij}^{B_i} = 1, & if\ x_i\ R_{B_i} x_j; \\ r_{ij}^{B_i} = 0, & if\ x_i\ \not R_{B_i} x_j. \end{cases} \tag{2}$$

Then, the relation matrix of composite relation CR_B is $M_{CR_B} = [z_{ij}]_{n \times n}$, where

$$z_{ij} = \bigwedge_{i=1}^{k} r_{ij}^{B_i}. \tag{3}$$

For convenience, three kinds of binary relations used in Example 21, e.g., neighborhood relation between numerical objects, partial relation between interval valued objects, and tolerance relation in between set-valued objects are introduced briefly as follows [6, 7, 11].

If $\forall v \in V_B$ ($B \subseteq C$), v is set-valued, then the tolerance relation between set-valued objects is

$$R_B^{\cap} = \{(y,x) \in U \times U | f(y,a) \cap f(x,a) \neq \emptyset (\forall a \in B)\}$$

If $\forall v \in V_B$, v is a set of interval numbers, $f(x,a) = [f^L(x,a), f^U(x,a)]$, where $f^L(x,a)$, $f^U(x,a) \in R$, $f^L(x,a)$, and $f^U(x,a)$ are lower and upper limits of the interval number, respectively. Then the partial order relation between interval valued objects is

$$\begin{aligned} R_{I_B}^{\geq} = \{(y,x) \in U \times U | f^L(y,a) \geq f^L(x,a), \\ f^U(y,a) \geq f^U(x,a), f^L(y,b) \leq f^L(x,b), \\ f^U(y,b) \leq f^U(x,b), a \in A_1, b \in A_2\} \end{aligned} \tag{4}$$

If $\forall v \in V_B$ ($B \subseteq C$), v is a numerical value, then $\forall x \in U$ and $B \subseteq C$, the neighborhood $\delta_B(x)$ of x in B is defined as:

$$\delta_B(x) = \{y | y \in U, \Delta^B(x,y) \leq \delta\} \tag{5}$$

where Δ is a distance function. The formula of Δ is

$$\Delta_P(x,y) = \left(\sum_{i=1}^{N} |f(x,a_i) - f(y,a_i)|^P \right)^{1/P} \tag{6}$$

In the following, we illustrate CRS by an example.

Example 21 Table 1 is an example of *MDTDIS*, where $U = \{x_i, 1 \leq i \leq 8\}$, $C = \{a_i, 1 \leq i \leq 4\}$, $D = \{d\}$, $V_{a_1} = \{1,2,3,4\}$ is nominal values, $V_{a_2} = \{0.2, 0.85, 0.31, \ 0.74, 0.82, \ 0.72, 0.6, 0.44\}$ is numerical values, $V_{a_3} = \{[2.17, 2.86], \cdots, [2.51, 4.04]\}$ is interval values, $V_{a_4} = \{\{0,1\}, \cdots, \{1,2\}\}$ is se-t valued. Suppose the relationship in different attributes are equivalence relation, neighborhood relation (δ= 0.26 , Euclidean distance) [11], partial relation [7], tolerance relation [6] on a_1 to a_6, respectively.

The relation matrixes of $R_{a_i} (1 \leq i \leq 4)$ are given as follows:

$$M_{R_{a_1}} = \begin{bmatrix} 1 & 0 & 0 & 0 & 0 & 1 & 0 & 0 \\ 0 & 1 & 0 & 0 & 0 & 0 & 0 & 0 \\ 0 & 0 & 1 & 0 & 0 & 0 & 0 & 1 \\ 0 & 0 & 0 & 1 & 1 & 0 & & 0 \\ 0 & 0 & 0 & 1 & 1 & 0 & 1 & 0 \\ 1 & 0 & 0 & 0 & 0 & 1 & 0 & 0 \\ 0 & 0 & 0 & 1 & 1 & 0 & 1 & 0 \\ 0 & 0 & 1 & 0 & 0 & 0 & 0 & 1 \end{bmatrix}, \quad M_{R_{a_2}} = \begin{bmatrix} 1 & 0 & 1 & 0 & 0 & 0 & 0 & 1 \\ 0 & 1 & 0 & 1 & 1 & 1 & 1 & 0 \\ 1 & 0 & 1 & 0 & 0 & 0 & 0 & 1 \\ 0 & 1 & 0 & 1 & 1 & 1 & 1 & 0 \\ 0 & 1 & 0 & 1 & 1 & 1 & 1 & 0 \\ 0 & 1 & 0 & 1 & 1 & 1 & 1 & 0 \\ 0 & 1 & 0 & 1 & 1 & 1 & 1 & 1 \\ 1 & 0 & 1 & 0 & 0 & 0 & 1 & 1 \end{bmatrix},$$

$$M_{R_{a_3}} = \begin{bmatrix} 1 & 1 & 1 & 1 & 1 & 1 & 1 & 1 \\ 0 & 1 & 0 & 1 & 1 & 0 & 0 & 0 \\ 0 & 1 & 1 & 1 & 1 & 0 & 0 & 0 \\ 0 & 0 & 0 & 1 & 0 & 0 & 0 & 0 \\ 0 & 0 & 0 & 1 & 1 & 0 & 0 & 0 \\ 0 & 1 & 1 & 1 & 1 & 1 & 0 & 1 \\ 0 & 1 & 1 & 1 & 1 & 1 & 1 & 1 \\ 0 & 1 & 1 & 1 & 1 & 0 & 0 & 1 \end{bmatrix}, \quad M_{R_{a_4}} = \begin{bmatrix} 1 & 0 & 1 & 1 & 1 & 1 & 1 & 1 \\ 0 & 1 & 1 & 0 & 0 & 0 & 1 & 1 \\ 1 & 1 & 1 & 1 & 1 & 1 & 1 & 1 \\ 1 & 0 & 1 & 1 & 1 & 1 & 0 & 1 \\ 1 & 0 & 1 & 1 & 1 & 1 & 0 & 1 \\ 1 & 0 & 1 & 1 & 1 & 1 & 0 & 1 \\ 1 & 1 & 1 & 0 & 0 & 0 & 1 & 1 \\ 1 & 1 & 1 & 1 & 1 & 1 & 0 & 1 \end{bmatrix}.$$

Then, $M_{CR_C} = \begin{bmatrix} 1 & 0 & 0 & 0 & 0 & 0 & 0 & 0 \\ 0 & 1 & 0 & 0 & 0 & 0 & 0 & 0 \\ 0 & 0 & 1 & 0 & 0 & 0 & 0 & 0 \\ 0 & 0 & 0 & 1 & 0 & 0 & 0 & 0 \\ 0 & 0 & 0 & 1 & 1 & 0 & 0 & 0 \\ 0 & 0 & 0 & 0 & 0 & 1 & 0 & 0 \\ 0 & 0 & 0 & 0 & 0 & 0 & 1 & 0 \\ 0 & 0 & 1 & 0 & 0 & 0 & 0 & 1 \end{bmatrix}$.

Then, $[x_1]_{CR_C} = \{x_1\}, [x_2]_{CR_C} = \{x_2\}, [x_3]_{CR_C} = \{x_3\}, [x_4]_{CR_C} = \{x_4\}, [x_5]_{CR_C} = \{x_4, x_5\}, [x_6]_{CR_C} = \{x_6\}, [x_7]_{CR_C} = \{x_7\}, [x_8]_{CR_C} = \{x_3, x_8\}$.

3 Probabilistic Composite Rough Set

In this section, we present a new extended rough set model, Probabilistic Composite Rough Set (PCRS), by introducing probability method in CRS.

$MDTDIS = (U, A, V, f)$ is an Multi-Data Decision Information system, $CR_B(x)$ is a composite class of x, for $\forall X \in U$, then $P(X, [x]_{CR_B}) = \frac{|X \cap [x]_{CR_B}|}{|[x]_{CR_B}|}$ is the conditional probability. Probabilistic Composite Rough Set (PCRS) is defined as follows:

Definition 31 Given a quadruple $MDTDIS = (U, A, V, f)$, $\forall X \in U$, $0 \leq \mu \leq l \leq 1$,

$$\underline{apr}^{l}_{CR_B}(X) = \left\{x | P(X, [x]_{CR_B}) \geq l\right\} \tag{7}$$

$$\overline{apr}^{\mu}_{CR_B}(X) = \left\{x | P(X, [x]_{CR_B}) \geq \mu\right\} \tag{8}$$

Definition 32 For any $X \subseteq U$, the approximation accuracy of X is defined as follows:

$$\alpha^{l,\mu}_{CR_B}(X) = \frac{\left|\underline{apr}^{l}_{CR_B}(X)\right|}{\left|\overline{apr}^{\mu}_{CR_B}(X)\right|} \tag{9}$$

The roughness of X is defined below:

$$\rho^{l,\mu}_{CR_B}(X) = 1 - \alpha^{l,\mu}_{CR_B}(X) \tag{10}$$

Property 31 For $\underline{apr}^{l}_{CR_B}(X)$ and $\overline{apr}^{\mu}_{CR_B}(X)$, we have

1. If $l_1 \leq l_2$, then $\underline{apr}^{l_1}_{CR_B}(X) \supseteq \underline{apr}^{l_2}_{CR_B}(X)$;
2. If $\mu_1 \leq \mu_2$, then $\overline{apr}^{\mu_1}_{CR_B}(X) \supseteq \overline{apr}^{\mu_2}_{CR_B}(X)$;
3. If $l_1 \leq l_2$, $\mu_1 \geq \mu_2$, then $\alpha^{l_1,\mu_1}_{CR_B}(X) \geq \alpha^{l_2,\mu_2}_{CR_B}(X)$,
4. If $l_1 \geq l_2$, $\mu_1 \leq \mu_2$, then $\alpha^{l_1,\mu_1}_{CR_B}(X) \leq \alpha^{l_2,\mu_2}_{CR_B}(X)$, $\rho^{l_1,\mu_1}_{CR_B}(X) \geq \rho^{l_2,\mu_2}_{CR_B}(X)$;

Example 31 For Table 1, R_d is an equivalence relation on decision attributes. $U/R_d = \{D_1, D_2, D_3\}, D_1 = \{x_1, x_4, x_6, x_7\}, D_2 = \{x_2\}, D_3 = \{x_3, x_5, x_8\}$.*Then*, $P(D_1, [x_1]_{CR_B}) = 1, P(D_1, [x_2]_{CR_B}) = 0, P(D_1, [x_3]_{CR_B}) = 0, P(D_1, [x_4]_{CR_B}) = 1,$ $P(D_1, [x_5]_{CR_B}) = 0.5, P(D_1, [x_6]_{CR_B}) = 1, P(D_1, [x_7]_{CR_B}) = 1, P(D_1, [x_8]_{CR_B}) = 0.$

If $l = 0.6$, $\mu = 0.3$, then $\underline{apr}^{l}_{CR_B}(D_1) = \{x_1, x_4, x_6, x_7\}$, $\overline{apr}^{\mu}_{CR_B}(D_1) = \{x_1, x_4, x_5, x_6, x_7\}$, $\alpha^{l,\mu}_{CR_B}(D_1) = 0.8$, $\rho^{l,\mu}_{CR_B}(D_1) = 0.2$.

4 Attribute Reduction in Probabilistic Composite Rough Set

Attribute reduction is an important task in data mining. RST has been applied successfully in attribute reduction [24]. In this section, distribution reduction method in [25] is extended to CRS. R_D is an equivalence relation on the decision attribute D. Then ${}^{U}/_{R_d} = \{D_1, D_2, \ldots, D_j, \ldots, D_r\}$ forms a partition on the universe U. $[x]_{R_D} = \{y | yR_D x, x, y \in U\}$. If $[x]_{CR_B} \subseteq [x]_{R_D}$, then $MDTDIS$ is a consistent information system; If $[x]_{CR_B} \not\subset [x]_{R_D}$, then *MDTDIS* is an inconsistent information system.

Let $P(D_j, [x]_{CR_B}) = \frac{|D_j \cap [x]_{CR_B}|}{|[x]_{CR_B}|}$ be the conditional probability of $[x]_{CR_B}$ on D. Let $\mu_B(x) = (P(D_1, [x]_{CR_B}), P(D_2, [x]_{CR_B}), \ldots, P(D_r, [x]_{CR_B}))$ denote the distribution function of x on B. Then the distribution reduction on *MDTDIS* is given.

Definition 41 In $MDTDIS = (U, C \cup D, V, f)$, $B \subseteq C$, if $\forall x \in U$, $\mu_B(x) = \mu_C(x)$, and $\neg \exists E \subset B$, $\mu_E(x) = \mu_C(x)$, then B is the distribution reduction of C.

Definition 42 The discernibility matrix of distribution reduction $M_{Dis} = [m_{ij}]_{n \times n}$ is

$$m_{ij} = \begin{cases} DisAtr_{ij}\left([x_i]_{CR_B}, [x_j]_{CR_B}\right) \in D^* \\ C\left([x_i]_{CR_B}, [x_j]_{CR_B}\right) \notin D^* \end{cases} \tag{11}$$

where $D^* = \left\{([x_i]_{CR_B}, [x_j]_{CR_B}) : \mu_C(x_i) \neq \mu_C(x_j)\right\}$, $DisAtr_{ij} = a_k$, $a_k \in B_i : x_i R_{B_i} x_j$.

Definition 43 The discernibility formula of distribution reduction is

$$M = \wedge\{\vee\{a_k : a_k \in DisAtrij\}\}(i, j \leq n) \tag{12}$$

The minimum disjunction form is

$$M_{min} = \bigvee_{k=1}^{p} \left(\bigwedge_{i=1}^{q_k} a_i\right) \tag{13}$$

Let $B_k = \{a_s : s = 1, 2, \cdots, q_k\}$. Then $RED = \{B_k : k = 1, 2, \cdots, p\}$ is the set of distribution reduction.

Example 41 (Continuation of Examples 2.1 and 3.1) We compute the reducts of the information system.

1. Firstly, the distribution function of $x_i (1 \leq i \leq 8)$ is $\mu_C(x_1) = \{1, 0, 0\}$, $\mu_C(x_2) = \{0, 1, 0\}$, $\mu_C(x_3) = \{0, 0, 1\}$, $\mu_C(x_4) = \{1, 0, 0\}$, $\mu_C(x_5) = \{0.5, 0, 0.5\}$, $\mu_C(x_6) = \{1, 0, 0\}$, $\mu_C(x_7) = \{1, 0, 0\}$, $\mu_C(x_8) = \{0, 0, 1\}$.
2. Next, by Definition 4.3, the discernibility matrix M_{Dis} is
3. Then, $M = \{a_1\} \wedge \{a_3\}$. $M_{min} = \{a_1, a_3\}$.
4. There is a reduct for the information system, i.e., $B_1 = \{a_1, a_3\}$.

$$M_{Dis} = \begin{bmatrix} C & \{a1,a2,a4\} & \{a1\} & C & \{a1,a2\} & C & C & \{a1\} \\ \{a1,a2,a3,a4\} & C & \{a2,a3\} & \{a1,a4\} & \{a1,a4\} & \{a1,a3,a4\} & \{a1,a3\} & \{a1,a3\} \\ \{a1,a3\} & \{a1,a2\} & C & \{a1,a2\} & \{a1,a2\} & \{a1,a2,a3\} & \{a1,a2,a3\} & C \\ C & \{a1,a2,a3,a4\} & \{a1,a3\} & C & \{a3\} & C & C & \{a1,a2,a3\} \\ \{a1,a2,a3\} & \{a1,a3,a4\} & \{a1,a2,a3\} & C & C & \{a1,a3\} & \{a3,a4\} & \{a1,a2,a3\} \\ C & \{a1,a4\} & \{a1,a2\} & C & \{a1\} & C & C & \{a1,a2\} \\ C & \{a1\} & \{a1,a2\} & C & \{a1\} & C & C & \{a1\} \\ \{a1,a3,a4\} & \{a1,a2\} & C & \{a1,a2\} & \{a1,a2\} & \{a1,a2\} & \{a1,a3,a4\} & C \end{bmatrix}$$

5 Conclusions

In this paper, we proposed the PCRS which aims to deal with multiple binary relations in a *MDTDIS* in the framework of probabilistic method. Then, the distribution attribute reduct is investigated. Examples are given to illustrate the method proposed in the paper. In the future work, we will develop algorithms to verify the efficiency of the approach in real-life applications.

References

1. Pawlak Z (1982) Rough sets. Int J Inf Comput Sci 11(5):341–356
2. Pawlak Z, Skowron A (2007) Rough sets: some extensions. Inf Sci 177(1):8–40
3. Li TR, Ruan D, Geert W, Song J, Xu Y (2007) A rough sets based characteristic relation approach for dynamic attribute generalization in data mining. Knowl Based Syst 20(5):485–494
4. Greco S, Matarazzo B, Slowinski R (1999) Rough approximation of a preference relation by dominance relations. Eur J Oper Res 117(1):63–83
5. Guan YY, Wang HK (2006) Set-valued information systems. Inf Sci 176:2507–2525
6. Qian YH, Dang CY, Liang JY, Tang DW (2009) Set-valued ordered information systems. Inf Sci 179:2809–2832
7. Qian YH, Liang JY, Dang CY (2008) Interval ordered information systems. Comput Math Appl 56:1994–2009
8. Hu QH, Yu DR, Guo MZ (2010) Fuzzy preference based rough sets. Inf Sci 180:2003–2022
9. Kryszkiewicz M (1998) Rough set approach to incomplete. Inf Sci 112:39–49
10. Wang GY (2002) Extension of rough set under incomplete information systems. Comput Res Dev 39:1238–1243
11. Hu QH, Liu JF, Yu DR (2008) Mixed feature selection based on granulation and approximation. Knowl Based Syst 21(4):294–304
12. Hu QH, Yu DR, Liu JF, Wu CX (2008) Neighborhood rough set based heterogeneous feature subset selection. Inf Sci 178:3577–3594
13. Hu QH, Xie ZX, Yu DR (2007) Hybrid attribute reduction based on a novel fuzzy-rough model and information granulation. Pattern Recognit 40:3509–3521
14. Wei W, Liang JY, Qian YH (2012) A comparative study of rough sets for hybrid data. Inf Sci 190(1):1–16
15. An LP, Tong LY (2010) Rough approximations based on intersection of indiscernibility, similarity and outranking relations. Knowl Based Syst 23:555–562
16. Abu-Donia HM (2012) Multi knowledge based rough approximations and applications. Knowl Based Syst 26:20–29
17. Zhang JB, Li TR, Chen HM (2012) Composite rough sets. In: proceedings of the 2012 international conference on artificial intelligence and computational intelligence (AICI'12), pp 150–159
18. Yao YY (2008) Probabilistic rough set approximations. Int J Approx Reason 49(2):255–271
19. Ziarko W (1993) Variable precision rough set model. J Comput Syst Sci 46(1):39–59
20. Ziarko W (1993) Analysis of uncertain information in framework of variable precision rough set. Found Comput Decis Sci 18:381–396
21. Yao YY, Wong SKM (1992) A decision theoretic framework for approximatingconcepts. Int J Man Mach Studies 37(6):793–809
22. Slezak D, Ziarko W (2005) The investigation of the Bayesian rough set model. Int J Approx Reason 40(1–2):81–91

23. Chen HM, Li TR, Ruan D (2012) Dynamic maintenance of approximations under a rough-set based variable precision limited tolerance relation. J Multple Valued Logic Soft Comput 18:577–598
24. Thangavel K, Pethalakshmi A (2009) Dimensionality reduction based on rough set theory: a review. Appl Soft Comput 9:1–12
25. Mi JS, Wu WZ, Zhang WX (2004) Approaches to knowledge reduction based on variable precision rough set model. Inf Syst 159(2):255–272

AM-FM Signal Modulation Recognition Based on the Power Spectrum

Sen Zhao, Liangzhong Yi, Zheng Pei and Dongxu Xu

Abstract Modulation recognition in frequency domain is a complex problem. This paper deals with classification for analog modulated signals based on the frequency domain information. Here, two feature parameters are designed to describe the spectrum property of the modulated signals, one is the ratio of the square of the mean value of the spectrum amplitude to the variance, and the other is the kurtosis of the normalized spectrum amplitude. Modulation recognition rules based on the two feature parameters are extracted. The rules are used to recognize AM-FM analog modulated signals, which are actual spectrum data by Agilent E4407 Spectrum Analyzer. The test results show the effectiveness of our method in this paper.

Keywords AM · FM · Modulation recognition · Spectrum feature · Discrimination rule

1 Introduction

Nowadays, Automatic Modulation Recognition has become an important topic in many applications for several reasons. The first reason is to select an appropriate demodulator to unknown modulation type. This state prevents partially or

---

S. Zhao (✉) · L. Yi · Z. Pei · D. Xu
Xihua University, the School of Mathematics and Computer Engineering,
610039 Chengdu, China
e-mail: zhaosen520@126.com

L. Yi
e-mail: yiliangzhong@x263.net

Z. Pei
e-mail: zhengpei@mail.xhu.edu.cn

F. Sun et al. (eds.), *Knowledge Engineering and Management*,
Advances in Intelligent Systems and Computing 214,
DOI: 10.1007/978-3-642-37832-4_18,

completely damages of the communication signal information content [1]. Thus, this signal information content is correctly obtained from intercepted signal. The second reason is to know the correct modulation type helps to recognize the threat and determine the suitable jamming waveform [2].

To improve the probability of correct modulation recognition, from 1995 to 2007, Nandi and Azzouz proposed a series of algorithms for automatic modulation recognition to identify digital and analog signals [1, 3–5]. They proposed nine parameters based on instantaneous information for recognition by method of decision theory, neural networks, and neural network cascade, which became quite classical. The literature [6] was first proposed modulation recognition method of the satellite signals based on the characteristics of the power spectrum. Wu studied the automatic modulation recognition of digital communication signals using statistical parameters methods [7]. The modulation recognition algorithm based on combined feature parameter and modified probabilistic neural network is proposed to identify AM, VSB, 8QAM, FM, ASK, FSK, and MSK [8–10].

AM is Amplitude Modulation, and FM is Frequency Modulation. Generally, the medium-wave broadcast and the international shortwave broadcast in the high frequency (3–30 MHz) use Amplitude Modulation (AM). Even Air Navigation Communication which has the higher frequency than the FM radio also uses AM. Moreover, Frequency Modulation is most widely used in the FM radio (88–108 MHz in China), while, Amateur radio, the application band of satellite communication in the shortwave range of 27–30 MHz, and the intercom and paging (137–167 MHz) use FM [11]. In recent years, it is worsening that Aviation band voice communications business is interfered, which affect the normal air-ground voice communications. Among them, the interference are the most prominent which are caused by the frequency modulation radio stations. Since, the modulation type is an important parameter of the frequency modulation radio and the normal air-ground voice communications. However, Modulation Recognition based on time domain usually does not apply to the radio monitoring based on frequency domain.

The signal spectrum is a description of the frequency domain for a signal, and then different modulation signals have different manifestations in the frequency domain. The signal power spectrum directly reflects the distribution of the power of the modulation signal frequency components, which can better reflect the characteristics of the different modulation. In addition, the radio managers frequently use the signal power spectrum in the practical applications. Therefore, the modulation recognition has a practical significance by utilizing the spectrum data. In this paper, we use two feature parameter to describe the spectrum property. In order to achieve the identification of the AM and FM modulation, we use the rule of discrimination classifier which has a simple calculation and a good real-time.

The remaining of this paper is organized as follows. The Amplitude Modulation and Frequency Modulation are derived in the next section. Section 3 presents the data preprocessing and the feature extraction. The Simulation result is introduced in Sect. 4. Finally, the paper is concluded in Sect. 5.

2 Amplitude, Frequency Modulation

In this study, two analog modulation types are used. These analog modulation types are double side band with transmission carrier amplitude modulation (AM) and frequency modulation (FM) [12]. A modulated signal $c(t)$ can be given as below:

$$c(t) = b_c r(t)\cos[2\pi f_c(t)] + \psi(t) + \theta_0$$

where $r(t)$ is the signal envelope, $f_c(t)$ is the carrier frequency and $\psi(t)$ is the phase and θ_0 is initial phase and b_c controls the carrier power. And, AM and FM signals are given as below:

$$c(t) = [1 + mf(t)]\cos(2\pi f_c(t)) \quad (1)$$

$$c(t) = \cos[2\pi f_c(t) + K_f \int_{-\infty}^{t} x(\tau)d\tau] \quad (2)$$

where m is modulation index, $f(t)$ is the message signal and $f_c(t)$ is the carrier frequency and K_f is the frequency deviation coefficient of FM signal.

Fast Fourier Transform (FFT) is one of the methods for signal. FFT decomposes a signal defined on infinite time interval into a λ-frequency component where λ can be either real or complex number [13]. FFT is actually a continuous form of Fourier series. The definition for DFT (Discrete Fourier Transform) of an N-point discrete-time signal $c(t)$ is:

$$x_k = \sum_{n=0}^{N-1} c(t)e^{-i2\pi k\frac{n}{N}}, k = 0, \ldots, N-1 \quad (3)$$

We can compress a signal by taking its FFT and then discard the small Fourier coefficients. We can apply the FFT to transform the time domain signal into frequency domain to obtain the power spectrum data.

In addition, we can intuitively feel the performance by studying the energy distribution for the signals in different frequency bands. Since, the radio managers more frequently contact frequency-domain signals by using the signal power spectrum in the practical applications. In this paper, we directly use the power spectrum data to carry out the identification of the modulation types. The spectrum descriptions of the frequency modulation radio and the normal air-ground voice communications are shown in Figs. 1 and 2.

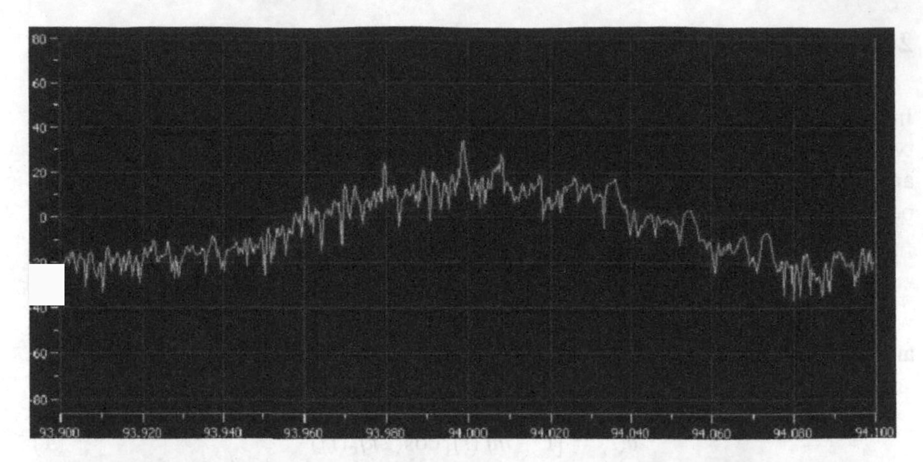

Fig. 1 The spectrum of the frequency modulation radio

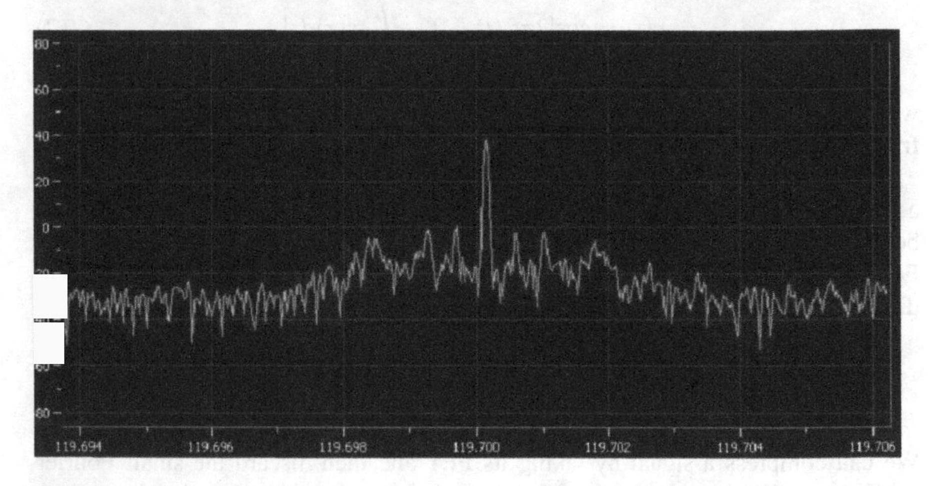

Fig. 2 The spectrum of the normal air-ground voice communications

3 Pre-Process Data and Feature Extraction

3.1 Pre-Process Data

In this paper, we consider the complexity of radio environment, then pre-process the AM and FM spectrum data which are received by the spectrum analyzer. Our purpose is to eliminate the difference of the background noise level, and enhance the useful information [14, 15]. The background noise is complex and uncertainty which mainly displays in energy so that it is difficult for us to accurately eliminate the difference by the classical electromagnetic theory. In order to eliminate this difference to a certain extent, we take the fuzzy c-means (FCM) clustering algorithm into account.

The FCM clustering algorithm is proposed by Bezdek, which is an improvement of the hard k-means algorithm [16]. It is determined by the degree of each data point belongs to some cluster with the degree of subjection. FCM algorithm can divide the fuzziness of data that make it more objective and reality. In our research, the background noise is clustered into three clusters by using FCM. It goes without saying that three cluster centers are obtained. So the fourth largest value of the smallest cluster center will be viewed as the background noise threshold. The pre-process data is that every spectrum data point minus the background noise threshold of each group data. Specifically, there is that $X = \{X_1, X_2, \ldots, X_n\}$ is the n groups spectrum data which will be analyzed, and $X_i = \{x_{1i}, x_{2i}, \ldots, x_{ki}\} (1 \le i \le n)$ is the ith group data, and $X_i^{'} = \{x_{1i}^{'}, x_{2i}^{'}, \ldots, x_{ki}^{'}\}$ $(1 \le i \le n)$ is the processed data, where k is the sample number. Moreover, we can define that D_{level} is the background noise threshold of X. So,

$$x_{ki}^{'} = \begin{cases} x_{ki} - D_{level}, & \text{if} \quad x_{ki} > ; D_{\text{level}}, \\ 0, & \text{if} \quad x_{ki} \le D_{\text{level}}. \end{cases} \tag{4}$$

The physical significance of this process is that the spectrum signal data is intercepted by the background noise threshold, which is equivalent to retain the points above the threshold and move up the height of the noise floor level. It can effectively eliminate differences and preserve the useful information.

3.2 Feature Extraction

The two key feature parameters (R, u_{42}^x) are utilized to identify the modulation signals. Considering computation complexity and recognition efficiency, the following two key feature parameters are selected as the decision criterion for recognition [17].

The first key feature parameter, R, is defined by:

$$R = \frac{u^2}{d} \tag{5}$$

where u is the mean value of spectrum amplitude $x(i)$ and d is the variance of spectrum amplitude $x(i)$.

Thus, R is the ratio of the square of the mean value of the spectrum amplitude to the variance. This parameter is improvements to the parameter R proposed by Chan [18].

The other spectrum feature parameter u_{42}^x which is the kurtosis of the normalized spectrum amplitude for the modulation recognition. The u_{42}^x is defined as follows:

$$u_{42}^{x} = \frac{E\{x_{cn}^{4}(i)\}}{\{E\{x_{cn}^{2}(i)\}\}^{2}} \tag{6}$$

where $x_{cn}(i) = x_n(i) - 1, x_n(i) = x(i)/m_x$, $m_x = \frac{1}{N}\sum_{i=1}^{N} x(i)$, $x_{cn}(i)$ is the value of the normalized-centered spectrum amplitude $(i = 1, 2, \ldots, N)$, N is the sample number and $x(i)$ is the spectrum amplitude.

The design of u_{42}^{x} as the feature for the proposed algorithm is based on the following fact: This feature as defined by Eq. 6 is used to measure the compactness of the spectrum amplitude distribution, so it can be used to discriminate between the signals. For example, the spectrum amplitude of the AM signals has high compact distribution ($u_{42}^{x} > t_{u_{42}^{x}}$) and the FM signals' have less compact distribution ($u_{42}^{x} \leq t_{u_{42}^{x}}$).

Because there are only AM and FM modulation types to be identified, we can obtain the discrimination rule based on the above two feature as follows:

(I) If a signal $u_{42}^{x} \geq t_{u_{42}^{x}}$ and $R \leq t_R$, then a modulation signal is a AM signal.
(II) If a signal $u_{42}^{x} < t_{u_{42}^{x}}$ and $R > t_R$, then a modulation signal is a FM signal.
(III) If a signal $u_{42}^{x} \geq t_{u_{42}^{x}}$ and $R > t_R$, then a modulation signal is a AM signal.
(IV) If a signal $u_{42}^{x} < t_{u_{42}^{x}}$ and $R \leq t_R$, then a modulation signal is a AM signal.

By analyzing the above rules, we summarize and get new rules:

(1) If a signal $u_{42}^{x} < t_{u_{42}^{x}}$ and $R \leq t_R$, then a modulation signal is a AM signal.
(2) If a signal $u_{42}^{x} < t_{u_{42}^{x}}$ and $R > t_R$, then a modulation signal is a FM signal.
(3) If a signal $u_{42}^{x} \geq t_{u_{42}^{x}}$, then a modulation signal is a AM signal.

So, we can recognize the AM-FM modulation by using the above rules.

4 Simulation Result

The implementation of the algorithm requires the determination of the feature thresholds t_R and $t_{u_{42}^{x}}$. The determination of the optimum feature threshold, t_{opt} has the largest average probability (approach 1) for the following formula [5]:

$$P_{av}[t_{opt}] = \frac{P\{AM[t_{opt}]/AM\} + P\{FM[t_{opt}]/FM\}}{2} \tag{7}$$

where, $P\{AM[t_{opt}]/AM\}$ is the correct probability that a signal is judged to belong to AM with the threshold t_{opt} under the condition that a signal is belong to AM signal, and $P\{FM[t_{opt}]/FM\}$ is the correct probability that a signal is judged to belong to FM with the threshold t_{opt} under the condition that a signal is belong to FM signal.

In this paper, all experimental data which is the actual spectrum data is collected by Agilent E4407 Spectrum Analyzer. Frequency Modulation radio signal

Table 1 Feature parameters of modulation signal

Modulation\Feature	R	u_{42}^{x}
AM	0.2153	21.397
FM	0.8373	1.979

Table 2 The identification rete (%)

Modulation	FM	AM
Identification rate	96.3	98.9

(88–108 MHz) is chosen as the FM signal, in addition the air-ground voice communications signal (108–128 MHz) are viewed as AM signal. One class of modulation signal has 1500 samples. Five hundred of them are to test the performance of the proposed feature parameters and used to calculate the threshold. The other signals are used to determine the identification rate. Table 1 give the mean value of two feature parameters for one hundred samples. It is obvious from the Table 1 that the value u_{42}^{x} of AM is larger than the FM's and the value R of AM and FM are not much different. However, the value u_{42}^{x} of AM should be the largest than the FM's for based on the theoretical basis. This situation may be caused by terrain obstacles, the electromagnetic interference, or Multi-path Effect, which result in the faster decline of the amplitude.

The optimum features threshold values, t_R and $t_{u_{42}^{x}}$ are chosen to be 0.43 and 3.1. The identification rate of the actual signals is presented in Table 2.

5 Conclusions

In this paper, we study Automatic Modulation Recognition for AM and FM signal with the spectrum analyzer in the actual environment. Taking the decline of the amplitude caused by the environmental impact into account, we design two new parameters and give the formula. The test shows that the recognition result has the practical significance by using the spectrum data to identify the modulation types of the frequency modulation radio and the air-ground voice communications. The future work is focused on how to recognize other analog modulation types by utilizing the spectrum data.

Acknowledgments This work is partially supported by the research fund of Sichun Key Laboratory of Intelligent Network Information Processing (SGXZD1002-10) , the National Natural Science Foundation (61175055, 61105059), Sichuan Key Technology Research and Development Program (2012GZ0019, 2011FZ0051), and the Innovation Fund of Postgraduate (Xihua university) (YCJJ201230), and the research fund of education department of Sichuan province (10ZC058).

References

1. Nandi AK, Azzouz EE (1995) Automatic identification of digital modulation types. Signal Process 46(2):211–222
2. Avci D (2010) An intelligent system using adaptive wavelet entropy for automatic analog modulation identification. Digit Signal Process 20:1196–1206
3. Azzouz EE, Nandi AK (1997) Modulation recognition using artificial neural networks. Signal Process 56(3):166–175
4. Wong MLD, Nandi AK (2004) Automatic digital modulation recognition using artificial neural network and genetic algorithm. Signal Process 84:354–356
5. Azzouz EE, Nandi AK (1997) Automatic Modulation Recognition-I. J Franklin Inst 334B(2):241–273
6. Fan HB, Yang ZD, Cao ZG (2004) Automatic recognition for comlllon used modulations in satellite communication. J China Inst Commun 25(1):140–149
7. Wu JP, Han YZ, Zhang JM (2007) Automatic modulation recognition of digital communication signals using statistical parameters methods. In: Proceedings of 2007 IEEE international conference on communications, Kokura, Japan, pp 697–700, 2007
8. Gao YL, Zhang ZZ (2006) Modulation recognition based on combined feature parameter and modified probabilistic neural network. In: Proceedings of the 6th world congress on intelligent control and automation, Dalian, China, pp 2954–2958, 2006
9. Wei PM, Ye J (2010) Improved intuitionistic fuzzy cross-entropy and its application to pattern recognitions. In: Intelligent systems and knowledge engineering (ISKE), 2010 international conference, pp 114–116, 2010
10. Wei CF, Yuan RF (2010) A decision-making method based on Linguistic Aggregation operators for coal mine safety evaluation. In: Intelligent systems and knowledge engineering (ISKE), 2010 international conference, pp 17–20, 2010
11. National Spectrum Management Manual (2005) International Telecommunications Union. Switzerland, Geneva
12. Roder H (1931) Amplitude, phase, and frequency modulation. Proc Inst Radio Eng 19(12):2146–2148
13. Boggess A, Narcowich FJ (2009) A first course in wavelets with Fourier analysis, 2nd ed. Wiley, Hoboken
14. Bouchon-Meunier B, Marsala C (2008) Fuzzy inductive learning: principles and applications in data mining. In: Intelligent systems and knowledge engineering (ISKE), 2010 international conference, pp 1–4, 2008
15. Martinez L (2010) Computing with words in linguistic decision making: analysis of linguistic computing models. In: Intelligent systems and knowledge engineering (ISKE), 2010 international conference, pp 5–8, 2010
16. Bezdek JC, Ehrlich R (1981) FCM: the fuzzy c-means clustering algorithm. Comput Geosci New 10(2):191–203
17. Wang J, Liu YH, Zhang J, Zhu JQ, Dong TZ (2008) Research on hiberarchy trusted network based on the grade division. In: Intelligent systems and knowledge engineering (ISKE), 2010 international conference, pp 19–25, 2008
18. Chan YT, Gadbois LG (1989) Identification of the modulation type of a signal. Signal Process 16(2):149–154

Model-Based Approach for Reporting System Development

Jinkui Hou

Abstract From the viewpoint of software engineering implementation, a model-based development approach for reporting systems is proposed systematically based on the concern of separating application descriptions and UI designs. The development process consists of four steps: data modeling, report modeling, model transformation, and code generation. The experiment shows that this approach enhances the efficiency and quality of reporting system development, which can be well combined together with other application development frameworks, and thus can support model-driven software engineering effectively.

Keywords Software engineering · Model-driven development · Reporting system · Modeling approach

1 Introduction

Reporting system is a very important subsystem in the business application system, which is used frequently. It is a heavy task to develop embedded reporting system, and the product is with vulnerability of short life cycle. Therefore, research of automatic code generation for reporting system can reduce the workload of system development, and enable the system to meet the complex application environments. The current Web-based applications are gradually replacing the traditional C/S mode software, which becomes the mainstream of application software. It is also an urgent need for reporting system to adapt to this situation. Most of the existing reporting tools do not have the learning function, in which the versatility is not enough [1]. It generally cannot generate reports of different styles in the

J. Hou (✉)
School of Computer Engineering, Weifang University, 261061 Weifang, China
e-mail: jkhou@163.com

F. Sun et al. (eds.), *Knowledge Engineering and Management*,
Advances in Intelligent Systems and Computing 214,
DOI: 10.1007/978-3-642-37832-4_19,

same run-time, and cannot meet the needs of generating reports of real-time based on user requirements. When the report format change greatly, it is difficult to add new reports to meet user requirements dynamically.

Model-driven development has become a hot topic and main trends of software engineering, which enhances abstraction level to deal with the complexity of software development through the application of models and modeling techniques. OMG's model-driven architecture (MDA) [2] provides theoretical support for automatic transformation between models. Through in-depth understanding of traditional approaches of report generation and basic requirements of reporting system, this paper provides a MDA-supported development model for Web reporting system. It can be well combined together with other application development frameworks. Thereby, it is a good application development mode with many merits, such as simple, intuitive, automatic generation of target code, and so on.

2 MDA-Supported Development Model for Web Reporting System

On the basis of the model-driven development model named ASLP [3] proposed in our previous study, a development model for Web reporting system is proposed in this paper based on the theory of MDA [4, 5]. This model support model-driven software development, which is shown in Fig. 1.

The model mainly comprises four parts: data model, report model, model transformation engine, and code generator. Data model is the basis for the whole framework of the model, which separate the report module from the actual data source. It reduces the coupling degree of the report module and the actual data source, and ensures the continuity of the knowledge of specific report definitions. It also makes the information of report definition to be no longer depended on the actual database table. Report model describes the realization of the content and format of the report which is customized by the user, which provide all the parameter information for target code generation. Model transformation engine is used to achieve model transformation from abstract model to specific platform model, while code generator generates final source codes of the report system.

2.1 Data Model

Data model comprises two parts: functional view and data object view. Functional view is used to determine the requirements of report pages on the inner model and relationships between interfaces through the analysis of user's needs. Data object view is used for modeling domain concepts, which describe the object class used

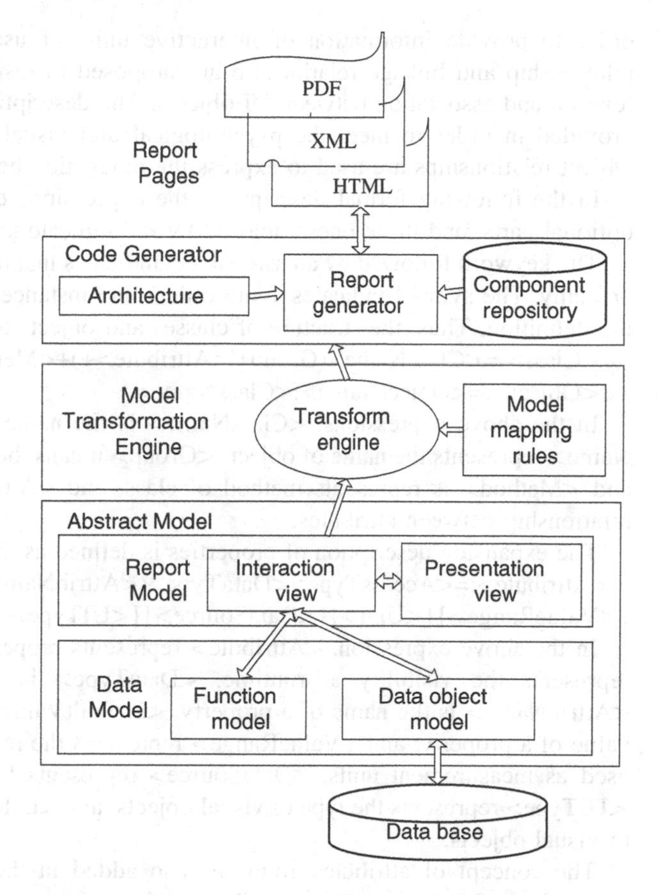

Fig. 1 Model-driven development model for Web reporting system

in application systems. It describes the data objects and their relationships required by the reports from static aspects.

The extended use case diagrams in UML are used in functional view to describe user requirements. Through analyzing user requirements, functional view can be used to determine the functions and the framework of user interfaces, the relationship between user interfaces as well as the requirement on inner models of interface presentation.

Static view of ASLP describes the composition and behavior of objects from the view of abstract calculation relationships, which cannot meet the requirements of establishing interface structure. In order to meet the requirement of the customization of report interface, object view is built by expanding static view of ASLP. In addition to considering the general composition and behavior of objects, object properties used for interface presentation and code generation is introduced and the constraints on the realization of specific interactions is minimized to the maximum extent. Expansion of the data object view includes the following aspects. Object properties are expanded by adding UI type, default values, labels, and units of measurement, and the description of data types is also extended in order to provide information of interactive units of user interface. Deduction relationship and linkage relationship are proposed to respectively represent correlation and association between UI objects. The description of property group is provided in order to meet the psychological and visual requirements of users. Object relationships are used to express the navigation between use interfaces.

In the following formal description, the expressions enclosed by {} indicate optional parts, and those ones enclosed by <> indicate keywords.

The keyword followed by an asterisk (*) indicates that the object may have zero or many. The symbol indicates + more than one instances. The symbol = means the definition. Thus, the structure of classes and objects are defined as follows:

<Class>:=<ClassName>(Group+|<Attribute>+)+<Method>*<Attriblink>*;

<Object>:=<ObjectName>:<ClassName>

In the above expressions, <ClassName> is the name of class, and <ObjectName> represents the name of object. <Group> means the set or group of objects, and <Method> is represents method of class, and <AttribLink> is the linkage relationship between attributes.

The expansion description of properties is defined as the following expression:

<Attribute>:=<AccessType><DataType>(<AttribName>{=<DefaultValue>}) {<ValueRange>}{<Unit>}{<DataSource>}{<UIType>}{<Label>}

In the above expression, <Attribute> represents property, and <AccessType> represents the visibility at runtime. <DataType> is the type of data, and <AttribName> is the name of a property. <DefaultValue> represents the default value of a property, and <ValueRange> represents the range of value. <Unit> is used as measurement units. <DataSource> represents for the source of values. <UIType> represents the type of visual objects, and <Label> is the label attached to visual objects.

The concept of attribute groups is also added in the model to support the generation of user interface. Attributes of the same group can be combined together by a frame and used as a whole. In complicated situation, the group is divided into sub-groups, and forms a nested relationship.

Attribute group is a gathering of attributes of sub-groups or properties, which is depicted with the following manner.

<Group>:=[<GroupName><Attribute>+]|[<GroupName>{<Attribute>+ |<Group>+}+].

In the above expression, <GroupName> represents of the group name, and <Attribute> represents its properties.

2.2 Report Model

Report model consists of interaction view and presentation view. Interaction view is used to describe the system from dynamic aspect, which is also an abstract description of behavior of user interface and provides the internal association between UI behaviors and system functions. The function of UI presentation view is to show the layout and forms of the interface according to the inner models (data object view and interaction view) and user requirements for data presentation. It provides a full description for the intuitive presentation of user interface, and provides the binding relationship between the interface elements and the visible elements of interaction view.

Interaction view plays a key role of connecting link between the preceding and the following in the whole framework, which is shown in Fig. 2. On the one hand, interaction view, function view, and data object view are organically combined together to form abstract outline of the interface. On the other hand, the classification of interaction views is the basis of the retrieve of interface template in presentation model. That is to say, the type of interface templates is determined by the interaction relationship between objects of interaction views. It provides constraints for the users to select the appropriate interface and ensures the correctness of UI generation. In addition, the degree of abstraction of interactive view is lower than that of the functional model and object model. Compared with them, interaction model is more close to the interface.

Data objects of interactive view is the instance of the data objects defined in object view. Data object of object view is the concept of class, and data object of interactive view is the instance of class applied in specific user interface. Data objects of object view are instantiated in interaction view, which can be further subdivided to determine the appropriate presentation form according to its role in the description of different interfaces at the same time. Interactive view is defined as a triple $<V(G), E(G), \varphi_G>$, where $V(G)$ is a collection of interacting objects, and $E(G)$ is the set of interactive relations, and φ_G is a function from the set of interactions to the set of ordered pair of interactive objects.

Interactive objects is entity objects in UI which can interact with other objects, such as data object, collection, user, use case, page reference, and so on. Data objects of interactive view come from data object view and serve for the corresponding user interface. After being presented in UI, the corresponding interface elements can be in various forms of navigation links and a variety of forms, such as text box, password boxes, radio buttons, check boxes, and so on. Interactive relationship is interactions between interactive objects, which emphasize interactive behavior between objects and use cases as well as the influence to the relationship between UI objects, which including calls, participation, information access, and navigation. Invoking relationship refers that one object calls the method of another object and returns the results of the function.

In interactive view, the relationship between user to use case is use case calling, which reflects on UI is the click on buttons. The calling relations between use cases and data objects are trigged by causing the behavior of data object or data collection. Participation relationship is used to express data providing, where the general pointer is from the object or collection to use cases, which indicates that the users provide data or parameters. Message connection is used to express message transfer among objects and visible data objects or data collection. It can also be message transfer from an event trigger to component, or message transfer

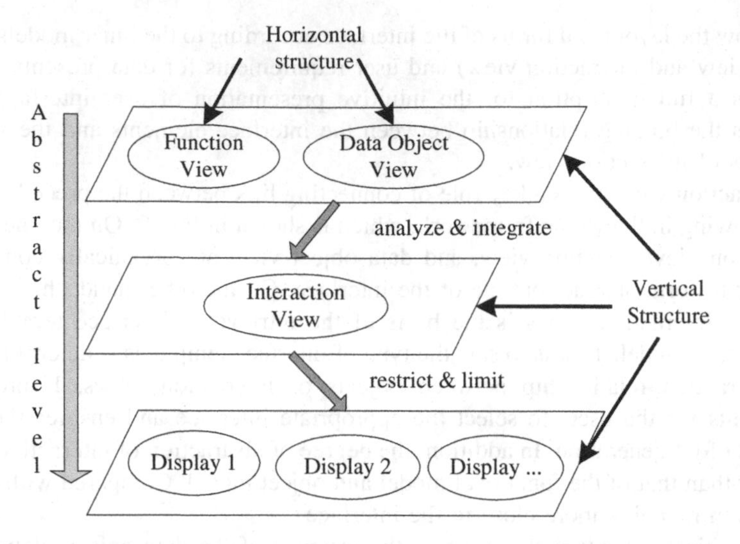

Fig. 2 The role of interaction view

from visible objects or data collection to component. Navigation relation refers page forwarding operation inspired by use case.

UI presentation view provides a constraint environment to show the layout of user interface. Interactive view is the inner basis of UI presentation view, while UI presentation view is the external depiction of interactive view. UI presentation view is used to describe the appearance of UI on the basis of interaction view, which mainly deals with macro layout of UI and display-control, and provide service for automatic code generation of graphical user interface. It embodies abstract interface based on the inner model of UI, thus solve the problem of the overall layout of user interface. UI presentation mainly includes template object, area object, dividing line object, and template interactive objects. Its comprehensive introduction can be seen in [3], and here we are no longer on it.

3 Model Transformation and Code Generation Based on ASP.NET

ASP.NET is a framework widely used for Web application development [6]. C# is used as the target code in the experiment introduced in this paper.

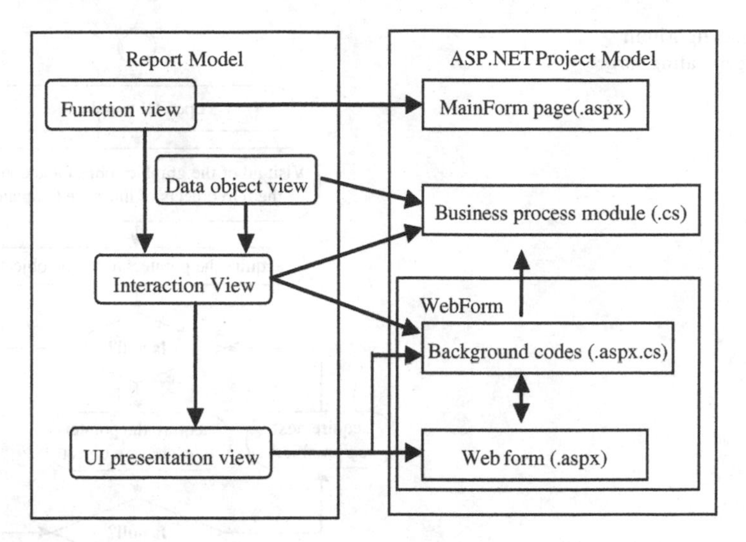

Fig. 3 Mapping relations from report model to target model

3.1 Model Mapping Relations

The mapping relations from report model to ASP.NET project model are shown in Fig. 3, in which the corresponding generated codes mainly include project information, business processing module, Web forms, and background codes.

Object view is mapped to the business processing module in the generated ASP.NET project, which provides corresponding support for interactive view and UI presentation view. The compound use case of functional view is mapped to functional selection items of the target project. Interaction view combined with module processing information of UI presentation view are mapped to the background codes of the generated Web pages (**.aspx.cs* files). Use cases of interaction view are mapped to operations of corresponding UI control objects, such as menu, button or hyperlink, and so on. Method invoking relations of the source model are mapped to operation call of the corresponding object. The navigation relations are mapped to display operations of the target pages. The template object of UI presentation view are mapped to Web pages, in which object information are mapped to attribute information of presentation elements of Web forms, such as the type, location, size, color, and other information (*. *aspx* files).

3.2 Code Generation

The algorithm of target code generation is mainly responsible for the generation of reporting framework, interface elements, the target program for print, and preview of reports. The algorithm flow is shown in Fig. 4.

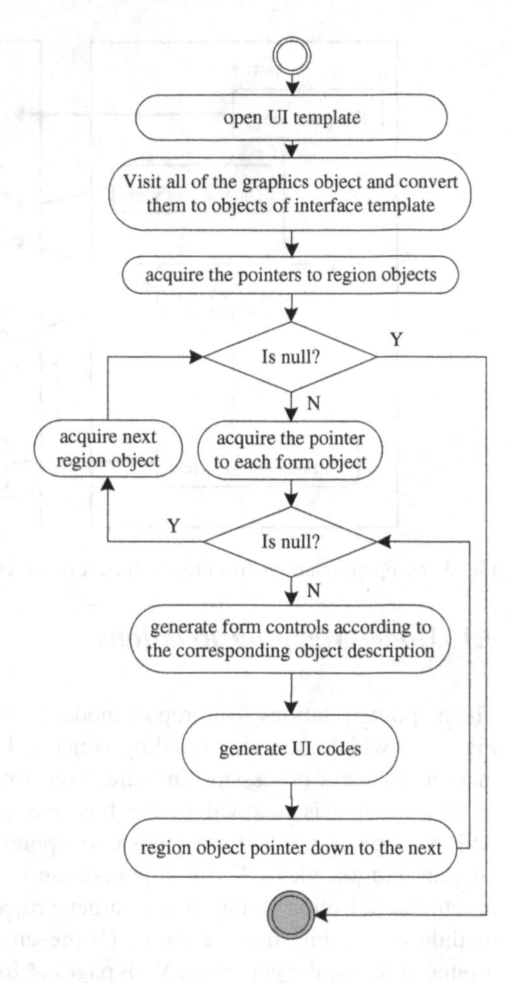

Fig. 4 The algorithm flow for generating target codes

The algorithm for generation of reporting framework is the entrance to the whole automatic code generation, which is most upper algorithm of the UI generation for the reporting system. The user interface, interface elements, and the layout of interface elements are generated according to the information provided by interface templates. The appropriate background codes are generated on the basis of interaction view corresponding to UI template. In the algorithm flow, each region of object UI template is visited and the corresponding code generation algorithm is invoked according to the data types and presentation form of each display unit.

The generation algorithm of interface elements is the algorithm for generation of the basic elements of UI, which is mainly used to generate specific interface controls. It is a recursive algorithm because some of the controls may also contain sub-controls, such as *Frame*, *Tabstrip*, and so on.

The generation algorithm for report printing and preview is mainly used to generate the print and preview program codes. The print and preview module is a core module of the reporting system, which is output part of the whole system. The basic process to achieve report printing and preview on .NET platform is described as follows: we should first define the layout of the printed page, and then define each print area which includes the size of the print area, and the text, graphics, and images needing to be printed. Finally we call the system methods (e.g., *DrawString*() and *DrawRectangle*()) to output the corresponding text, graphics, and images.

4 Conclusion and Future Work

A model-based development approach for reporting systems is proposed systematically in this paper, which can be well combined together with other application development frameworks, and thus can provides reporting capabilities for application system and has a certain capacity of software reuse. It can dynamically add new reports to meet the requirements of users. This approach follows the essence, process, and requirements of model-driven software development, which can make an effective support for model-driven software engineering.

Future works are as follows: (1) to further improve the description of interaction view, and enhance its ability of semantic interpretation; (2) to fully abstract and describe the UI presentation view, and enhance visual attractiveness of the generated pages; (3) to diversify target platform in order to verify the practicability of this approach.

Acknowledgments The author is most grateful to the anonymous referees for their constructive and helpful comments on the earlier version of the manuscript that helped to improve the presentation of the paper considerably. This research was supported by the foundation of science-technology development project of Shandong Province of China under Grant No. 2011YD01042 and No. 2011YD01043.

References

1. Hailpern B, Tarr P (2010) Model-driven development: the good, the bad, and the ugly. IBM Syst J 45(3):451–461
2. Miller J, Mukerji J (2011) MDA guide version 1.0.1 (document number omg/20011-06-01). http://www.omg.com/mda
3. Hou J, Wan J, Yang X (2006) MDA-based modeling and transformation approach for WEB applications. In: Proceeding of the sixth international conference on Intelligent System Design and Applications (ISDA), pp 867–812. IEEE Computer Society, New York
4. Kleppe A, Warmer J, Bast W (2009) MDA explained, the model driven architecture: practice and promise. Addison-Wesley, Boston
5. Thomas D (2009) MDA: revenge of the modelers or UML utopia? IEEE Softw 21(3):15–17
6. Jeffrey R, Francesco B (2009) Applied Microsoft.NET framework programming. Microsoft Press, Washington

A Time Dependent Model via Non-Local Operator for Image Restoration

Zhiyong Zuo, Xia Lan, Gang Zhou and Xianwei Liu

Abstract Image is ubiquitous in modern communication. However, during the process like image acquisition and transmission, blur will appear on the reproduced image. This paper presents a time dependent model for image restoration based on the non-local total variation (TV) operator, which is robust to the noise and makes full use of the spatial information distributed in the different image regions. Experiment results demonstrate that the proposed model produces results superior to some existing models in both visual image quality and quantitative measures.

Keywords Image restoration · Inverse problem · Variational model · Nonlinear diffusion

1 Introduction

In many applications, such as microscopy imaging, remote sensing, and astronomical imaging, observed images are often degraded by blurring. Examples of the most common sources of blur are atmospheric turbulence, relative motion between a camera, and an object or wrong focus [1]. Restoration of the degraded image is an important problem because it allows the recovery of the lost information from the observed degraded image data. To this day, the restoration of

Z. Zuo (✉) · G. Zhou · X. Liu
National Key Laboratory of Science and Technology
on Multispectral Information Processing,
Huazhong University of Science and Technology,
Wuhan 430074, China
e-mail: zzy.iprai@gmail.com

X. Lan
School of Mathematics and Statistics,
Wuhan University, Wuhan 430072, China

F. Sun et al. (eds.), *Knowledge Engineering and Management*,
Advances in Intelligent Systems and Computing 214,
DOI: 10.1007/978-3-642-37832-4_20,

degraded images has been actively studied [2]. In most instances, the degraded process is assumed to be a linear shift invariant as follows:

$$g = h * f + n \quad (1)$$

where g, h, f, and n are the observed image, the point spread function (PSF) or blur kernel, the original image, and the zero mean Gaussian white noise with standard variance σ, respectively, and $*$ denotes 2-D convolution operator. It is well known that recovering the image with known PSF is a mathematically ill-posed problem [3]. A classical way to overcome this ill-posed problem is to add a regularization term. This idea was introduced in 1977 by Tikhonov and Arsenin [4]. The authors proposed the following deblurring and denoising model:

$$f_t = \Delta f - \lambda h(-x, -y) * (h * f - g) \quad (2)$$

where Δ is the Laplacian operator, and $\lambda > 0$ is the Lagrange parameter will control the trade-off between the smoothness of f and the goodness of fit-to-the-data.

Since the Laplacian operator has very strong isotropic smoothing properties and does not preserve edges, one should then propose anisotropic regularization in order to preserve the edges as much as possible. The popular and effective regularization approach is the total variation, which was first given by Rudin, Osher, and Fatemi in [5] as follows:

$$f_t = div\left(\frac{\nabla f}{|\nabla f|}\right) - \lambda h(-x, -y) * (h * f - g) \quad (3)$$

with $f(x, y, 0)$ given as initial data (used as initial guess the original blurred and the noisy image g) and homogenous Neumann boundary conditions. Here div and ∇ denote the divergence and the gradient, respectively. However, since the formula (3) is not well defined at points where $|\nabla f| = 0$, due to the presence of the term $\frac{1}{|\nabla f|}$, it is common to slightly perturb the TV functional to become:

$$f_t = div\left(\frac{\nabla f}{\sqrt{|\nabla f|^2 + \beta}}\right) - \lambda h(-x, -y) * (h * f - g) \quad (4)$$

with $f(x, y, 0)$ given as initial data and homogenous Neumann boundary conditions as above, and β is a small positive parameter. However, this method becomes highly ill-conditioned for deblurring case where the computational cost is very high and parameter dependent. Furthermore, this method also suffers from the undesirable staircase effect. In order to regularize the parabolic term, Marquina and Osher [6] proposed a new time dependent model as follows (denoted by NTV):

$$f_t = |\nabla f| div\left(\frac{\nabla f}{\sqrt{|\nabla f|^2 + \beta}}\right) - \lambda |\nabla f| h(-x, -y) * (h * f - g) \quad (5)$$

with $f(x, y, 0)$ given as initial data and homogenous Neumann boundary conditions as above.

In order to eliminate some noise before solving the above formula, Shi and Chang [7] proposed an improved time dependent model as follows (denoted by INTV):

$$f_t = |\nabla G_\sigma * f| div\left(\frac{\nabla G_\sigma * f}{\sqrt{|\nabla G_\sigma * f|^2 + \beta}}\right) - \lambda |\nabla G_\sigma * f| h(-x, -y) * (h * \nabla G_\sigma * f - g) \tag{6}$$

with $f(x, y, 0)$ given as initial data and homogenous Neumann boundary conditions as above. $G_\sigma(x, y) = \frac{1}{2\pi\sigma} e^{-(x^2+y^2)/2\sigma}$ and σ have the same definition as (1). However, these methods do not preserve the fine structures, details, and textures.

Recently, Lou et al. [8] proposed a non-local total variation operator. Since the nonlocal total variation exploits the correlation in the image, it has shown great promise and obtained the state-of-the-art results. Meantime, the abundant experimental results also show the non-local total variation operator outperforms most previous ones that it can suppress noise, while preserving the edge and fine details.

In this paper, based on the non-local total variation operator, we present an improved time dependent model for image restoration, and then three image quality assessment indices are applied to give an objective assessment of the restoration results.

The rest of this paper is organized as follows. In Sect. 2, we review the related work of nonlocal regularization and present the proposed time dependent model. Section 3 gives some experimental results and discussion. Finally, we conclude our work in Sect. 4.

2 The Proposed Model Via Non-Local Operator

Generally speaking, the information in a natural image is redundant to some extent. Based on this observation, Buades [9] initially developed a Non-Local means (NL means) image denoising algorithm which takes full advantage of image redundancy as follows:

$$NLf(x) = \int_\Omega w(x, y) f(y) dy \tag{7}$$

where the weight function $w(x, y)$ has the form

$$w(x, y) = \frac{1}{C(x)} \exp\left(-\frac{(G_a * |f(x + \cdot) - f(y + \cdot)|)(0)}{h^2}\right) \tag{8}$$

$$C(x) = \int_{\Omega} w(x,y)dy \tag{9}$$

where $C(x)$ acts as a normalization to ensure that a constant f is mapped to itself, and G_a is the Gaussian kernel with standard deviation a and h is a filtering parameter. In general, h corresponds to the noise level; usually we set it to be the standard deviation of the noise [9].

Due to its virtue of reducing the noise and preserving edge, Kindermann [10] incorporated the non-local means filter into the variational framework and formulated the non-local regularization function. However, these functions generally are not convex. Thus, Gilboa and Osher formalized the convex non-local function inspired from graph theory

$$U(f) = \frac{1}{2}\int_{\Omega\times\Omega} \phi(|f(x) - f(y)|)w(x,y)dxdy \tag{10}$$

where ϕ is a convex, positive function and the weight function $w(x,y)$ is non-negative and symmetric [11]. Moreover, based on the gradient and divergence definitions on graphs, Gilboa and Osher derived the non-local operators. Here, we review some definitions. Let $\Omega \subset \mathrm{R}^2$ and $x, y \in \Omega$., we define a local gradient $\nabla_w f : \Omega \to \Omega \times \Omega$

$$(\nabla_w f)(x,y) = (f(y) - f(x))\sqrt{w(x,y)} \tag{11}$$

Non-local divergence $div_w \vec{f} : \Omega \times \Omega \to \Omega$

$$\left(div_w \vec{f}\right)(x) = \int_{\Omega} (f(x,y) - f(y,x))\sqrt{w(x,y)}dy \tag{12}$$

Non-local Laplacian $\Delta_w f : \Omega \to \Omega$

$$(\Delta_w f)(x) = \frac{1}{2} div_w(\nabla_w f(x)) = \int_{\Omega} (f(y) - f(x))w(x,y)dy \tag{13}$$

Note that a factor $\frac{1}{2}$ is used to get the related standard Laplacian definition.

Then Lou [8] proposed the non-local TV operator for image recovery as follows:

$$J_{NL/TV}(f) = |\nabla_w f|_{L^1} = \int_{\Omega} \sqrt{\int_{\Omega} (f(x) - f(y))^2 w(x,y)dydx} \tag{14}$$

where $w(x,y)$ is the weight function shown in (8).

In this paper, based on the non-local TV operator, we presented a time dependent model as follows:

$$f_t = |\nabla G_\sigma * f| div_w \left(\frac{\nabla_w f}{\sqrt{|\nabla_w f|^2 + \beta}} \right) - \lambda |\nabla G_\sigma * f| h(-x,-y) * (h * f - g) \tag{15}$$

with $f(x, y, 0)$ given as initial data and homogenous Neumann boundary conditions as above, and $\lambda > 0$ is the Lagrange parameter will control the trade-off between the smoothness of f and the goodness of fit-to-the-data, and the G_σ has the same definition as (6).

3 Experiment Results

In this section, simulations are carried out to verify the performance of the proposed method (denoted by NNLTV). Three test images with size of 256×256, as shown in Fig. 1, are selected for the simulated experiments. The Gaussian blur, the Gaussian noise is generated using the Matlab "fspecial" command, and the Matlab "randn" command, respectively.

In order to save the computational time and improve the storage efficiency, we only compute the "best" neighbors and "best" patches, that is, for each pixel x, we only take a similarity window of size 3×3 and a semi-local searching window of size 21×21 centered at x in the formula (8).

The peak signal-to-noise ratio (PSNR), the structural similarity (SSIM) [12], and a promising recently proposed full reference image quality assessment index called FSIM [13] are used to evaluate the restoration results, which are considered very important because the quality assessments, just from the visual aspect, are used to illustrate the worth of the restoration results. To assess the relative merits of the proposed method, it is compared with TV [5], NTV [6], INTV [7], and NLTV [11] in the experiments.

The criterion $\|f^i - f^{i-1}\|_{L^2} / \|f^{i-1}\|_{L^2} < 0.001$ and the max iterative number set to 100 are used to terminate iteration. Since we try to get the larger values of PSNR, the iterative numbers may be different. The PSNR values, SSIM values, FSIM values, and the iterative numbers of the different algorithms are presented in Table 1. In order to give a visual impression about performances of the methods included in the comparison, the detailed regions cropped from the results of "Saturn" image are presented in Fig. 2.

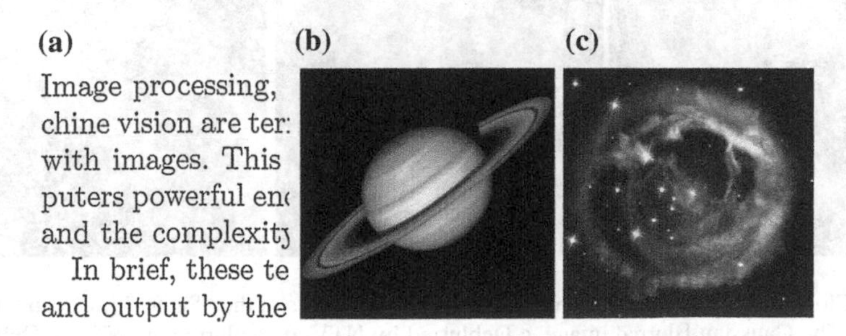

Fig. 1 The test images. **a** Word, **b** Saturn, **c** Light echoes

Table 1 The PSNR, SSIM, and FSIM values of different restoration methods in the experiments

	Assessment index	TV	NTV	INTV	NLTV	NNLTV
Word	PSNR	16.5631	16.5796	16.5816	20.0654	20.1515
	SSIM	0.6449	0.6530	0.6541	0.8573	0.8598
	FSIM	0.6280	0.6349	0.6373	0.8386	0.8410
	Iterative number	100	19	17	13	11
Saturn	PSNR	31.9487	32.8451	32.9815	35.8810	36.0650
	SSIM	0.9016	0.9124	0.9184	0.9667	0.9668
	FSIM	0.8974	0.9019	0.9073	0.9698	0.9705
	Iterative number	100	17	16	8	8
Light echoes	PSNR	31.1604	31.5782	31.5941	33.6707	33.6820
	SSIM	0.9096	0.9101	0.9128	0.9361	0.9372
	FSIM	0.9420	0.9439	0.9441	0.9587	0.9598
	Iterative number	100	14	13	9	8

Among the four restoration results in Fig. 2, it is shown that the proposed NNLTV method produces a better restoration result. Not only the detailed information is well preserved, but also the high-intensity noise is well suppressed in the homogenous regions. However, for the NTV and INTV methods, because the spatial information is not considered in the restoration process, they cannot suppress the noise and preserve the edge well.

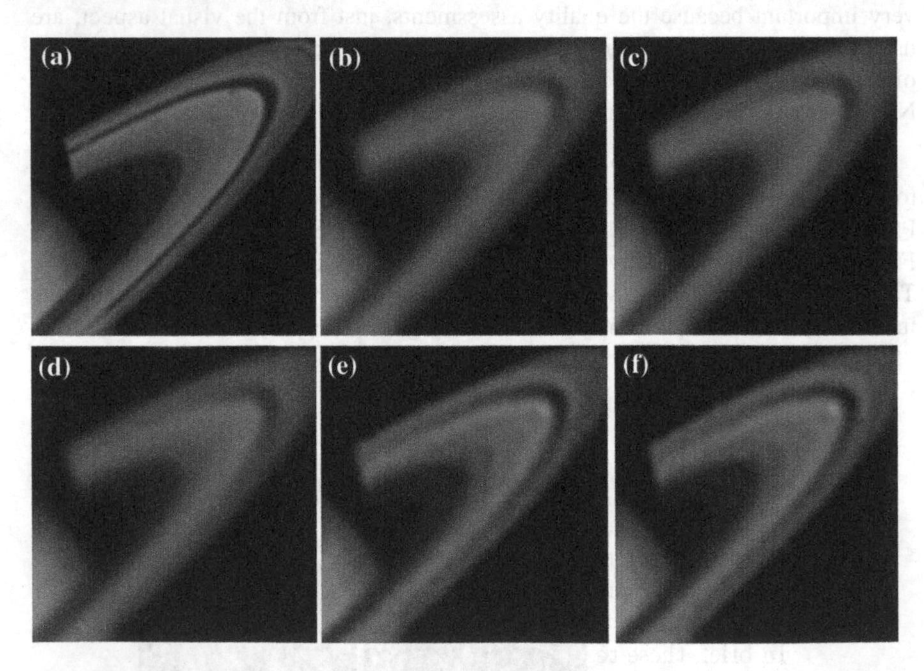

Fig. 2 Close-ups of selected sections of the restoration results of the "Saturn" image. **a** Original image, **b** The Gaussian Blurred image, **c** Deblurred by NTV, **d** Deblurred by INTV, **e** Deblurred by NLTV, **f** Deblurred by NNLTV

The good performance of the proposed NNLTV method can also be illustrated by PSNR, SSIM, and FSIM values presented in Table 1. It is shown that the proposed NNLTV method need the least iterative number except the Saturn image, but produces the highest PSNR value, and also has the highest SSIM and FSIM values, which illustrates that the proposed NNLTV method produces a better restoration result, close to the original image, both from gray value and image structure aspects.

4 Conclusion

In this paper, we have proposed a time dependent model to solve the image restoration problem based on the non-local total variation operator, which makes full use of the spatial information distributed in the different image regions. For the flat region, large total variation regularization is enforced to suppress noise; for the edge area, small total variation regularization is enforced to preserve the edge and detailed information. And three image quality assessment indices are given an objective assessment of the restoration results. The numerical experiment demonstrates the proposed time dependent converges quickly to the steady state solution, and achieves a better restoration effect in both visual image quality and quantitative measures compared with some of the state-of-the-art methods.

Acknowledgments The authors would like to thank Yifei Lou and Xiaoqun Zhang for supplying the Matlab implementation of their algorithm. This work was supported by the Project of the key National Natural Science Foundation of China under Grant No.60736010, No.60902060, and the Defense Advanced Research Foundation of the General Armaments Department of the PLA under Grant No.9140A01060110 JW0515.

References

1. Sroubek F, Flusser J (2005) Multichannel blind deconvolution of spatially misaligned images. IEEE Trans Image Process 14:874–883
2. Yan LX, Jin MZ, Fang HZ, Liu H, Zhang TX (2012) Atmospheric-turbulence-degraded astronomical image restoration by minimizing second-order central moment. IEEE Geosci Remote Sens Lett 9:672–676
3. Zhang K, Zhang TX, Zhang BY (2006) Nonlinear image restoration with adaptive anisotropic regularizing operator. Opt Eng 45:127004
4. Tikhonov AN, Arsenin VY (1977) Solutions of ill-posed problems. Winston and Sons, Washington, D.C
5. Rudin L, Osher S (1994) Total variation based image restoration with free local constraints. In: Procedings of the first IEEE ICIP, Austin, USA, pp 31–35
6. Marquina A, Osher S (2000) Explicit algorithms for a new time dependent model based on level set motion for nonlinear deblurring and noise removal. SIAM J Sci Comput 22:387–405
7. Shi YY, Chang QS (2006) New time dependent model for image restoration. Appl Math Comput 179:121–134

8. Lou YF, Zhang XQ, Osher S, Bertozzi A (2009) Image recovery via nonlocal operators. J Sci Comput 42:185–197
9. Buades A, Coll B, Morel JM (2005) A review of image denoising algorithms, with a new one. Multiscale Model Simul 4:490–530
10. Kindermann S, Osher S, Jones PW (2005) Deblurring and denoising of images by nonlocal functional. Siam Multiscale Model Simul 4:1091–1115
11. Gilboa G, Osher S (2008) Nonlocal operators with applications to image processing. Multiscale Model Simul 7:1005–1028
12. Wang Z, Bovik A (2009) Mean squared error: love it or leave it?—A new look at signal fidelity measures. IEEE Signal Process Mag 26:98–117
13. Zhang L, Zhang L, Mou X, Zhang D (2011) FSIM: a feature similarity index for image quality assessment. IEEE Trans Image Process 20:2378–2386

CBM: Free, Automatic Malware Analysis Framework Using API Call Sequences

Yong Qiao, Yuexiang Yang, Jie He, Chuan Tang and Zhixue Liu

Abstract Classic static code analysis for malware is ineffective when challenged by diverse variants. As a result, dynamic analysis based on malware behavior is becoming thriving in malware research. Most current dynamic analysis systems are provided as online services for common users. However, it is inconvenient and ineffective to use online services for the analysis of a big malware dataset. In this paper, we propose a framework named CBM enabling tailored construction of an automated system for malware analysis. In CBM, API call sequences are extracted as malware behavior reports by dynamic behavior analysis tool, and then API calls will be transformed to byte-based sequential data for further analysis by a novel malware behavior representation called BBIS. The peculiar characteristic of CBM is that it can be customized freely, contrary to current online systems, which supports local deployment and runs mass malware analysis automatically. Experiments were carried out on a large-scale malware dataset, which have demonstrated that CBM is more efficient in reducing storage size and computation cost while keeping a high precision for malware clustering.

Keywords Automatic malware analysis · Open-source · API-call sequences · Clustering · API-Hook

Y. Qiao (✉) · Y. Yang · J. He · C. Tang
National University of Defense Technology, Changsha 410073, China
e-mail: qiaoyong10@nudt.edu.cn

Y. Yang
e-mail: yyx@nudt.edu.cn

J. He
e-mail: hejie@nudt.edu.cn

C. Tang
e-mail: tangchuan@nudt.edu.cn

Z. Liu
China Navy Equipment Academy, Beijing 100161, China
e-mail: lstprince@163.com

F. Sun et al. (eds.), *Knowledge Engineering and Management*, Advances in Intelligent Systems and Computing 214,
DOI: 10.1007/978-3-642-37832-4_21, © Springer-Verlag Berlin Heidelberg 2014

1 Introduction

Malware, as malicious software, is the basement for attackers to implement intrusions and maintain them. Conventional methods disassemble the malware to carry out detailed analysis on the malware at the assembly code level. There are two prerequisites in such situation: one is that the researchers should have deep technical insights of assembly, the other is that the analysis processes should be efficient enough to cope with the constant renewal and variation of the malware. However, the widespread packing and obfuscation technology utilized by malware make it even more challenging.

Fortunately, dynamic analysis based on behaviors provides a new perspective to analyze malware, different from static code analysis, it runs malware in a controlled environment called *sandbox* and captures the behaviors triggered upon operation systems. With such technique, we can perform the malware analysis automatically at a large scale. Several systems have been implemented, such as *CWSandbox* [1], Anubis [2], Norman,[1] ThreatExpert,[2] et al. Usually, most of them provide free service for the submission and online analysis of malware binaries.

Unfortunately, source codes or packages for local installation are not available for those systems. Moreover, above tools have limitations on the number of submissions and the size of executable applications, which limits their usage in large-scale analysis. Therefore, in this paper we will introduce a locally deployable system for automatic malware analysis. This framework would enable a fully controllable analysis procedure under your control as our framework is based on two open source systems: *Cuckoo sandbox*[3]and *Malheur* [3]. Our contributions can be listed as follows:

- *Reasonable integration solution.* We proposed an automatic malware analysis framework CBM based on the improvement and integration of the existed open source system *Cuckoo sandbox* and *Malheur*. CBM can abstract the analysis reports from *Cuckoo sandbox* and encode the reports to sequential data for *Malheur* to perform clustering and classification analysis.
- *BBIS* (Byte-based Behavior Instruction Set) and CARL (Compression Algorithm of high Repeatability in Logarithmic level). We designed BBIS to transform *Cuckoo sandbox*'s analysis reports and make them recognizable by *Malheur*. In theory, BBIS can maintain a minimum size of reports while keeping the full messages needed. CARL can further compress the reports by means of reducing the high repeatability in API calling while keeping or improving the malware clustering performance.

[1] http://www.norman.com/

[2] http://threatexpert.com/

[3] http://www.cuckoobox.org/

- *Evaluation on a large-scale malware data set.* We have achieved bigger then 90% precision of clustering while using less computation time and less storage size.

2 Related Works

One of the first approaches for analysis and detection of malware was introduced by Raymod [4] in 1995. Malware binaries are manually analyzed by extracting static code signs, indicative for malicious activity. Those features are then applied for detection of other malware samples. This approach has been improved further by Christodorescu [5] and Preda [6] et al. and became a semantics-aware analysis method. On the attackers' side, Popov [7], Ferrie [8] et al. proposed obfuscation techniques to thwart static analysis. Although Martignoni [9],Sharif [10] have proposed several systems to generically unpack malware samples, human intervention is still needed.

Dynamic analysis of malware has attracted lots of attention recently. Multiple systems have been proposed, such as *CWSandbox* [1], *Anubis* [2], *BitBlaze* [11]. Those systems can execute malware binaries within an instrumented environment and monitor their behaviors for analysis and development of defense mechanisms. For further analysis, Konard [3] developed *Malheur* to cluster and classify malware by processing the malware behaviors, he employed the CWSandbox for monitoring malware behavios and represented the results in MIST [12] format, by means of n-grams algorithm and several related approaches. Malheur can classify the malware to a predefined set of classes and find novel classes by clustering. Unfortunately *CWSandbox* and MIST are not open-source, so we use *Cuckoo sandbox* and BBIS as the replacement.

Mamoun [13] and Xiaomei [14] adopted API call sequences to reflect malware system behaviors. We adopt the same methodology and it turns out to be feasible and efficient.

3 CBM: Build your Own System for Automatic Malware Analysis

Our system is named CBM since it consists of three major components: *Cuckoo Sandbox*[4], *BBIS* and *Malheur*. The relationship among the three modules is demonstrated in Fig. 1. The workflow is summarized below:

[4] We will use *Cuckoo* as the shorthand of *Cuckoo Sandbox* later in this article.

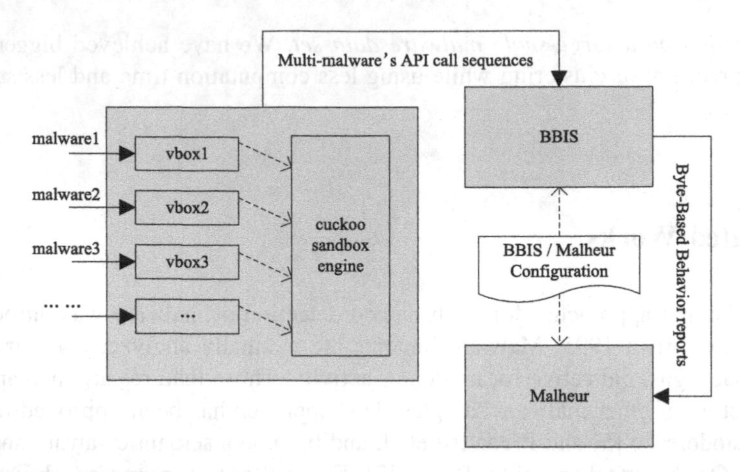

Fig. 1 The framework of CBM

1. CBM first executes and monitors multi malware binaries in Cuckoo simultaneously. Based on the analysis results, CBM extracts the API call sequences as each binary's behavior report.
2. CBM encodes the API call sequences to byte-based behavior reports using BBIS and CARL algorithms.
3. CBM uses Malheur to embed the sequential data in a high-dimensional vector space by n-gram algorithm, and then calculates the similarity of vectors, at last machine learning techniques for clustering and classification are applied on the vectors.

In the following sections, we will discuss each step in detail, including: how to improve Cuckoo's monitoring capabilities by adding API hooks, the design of BBIS and CARL, and the extensions in Malheur.

3.1 The Use and Improvement of Cuckoo

Cuckoo can deploy multi-VMs on one single host and run them simultaneously. This guarantees the efficiency of Cuckoo to analyze massive malware binaries quickly. Another advantage is that Cuckoo's new hook engine *cHook.dll*[5] embedded in *cmonitor.dll* has implemented a new technique called *trampoline* which can make the monitoring upon malware more difficult to be discovered. CBM use Cuckoo to obtain the malware binaries's behavior reports composed of API call sequences. A number of API hooks are needed to fetch the individual

[5] http://honeynet.org/node/755

Table 1 The differences between cuckoo and CWSandbox's hooking objects

	Cuckoo	CWSandbox
API Hooks	42	120
Categories	11	14
Winsock API	No	Yes
ICMP	No	Yes
SystemInfo	No	Yes

behaviors. We compared Cuckoo with CWSandbox and found that Cuckoo's monitoring scope is not comprehensive and should be improved.

Improvement in Cuckoo by adding Hooks. Cmonitior.dll is a kernel component of Cuckoo, which will be injected into malware's memory space to hook the original API calls for tracking. Herein, we improve Cuckoo's monitoring capabilities by adding API hooks in *cmonitior.dll. cmonitior.dll* is written in *CPP* language and compiled by *VisualStudio 2010*, in which 42 windows API functions were hooked and tracked. Thus, we can extract malware behaviors brought about by these API calls. Table. 1 lists the differences between Cuckoo and CWSandbox's hooking number and categories.

The number of hooks in Cuckoo are much less than that in CWSandbox with an approximate number of categories, which means that Cuckoo can hold the basic monitoring ability from 11 categories. However, Table. 1 also shows that Cuckoo does not set hooks in *Winsock API*, *ICMP API* and *SystemInfo API*. Since the malware like *bots*, *worms* et al. always have similar behavior in network communication, therefore, it is not wise to ignore those related API calls. In CBM, we have selected 18 additional API functions in these three aspects to be hooked.

To add hooks on API functions, specific hooking processes in *cmonitor.dll* need to clarified. Figure. 2 demonstrates the differences between regular API calls and API calls hooked by *comnitor.dll*.

We can see from Fig. 2 that *chook.dll* is responsible for the preservation of specific API functions' pointer, the transfer of the calls to new functions, and the return of the original results to invoker. We can add hooks in three steps: (1) Append real API function pointers for applying virtual memory space; (2) append customized functions to call the real API functions inside and record the corresponding message including timestamp, operation parameters, and returning results; (3) install hooks by invoking the output function *Hookattach()* from chook.dll. The space is limited to give adequate coverage of these detailed processes, we will publicize a full version report including the comprehensive solution.

Obtaining API call sequences. In Cuckoo's results folder, each process of malware was assigned a *csv* file that contains the detailed API call messages. However, CBM does not care about intricate things like the arguments, the process ID, the return value, and so on. CBM only extracts the lists of the API calls from all of the *csv* files that correspond to the process and concatenates them to be one sequence. CBM employs three rules to make the concatenation. First, the API call list within the same process are ordered by timestamp, with the earlier one coming

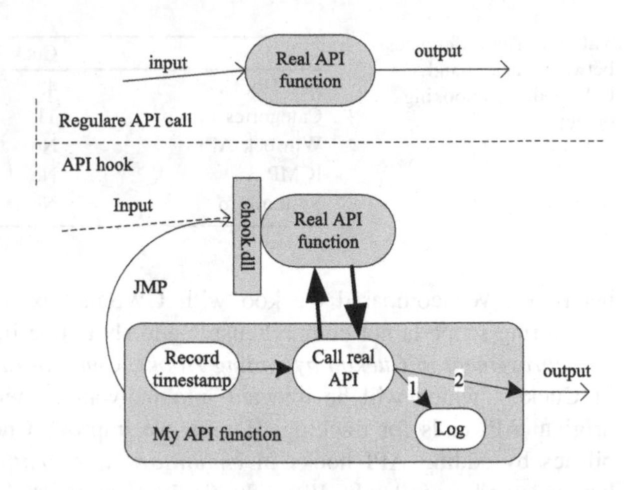

Fig. 2 Hooking processes in cmonitor.dll

first; second, API call lists from different processes will be connected from head to tail in the whole sequence; finally, different lists of multi-process will be ordered by the first timestamp of each API call list, with the earlier one coming first.

3.2 The Design of BBIS and CARL

BBIS is designed to transfer API call sequences to byte-based sequential data recognizable by Malheur in order to do clustering and classification.

Currently, Malheur only supports MIST [12] and byte-based sequential data. MIST can transfer the Malware original behavior reports to multi-level instructions, reflecting behaviors with different degree of granularity. For example, the category and operation are classified into the level 1 in MIST, followed by the arguments. Those considerations are good at recoding the whole messages of behaviors. However, testing results of Malheur showed that MIST with more levels consistently got worse results than with only one level in clustering and classification computations. Therefore, in CBM we only extract the API names and in BBIS we do not set multi-levels. We only need to build a suitable mapping table to change the API call sequence to a byte-based sequence. Byte-based here means each byte or fixed length of bytes will reflect one feature in the sequence. For instance, *abc* can reflect three API calls. In CBM, we utilize the following rules to build the mapping table.

Mapping rules of BBIS. There are two parameters that should be set up first. One is *IsVisible*, which indicates whether visible characters are used or not in the mapping table. The other *UpFeatures* is the upper limit of amount of unique features in all of the sequences. For example, Cuckoo has hooked 42 API functions by default, so the *UpFeatures* here is 42, CBM has improved the number of API hooks to 60, so the *UpFeatuers* will be 60 in CBM. Once the two parameters are specified, BBIS can build the mapping table in following way:

Table 2 Part of BBIS mapping table

API Function	BBIS-1	MIST	BBIS-2
LoadLibraryA	a	02 02 | ...	06
CreateFileW	b	03 01 | ...	08
RegQueryValueExW	p	09 05 | ...	2D
bind	#	12 06 | ...	68

(1) If *IsVisible* is TRUE, visible characters will be chosen to build the table. In the ASCII coding table, one byte can represent 94 different visible characters (from 33 to 126 in decimal system). When $UpFeatures <= 94$, we can select characters randomly with the number of *UpFeatures* to map the features in sequences. If $UpFeatures > 94$, BBIS need more bytes to map one feature, the number of bytes needed can be calculated by Eq. (1):

$$94^{n-1} < UpFeatures \leq 94^{n} \quad (1)$$

(2) If *IsVisible* is FALSE, It is not necessary to choose the visible characters to map the features. In such case, one byte can represent $2^8 = 256$ features. Consequently, in the map table, we can use the hexadecimal characters (from 00 to FF) to map the features and write the hexadecimal sequence to report file in binary mode. If $UpFeatures > 256$, BBIS also needs more bytes to map one feature by calculating Eq. (2):

$$256^{n-1} < UpFeatures \leq 256^{n} \quad (2)$$

Since in CBM, the $UpFeatures = 60$, which is less then 94, we can select the visible mode to build the mapping table. For example, the left column of Table. 2 demonstrates the part of mapping table we used in CBM. BBIS is a common way to represent features, not only designed for CBM. For example, MIST has 120 features, since $120 > 94$, thus we can use 2 bytes to represent MIST code, just like the right column of Table. 2 which uses the 2 hexadecimal characters to map the MIST code, it is interesting that the 2 hexadecimal characters can also be written to file in binary mode within one byte size if *isVisible* is set to *FALSE*.

CARL - Compression Algorithm of high Repeatability in Logarithmic level. During the process of dynamic analysis of malware, one notable thing is that some API functions can be continuously invoked thousands of times in a single analysis. Usually, this can be attributed to mistakes in coding or unsuitable execution environment for the malware. The API functions like *ReadFileW*, *VirtualAllocEx*, *RegOpenKeyW* are found many times with such situations. However, if parts of malware in the family have high repeatability problem, the similarity between malware will be greatly affected and much more redundant computation might be introduced. Therefore, we proposes an algorithm called CARL in CBM to reduce the repeatability.

Definition 1 REPEAT. For a sequence *Seq* and a subsequence *SSeq*, if *SSeq* is composed of consistent characters and the length of it is more than one, we define the length as the REPEAT of the *SSeq*.

Definition 2 SN and NL. For a sequence *Seq* and a subsequence *SSeq*, *SN* is a start number to run the CARL process, *NL* is the new length of *SSeq*. If the *REPEAT* of *SSeq* is bigger then *SN*, the length of *SSeq* will be reduced to *NL* by Eq. (3).

The calculation of CARL is illustrated in the following Equation:

$$NL = SN + \beta \cdot Round(log_{\gamma}(REPEAT - SN)) \tag{3}$$

In Eq. (3), β and γ are used as the adjustment coefficients. Usually, γ is set as 2, β is set as 1. For example, if $S = \{aQcQQQQQQQQQQQQQddhhhdddd\}$, we can get sub-sequences:$SSeq1 = \{QQQQQQQQQQQQQ\}$, $SSeq2 = \{dd\}$, $SSeq3 = \{hhh\}$, $SSeq4 = \{dddd\}$, which REPEAT are bigger then 1. If we define $SN = 5$ here, the only subsequence that needs to be reduced is $SSeq1$, we can see the REPEAT of $Seq1$ is 13, so we bring the parameters to Eq. (3) and get the new length of $Seq1$ as: $NL\_SSeq1 = 5 + Round(log_2(13 - 5)) = 5 + 3 = 8$. In this case, the length of this subsequence has changed from 13 to 8, which is not an obvious reduction. However, if the REPEAT is a big number like 1000, the new length of the subsequence will be $NL\_SSeq = 5 + Round(log_2(1000 - 5)) = 5 + 10 = 15$ with a huge amount of reduction.

3.3 The Use and Improvement of Malheur

CBM uses Malheur to perform clustering and classification analysis. Malheur can embed the byte-based sequential data in a high-dimensional vector space using n-gram algorithm, and then extract the prototypes for clustering and classification. Konrad et al. [3] has introduced brief details of the algorithms related to Malheur. Here we focus on the modification of Malheur. CBM has made two changes in Malheur. First, to accommodate the features of multi-byte, CBM improves the function in Malheur to parse the contents from sequential data. Second, in order to use historical data, CBM adds database function to Malheur for analysis upon global malware reports. Once the database system is set up, Malheur can analyze single behavior based on historic data. Limited to the space, we will talk about this part in detail in our public report.

4 Empirical Evaluation

We carried out a systematic evaluation of CBM. For this evaluation, we consider a total number of 3131 malware binaries obtained from the website http://pi1.informatik.uni-mannheim.de/Malheur/. The malware binaries have been collected over a period of three years via various sources, including honeynets, spamtraps, anti-malware vendors and security researchers. Those binaries have also been assigned a known class of malware by the majority of six-independent anti-virus products.

In the following experiments, the 3131 malware binaries are executed and monitored using Cuckoo, CuckooEx[6] individually. We compare CBM to the state-of-the-art analysis framework CMM(*CWSandbox/MIST/Malheur*) throughout our experiments.

In the following experiments, we will evaluate clustering results of CBM. To assess the performance of clustering, we employ the evaluation metrics of *precision* and *recall* [3]. The precision P reflects how well individual clusters agree with malware classes and the *recall* R measures to what extent classes are scattered across clusters. Formally, we define precision and recall for a set of clusters C and a set of malware classes Y as:

$$P = \frac{1}{n}\sum_{c \in C} \#c \qquad \textit{and} \qquad R = \frac{1}{n}\sum_{y \in Y} \#y \tag{4}$$

where $\#c$ is the largest number of reports in cluster c sharing the same class and $\#y$ the largest number of reports labelled y within one cluster. Consequently, the goal is to seek an analysis setup which maximizes precision and recall. An aggregated performance score is adopted for our evaluation, denoted as *F-measure*, which combines precision and recall. A perfect discovery of classes yields $F = 1$, while either a low precision or recall results in a lower *F-measure*.

$$F = \frac{2 \cdot P \cdot R}{P + R} \tag{5}$$

Experiment 1: Comparisons of CBM and CMM

Results for the evaluation of clustering are presented in Fig. 3, with CBM using Cuckoo yields an best F-measure 88.4% corresponding to a discovery of 28 known malware classes in the malware data set while using CuckooEx yields a best F-measure 90.9% corresponding to a discovery of 25 known malware classes very close to the real 24 classes. Compared with CBM, CMM performs better with a best F-measure 95.0% corresponding to a discovery of 24 malware classes. But we find that CBM using CuckooEx can achieve a better clustering performance even

[6] In the following experiments,we use *CuckooEx* to represent the improved *Cuckoo*.

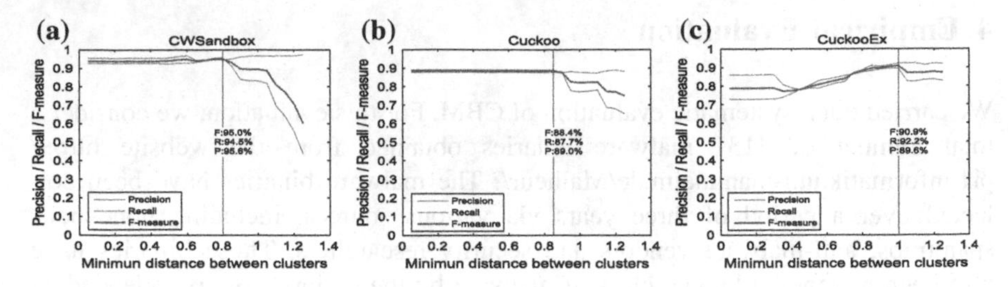

Fig. 3 Clustering result

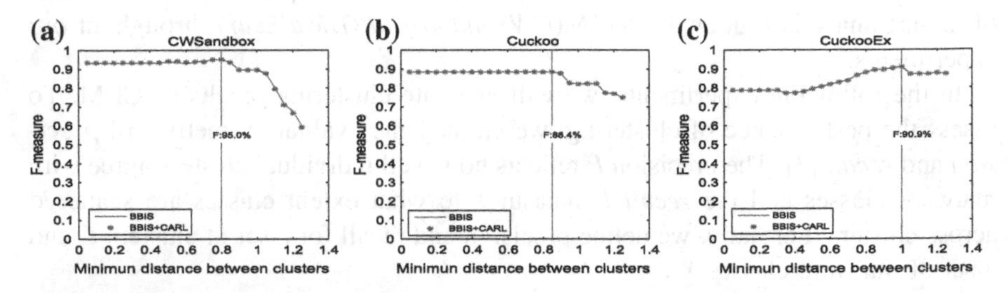

Fig. 4 Comparisons of BBIS and (BBIS + CARL) for clustering

the minimum distance between clusters is bigger than 1. Overall, CBM can achieve a competitive performance compared with CMM and it is more robust.

Experiment 2: Comparisons of BBIS and (BBIS + CARL) for clustering

In this experiment we evaluate the performance of clustering by using BBIS with or without CARL algorithm. Figure. 4 shows that all three systems have achieved nearly the same results by using CARL to compress the behavior reports. It seems that CARL is an effective way to reduce the computation cost by reducing the size of reports while keeping the performance of clustering.

Experiment 3: Time consumption

From the experiments, CBM uses 8.9 s per clustering on average while 11.6 s for CMM, which demonstrates that CBM is more efficient then CMM at time consumption.

Experiment 4: The storage size of Reports

From the statistic of the storage size of reports from above experiments, we get that BBIS can approximately reduce the MIST reports size to a 15% proportion and CARL algorithm can further reduce the size by half, which demonstrates that BBIS can reduce the size of behavior reports to a large extent, which is a huge advantage over MIST in terms of storage and computation. Moreover, CARL is useful to further reduce the size while keeping the inherent features of reports as testified by the clustering results illustrating in experiments 2.

5 Conclusion

In this article, we have introduced a framework named CBM, which utilizes and connects the open-source software tools to construct an automated malware analysis system. In CBM, we use Cuckoo to extract the API call sequences obtained from monitoring malware dynamic execution and create a new representation method of malware behaviors called BBIS to convert the API calls to byte-based sequential data. We have introduced how to improve Cuckoo's monitoring capability by appending new API hooks. The CARL algorithm proposed for reducing the high repeatability in API call sequences can effectively reduce the computation cost without a significant loss in performance. Serious experiments demonstrate that our framework is a competitive alternate to the state-of-the-art analysis framework CMM and easy to be realized. However, the main advantage of CBM is its ability in supporting local deployment. We hope CBM can replace the non-open source online services in large-scale malware dynamic analysis. CBM needs to be improved in several aspects including the monitoring range, the stability to various malware, and so on.

Acknowledgments This work was supported by NSFC under grants No. 61170286 and No.61202486.

References

1. Willems C, Holz T, Freiling F (2007) Toward automated dynamic malware analysis using cwsandbox. IEEE Security and Privacy, IEEE Computer Society, pp 32–39
2. Bayer U, Moser A, Kruegel C, Kirda E (2006) Dynamic analysis of malicious code. J Comp Virol, Springer, 2(1): 67–77
3. Rieck K, Trinius P, Willems C, Holz T (2011) Automatic analysis of malware behavior using machine learning. J Comp Virol, IOS Press, 19(4): 639–668
4. Lo RW, Levitt KN, Olsson RA (1995) MCF: A malicious code filter. Computers and Security, Elsevier, 14(6): 541–566
5. Christodorescu M, Jha S (2006) Static analysis of executables to detect malicious patterns. DTIC Document
6. Preda MD, Christodorescu M, Jha S, Debray S (2007) A semantics-based approach to malware detection. ACM SIGPLAN Notices, ACM, vol 24. pp 377–388
7. Popov IV, Debray SK, Andrews GR (2007) Binary obfuscation using signals. Proceedings of 16th USENIX Security Symposium on USENIX Security Symposium, USENIX Association, pp 19
8. Ferrie P (2009) Anti-unpacker tricks 2 part seven. June
9. Martignoni L, Christodorescu M, Jha S (2007) Omniunpack: Fast, generic, and safe unpacking of malware. Computer Security Application Conference, 2007. ACSAC 2007. Twenty-Third Annual, IEEE, pp 431–441
10. Sharif M, Lanzi A, Giffin J, Lee W (2009) Automatic reverse engineering of malwar emulators. Security and Privacy, 2009 30th IEE Symposium on, IEEE, pp 94–109
11. Song, D, Brumley D, Yin H, Caballero J, Jager I, Kang M, Liang Z, Newsome J, Poosankam P, Saxena P (2008) BitBlaze: A new approach to compute security via binary analysis. Information Systems Security, Springer, pp 1–25

12. Trinius P, Willems C, Holz T, Rieck K (2009) A malware instruction set for behavior-based analysis. Technical Report TR-2009-005, University of Mannheim
13. Alazab M, Venkataraman S, Watters P (2010) Towards understanding malware behaviour by the extraction of API calls. Second cybercrime and trustworthy computing workshop, pp 52–59
14. Dong X, Zhao Y, Yu X (2012) A Bot Detection Method Based on Analysis of API Invocation. Recent Advances in Computer Science and Information Engineering, Springer, pp 603–608

Shop&Go: The Routing Problem Applied to the Shopping List on Smartphones

Inmaculada Pardines and Victoria Lopez

Abstract Mobile applications have achieved a great development in the past years. Some of the existing tools are useful for carrying out daily tasks. In this regard, we propose in this paper a mobile application, called Shop&Go, to do the shopping in a supermarket covering the shortest distance. This kind of problems is a classic paradigm from the field of operational research. We propose two heuristics to obtain the optimum path to do the shopping. The distances and the products layout are obtained from a map previously provided by the supermarket. We have compared both heuristics, proving that any of them can achieve the best solution depending on which products are in the shopping list. Therefore, we have decided to include both algorithms in our tool, leaving to the user the decision of which of the solution paths to take. The application has also other functional requirements which are described in the paper. Simulations of how the tool is working over an HTC Desire mobile are shown.

Keywords Mobile applications · Routing problems · Android operating system · Shopping list · Optimum path

1 Introduction

Nowadays mobile technology is under constant development. There are a lot of applications that allow users to carry out different tasks. Smartphones have become an essential tool in our lives. It is not only used to communicate by a voice

---

I. Pardines (✉) · V. Lopez
Mobile Technologies and Biotechnology Research Group (G-TeC),
Department of Computer Architecture, Universidad Complutense,
Profesor José García Santesmases s/n 28040 Madrid, Spain
e-mail: inmapl@dacya.ucm.es

V. Lopez
e-mail: vlopez@fdi.ucm.es

F. Sun et al. (eds.), *Knowledge Engineering and Management*,
Advances in Intelligent Systems and Computing 214,
DOI: 10.1007/978-3-642-37832-4_22,

call or a SMS, but for executing a great number of applications such as: games, music, e-mail, MMS, PDA, digital photograph and digital video, video call, Internet, and even digital television.

In the beginning, these applications were made for entertainment whereas in the present they develop more useful services as street maps, product guides, latest news, etc. The most innovate applications are related to Augmented Reality [1, 2]. This technology defines a vision of the world as a combination of the real scene viewed by the user and a virtual scene generated by the computer. Moreover, all elements of the scene are augmented with additional information (sound, video, graphics, or GPS data).

Due to the great success of mobile applications, the main mobile phone companies have created portals where users can buy or download any kind of tools. Among the most important ones are the iTunes Store from Apple [3] and the Android Market from Google [4], but there are also others like Ovi Store from Nokia [5] or App Place from Toshiba [6].

In our working group, we develop mobile applications that try to make people's life easier with the aid of a mobile. The work presented in this paper is focused in this idea. We propose a new mobile tool (called Shop&Go) which optimizes the route, and so on, the time necessary to do the shopping in a supermarket. There were some studies about this subject but not applied to mobiles [7]. Moreover, there are Android applications related to the shopping list. Some examples are Hungry! Shoplist [8] and ShopSavvy Barcode Scanner [9]. The first one allows to write and to save the shopping list, to organize the different products by colors, and to send this list by e-mail or SMS. The latter one is able to identify a barcode using the mobile camera and to provide a price, and store lists.

Our application has been developed for the Android operating system [10, 11] in the Java programming language [12]. The Shop&Go tool executes more of the tasks of the Hungry! and ShopSavvy Barcode Scanner applications and also optimizes the shopping route in a supermarket. The idea is that the supermarket provides the Shop&Go application to their customers. They download the tool in their mobile with a distribution map of the products. In this way the customer will be able to do the shopping in a simpler way and in less time.

The paper is organized as follows. In Sect. 2 the routing problem optimization and the method used to solve it are described. The tool requirements are specified in Sect. 3. Different examples to show how works our mobile application is shown in Sect. 4. Finally, the main conclusions from this work are discussed in Sect. 5.

2 Routing Problems: The Shopping List

The routing problems belong to the combinatorial optimization area [13], and they can be applied to solve many problems. They are usually modeled by graphs that are made up of both a finite vertices (nodes) and edges set. The objective of these

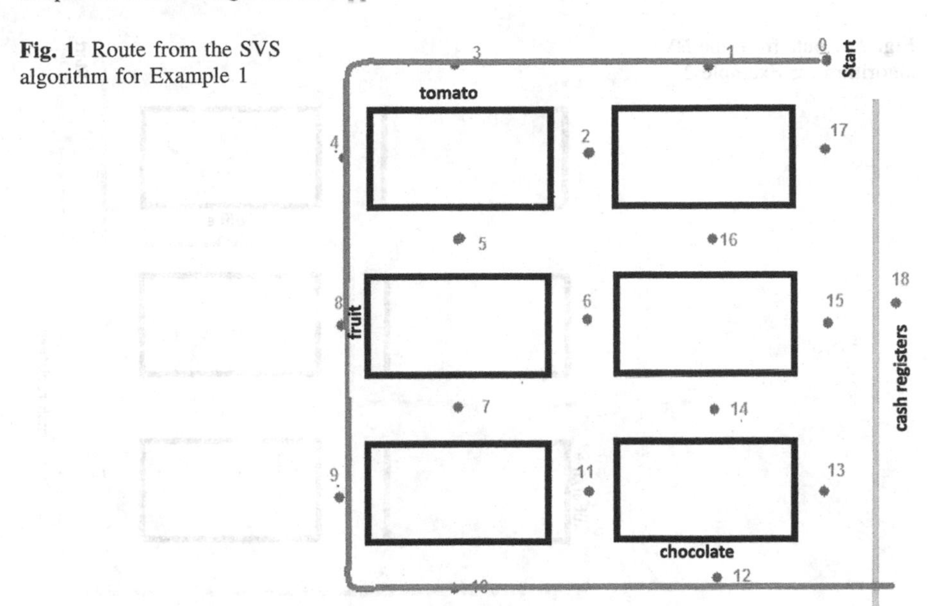

Fig. 1 Route from the SVS algorithm for Example 1

problems is to cover part or all the vertices of the graph using the minimum possible path.

The problem studied in this paper is a routing problem that tries to optimize doing the shopping in a supermarket. Therefore, we have simulated a supermarket map based on a real map supplied by El Corte Inglés, an important Spanish chain of supermarkets [14]. The map is made up of six blocks with shelves. Three vertical corridors and four horizontal ones separate the six blocks as we can see in Figs. 1 and 2. In the sketch we represent the corridors as numbered nodes and the blocks as rectangles. We have decided to number the vertices in a way that the lower ones are assigned to the shelves sited far away from the cash registers.

Following we describe the two heuristic methods proposed to solve the shopping problem. Both of them are based in the Dijkstra algorithm [15].

2.1 *Sorted Vertex Search Algorithm (SVS Algorithm)*

This algorithm is based on the nodes numbering described before. The idea is that the distance covered by the customer to make the shopping is the shortest if the products close to the cash registers are taken as the last ones.

The proposed method can be considered as a change of the Dijkstra algorithm. This algorithm obtains the shortest path from an origin vertex to every vertex of a graph with nonnegative path costs. Our algorithm sorted the vertices to cover from the lowest to the highest. After that it uses the Dijkstra matrix to find the shortest path to go from a node to the next one. The proposed heuristic used two sets:

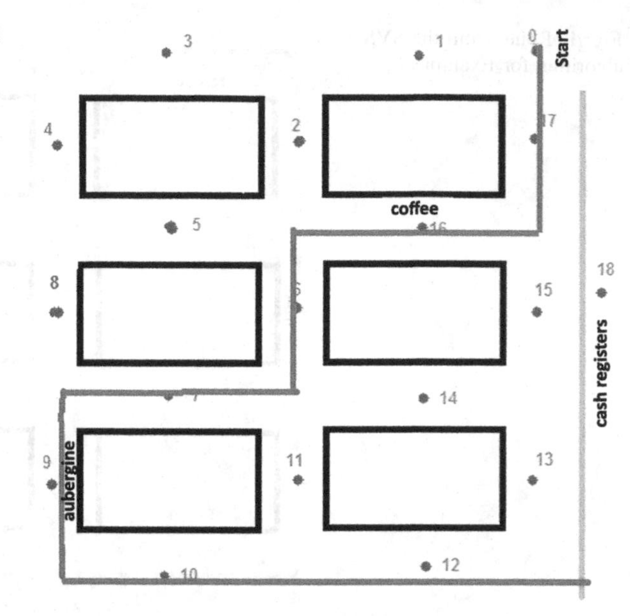

Fig. 2 Route from the NVS algorithm for Example 2

- *Solution* (*S*): Contains vertices that the customer has to go to do the shopping.
- *List* (*L*): Stores nodes that have products from the shopping list.

Once defined these sets, we can describe the steps followed by our algorithm:

1. The S set is initialized with the origin node 0.
2. The L set is created by searching in a database which node is associated to each product of the shopping list. It is necessary to include in the set the cash register node. Then, these vertices are sorted from low to high.
3. This loop is executed until the L set is empty: the heuristic finds the shortest path between the last node of S and the first node of L using the Dijkstra matrix. This node of L is removed and the nodes that belong to this path are included in the S set.

Following we propose the Example 1 which emphasizes the heuristic behavior. The shopping list is made up of tomatoes, fruit, and chocolate. These products are arranged in shelves as shown in Table 1.

The heuristic starts from the sets: $S = \{0\}$ and $L = \{3,8,12,18\}$, where the node 18 corresponds to the cash registers.

Table 1 Shopping list of Example 1

Product	Vertex
Tomato	3
Fruit	8
Chocolate	12

First, the shortest path between nodes 0 and 3 is searched. This path is through node 1, so $S = \{0,1,3\}$. Node 3 is removed from L. Then, the optimum path between nodes 3 and 8 is found through node 4. The process continues in this way until L is empty. Finally, the Solution Set is $S = \{0,1,3,4,8,9,10,12,18\}$ and the obtained route is shown in Fig. 1.

2.2 *Nearest Vertex Search Algorithm (NVS Algorithm)*

The decision to establish the optimum path in this heuristic is based on the idea that the customer moves always from a node to the closest one which is related to a product of the shopping list.

The decision to establish the optimum path in this heuristic is based on the idea that the customer moves always from a node to the closest one which is related to a product of the shopping list.

We use the same sets defined in the previous subsection. The first two steps of the algorithm can be described by the following rules:

1. The S set is initialized with the origin node 0.
2. The L set is created by searching in a database which node is associated to each product of the shopping list. It is necessary to include in the set the cash register node.
3. This loop is executed until the L set is empty: the distance from the current vertex to every node of L is found. The node with the shortest distance is selected as the next one to move. This node is included in the Solution set and removed from the List set.

As in the SVS algorithm, we propose Example 2 to describe the heuristic. The shopping list is made up of coffee and aubergines. These products are arranged in shelves as Table 2 shows.

The heuristic starts from the sets: $S = \{0\}$ and $L = \{16,9\}$. In this case, the L elements are not numbering sorted; they are saved as the product appears in the shopping list.

Initially, the distance between node 0 and every node of L is calculated. For doing this we use the distance between shelves from the map provided by the supermarket. In our example node 0 is 5 m apart from node 16 and 13.5 m apart from node 9. So, the heuristic selects vertex 16 as the next one to go, adding it to

Table 2 Shopping list of Example 2

Product	Vertex
Coffee	16
Aubergines	9

the S set and removing it from L set. As in the SVS algorithm, the Dijkstra matrix is used to calculate the minimum path between two vertices.

Then, the selected node in the current iteration is used as the starting point of the next one, calculating the distances from every node of L to it. The vertex with the low distance will be the next one to move. The process continuous in this way until L is empty. When all the nodes of the shopping list have been covered, the algorithm searches for the best path from the last node of S to the cash registers. Finally, the Solution Set is $S = \{0,17,16,6,7,9,10,12,18\}$ and the obtained route is shown in Fig. 2.

2.3 Comparative Between the SVS and the NVS Algorithms

The SVS and the NVS algorithms are heuristics, so it cannot assure that the best solution is achieved in all cases. Following, we describe two examples to show that this assertion is right (see Table 3).

Applying the SVS algorithm to the example 3, the customer will have to pass the nodes 0, 1, 3, 2, 5, 8, 9, 10, 12, 11, 6, 16, 18, in this order. The total distance associated with this path is 36 m. For this example the NVS heuristic achieves a better solution with a total distance of 27.5 m and a route which the vertices follows the order 0, 17, 16, 5, 8, 4, 3, 2, 6, 11, 12, 18.

However, for the example 4 the SVS heuristic achieves the best solution with a total distance of 22 m ($S = \{0,1,3,4,5,16,18\}$) whereas the NVS algorithm obtains a path of 24.5 m long ($S = \{0,17,16,2,3,4,5,16,18\}$).

As we can see the selected combination of the products in the shopping list will determine which heuristic will be the optimum one. So, we have decided to include both algorithms in the Shop&Go mobile application. Moreover, knowing two possible paths a priori the customer can decide which is the most interesting route for him. Perhaps there were in any path other products, which being out of the shopping list, he likes.

Table 3 Shopping list of Examples 3 and 4

List of Example 3		List of Example 4	
Product	Vertex	Product	Vertex
Salad	3	Cereals	16
Rice	5	Legumes	4
Fruit	8	Tomatoes	3
Chocolates	12		
Honey	16		

3 Shop&Go Requirements Specification

The Shop&Go mobile application offers to the user a great variety of functional requirements which are described as follows:

- **Search a product in the supermarket.** The customer can select a product looking for it in a shopping list or writing their name in the corresponding field. Then, the application shows the product location pointing out in a map.
- **List creation and edition.** Users can create lists, give then a name, and add them a description. The new lists will be permanent. Once they have been created, they are saved for next searches in a new execution of the application in the future.
- **List removing.** The user can remove an existing shopping list at any time.
- **Search the price of a product.** After the customer selects the products of his shopping list, the application shows the prices of all these products.
- **Send a shopping list by e-mail.** Once a shopping list is created, it can be sent by e-mail and another person can do the shopping.
- **Selection between two heuristics.** The user can select between the solution offered by the SVS heuristic and the solution achieved by the NVS algorithm. The routes provided by this two heuristics are usually different. Our mobile application displays the path to cover and its total distance for both heuristics. Then, it is the customer according to his preferences who decides which route is he going to select.
- **Display the route in a map.** The user can see in a map the route which has to cover to purchase all the products of a shopping list. Pressing a product after taking it, removes it from the list.
- **Show the covered distance.** The map provides additional information about the distance to cover.
- **Display the route in text.** The route can also be seen in text format, with simple instructions about how to advance, straight ahead, turn to the right or to the left, for example.
- **Voice routing.** The Shop&Go mobile application provides the possibility of listening the routing instructions through the mobile loudspeakers. In this way the application is very useful for people with visual discapacity.

Apart from the previous requirements, related to the application functionality, there are others related to evaluate the tool quality from a technical point of view. In this regard, we expect that our application will be scalable, with high performance, very intuitive to use, and small. These objectives seem to be achieved from the obtained results of our tests.

4 Simulations

We have tested the Shop&Go application in a HTC Desire mobile with an operating system Android 2.1 *Éclair*. It is a powerful Smartphone with a 1 GHz *Snapdragon* processor, 512 MB of ROM and 576 MB of RAM. It can execute several applications simultaneously and it can browse the Internet very fast. Moreover, it has a 3.7 inch screen, which allows a good display of the shopping routes created by our application.

At present the Shop&Go can be installed from a CD, but in the future will be downloaded from the Web page of the supermarket or from a bidi code. Once it is installed in our mobile for executing the application it is enough to click on its icon. Figure 3 shows the welcome screen of the tool.

The Shop&Go application has implemented all the functionalities defined in Sect. 3. Among then the most important ones are the search of a certain product and the display of the shortest path to do the shopping. Therefore, we show the behavior of our tool in these cases.

To search a product we have to press the icon *Search a product* in the initial screen (see Fig. 3). Then, the interface of Fig. 4a will appear in the mobile. We have to introduce the name of the product and press the icon *Display Map*. Following, the supermarket map is displayed with a thick dot point to the shelves where the product is (Fig. 4b).

Fig. 3 Initial screen of Shop&Go

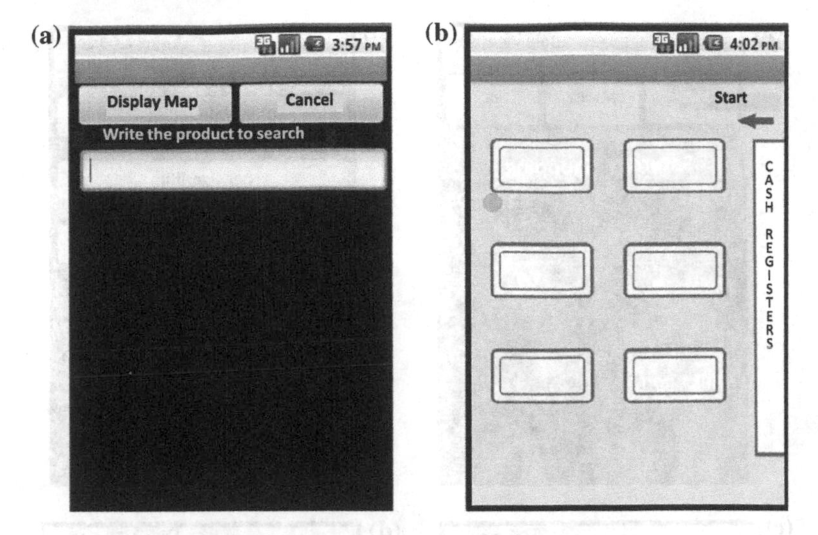

Fig. 4 Search and localization of a product in the supermarket map

In the case that we wish to know which one is the shortest path to do the shopping, we press the *Lists* icon from the initial screen. If the list is already saved in the mobile, we select it, but if not we have to create a new list. Once the content of the list is displayed, the *Display Route* icon must be clicked (see Fig. 5a). Then, it appears as the interface shown in Fig. 5b. The user can choose how the path will be computed. He can see the solutions of the SVS and the NVS heuristics and after that he decides which option is the best for him, or directly selects the best route to do the shopping.

Pressing one of these three options, the application displays a map. The path to take is pointed out as we can see in Fig. 5c. The products to buy are also indicated in the screen together with the order in which they must be taken. Finally, the total distance that the user has to cover is shown in the bottom left on the screen.

The *Menu* button of the mobile opens the popup of Fig. 5d. If the user wishes to exit he must press the corresponding icon. He has also the possibility of asking for the instructions about how to do the shopping, either by text or by voice.

These simulations prove that the Shop&Go application is very friendly and easy to use. Moreover, it works quickly, which is important for its successful marketing.

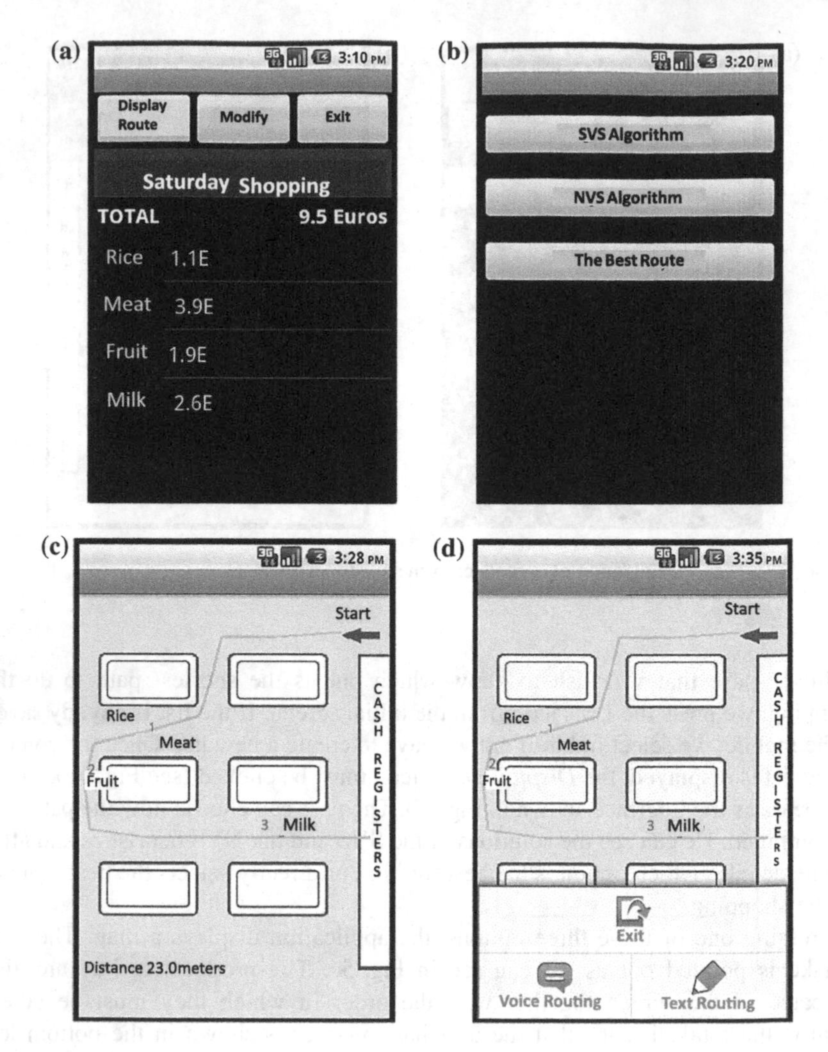

Fig. 5 Different stages in the search of the shortest path to do the shopping

5 Conclusions

A mobile application to do the shopping in an efficient way is presented in this paper. Two heuristics have been proposed to find the shortest route for taking the products of a shopping list. We have demonstrated that it is impossible to assure which heuristic is the best one. Therefore, we have decided to include the two algorithms in the Shop&Go application. Our tool has also other functional requirements such as creating or deleting lists, searching for a product price, locating a product in the supermarket map, and so on. Several simulations that prove the high performance of the proposed application are shown. However, despite its good quality, this application has a difficult marketing. The idea of doing the shopping in a short time benefits the customers but it is opposed to the interest of the supermarkets. These businesses frequently change the products layout to force the customer to go through many corridors, increasing the probability that the customer buys more products than he needs.

References

1. Wagner D, Reitmayr G, Mulloni A, Drummond T, Schmalstieg D (2010) Real-time detection and tracking for augmented reality on mobile phones. IEEE Trans Visual Comput Graph 16(3):355–368
2. Henrysson A, Ollila M (2004) UMAR: ubiquitous mobile augmented reality. In: Proceedings of 3rd international conference on mobile and ubiquitous multimedia (MUM '04). ACM, New York, pp 41–45
3. Itunes Store de Apple. http://www.apple.com/es/itunes/
4. Ganapati P (2010) Independent app stores take on Google's Android Market. Wired News. Available at SSRN: http://www.wired.com/gadgetlab/2010/06/independent-app-stores-take-on-googles-android-market/
5. Ovi Store de Nokia. http://www.store.ovi.com
6. App Place de Toshiba. http://apps.toshiba.com/
7. Hui SK, Fader P, Bradlow E (2008) The traveling salesman goes shopping: the systematic deviations of grocery paths from TSP-optimality. Available at SSRN: http://ssrn.com/abstract=942570
8. Hungry!. http://uk.androlib.com/android.application.com-xta-foodmkt-xzz.aspx
9. ShopSavvy Barcode Scanner. http://shopsavvy.softonic.com/android
10. Haseman C (2008) Android essentials. Books for professionals by professionals. First Press, New York
11. Gargenta M (2011) Learning android. O'Really Media, Inc., Sebastopol
12. Arnold K, Gosling J, Holmes D (2005) Java programming language. Java series. Addison Wesley Professional, Boston
13. Schrijver A (2002) Combinatorial optimization-polyhedra and efficiency. Springer, Berlín
14. El Corte Inglés Supermarket. https://www.elcorteingles.es/supermercado/sm2/login/login.jsp
15. Cormen TH, Leiserson CE, Rivest RL, Stein C (2001) Dijkstra's algorithm. In: Cormen TH, Leiserson CE, Rivest RL, Stein C (eds) Introduction to algorithms. MIT Press and McGraw-Hill, Cambridge, pp 595–601

A Node Importance Based Label Propagation Approach for Community Detection

Miao He, Mingwei Leng, Fan Li, Yukai Yao and Xiaoyun Chen

Abstract Community detection provides an important tool to get a deeper insight about the various existing real-world social networks. A large amount of algorithms for detecting community in social networks have been developed in recent years. Most of these algorithms have high computational complex and expensive time-consuming which result that they are not scalable for large-scale networks, and some of them cannot find stable communities. In this paper, we take the importance of node into consideration and give a score to quantize the importance of nodes. Based on this importance quantification, we propose a novel node importance based label propagation algorithm for community detection. We implement our algorithm and other compared algorithms on several benchmark networks. And the experimental results show that our algorithm performs better than others.

Keywords Community detection · Social network · Label propagation approach · Node importance

1 Introduction

When we take the notion of network where individuals are described as nodes and their interactions as edges directed or undirected, we can treat a wide variety of now existing systems as networks. Due to the networks complexity and large scale, more and more researchers in recent years focus on understanding the evolution

M. He · M. Leng · F. Li · Y. Yao · X. Chen (✉)
School of Information Science and Engineering, Lanzhou University,
730000 LanZhou, China
e-mail: chenxy@lzu.edu.cn

M. He
e-mail: hem2010@lzu.edu.cn

F. Sun et al. (eds.), *Knowledge Engineering and Management*,
Advances in Intelligent Systems and Computing 214,
DOI: 10.1007/978-3-642-37832-4_23,

and organization of such networks and the effect of network topology on dynamics and behaviors of the system [1]. Uncovering the community structure in networks is an important method for getting more information which we do not know the networks before. In general, the notion of community was proposed by Girvan and Newman in the physics literature [2]. An intuitive feature of the community in network is that communities are groups of nodes in which nodes are densely connected and between which nodes are sparsely connected. The goal of community detection is to find out these groups of nodes in network. In this paper, we introduce the importance of node in the network into community detection and propose a novel community detection algorithm. Each node in the network plays a different role and their degree of importance is also different. Taking the router network for example, some key routers are more important than others as these critical routers crash will result in the throughput of the entire network sharply decrease. Thus, we should not treat them equally as most other community detection algorithms. In our algorithm, we give each node a score to quantify the importance of nodes in the network. During the label propagation course, we take both the importance of each node and the frequencies of neighbor community labels into consideration. We experiment out algorithm on some network data and illustrate the results of our algorithm, and compare with other community detecting algorithms. The structure of the paper is as follows: In Sect. 2, we give a brief introduction of the various related work. Section 3 discusses our approach of community detection in detail. Experiments results and discussion are given in Sect. 4 and conclusion in Sect. 4.1.

2 Related Work

In [1], the authors invented a near linear time community detection algorithm LPA (label propagation algorithm). According to the paper, each node is initialized with a unique label. During the course of the label propagation, each node adopts the label which is owned by most of its neighbors. The community label updating rule of each node can be expressed in formula (1),

$$c_x = \arg\max_l \left|N^l(x)\right| \quad (1)$$

where $N^l(x)$ indicates the set of neighbors of node x that owns the label l. If there exists multiple most frequent neighbor label, the new label is chosen at random from them. In theory, the label propagation course executes iteratively until each node does not change its community label. Eventually, the nodes which share the same label are seen as a community. The main advantage of LPA is near linear time complexity [1], which is superior to most other community detection algorithms. The disadvantage of LPA intuitively is its random characteristics, besides the communities detected by LPA are not stable. Based on the advantage and disadvantage of LPA, a large number of its varieties have been proposed, like

[3–6]. Newman developed a fast algorithm in [7]. This algorithm is based on modularity value Q which is a measure of the community division. The author defined the modularities computing formula as follows:

$$Q = \sum_{i} \left(e_{ii} - a_i^2\right) \tag{2}$$

where e_{ii} represents the fraction of edges that fall within group i and a_i is the fraction of edges which have a vertex in group i. According to Newmans definition, the higher value of Q is, the better division of network is. At first, we take each vertex as an individual community with a unique label; then combine communities in pairs which will cause the greatest increase or smallest decrease of Q value. The combining course of the algorithm can be viewed as a dendrogram building course. Cutting the dendrogram at different level results different community division. The author chooses the level which will generate the maximum Q to cut the dendrogram. The change of Q after combining two communities can be computed by

$$\Delta Q = e_{ij} + e_{ji} - 2\,a_i\,a_j = 2\left(e_{ij} - a_i\,a_j\right) \tag{3}$$

where e_{ij} denotes one-half of the edge fraction in the network that one node in group i and another node in group j. a_i is the same with formula (2). By means of modifying the rule of label updating for maximal modularity, Barber and Clark create a new algorithm named LPAm in [8]. LPAm remains the benefit of fast processing speed of LPA, and simultaneously avoids the trivial solution of LPA. But, the maximum modularity is just local maximum, not global maximum.

3 Proposed Algorithm

In this section, we first present two drawbacks of the LPA. First, the sequence of nodes to be updated is random, and if there is more than one the most frequent neighbor label, LPA chooses one from them randomly. This reduplicative randomicity make that LPA is sensible to the updating order of nodes. Hence, the result of LPA is not stable. Second, considering a network in Fig. 1. At time t, it is the order of d to update the community label. From all neighbors label, node d chooses the most frequent one. Assuming there is no identical community label in a, b, c and e at this moment. LPA will choose a label from them arbitrarily. But taking the importance of nodes into consideration, it is evident from the figure that node e is more important than other neighbors, so we should use e's label to update d.

To solve the problems mentioned above, we introduce some innovations to LPA by bringing in the importance of node and get rid of the randomicity. Thus, the first thing we should overcome is how to find a way to measure and quantify the importance of each node. In the field of information retrieval, Sergey Brin and Lawrence Page in [9] invent a method called PageRank to estimate the importance

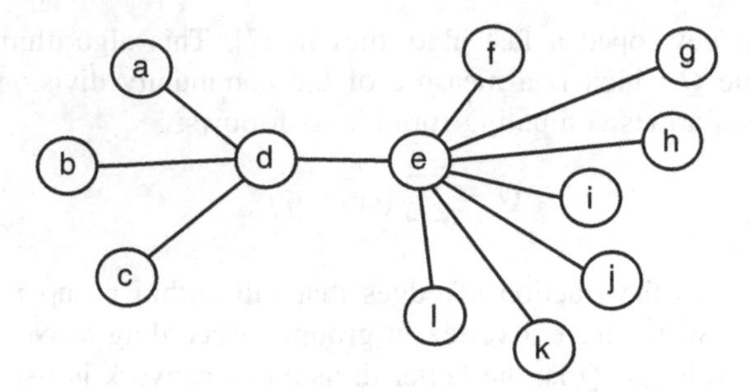

Fig. 1 A simple network diagram

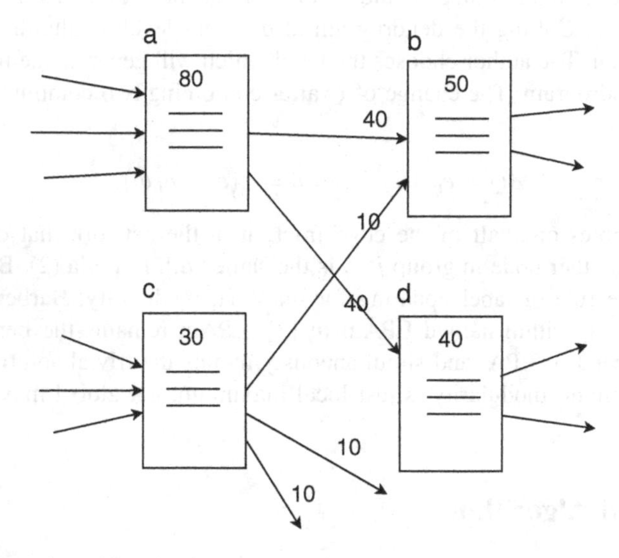

Fig. 2 An illustration of PageRank computing

of the Web page. The basic notion of the PageRank is that: (a) page i conveys some authority to page j if there is a hyperlink from i to j. The prestige of page j is proportional to the number of hyperlinks that page j receives; (b) page i that targets at page j maintains an authority score for itself. The higher prestige score that a page owns, the more prestige it can contributes to other pages that it can reach directly. But the score a page can contribute is inversely proportional to the number of hyperlinks which start from the page. We illustrate the course of PageRank score computing in Fig. 2.

Treating the Web system as a directed graph $G = (V, E)$, V is the set of vertices represented by web pages, and E is the set of edges represented by hyperlinks. $|V| = n$, we use $P(i)$ to denote the PageRank score of page i, the formulation of $P(i)$ is defined as:

$$P(i) = \sum_{(j,i)\in E} \frac{P(j)}{O_j} \tag{4}$$

where O_j is the number of hyperlinks that start from page j.

We use PageRank to measure the importance of a node in the network on account of the importance of a node is also determined by the importance of its neighbors and the number of edges that attached to it. The Web system can be abstracted as a set of vertexes and edges; at this point, it is almost identical with social network. After obtaining the score of each nodes importance, we sort the nodes in descending order according to their scores. Then, we select several core nodes. We describe the pseudo-code of core nodes selecting method in Algorithm 1.

Algorithm 1 SelectCore

1. Input: Graph $G = (V, E)$, δ
2. Output: core nodes list K
3. $K \leftarrow \emptyset$
4. Compute the score of each node using PageRank algorithm
5. Compute the average value of all PageRank scores, let it be *avg*
6. Sort the nodes in the descending order by the PageRank score, and get the sorted nodes list L
7. **for** *node* **in** L:
8. **if** *node* does not appear in δ neighborhood of all nodes in K **and** the PageRank score of *node* is not less than *avg*, **then** appends *node* into K
9. end for
10. return K

After obtaining the core nodes, we assign a unique community label to them. Then, we can change the community label of thc rest nodes. The rule when updating a node label is defined by

$$c_x = \arg\max_{l} \sum_{i\in N^l(x)} P(i) \tag{5}$$

where $N^l(x)$ is the neighbors of node x that owns community label l, and $P(i)$ is the PageRank score of node i. The basic ideal of our algorithm is: we recursively select one node from the core nodes and use (5) to update its neighbors, but each core expands outward only one layer at a time; then, it is the turn of other cores. After all cores have expanded, the whole expanding process starts again from the first core node. In other words, the whole updating process is a process that each core occupies more nodes layer by layer. When encountering the situation that the label of a node x has been updated by previous step, assuming the old community label is l_{old} and the new label is l_{new}, we should compare $\sum_{i\in N^{l_{new}}(x)} P(i)$ and $\sum_{i\in N^{l_{old}}(x)} P(i)$, only when the previous value is greater than the latter one, l_{new} is adopted, otherwise, do nothing. We give the pseudo-code of our algorithm in

Algorithm 2.

Algorithm 2 LPAp

1. Input: Graph $G = (V, E)$
2. Output: Communities C
3. $K \leftarrow SelectCore\,(G)$
4. Assign each core node a unique community label
5. Let *dict* be a hash table, the keys are the cores and the values are the neighbors of the cores.
6. **while** there is a node not updated **do**
7. **for** *core* **in** K:
8. $nodes \leftarrow dict\,[core]$
9. $NextExpanding \leftarrow \emptyset$
10. **for** *node* **in** *nodes* **do**
10. $NextExpanding+ = neighbors\,(node)$
11. **if** *node* not updated **then**
12. use the most frequent and important neighbor label to update *node*
13. **else**
14. **if** the score of l_{new} greater than l_{old} **then**
15. use l_{new} to update *node*
16. **end if**
17. **end for**
18. $dict\,[core] = NextExpanding - dict\,[core]$
19. **end for**
19. **end while**

Each iteration of LPA takes linear time in the number of edges ($O(m)$). Though experiments, the authors found that irrespective of *n*, 95 % of the nodes or more are classified correctly by the end of 5 iterations. In our algorithm, we remove the randomicity of LPA, so there is no need for multiple iteration process. The time complexity in Algorithm CoreSelect is the sum of PageRank Algorithm's time complexity and the node filter process's time complexity $O(n)$. And that, The time consumption in label spreading course of Algorithm LPAp is the same as one iteration of LPA. Thus, the total time complexity of our algorithm is $O(n^2 + n + m)$.

4 Experimental

In this section, we implement our algorithm on several real-world networks which are commonly used for evaluating community detection algorithm. As a comparison, we test several other community detection algorithms on the same data. We carried out all our experiments on a Window7 notebook computer with four 2.3 GHz processors and 4 GB of RAM.

4.1 Dataset and Their Detection Results

Zacharys Karate club network. The Zacharys Karate Club [10] consist of 34 members of a karate club in the USA for a 2-year period. During the observation period, the club split up into two groups because of a conflict and the two separated groups can be thought as the right community division. Figure 3a shows the detected result. The graph with the same color and the same shape represents an individual community. All nodes are divided correctly by our algorithm.

Dolphins association network. The dolphins association network [11] consists of 62 bottlenose dolphins living in Doubtful Sound, New Zealand. Lusseau collected the network after 7 years of field studies of the dolphins. The relationship between dolphin pairs is built by the observation of statistically substantial frequent association. For a period of 2 years, Lusseau observed that the dolphins can be separated into two groups. Figure 3b shows the community structure detected by our algorithm. There is no node with wrong community label.

Collaboration network. This dataset is a network of collaborations of scientists working at the Santa Fe Institute (SFI). This network was constructed by taking names of authors from all the journals and book publications by the

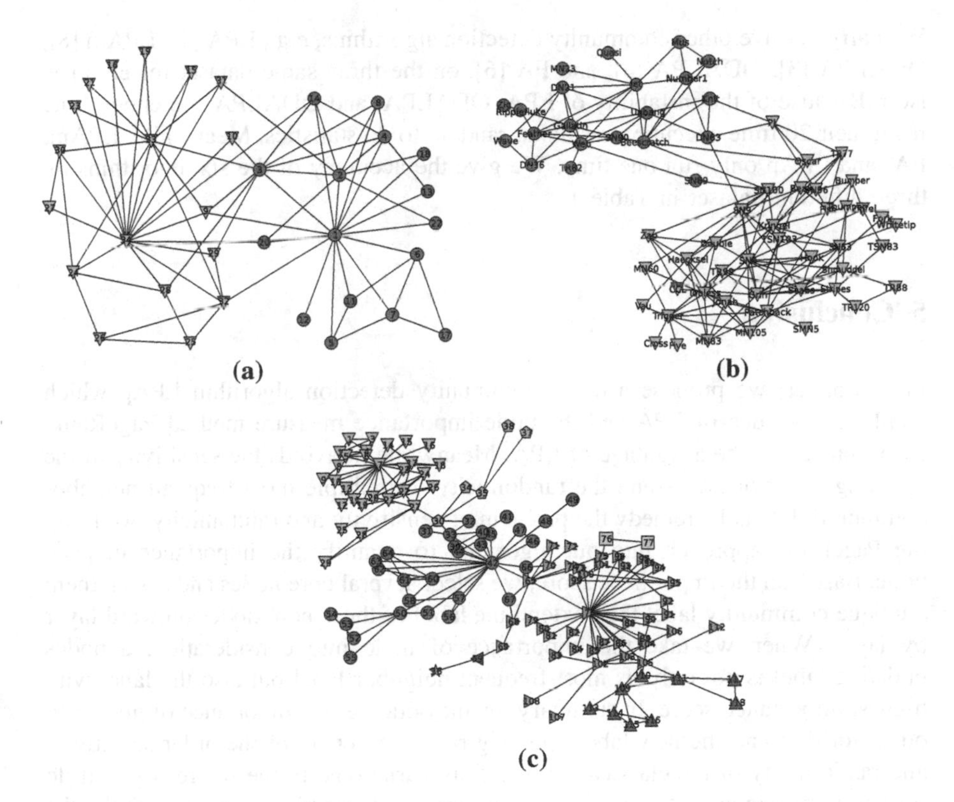

Fig. 3 The detected communities

Table 1 The accuracy of the algorithms on different networks

	LPA	LPAm	ODALPA	DDALPA	FA	LPAp
Karate	0.676	0.971	0.559	0.912	0.618	1
Dolphins	0.710	0.968	0.677	0.516	0.677	1
Collaboration	0.576	0.432	0.415	0.576	0.644	0.695

scientists involved, along with all papers which appeared in the institutes technical reports series. Vertices in the network denote the scientists in residence at SFI and their collaborators. Edges denote the co-authorship between scientists. Unlike the previous dataset, we do not know the real community division of collaboration network. However, Girvan and Newman in [2] use the proposed algorithm and divided this network into four communities. You can see the community division in Fig. 3c. There exist some nodes with wrong community label, but we can see the general structure of the four communities.

4.2 Comparisons with Other Algorithms

We carry out five other community detection algorithms, e.g., LPA [1], LPAm [8], DDALPA [3], ODALPA [3], and FA [6], on the three same dataset for comparison. Because of the instability of LPA, ODALPA, and DDALPA, we choose one from their 20 time executing results at random to do statistics. Meanwhile, LPAm, FA, and LPAp only run one time. We give the accuracy of the six algorithms on three different dataset in Table 1.

5 Conclusion

In this paper, we propose a novel community detection algorithm LPAp which combines the idea of LPA and the node importance measure method PageRank. LPAp preserves the advantage of LPA. Meanwhile, it avoids the sensitivity of the updating order of nodes and the randomicity of multiple most frequent neighbor community label. To remedy the problem of sensitivity and randomicity, we bring the PageRank approach into our algorithm to quantify the importance of each node. Based on the importance score, we select several core nodes and assign them a unique community label then extend the label of these core nodes outward layer by layer. When we take the importance of node into consideration, a nodes updating label is not only its most frequent neighbor label but also the label with highest importance score. In summary, by introducing the importance of node into our algorithm and the new label updating rule, we get rid of the order sensitivity and randomicity of the classical LPA and its variations. In the future, we will do further improvements to the algorithms proposed in this paper, especially the

SelectCore algorithm in respect that the communities are affected very much by the core nodes. We expect the future algorithms can perform well on a variety of network data set with different structure.

Acknowledgments This paper is supported by the Fundamental Research Funds for the Central Universities (lzujbky-2012-212).

References

1. Raghavan UN, Albert R, Kumara S (2007) Near linear time algorithm to detect community structures in large-scale networks. Phys Rev E 76:036106
2. Girvan M, Newman MEJ (2002) Community structure in social and biological networks. Proc Natl Acad Sci 99:7821–7826
3. Šubelj L, Bajec M (2011) Unfolding communities in large complex networks: combining defensive and offensive label propagation for core extraction. Phys Rev E 83(3):036103
4. Liu X, Murata T (2010) Advanced modularity-specialized label propagation algorithm for detecting communities in networks. Physica A 389:1493
5. Pang S, Chen C, Wei T (2009) A realtime clique detection algorithm: time-based incremental label propagation. In: Proceedings of the international conference on intelligent information technology application, vol 3, pp 459–462
6. Gregory S (2010) Finding overlapping communities in networks by label propagation. New J Phys 12:103018
7. Newman MEJ (2004) Fast algorithm for detecting community structure in networks. Phys Rev E 69:066133
8. Barber MJ, Clark JW (2009) Detecting network communities by propagating labels under constraints. Phys Rev E 80(2):026129
9. Brin S, Page L (1998) The anatomy of a large-scale hypertextual web search engine. In: Proceedings of the seventh international conference on world wide web 7, April 1998, pp 107–117, Brisbane
10. Zachary W (1977) An information flow model for conflict an fission in small groups. J Anthropol Res 33:452–473
11. Lusseau D, Schneider K, Boisseau OJ, Haase P, Slooten E, Dawson SM (2003) The bottlenose dolphin community of doubtful sound features a large proportion of long-lasting associations. Behav Ecol Sociobiol 54:396–405

An Object-Oriented and Aspect-Oriented Task Management Toolkit on Mobile Device

Yongsheng Tian, Qingyi Hua, Yanshuo Chang, Qiangbo Liu and Jiale Tian

Abstract In order to provide highly available and effective task management operations on mobile devices, an object-oriented and aspect-oriented task management toolkit on mobile device called OATMTM is designed. It uses object-oriented (OO) technology and aspect-oriented (AO) technology to provide a set of operation objects and aspect objects, which makes the developers only need to define objects without considering the detail of task management. The operating process is simplified. And it also provides mobile task management architecture for developer to reduce the effort of development.

Keywords Task management · Object oriented · Aspect oriented · Mobile handheld device

1 Introduction

Nowadays, as a diversity of mobile handheld devices has been entering the customer market, it plays an increasingly important role in people work and life. But task management developers are facing a challenge, that is, highly available and effective task management operations on mobile handheld devices. Restrictions by its own physical characteristics, the user interface has to fall back to the era of full-screen editing, so that the mobile interaction is very different from PC [1]. And with the emergence of different handheld mobile operating system, mobile development will no longer just for the experts in the field, but more for the general public. However, there is still no standard for task management on mobile handheld devices.

Y. Tian (✉) · Q. Hua · Y. Chang · Q. Liu · J. Tian
Department of Information Science and Technology, Northwest University, Xi'an, China
e-mail: jinmuzhao@163.com

F. Sun et al. (eds.), *Knowledge Engineering and Management*,
Advances in Intelligent Systems and Computing 214,
DOI: 10.1007/978-3-642-37832-4_24,

According to different mobile handheld devices operating system, each manufacturer development task management application is also different. But summarized up nothing more than a little: directly call their operating system code library to write. But this is time consuming and not conducive to the code reuse. In addition, it goes against the principle of separation of interface and application [2], if the interface design changes, the application code must be changed accordingly, is not conducive to the interface prototyping.

OATMTM design goal is to help the development of users to establish task management operation object by using user-centered [3] thought, and maintain the relationship between them and application. OATMTM provides a mobile handheld task management system structure for developer, and use of object-oriented (OO) and aspect-oriented (AO) technologies to provide a flexible and extensible task management operation processing objects set.

2 OATMTM System Structure

OATMTM provides a task management system structure for developer. OATMTM establishes a connection between task management operation object and particular application entity, so that the developer can call the interface provided by the OATMTM to complete specific task management operations. During operation process, OATMTM converts the lower event from mobile handheld device operating system to abstract events after computing, and then polymerization for task management operation, and process by the event processor.

2.1 System Structure Overview

According to the process of the task management operation, OATMTM system structure is divided into five modules, as showed in Fig. 1:

Virtual Task Manager Application Interface: This module, on the one hand, is responsible for render event handing results for the specific task management application calls; on the other hand, it is responsible for receiving the user's task management operation, for example: the task view, process view, switching, close a task, etc. Virtual task manager application interface send users direct manipulation of low-level task management operations event-to-event converter.

Event converter: This module is responsible to convert the low-level task management operations event to the corresponding abstract event object. It can support more task management types of input devices by expansion of the abstract event object. Event converter send abstract event object to the event interpreter.

Event interpreter: This module is responsible for interpreting the abstract task management operation events as specific mobile handheld device-independent task

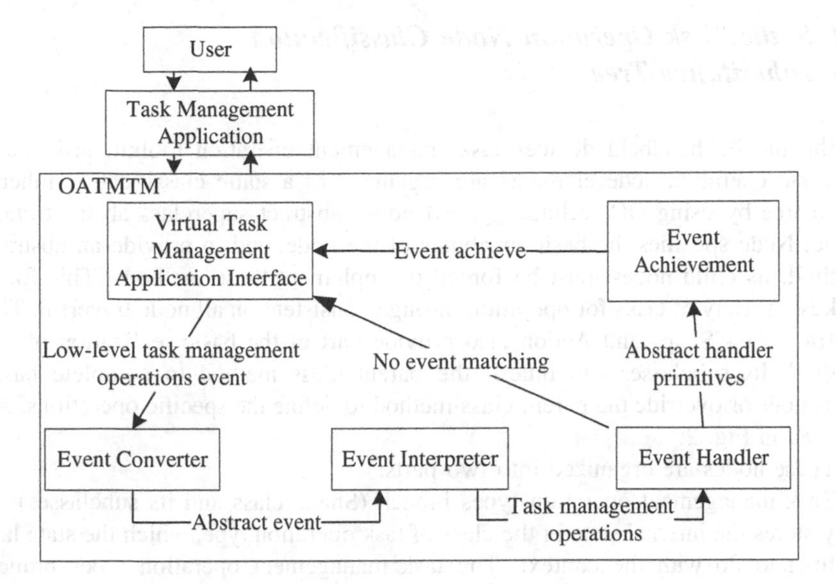

Fig. 1 The system structure of OATMTM

management operation object. These objects are more abstract and greater granularity interactive primitive. And a class represents an event action.

Event handler: This module is responsible for handling the abstract task management operation events sent by the event interpreter. When abstract task management operation events into this module, the event requires loop cycle in the event handler will deal with it. Generally, the event requires loop cycle press the abstract task management operation events into a queue. And when there is an event in the queue, the event handle loop cycle will handle the event through traversal the event require loop. When event handler action is executed, the abstract task management operation object only traverse the nodes of each task operate in accordance with the traversal strategy. At the same time, send the appropriate execute message to the node. And it is the various object in the queue which deals with the implementation of specific operational action. Generally, task management operation object registered the abstract operation primitives related to the application. At the same time, it will trigger log processing, exception handling, etc.

Event achievement: This module is responsible for receiving the abstract task operation primitives sent by the event handler. And its calculation associated with the context of the specific mobile handheld devices operating system information, and then handle the task operations by calling the specific mobile device operating system low-level code. And operating results are sent to the virtual task management application interface.

2.2 Static Task Operation Node Classification Inheritance Tree

In the mobile handheld devices task management operation toolkit, task management operation node elements are organized as a static classification inheritance tree by using OO technology. All nodes abstract superclass abstract class Node. Node specifies the basic attributes of the node, and to provide an abstract method. Its child nodes must be forced to implement these methods. This force makes a variety of tasks for operation through consistent on all node traversal. The abstract class Shape and Action also provide part of the basic realization of the method. Its subclasses can inherit the parent class method to complete basic operations or override the parent class method to define the specific operations. As showed in Fig. 2:

These nodes are organized into two parts:

Task management operation types branch (Shape class and its subclasses): It only stores the internal state in the class of task operation type, which the state has nothing to do with the context. The task management operation types branch responsible for describing the type of the task management operations. Shape is the abstract superclass of this branch, and its subclass is specific description of the type of operation, such as: task (Task), service (Service), process (Process), etc. In the design and development process, developers can direct instance of the kind of operation according to the actual needs or first specialized and then instantiate operation types.

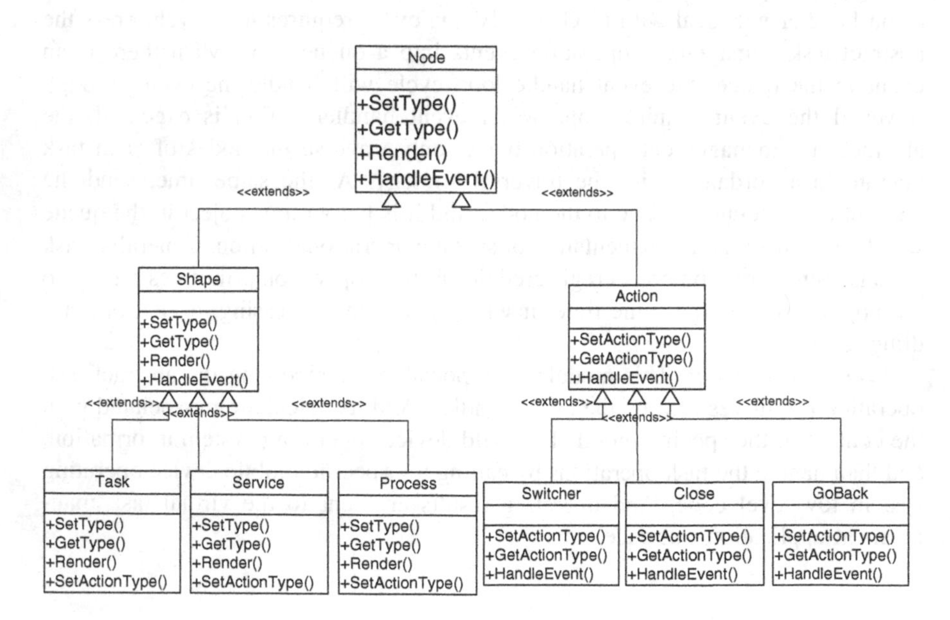

Fig. 2 Static task operation node classification inheritance tree

Task management operation action branch (class Action and its subclasses): This branch is responsible for describing the task management operation action, where the action object is the main implementation class in task management operations. Action is the abstract superclass of this branch, and its subclasses are responsible for describing various specific task management operations, such as: switch (Switch), background processing (Go Back), close (Close), etc. The most commonly used functions are interface scheduling in mobile task management. So it can provide more task management operation action through expanding this branch.

2.3 Apply Aspect-Oriented Technologies to Improve OATMTM Availability

In order to enhance the toolkit availability [4], OATMTM makes the availability of concerns of abstraction in the module by using AO technology [5, 6], which the concerns are dispersion in multiple modules or class when using OO to development, but having the modularity in logic, such as: the log processing module, the exception handing module, the return module, the close module, etc. During the task management operation, it may be an "aspect" or a plurality of interacting "aspect", it also can contain more than one class or interface class. And it is the communication between the "aspect" and class and AO intercept mechanism and woven into mechanism that to meet the usability of the modular of the system. The code for toolkit usability can be woven into the connection point by using AO weaving mechanism as showed in Fig. 3.

In the task management toolkit, aspects member can be divided into three parts: abstract aspects, interface, and specific aspects. Primarily responsible for: exception handling, log processing, return processing, and close processing.

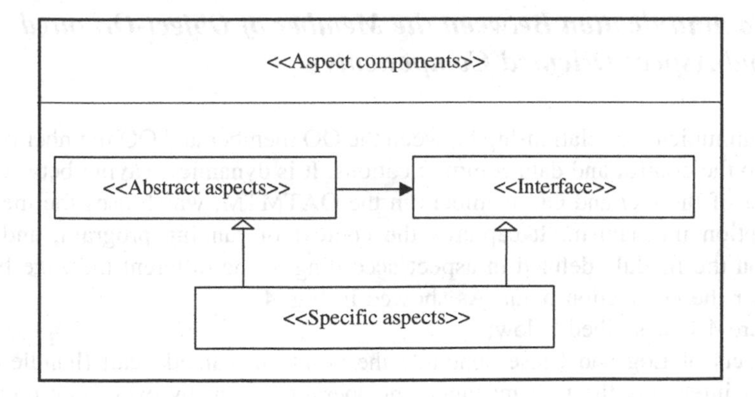

Fig. 3 Aspect components

Abstract aspects: It is mainly to define the action and behavior in the weave, which to define the notice entity. In order to determine the position of the weaving action, it defined the abstract point of tangency. It also defined the abstract methods depending on the context of application. And static weaving mechanism is also defined. What is defined in the abstract aspects including: abstract aspect log, abstract aspect close, abstract aspect return, and abstract aspect exception. The abstract aspect log is mainly for logging applications, including the abstract cut point call () and notification check (), do-Log (). The abstract aspect close is used to implement undo in the application, including the abstract point-cut call () and notification check (), do-return (). The abstract aspect return is used to implement return to the previous step, including the abstract point cut call () and notification save scene (), restore scene (). The abstract aspect exception is used to implement the abnormal reminder when application is wrong, including abstract point cut call () and notification check (), do-exception ().

Interface: The main duty of the interface is to provide service when the abstract aspects and specific aspects have announcement, and to achieve a unified interface.

Specific aspects: It is inherited the abstract aspects, and the role is to define the weave behavior, namely how to weave and where to weaving. The specific aspects inherited the weaving position from abstract tangent, and it can override the parent method or redefine the terms of the behavior. Specific aspect can also inherit interface class. And it implements the interface in the interface class. It completes the specific aspects through implementing the interface class or inherit the notification entity defined in the parent abstract aspect. The specific aspect logs inherit abstract aspect log, it redefines the point cut check (), do-log (), and the notification call (). The specific aspect close inherited the abstract aspect close. It redefines the notification call () and the point cut check (), do-return (). The specific aspect return inherited the abstract aspect return. It redefines the definition of the notification call () and point cut save scene (), restore scene (). The specific aspect exception inherited the abstract aspect exception. It redefines the definition of the notification call () and point cut check (), do-exception ().

2.4 Communication Between the Member of Object-Oriented and Aspect-Oriented Components

The communication relationship between the OO member and OO member mainly refers to the control and data communications. It is dynamic weaving between the member of the OO and OO members in the OATMTM, which uses the message interception mechanism. It captures the context of running program, and then injection the module defined in aspect according to the different message before and after the connection point. As showed in Fig. 4:

Figure 4 is described below:

Aspects of Log and Close conclude the point cut named "call (handle-event (*))". It intercepts the task management operations sent by event interpreter. It

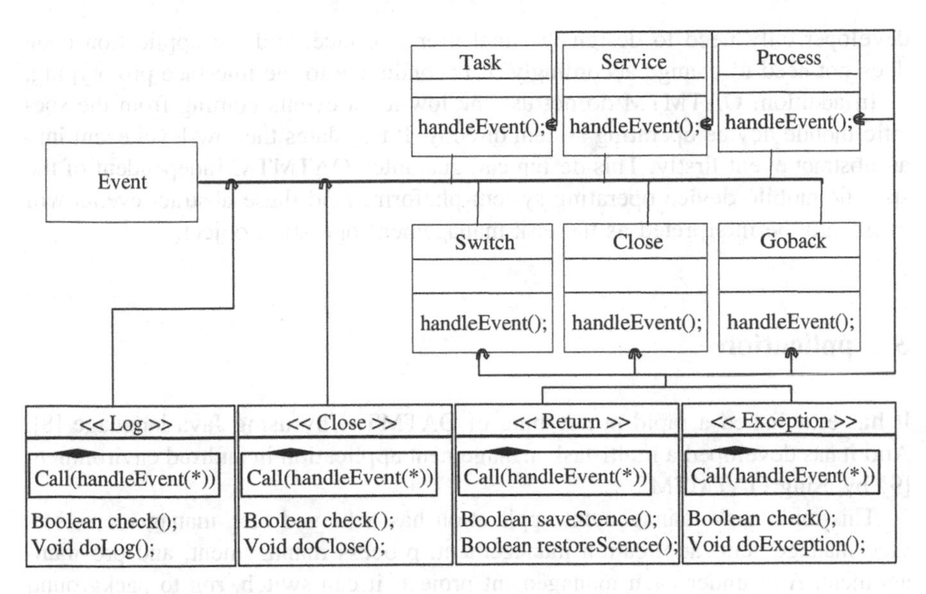

Fig. 4 Communication between member of object-oriented and aspect-oriented components

means it will trigger an action defined in notification written in aspect of log when Event call the handle event () in task management operations. The method check () will check whether there need to be woven into the task management operations. And it accordance with the context of running to judge whether woven the action declared in the notification, that is the action defined in method do ().

Aspect of Return concludes the point cut named "call (handle-event (*))". It intercepts the task management operations event. It means it will trigger the action defined in notification written in aspect of Return when task management operations event call the method named handle event () by itself. And Return defined two methods: save scene () and restore scene (). The save scene () is mainly to save current state site, and restore scene () is mainly to reply the scene.

Aspect of Exception will trigger when there is something wrong or exception in task management operations running state. It means it will trigger the action defined in notification written in the Exception, namely the action in do-exception ().

2.5 Abstract Event Model

OATMTM provides a virtual task manager application interface that not provides a specific user interface for developer directly. This virtual task manager application interface is an abstract interface model [7]. It uses abstract interaction object (AIO) to develop interactive structure that has nothing to do with the technical environment. But the final user interface is a specific platform interface. This implements the principle of separation of interface and application. It makes the developer only need to design the final user interface, and the application code does not need to change accordingly. It is conducive to the interface prototyping.

In addition, OATMTM do not use the low-level events coming from the specific mobile device operating system directly. It translates the low-level event into an abstract event firstly. This design can guarantee OATMTM independent of the specific mobile device operating system platform. And these abstract events will eventually be interpreted as the task management operation object.

3 Application

It has established a rapid prototyping of OATMTM by using Java language [8]. And it has developed a multi-task management application in android environment [9] by using OATMTM.

This multi-task management application has achieved task management, service management, application management, process management, and pro management. And under each management project, it can switch, run to background and close operations. As showed in Fig. 5:

The application presents the effect of use and the uninstall task management in task management application under android2.3.3 simulator.

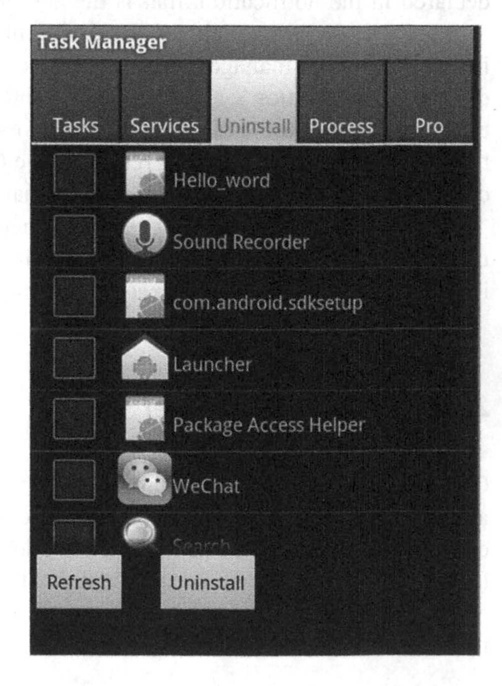

Fig. 5 Application of task manager

4 Conclusion

The goal of OATMTM is to help developers quickly build mobile task management operations which are user centric. It is independent of specific applications, and is beneficial for reuse and stratified in logical layer. OATMTM provides a flexible and extensible task management operation processing object sets. These objects are easily expanded to adapt to different applications. And it uses AO technology to improve the usability of the application.

Acknowledgments This research is supported by the National Natural Science Foundation of China (Granted No. 61272286).

References

1. Yue W, Dong S, Wang Y, Wang G, Wang H, Chen W (2004) Study on human-computer interaction frame work of pervasive computing. J Softw 27:1657–1664
2. Hua Q (2003) From conceptual modeling to architectral modeling—A UCD method for interactive systems. J Asian Sci Inf Life 3:227–234
3. Kankainen A (2003) UCPCD:user-centered product concept desing. In: Designing for user experiences. Proceedings of the 2003 conference on designing for user experiences, ACM Press, New York, pp 1–13
4. Dix A, Finlay J, Abowd G, Beale R (2003) Human-computer interaction. Publishing House of Electronics Industry, Beijing
5. Filman RE, Rlrad T, Clarke S, Aksit M (1997) Aspect-oriented programming. Springer, Finland
6. Soule P (2010) Autonomics development: a domain-specific aspect language approach. Auton Syst 0:1–6
7. Calvery G, Coutaz J, Thevenin D, Limbourg Q, Bouiloon L, vanderdonckt J (2003) A unifying reference framework for multi-target user interfaces. Interact Comput 15:289–308
8. Eckel B (2006) Thing in java fourth edition. Prentice Hall, London
9. Felker D (2010) Android application development for dummies. Hungry Minds, Hoboken

4 Conclusion

The goal of OATMTM is to help developers quickly build mobile task management operations which are user-centric. It is independent of specific applications and is beneficial for reuse and simplified in design. Besides, OATMTM provides flexible and extensible task management operation processing objects. These objects are easily extended to adapt to different applications. And it uses AO technology to improve the usability of the application.

Acknowledgments. This research is supported by the National Natural Science Foundation of China (Grant No. 61170[illegible]).

References

1. Yue W, Hong S, Wang Y, Wang G, Wang H, Chen Y (2008) Service in human computer interaction: the work of pervasive computing. [illegible] 2:165–[illegible]
2. Hua Q (2003) From conceptual modeling to behavioral modeling—A [illegible] method for interactive systems. [illegible] Inf [illegible] 972–[illegible]
3. Sankaran A (20[illegible]) User-centered product concept design. In: Designing for user experiences. Proceedings of the 20[illegible] conference on designing for user experiences. ACM Press, New York, pp 1–[illegible]
4. Dix A, [illegible] J, Abowd G, Beale R (2003) Human computer interaction. Publishing House of Electronics Industry, Beijing
5. Tolman P, Kiczales G, Clarke S, [illegible] M (1997) Aspect-oriented programming. Springer, Finland
6. Soule P (2010) Autonomous development: a domain-specific aspect language approach. Autom Syst 0:1–5
7. Calvary G, Coutaz J, Thevenin D, Limbourg Q, Bouillon L, Vanderdonckt J (2003) A unifying reference framework for multi-target user interfaces. Interact Comput 15:289–[illegible]
8. Eckel B (2006) Thinking in Java fourth edition. Prentice Hall, London
9. [illegible] Android application development for dummies. [illegible] Hoboken

The Heuristic Methods of Dynamic Facility Layout Problem

Lianghao Li, Bo Li, Huihui Liang and Weiwei Zhu

Abstract The hybrid heuristic simulation algorithms for dynamic facility layout problem (DFLP) have been proposed in this paper. By the combination of genetic algorithm and simulated annealing, we develop two heuristic methods and design the process of the algorithms with the forecasting and backtracking strategies in order to obtain the optimal solutions. Finally, a lot of experiments have been performed to reduce the random property of the algorithms. The simulation experiments have shown that the two combined heuristic algorithms proposed in this paper have much better performance than the other algorithms proposed in the reference when solving the sequential coding problem, especially in solving large-size problems.

Keywords Dynamic facility layout problem (DFLP) · Hybrid heuristic algorithms · Forecasting and backtracking strategies

1 Introduction

Recently some change over a time period can be significant for facility layout problems. For instance, the demand from a customer may change as a result of a trend, or the cost of delivery may change as the transportation mode changes. Even the cost of operation may vary from period to period in each location with changes in economics conditions that affect the regional operating. Thus, the problem is to plan a schedule layout optimally for a predetermined time period, over long-time layout problems that is optimum in one period may or may not be optimum in the following period. This is especially true when there are limitations on the number

L. Li · B. Li · H. Liang (✉) · W. Zhu
College of Management and Economics, Tianjin University, 300072 Tianjin, China
e-mail: libo0410@tju.edu.cn

F. Sun et al. (eds.), *Knowledge Engineering and Management*, Advances in Intelligent Systems and Computing 214,
DOI: 10.1007/978-3-642-37832-4_25,

and capacities of facilities that we may have. We call this kind of problem Dynamic Facility Layout Problem (DFLP).

DFLP involves material transportation and facility rearranging schedules in several production periods, and solving DFLP is a little complex. Many researchers have done a lot of research work on DFLP. Rosenblatt [1] is the first person who has described DFLP completely, he promoted to apply dynamic programming for solving the DFLP, but the proposed methods just handled the small layout problems. Further, he has put forward some heuristic ideas and relative strategies based on dynamic programming to speed the solving progress. After that, it is popular to study the heuristic methods for solving DFLP. For examples, Venkatamanan [2] used genetic algorithm to solve DFLP, and Balakrishnan [3–5] developed two heuristic algorithms to improve the downhill pair exchanged heuristic algorithm. The first algorithm is the same heuristic algorithm that Urban [6] used to generate the feasible solutions for DFLP, but he proposed a tracing pairing exchanged heuristic algorithm to improve the solutions, which generated by the forecast window, then choose the best solution; the second heuristic algorithm combined Urban's algorithm with dynamic programming. Simultaneity, Baykasoglu and Gindy etc. [7, 8] also discussed using simulated annealing algorithm to solve DFLP and Conway [2] addressed a mixed genetic algorithm to solve DFLP. Sadan [9] gave a three-stage heuristic algorithm to study DFLP. The first Stage collected all the logistics data from T stages, and then each production period was looked as a static facility layout problem. By a weight scheme, some feasible layout planning could be obtained. At the second stage, by improving the solutions obtained in the first stage using dynamic programming, and then some optimal solutions for dynamic facility layout problem was obtained. Last, a random cooling exchange strategy was be further improved the solutions obtained in the second stage. McKendall and Shang [10] explored the hybrid ant colony algorithm to solve the dynamic facility layout problem, and they used simulated annealing heuristic algorithm to improve this problem. McKendall etc. [11] proposed the Heuristics for the dynamic facility layout problem with unequal-area departments. A boundary search (construction) technique, which placed departments along the boundaries of already placed departments, was developed for the DFLP and the solution was improved using a tabu search heuristic.

Based on the results proposed in the above references, multi-line DFLP has been discussed with the aid of the hybrid thought in this paper. Two heuristics, named BGSA_GA_D I and BGSA_GA_D II, respectively, have been proposed and studied through the combination of genetic algorithm and simulated annealing with the forecasting and backtracking strategy. Finally, a series of simulation experiments have been performed, and the result have been analyzed and compared to the optimal results from the above references. The results have showed the effectiveness of the method proposed in this paper.

2 The Dynamic Facility Layout Model

In order to simplify the problem, while a dynamic facility layout problem is modeled, the assumptions will be considered as [12]: (1) Each facility has the same area; (2) The number of the facilities is less than or equal to the number of the location area(if less than, assuming the virtual facilities); (3) The same cost of the unit transportation between the facilities; (4) The facility moving cost is irrelative with the layout (assuming a constant); (5) The distances between the facilities in the same row or column are calculated by measuring their centers of these facilities, else the rectangular distances.

Referring to Koopmans and Beckman's QAP model, a DFLP mathematical model is given as:

$$\mathrm{Min}Z = \sum_{t=1}^{P}\sum_{i=1}^{n}\sum_{j=1}^{n}\sum_{k=1}^{n}\sum_{l=1}^{n} f_{tik} d_{tjl} X_{tij} X_{tkl} + \sum_{t=2}^{P}\sum_{i=1}^{n}\sum_{j=1}^{n}\sum_{k=1}^{n}\sum_{l=1}^{n} A_{tijl} Y_{tijl}. \quad (1)$$

Subject to:

$$\sum_{i=1}^{n} X_{tij} = 1, \qquad j = 1, \ldots, n, \qquad t = 1, \ldots, P. \quad (2)$$

$$\sum_{j=1}^{n} X_{tij} = 1, \qquad i = 1, \ldots, n, \qquad t = 1, \ldots, P. \quad (3)$$

$$Y_{tijl} = X_{(t-1)ij} X_{til}, \quad i, j, l = 1, \ldots, n, \quad t = 2, \ldots, P. \quad (4)$$

where the total periods P; the number of the facilities n; the facility indexes i, k; the location j; f_{tik} is the transportation cost from facility i to k, assuming f_{tik} is the logistics capacity between the facilities and assuming the unit cost between them is 1; d_{tjl} is the distance from location l to location j; and in period t, define

$$X_{tij} = \begin{cases} 1 & facility \quad i \quad in \quad location \quad j \\ 0 & else \end{cases} \quad (5)$$

$$Y_{tijl} = \begin{cases} 1 & moving \quad facility \quad i \quad from \quad location \quad j \\ 0 & else \end{cases}. \quad (6)$$

$$A_{tijl} = \begin{cases} 1 & moving \quad \cos t \quad of \quad facility \quad i \quad from \quad j \quad to \quad l \\ 0 & else \end{cases} \quad (7)$$

The model is to minimize the total material transportation cost and the facility moving cost in the planning time in formula (1); Constraints (2) and (3) show that every facility has a location and each location has a facility; Constraint (4) means only when the facility location has changed in the continuous two periods, the value of $Y_{tijl} = 1$.

3 The Hybrid Heuristic Algorithm

This paper puts forward two combined heuristic algorithm to solve the dynamic facility layout problem that modeled above; the first one uses BGSA [12] in the literature, and then develops the single parent genetic algorithm to get the initial solution as the input of BGSA, that is BGSA_GA_D I; the second improves the forecasting and backtracking strategy in simulated annealing algorithm based on the first algorithm, this new heuristic algorithm is called BGSA_GA_D II.

3.1 BGSA_GA_D I

From Fig. 1, it is obvious that BGSA_GA_D I combined single parent genetic algorithm and simulated annealing algorithm. The main idea is: First, use PGA to generate the initial solution, and then change the multi-phases problem into the single-phase static layout problem. By applying the procedure of PGA, all optimal solutions of every period will be obtained. Therefore, the layout of each period forms the initial solution of the DFLP, that is the initial layout of the BGSA_GA_D I. Then, simulated annealing, which is the core algorithm of BGSA_GA_D I, will be adopted to optimize the entire process. The cooling schedule and accept rule involved in SA have not improved; but in BGSA_GA_D I, the random downhill pair exchange is applied to generate neighbor solution.

3.2 BGSA_GA_D II

BGSA_GA_D II is an innovation of the combined heuristic algorithm based on BGSA_GA_D I in this paper. The forecasting and backtracking strategies have been developed based on the simulated annealing to search the global optimization.

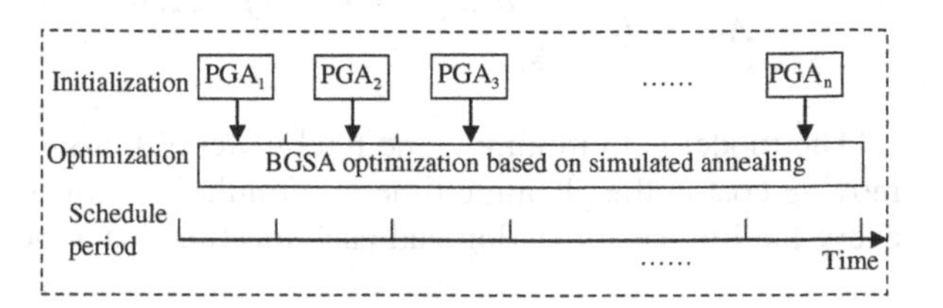

Fig. 1 The ideas of BGSA_GA_D I

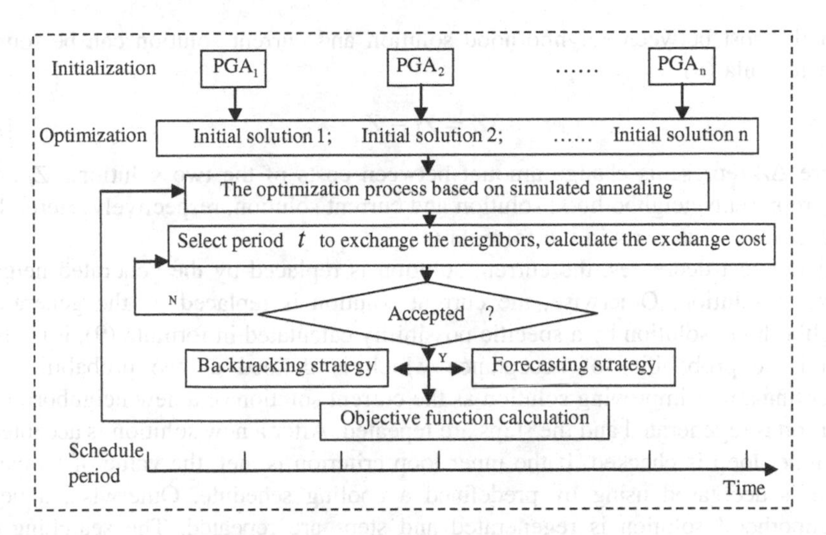

Fig. 2 The ideas of BGSA_GA_D II

Figure 2 gives the combined heuristic idea of BGSA_GA_D II. The idea to initialize the solution is the same as that in BGSA_GA_D I, and the neighbor structure exchanging arrangement using simulated annealing to optimize the whole problem is the same with that in BGSA_GA_D I. But when the neighbor solution is accepted, the new two heuristic algorithms, the backtracking strategy and the forecasting strategy are developed to improve the global solution. That is, once the layout of the period t is exchanged, the same two locations in period $t-1$, $t-2$,... and period $t+1$, $t+2$,..., n will also be calculated and then decided whether it is acceptable. If the neighbor solution cannot be accepted by computing them, for example, the solution for period $t-2$ is acceptable, but solution for period $t+2$ is not acceptable. In the following, the backtracking strategy and the forecasting strategy, the solution of period $t-3$ and periods before will be considered and be evaluated until the first period or the last period is acceptable or unaccepted. Finally, summarize the exchange of the global objective because of the period t and the influence of the preceding periods and following periods drawn from period t. The realization detail of the two strategies can be referred to the fourth section in this paper.

The main procedures are described as follows:

(1) Generate the initial population

The initial population is generated as following: by transferring the PGA procedure, the server static deployment of each period t is obtained, and then the initial solution in all of the periods can be given by combining all results of each period. The detail algorithm is shown in the Ref. [12].

(2) Improving method based on simulated annealing

SA is a search algorithm to find a possible best solution of a combinatorial optimization problem. The searching process starts with one initial population. A neighborhood of this solution is generated using any neighborhood move rule and then the cost between neighborhood solution and current solution can be found with formula (8).

$$\Delta Z = Z_i - Z_{i-1} \tag{8}$$

where ΔZ represents change amount between costs of the two solutions. Z_i and Z_{i-1} represents neighborhood solution and current solution, respectively. Here, the cost includes two parts in the formula (1).

If the cost decreases, the current solution is replaced by the generated neighborhood solution. Otherwise, the current solution is replaced by the generated neighborhood solution by a specific possibility calculated in formula (9), it is also called the probability of acceptance which is defined as the probability of accepting a non-improving solution as the current solution or a new neighborhood solution is regenerated and the steps are repeated. After a new solution is accepted, the inner loop is checked. If the inner loop criterion is met, the value of temperature is decreased using by predefined a cooling schedule. Otherwise, a new neighborhood solution is regenerated and steps are repeated. The searching is repeated until the termination criteria are met or no further improvement can be found in the neighborhood of the current solution. The termination criterion (outer loop) is predetermined.

$$e^{(-\Delta Z/T)} > R \tag{9}$$

where T temperature is a positive control parameter. R is a uniform random number between 0 and 1.

(1) Swapping move: this operator is used to produce a near solution to current solution in search space. Swapping move randomly selects two genes in solution and swaps the positions of these genes. Then a neighborhood solution is produced. In shifting move, randomly selects two genes in solutions and the second gene is put in front of another genes. Thus a new solution is obtained.

Note the second heuristic considers the forecasting and backtracking strategies, after performing a swapping move, the next step is modified to include the forecasting and backtracking strategies.

(2) Cooling schedule: the proportional decrement schedule is applied and two temperature values, which are in ith and $i - 1$th iterations, are computed by formula (10).

$$\mathrm{T}_i = \alpha \mathrm{T}_{i-1} \tag{10}$$

where T_i and T_{i-1} are the temperatures in ith and $i - 1$th iterations, respectively. α is the cooling rate between two temperatures and varies between 0 and 1, and it is usually set at 0.90.

Here, the changes in the total cost of exchanging the locations of two facilities in the preceding $(t - 1)$ and succeeding $(t + 1)$ periods are studied. If the total cost holds for both periods, update the current solution and consider the changes in the total costs of exchanging the locations of the two facilities in the preceding $(t - 2)$ and succeeding $(t + 2)$ periods. If the exchange is accepted for only one of the periods, such as ($t + 2$), then update the current solution, and then consider the exchange in the succeeding ($t + 3$) period, and reject the exchange in period ($t-2$). Continue considering the exchange in the succeeding periods until $t = n$ or the exchange is not accepted. Then go to the next step.

(3) Inner loop and outer loop criterion: Inner loop criterion decides how many possible new solutions to produce in every temperature. Outer loop criterion is used to stop the search process. In this study, inner loop criterion is set to 3, 5, and 10. Outer loop criterion is 1,000.

4 Simulation Experiments

To prove the algorithms proposed in this paper, the simulation experiments have been performed and the results have been compared with those in the other research. The simulation data comes from the reference. The 16 problems of 48 problems from Balakrishman and Cheng [4], include: (1) 6 facilities and 10 periods, (2) 15 facilities and 5 periods, and (3) 15 facilities and 10 periods; the total experiment number is 16. Matlab for programming is applied to run these programming.

For different problems, the parameters of the simulated annealing in the three simulation experiments are as Table 1. The parameters of PGA which are used to generate initial solutions follow the parameters given in the Ref. [12], that is: popsize = 50, GEN = 50, where popsize is the size of the population, and GEN is the number of the evolution generation. The experiment involves four dynamic facility layout problems with different sizes. For each size, the experiment has the respective paramcters,shown as Table 1.

In the simulation, we test five times with both BGSA_GA_D I and BGSA_-GA_D II for the first cight problems, and for the last eight problems, we have three tests for each problem. The results are in Tables 2, 3, 4, and 5.

Where n is the number of facility, t is the period number; Tin is the initial temperature, elmax is the external iterative number, α is the cooling rate in simulated annealing.

From Tables 2, 3, 4, and 5, it shows that for dynamic facility layout problem with different sizes, BGSA_GA_D II can get the better results than BGSA_GA_D I in most cases. So the algorithm that combines the heuristic strategy has the better performance.

Table 1 The parameters of the simulated annealing

Test number	n	t	Tin	Elmax	α
1–4	6	10	16671	3000	0.996
5–8	15	5	37164	3200	0.996
9–12	15	10	16572	4000	0.997

Table 2 The cost results of 6 facilities and 10 periods

Times	Problem1		Problem 2		Problem 3		Problem 4	
	I	II	I	II	I	II	I	II
1	215931	214313	212238	214128	208605	208786	212949	213561
2	214313	217691	214627	213307	208791	209273	214961	214387
3	216452	214313	214029	212943	208581	209045	214626	**212747**
4	216063	215723	213161	212293	209452	209045	214054	213287
5	215200	215351	213240	213355	210815	209839	213303	213185
Average	215592	215478	213459	213205	209249	209198	213979	**213433**

Unit: Yuan

Table 3 The cost results of 15 facilities and 5 periods

Times	Problem5		Problem 6		Problem 7		Problem 8	
	I	II	I	II	I	II	I	II
1	483845	486242	488637	485205	495358	492598	491701	488124
2	483716	487230	491337	489431	493905	494583	491895	491515
3	486109	485424	490001	487024	492665	493703	485928	490076
4	485716	485714	487191	489364	497900	494650	489953	488038
5	486521	485833	489348	490962	495176	494600	489616	488454
Average	485181	486089	489303	488397	495001	494027	489819	489241

Unit: Yuan

Table 4 The cost results of 15 facilities and 10 periods

Times	Problem 9		Problem 10		Problem 11		Problem 12	
	I	II	I	II	I	II	I	II
1	**987536**	991723	**986311**	**985114**	995404	991719	988714	**981246**
2	989969	988590	989506	985259	997734	**991055**	983858	982984
3	988763	**987437**	989691	984966	**993290**	993372	**979675**	**981246**
Average	**988756**	989250	988503	**985113**	995476	**992049**	984082	**981825**

Unit: Yuan

The data from Balakrishman and Cheng [4] has been used by many literatures to do simulation experiments. We have obtained some results of the other algorithms in the reference, and compared them to the 16 problems in this paper to prove the performance of the two hybrid algorithms proposed in this paper.

Table 5 gives best optimal objectives of the 16 problems in Ref. [4], which have used the other algorithms proposed in other literatures, like CVGA, BGSA, and PGA. The last two columns are the optimal results of BGSA_GA_D I and BGSA_GA_D II in this paper.

From Table 5, the two hybrid intelligent algorithms in this paper show better performance than the other algorithms proposed before. In the simulation, BGSA_GA_D II have obtained the best optimal solutions 13 times, it is 81.25 %

Table 5 The cost comparisons of the five different algorithms [4]

Problems	CVGA	BGSA	PGA	BGSA_GA_D I	BGSA_GA_D II
1	504759	501447	493587	483716	485424
2	514718	506236	502934	488637	485205
3	516063	512886	510955	492665	492598
4	508532	504956	500679	485928	488038
5	1055536	1017741	1044941	987536	987437
6	1061940	1016567	1030050	986311	985114
7	1073603	1021075	1037809	993290	991055
8	1060034	1007713	1026075	979675	981246
9	632737	604408	627750	581512	581315
10	647585	604370	601391	577569	576749
11	642295	603867	602613	580713	580136
12	634626	596901	593872	572825	572503
13	1362513	1223124	1273014	1176115	1175875
14	1379640	1231151	1273060	1177859	1177455
15	1365024	1230520	1258018		1173231
16	1367130	1200613	1251336		1157594

Unit: Yuan

of the total number of problems, BGSA_GA_D I have obtained the best optimal solution three times, while the other three algorithms do not get the best optimal solution. Obviously, we can see, the hybrid intelligent algorithms proposed in this paper have the good performances in solving the dynamic facility layout problems of different sizes, especially for large-scale multi-phase facility layout problems.

5 Conclusion

This paper puts forward two combined heuristic algorithm for dynamic facility layout problem, and explains the principle and the process of the algorithms. Finally, the paper designs three groups, summed 16 problems, and compiles the procedure and then do a lot of simulation computing. In order to reduce the randomness of the algorithm, every experiment runs 3 or 5 times, and then gets the average results. By studying the algorithm and the simulation experiment, it shows that the two combined heuristic algorithms are much better than other simplex algorithm when solving sequential coding problem, especially in solving large problems.

Acknowledgments This work was supported by Higher education specialized research fund for the doctoral program funding issue, China (No. 20100032110034).

References

1. Rosenblatt MJ (1986) The dynamics of plant layout. Manag Sci 37:272–286
2. Conway DG, Venkataramanan MA (1994) Genetic search and the dynamic facility layout problem. Comput Oper Res 21(8):955–960
3. Balakrishnan J, Cheng CH (2000) Genetic search and the dynamic layout problem: an improved algorithm. Comput Oper Res 27:587–593
4. Balakrishnan J, Cheng CH (2003) A hybrid genetic algorithm for the dynamic plant layout problem. Int J Prod Econ 86:107–120
5. Balakrishnan J, Cheng CH (2009) The dynamic plant layout problem: incorporating rolling horizons and forecast uncertainty. Omega 37:165–177
6. Urban TL (1998) Solution procedures for the dynamic facility layout problem. Anneals Oper Res 76:323–342
7. Baykasoglu A, Dereli T, Sabuncu I (2006) An ant colony algorithm for solving budget constrained and unconstrained dynamic facility layout problems. Omega 34:385–396
8. Baykasoglu A, Gindy NNZ (2001) A simulated annealing algorithm for dynamic facility layout problem. Comput Oper Res 28:1367–1460
9. Kulturel-Konak S (2007) Approaches to uncertainties in facility layout problems: perspectives at the beginning of the 21st century. J Intell Manuf 18:273–284
10. McKendall AR, Shang J (2006) Hybrid ant systems for the dynamic facility layout problem. Comput Oper Res 33:790–803
11. McKendall AR Jr, Hakobyan A (2010) Heuristics for the dynamic facility layout problem with unequal-area departments. Eur J Oper Res 201:171–182
12. Drira A, Pierreval H, Hajri-Gabouj S (2007) Facility layout problems: a survey. Annu Rev Control 31(2):255–267

Attitude-Based Consensus Model for Heterogeneous Group Decision Making

R. M. Rodríguez, I. Palomares and L. Martínez

Abstract Usually, human beings make decisions in their daily life providing their preferences according to their knowledge area and background. Therefore, when a high number of decision makers take part in a group decision-making problem, it is usual that they use different information domains to express their preferences. Besides, it might occur that several subgroups of decision makers have different interests, which may lead to situations of disagreement amongst them. Therefore, the integration of the group's attitude toward consensus might help optimizing the consensus reaching process according to the needs of decision makers. In this contribution, we propose an attitude-based consensus model for heterogeneous group decision-making problems with large groups of decision makers.

Keywords Group decision making · Heterogeneous information · Attitude · Consensus reaching

1 Introduction

Decision making is a usual process for human beings in their daily life. In group decision-making (GDM) problems, a group of decision makers try to reach a solution to a problem that consists of a set of possible alternatives, providing their preferences [3]. An important aspect in GDM problems is to achieve a common solution which is accepted by all decision makers involved in the problem.

R. M. Rodríguez (✉) · I. Palomares · L. Martínez
Department of Computer Science, University of Jaén, Campus Las Lagunillas, s/n. 23071, Jaén, Spain
e-mail: rmrodrig@ujaen.es

I. Palomares
e-mail: ivanp@ujaen.es

L. Martínez
e-mail: martin@ujaen.es

F. Sun et al. (eds.), *Knowledge Engineering and Management*, Advances in Intelligent Systems and Computing 214,
DOI: 10.1007/978-3-642-37832-4_26,

Usually, GDM problems have been solved by applying approaches that do not guarantee to reach such a collectively accepted solution. Hence, Consensus Reaching Processes (CRPs) become necessary to obtain accepted solutions by all decision makers participating in the GDM problem [8].

Classically, consensus models proposed in the literature to deal with CRPs focused on the resolution of GDM problems where a low number of decision makers take part. However, nowadays new trends like social networks [9] and e-democracy [4] imply the participation of larger groups of decision makers in discussion processes. When large groups of decision makers are involved in a GDM problem, it is usual that each one expresses her/his preferences in different information domains, such as, numerical, interval valued or linguistic values, according to their profile, the area of knowledge they belong to, and the nature of alternatives. When alternatives are quantitative in nature, they are normally assessed by means of numerical or interval-valued values; however, when their nature is qualitative, the use of linguistic information might be more suitable [1]. In such cases, the GDM problem is defined in an heterogeneous framework. Different approaches to deal with heterogeneous information have been presented [2, 5].

Another issue in GDM problems with a large number of decision makers is that there might exist several subgroups of decision makers with conflicting interests, which may lead to situations of disagreement amongst such subgroups, thus making it hard to achieve an agreed solution and delaying the decision process. Therefore, the integration of the group's attitude toward consensus, i.e., the capacity of decision makers to modify their own preferences during the CRP, becomes an important aspect to consider in any CRP involving large groups [6].

The aim of this paper is to propose a new consensus model for GDM problems defined in heterogeneous contexts, which is able of integrating the attitude of decision makers toward consensus in CRPs involving large groups.

This paper is structured as follows: Sect. 2 revises some preliminary concepts about management of heterogeneous information and attitude integration in CRPs. Section 3 presents the heterogeneous consensus model that integrates the group's attitude toward consensus. Section 4 shows an illustrative example of the proposed model, and Sect. 5 points out some conclusions.

2 Preliminaries

In this section, some concepts about managing heterogeneous information in GDM and integrating attitudes in CRPs are briefly reviewed.

2.1 Heterogeneous Information in GDM

In GDM problems, where a large number of decision makers are involved, it is frequent that they have different background or they have different degrees of knowledge about the problem. For these reasons, the use of different information domains might allow decision makers to express their preferences in a more suitable way, thus leading to better results than those obtained if they had to express such preferences in a single domain imposed for the whole group. In this paper, we consider an heterogeneous framework compound by the following domains:

- *Numerical*: $N = \{v | v \in [0, 1]\}$.
- *Interval-valued*: $I = P([0, 1]) = \{[l, u] | l, u \in [0, 1] \wedge l \leq u\}$.
- *Linguistic*: $S = \{s_0, \ldots, s_g\}$, where S is a linguistic term set defined in the unit interval. It is assumed that each linguistic term $s_j \in S, j \in \{0, \ldots, g\}$, has associated a fuzzy membership function, denoted as $\mu_{s_j}(y), y \in [0, 1]$ [12].

In spite of the different approaches to deal with heterogeneous information in the literature [2, 5], each one is based on different features. Here, we use the method proposed by Herrera et al. [2] to unify assessments expressed in different domains into fuzzy sets $F(S_T)$, in a common linguistic term set $S_T = \{s_0, \ldots, s_g\}$ chosen according to the rules introduced in [2], by means of the transformation functions defined below for each type of information.

Definition 1 [2] Let $v \in [0, 1]$ be a numerical value, the function $\tau_{NS_T} : [0, 1] \rightarrow F(S_T)$ transforms a numerical value into a fuzzy set in S_T.

$$\tau_{NS_T}(v) = \{(s_0, \gamma_0), \ldots, (s_g, \gamma_g)\} \quad s_k \in S_T, \gamma_k \in [0, 1]$$

$$\gamma_k = \mu_{s_k}(v) = \begin{cases} 0 & if \quad v \notin support\ (\mu_{s_k}(x)), \\ \frac{v-a_k}{b_k-a_k} & if \quad a_k \leq v \leq b_k, \\ 1 & if \quad b_k \leq v \leq d_k, \\ \frac{c_k-v}{c_k-d_k} & if \quad d_k \leq v \leq c_k. \end{cases}$$

being $\mu_{s_k}(\cdot)$ a membership function for linguistic term $s_k \in S_T$, represented by a parametric function (a_k, b_k, d_k, c_k).

Definition 2 [2] Let $I = [l, u]$ be an interval valued in [0, 1], the function $\tau_{IS_T} : I \rightarrow F(S_T)$ transforms an interval valued into a fuzzy set in S_T.

$$\tau_{IS_T}(I) = \{(s_k, \gamma_k)/k \in \{0, \ldots, g\}\}\}$$
$$\gamma_k = max_y min\{\mu_I(y), \mu_{s_k}(y)\}$$

where $\mu_I(\cdot)$ and $\mu_{s_k}(\cdot)$ are the membership functions of the fuzzy sets associated with the interval valued I and linguistic term s_k, respectively.

Definition 3 [2] Let $S = \{l_0, \ldots, l_p\}$ and $S_T = \{s_0, \ldots, s_g\}$ be two linguistic term sets, such that $g \geq p$, a linguistic transformation function $\tau_{SS_T} : S \rightarrow F(S_T)$ transforms a linguistic value l_i into a fuzzy set in S_T.

$$\tau_{SS_T}(l_i) = \{(s_k, \gamma_k^i)/k \in \{0, \ldots, g\}\} \qquad \forall l_i \in S$$
$$\gamma_k^i = max_y min\{\mu_{l_i}(y), \mu_{s_k}(y)\}$$

where $i \in \{0, \ldots, p\}$. $\mu_{l_i}(\cdot)$ and $\mu_{s_k}(\cdot)$ are the membership functions of the fuzzy sets associated with the terms l_i and s_k, respectively.

2.2 Attitude Integration in CRPs: Attitude-OWA

CRPs in GDM problems attempt to find a collective agreement amongst decision makers before making a decision, so that a more accepted solution by the whole group is achieved [8]. Although a large number of consensus models have been proposed to support groups in CRPs, most of them do not consider the integration of the group's attitude toward consensus in situations where several subgroups of decision makers, with different (and often conflicting) interests and attitudes, take part in the GDM problem. Since this aspect might help optimizing the CRP according to the needs of decision makers and each particular problem, a consensus model that integrates such an attitude was recently proposed in [6], where the different types of group's attitudes to be considered were also introduced:

- *Optimistic attitude*: Achieving an agreement is more important for decision makers than their own preferences. Therefore, more importance is given to positions in the group with higher agreement.
- *Pessimistic attitude*: Decision makers consider more important to preserve their own preferences. Therefore, positions in the group with lower agreement are given more importance.

In the following, we briefly review the definition of an aggregation operator, so-called Attitude-OWA, that will be used to integrate the attitude of decision makers in CRPs in the proposed consensus model. Such an operator extends OWA aggregation operators [10], and it is specially suitable for dealing with large groups of decision makers [6].

Definition 4 [10] An OWA operator on a set $A = \{a_1, \ldots, a_h\}$, $a_i \in R$ is a mapping $F : R^h \rightarrow R$, with an associated weighting vector $W = [w_1 \ldots w_h]^\top$:

$$F(a_1, \ldots, a_h) = \sum_{j=1}^{h} w_j b_j \tag{1}$$

with $w_i \in [0, 1]$, $\sum_i w_i = 1$. b_j is the jth largest a_i value.

The Attitude-OWA operator extends the OWA operators by introducing two *attitudinal parameters* that must be provided by the decision group, $\vartheta, \varphi \in [0, 1]$:

- ϑ represents the group's attitude, which can be optimistic ($\vartheta > 0.5$), pessimistic ($\vartheta < 0.5$), or indifferent ($\vartheta = 0.5$). It is equivalent to the measure of optimism (*orness*) that characterizes OWA operators [10].
- φ indicates the amount of agreement positions that are given nonnull weight in the aggregation process. The higher φ, the more values are considered.

Attitude-OWA operator is then defined as follows:

Definition 5 [6] An Attitude-OWA operator of dimension h on a set $A = \{a_1, \ldots, a_h\}$, is an OWA operator based on attitudinal parameters ϑ, φ given by a group of decision makers to indicate their attitude toward consensus,

$$Attitude\text{-}OWA_W(A, \vartheta, \varphi) = \sum_{j=1}^{h} w_j b_j \tag{2}$$

where b_j is the j-th largest of a_i values and A is the set of values to aggregate.

Attitude-OWA is characterized by a weighting vector W, computed according to attitudinal parameters, so that weights reflect an specific attitude adopted by decision makers. The following scheme was proposed in [6] to compute Attitude-OWA weights.

(i) The group provides values for ϑ, φ, based on their interests and/or the nature of the GDM problem.
(ii) A RIM (Regular Increasing Monotone) linguistic quantifier with membership funcion $Q(r)$, $r \in [0, 1]$:

$$Q(r) = \begin{cases} 0 & \text{if } r \leq \alpha, \\ \frac{r-\alpha}{\beta-\alpha} & \text{if } \alpha < r \leq \beta, \\ 1 & \text{if } r > \beta. \end{cases} \tag{3}$$

is defined upon ϑ, φ, by computing $\alpha = 1 - \vartheta - \frac{\varphi}{2}$ and $\beta = \alpha + \varphi$.
(iii) The following method proposed by Yager in [11] is applied to compute weights w_i:

$$w_i = Q\left(\frac{i}{h}\right) - Q\left(\frac{i-1}{h}\right), i = 1, \ldots, h \tag{4}$$

3 Attitude-Based Consensus Model for GDM in Heterogeneous Contexts

This section presents a consensus model for GDM problems that deals with heterogeneous information and integrates the attitude in CRPs when there are large groups of decision makers.

Classically, consensus models proposed in the literature consider the existence of a human figure so-called moderator, who is responsible for coordinating the overall CRP [8]. Nevertheless, this approach facilitates the automation of his/her tasks, by implementing such a model into a *Consensus Support System* based on intelligent techniques [7].

GDM problems considered in this model are formed by a set $E = \{e_1, \ldots, e_m\}$, $(m \geq 2)$, of decision makers who express their preferences over a set of alternatives $X = \{x_1, \ldots, x_n\}$, $(n \geq 2)$ by using a preference relation P_i:

$$P_i = \begin{pmatrix} - & \ldots & p_i^{1n} \\ \vdots & \ddots & \vdots \\ p_i^{n1} & \ldots & - \end{pmatrix}$$

Each assessment $p_i^{lk} \in D_i$ represents the degree of preference of the alternative x_l over x_k, $(l \neq k)$ for the decision maker e_i, expressed in an information domain, $D_i \in \{N, I, S\}$ (see Sect. 2.1).

A scheme of the consensus model is depicted in Fig. 1, and its phases are described in detail below:

1. *Determining Group's Attitude*: The first phase consists of determining the group's attitude towards the measurement of consensus, gathered by means of the attitudinal parameters ϑ, φ.
2. *Gathering Preferences*: Given that the GDM problem is defined in an heterogeneous framework, each e_i provides his/her preferences on X by means of a preference relation P_i, consisting of a $n \times n$ matrix of assessments $p_i^{lk} \in D_i = \{N, I, S\}$.
3. *Making Heterogeneous Information Uniform*: Preferences expressed by decision makers in different information domains are unified by applying the approach proposed in [2], that unifies heterogeneous information into fuzzy sets

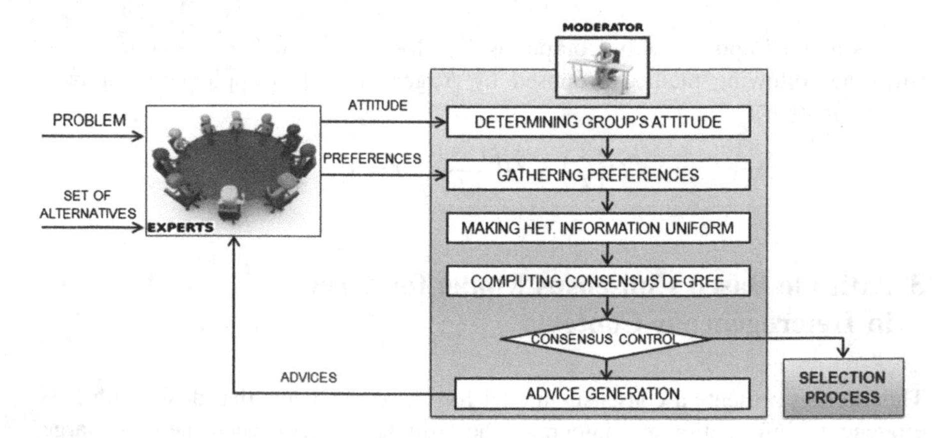

Fig. 1 Consensus model scheme

in a common linguistic term set (see Def. 1, 2, 3). Assuming that each unified assessment is represented by $p_i^{lk} = (\gamma_{i0}^{lk}, \ldots, \gamma_{ig}^{lk})$, each decision maker's preference relation is represented as follows:

$$P_i = \begin{pmatrix} - & \cdots & (\gamma_{i0}^{1n}, \ldots, \gamma_{ig}^{1n}) \\ \vdots & \ddots & \vdots \\ (\gamma_{i0}^{n1}, \ldots, \gamma_{ig}^{n1}) & \cdots & - \end{pmatrix}$$

4. *Computing Consensus Degree*: It computes the degree of agreement amongst decision makers [3], measured as a value in [0, 1]. The group's attitude toward consensus is integrated in step c) during this phase.
   (a) For each unified assessment (fuzzy set) $p_i^{lk} = (\gamma_{i0}^{lk}, \ldots, \gamma_{ig}^{lk})$, a *central value* $cv_i^{lk} \in [0, g]$ is obtained to facilitate further computations, as follows:

$$cv_i^{lk} = \frac{\sum_{j=0}^{g} index(s_j) \cdot \gamma_{ij}^{lk}}{\sum_{j=0}^{g} \cdot \gamma_{ij}^{lk}}, \quad s_j \in S_T \tag{5}$$

where $index(s_j) = j \in \{0, \ldots, g\}$.
   (b) For each pair of decision makers e_i, e_t, $(i<t)$, a *similarity matrix* $SM_{it} = (sm_{it}^{lk})_{n\times n}$ is computed. Each similarity value $sm_{it}^{lk} \in [0, 1]$ represents the agreement level between e_i and e_t in their opinion on (x_l, x_k), computed as:

$$sm_{it}^{lk} = 1 - \left| \frac{cv_i^{lk} - cv_t^{lk}}{g} \right| \tag{6}$$

   (c) A *consensus matrix* $CM = (cm^{lk})_{n\times n}$ is obtained by aggregating similarity values, by means of an Attitude-OWA operator defined upon ϑ, φ (see Sect. 2.2) to reflect the group's attitude [6]:

$$cm^{lk} = Attitude - OWA_W(SIM^{lk}, \vartheta, \varphi) \tag{7}$$

$SIM^{lk} = \{sm_{12}^{lk}, \ldots, sm_{1m}^{lk}, \ldots, sm_{(m-1)m}^{lk}\}$ is the set of all pairs of decision makers' similarities in their opinion on (x_l, x_k), with $|SIM^{lk}| = \binom{m}{2}$, being cm^{lk} the degree of consensus achieved by the group in their opinion on (x_l, x_k).
   (d) Consensus degrees ca^l on each alternative x_l, are computed as

$$ca^l = \frac{\sum_{k=1,k\neq l}^{n} cm^{lk}}{n-1} \tag{8}$$

   (e) Finally, an overall consensus degree, cr, is obtained as follows:

$$cr = \frac{\sum_{l=1}^{n} ca^l}{n} \tag{9}$$

5. *Consensus Control*: Consensus degree cr is compared with a consensus threshold $\mu \in [0, 1]$, established a priori by the group. If $cr \geq \mu$, the CRP ends and the group moves on the selection process; otherwise, the process requires further discussion. A parameter $Maxrounds \in \mathbb{N}$ can be used to control the maximum number of discussion rounds.

6. *Advice Generation*: When consensus required is not achieved, $cr < \mu$, decision makers are advised to modify their preferences to make them closer to each other and increase the consensus degree in the following CRP round. As stated above, despite a human moderator has been traditionally responsible for advising and guiding decision makers during CRPs, the proposed model allows an automation of his/her tasks [7], many of which are found in this phase of the CRP. The following steps are conducted in this phase (based on central values cv_i^{lk}):

   (a) Compute a collective preference and proximity matrices: A collective preference $P_c = (p_c^{lk})_{n \times n}$, $p_c^{lk} \in [0, g]$, is computed for each pair of alternatives by aggregating preference relations:

$$p_c^{lk} = v(cv_1^{lk}, \ldots, cv_m^{lk}) \tag{10}$$

Afterwards, a proximity matrix PP_i between each e_i's preference relation and P_c is obtained:

$$PP_i = \begin{pmatrix} - & \ldots & pp_i^{1n} \\ \vdots & \ddots & \vdots \\ pp_i^{n1} & \ldots & - \end{pmatrix}$$

Proximity values $pp_i^{lk} \in [0, 1]$ are obtained for each pair (x_l, x_k) as follows:

$$pp_i^{lk} = 1 - \left| \frac{cv_i^{lk} - p_c^{lk}}{g} \right| \tag{11}$$

Proximity values are used to identify the furthest preferences from the collective opinion, which should be modified by some decision makers.

   (b) Identify preferences to be changed (CC): Pairs of alternatives (x_l, x_k) whose consensus degrees ca^l and cp^{lk} are not enough, are identified:

$$CC = \{(x_l, x_k) | ca^l < cr \wedge cp^{lk} < cr\} \tag{12}$$

Afterwards, the model identifies decision makers who should change their opinions on each of these pairs, i.e. those e_is whose assessment p_i^{lk} on $(x_l, x_k) \in CC$ is such that cv_i^{lk} is furthest to p_c^{lk}. To do so, an average proximity $\overline{pp}^{lk}$ is calculated, by using an aggregation operator λ:

$$\overline{pp}^{lk} = \lambda(pp_1^{lk}, \ldots, pp_m^{lk}) \tag{13}$$

As a result, decision makers e_i whose $pp_i^{lk} < \overline{pp}^{lk}$ are advised to modify their assessments p_{ij}^{lk} on (x_l, x_k).

(c) Establish change directions: Several direction rules are applied to suggest the direction of changes proposed to decision makers, in order to increase the level of agreement in the following rounds. An acceptability threshold $\varepsilon \geq 0$, which should take a positive value close to zero is used to allow a margin of acceptability when cv_i^{lk} and p_c^{lk} are close to each other.

- DIR.1: If $(cv_i^{lk} - p_c^{lk}) < -\varepsilon$, then e_i should *increase* his/her assessment p_i^{lk} on (x_l, x_k).
- DIR.2: If $(cv_i^{lk} - p_c^{lk}) > \varepsilon$, then e_i should *decrease* his/her assessment p_i^{lk} on (x_l, x_k).
- DIR.3: If $-\varepsilon \leq (cv_i^{lk} - p_c^{lk}) \leq \varepsilon$, then e_i should not modify his/her assessment p_i^{lk} on (x_l, x_k).

4 Application Example

In this section, a real-life GDM problem is solved by using a Web-based Consensus Support System that facilitates the implementation of the previous consensus model [7].

Let us suppose a city council compound by 40 politicians with different background $E = \{e_1, \ldots, e_{40}\}$, must make an agreed decision about defining a budget allocation related to a recent income from the national government. The investment options proposed are, X = $\{x_1$:Introduction of a new tram line in the city center,x_2:Construction of an indoor shopping center,x_3:Expand green areas and parks,x_4:Improve leisure centers and sports facilities$\}$: The information domains defined are:

- Numerical: $[0, 1]$
- Interval-valued: $I([0, 1])$
- Linguistic: $S = \{s_0 : null\ (N), s_1 : very\_low\ (VL), s_2 : low\ (L), s_3 : medium\ (M), s_4 : high\ (H), s_5 : very\_high\ (VH), s_6 : perfect\ (P)\}$.

Without loss of generality, the term set S is chosen as the common linguistic term set used to unify heterogeneous information, i.e. $S_T = S$. The group's attitude given by ϑ, φ and other CRP parameters are summarized in Table 1.

Once the problem is defined, the CRP begins, following the phases shown in Sect. 3 and Fig. 1:

Table 1 Parameters defined at the beginning of the CRP

Attitudinal parameters	Consensus threshold	Maximum #rounds	Accept. threshold
$\vartheta = 0.35$, $\varphi = 0.6$	$\mu = 0.85$	*Maxrounds* = 10	$\varepsilon = 0.1$

(1) *Determining group's attitude*: The group's attitude toward consensus is gathered by means of $\vartheta, \varphi \in [0,1]$ (see Table 1). Based on them, a RIM quantifier's parameters are computed as: $\alpha = 1 - \vartheta - \frac{\varphi}{2} = 0.35$ and $\beta = \alpha + \varphi = 0.95$, respectively, thus obtaining the following quantifier $Q(r)$ (see Fig. 2a):

$$Q(r) = \begin{cases} 0 & \text{if } r \leq 0.35, \\ \frac{r-0.35}{0.6} & \text{if } 0.35 < r \leq 0.95, \\ 1 & \text{if } r > 0.95. \end{cases} \tag{14}$$

Eq. (4) is then used to compute a weighting vector W of dimension $\binom{40}{2} = 780$, thus defining an $Attitude - OWA_W(SIM^{lk}, 0.35, 0.6)$ operator.

(2) *Gathering preferences*: Decision makers provide their preferences, expressed in their domains. An example of preference relations expressed by three experts, e_s, e_r, e_t in different information domains is given below:

$$P_s = \begin{pmatrix} - & [.6,.8] & [.5,.7] & [.1,.4] \\ [.2,.4] & - & [.7,.9] & [0,.2] \\ [.3,.5] & [.1,.3] & - & [0,0] \\ [.6,.9] & [.8,1] & [1,1] & - \end{pmatrix}, \; P_r = \begin{pmatrix} - & .9 & .8, & .5 \\ .1 & - & .4 & 0 \\ .2 & .6 & - & .2 \\ .5 & 1 & .8 & - \end{pmatrix},$$

$$P_t = \begin{pmatrix} - & P & P & H \\ N & - & M & L \\ N & M & - & VL \\ L & H & VH & - \end{pmatrix}$$

(3) *Making Heterogeneous Information Uniform*: The unification scheme described in Sect. 2.1 is applied to decision makers' preferences obtaining fuzzy sets

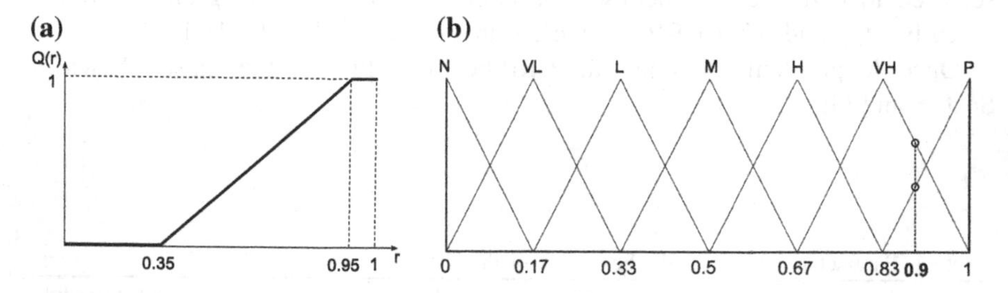

Fig. 2 **a** RIM quantifier defined upon attitudinal parameters $\vartheta = 0.35$, $\varphi = 0.6$. **b** Unification of heterogeneous information into fuzzy sets in S_T

Table 2 Global consensus degree for each round

Round 1	Round 2	Round 3	Round 4	Round 5
0.616	0.693	0.762	0.821	0.870

in S_T. For example, assessment $p_r^{12} = 0.9$ on (x_1, x_2) in the numerical preference shown above is unified into $(\gamma_{r0}^{12}, \ldots, \gamma_{r6}^{12})$=$(0,0,0,0,0,.59,.41)$ (see Fig. 2b).

(4) *Computing consensus degree*: The level of agreement is computed taking into account the Attitude-OWA operator defined above. A central value must be previously computed upon each unified assessment (see Eq. (5)).
   Pairwise similarities are computed and aggregated, taking into account the Attitude-OWA operator defined above. Afterwards, the overall consensus degree is obtained as $cr = 0.616$.

(5) *Consensus control*: The global consensus degree, $cr = 0.616 < 0.85 = \mu$, therefore consensus achieved is not enough and the CRP must continue.

(6) *Advice generation*: Some recommendations are generated for each politician to modify his/her preferences and increase the level of collective agreement. Afterwards, the second CRP round begins.

In this problem, due to the moderately pessimistic attitude provided by the group, it was necessary to carry out a total of five rounds of discussion (see Table 2) to reach the consensus threshold $\mu = 0.85$.

The proposed consensus model allowed us to solve the GDM problem taking into account the attitude of decision makers toward consensus and giving them the possibility to use different information domains to express their preferences, which lead to make an agreed and highly accepted solution by the whole group.

5 Conclusions

Nowadays, new trends like e-democracy and social networks imply the participation in discussion processes of large groups of decision makers, who might have different backgrounds. Therefore, the use of heterogeneous information is common in GDM problems, where a high number of decision makers take part. In this contribution, we have presented a consensus model that deals with heterogeneous information and integrates the attitude of decision makers to achieve the consensus.

Acknowledgments This work is partially supported by the Research Project TIN-2009-08286 and FEDER funds.

References

1. Herrera F, Herrera-Viedma E (2000) Linguistic decision analysis: steps for solving decision problems under linguistic information. Fuz Sets and Sys 115:67–82
2. Herrera F, Martínez L, Sánchez PJ (2005) Managing non-homogeneous information in group decision making. Eur J Oper Res 166(1):115–132
3. Kacprzyk J (1986) Group decision making with a fuzzy linguistic majority. Fuzzy Sets Syst 18(2):105–118
4. Kim J (2008) A model and case for supporting participatory public decision making in e-democracy. Group Decis Negot 17(3):179–192
5. Li DF, Huang ZG, Chen GH (2010) A systematic approach to heterogeneous multiattribute group decision making. Comput Ind Eng 59(4):561–572
6. Palomares I, Liu J, Xu Y, Martínez L (2012) Modelling experts' attitudes in group decision making. Soft Comput 16(10):1755–1766
7. Palomares I, Rodríguez RM, Martínez L (2013) An attitude-driven web consensus support system for heterogeneous group decision making. Expert Syst Appl 40(1):139–149
8. Saint S, Lawson JR (1994) Rules for reaching consensus. A modern approach to decision making. Jossey-Bass, San Francisco
9. Sueur C, Deneubourg JL, Petit O (2012) From social network (centralized vs. decentralized) to collective decision-making (unshared vs. shared consensus). PLoS one 7(2):1–10
10. Yager RR (1988) On orderer weighted averaging aggregation operators in multi-criteria decision making. IEEE Trans Syst Man Cyber 18(1):183–190
11. Yager RR (1996) Quantifier guided aggregation using OWA operators. Int J Intell Syst 11:49–73
12. L.A. Zadeh (1975) The concept of a linguistic variable and its applications to approximate reasoning. Inf Sci Part I, II, III, 8,8,9:199–249,301–357,43–80

An Application of Soft Computing Techniques to Track Moving Objects in Video Sequences

Luis Rodriguez-Benitez, Juan Moreno-Garcia, Juan Giralt, Ester del Castillo and Luis Jimenez

Abstract In this paper, an approach for tracking objects using motion information over compressed video is proposed. The input data is a set of blobs or regions obtained from the segmentation of H264/AVC motion vectors. The tracking algorithm establishes correspondences between blobs in different frames. The blobs belonging to the same object must satisfy a set of constraints between them: continuity, temporal coherence, and similarity in attributes like position and velocity. The uncertainty and dispersion inherent in motion data available from compressed video makes fuzzy logic a suitable technique to achieve good results. Then, the blobs are represented as sets of linguistic data called linguistic blobs and all the operations are performed over this structure. A major contribution of this work is the design of a tracking process that is able to operate in real-time because the size of input data is very small and the computational cost of operations is low.

This work has been funded by the Regional Government of Castilla-La Mancha under the Research Project PII1C09-0137-6488, and by the Spanish Ministry of Science and Technology under the Research Project TIN2009-14538-C02-02.

L. Rodriguez-Benitez (✉) · J. Moreno-Garcia · J. Giralt · E. del Castillo · L. Jimenez
Information Systems and Technologies Department,
University of Castilla-La Mancha, Paseo de la Universidad,
4, Ciudad Real, Spain
e-mail: luis.rodriguez@uclm.es
URL: http://oreto.esi.uclm.es

J. Moreno-Garcia
e-mail: juan.moreno@uclm.es

J. Giralt
e-mail: juan.giralt@uclm.es

E. del Castillo
e-mail: ester.castillo@uclm.es

L. Jimenez
e-mail: luis.jimenez@uclm.es

F. Sun et al. (eds.), *Knowledge Engineering and Management*,
Advances in Intelligent Systems and Computing 214,
DOI: 10.1007/978-3-642-37832-4_27,

The main calculation cost is reduced by using simple mathematical computations over the linguistic variables.

Keywords H264/AVC compressed video · Motion vectors · Linguistic objects · Approximate reasoning

1 Introduction

The tasks involved in video surveillance usually require moving objects to be tracked because information about their behavior can be obtained from characteristics of their trajectories and the interaction between them, as summarized in Fuentes and Velastin [2].

Typically, the tracking process involves the matching of image features for non-rigid objects such as people, or correspondence models, widely used with rigid objects like cars. Tracking is usually performed by matching boxes from two consecutive frames. The matching process uses the information of 3D boxes [5], color histogram back projection, or different blob features such as color [1] or distance between blobs. In function of the features exploited, the algorithms can be grouped into four approaches: region based, active contour based, feature based, and model based [4]. Several techniques are described in different reviews [4, 12, 13] in which can be deduced that fuzzy and clustering algorithms are widely used.

In fact, most current tracking algorithms build correspondences among the objects across different video frames using models such as Kalman filters [6] or particle filters [8]. In Motamed et al. [7], the tracking algorithm uses globally the Nearest Neighbor strategy. In that work, for each detected region are computed three features: center of gravity, bounding box, and color histogram. It uses a belief indicator representing the tracking consistency for each object allows solving defined ambiguities at the tracking level. Zhou [14] propose a new tracking algorithm using probabilistic fuzzy c-means and Gibbs Random Fields to associate the segmented regions to form video objects.

Due to more reliable features that can be extracted from the pixel data, the majority of object trackers require full decoding of the video stream. However, a number of publications have demonstrated good results of the compressed domain approach to video analysis. Concretely, during the last years there have been developed a considerable amount of tracking algorithms based on the video compressed domain, for example over MPEG [10].

In this work, a new H.264 [9] tracking algorithm is proposed in which the tracking process associates blobs or regions detected in different frames to objects. The blobs are located by means of a segmentation algorithm [11] that, using fuzzy techniques, obtains linguistic representations of blobs called linguistic blobs from H264/AVC motion vectors. The proposed technique is based on managing linguistic components using similar representations to the 2-tuple fuzzy linguistic representation model for computing with words [3]. Then, first in Sect. 2, a set of fuzzy elements used in this work is described. Later, in Sect. 3 is presented the tracking algorithm in detail and finally the experiments on several real traffic video sequences and the conclusions of this work are exposed in Sects. 4 and 5, respectively.

2 Linguistic Elements used in the Problem Domain

Before starting with the description of the tracking method it is necessary to define the set of fuzzy elements used in this work.

Let $A_j = \{A_j^1, A_j^2 \ldots A_j^{i_j}\}$ be an ordered set of linguistic labels over a continuous variable X_j. An **ordered linguistic interval** $I_j^{p,q}(x)$ is the result of the fuzzification of the real value x over X_j and is represented as:

$$I_j^{p,q}(x) = \begin{cases} [A_j^p : \mu_{A_j^p}(x);\ A_j^q : 1 - \mu_{A_j^p}(x)] & \textit{if } p \neq q \\ [A_j^p : 1] & \textit{if } p = q \end{cases} \tag{1}$$

where p and q are the positions in A_j of the two consecutive linguistic labels in the interval, and the value of $\mu_{A_j^p}(x)$ is not 0.

A **linguistic blob** (***LB***) [11] is defined as a 6-tuple composed by one or more conceptually similar motion vectors that describe a motion region in the image in a linguistic way:

$$LB = (FrameNumber, Size, I_{HV}, I_{VV}, I_{VP}, I_{HP})$$

where *FrameNumber* is the frame number where the blob is located, *Size* is the number of linguistic motion vectors grouped in the blob, and the last four elements are the ordered linguistic intervals that represent the position and velocity of the blob: *Horizontal Position*, *Vertical Position*, *Horizontal Velocity*, and *Vertical Velocity*. An example of *LB* is shown in Table 1.

The formal representation of a target object is called **Linguistic Object** (*LO*) and is defined as:

$$LO = (FirstFrame, LastFrame, Size, Blobs)$$

Table 1 Representation of a linguistic blob

Attribute	Value
FrameNumber	3252
Size	11
I_{HV}	[Fast Left:1]
I_{VV}	[No Motion:0.7; Slow Down:0.3]
I_{VP}	[Up:1]
I_{HP}	[Right:0.75; Very Right:0.25]

Table 2 Representation of a linguistic object

FirstFrame: 2
LastFrame: 9
Size: 2.82
Blobs: $\{LB_1, LB_5, LB_6, LB_{12}, LB_{23}, LB_{27}\}$
LB_1: $(2, 3, [SR : 1], [NM : 1], [DW : 1], [CH : 1])$
LB_5: $(3, 4, [SR : 1], [NM : 1], [DW : 1], [CH : 1])$
LB_6:$(5, 1, [SR : 1], [NM : 1], [DW : 1], [CH : 1])$
LB_{12}:$(6, 3, [SR : 1], [NM : 1], [DW : 1], [CH : 1])$
LB_{23}:$(8, 2, [SR : 1], [NM : 1], [DW : 1], [CH : 0.7;\ R : 0.3])$
LB_{27}:$(9, 4, [SR : 1], [NM : 1], [DW : 1], [CH : 0.7;\ R : 0.3])$

where *FirstFrame* and *LastFrame* correspond with frames the moving object appears and disappears, respectively. Attribute *Size* corresponds with the average size of every blob in *Blobs*. Finally, *Blobs* contains linguistic blobs from different frames associated with this *LO*. If *Blobs* is $\{LB_i, LB_{i+j} \ldots, LB_m\}$ next conditions must be satisfied:

$$LB_i(FrameNumber) = LO(FirstFrame) \tag{2}$$

$$LB_m(FrameNumber) = LO(LastFrame) \tag{3}$$

In Table 2 an example of *LO* is shown. The moving object appears during 6 frames and Eqs. 2 and 3 are satisfied:

$$LB_1(FrameNumber) = 2 = LO(FirstFrame) = 2$$
$$LB_{27}(FrameNumber) = 9 = LO(LastFrame) = 9$$

Finally, *Size* is computed by means of Eq. 4.

$$\frac{3+4+1+3+2+4}{6} = 2.82 \tag{4}$$

From the information contained in the linguistic intervals, it can be deduced that the object is moving slowly to the right (*Slow Right*), it has no vertical motion (*No Motion*), its horizontal position changes from the center of the image to the right (*CH and CH, R*), and its vertical position is *Down*.

3 Tracking of Motion Regions

The tracking technique consists in an iterative method repeated for every frame in the video sequence. For our purpose, a frame ***F*** is defined as a set of *LBs* detected during the segmentation process of each frame:

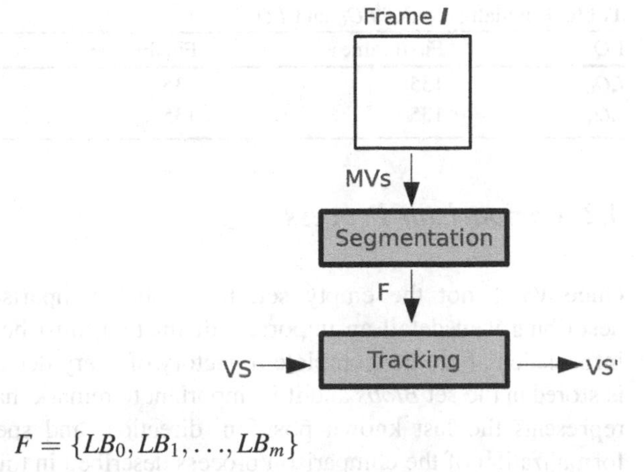

Fig. 1 Operation of the tracking process in every frame of the video

$$F = \{LB_0, LB_1, \ldots, LB_m\}$$

The information obtained by the tracking process is stored in a set called Video Sequence (*VS*). As the tracking process obtains *LOs*, **VS** is defined as:

$$VS = \{LO_0, LO_1, \ldots, LO_n\}$$

Figure 1 shows the set of actions to be executed when a frame *F* is processed. It can be observed how the H264/AVC motion vectors (*MVs*) are the input of the segmentation stage and *F* is the output. *F* and *VS* are the inputs of the tracking process and *VS'* is the output. In *VS'* new *LOs* are added or the attributes of the existing *LOs* are modified. Once all the frames of the video have been studied, the set *VS* contains all the detected objects, their positions, and trajectories. Now, the tracking process will be described in detail.

3.1 Initialization

At the beginning of tracking the set *VS* is the empty set (∅). Then, when first *LBs* are detected in the video sequence such set must be initialized. For example, if Table 3 represents the set *F* at frame 135, two *LOs* are created from *LBs* values in this way:

- LO(FirstFrame) ← LB(FrameNumber)
- LO(LastFrame) ← LB(FrameNumber)
- LO(Size) ← LB(Size)
- LO(Blobs) ← {LB}

In Table 4 the result of the initialization of LO_0 and LO_1 from LB_0 and LB_1 is shown.

Table 3 *LBs* used in initialization process

F at frame 135
LB_0: $(135, 3, [SR:1], [NM:1], [DW:1], [CH:1])$
LB_1: $(135, 4, [SL:1], [NM:1], [UP:1], [R:1])$

Table 4 Initialization of LO_0 and LO_1

LO	FirstFrame	FinalFrame	Size	Blobs
LO_0	135	135	3	$\{LB_0\}$
LO_1	135	135	4	$\{LB_1\}$

3.2 Comparison Process

Once *VS* is not the empty set, the global comparison process starts. Before describing it in detail an important distinction must be emphasized: *LOs* contain information about the complete trajectory of every detected object. This trajectory is stored in the set *Blobs* and it is important to remark that the last element in *Blobs* represents the last known position, direction, and speed of that object. In the formalization of the comparison process described in this section, this last element is referred as LB_{LO} while any element in *F* is named LB_j.

The comparison process is divided into two main stages. The first consists in checking three conditions between all the *LOs* in *VS* and all the *LBs* in *F*. These conditions or constraints are typically used in other tracking algorithms (Figure 2). From this comparison, a set of pairs (LO_i, LB_j) fulfilling such three conditions are obtained. In the second stage, a selection criteria is used to determine the final set of correspondences object-blob. This criteria is based on a value stored in **Decision Matrix** (*DM*). In this matrix, the result of the comparison between $LO_i \in VS$ and $LB_j \in F$ is stored in the row number *i* and the column number *j*. This position is identified as $DM[i,j]$ and the matrix dimension is $|VS| \times |F|$, being $|\ |$ the number

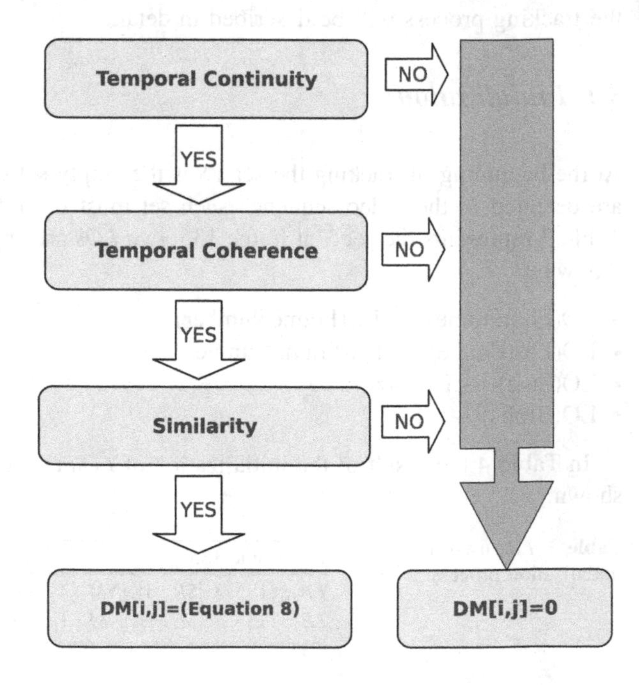

Fig. 2 Decision matrix construction

of elements of each set. As it can be observed in Figure 2, if at least one of the conditions are not satisfied $DM[i,j]$ takes the value 0. This means that LB_j is not considered a valid representation for the object LO_i in the processed frame. Now, each condition is studied.

Temporal continuity The first condition guarantees the assumption of continuity of motion along the time and more precisely tries to avoid confusions between moving objects in nearby regions of the image at different times. Then, an object LO_i and a blob LB_j satisfies the temporal continuity condition, if the result in Eq. 5 is *TRUE* (**Continuity**(LO_i, LB_j)).

$$LB_j(FrameNumber) - LO_i(FinalFrame) \leq DF \quad (5)$$

where DF is a configuration variable.

For example, if the parameter DF is equal to 10 and LB_{13}:

$$LB_{13} = (44, 5, [NM:1], [SD:1], [DW:1], [VL:1])$$

this LB can be only compared with those LOs whose parameter $FinalFrame$ is equal or greater than 34 ($LO(FinalFrame) \geq 34$).

Temporal coherence The next condition to satisfy is known as **temporal coherence** and it exploits the temporal and spatial consistency of appearance from frame to frame. If motion characteristics of a LO are known, a prediction about the position of the LO in a later frame can be made. For example, if intervals I_{HV} and I_{HP} of LB_{LO} have values *[Slow Right: 1]* and *[Left: 1]* and the value of I_{HP} in LB_j is *[Very Left: 1]*, the condition of temporal coherence between LO_i and LB_j is not fulfilled. That is because the motion direction is to the right and the new position of an object cannot be placed more to the left than the last position known. Now this condition is formalized. First, two preliminary definitions are given.

Let be $I_j^{a,b}(y)$ an ordered linguistic interval over the linguistic variable X_j, the **Position of the Interval in the Variable** (PIV) is a numerical value obtained from Eq. 6.

$$PIV(I_j^{p,q}(y)) = \begin{cases} \frac{p+q}{2}, & if\, p \neq q \\ p, & if\, p = q \end{cases} \quad (6)$$

For example, next linguistic interval is defined over the variable X_{HP} (Fig. 3): $I_{HP}^{2,3}(7) = [Left: 0.5;\ Center\ Horizontal: 0.5]$ and PIV is $PIV(I_{HP}^{2,3}(7)) = (2+3)/2 = 2.5$.

Let be X_j a linguistic variable, the **Position of the Central Label of a Variable** ($PCLV$) is a numerical value obtained from Eq. 7.

$$PCLV(X_j) = \frac{|X_j| + 1}{2} \quad (7)$$

being $|X_j|$ the number of labels of the variable.

As it can be observed in Fig. 4, $|X_{HV}|$ is 7 and $PCLV(X_{HV})$ is 4 ($(7+1)/2$).

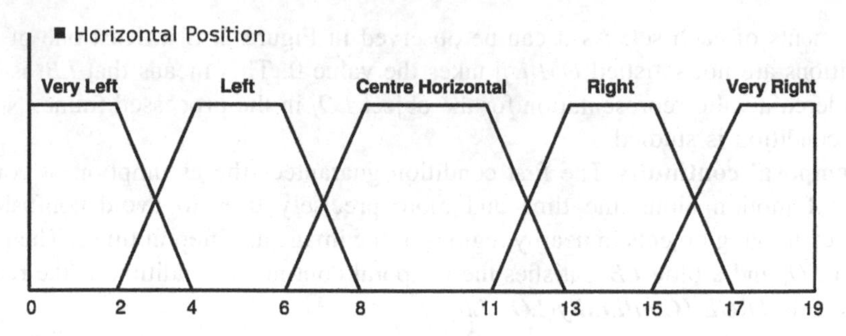

Fig. 3 Linguistic variable Horizontal Position (*HP*)

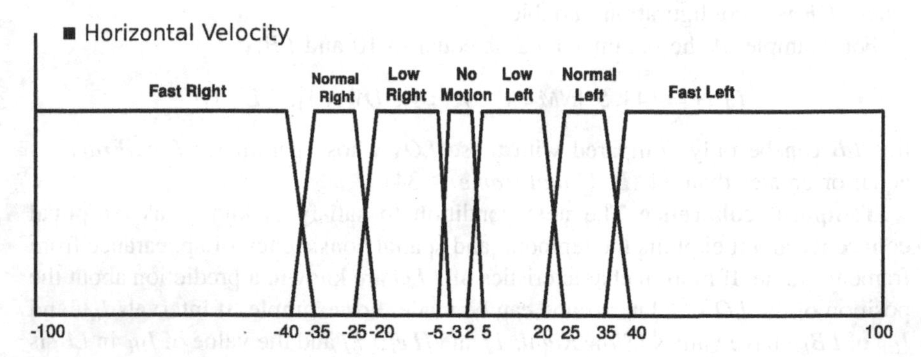

Fig. 4 Linguistic variable Horizontal Velocity (*HV*)

The design of the linguistic variables provides additional information that allows to give the next set of axioms:

1. $A_{HV}^{PCLV(X_{HV})}$ is the numeric position of the label *No Motion*.
2. From every interval I_{HV}, it can be determined the direction of displacement of an object:

   (a) $PIV(I_{HV}) > PCLV(X_{HV})$ represents motion to the left.
   (b) $PIV(I_{HV}) = PCLV(X_{HV})$ represents no motion.
   (c) $PIV(I_{HV}) < PCLV(X_{HV})$ represents motion to the right.

3. $A_{VV}^{PCLV(X_{VV})}$ is the numeric position of the label *No Motion*.
4. From every interval I_{VV}, it can be determined the direction of displacement of every object:

   (a) $PIV(I_{VV}) > PCLV(X_{VV})$ represents motion to the bottom of the image.
   (b) $PIV(I_{VV}) = PCLV(X_{VV})$ represents no motion.
   (c) $PIV(I_{VV}) < PCLV(X_{VV})$ represents motion to the top.

The temporal coherence between two *LBs* (**TCoherence**(LB_{LO}, LB_j)) is satisfied if at least one of next conditions is fulfilled for every pair $\{z = HV, y = HP\}$ and $\{z = VV, y = VP\}$:

- **Condition 1**: If $PIV(LB_{LO}(I_z)) > PCLV(X_z)$ then $PIV(LB_{LO}(I_y)) \geq PIV(LB_j(I_y))$.
- **Condition 2**: If $PIV(LB_{LO}(I_z)) < PCLV(X_z)$ then $PIV(LB_{LO}(I_y)) \leq PIV(LB_j(I_y))$.
- **Condition 3**: $PIV(LB_{LO}(I_z)) = PCLV(X_z)$.

Now an example is shown where LB_{LO} and LB_j are: $(26, 1, [SR : 0.75; NM : 0.25], [NM : 1], [DW : 1], [VL : 1])$ and $(27, 2, [S : 1], [NM : 1], [DW : 0.75; VDW : 0.25], [VL : 0.9; L : 0.1])$, respectively.

The condition for the first pair $\{z = HV, y = HP\}$ is evaluated:

$$PIV(LB_{LO}(I_{HV})) = PIV([SR : 0.75; NM : 0.25]) = (3 + 4)/2 = 3.5$$
$$PCLV(X_{HV}) = (7 + 1)/2 = 4$$

Being $PIV(LB_{LO}(I_{HV}))$ less than $PCLV(X_{HV})$, $PIV(LB_{LO}(I_{HP}))$ must be less than or equal to $PIV(LB_j(I_{HP})$.

Now, these values are calculated and it can be observed how the **Condition 2** is satisfied:

$$PIV(LB_{LO}(I_{HP})) = PIV([VL : 1]) = 1$$
$$PIV(LB_j(I_{HP})) = PIV([VL : 0.9; L : 0.1]) = (1 + 2)/2 = 1.5$$

Now, the condition for the pair $\{z = VV, y = VP\}$ is evaluated:

$$PCLV(X_{VV}) = (7 + 1)/2 = 4$$
$$PIV(LB_{LO}(I_{VV})) = PIV([NM : 1]) = 4$$

Being $PCLV(X_{VV})$ equal to $PIV(LB_{LO}(I_{VV}))$ **Condition 3** is fulfilled.

Then, as one condition for each one of the pairs $\{z = VV, y = VP\}$ and $\{z = HV, y = HP\}$ is satisfied, there exists temporal coherence between LO_i and LB_j.

Similarity Now, the similarity between (LO_i, LB_j) is measured using Eq. 8.

$$TD(LO_i, LB_j) < \gamma \tag{8}$$

where *TD* [11] is a distance measure between blobs indicating how different are their velocities and their positions in the scene and γ is a configuration variable. ($\gamma \in [0, 1]$)

Decision Matrix filling Finally, a value that discriminates between all the pairs satisfying the three mandatory conditions is stored in each position in *DM*. Greater values correspond to those pairs where (a) $LB \in (LO_i, LB)$, *LB* must be the blob that best represents LO_i in the current frame and (b) $LO \in (LO, LB_j)$, *LO* must be a real object in the sequence once the video is fully processed.

To compute this value, information obtained from the representation of *LOs* is used. More concretely the total number of frames in which the *LO* appears (*NTF*). Then:

$$DM[i,j] = NTF - TD(LO_i, LB_j) \tag{9}$$

being $NTF = LO_i(LastFrame) - LO_i(FirstFrame)$.

NTF favors the union of *LBs* with *LOs* with stable trajectories against objects appearing in a small number of images. That is because the first set of elements used to correspond with real objects in the video sequence, while the second set that corresponds to blobs is not correctly segmented. The negative value of the distance *TD* (Eq. 8) discriminates between objects with the same duration of appearance.

3.3 Selection of Correspondences and Aggregation of Blobs into Objects

Once the matrix *DM* is filled, the final correspondences between *LBs* and *LOs* are selected. This procedure is an iterative process where, in each iteration step, the maximum value in the matrix is selected taking into account that the relation between *LBs* and *LOs* in a frame is univocal: One *LB* only can be assigned to one *LO* and One *LO* only can receive the assignment of one *LB*. Now, an example of operation of the selection process is given taken as input Table 5, steps 1 to 3 of the iterative process are shown.

In this matrix, the maximum value is 24. Then (LO_6, LB_0) is a valid correspondence. Now, to meet the univocality constraint value 0 is assigned to every element at row 7 and column 0. The new maximum is 14 and the correspondence (LO_7, LB_1) is added. Row 8 and column 1 take a value of 0 and after this modification all the positions in the array store the value 0, then, this process is finished.

Now, for each pair (LO_i, LB_j) selected, the **aggregation operation** of LB_j into LO_i is done consisting of these operations:

Table 5 Decision matrix

1	LB_0	LB_1	2	LB_0	LB_1	3	LB_0	LB_1
LO_0	0	0		0	0		0	0
LO_1	0	0		0	0		0	0
LO_2	0	0		0	0		0	0
LO_3	0	0		0	0		0	0
LO_4	0	0		0	0		0	0
LO_5	14.5	0		0	0		0	0
LO_6	24	0.5		0	0		0	0
LO_7	0	14		0	14		0	0
LO_8	0	5.5		0	5.5		0	0

1. $LO_i(LastFrame) \leftarrow LB_j(FrameNumber)$
2. $LO_i(Size)$ is modified by Eq. 10.

$$LO_i(Size) \leftarrow \frac{LO_i(Size) \cdot |Blobs|}{|Blobs|+1} + \frac{LB_j(Size)}{|Blobs|+1} \tag{10}$$

3. $LO_i(Blobs) \leftarrow LO_i(Blobs) \bigcup LB_j$

4 Experimental Results

The two scenarios selected for experimentation are shown in Figs. 5 (groups of persons walking are observed) and 6 (a van and a bicycle are shown). The characteristics of each video sequence can be observed in Table 6.

The parameters used to evaluate the performance of the tracking technique are: *True-Positive (TP)*: the result provided by the system is a real object in the scene; *False Positive (FP)*: the system solution is not an object. Using these values two quality measures, *Detection Probability (DP=TP/Real Objects)* and *Precision (P=TP/(TP+FP))* (Table 7).

Fig. 5 Pictures of first set of experiments

Table 6 Experiments characteristics

Experiment	Time	Frames	Resolution
1	5.2	9360	320 × 240
2	3.5	6300	320 × 240

Fig. 6 Pictures of second set of experiments

Table 7 Tracking evaluation results

Experiment	Objects	TP	FP	DP (%)	P (%)
1	98	91	62	92	59
2	102	102	11	100	90

5 Conclusions

In this work, a novel approach for tracking objects directly on the compressed video data is presented. The average processing time of each frame is between 10 and 15 milliseconds. These values are less than 40 milliseconds and this method can be considered that works in real time. This is possible because complex pixel algorithms are discarded and replaced by new ones based on computations on the integers used to define and represent the order of fuzzy sets over fuzzy variables. Another contribution is the linguistic description of the trajectory of an object stored in the *Linguistic Objects*. Furthermore, the proposed design of fuzzy variables includes the representation of the scenario and information about the motion in the scene that can be easily recovered and processed. As future works, it is very important to improve precision results in videos with small motion objects.

For example, updating dynamically the configuration variables of the system used in Eqs. 5 and 8. Such modification should be conditioned by the level of motion in the scene or the number of objects in the same frame.

References

1. Brox T, Rousson M, Deriche R, Weickert J (2010) Colour, texture, and motion in level set based segmentation and tracking. Image Vis Comput 28(3):376–390
2. Fuentes LM, Velastin SA (2006) People tracking in surveillance applications. Image Vis Comput 24:1165–1171
3. Herrera F, Martinez L (2000) A 2-tuple fuzzy linguistic representation model for computing with words. IEEE Trans Fuzzy Syst 8:746–752
4. Hu W, Tan T, Wang L, Maybank S (2004) A survey on visual surveillance of object motion and behaviours. IEEE Trans Syst Man Cybern 34(3):334–352
5. Johansson B, Wiklund J, Forssn P, Granlund G (2009) Combining shadow detection and simulation for estimation of vehicle size and position. Pattern Recognit Lett 30(8, 1):751–759
6. Kalman RE (1960) A new approach to linear filtering and prediction problems. Trans ASME J Basic Eng 82:35–45
7. Motamed C (2006) Motion detection and tracking using belief indicators for an automatic visual-surveillance system. Image Vis Comput 24:1192–1201
8. Muoz-Salinas R, Medina-Carnicer R, Madrid-Cuevas FJ, Carmona-Poyato A (2009) People detection and tracking with multiple stereo cameras using particle filters. J Vis Commun Image Represent 20(5):339–350
9. Richardson IEG (2003) H.264 and MPEG-4 Video compression. Willey, New York
10. Rodriguez-Benitez L, Moreno-Garcia J, Castro-Schez JJ, Jimenez L (2008) Fuzzy logic to track objects from MPEG video sequences. In: 12th international conference on information processing and management of uncertainty in knowledge-based systems, IPMU'08, pp 675–681
11. Solana-Cipres C, Fernandez-Escribano G, Rodriguez-Benitez L, Moreno-Garcia J, Jimenez-Linares L (2009) Real-time moving object segmentation in H.264 compressed domain based on approximate reasoning. Int J Approx Reason 51(1):99–114
12. Valera M, Velastin SA (2005) Intelligent distributed surveillance systems: a review. In: IEEE proceedings of vision, image and signal processing, pp 192–204
13. Yilmaz A, Javed O, Shah M (2006) Object tracking: a survey. ACM Comput Surv 38(4):1–45 (Article 13)
14. Zhou J, Zhang XP (2005) Video object segmentation and tracking using probabilistic fuzzy C-means. IEEE workshop on machine learning for signal processing, pp 201–206

Key Technologies and Prospects of Individual Combat Exoskeleton

Peijiang Yuan, Tianmiao Wang, Fucun Ma and Maozhen Gong

Abstract With the development of modern warfare, the load-carrying of the soldier is more and more heavy. The overload affects the soldier's ability and readiness and causes acute and chronic musculoskeletal injuries. Exoskeleton can greatly reduce the oxygen consumption of the soldiers and support energy for transferring, running, and jumping, and enhance locomotor and operational capability of the soldiers. The Berkeley Lower Extremity Exoskeleton (BLEEX), Raytheon XOS, Human Universal Load Carrier (HULC), and Hybrid Assisted Limb (HAL) are the most typical exoskeleton robots. The first three are individual combat exoskeletons in support of U.S. Defense Advanced Research Projects Agency (DARPA). The HAL is mainly used for civilian. We research and analyze the structural characteristics and joints movement of the lower limb and structural design, power system, control system, and so on key technologies of those four exoskeletons. At last, we predict the trend of prospective individual combat exoskeleton.

This work is partially supported by National High Technology Research and Development Program (863 Program) of China under Grant No. 2011AA040902, National Natural Science Foundation of China under Grant No. 61075084, and Fund of National Engineering and Research Center for Commercial Aircraft Manufacturing (Project No. is SAmc12-7s-15-020).

P. Yuan (✉) · T. Wang · F. Ma · M. Gong
School of Mechanical Engineering and Automation, Beihang University,
37 Xueyuan Road, Beijing 100191, China
e-mail: itr@buaa.edu.cn

T. Wang
e-mail: itm@buaa.edu.cn

F. Ma
e-mail: giantmfc@163.com

M. Gong
e-mail: gongmaozhen@126.com

F. Sun et al. (eds.), *Knowledge Engineering and Management*,
Advances in Intelligent Systems and Computing 214,
DOI: 10.1007/978-3-642-37832-4_28, © Springer-Verlag Berlin Heidelberg 2014

Keywords Individual combat · Exoskeleton · Power · Control · Prospect

1 Introduction

Individual combat is that physical quality of the soldier is eminent, and a soldier has the combat capability to defeat doses of enemy soldiers. With the development of modern warfare, the load-carrying of the soldier is more and more heavy. For instance, the load-carrying including weapons and protections of the soldier is 20 kg in World War II and the load is 90 kg in Vietnam War, while the load is up to 200 kg in Gulf War [1]. Excessive weight-bearing will increase oxygen consumption and cause physical premature decline, and finally will lead to the decline of soldier's combat capability. Especially in the harsh environment of the jungle, highlands, and street fighting, the physical consumption of the soldier is greater. Moreover, the overload causes acute and chronic musculoskeletal injuries at the lower limb joints and the back leading to related physiological and psychological health problems [2]. In this case, the individual combat exoskeleton robot emerged as soldiers' auxiliary equipment. It can greatly reduce the oxygen consumption of the soldiers and support energy for transferring, running and jumping, and enhance locomotor and operational capability of the soldiers. It also obviously reduces the risk of wound for soldiers (Fig. 1). The exoskeleton robot, also called wearable robot, is a mechanical means that human can wear. It integrates with human intelligence and robot physical. The exoskeleton robot can complete the task that soldier cannot complete alone by relying on human intelligence to control the robot. The concept of exoskeleton robot derives from the term of biology exoskeleton. The concept was introduced 60 years ago when the company General Electric crafted prototypes of autonomous manipulators and walking structures with the purposes of duplicating the biomechanical strength of the human body [2]. The advantages of exoskeleton have caused

Fig. 1 Four typical individual combat exoskeletons

great interest of scientists who began to study artificial exoskeleton from the point of view of bionics one after another [3].

The exoskeleton robot technology is the integration of sensing, control, information, mobile computing, and other disciplines technology. The leading countries on exoskeleton research are United States and Japan. The U.S. Defense Advanced Research Projects Agency (DARPA) began the project of Exoskeleton for Human Performance Augment (EHPA) in 2000. In support of this project, the University of California at Berkeley developed the Berkeley Lower Extremity Exoskeleton (BLEEX), and Raytheon company developed XOS (XOS is short for exoskeleton) series exoskeletons. Over the same period, the U.S. Lockheed Martin company developed the Human Universal Load Carrier (HULC). Those exoskeleton robots are shown as Fig. 1. The BLEEX, the first load-bearing and energetically autonomous exoskeleton, can bear 45 kg, and the battery for control system can be maintained more than 20 h [4]. In order to address issues of field robustness and reliability, the BLEEX is designed such that, in the case of power loss, the exoskeleton legs can be easily removed and the remainder of the system can be carried like a standard backpack [5–8]. The XOS-2 can complete thousands of push-ups, easily lift 200 pound, and split 3 inches thick planks by one-handed. But the energy problem has not been resolved, the XOS-2, whose weight is 68 kg, can only work 40 min relying on its own internal battery. It must rely on external hydraulic lines to test or performance for long time [4, 9]. The HULC is made of titanium and casting carbon fiber materials and its shape is simplest. It can bear 90.7 kg. The soldier whose load is 90 kg can run one hour under the speed of 4.8 km/h, while the sprint speed can reach 16 km/h [10]. The Cyberdyne Company founded by the Yoshiyuki Sankai Professor of the University of Tsukuba developed hybrid assisted limb (HAL) exoskeleton robot. The HAL-5 is mainly used for civilian, its structure is simple and the weight of the whole body is only 23 kg. In addition to several above-mentioned typical exoskeleton robots, HERCULE (France) and the NTULEE (Singapore) exoskeleton robots have obtained a large amount of research achievements [11].

The research and development of exoskeleton robot has become a new research focus in the field of robotics, electromechanical engineering, automatic control, bio-engineering, and artificial intelligence disciplines in recent years and gradually been widely used in military research, industrial production, and daily life [12]. From a functional point of view, the exoskeleton robot is mainly composed of the mechanical structure system, the information processing system, control system, and power system. Figure 2 researches exoskeleton robot by several key technologies and predicts the future development trend.

2 Structure Research of Human Lower Limb

To optimize the gait of the exoskeleton, we must understand the structural characteristics and joints movement of the lower limb. Through the research and analysis of the lower limb structure of clinical medicine, sports science, and

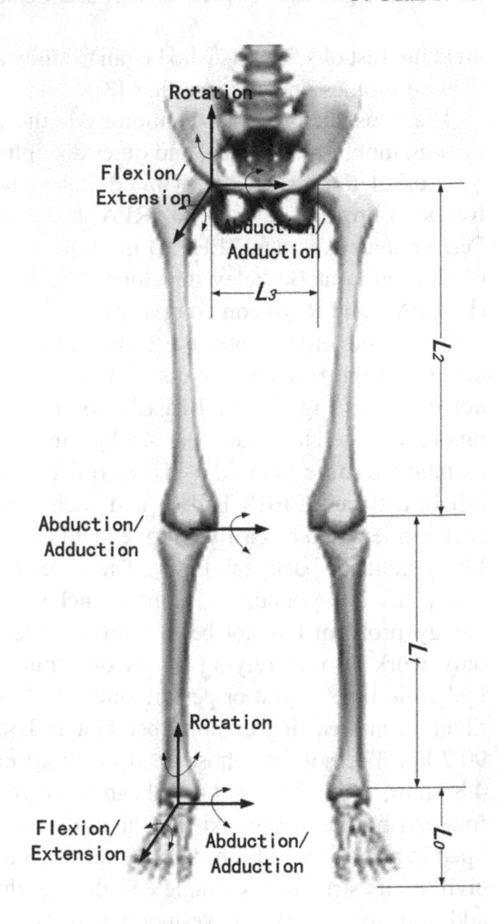

Fig. 2 The seven-degree-of-freedom model of lower limb

kinesiology, hip joint is a ball-and-socket joint which consists of spherical femoral head and the concave acetabula component. It is a typical triaxial joint and has 3 degrees-of-freedom (DOF). It can bend, extend, and rotate, and so on multi-directional movement in three mutually perpendicular motion axes. It is the deepest joints of the human body. Due to supporting the upper body weight, the joint is both solid and flexible. Knee joint is the most complex joint of the body. It is a hinge joint and can bend and extend. It also can do tiny spin movement under the knee bending state. But in general, the range of spin is so slightly that is ignored in modeling in order to simplify the design. It is designed as hinge joint and has only 1DOF. The ankle joint is also a ball-and-socket joint and has 3DOFs. The feet can do flexion/extension, abduction/adduction, and rotation movement around the ankle joint. Human lower limb is linked by joints and ligaments to achieve a flexible, efficient, and stable movement. Due to the lower limb of the human body is too complex, the movement of joints occurred in any one direction is inevitably accompanied by the movement of the other two directions. It is the manifestation of motion complexity and is a difficult problem of bionics. Now human lower limb is simplified to a 7DOFs. 3DOFs is located in hip joint, 1DOF is located in knee joint, and 3DOFs is located in ankle joint.

3 Comparison of the Key Technologies of Typical Exoskeleton Robot

3.1 The Mechanical Structural Design

The mechanical structural design should meet the requirements of ergonomics. Ergonomics notable feature is that it takes Man–Machine–Environment as a single system in which the three elements are interaction and interdependence among man, machine, and environment. According to the studies of probability distribution of the ergonomic measurement, there is a directly proportional relationship between the sizes of various parts of the lower limb for Chinese young people. According to above-mentioned study, we can confirm the size between the centers of the lower limb skeleton model joints.

Exoskeleton robots can be classified according to muscle strength supporting parts: upper limb exoskeleton, lower limb exoskeleton, whole-body exoskeleton, and specific joint exoskeleton [10]. The BLEEX is a lower limb exoskeleton and the mechanical structure of each leg has 7DOF which are three actuated DOFs in Hip, knee, ankle joints driven by three pieces of hydraulic mechanisms. There is each of the two following DOFs in the hip and ankle joints which are spring-loaded and free spinning. Moreover, there is a following DOF in the center of waist to adjust motion between two legs [4]. The BLEEX has 15 DOFs that is similar to the quantity of DOF of the human body. Although a seemingly complicated device, the exoskeleton does not require any special training to use it [13]. The vest is designed from several hard Polycarbonate surfaces connected compliantly together to conform to the pilot's chest, shoulders, and upper back, thereby preventing concentrated forces between the exoskeleton and the wearer [5]. It only needs to bundle in the shank and feet. It is very ergonomic and convenient [11, 14–17]. The XOS-2 is a whole-body exoskeleton and not entirely imitates the shape of human. The size of exoskeletons is much larger than human body's. Each leg only has 3DOF and moves objects through auxiliary fixture. Due to the larger size, the target is not conducive to covert. The XOS-2 must simplify the structure and reduce the size in order to equip to U.S. troops. The shape of the HULC is the simplest among those exoskeletons. The HULC is a lower limb exoskeleton. It is made of titanium metal and casting carbon fiber material in order to reduce the weight of the equipment. Each leg has only one actuated DOF and power plant is placed on the back. The HAL is also a whole-body exoskeleton and each leg has only two actuated DOFs. As a result of DC motors with harmonic drive placed directly on the joints, so hip and knee joints have a protruding part [18]. With a constantly updated version of HAL series robots, structure of the HAL-5 is very simple and its weight is 21 kg and it can enhance the bearing capacity to more than ten times [19]. Because of the limited power of the motor, it is unable to meet the requirements of individual combat. The structure of future exoskeleton should be simple, lightweight, wearable, and can be put on and take off smoothly without professional training. It should be able to adapt to various soldiers, the size of length and width direction should be able to adjust to meet different soldiers.

3.2 The Power System

The soldiers go to the regions of hundreds of kilometers away and often climb the mountains or wading, so requirements for the power system are particularly rigorous. However, the development of the power mechanism cannot meet the requirements of individual combat. Currently, the most commonly used driving systems are electrical, pneumatic, and hydraulic devices. A study determined that electric motor actuation significantly decreased power consumption during level walking in comparison to hydraulic actuation. However, the weight of the implementation of the electrically actuated joint was approximately twice that of their hydraulic actuated joint (4.1 vs. 2.1 kg) [4]. The power of pneumatic driving system is much smaller than hydraulic driving system. Most of the military exoskeletons use hydraulic device to drive mechanical system. The BLEEX applies hybrid power union; the hydraulic power is used to drive the mechanical movement of the exoskeleton while battery is used to drive the computer system. Hydraulic power is driven by three hydraulic cylinders and the maximum load is 45 kg. The longest work time of battery is 20 h [20, 21].

The XOS-2 is also driven by hydraulic power. However, the XOS-2 adapts rotary hydraulic actuators located directly on the powered joints of the device instead of linear hydraulic actuators [4]. The hydraulic actuators in the XOS-2 cannot act by themselves. In order for the actuators to work properly, a torque pump in the valve must be used to add high pressure to the hydraulic fluid that runs through the hydraulic lines and ultimately into the actuators. The company designs 30 hydraulic components required itself. But the energy problem has not been resolved, the XOS-2, whose weight is 68 kg, can only work 40 min relying on its own internal battery. It must rely on external hydraulic lines to test or performance for a long time. It greatly limits application of the XOS-2. Due to the larger size, it cannot be continuously worn when the battery is exhausted. It must be given up or find charging power before the battery is exhausted, so operational cost is too high. The HULC applies single hydraulic cylinder system and uses two lithium polymer batteries weighted 3.6 kg to provide power for the control system and mechanical structure. Its maximum load is up to 90.7 kg. After a full charge, the wearer whose load is 90 kg can run 1 hour under the speed of 4.8 km/h, while the sprint speed can reach 16 km/h. The HAL is driven by the motor and harmonic reducer. It uses lithium battery power located in the waist to supply for the control system. It can easily lift a common adult. Due to the power limit of the motor, the HAL cannot be used for the military and it has been used for civilian that need not too large power. The prospective exoskeleton power system should be able to adapt to different environments, for example, plateau, rainforest, desert, and wading. It should provide adequate power at least 24 h. When the battery is exhausted, the equipment can easily be brought back. It should be able to easily destroy the core components of the exoskeleton or power system to prevent enemy using when the soldiers have to give up.

3.3 The Control System

The control system is the core of the exoskeleton robot. If control system cannot fleetly process information transmitted by sensors and drive power system, man and exoskeleton cannot match up well. The BLEEX installed acceleration sensors, force sensors, encoders, and so on a large number of sensors can real-time monitor of the state of motion of the human body [4]. The BLEEX uses high-speed synchronous ring network topology to achieve the communication of control information. First, the sensor system detects the motion of independent and subconscious of soldier. Then control algorithm disposes information transmitted by sensors and passes drive signal to all actuators to save the soldier's strength. The XOS-2 control principle is similar to the aircraft linear conduction control. The designer provides force sensors for each joint. When the wearer want to move the limb to do a certain action, all the force sensor immediately transmit motion data to the central control computer. Then exoskeleton takes action to help users by high-speed calculation of central control computer. According to the calculations, the central control computer indicates the proper position of the hydraulic components to move piston. The control principle of XOS-2 is that the forces of the human body received from the force sensors are minimized. In other words, because the exoskeleton bears the load instead of soldier, the soldier can get help without having to contribute at the beginning. The sensors of relevant position will not continue to perceive the force applied by the human body. It is shown as Fig. 3. Jack Dobson, the XOS-2 inventor, called this control principle "get out of the way." The control system is designed to counteract the force on the sensors and take the force on the sensors to zero by opening the valves, so that the soldier does not feel any force on their body [9]. This control principle looks simple, but the actual implementation is not easy. Sensors detect human force status hundreds of times per second all the time and transmit to the central control computer immediately. Central control computer must immediately complete the operation and give commands to the hydraulic components to finish the action. If the process is not quick enough, users will feel like walking in the water. The real-time control is a big challenge for research personnel. The quantity of sensors of HULC is fewer than other exoskeleton. It only equips with two pressure sensors which are in charge of sending traveling speed and walking posture of soldiers to control

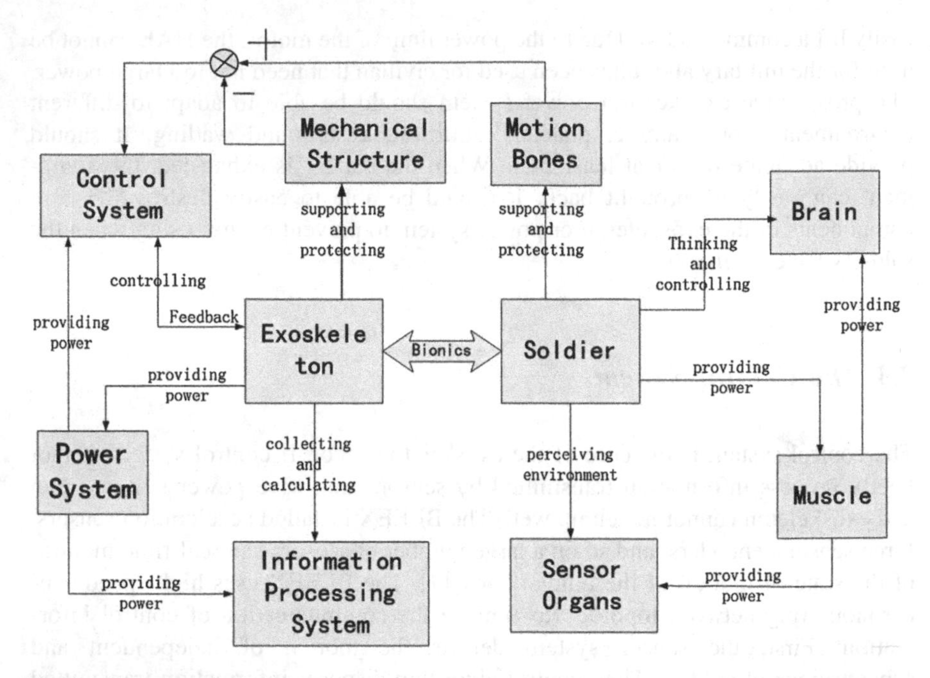

Fig. 3 The control principle of exoskeleton

computer in the foot pad of the internal. The portable microcomputer is responsible for controlling the hydraulic system to provide the pressure to reduce the force contributed [22]. There is a poor synchronization with human motion because of being unable to obtain real-time movement data of the soldiers. It is worth noting that the HAL system does not transfer the load to the ground surface in distinction to the other three exoskeletons. So it cannot bear too heavy load. The HAL-5 has two sets of control systems: EMG-based system and walking pattern-based system. The former uses electrode stuck on the human skin to test weak bioelectrical to determine contribute way without using the force sensors [23]. The HAL perceives action intention of the operator by collecting the EMG of the lower part of the hip joint and the upper part of the knee joint and control the movement of the motor in the hip and knee joint. The latter can remember and imitate the characteristics of user action. For example, user will more closely match with a weak leg; it can also solve the problem of some biological current weak. However, EMG signals are easily interference by electromagnetic noise. The EMG devices need to be connected with a special adhesive on the skin surface in order to accurately measuring. It is very inconvenient for soldiers. Prospective individual combat exoskeleton could be introduced in default parameters and self-learning mechanism. Because of personalized difference of human walking and the use of walking motion is random according to the road conditions. It is difficult to obtain a fixed output setting. Default parameter settings of the walking motion are vital and must use this parameter to control the operation of various circumstances, in order to achieve the purpose of synchronized movement coordinated with the soldiers. Exoskeleton can remember movement parameter of soldiers by using self-learning mechanism. The control system determines the soldier's movement trends and provides the impetus for the soldiers in advance. This will better match with the body's movement.

4 The Prospects of the Future Individual Combat Exoskeleton Robot

In the recent decade, many countries extravagantly develop individual combat. Some U.S. troops have equipped the Land Warrior system which is including five subsystems, weapons subsystem, protective clothing and individual equipment subsystem, computer/radio subsystem, helmets component subsystem, and software subsystems [24]. The total weight of the Land Warrior system is 45 kg. Moreover, the total weight of weapons system of Project-Wolf/soldier2000 (Russian) and FELIN (France) is more than 50 kg [25, 26]. This overload causes acute and chronic musculoskeletal injuries at the lower extremities joints and the back leading to related physiological and psychological health problems. These injuries often compromise the soldier's ability and readiness for his current and future missions.

After comparing the structural design, power system, and control system of four typical exoskeletons, we realize the development situation of individual combat exoskeleton. We predict the trend of prospective individual combat exoskeleton. It is shown as Fig. 4.

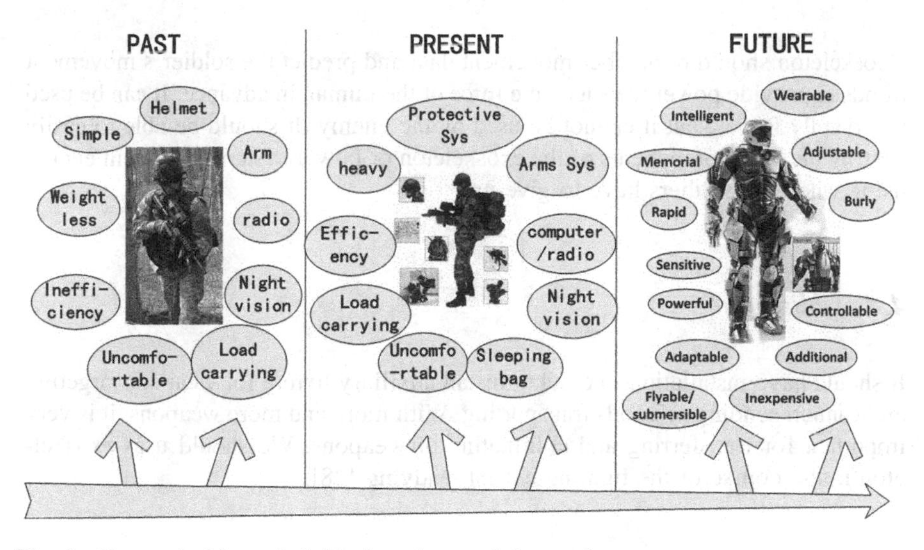

Fig. 4 The trend of future individual combat exoskeleton robot

4.1 Wearable, Burly, and Adjustable

The future individual combat exoskeleton robot should be very easy to put on and take off without man–machine training. The man–machine interface should be comfortable and effective [27]. It should be able to withstand the weight of several of military supplies of the back. Institutional strength should meet any impact load bought by tactical action of the soldiers. We should develop new materials such as artificial muscle fibers to reduce the weight of the exoskeleton. The size of exoskeleton should be suitable for the size of soldier. When a soldier wears the exoskeleton, he can finish run, squat, crawl, and crossing and so on actions that he want to do. The exoskeleton should be able to adapt to various soldiers, the size of the length and width direction should be able to adjust.

4.2 Sensitive, Rapid

Varies sensors can obtain movement data of soldier and transits to control system without any delay. Ideally, the sensors will be part of a computerized system that feels what the user is doing and initiates the machine to do the work without the user straining [13]. The control computer must immediately complete the data that detected by sensors and give commands to the hydraulic components to finish the operation without dragging.

4.3 Controllable, Intelligent, and Memorial

Exoskeleton should remember movement data and predict the soldier's movement trends to provide power to reduce the force of the human in advance. It can be used by friendly forces, but it cannot be used by the enemy. It should be able to easily destroy the core components of the exoskeleton or power system to prevent enemy using when the soldiers have to give up.

4.4 Additional

It should have installation to quickly install auxiliary fixture for weapons targeting and cannon sending or goods transporting. With more and more weapons, it is very important for transferring and collimating of weapons. We should uses exoskeleton in the course of the fighting except readying [28].

References

1. Zhang K (2011) The structure and motion analysis of individual soldier exoskeleton. J Sci Technol Innov Her 13:224–226
2. Knapik Maj(ret) J (2000) Physiological, biomechanical and medical aspects of soldiers load carriage. In: Meeting on soldier mobility: innovations in load carriage system design and evaluation, Kingston, pp 27–29
3. Yang Z, Zhang J, Gui L, Zhang Y (2009) Summarize on the control method of exoskeleton robot. J Nav Aeronaut Astronaut Univ 24:520–525
4. Dollar AM, Herr H (2008) Lower extremity exoskeletons and active orthoses: challenges and state-of-the-art. J IEEE Trans Robotics 24:1–15
5. Kazerooni HW (2005) Exoskeletons for human power augmentation. In: 2005 IEEE/RSJ international conference on intelligent robots and system. IEEE Press, New York, pp 3120–3125
6. Zoss A, Kazerooniw H (2006) Design of an electrically actuated lower extremity exoskeleton. J Adv Robotics 20:967–988
7. Zoss A, Kazerooni HW, Chu A (2005) On the mechanical design of the berkeley lower extremity exoskeleton (BLEEX). In: 2005 IEEE/RSJ international conference on intelligent robots and system. IEEE Press, New York, pp 3120–3125
8. Kazerooni HW (2005) The berkeley lower extremity exoskeleton. In: International conference on field and service robotics, Port Douglas, pp 17–20
9. Dombrowski P, Coval D (2012) Private first class iron man. J Swanson Sch Eng 04:149–157
10. Lee H, Kim W, Han C (2012) The technical trend of the exoskeleton robot system for human power assistance. Int J Precis Eng Manuf 13:1491–1497
11. Ferris PD (2010) Robotic lower limb orthosis: goals obstacles and current research. In: the 34th annual meeting of the american society of biomechanics, symposia: robotic lower limb ortheses and prostheses, San Diego, pp 324–332
12. Yin J (2010) Analysis and design of wearable lower extremity exoskeleton. J Syst Simul 12:7–18
13. Hirsh C, Karloski D (2009) Design and implementation of mechanized exoskeletons in the armed forces. In: the ninth annual freshman conference, Pittsburgh, pp 301–307
14. Raade J (2004) Monopropellant-driven free piston hydraulic pump for mobile robotic systems. J Dyn Syst Meas Control 126:75–81
15. Kazerooni H, Steger W, Huang C (2006) Hybrid control of the berkeley lower extremity exoskeleton (BLEEX). Int J Robotics Res 25:561–573
16. Kazerooni D, Racine W (2005) On the control of the berkeley lower extremity exoskeleton (BLEEX). In: Proceedings of the 2005 IEEE international conference on robotics and automation, Pittsburgh, pp 4364–4371
17. Chu A, Kazerooni H, Zoss A (2005) On the biomimetic design of the berkeley lower extremity exoskeleton (BLEEX). In: Proceedings of the 2005 IEEE international conference on robotics and automation, Pittsburgh, pp 4356–4363
18. Maeshima S, Osawa A, Nishio D, Hirano Y (2011) Efficacy of a hybrid assistive limb in post-stroke hemiplegic patients: a preliminary report. J BMC Neurol 07:96–101
19. Suzuki S, Kawamoto A, Hasegawa D (2012) Intention-based walking support for paraplegia patients with robot suit HAL. J Climb Walk Robots 07:383–408
20. Ghan J, Kazerooni H (2006) System identification for the berkeley lower extremity exoskeleton (BLEEX). In: Proceedings of the 2006 IEEE international conference on robotics and automation, Pittsburgh, pp 3477–3484
21. Steger R, Kim SH, Kazerooni H (2006) Control scheme and networked control architecture for the berkeley lower extremity exoskeleton (BLEEX). In: Proceedings of the 2006 IEEE international conference on robotics and automation, Pittsburgh, pp 3469–3476
22. Okamura S, Tanaka A (2011) Exoskeletons in neurological diseases—current and potential future applications. J Adv Clin Exp Med 06:227–233

23. Steger R, Kim SH, Kazerooni H (1999) EMG-based prototype powered assistive system for walking aid. In: Proceedings of the Asian symposium on industrial automation and robotics (ASIAR'99), Bangkok, pp 229–234
24. Cao X (2010) The analysis of the development of American soldiers system. J Mod Mil 02:50–54
25. Li H, Wang H, Zhang P (2012) Application prospect of exoskeleton equipment in future individual soldier system. J Mach Des Manuf 03:275–276
26. Li H (2010) The advanced soldiers' equipment system (FELIN) of French army finished the experiment evaluation. J Arms Equip 01:67–69
27. Herr H (2009) Exoskeletons and orthoses: classification, design challenges and future directions. J Neuro-Eng Rehabil 06:1743–1751
28. Baklouti M, Saleh JA, Monacelli E, Couvet S (2009) Human machine interface in assistive robotics: application to a force controlled upper-limb powered exoskeleton. J Adv Robotics 12:211–221

Adapting Real-Time Path Planning to Threat Information Sharing

Zheng Zheng, Xiaoyi Zhang, Wei Liu, Wenlong Zhu and Peng Hao

Abstract Information sharing is an important characteristic of cooperative flights, bringing forward new challenges to real-time path planning approaches. In this paper, problems brought out by threats information sharing are proposed, while distinguishing characteristics of threats environment with information sharing are analyzed. Based on these, solutions to these characteristics and a new path planning approach based on three strategies are advanced. At last, to test the applicability of the proposed approach, it is implemented to the improvement of a real-time Rapidly-exploring Random Tree (RRT) algorithm. The effectiveness and efficiency of the resultant algorithm are verified by stochastic and representative scenarios, which show that the new approach can be more adaptive to the threat information sharing environments.

Keywords Path planning · Real-time path planning · Multi-UAVs · Information sharing

1 Introduction

Cooperation of multiple unmanned aerial vehicles (UAVs) in complicated environment is necessary and inevitable. Rather than operate separately, UAVs cooperate in two ways: (1) by sharing information among the team; and (2) by

This research work was supported by National Natural Science Foundation of China (Grant no. 60904066).

Z. Zheng (✉) · X. Zhang · W. Liu · W. Zhu · P. Hao
Department of Automatic Control, Beijing University of Aeronautics and Astronautics, Beijing, China
e-mail: Zhengz2011@gmail.com

F. Sun et al. (eds.), *Knowledge Engineering and Management*,
Advances in Intelligent Systems and Computing 214,
DOI: 10.1007/978-3-642-37832-4_29,

coordinating their tasks [1], in which information sharing is central to cooperation and coordination [2]. In real-time path planning problems, where the location of threats are uncertain and changed in real time, the impacts of information sharing on single UAVs path planning are especially great. This paper concentrate on the impacts of threat information sharing on real-time path planning.

Real-time path planning for UAVs has been considered as a fundamental and necessary mission in diverse military applications, such as reconnaissance, rescue, navigation, and guidance of air and precision strike [3]. However, case studies indicate that many real-time path planning approaches does not come up to the expectations. For example, for MILP-based methods [4] and bouncing-based A2D/A3D methods [5], usually no effects of sharing threat information on resulting path are discovered; for RRT methods, sharing threat information may results in a longer path length and computational times; for the behavior-based approach, such as the FBCRI-based method [6] and BCV method [7], if considering the shared information as theirs inputs, the path obtained even gets worse. In these cases, the approaches tend to fall into iterations of previous trajectory, failing to converge to the goals as a result.

Based upon the findings, two questions arise:

(1) What are the characteristics of the shared threat information causing the results not coming up to the expected?
(2) Can we embed techniques into existing methods to improve their performance under threat information sharing environment?

In this paper, both questions will be explored. Characteristics of shared threat information, including the increasing amount, unconnectedness and imbalance, and their effects on real-time path planning are analyzed in detail. Furthermore, the improving strategies for eliminating negative effects are put forward. Based on the analysis, a new real-time path planning approach, embedding the strategies, is proposed. The effectiveness and efficiency of the approach are verified by its realization on a RRT algorithm [8] and representative test examples, which show that embedding the strategies into a real-time path planning approach can really make it more adaptive to threat information sharing and improve the quality of flight paths.

2 Problem Formulation

The mission scenario considered throughout this paper comprises the following elements:

- A bounded $L_x \times L_y$ mission environment in which the UAVs operate, denoted as E. It is represented in the UAVs information bases as a grid of cells: $E = \{w|x = 1...L_x, y = 1,...L_y, w = (x, y)\}$.

- M threat sources, denoted as $T_k, k = 1, \ldots, M$. The position of T_k is denoted as $w_{T_k} = (x_{T_k}, y_{T_k})$, in which $(x_{T_k}, y_{T_k}) \in E$ and $k = 1, \ldots, M$. This paper applies the description method of obstacles distribution based on probabilistic risk presented in [9].
- N UAVs, $\{u_k | k = 1, \ldots, N\}$. The parameters of them will be described later.
- Target of the kth UAV: $P_k^{tar} = (x_k^{tar}, y_k^{tar})$, in which $(x_k^{tar}, y_k^{tar}) \in E$ and $k = 1, \ldots, N$.
- For the UAVs, the threat information is unknown before their flight. Sensors equipped on the UAVs have limited ranged, and each UAV shares its new obtained information with others. The characteristics of information sharing will be presented in Sect. 3.

The mission of the UAV team is to escape from obstacles and move to targets. Note that, given the focus in this paper, we suppose that the UAVs cooperation only comes from their threat information sharing. Besides, to guarantee the existence of feasible paths and the precondition of threat information sharing, assume that the targets of the UAV team are in the supplement of threat area.

The state of UAV u_k at time t includes the following parameters:

- A unique ID, u_k, used to tag information sent out to other UAVs.
- Position$(x^{u_k}(t), y^{u_k}(t))$, denoted as $w^{u_k}(t)$, in which $x^{u_k}(t) = 1, \ldots, L_x, y^{u_k}(t) = 1, \ldots, L_y$.
- Speed $v^{u_k}(t)$, which is constrained by $v_{min} \leq v^{u_k}(t) \leq v_{max}$;
- Heading angle $\psi^{u_k}(t)$;
- The turning radius $r^{u_k}(t)$, which is constrained by $r^{u_k}(t) > r_{min}$;
- At any time t, the detection radius of sensor is R_k, and the sensory region is defined by the positions whose distances with $(x^{u_k}(t), y^{u_k}(t))$ are not more than R_k. Note, assume that R_k is larger than that of a separate threat region, so that the UAV can sense and calculate the threat region before stepping into it.
- At any time t, the communication radius of u_k is C_k, and the communication region is defined by the positions whose distance with $(x^{u_k}(t), y^{u_k}(t))$ is not more than C_k. If $C_k = 0$, there is no information sharing among UAVs.
- The period of the data communication is T.

3 Threat Information Base and Its Characteristics

3.1 Treat Information Base

As mentioned earlier, each UAV in threat information sharing environments carries its own subjective information base, representing its view of the mission status. The threat information base of a UAV comes from two sources: (1) Information generated by the UAVs own actions; and (2) Information received from other UAVs. At time t, the threat information base (*TIB*) of a UAV can be denoted as $\Xi^{u_k}(t)$, in which u_k is the ID of UAV. $\Xi^{u_k}(t)$ is composed of a lot of elements (called threat elements), each of which corresponds to a cell c. A threat element I is defined by a triple (w_I, d_I, t_I), in which:

- w_I denotes the position of c;
- d_I denotes the threat degree of threat source in cell c. Note, if there is no threat source in c, set $d_I = 0$;
- t_I denotes the time detecting the threat element.

During the sharing of threat information, $\Xi^{u_k}(t)$ comprises three subsets:

- $\Xi^{u_k}(t-1)$, the threat information base at time $t-1$;
- $I_S^{u_k}(t)$, containing the threat information in the sensory range of u_k at t.

The probabilistic risk map $M_k^u(t)$ can be calculated as follows:

$$M_t^{u_k}(x,y) = 1 - \prod_{i=0}^{n^{u_k}(t)} 1 - P_i(x,y) \tag{1}$$

where $P_i(w)$ is the probabilistic risk of point (x,y) to threat source $T_i(i = 1, 2, \ldots, n^{u_k}(t))$, and $n^{u_k}(t)$ the threat sources detected at time t, i.e., $n^{u_k}(t) = |\{I | I \in \Xi^{u_k}(t) \text{ and } d_I > 0\}|$ in which $|\cdot|$ denotes the size of a set.

3.2 characteristics of TIB

In this section, the characteristics of *TIB* will be presented and analyzed. Illustrative examples are presented to show the impacts of them on real-time path planning. The measures for the characteristics and their handling strategies will be

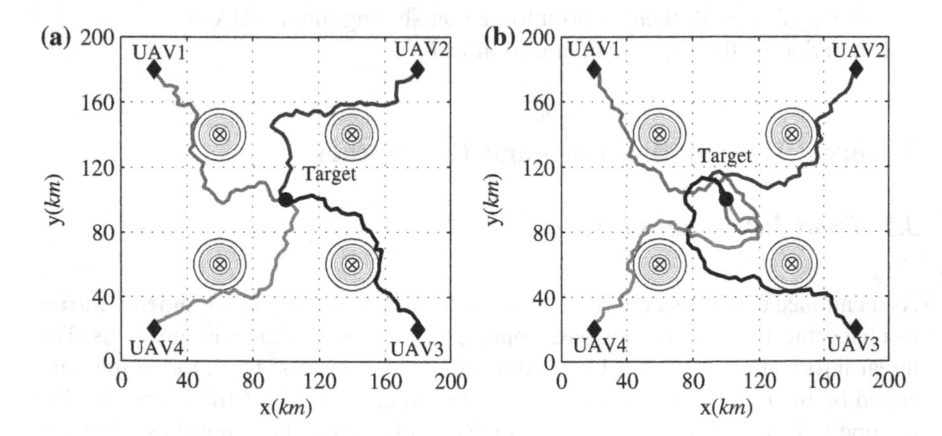

Fig. 1 Impacts of characteristic 1 on path planning

introduced in Sect. 4.1. To be brief, we call the three characteristics as CIT 1, CIT 2, and CIT 3, respectively.

(1) *CIT 1: Increasing amount of threat information.* In the scene illustrated by Fig. 1, which has four threat sources, UAVs are coordinated flying from four start positions at different sides of the scene to the central target. It is shown that the path planned by each UAV deteriorates when sharing threat information with each other. The reason comes from the unnecessary sharing of obstacles, i.e., the threat information detected by others is redundant for a UAV. Figure 1 shows the path planning results in the scene according to a real-time RRT method [8].

The increasing of information is a dichotomy for UAV path planning. On one side, the increasing of threat information extends the ability of threat detection for each UAV. If threat information acquired from other UAVs can be utilized effectively, the path planned by each UAV would be closer to the global optimization. On the other side, threat information sharing brings about negative impacts on path planning of UAVs. First, a real-time path planning algorithm will be more time-consuming because of the additional information to be processed. Second, the negative influence of threat information sharing comes from the information redundancy. Some shared information may be redundant for current decisions of a UAV.

(2) *CIT:2 Unconnectedness of threat elements*

In the scene illustrated by Fig. 2, which has seven threat sources $T_1 - T_4$, three UAVs are coordinated flying from their start positions to a same target. At the start time, UAV1 detects threat T_1, and shares the information with UAV3. Note, at this time the threat sources $T_2 - T_5$ are undetected. After sharing the information of T_1 from UAV1, UAV3 has more probability to turn to left when confounding T_1, because it thinks the left side is more safety than the right. Unfortunately, it makes a wrong decision. As a result, the path planned by UAV3 deteriorates when sharing threat information with UAV1.

The reason comes from the undetected region (marked by a rectangle in Fig. 2) at the start time. Although T_1 is known by UAV1, it is not suitable for UAV3 to use because a large region has not been detected between T_1 and UAV3. We call this characteristic as unconnectedness in *TIB*, which will be formalized in Sect. 4.1. The characteristics is not in accord with the fundamental assumption of many real-time path planning approach, such as graph-based approach, artificial potential field approach, fuzzy control based approach, and etc. As a result, the characteristic may influence the results of these kinds of approaches. Therefore, how to effectively utilize the threat information, considering the new characteristic need to be considered for real-time path planning algorithms.

(3) *CIT 3: Imbalance of shared threat regions*

As illustrated by Fig. 3a, the white part is the region having been detected, in which there are threat sources in the cells or not. The gray part is the region whose threat information is still unknown. In this case, although the detected threat elements are connective and not redundant, they are distributed in one side of the

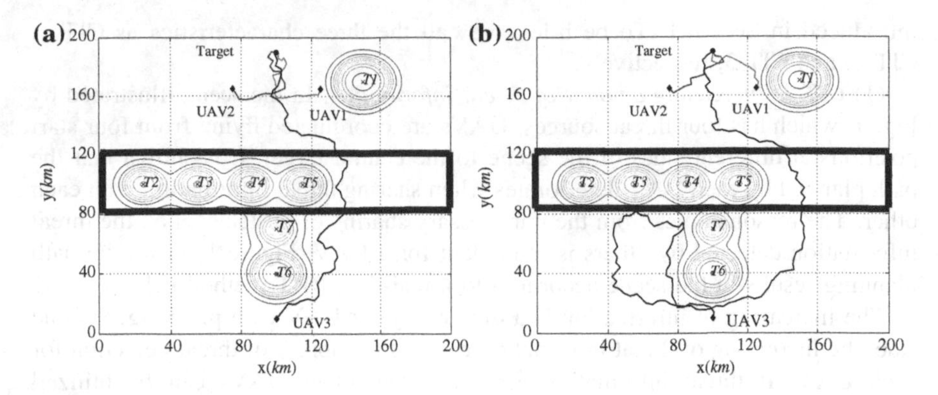

Fig. 2 Impacts of characteristic 2

line between UAV1 and its target. This leads to the unbalanced influence from the threats to UAV1. In that way, UAV1 has more possibility to turn left because the right side seems to be more dangerous. However, most threat information in left side is still undetected, flying to which is actually blind. In Fig. 3b, T_6 is detected by UAV2 and UAV3 and threat sources $T_1 - T_5$ are still unknown. In this case, UAV1 tends to turn left at first to steer clear of the known threat region, which is actually unreasonable because it will retrace the steps to the start point.

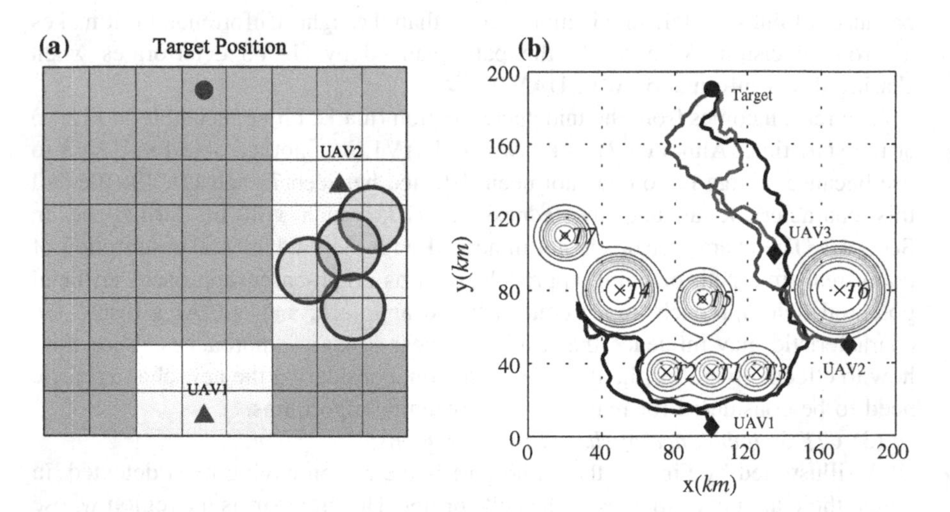

Fig. 3 Impacts of characteristic 3

4 Real-Time Path Planning in Threat Information Sharing Environments

Intuitively, more sufficient threat information is obtained, more efficient a real-time path planning will be. However, experiment results do not support the intuition. For example, for MILP based methods and bouncing-based A2D/A3D methods, usually no effect of sharing information on resulting path are discovered; for RRT methods, sharing threat information may results in a longer path length and computational time; for the behavior-based approach, such as the FBCRI-based method [6] and BCV method [7], if considering the shared information as theirs inputs, the path obtained even gets worse. The approach tends to fall into iterations of previous trajectory, failing to converge to the goals as a result. Based upon the findings, two questions arise:

(1) What are the characteristics of the shared threat information causing the results not coming up to the expected?
(2) Can we embed techniques into existing methods to improve their performance under threat information sharing environment?

For the first question, we have started to explore it in previous section, and will formalize the characteristics (CIT 1 to CIT 3) soon. For the second one, the handling strategies for CITs are proposed in this section, and the strategies will be embedded into a RRT algorithm to test their effectiveness.

4.1 Improving Strategies

(1) *Strategy 1: Adapting CIT 1*

In the environments without threat information sharing, the threat sources detected in real time are usually around the UAV. Therefore, they are useful for both local and global decisions. However, in the threat information sharing environment, threat elements (refer to Sect. 3.1) detected by other UAVs may be redundant for a specific one. Thus, mechanisms are needed to measure the redundancy degrees of new threat elements. In this section, a measure of redundancy is proposed.

Definition 1 (*Redundancy measure*) Given the current position and the target of a UAV as (x_u, y_u) and (x_t, y_t) respectively, the redundancy degree of any threat element $I = (w_I, d_I, t_I)$ is measured by

$$d_r(I) = \left(v_{ut} \cdot v_{uI}^T\right) \times \left(v_{ut} \cdot v_{tI}^T\right) \quad (2)$$

in which $v_{ut} = [x_t - x_u, y_t - y_u]$, $v_{uI} = [x_I - x_u, y_I - y_u]$, $v_{tI} = [x_t - x_t, y_I - y_t]$.

In above definition, v_{ut} denotes the vector from the current position of UAV to its target, v_{uI} denotes the vector from the current position of UAV to the obstacle, and v_{ul} denotes the vector from the target to the obstacle. For an element, the higher the value of $d_r(I)$, the more redundancy the element is deemed to be. The range of $d_r(I)$ is $[-\infty, +\infty]$.

The basic idea of the measure is to judge whether the position of an element is within the region between current position of the UAV and its target. If not, the element is deemed to be redundant for the planning of the UAV at the current time. In this paper, we set the threshold as 0. If $d_r(I) < 0$, I is not redundant; otherwise, I is redundant.

(2) *Strategy 2: Adapting CIT 2*

In the environment without threat information sharing, all elements in $\Xi^{u_k}(t)$ are connected. However, in the environments with threat information sharing, $\Xi^{u_k}(t)$ is composed of not only $I_S^{u_k}(t)$ but $I_S^{u_s}(t)$, in which $k \neq s$. Thus, the elements in $\Xi^{u_k}(t)$ may be unconnected. To measure the characteristic quantitatively, we propose the following definition.

Definition 2 (*Unconnectedness measure of threat elements*) Given a threat element I_1 and a set of threat elements $S \subseteq \xi^{u_k}(t)$, the unconnectedness of I_1 on S is defined as $d_u(I_1, S) = \min\{\|w_I - w_{u_k}\| \,|\, \forall I \in S$ and I is in the line connecting w_{I_1} and $w_{u_k}\}$, in which w_{u_k} is the current position of UAV u_k.

The unconnected degree of a threat element is measured by the minimal distance between u_k and any element in the line connecting I_1 and current position of u_k.

(3) *Strategy 3: Adapting CIT 3*

In this section, the imbalance measure of shared threat regions (CIT 3) will be proposed step by step.

Definition 3 (*Gravity center of threat region*) Given a set of threat elements, its gravity is defined as:

$$w_o = (x_{w_o}, y_{w_o}), \tag{3}$$

where

$$x_{w_o} = \left(\sum_{\forall I \in S} x_{w_I}\right) / |S|; \, y_{w_o} = \left(\sum_{\forall I \in S} y_{w_I}\right) / |S|. \tag{4}$$

Definition 4 (*Imbalance Measure for threat elements*) Suppose $S \in \xi^{u_k}(t)$ is a set of threat elements, $w_T = (x_{w_T}, y_{w_T})$ the target of the UAV and $w_k = (x_{w_k}, y_{w_k})$ the current position of the UAV, the imbalance degree of S can be measured by the distance from the gravity center of the elements in S to the line between the UAV and its target, which is formalized as

$$d_i(S) = \frac{|(y_{w_k} - y_{w_T}) \cdot x_{w_o} - (x_{w_k} - x_{w_T}) \cdot y_{w_o} + x_{w_k} y_{w_T} - x_{w_T} y_{w_k}|}{\sqrt{(y_{w_k} - y_{w_T})^2 + (x_{w_k} - x_{w_T})^2}}. \tag{5}$$

in which $w_o = (x_{w_o}, y_{w_o})$ is the gravity center of the elements in S.

Definition 5 (*Relative imbalance measure for threat elements*) Given two sets of threat elements $S_1, S_2 \in \xi^{u_k}(t)$, the relative imbalance measure of S_2 to S_1 is defined as

$$d_{i_r}(S_1, S_2) = d_i(S_1 \cup S_2) - d_i(S_1). \tag{6}$$

Obviously, the longer the distance, the more serious the imbalance would be. Actually, the relative imbalance measure considers the degree of deterioration combining S_2 to S_1. If the degree is too great, the combination will be forbidden.

4.2 Improved RRT Algorithm

According to above strategies, a threat information filtering mechanism is developed. The mechanism first measures the redundancy of each threat element and filters those whose redundancy degrees are greater than zero. For remaining elements, the unconnected degrees (Definition 2) are measured and filters those whose degrees are greater than μ_c. At last, the elements, which are not redundant and whose unconnected degrees are acceptable, are measured by its imbalance degrees (Definition 5). The elements which make little deterioration on the imbalance degree of a threat region are kept. The detail of the information filtering mechanism is described in Table 1. In the function, *IS* denotes the set of threat elements including detected elements at the previous time and the elements detected by the UAV itself at this time. *SHAR_IS* includes all shared threat elements at this time and those filtered before. The output of the function is a reduction of *IS*.

In this paper, we embed the threat information filtering mechanism into a RRT algorithm, and an improved RRT algorithm, named as SRRT in short, is developed to confront the problems caused by threat information sharing. RRT algorithms are introduced in [10] as an efficient data structure and sampling scheme to quickly search high-dimensional spaces that have both algebraic constraints (arising from obstacles) and differential constraints (arising from nonholonomy and dynamics). The key idea is to bias the exploration toward unexplored portions of the space by randomly sampling points in the state space, and incrementally pulling the search tree toward them.

5 Experiments

In order to validate the new approach and its implementation algorithm, we perform simulations in stochastic scenarios with various obstacle distributions.

In this section, to test the validity of SRRT algorithm, we generate ten groups of scenarios in which the numbers of threat sources range from 1 to 10. Each group, denoted as S_k, includes 100 stochastic scenarios, where $k = 100(i-1)+j$;

Table 1 Theat information filtering function

	Function: InformationFilter(*IS, SHAR_IS*)
1	*IS'=SHAR_IS*
2	**for each** $I \in$ SHAR_IS
3	**if** $\sqrt{(x_I - x_{current})^2 + (y_I - y_{current})^2} > R_k$ **then**
4	$w_I = (x_I, y_I)$;//The candidates are out the sensory region of the UAV itself
5	**if** $d_r(I) \geq 0$ **then**
6	*IS'=IS'*-$\{I\}$; //Redundancy measure
7	**else**
8	**if** $d_u(I, IS) \geq \mu_c$ **then**
9	*IS'=IS'*-$\{I\}$; //Unconnectedness measure
10	**else**
11	$d_{i_r}(IS, \{I\}) > 0$,*IS'=IS'*-$\{I\}$;//Imbalance measure
12	**endif endif endif enfor**
13	**if** $(M' == M)$ **then**
14	*ifreplan*=0;
15	**else**
16	*ifreplan*=1;
17	**endif**
18	**return** *IS*';

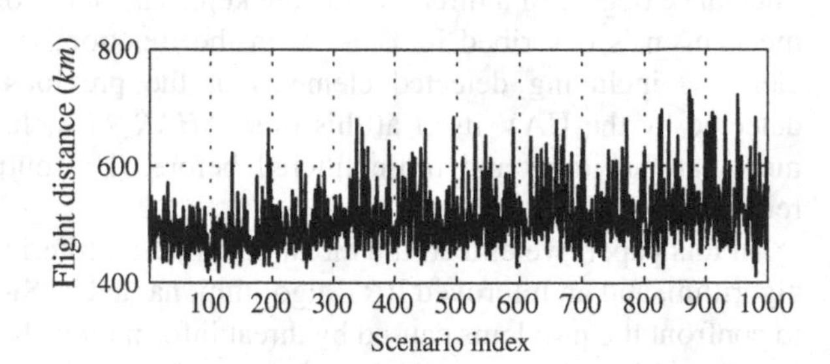

Fig. 4 Distances in stochastic scenarios

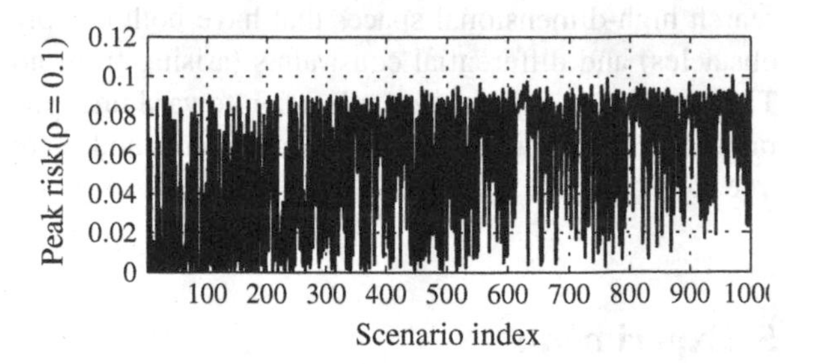

Fig. 5 Peak risks in stochastic scenarios

$i = 1, 2, \ldots, 10$; $j = 1, 2, \ldots, 100$. All the scenarios are set in a large operational space [0, 200] × [0, 200] km deployed with two UAVs. The two UAVs start points are [20, 20, 8] and [60, 160, 8] km, respectively, and both their target points are [180, 180, 8] km. The simulation results in 1,000 scenarios are illustrated in Figs. 4 and 5.

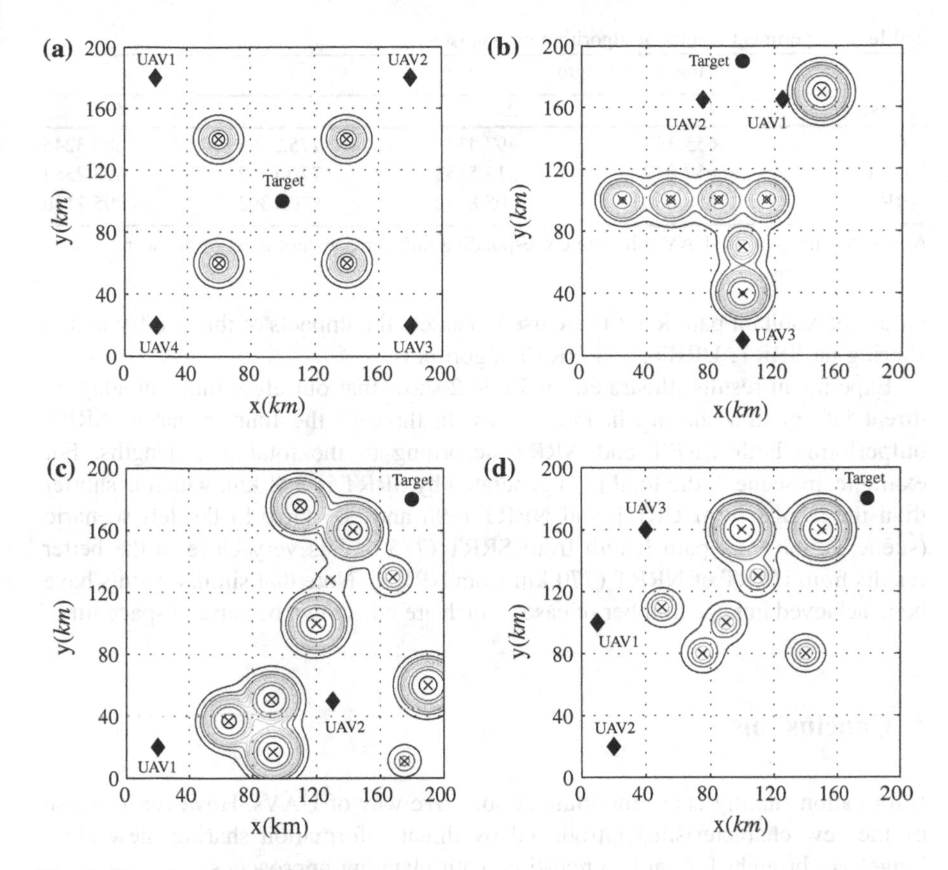

Fig. 6 Scenarios for experiments

As shown in Figs. 4 and 5, the SRRT algorithm is capable of providing convergent and safe flight paths in all the 1,000 scenarios. In Fig. 4, the two UAVs total flight distances are ranged from 411.0 to 728.2 km (the average flight distance is 507.1 km), indicating relative short flight paths in cooperative missions. In Fig. 5, let the peak risk represent the maximal threat degree during the course of a flight simulation. It is evident that the peak risks of the 1,000 flights are all less than 0.1, which is set as the threshold of flight safety, i.e., the paths generated by our algorithm can absolutely guarantee safety to UAVs in various scenarios.

The following test examples are used to compare the feasibility and efficiency of the proposed algorithm (SRRT) with two algorithms, which are RRT algorithm with unconditional threat information sharing (URRT in brief) and RRT algorithm without threat information sharing (NRRT in brief). These test examples are based on four representative scenarios with obstacles as shown in Fig. 6a–d. Table 2 shows the experiment results.

Because of the random sampling methods used in RRT algorithm to generate extension trees, we make experiments in each scene for 1,000 times. Average

Table 2 Experiment results of algorithm comparisons

	Total path length			
Alg./scene	1	2	3	4
SRRT	655.3387	497.419	775.3368	888.3245
URRT	659.0491	513.5153	770.1302	898.2384
NRRT	658.6863	505.8142	870.1392	905.7136

Note– denotes that the UAV with the corresponding number does not exist in the scene

values of resultant path lengths are used to assess the impacts of threat information sharing on SRRT, URRT, and NRRT algorithms.

Experiment results illustrated by Table 2 show that our algorithm can adapt to threat information sharing in most cases. In three of the four scenarios, SRRT outperforms both URRT and NRRT according to the total path lengths. For example, in scene 4, the total path generated by SRRT is 888 km, which is shorter than the results from URRT and NRRT (898 and 905 km). In the left scenario (scene 3), the total path length from SRRT (775 km) is very close to the better results from URRT or NRRT (770 km from NRRT). Note that similar results have been achieved in huge number of cases which are not shown because of space limit.

6 Conclusions

Information sharing is an important cooperative way of UAVs. However, because of the new characteristics introduced by threat information sharing, new challenges are brought forward to real-time path planning approaches.

The paper makes three contributions. First, we found that the sharing of threat information may have negative effects on path planning algorithms. Second, three characteristics are explored to analyze the effects. Third, a new approach and its implement on RRT algorithm are proposed showing that the negative effects are eliminated effectively by embedding useful strategies. The effectiveness of the algorithm is verified by random testing and representative scenarios.

References

1. Liao Y, Jin Y, Minai AA, Polycarpou MM (2005) Information sharing in cooperative unmanned aerial vehicle teams. In: Proceedings of IEEE conference on decision and control, IEEE Press, Seville, pp 90–95
2. Ren W, Beard RW, McLain TW (2004) Coordination variables and consensus building in multiple vehicle systems. In: Kumar V, Leonard N, Morse AS (eds) Cooperative control. Springer, Heidelberg, pp 171–188
3. United States Air Force (2009) Unmanned aircraft systems flight plan 2009–2047. Technical report, Headquarters, United States Air Force

4. Kim Y, Gu DW, Postlethwaite I (2007) Real-time optimal mission scheduling and flight path selection. IEEE Trans Autom Control 52:1119–1123
5. Kim Y, Gu DW, Postlethwaite I (2008) Real-time path planning with limited information for autonomous unmanned air vehicles. Automatica 44:696–712
6. Zheng Z, Wu SJ, Liu W, Cai KY (2011) A feedback based CRI approach to fuzzy reasoning. Appl Soft Comput 11:1241–1255
7. Wu SJ, Zheng Z, Cai KY (2011) Real-time path planning for unmanned aerial vehicles using behavior coordination and virtual goal. Control Theory Appl 28:131–136
8. LaValle SM (2006) Planning algorithms. Cambridge University Press, Cambridge
9. Gu DW, Kamal W, Postlethwaite I (2004) A UAV waypoint generator. In: Proceedings of AIAA 1st Intelligent systems technical conference. AIAA Press, Chicago, pp 1–6
10. LaValle SM, Kuffner JJ (2001) Rapidly-exploring random trees: progress and prospects. In: Donald BR, Lynch KM, Rus D, Wellesley E (eds) Algorithmic and computational robotics: new directions. A K Peters, pp 293–308

A Femtocell Self-Configuration Deployment Scheme in Hierarchical Networks

Peng Xu, Fangli Ma, Jun Wang, Yang Xu, Xingxing He and Xiaomei Zhong

Abstract More Femtocells will be deployed in existing cell networks, improving system capacity and enhancing indoor coverage. However new Femtocells will introduce a series of several interferences, whether which are well dealt with or not is the key to widely deploy for Femtocells. A Femtocell self-configuration deployment scheme is proposed, in which an optimal power issue is formulized and solved based on various interferences that Femtocell base station may receive from other base stations. And that is supposed to maximize the Femtocell system capacity under the condition guaranteeing the original users' usual communication. The analysis and simulation results show that the proposed scheme cuts down the transmit power of Femtocell, increases the throughput.

Keywords Femtocells · Transmit power · Self-configuration · Signal to interference plus noise ratio

1 Introduction

The rapid increase in mobile data activity has raised the stakes on developing innovative new technologies and cellular topologies in an energy efficient manner. One of most interesting trends to emerge from cellular evolution is femtocells.

P. Xu (✉) · Y. Xu · X. He · X. Zhong
School of Mathematics, Sowthwest Jiaotong University, Chengdu 610031, China
e-mail: pengxup@my.swjtu.edu.cn

P. Xu · F. Ma · J. Wang · Y. Xu · X. He · X. Zhong
Intelligent Control Development Center, Southwest Jiaotong University, Chengdu 610031, China

F. Ma
Sichuan Provincial Radio Monitoring Station, Chengdu 610016, China

J. Wang
Radio Manage Office, Sichuan Province, Chengdu 610017, China

F. Sun et al. (eds.), *Knowledge Engineering and Management*, Advances in Intelligent Systems and Computing 214,
DOI: 10.1007/978-3-642-37832-4_30,

Femtocells are small, inexpensive, low-power base stations that are generally consumer-deployed and connected to their own wired backhaul connection [1]. The main benefit of these small base stations is the improved indoor coverage and offloading of traffic from macro base stations [2]. As a result, not only do users enjoy better coverage due to the close vicinity to BSs, the network operators also benefit from a reduced demand for constructing macrocell towers. Due to the "win–win" situation, several standard bodies, such as 3GPP, 3GPP2, and WiMAX Forum, currently have started to standardize WCDMA, LTE, and WiMAX femtocells [3].

Femtocell networks are largely installed by customers or private enterprises often in an ad hoc manner without traditional Radio Frequency planning, site selection, deployment, and maintenance by the operator. In addition, as the number of femtocells is expected to be orders of magnitude greater than macrocells, manual network deployment and maintenance is simply not scalable in a cost-effective manner for large femtocell deployments. Femtocells must therefore support an essentially plug-and-play operation with automatic configuration and network adaptation, as is referred to a self-organizing network (SON) [2]. A self-organizing network, defined as a network that requires a minimal human involvement due to the autonomous and/or automatic nature of its functioning, will integrate the processes of planning, configuration, and optimization in a set of autonomous/automatic functionalities. These functionalities will allow femtocells to scan the air interface, and tune their parameters according to the dynamic behavior of the network, traffic, and channel [4]. The need for self-organization in femtocell networks is driven by achieving a substantial reduction in the Capital Expenditures (CAPEX) and Operational Expenditure (OPEX) of the network by reducing the human involvement, optimizing the performance of the network in terms of coverage, capacity, or QoS, and allowing the deployment of a larger number of femtocells. One aspect of SON that has attracted considerable research attention is automatic channel selection, power adjustment, and frequency assignment for autonomous interference coordination and coverage optimization. Such problems are usually formulated as mathematical optimization problems.

Self-configuration of femtocells, as an important part of SON, is aiming to minimize the impact on existing base stations and user equipments. Likewise transmit power self-configuration of femtocell base stations is the key that whether the femtocells can be successfully deployed in macrocell networks. A reducing interference method for UMTS networks has been proposed in [5], which adopts power control on pilot and data channel guaranteeing coverage, also analyzes how to set downlink and uplink power. The information of terminal mobility used to adaptively adjust the coverage of femtocell is studied in [6]. In [7], the uplink interference is solved through adjusting noise threshold. In [8], downlink power control is used to make the users of macrocell and femtocells achieve predefined Signal to Interference plus Noise (SINR). However methods above do not consider the interference on existing users. The impact on power setting of femtocell base station, which includes downlink signal of macrocell base station and uplink signal of macrocell users, is analyzed in [9]. But the interference on macrocell from femtocell is neglectful. In [10], a new cross-tier interference avoid scheme is proposed, in which power and sub-channel among macrocell and femtocell are reassigned based on information of cross-tier interference. The co-channel femtocell deployment in two-tier networks is investigated in [11], while considering cellular geometry and cross-tier interference in downlink. The issue on uplink power control in LTE hierarchical networks is researched thoroughly in [12]. In [13], the joint power and sub-channel allocation scheme is analyzed when large dense femtocells are deployed, and a binary power allocation form of power allocation is proposed. On-demand resource sharing and femtocell access control in OFDMA femtocell networks are studies in [14], also a more comprehensive perspective on self-organizing femtocell networks where users optimize their performance in a distributed manner is provided. However all above references are still not comprehensive, such as choosing excessive femtocells for analyzing interference or ignoring the impact of new added femtocell.

A femtocell self-configuration deployment scheme is proposed, which not only reduces the number of referred femtocells but also restricts the interference on original users. The rest is organized as follows: The interference model is given in Sect. 2; The proposed scheme is given in Sect. 3; The numerical results and conclusions are given in Sect. 4 and in Sect. 5, separately.

2 Interference Model

Generally hierarchical networks can be described as Fig. 1. When a new femtocell is deployed by a user, how to configure transmit power of the femtocell becomes a pressing issue. With 3GPP standard, power self-configuration means measuring around radio environment at the first stage, including various received signal and interference, then optimizing the transmit power at the self optimization stage [4].

The various received signal and interference can be represented with Fig. 2. It mainly contains signal from every macrocell base station (m-BS) and adjacent femtocell base station(f-BS). In addition because the huge difference (13dBm ~ 33dBm) in the transmit power between m-BSs and f-BSs [15], the impacts of both m-BSs and adjacent f-BSs are considered when setting transmit power of f-BS. Assuming transmit power and radius of f-BS located in the center of room is 15 dBm and 10 m, separately.

The signals crossing one wall and two walls can be obtained as Fig. 3 [16]. From Fig. 3, the conclusion that adjacent femtocells in which only the number of walls is less than two need to be considered, can be obtained.

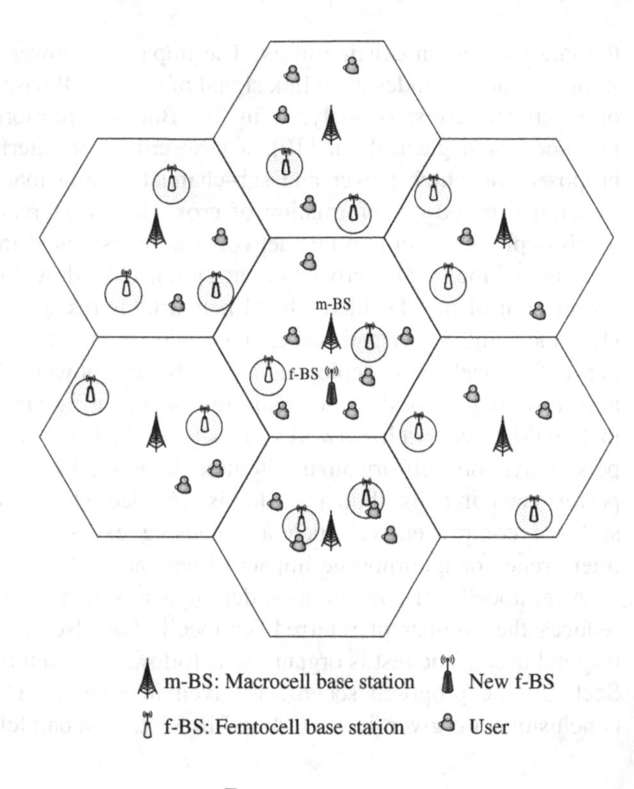

Fig. 1 Hierarchical networks

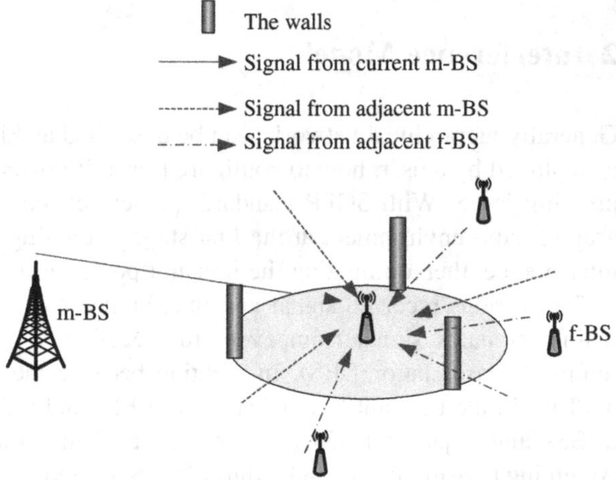

Fig. 2 Received signal and interference of new femtocell

3 Proposed scheme

Assuming the new f-BS can work as the user terminal [17], which can detect around radio environment and obtain network parameters through wired backhaul. Then the problem can be described as:

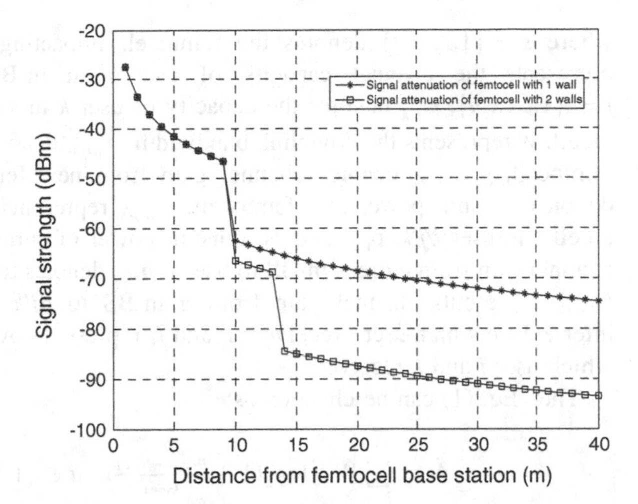

Fig. 3 Signal attenuation of femtocell

$$
\begin{aligned}
& Max \sum_{i=1}^{l} C_{f\_new,i} && i=\{1,2,\ldots,l\} \\
& s.t.\ SINR_{m,j} \geq SINR_{m,thr} && j=\{1,2,\ldots,n\} \\
& SINR_{f,k}^{s} \geq SINR_{f,thr} && k=\{1,2,\ldots,r\}, \\
& && s=\{1,2,\ldots,t\}
\end{aligned}
\tag{1}
$$

where $C_{f\_new,i}$ denotes the throughput of user i, and $i = \{1,2,\ldots,l\}$ denotes the users in the new femtocell. $SINR_{m,j}$ denote the Signal to Interference plus Noise(SINR) of user j in adjacent m-BS, and $SINR_{f,k}^{s}$ denotes SINR of user k in s femtocell. $SINR_{m,thr}$, $SINR_{f,thr}$ denotes the related SINR threshold.

The throughput can be computed[18] as:

$$
\begin{aligned}
C_{f\_new,i} &= W * \log_2\left(1 + \frac{p_{f,new} * g_{new,i}}{I_{new,i}}\right) \\
I_{new,i} &= \sum_{a=1}^{q} p_{m,a} * h_{a,i} + \sum_{s=1}^{t} p_{f,s} * g_{s,i}
\end{aligned}
\tag{2}
$$

$$
\begin{aligned}
C_{m,j} &= W * \log_2\left(1 + \frac{p_m * h_{m,j}}{I_j + p_{f,new} * g_{new,j}}\right) \\
I_j &= \sum_{a=1}^{q-1} p_{m,a} * h_{a,j} + \sum_{s=1}^{t} p_{f,s} * g_{s,j}
\end{aligned}
\tag{3}
$$

$$
\begin{aligned}
C_{f,k}^{s} &= W * \log_2\left(1 + \frac{p_{f,s} * g_{s,k}}{I_k + p_{f,new} * g_{new,k}}\right) \\
I_k &= \sum_{a=1}^{q} p_{m,a} * h_{a,k} + \sum_{s=1}^{t-1} p_{f,s} * g_{s,k}
\end{aligned}
\tag{4}
$$

where $s = \{1,2,\ldots,t\}$ denotes the femtocell impacting on power setting, $C_{m,j}$ represents the channel capacity of user j in m-BS near new femtocell, $j = \{1,2,\ldots,n\}$. $C^{s}_{f,k}$ denotes the capacity of user k in s femtocell near new femtocell, W represents the downlink bandwidth. $p_{f,new}$ denotes transmit power of new femtocell, $g_{new,i/j/k}$ denotes channel gain from new femtocell to user $i/j/k$. $p_{f,s}$ denotes transmit power of s femtocell, $g_{s,i/j/k}$ represents channel gain from femtocell s to user $i/j/k$. $\mathrm{p_m}$ denotes transmit power of current m-BS, $\mathrm{h_{m,j}}$ represents channel gain from current m-BS to user j. $p_{m,a}$ denotes transmit power of a m-BS, $h_{a,i/j/k}$ represents channel gain from a m-BS to $i/j/k$. $I_{new,i}$ denotes downlink interference which user i received. I_j and I_k represent downlink interference signal which user j and k received.

Then Eq. (1) can be changed as:

$$\begin{array}{ll} Max \sum_{i=1}^{l} W * \log_2(1 + \frac{p_{f,new} * g_{new,i}}{I_{new,i}}) & i \in \{1,2,\ldots,l\} \\ s.t.\, p_{f,new} * g_{new,j} \le (\frac{p_m * h_{m,j}}{SINR_{m,thr}} - I_j) & j \in \{1,2,\ldots,n\} \\ p_{f,new} * g_{new,k} \le (\frac{p_{f,s} * g_{s,k}}{SINR_{f,thr}} - I_k) & k \in \{1,2,\ldots,r\}, \\ p_{f,new} > 0 & s \in \{1,2,\ldots,t\} \end{array} \quad (5)$$

For convenience some transforms are needed as:

$$\begin{aligned} \alpha &= \min(\frac{1}{g_{new,j}} * (\frac{p_m * h_{m,j}}{SINR_{m,thr}} - I_j))\, j \in \{1,2,\ldots,n\} \\ \beta &= \min(\frac{1}{g_{new,k}} * (\frac{p_{f,s} * g_{s,k}}{SINR_{f,thr}} - I_k))\, k, s \\ \gamma &= \min(\alpha, \beta) \end{aligned} \quad (6)$$

Lagrange multiplier method [19] can be used in Eq. (5),

$$\begin{aligned} L(p_{f,new}, \lambda, \mu) = & -(\sum_{i=1}^{l} \log_2(1 + \frac{p_{f,new} * g_{new,i}}{I_{new,i}})) \\ & + \lambda * (p_{f,new} * g_{new,x} - \gamma * g_{new,x}) \\ & + \mu * (-p_{f,new}) \end{aligned} \quad (7)$$

$$\begin{aligned} \frac{\partial(p_{f,new}, \lambda, \mu)}{\partial p_{f,new}} = & \frac{-1}{(1 + \frac{p_{f,new} * g_{new,i}}{I_{new,i}}) * \ln 2} * \frac{g_{new,i}}{I_{new,i}} \\ & + \lambda * g_{new,x} - \mu = 0 \end{aligned} \quad (8)$$

With Kuhn-Tucker condition, we have:

$$\begin{cases} \lambda * (p_{f,new} * g_{new,x} - \gamma * g_{new,x}) = 0 \\ \mu * p_{f,new} = 0 \end{cases} \quad (9)$$

$$\begin{cases} p_{f,new} = \gamma \\ \lambda = \dfrac{1}{\ln 2 * g_{new,x}} * (\dfrac{I_{new,i}}{g_{new,i}} + \gamma)^{-1} \\ \mu = 0 \end{cases} \quad (10)$$

Then optimal power $p^*_{f,new}$ can be obtained if we define x^+:max(x,0).

$$p^*_{f,new} = [\gamma]^+ \quad (11)$$

From Eq. (11), the optimal power is related with original user's location and radio environment.

4 Numerical Results

In this part, the proposed scheme is compared with scheme-RSS [17]. The parameters setting is listed in Table 1. Assuming some femtocells distributed uniformly in each macrocell, the number of femtocells affecting new femtocell is 1 ~ 4, and there are 2 users in each femtocells.

From Fig. 4, the power is reducing with apart from m-BS for scheme-RSS. When distance is more than 450 m, the change is small. That is because the signal from f-BS is larger than from m-BS. When distance is less than 450 m, f-BS has to increase the power to improve users' SINR. While the power is more large, it may affect both macro users and femto users. In addition, the power of proposed scheme is slowly changing, that is because of the constraint on users' SINR. That is to say, proposed scheme has minimal impact on original users.

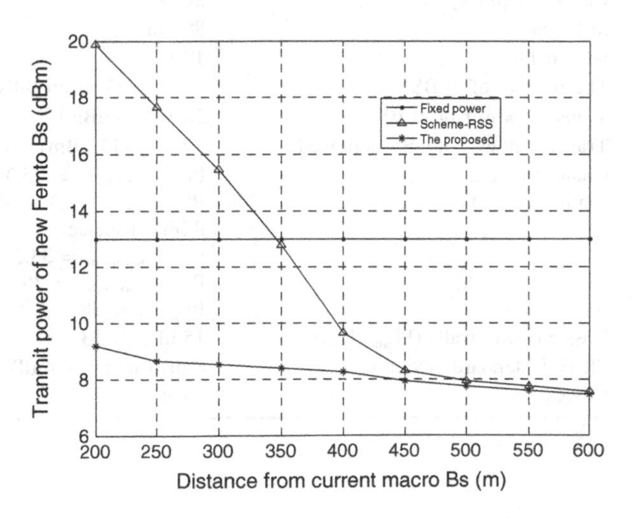

Fig. 4 New femtocell power setting

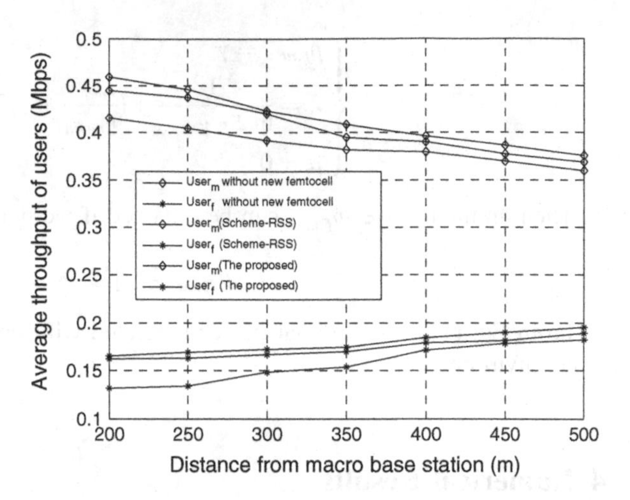

Fig. 5 Users' throughput

From Fig. 5, the scheme-RSS reduces original user's throughput because of setting high power when distance from m-BS is less than 400 m. While distance is more than 400 m, power setting accords with radio environment, which leads to slowdown throughput decrease. The proposed scheme has no difference in throughput, because of considering the impact on original users. That is the scheme make the interference introduced minimized (Table 1).

Table 1 Parameter setting

Parameters	Setting
Bandwidth	5 MHz
Carrier frequncey	2Ghz
m-BS radius	800 m
f-BS radius	10 m
The number of m-BS	One tier (7 macrocells)
Transmit power of m-BS	20 W (43 dBm)
Transmit power of f-BS deployed	20 mW (13 dBm)
Channel model	$PL_{outdoor\text{-}indoor} = 15.3 + 37.6 \log_{10}(d) + PL_{out}[dB]$
(Path loss, PL)	$PL_{outdoor\text{-}} = 15.3 + 37.6 \log_{10}(d)$
	d:[m], distance
	$PL_{indoor\text{-}indoor} = 38.5 + 20 \log_{10}(d) + 1.5{*}PL_{in}[dB]$
	$PL_{indoor\text{-}outdoor} = 38.5 + 20 \log_{10}(d) + PL_{out}[dB]$
	$PL_{indoor} = 38.5 + 20 \log_{10}(d)$
Loss crossing walls (PL_{in},PL_{out})	15 dB, 30 dB
Shadow standard deviation	8 dB (outdoor), 4 dB (indoor)
$SINR_{th}$	1 dB

5 Conclusion

According to above analysis, the proposed scheme can efficiently reduce the impact on original users, simultaneously guaranteeing avail communication between users. Although the analysis is only as to one femtocell, it is not limited and can be expanded to more femtocells. Based on the same mechanism, the proposed scheme will obviously perform as good as the state with one femtocell, and the similar results will be obtained in the same way.

Acknowledgments This work is supported by National Science Foundation of China (Grant No.61175055),Sichuan Key Technology Research and Development Program (Grant No.2011FZ0051), Radio Administration Bureau of MIIT of China (Grant No.[2011]146), China Institution of Communications (Grant No.[2011]051). The Fundamental Research Funds for the Central Universities (Grant No. A0920502051305-25)

References

1. Andrews JG, Claussen H, Dohler M, Rangan S, Reed MC (2012) Femtocells: Past, Present, and Future. Selected Areas in Communications, IEEE Journal on 30(3):497–508
2. Barbieri, A.; Damnjanovic, A.; Tingfang Ji; Montojo, J.; Yongbin Wei; Malladi, D.; Osok Song; Horn, G.;, "LTE Femtocells: System Design and Performance Analysis," Selected Areas in Communications, IEEE Journal on, vol.30, no.3, pp.586-594, April 2012
3. Jia Liu; Tianyou Kou; Qian Chen; Sherali, H.D.;, "Femtocell Base Station Deployment in Commercial Buildings: A Global Optimization Approach," Selected Areas in Communications, IEEE Journal on, vol.30, no.3, pp.652-663, April 2012
4. ZHANG J, GUILLAUME R. Femtocells : Technologies and Deployment[M]. Singapore: Wiley, 2010:225-229
5. Claussen H, Ho L, Samuel L (2008) An overview of the femtocell concept[J]. Bell Labs Technical Journal 3(1):221–245
6. CLAUSSEN H, HO L, SAMUEL L. Self-optimization coverage for femtocell deployments[C].Wireless Telecommunications Symposium, California, USA, Apr. 24–26, 2008, 278–285
7. 3GPP TR25.967 V8.0.1. FDD Home NodeB (HNB) RF Requirements[S]. 2009
8. LI X, QIAN L, KATARIA D. Downlink Power Control in Co-Channel Macrocell Femtocell Overlay[C]. Conference on Information Sciences and Systems (CISS), Baltimore, USA, Mar.18-20, 2009,383 – 388
9. MORITA M, MATSUNAGA Y, HAMABE K. Adaptive Power Level Setting of Femtocell Base Stations for Mitigating Interference with Macrocells[C]. Vehicular Technology Conference Fall (VTC 2010-Fall), Ottawa, Canada, Sept.6-9, 2010, 1-5
10. LOPEZ D, LADANYI A, JUTTNER A, et al. OFDMA Femtocells: Intracell Handover for Interference and Handover Mitigation in Two-Tier Networks[C]. Wireless Communications and Networking Conference (WCNC), Sydney, Australia, Apr.18-21, 2010, 1-6
11. Kim Y, Lee S, Hong D (2010) Performance Analysis of Two-Tier Femtocell Networks with Outage Constraints[J]. IEEE Trans Wireless Commun 9(9):2695–2700
12. GORA J, PEDERSEN K, SZUFARSKA A, et al. Cell-Specific Uplink Power Control for Heterogeneous Networks in LTE[C]. Vehicular Technology Conference Fall (VTC 2010-Fall), Ottawa, Canada, Sept.6-9, 2010, 1-5

13. Kim J, Cho D (2010) A Joint Power and Subchannel Allocation Scheme Maximizing System Capacity in Indoor Dense Mobile Communication Systems[J]. IEEE Trans Veh Technol 59(9):4340–4353
14. Chun-Han Ko; Hung-Yu Wei;, "On-Demand Resource-Sharing Mechanism Design in Two-Tier OFDMA Femtocell Networks," Vehicular Technology, IEEE Transactions on, vol.60, no.3, pp.1059-1071, March 2011
15. 3GPP. TS 25.104 V9.5.0, Base Station (BS) radio transmission and reception (FDD)[S]. 2010
16. NAMGEOL O, HAN S, KIM H. System Capacity and Coverage Analysis of Femtocell Networks[C]. Wireless Communications and Networking Conference (WCNC), Sydney, Australia, Apr.18-21, 2010, 1-5
17. Madan R, Borran J, Sampath A et al (2010) Cell Association and Interference Coordination in Heterogeneous LTE-A Cellular Networks[J]. IEEE J Sel Areas Commun 28(9):1479–1489
18. Shannon C (1948) A mathematical theory of communication[J]. Bell Syst Tech J 27:379–423
19. KUHN H, TUCKER A. Nonlinear programming[C]. Proceedings of 2nd Berkeley Symposium. Berkeley, USA, Jul.31-Aug.12, 1950, 481–492

Programming Global and Local Sequence Alignment by Using R

Beatriz González-Pérez, Victoria López and Juan Sampedro

Abstract R [2] is a programming language primarily oriented to statistical and graphical analysis. Since R is an open source language, new functions of very different fields are continuously appearing all around the world. Operations Research is a multidisciplinary science and given the interest that exists between teachers and researchers to develop procedures that can be applied across the board by students, professionals and scientists, the use of R is needed to solve problems related to the optimization of a system. Dynamic Programming algorithms are essential basis for the development of algorithms that solve other problems. One of the main research areas in Bioinformatics is Sequence Alignment of nucleotide or amino acid residues to identify regions of similarity. The Bioconductor project [3] provides R packages for the analysis of genomic data. This work focuses on alignment of pairs. We develop two functions with R code: localAlignment and globalAlignment. These functions solve standard problems of Local and Global Sequence Alignment by using Dynamic Programming.

Keywords R · Global and local sequence alignment · Dynamic programming · Bioinformatics

B. González-Pérez (✉) · V. López · J. Sampedro
Mobile Technologies and Biotechnology Research Group (G-TeC),
Universidad Complutense de Madrid, 28040 Madrid, Spain
e-mail: beatrizg@mat.ucm.es

V. López
e-mail: vlopez@fdi.ucm.es

F. Sun et al. (eds.), *Knowledge Engineering and Management*,
Advances in Intelligent Systems and Computing 214,
DOI: 10.1007/978-3-642-37832-4_31,

1 Introduction

Bioinformatics is the application of computer technology to storing, retrieving, and analyzing of biological data. The countless biology experiments produce large amounts of information, such as large nucleic acid (DNA/RNA) or protein sequences, that can be stored and reused. To do this, databases are created to store all this information in order to be accessed quickly and easily.

One of the main research areas in Bioinformatics is Sequence Alignment of most simple units sequences as nucleotide or amino acid residues to identify regions of similarity that may be a consequence of functional, structural, or evolutionary relationships.

This work focuses on Pairwise Alignment between two sequences at a time. They are efficient to compute and are often used when extreme precision does not needed, such as searching a database for sequences with high similarity to a query. A Multiple Sequence Alignments between three or more biological sequences at a time are often used to identify conserved regions among a group of sequences which are supposed to have a common ancestor. The three primary methods of producing Pairwise Alignments are Dot-Matrix Methods, Dynamic Programming, and Word Methods. Multiple Alignments in many cases lead to combinatorial optimization problems of the complexity class NP-complete.

Global Alignments align every residue in every sequence by searching the most large common subsequence of two allowing gaps and forcing the alignment to span the entire length of the query sequence. These alignments are used when the sequences to compare are similar and of roughly equal size. A general Global Alignment technique based on Dynamic Programming is the Needleman-Wunsch algorithm.

By contrast, Local Alignments identify regions of similarity within long sequences that are often widely divergent overall. Local Alignments are often preferable with distantly related biological sequences where mutations have generated too much noise in some regions to conduct a meaningful comparison. Local Alignments avoid these areas and concentrate on those with conserved similarity signals. A general Local Alignment technique based on Dynamic Programming is the Smith–Waterman algorithm.

When given a subsequence of solutions, it is possible to safely always know the next decision that leads to the optimal solution, the problem can be solved by an appropriate selection criterion in each step which is called Greedy Strategy. Although such Greedy Strategy does not always exist, sometimes it is possible to verify the Bellman's Optimality Principle. Dynamic Programming is specially designed for this kind of optimization problems. This Principle is true in those problems where any subsolution of the optimal solution is the optimal solution of the corresponding reduced problem. This Principle is easily recognizable in Shortest Path Problems. When the shortest path between two points A and B is known and goes through a point C, it is true that the shortest path between A and C is precisely the initial piece of the shortest path between A and B. This Principle is closely linked with resolving problems that can be modeled by using recursive equations.

The Dynamic Programming method is guaranteed to find an optimal alignment given a particular scoring function. Scoring systems are particularly interesting in the case of amino acid sequence alignment where to be replaced one amino acid by a mutation in a protein, all amino acids are not equally likely substitute. This is because the substitution of one amino acid for another with very different properties could affect protein folding and thus its functionality. PAM and BLOSUM are scoring matrices widely used. In this work, we consider the scoring function based on the identity matrix and without penalty by gap.

2 Global Alignment

Global Sequence Alignment is based on constructing the best alignment over the entire length of two sequences. This kind of alignment is suitable when the two sequences have a significant degree of similarity throughout and similar length.

We develop a function called globalAlignment with R code for seeking an alignment between two sequences to maximize the number of residue-to-residue matches. The problem can be modeled by using the following recursive equations

$$t_{i,j} = \begin{cases} t_{i-1,j-1} + 1 & \text{if} \quad x_i = y_j \\ \max\{t_{i-1,j}, t_{i,j-1}\} & \text{otherwise} \end{cases}$$

$$t_{i,j} = 0 \quad \text{if} \quad i = 0 \quad \text{or} \quad j = 0$$

where $t_{i,j}$ is the length of the longest common subsequence among the subsequences with the i-first characters of the sequence x and the j-first of the sequence y.

This algorithm is of $O(nm)$ complexity order, where n and m are respectively the length of the initial two sequences to compare.

2.1 R Code GlobalAlignment

```
globalAlignment<-function(x,y){
N=length(x)
M=length(y)
t=matrix(0,N+1,M+1)
for(i in 1:N){
for(j in 1:M){
if(x[i]!=y[j])
t[i+1,j+1]<-max(t[i,j+1],t[i+1,j])
else
t[i+1,j+1]<-t[i,j]+1
}
}
t<-t[2:(N+1),2:(M+1)]
long<-t[N,M]
i<-N
j<-M
cont<-long
xIndex<-c()
yIndex<-c()
seq=c()
error=FALSE
while(i>0 && j>0 && cont>0){
if(i>1 && t[i,j]==t[i-1,j]){
i<-i-1
}else if(j>1 && t[i,j]==t[i,j-1]){
j<-j-1
}else {
seq<-c(x[i],seq)
xIndex<-c(i,xIndex)
yIndex<-c(j,yIndex)
cont<-cont-1
i<-i-1
j<-j-1
}
}
if(cont!=0)
error=TRUE
alignment<-matrix(c(" ","-"," "),3,M+N-long)
currPosX<-1
currPosY<-xIndex[1]-yIndex[1]+1
currMatchX<-1;
currMatchY<-1;
if(xIndex[1]<yIndex[1]){
currPosX<-yIndex[1]-xIndex[1]+1
currPosY<-1
}
```

```
for(i in 1:N){
if(currMatchX<=length(xIndex) && i==xIndex[currMatchX]){
m<-max(xIndex[currMatchX],yIndex[currMatchX])
alignment[1,m]<-x[i]
alignment[2,m]<-"+"
currPosX<-m+1
currMatchX<-currMatchX+1
}else{
alignment[1,currPosX]<-x[i]
currPosX<-currPosX+1
}
}
for(j in 1:M){
if(currMatchY<=length(yIndex) && j==yIndex[currMatchY]){
m<-max(xIndex[currMatchY],yIndex[currMatchY])
alignment[3,m]<-y[j]
currPosY<-m+1
currMatchY<-currMatchY+1
}else{
alignment[3,currPosY]<-y[j]
currPosY<-currPosY+1
}
}
alignment<-alignment[,1:max(currPosX-1,currPosY-1)]
k<-ncol(alignment)
numGaps<-0
for(i in 1:k){
if(alignment[1,i]==" " || alignment[3,i]==" ")
numGaps<-numGaps+1
}
list(long=long,xIndex=xIndex,yIndex=yIndex,sequence=seq,
      alignment=alignment,numGaps=numGaps,error=error)
}}
```

2.2 Example

```
> seq1
 [1] "A" "T" "C" "G" "T" "A" "T" "T" "C" "G" "G" "T" "C" "A" "A" "C" "T"
> seq2
 [1] "G" "T" "A" "T" "C" "A" "A" "T" "T" "G" "C" "T" "A" "C" "C"

> globalAlignment(seq1,seq2)
$long
[1] 10

$xIndex
 [1]  1  2  3  6  7  8  9 12 14 16

$yIndex
 [1]  3  4  5  6  8  9 11 12 13 14

$sequence
 [1] "A" "T" "C" "A" "T" "T" "C" "T" "A" "C"

$alignment
     [,1] [,2] [,3] [,4] [,5] [,6] [,7] [,8] [,9] [,10] [,11] [,12] [,13]
[1,] " "  " "  "A"  "T"  "C"  "A"  "T"  "T"  "T"  " "   "C"   "T"   "C"
[2,] "-"  "-"  "+"  "+"  "+"  "+"  "-"  "+"  "+"  "-"   "+"   "+"   "-"
[3,] "G"  "T"  "A"  "T"  "C"  "A"  "A"  "T"  "T"  "G"   "C"   "T"   " "
     [,14] [,15] [,16] [,17]
[1,] "A"   "A"   "C"   "T"
[2,] "+"   "-"   "+"   "-"
[3,] "A"   " "   "C"   "C"

$numGaps
[1] 5

$error
[1] FALSE
```

We observe that a total of 10 nucleotides have been aligned by using 5 gaps (the initial sequences were formed respectively by 17 and 15).

3 Local Alignment

Local Sequence Alignment involves stretches that are shorter than the entire sequences, possibly more than one. This kind of alignment is suitable for comparing substantially different sequences, which possibly differ significantly in length, and have only a short patches of similarity.

We develop a function called localAlignment with R code for seeking all maximum length common fragment without gaps to two sequences. The problem can be modeled by using the following recursive equations

$$t_{i,j} = \begin{cases} t_{i-1,j-1} + 1 & \text{if} \quad x_i = y_j \\ \\ 0 & \text{otherwise} \end{cases}$$

$$t_{i,j} = 0 \quad \text{if} \quad i = 0 \quad \text{or} \quad j = 0$$

where $t_{i,j}$ is the maximum length of the common fragments to both chains ending in the position x_i, y_j.

This algorithm is of $O(nm)$ complexity order, where n and m are respectively the length of the initial two sequences to compare.

3.1 R Code LocalAlignment

```
localAlignment<-function(x,y,all){
N=length(x)
M=length(y)
t=matrix(0,N+1,M+1)
for(i in 1:N){
for(j in 1:M){
if(x[i]==y[j])
t[i+1,j+1]<-t[i,j]+1
else
t[i+1,j+1]<-0
}
}
t<-t[2:(N+1),2:(M+1)]
max<-0
fin1<-0
fin2<-0
for(i in 1:N){
for(j in 1:M){
if(t[i,j]>=max){
max=t[i,j]
fin1=i
fin2=j
}
}
}
long<-max;
if(all==FALSE){
in1<-fin1-(long-1)
in2<-fin2-(long-1)
if(in1<in2){
alignment<-matrix(c(" ","-"," "),3,max(N+in2-in1,M))
alignment[1,(in2-in1+1):(in2-in1+N)]<-x
alignment[2,in2:fin2]<-array("+",fin2-in2+1)
alignment[3,1:M]<-y
} else{
```

```
alignment<-matrix(c(" ","-"," "),3,max(N,M+in1-in2))
alignment[1,1:N]<-x
alignment[2,in1:fin1]<-array("+",fin1-in1+1)
alignment[3,(in1-in2+1):(in1-in2+M)]<-y
}
list(long=long,in1=in1,in2=in2,fin1=fin1,fin2=fin2,
                    sequence=c2s(x[in1:fin1]),alignment=alignment)
}else{
fin1=c()
fin2=c()
for(i in 1:N){
for(j in 1:M){
if(t[i,j]==long){
fin1=c(fin1,i)
fin2=c(fin2,j)
}
}
}
in1<-fin1-(long-1)
in2<-fin2-(long-1)
k=length(fin1)
alignment=list()
sequence=list()
for(i in 1:k){
currIn1=in1[i]
currIn2=in2[i]
sequence[[i]]<-x[currIn1:fin1[i]]
if(currIn1<currIn2){
currAlignment<-matrix(c(" ","-"," "),3,max(N+currIn2-currIn1,M))
currAlignment[1,(currIn2-currIn1+1):(currIn2-currIn1+N)]<-x
currAlignment[2,currIn2:fin2[i]]<-array("+",fin2[i]-currIn2+1)
currAlignment[3,1:M]<-y
} else{
currAlignment<-matrix(c(" ","-"," "),3,max(N,M+currIn1-currIn2))
currAlignment[1,1:N]<-x
currAlignment[2,currIn1:fin1[i]]<-array("+",fin1[i]-currIn1+1)
currAlignment[3,(currIn1-currIn2+1):(currIn1-currIn2+M)]<-y
}
alignment[[i]]<-currAlignment
}
list(long=long,in1=in1,in2=in2,fin1=fin1,fin2=fin2,
          sequence=sequence,alignment=alignment)
}
}
```

3.2 Example

```
> seq1
 [1] "A" "T" "C" "G" "T" "A" "T" "T" "C" "G" "G" "T" "C" "A" "A" "C" "T"
> seq2
 [1] "G" "T" "A" "T" "C" "A" "A" "T" "T" "G" "C" "T" "A" "C" "C"
> localAlignment(seq1,seq2,'TRUE')
$long
[1] 4

$in1
[1]  4 12

$in2
[1] 1 4

$fin1
[1]  7 15

$fin2
[1] 4 7

$sequence
$sequence[[1]]
[1] "G" "T" "A" "T"

$sequence[[2]]
[1] "T" "C" "A" "A"


$alignment
$alignment[[1]]
     [,1] [,2] [,3] [,4] [,5] [,6] [,7] [,8] [,9] [,10] [,11] [,12] [,13]
[1,] "A"  "T"  "C"  "G"  "T"  "A"  "T"  "T"  "C"  "G"   "G"   "T"   "C"
[2,] "-"  "-"  "-"  "+"  "+"  "+"  "+"  "-"  "-"  "-"   "-"   "-"   "-"
[3,] " "  " "  " "  "G"  "T"  "A"  "T"  "C"  "A"  "A"   "T"   "T"   "G"
     [,14] [,15] [,16] [,17] [,18]
[1,] "A"   "A"   "C"   "T"   " "
[2,] "-"   "-"   "-"   "-"   "-"
[3,] "C"   "T"   "A"   "C"   "C"

$alignment[[2]]
     [,1] [,2] [,3] [,4] [,5] [,6] [,7] [,8] [,9] [,10] [,11] [,12] [,13]
[1,] "A"  "T"  "C"  "G"  "T"  "A"  "T"  "T"  "C"  "G"   "G"   "T"   "C"
[2,] "-"  "-"  "-"  "-"  "-"  "-"  "-"  "-"  "-"  "-"   "-"   "+"   "+"
[3,] " "  " "  " "  " "  " "  " "  " "  " "  "G"  "T"   "A"   "T"   "C"
     [,14] [,15] [,16] [,17] [,18] [,19] [,20] [,21] [,22] [,23]
[1,] "A"   "A"   "C"   "T"   " "   " "   " "   " "   " "   " "
[2,] "+"   "+"   "-"   "-"   "-"   "-"   "-"   "-"   "-"   "-"
[3,] "A"   "A"   "T"   "T"   "G"   "C"   "T"   "A"   "C"   "C"
```

We can observe that the optimum length is 4 and reached for two different fragments *GTAT* and *TCAA*.

4 Generalizations

4.1 Needleman–Wunsch Algorithm

The Needleman–Wunsch Algorithm performs a global alignment between two sequences for any scoring system. The algorithm was published in 1970 by Saul B. Needleman and Christian D. Wunsch [4]. The problem can be modeled by using the following recursive equations

$$\begin{aligned} F_{i,j} &= \max\left\{F_{i-1,j-1} + S(x_{i-1}, y_{j-1}), F_{i,j-1} + d, F_{i-1,j} + d\right\} \\ F_{0,j} &= dj \\ F_{i,0} &= di \end{aligned}$$

where $F_{i,j}$ is the maximum score of alignment between the i-first characters of the sequence x and the j-first of the sequence y, $S(r, s)$ is the scoring system of matching the residue r with the residue s and d is the gap penalty.

This algorithm is of $O(nm)$ complexity order, where n and m are respectively the length of the initial two sequences to compare.

4.2 Smith–Waterman Algorithm

The Smith–Waterman algorithm is a variation of the Needleman–Wunsch Algorithm that performs a local sequence alignment between two sequences for any scoring system by means of comparing segments of all possible lengths and optimizing the similarity measure. The algorithm was proposed by Temple F. Smith and Michael S. Waterman in 1981 [5]. The problem can be modeled by using the following recursive equations

$$H_{i,j} = \max\begin{cases} 0 & \\ H_{i-1,j-1} + S(x_{i-1}, y_{j-1}) & \text{if} \quad x_i y_j \\ H_{i-1,j} - d & \text{if} \quad x_i - \\ H_{i,j-1} - d & \text{if} \quad - y_j \end{cases}$$

$$\begin{aligned} H_{0,j} &= 0 \\ H_{i,0} &= 0 \end{aligned}$$

where $H_{i,j}$ is the maximum Similarity Score between a suffix of $x[1...i]$ and a suffix of $y[1..j]$.

5 Introduction to Word Methods

Theoretically, adapting a Global or Local Alignment between a query sequence and each protein or DNA sequence in a database is possible. Those sequences with higher scores would be possible to homologous traits. For example, for the human insulin molecule as a query, a search along a protein database should detect as homologous insulin molecules from other species next to humans. However, such an approach is in practice very computationally unaffordable.

Word methods, also known as *k*-tuple methods, are heuristic methods that are not guaranteed to find an optimal alignment solution, but are significantly faster than Dynamic Programming. These methods increase the specificity of searching, but reduce the sensitivity. This allows better detecting remote homologies, but missing some homology relations.

FASTA and BLAST are well-known implementations that can be found via Web through portals as EMBL FASTA [6] or NCBI BLAST [7]. The underlying idea is to use a word search of length *k*. FASTA is a DNA and protein sequence alignment software package first described (as FASTP) by David J. Lipman and William R. Pearson in 1985 [8]. The BLAST program was designed by Stephen Altschul, Warren Gish, Webb Miller, Eugene Myers, and David J. Lipman at the NIH and was published in the Journal of Molecular Biology in 1990 [9].

In order to compare Words Methods and Dynamic Programming we have develop a R function for computing the *k* best global alignments between a query sequence and each sequence in a database. Previously, it is needed downloading a database in FASTA format by using the package ff [10]. Then, we chose a query sequence to compare with a total of thousand (although the original dataset had more than a half million). The algorithm took 25 minutes. Both FASTA as BLASTA needed less than a minute with the whole database.

6 Conclusions

When talking about Bioinformatics and Biostatistics, it is needed to use the programming language R and free software projects as Bioconductor that provides R packages for the analysis of genomic data.

One of the main research areas in Bioinformatics is Sequence Alignment of nucleotide or amino acid residues to identify regions of similarity. Two functions have been implemented with R code that solve standard Sequence Alignment. The function globalAlignment for Global Sequence Alignment, focused on searching the best alignment over the entire length of two sequences and the function localAlignment for Local Sequence Alignment, based on aligning segments of two sequences with the highest density of matches. These two functions are applications of Dynamic Programming for solving complex problems by breaking them down into simpler subproblems.

These functions illustrate how Dynamic Programming methods guarantee the optimal solution, and although we can extend a Global or Local Alignment to compare a query sequence with each protein or DNA sequence in a database, such approach would be very computationally intensive. Word methods are heuristic methods that do not guarantee to reach an optimal alignment, but are significantly more efficient than Dynamic Programming by increasing the specificity of searching, but reducing the sensitivity. Heuristic methods are better detecting remote homologies, but missing some homology relations that perhaps only can be found by Dynamic Programming.

References

1. http://www.tecnologiaUCM.es
2. http://www.r-project.org/
3. http://www.bioconductor.org/
4. Needleman SB, Wunsch CD (1970) A general method applicable to the search for similarities in the amino acid sequence of two proteins. J Mol Biol 48(3):443–453
5. Smith TF, Waterman MS (1981) Identification of common molecular subsequences. J Mol Biol 147:195–197
6. http://www.ebi.ac.uk/Tools/sss/fasta/
7. http://blast.ncbi.nlm.nih.gov/
8. Lipman DJ, Pearson WR (1985) Rapid and sensitive protein similarity searches. Science 227(4693):1435–1441
9. Altschul S, Gish W, Miller W, Myers E, Lipman D (1990) Basic local alignment search tool. J Mol Biol 215(3):403–410
10. http://cran.r-project.org/web/packages/ff/index.html

Reliability and Quality in the Area of Statistical Thinking in Engineering

Raquel Caro, Victoria López, Jorge Martínez and Guadalupe Miñana

Abstract There is remarkable growing concern about the quality control at the time, which has led to the search for methods capable of addressing effectively the reliability analysis as part of the Statistic. Managers, researchers and Engineers must understand that 'statistical thinking' is not just a set of statistical tools. They should start considering 'statistical thinking' from a 'system', which means, developing systems that meet specific statistical tools and other methodologies for an activity. The aim of this article is to encourage them (engineers, researchers, and managers) to develop a new way of thinking.

Keywords Engineering · Quality · Reliability · Distribution models

1 Introduction

The complexity of industrial processes involves considerable risks [1]. If they are not analyzed and controlled, it may significantly affect the quality of the products and services we purchase.

R. Caro (✉)
Industrial Organization Department, Pontificial Comillas University, Madrid, Spain
e-mail: rcaro@doi.ucpomillas.es

V. López · G. Miñana
Department of Computer Arquitecture, Complutense University, Madrid, Spain
e-mail: vlopez@fdi.ucm.es

G. Miñana
e-mail: guamiro@fdi.ucm.es

J. Martínez
Industrial Engineering Department, Nebrija University, Madrid, Spain
e-mail: jmartine@nebrija.es

F. Sun et al. (eds.), *Knowledge Engineering and Management*,
Advances in Intelligent Systems and Computing 214,
DOI: 10.1007/978-3-642-37832-4_32, © Springer-Verlag Berlin Heidelberg 2014

Most goods and services are obtained and are forwarded to their destination by some production systems. Throughout their life cycle each production system (often of large size of both the number of people working in them as by the size and value of the facilities and equipment they use) goes through different phases. The final phase, called operation phase, involves the construction and commissioning of the system and it is the only truly productive.

At this stage the system can have failures that temporarily or permanently interrupt the operation phase. When an overtime is needed, this may cause some declines in their characteristics, qualities, and benefits. The maintenance intends to reduce the negative impact of such failures, by reducing their number or by reducing its consequences [2].

Both reliability and risk analysis are essential factors in the safety of a product, equipment, or system. It is said that a system or device fails when it ceases to provide us the service that should give us, or when undesirable effects occur according to design specifications with which it was built or installed the goods in question. The system failure will have repercussions that will depend on: the type of system, the type of mission that it plays, and the time when failure occurs. Since performance is running in a parallel way to the reliability of the system, it is desirable and sometimes essential that systems get an optimal performance and also optimal reliability in the sense that the user can work with them without a high risk of failure. The level of reliability, safety, or successful operation depends on the nature of the objective of the system [3].

Thus, system reliability and risk analysis will have some cost and effort associated, so that the requirement of reliability for a system may suit its purpose and significance.

2 Longevity of System

Any system consists of a series of interconnected devices in a manner capable of performing particular functions. These functional blocks may consist of a single component or complex subsystems, depending on the type of system and the interconnections on the same.

The state of the components and structure of the system determine whether a computer is working or not. In short, quantify the reliability of a system generally requires the system to consider the structure and the reliability of its components [4, 5].

The reliability engineering is the study of longevity and failure of products, equipment, and systems. To investigate the causes of failing it is applied scientific and mathematical principles. This type of research aims to gain a better understanding of device failure to identify where improvements can be made to the design of products to enhance its life or at least to limit the adverse consequences of failure.

3 Quality Versus Reliability

3.1 Quality

The quality of products and services has become a major decision factor in most businesses in the world. Regardless of whether the consumer is an individual, a multinational, or a retail store where the consumer is making a purchase decision, it is possible to assign equal importance to the quality that the cost and delivery time. Therefore, improving the quality has become a major concern for many companies.

The field of Statistical Quality Control can be defined broadly as those statistical and engineering methods used to measure, monitor and control, and improve quality.

It has become increasingly clear that raising quality levels can lead to reduced costs, increased customer satisfaction, and therefore more reliability. This has resulted in a renewed emphasis on statistical techniques for designing quality products and to identify quality problems in various stages of production and distribution.

It is impractical to inspect quality into a product: the product must be done right the first time. Accordingly, the manufacturing process must be stable or repeatable and have the ability to operate with little variability around the target or nominal dimension. Statistical Process Control is a powerful tool for achieving process stability and to improve their capacity by reducing variability.

Even the smallest of the products or services you can offer or present with kindness and courtesy, increasing their value automatically. The overall customer experience has a decisive influence on the perception of quality, so a higher quality perception reduces the feeling of uncertainty or risk in the purchase decision, making it easier to sell.

The ISO 9000 Quality Standards are a set of rules that define the minimum requirements that are accepted internationally for the development and implementation of systems of quality management, which in the current context of high global competitiveness in the economy, have become standard indicators of the increasing quality requirements that customers demand.

This situation presents a great market with quality consciousness. Customers, once satisfied with the technical skills, are now demanding greater reliability and quality of providers who seek to secure it.

3.2 Reliability

Moreover, in all fields of engineering, it is essential to study the time until a failure occurs in a system. Reliability refers to the permanence of the quality of products or services over time. We say that a device or component is reliable if properly developed its work throughout its life. A reliable device that works properly during its life, while another does not will have many problems.

The study of quality, in a first stage, is limited to ensuring that the product is delivered in good conditions. Reliability seeks to ensure that the product will be in good conditions for a reasonable period of time.

In recent times, consumers demand quality/reliability when acquiring any product or service. In fact legislation evolves responsibility to manufacturers and builders during certain periods that should take care of the failures of the products for defects that may appear after the acquisition and use.

Competitiveness in the market is such that the output of goods or services of poor quality/reliability is becoming more difficult and long-term only survives those companies with excellent image quality and reliability.

Consistency and reliability metrics are starting to gain popularity since the widespread addition of mobile devices and embedded software is having an adverse effect on most network monitoring tools. There are a lot of research in aiming to improve stability and risk of the systems that depends on new network technologies [6–8].

The simplest concept of reliability is one that verifies that the product meets certain specifications, and when this happens, it is sent to the customer. This means that the product may fail over time, and in some cases the warranty period is a way to anticipate this possibility in the short term [9].

All this leads to the need to consider quality control based on time. The reliability is therefore an aspect of uncertainty in engineering, since that system is driven for a certain period of time, it can only be studied in terms of probability. In fact the word "reliability" has a precise technical definition:

Reliability is the ability of a system or component to perform its required functions under stated conditions for a specified period of time.

Summarizing, reliability means to estimate the life time of a product or system and the probability of failure.

Reliability engineering is closely related to safety engineering, in that they use common methods for their analysis and may require input from each other. Reliability engineering focuses on costs of failure caused by system downtime, cost of spares, repair equipment, personnel, and cost of warranty claims.

The focus of safety engineering is normally not on cost, but on preserving life and nature, and therefore deals only with particular dangerous system failure modes.

Reliability engineering for complex systems requires a different, more elaborate systems approach than reliability for non-complex systems. Reliability analysis has important links with function analysis, requirements specification, systems design, hardware design, software design, manufacturing, testing, maintenance, transport, storage, spare parts, operations research, human factors, technical documentation, training, and more.

Software reliability is a special aspect of reliability engineering. System reliability, by definition, includes all parts of the system, including hardware, software, supporting infrastructure (including critical external interfaces), operators and procedures.

Traditionally, reliability engineering focuses on critical hardware parts of the system. Since the widespread use of digital integrated circuit technology, software has become an increasingly critical part of most electronics and, hence, nearly all present day systems. For details, see [9].

4 Analysis of System Reliability

Any system (mechanical, electrical, etc.) is constituted by a set of interconnected functional blocks or devices so that they are capable of performing particular functions. These functional blocks may consist of a single component or complex subsystems, depending on the type of system and the interconnections on the same. The state of the components (operating, failure, malfunction, etc.) and the structure of the system determine if a system is working or not. The structure of the system is described by a logic diagram illustrating the relationship between components and satisfactory operation.

Ultimately, quantifying the reliability of a system or improving the reliability of a system generally requires the system to consider, the structure and the reliability of its components. Therefore, to study the reliability of a system is the first step in the analysis of all failure modes of the system components and their effect on the same.

This analysis is known as FMEA (Failure Mode and Effects Analysis) and FMEA (Analysis of Failure Modes and Effects). It was developed in the mid-twentieth century by engineers in weaponry. The FMEA requires a qualitative analysis of the system and its components, and therefore must be conducted by engineers during the design stage of the system.

Be especially careful when defining the failure, as ambiguity can appear. These failures must be always related to a parameter that can be measured or linked to a clear indication without subjective interpretations. To all this, it is inevitable subjective judgments (usually when the database is not controlled).

Environment specifications should include loads, temperature, humidity, vibration, and all the necessary parameters that may affect the probability of failure of the product or system. These requirements should be set so that they are verifiable and logical, and should be related to the corresponding probability distributions.

5 Modeling the Reliability by Probability Distributions

In principle, it can be used any probability distribution to model time to failure of equipment or systems. In practice, the monotonic distributions functions seem more realistic and, within this group there are a few distributions which are considered to provide the most reasonable reliability models of devices.

- Exponential distribution: failure rate constant.

It is the distribution used most frequently to model reliability is the exponential distribution because it is simple to treat algebraically and is considered adequate for modeling the functional life span of the device lifecycle.

- Weibull distribution: failure rates increasing and decreasing.

A large majority of the teams have no real failure rate constant is more likely to fail as they age. In this case the failure rate is increasing. In any case, it is possible to find computers with decreasing failure rates.

- Lognormal distribution.

Its hazard function is increasing and is often used to model the reliability of structural and electronic components. Its disadvantage is that it is quite difficult to treat algebraically, but its advantage is that it arises naturally as the convolution of exponential distributions.

5.1 Goodness of Fit Statistics

Reliability is very important in performance evaluation and that when configuring a computer system will be as strong as possible in terms of performance, within a limited budget, and it will have high life expectancy. It is not used as a component with great performance if it fails soon and need to be replaced. Each component will be associated with a probability distribution that will calculate its reliability, paying particular attention to the distributions presented above and already widely proven by these models.

However, it would be useful to analyze the empirical distributions of time to failure making adjustments to a given probability model, regardless of the specifications that we can take the appropriate supplier [10, 11].

A common problem in reliability engineering is fitting a probability distribution to a set of observations for a variable. One does this to be able to make forecasts about the future. The most common situation is to fit a distribution to a single variable (like the lifetime of a mechanical or electrical component).

The principle behind fitting distributions to data is to find the type of distribution (normal, exponential, lognormal, gamma, beta, etc.) and the value of the parameters (mean, variance, etc.) that give the highest probability of producing the observed data.

Usually, of course, we do not know that the data came from any specific type of distribution, though we can often guess at some good possible candidates by matching the nature of the variable to the theory on which the probability distributions are based. The normal distribution, for example, is a good candidate if the random variation of the variable under consideration is driven by a large number of random factors (none of which dominate) in an additive fashion, whereas the lognormal is a good candidate if a large number of factors influence the value of the variable in a multiplicative way [12].

There are a number of software tools on the market that will fit distributions to a data set, and most risk analysis tools incorporate a component that will do this.

Chi Squared, Kolmogorov-Smirnoff, and Anderson-Darling tests are three of the most commonly used to goodness of fit in order to check whether a given sample of data is drawn from a given probability distribution. For example, the Chi squared test uses a measure of the discrepancy between observed and expected values under the model and quantifies how much a small value of the measure supports the hypothesis that the fitted distribution appears to match the data.

5.2 Adjusting Distributions with Statistical Packages

There are statistical commonly used packages as SAS, the free project software R or SPSS [13] (Statistical Package for Social Sciences). All of these statistical softwares present a large number of windows from which, on the one hand, managing and analyzing database and on the other hand, access to different aspects of the handling of the results generated.

ARENA [11] is a discrete event simulation software simulation and automation software developed by Systems Modeling and acquired by Rockwell Automation in 2000. It uses the SIMAN processor and simulation language.

In Arena, the user builds an experiment model by placing modules (boxes of different shapes) that represent processes or logic. Connector lines are used to join these modules together and specify the flow of entities. While modules have specific actions relative to entities, flow, and timing, the precise representation of each module and entity relative to real-life objects is subject to the modeler. Statistical data, such as cycle time and WIP (work in process) levels, can be recorded and outputted as reports. Arena can be integrated with Microsoft technologies.

When an input process for a Simulation is stochastic, it must develop a probabilistic Model to characterize the process's behavior over time. Procedure of Input Distribution Modeling is the following:

1. Document the process to be modeled.
2. Develop a plan to collect the data.
3. Graphical and statistical analysis of the data.
4. Hypothesize possible distribution. For example, see Fig. 1.
5. Estimate distribution parameters.
6. Check for goodness of fit of hypothesized distributions.
7. Check sensitivity of inputs on simulation outputs.

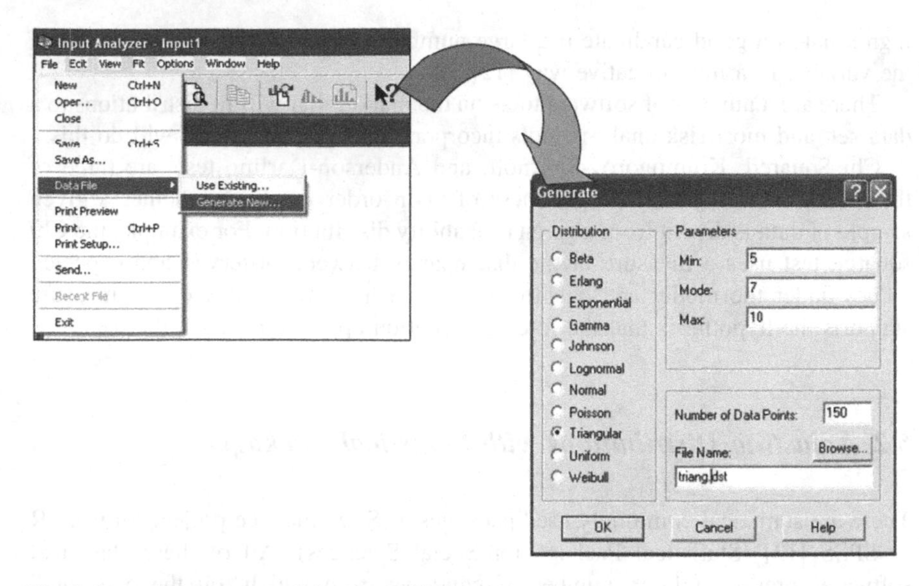

Fig. 1 Input distribution modeling in ARENA

5.3 An example: Overtime Production in a Make Rings Process

Table 1 shows the data specification of a problem in which an inspection process determines whether two rings fit or not (one inside the other) for an engine.

Table 2 shows the three possibilities after the analysis. With this information, if there are no problems, rings are sent to a packer whose time is lognormal of 7 min of mean and a standard deviation of 1 min. If there are problems, rings are sent to a rework craftsman that takes a time of 5 min plus a Weibull distribution with a

Table 1 Data specification of the problem

Data specification in a make rings process
100 Customer Calls with a success chance ~Beta (5, 1.5) with mean = 77 %
Each success call result in an order of a pair of rings
The number of rings is distributed by a binomial with a p given by the beta
The calling process is assumed time negligible
A ring is smaller than the other so that the smaller one must fit inside the larger one
The pair of rings is produced by a master ring maker which takes uniformly between 5 and 15 min
Rings are scrutinized by an inspector according to a triangular (2, 4, 7) min
The inspection determines whether the smaller ring is too big or too small when fit inside the bigger outer ring
The inside diameter of the bigger ring, Db, is normally distributed with a mean of 1.5 cm and a standard deviation of 0.002
The outside diameter of the smaller ring, Ds, is normally distributed with a mean of 1.49 and a standard deviation of 0.005

Table 2 Decision making

Three possibilities in making a decision
If $Ds > Db$, then the smaller ring will not fit
If $Db–Ds \leq tol = 0.02$ cm, then the rings will fit
If $Db–Ds > tol = 0.02$ cm, then the rings are considered too loose

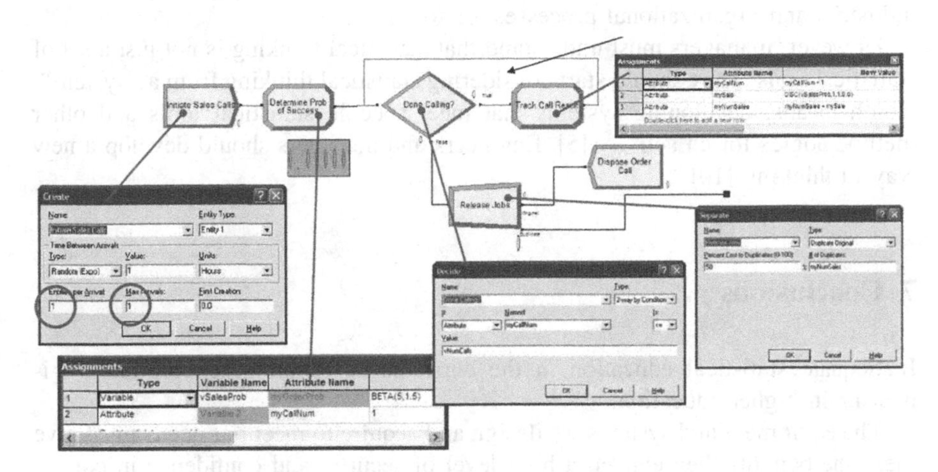

Fig. 2 Sales confirmation process

scale parameter of 15 and a shape of five. One problem is to know how about the probability of overtime production with two shifts of 480 min each. ARENA is used in this example to store and analyze this information and to get a solution. Figure 2 shows how to solve this particular issue.

6 New Functions for Engineers in the Area of Statistical Thinking

Inadequate statistical education in the curricula of business schools and engineering in higher education institutions has led many economists and engineers have completed their studies without understanding the value of statistics and its applications [9].

In Spain the subject Statistics is a course highly applied and it is taught students of Engineering or Business in the skills necessary to apply statistical techniques to enable them to understand and study phenomena nondeterministic.

Statistics is the part of Mathematics that is responsible for collecting, classifying, representing, and summarizing information from data and making inferences. An applied statistics course for students of Engineering and Business focuses in teaching skills necessary to apply statistical techniques to understand

and analyze random phenomena in order to enable them to design their own analysis.

Therefore, a statistics course for students of Engineering and Business should be designed taking into account the professional profile of students. The purpose of it is to train students in the application of statistical techniques in the engineering or management environment, to help them in decision-making and control of industrial and organizational processes, or so on.

However, managers must understand that statistical thinking is not just a set of statistical tools. They should start considering statistical thinking from a "system", which means, developing systems that meet specific statistical tools and other methodologies for an activity [5]. Engineers and managers should develop a new way of thinking [10].

7 Conclusions

Inadequate statistical education in the curricula of business schools and engineering in higher education.

The equipment and systems we design and acquire to meet our needs must give them the benefits they expect, a high level of security and confidence in correct operation. This will always depend both on the importance that we take the role of the equipment or system, and the consequences of failures that may occur. It is therefore necessary to consider reliability as a discipline in the design of any system, from the analysis of the identified need to dispose of the system designed, and integrated with other disciplines of logistics support.

A really effective reliability program can only exist in an organization where the fulfillment of the objectives of reliability is recognized as an integral part of corporate strategy.

Since the quality of the production will be the final determinant of reliability, quality control is an integral part of the reliability program. The quality control program should be based on reliability requirements and be directed not only to reduce production costs.

It is expensive to reach high goals of reliability, especially when the product or system is complex. But all this experience shows that all the efforts of well-managed program reliability are profitable because it costs less to discover and correct deficiencies in the design and development to correct the result of failures arising during operation of the product or system. Depending on the nature of the program we have the case of one or other type of cost.

Reliability Cost includes all costs charged during the design, production, warranty, etc. and it is based on the dual-user customer, while the life cycle cost comprises all costs charged by the system throughout its life, from conception to retirement at the end of its useful life, and this type of cost is based on manufacturer's perspective with a limited lifetime of the product. Therefore, there is a relationship between system reliability and cost of design and development.

It should be noted that reliability programs are usually constrained by the resources allocated to them during the design and development phases. The allocation of resources to the activities of a reliability program should be based on a consideration of the associated risks, being a subjective value based on experience.

Thus, the evaluation of reliability is a very important issue in organizations.

Studies of Engineering Degree tend to offer courses such as Statistics, Industrial Statistics, control and management of the quality, where it is studied the importance of the reliability and quality of systems.

References

1. Denney R (2005) Succeeding with use cases: working smart to deliver quality. Addison-Wesley Professional Publishing, Boston. Discusses the use of software reliability engineering in use case driven software development
2. Molero X, Juiz C, Rodeño MJ (2004) Evaluación y modelado del rendimiento de los sistemas informáticos. Pearson Prentice-Hall, Madrid
3. http://Informatica.Uv.Es/~Rmtnez/Ftf/Teo/ Tema01.Pdf
4. López V (2007) Performance and reliability of computer systems. EME-Editorial, Madrid
5. Puigjaner R, Serrano J, Rubio A (1995) Evaluación y explotación de sistemas informáticos. Ed. Síntesis. Madrid
6. Lopez V, Miñana G (2012) Modeling the stability of a computer system. Int J Uncertainty Fuzziness Knowl Based Syst 20:81–90
7. Lopez V, Santos M, Rodriguez T (2011) An approach to the analysis of performance, reliability and risk in computer systems. In: 6th international conference on intelligent systems and knowledge engineering, ISSN: 1867-5662, vol 2. Shanghai, pp 653–658
8. Gunther NJ (2005) Analyzing computer system performance with Perl: PDQ, Ed. Springer, Heidelberg
9. Caro R, López V, y Miñana G (2012) Análisis de fiabilidad de dispositivos según su sstructura para la herramienta Software EMSI. Actas Del XXXIII Congreso Nacional De Estadística E Investigación Operativa SEIO, Madrid
10. http://Www.Ucm.Es/Info/Tecnomovil/
11. Rossetti MD (2010) Simulation modeling and ARENA. Wiley, New York
12. http://www.vosesoftware.com/
13. www.ibm.com/software/analytics/spss

A Novel Fast Approach to Eye Location and Face Geometric Correction

Kejun Wang, Guofeng Zou, Lei Yuan and Guixia Fu

Abstract An effective method of eye location and face geometric correction is proposed in this paper. The classifier based on the Adaboost algorithm is used to quickly detect face region, then the eye candidate region is rapidly determined through the eyes detector. The positions of the eyes can be precisely located by using block integral projection according to the position and proportion of eyes in the candidate region. Especially in the case of eycs tilted and wearing glasses, the novel block integral projection approach can prevent the emergence of pseudo feature points, and the influences of the spectacle border and spectacle frame can be avoided. After eyes location, a novel calculation method of the rotation angle is proposed, then we give a detailed theoretical proof for this method, it is important for improving image rotation theory. Experimental results show that the proposed approach can realize the face geometric correction and also has better real time.

Keywords Eye location · The Adaboost algorithm · Block integral projcction · Face geometric correction

K. Wang (✉) · G. Zou · L. Yuan · G. Fu
College of Automation, Harbin Engineering University, 150001 Harbin, China
e-mail: wangkejun@hrbeu.edu.cn

G. Zou
e-mail: zgf841122@163.com

L. Yuan
e-mail: yuanlei8430@163.com

G. Fu
e-mail: fgx45101@163.com

F. Sun et al. (eds.), *Knowledge Engineering and Management*, Advances in Intelligent Systems and Computing 214,
DOI: 10.1007/978-3-642-37832-4_33,

1 Introduction

Face recognition has become an interesting research topic due to their enormously commercial and law enforcement applications, but the performance of many algorithms is reduced greatly when the face images have rotation change. Therefore, the study on effective face geometric correction method has important research value for improving the recognition effect. The eye precise automatic location is one of the absolutely necessarily key technologies for face geometric correction. The reason is that the distance between the eyes center is not subjected to the illumination and expression changes, the direction of the line between the eyes center would deflect as the deflection of the face, so it can be used as the basis for face rotation correction.

Generally, the eye location methods include two steps: (1) Estimation of the eye region. The most commonly used approaches in estimation of the eye region include: the template matching method [1], the threshold segmentation method [2], and the Adaboost algorithm [3]. (2) Precise location of the eyes. The precise location methods of eyes mostly include: the region segmentation method [4] and the projection method [5–7], etc.

The concept of image geometric correction refers to the process that the image rotates at an angle based on a fixed center point, so we need to determine the reference point, rotation angle and rotation direction. A fast face location and geometric correction method has been proposed by Hou [8], but it did not give the reference point selection method and rotation angle calculation formula. Su [9] defined rotation angle as an angle between the horizontal line and the line connecting the eyes center, while he did not specify how to select the reference point.

In this paper, we propose a novel method of precisely locating eyes and face geometric correction. The eye candidate regions can be detected in natural image by Adaboost cascade classifier. Then the candidate region is divided into three sub-districts, the eyes position is accurately located by gray integral projection in the left and right eye sub-district. Especially in the case of eyes tilted and wearing glasses, the proposed block integral projection is effective in preventing the emergence of pseudo feature points and avoiding the influences of the spectacle border and spectacle frame. After eyes location, the calculation method of the rotation angle is proposed when the center of the image is used as the reference point, and we give a detailed theoretical proof for this method, it is important for improving image rotation theory. On this basis, the face geometric correction can be quickly realized.

2 Estimation of the Eye Region

In 2001, Viola [10] achieved real-time processing of the face detection by combining the Adaboost algorithm [11] with Haar features based on the integral image, the algorithm has been applied in the real system. We use standard Adaboost

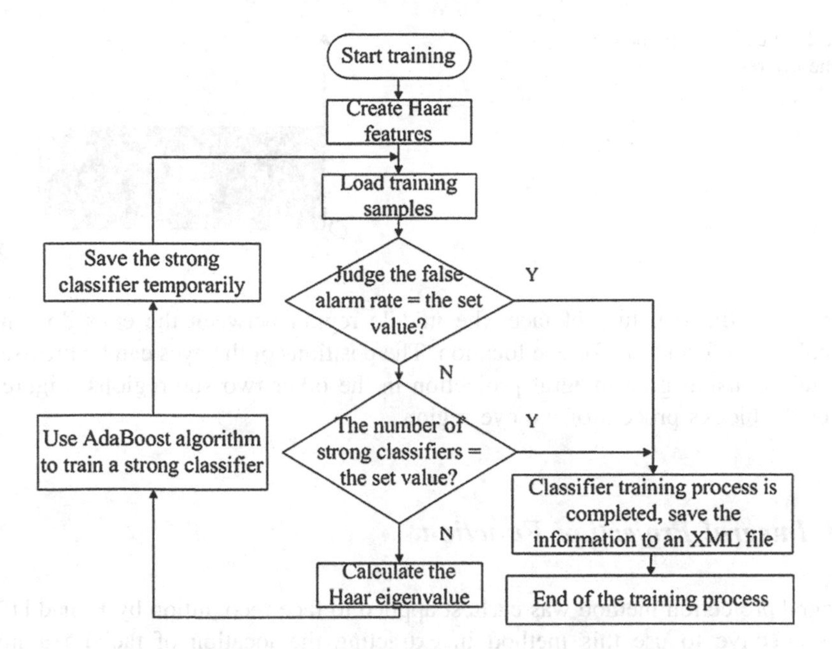

Fig. 1 The training process of cascade classifier

training methods combined with Viola's cascade approach to build eye classifier. After obtaining the classifier, the face region is rapidly detected based on the Adaboost face classifier, and then the eye region can be detected by using the eye classifier in face image, which is the estimation of the eyes accurate location. Figure 1 gives the training process of the eye classifier.

The eye classifier training process: There are in total 4,000 eye samples in the positive training set. All the eye samples are manually segmented from the JAFFE, the CAS-PEAL, and self-built face database. All the eye samples are processed with size normalization to pixels. The negative samples should choose the images which contain a variety of complex situation and are not supposed to include the eye area.

3 Precise Location of the Eyes

The gray integral projection is simple and faster, which is more accurate for eye location under ideal state. If the face is tilted, we will get four feature points by calculating traditional gray integral projection directly, because the vertical and horizontal projection appear two local minimum points respectively. Obviously, there must be two pseudo feature points (A, B) as shown in Fig. 2. In order to solve this problem, the novel block integral projection method is proposed, the eye region is divided into three same width sub-regions based on the shape of the eye

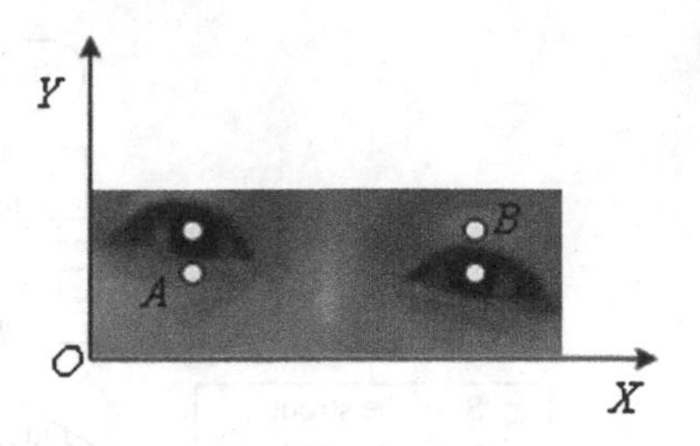

Fig. 2 Pseudo feature points of the eye region

region and the structure of face, the middle region between the eyes does not contain the information for eye location. The positions of the eyes can be precisely located by using gray integral projection in the other two sub-regions. Figure 3 gives the blocks process of the eye region.

3.1 Integral Projection Function

Integral projection method was earliest applied to face recognition by Kanad [12], it is effective to use this method in extracting the location of facial features. Suppose $I(x, y)$ is the intensity of a pixel at location (x, y), the vertical and horizontal integral projection in intervals $[y_1, y_2]$ and $[x_1, x_2]$ can be defined, respectively, as:

$$I_1(x_0) = \sum_{y=y_1}^{y_2} I(x_0, y) \; I_2(y_0) = \sum_{x=x_1}^{x_2} I(x, y_0) \tag{1}$$

where x_0 is the projection coordinate of x axis, y_0 is the projection coordinate of y axis. Usually the mean vertical and horizontal projections are used, which can be defined, respectively, as:

$$M_1(x_0) = \frac{1}{N_1} \sum_{y=y_1}^{y_2} I(x_0,\ y) \quad M_2(y_0) = \frac{1}{N_2} \sum_{x=x_1}^{x_2} I(x,\ y_0) \tag{2}$$

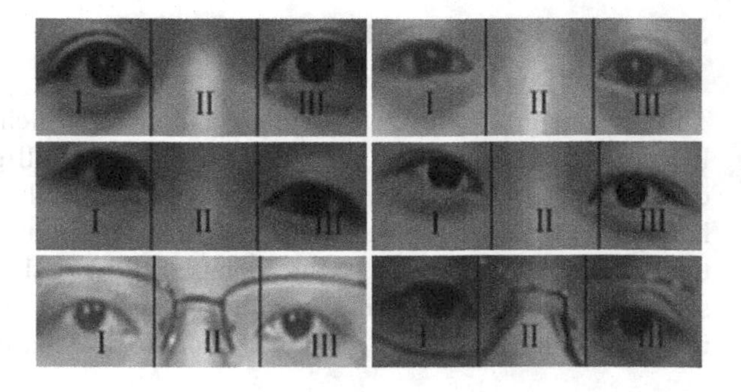

Fig. 3 The eye sub-regions

where N_1 and N_2 are the number of pixels in projection intervals for $x = x_0$ and $y = y_0$, respectively.

Ideally, the center coordinates of the eyes correspond to the local minimum points of vertical integral projection curve and horizontal integral projection curve.

3.2 Block Processing of the Eye Region

In order to avoid failure in locating the eyes because of the pseudo feature points, the eye region is divided into three sub-regions: the region I is left eye sub-region, the region II is middle sub-region, and the region III is the right eye sub-region. We calculate gray integral projection in vertical and horizontal directions for the left eye and right eye sub-region, respectively, in each sub-region, only one minimum point is determined, which is the coordinate of the eyes accurate position. The middle area does not contain useful information for locating the eyes, which could be ignored. As for wearing glasses, the removal of middle region can prevent thc interference of spectacles frame for the precise location of the eyes.

4 Face Geometric Correction

The accurate result of eyes location is shown in Fig. 4, the left eye coordinate $A(x_A, y_A)$ and the right eye coordinate $B(x_B, y_B)$. According to the definition of image geometric correction, it is necessary to ascertain the reference point, rotation angle, and rotation orientation to realize the image geometric correction. The center point of the image is selected as the reference point in this paper, and the distance between the center point and the other point of the image keeps invariant even if the image is rotated. The rotation orientation could be judged by the value of Y in the eyes coordinates $A(x_A, y_A)$ and $B(x_B, y_B)$. When $y_A < y_B$, the rotation orientation is clockwise, the rotation angle is $\theta = \alpha$, otherwise it is anticlockwise, the rotation angle is $\theta = -\alpha$.

Assume that XOY is the original coordinate system, $X'O'Y'$ is the new coordinate system after correction, coordinate (x, y) is a point in the image and the coordinate after rotation is (x', y'). The rotation formula is defined as

$$\begin{bmatrix} x' \\ y' \end{bmatrix} = \begin{bmatrix} \cos\theta & \sin\theta \\ -\sin\theta & \cos\theta \end{bmatrix} \begin{bmatrix} x \\ y \end{bmatrix} \tag{3}$$

Assume that the face image $I(x, y)$ with a resolution of $M \times N$ pixel is tilted. The left eye coordinate $A(x_A, y_A)$ and the right eye coordinate $B(x_B, y_B)$, then the rotation angle α is the function of the coordinate of point A and point B.

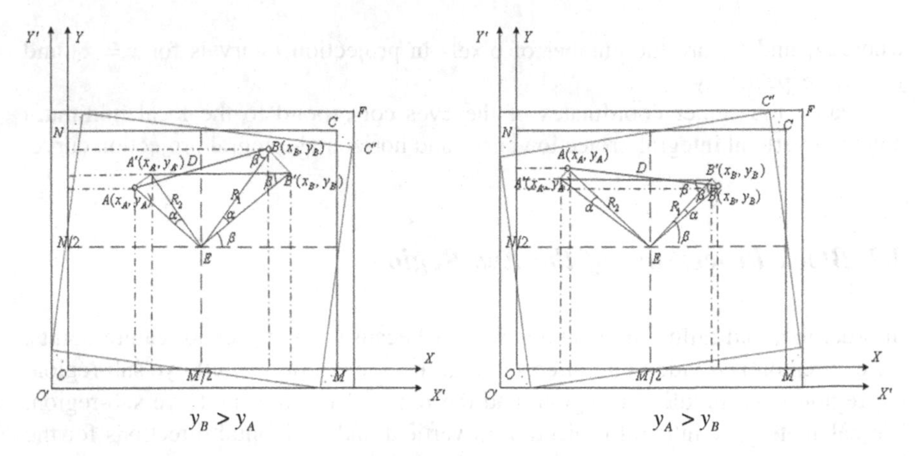

Fig. 4 The diagram of image rotation

(1) We assume $y_B > y_A$, it means the right eye position is higher than that of left eye. It is shown in Fig. 4.

The left eye coordinate $A(x_A, y_A)$ and right eye coordinate $B(x_B, y_B)$, the size of the image is $M \times N$ and the center point coordinate of face is E: $E(M/2, N/2)$.

The distance between $A(x_A, y_A)$ and $B(x_B, y_B)$ is defined as:

$$D = \sqrt{(x_A - x_B)^2 + (y_A - y_B)^2} \tag{4}$$

The distance between $B(x_B, y_B)$ and geometric center point is defined as:

$$R_1 = \sqrt{(x_B - M/2)^2 + (y_B - N/2)^2} \tag{5}$$

The distance between $A(x_A, y_A)$ and geometric center point is defined as:

$$R_2 = \sqrt{(x_A - M/2)^2 + (y_A - N/2)^2} \tag{6}$$

Then the point A, point B, and the center point E form a triangle ΔEAB.

We assume the rotation angle is α. After rotation, Point A corresponds to point A' and point B corresponds to point B', and the triangle $\Delta EA'B'$ is formed by point A', point B', and point E. Obviously, these two triangles are congruent triangles, and $\angle ABE = \angle A'B'E = \beta$.

According to the trigonometric relation:

$$\tan(\alpha + \beta) = \left(\frac{y_B - N/2}{x_B - M/2}\right) \tag{7}$$

According to the trigonometric cosine theorem:

$$D^2 + R_1^2 - 2DR_1 \cos(\beta) = R_2^2 \tag{8}$$

We get two formulas as follows:

$$(\alpha + \beta) = \arctan\left(\frac{y_B - N/2}{x_B - M/2}\right) \tag{9}$$

$$\beta = \arccos\left(\frac{D^2 + R_1^2 - R_2^2}{2DR_1}\right) \tag{10}$$

Obviously,

$$\alpha = \arctan\left(\frac{y_B - N/2}{x_B - M/2}\right) - \arccos\left(\frac{D^2 + R_1^2 - R_2^2}{2DR_1}\right) = f(x_A, y_A, x_B, y_B) \tag{11}$$

(2) We assume $y_A > y_B$, it means the position of right eye is lower than that of left eye.

The equation of solving rotation angle α is defined as:
The trigonometric relation:

$$\tan(\beta - \alpha) = \left(\frac{y_B - N/2}{x_B - M/2}\right) \tag{12}$$

The trigonometric cosine theorem:

$$D^2 + R_1^2 - 2DR_1\cos(\beta) = R_2^2 \tag{13}$$

Rotation angle is solved as:

$$\alpha = \arccos\left(\frac{D^2 + R_1^2 - R_2^2}{2DR_1}\right) - \arctan\left(\frac{y_B - N/2}{x_B - M/2}\right) = f(x_A, y_A, x_B, y_B) \tag{14}$$

5 Experiments and Analysis

Based on the above discussion, the procedure of the eyes rapid location and face geometry correction can be summarized as follows:

(1) Reading in face image and detecting the eye candidate region; (2) dividing the eye candidate region into three sub-regions and locating the precise positions of the eyes; (3) calculating the rotation angle and determining direction of image rotation; and (4) realizing the face geometric correction based on the image rotation formula.

The self-built face database is used in our experiments. This face database contains 20 persons, and everyone has 10 images with a resolution of 320 × 240 pixel. These images have posture change, distance change, expression variation, glasses and hair interference. Some images are shown in Fig. 5.

Fig. 5 Some images in the experimental database

Fig. 6 Eye region detection results

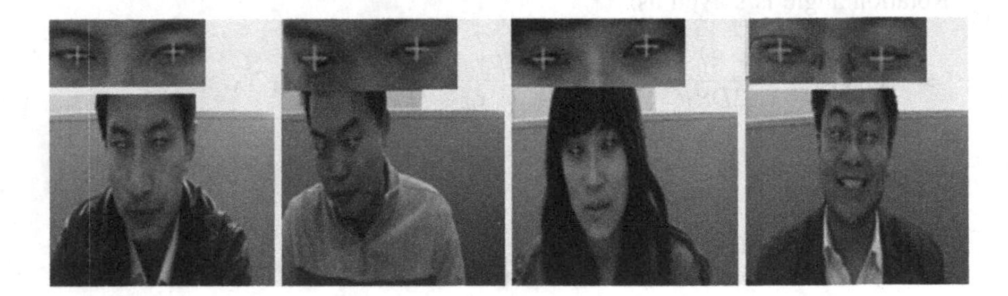

Fig. 7 Results of eyes accurate location

Figure 6 gives the detection results of the eye region in test image, obviously, the eye area rough estimation method based on Adaboost algorithm can obtain more accurate eye area, to some extent, this method overcomes the impact of the eyebrows and sideburns, and the detection is very fast to meet the requirements of real time.

Figure 7 gives some images of accurate locations, the results show that the presented block integral projection method can locate eyes accurately in complex situations even though the face is serious tilted or there are serious interference of glasses.

Table 1 Comparison of location accuracy our method with other methods

Method	Correct rate (%)	Detection time (s)
Method based on projection peak [14]	82.50	0.5301
Method based on block [15]	80.00	0.5532
Gabor method [16]	78.50	0.4701
Our method	89.50	0.1033

When the glasses frame keeps out the eye center position or there is serious uneven illumination, the interference factors would be more, which may result in the failure of eyes location. In order to verify the effectiveness of our method, we compare our method to other methods in experiment, the location correct rate are shown in Table 1.

The measurement criterion of Jesorsky [13] is used to judge the quality of eye location, let the C_l and C_r be the manually exacted left and right eyes positions, C_l' and C_r' be the detected positions, d_l be the Euclidean distance between C_l and C_l', d_r be the Euclidean distance between C_r and C_r'. The relative error of the detection is defined as:

$$err = \frac{\max(d_l, d_r)}{\|C_l - C_r\|} \tag{15}$$

If $err < 0.1$, we consider that the detection is accurate. Thus, for a database comprising N images, the location accuracy is defined as:

$$rate = (\sum_{i=1}^{N} 1/N) \times 100\ \% \quad (err_i < 0.1) \tag{16}$$

where err_i is error on the i-th image.

The method based on projection peak selected eye candidate window according to the proportion self-defined, and segmented the eyes through the histogram threshold segmentation. However, if there are greater pose changes, the method would not accurately exact the candidate region. What is more, the segmentation is vulnerable affected by illumination variation. Thus, the effect of location is not good. The block-based method is unable to overcome the affect of weak and side illumination, furthermore the computational complexity is higher and the processing time is longer. Gabor method is difficult to overcome the influence of side illumination and black frame spectacles, and may locate pseudo feature points in the situation of tilted face and greater pose changes. The accuracy of our method is better than the method based on projection peak; furthermore, it has good real-time performance, so the presented method has practical value.

The tilted face image should be corrected after the precise location of the eyes. Some geometric correction results are shown in Fig. 8, the red horizontal straight line in the figure is used to show that the eyes have been corrected to the horizontal position and the face images obtained an effective correction.

Fig. 8 Results of face geometric correction

6 Conclusions

It is an essential element of the face recognition to locate the eyes precise positions and correct tilted face. We propose an efficient eye location method, the Adaboost cascade classifier is use to quickly identify the eye candidate region, then the novel block integral projection is carried out in order to precisely locate the eyes. After the eyes' accurate location, through detailed analysis and theoretical proof, a novel calculation method of the rotation angle is presented when we rotate the image regarding the center of the image as the reference point, it is important for improving image rotation theory. Finally, the experimental results show that the proposed method can effectively rotate the face image to the horizontal direction and also has good real-time performance.

Acknowledgments This research is supported by National High Technology Research and Development Program of China (2008AA01Z148); National Natural Science Foundation of China (60975022); and National Research Foundation for the Doctoral Program of Higher Education of China (20102304110004).

References

1. Yuille AL, Hallinan PW, Cohen DS (1992) Feature extraction from faces using deformable templates. Int J Comput Vis 8(2):99–111
2. Song J, Liu J, Chi Z (2005) Precisely locating human eyes from front view face images. J Comput Aided Des Comput Graph 17(3):540–545
3. Tang X, Ou Z, Su T (2006) Fast locating human eyes in complex background. J Comput Aided Des Comput Graph 18(10):1535–1540

4. Shan S, Gao W, Chen X (2001) Facial feature extraction based on facial texture distribution and deformable template. J Softw 12(4):570–577
5. Feng GC, Yuen PC (1998) Variance projection function and its application to eye detection for human face recognition. Pattern Recogn Lett 19(9):899–906
6. Zhou Z, Geng X (2004) Projection functions for eye detection. Pattern Recogn 37(5):1049–1056
7. Zheng Y, Wang Z (2008) Minimal neighborhood mean projection function and its application to eye location. J Softw 19(9):2322–2328
8. Hou Y, Peng J, Li N (2005) A fast approach to face location and normalization. Acta Photonica Sin 34(12):1906–1909
9. Su G, Meng K, Du C et al (2005) Face recognition method based on the face geometric size normalized: China, 200510067962, X[P]
10. Viola P, Jones M (2001) Rapid object detection using a boosted cascade of simple feature. In: Proceedings of IEEE computer society conference on computer vision and pattern recognition. IEEE Press, Hawaii, pp 511–518
11. Freund Y, Schapire R (1997) A decision-theoretic generalization of online learning and an application to boosting. J Comput Syst Sci 55(1):119–139
12. Kande T (1973) Picture processing by computer complex and recognition of human faces. Kyoto University, Kyoto
13. Jesorsky O, Kirchberg KJ, Frischholz RW (2001) Robust face detection using the Hausdorff distance. In: Proceedings of international conference on audio and video-based biometric person authentication. LNCS, vol 2091. Springer, Halmstad, pp 90–95
14. Dai J, Liu D, Su J (2009) Rapid eye localization based on projection peak. Int J Pattern Recognit Artif Intell 22(4):605–609
15. Ai J, Yao D, Guo Y-F (2007) Eye location based on blocks. J Image Graph 12(10): 1841–1844
16. Sun D, Wu L (2001) Eye detection based on Gabor transforms. J Circu Syst 6(4):29–32

α-Generalized Lock Resolution with Deleting Strategies in $\mathscr{L}_n$F(X)

Xingxing He, Yang Xu, Jun Liu and Peng Xu

Abstract This paper focuses on refined non-clausal resolution methods in a Łukasiewicz first order logic $\mathscr{L}_n$F(X), i.e., α-generalized lock resolution with deleting strategies, which can further improve the efficiency of α-generalized lock resolution. First, the concepts of strong implication, weak implication, and α-generalized lock resolution with these two deleting strategies are given, respectively. Then the compatibilities of α-generalized lock resolution with strong implication deleting and weak implication deleting are shown in $\mathscr{L}_n$F(X), respectively. Finally, an algorithm for α-generalized resolution with these deleting strategies is given.

Keywords α-Generalized lock resolution · Deleting strategies · Non-clausal resolution · Łukasiewicz first order logic

1 Introduction

Semantics and syntactics are two important aspects for logic systems, which reflect properties of logical formulae from different views. For resolution-based automated reasoning [1, 7, 9] in classical logic, judging whether two literals are resolvable is simple, that is, they are complementary literals in form. This

X. He (✉) · Y. Xu · P. Xu
School of Mathematics, Southwest Jiaotong University, Chengdu 610031 Sichuan, People's Republic of China
e-mail: xinghe010@gmail.com

Y. Xu
e-mail: xuyang@swjtu.edu.cn

J. Liu
School of Computing and Mathematics, University of Ulster, Northern Ireland, UK
e-mail: j.liu@ulster.ac.uk

F. Sun et al. (eds.), *Knowledge Engineering and Management*, Advances in Intelligent Systems and Computing 214,
DOI: 10.1007/978-3-642-37832-4_34,

judgment can be realized in syntactic ways; hence, it can be implemented directly in computer for theorem proving. Computer language is developed based on a formal language of logic system, or to be precise, a class of formal languages, where it is a recursively defined set of strings on a fixed alphabet. Generally, the syntactic properties are easy to be judged in form, and can be applicable to realize on computer. However, they have some limitations, and are not general in theory [2, 3, 10, 11].

Uncertainty is often associated with human's intelligent activities, it is rather difficult to represent and reason it only by numbers or symbols in classical logic. α-Resolution principle [13, 14] in lattice-valued logic [15] based on lattice implication algebra (LIA) [12] is a great extension of Robinson's resolution principle [9], which can deal with uncertainty especially for incomparability in the intelligent information processing. Due to the extension of implication connective and language in lattice-valued logic, judging whether two g-literals are α-resolvable should be in both semantic and syntax ways, which is a relatively complex process. For improving efficiency of α-resolution, some refined strategies should be added. In α-lock resolution deduction [4, 5], we discussed α-lock resolution with deleting strategy [6] where this process is implemented only by a syntactic way, i.e., subsumed generalized clauses are deleted.

α-Generalized resolution [16, 17] is a non-clausal resolution method [8]. Compared with α-resolution, it has many advantages, for example, it can validate α-unsatisfiability of logical formulae without converting them to according generalized clausal forms, it is a dynamic resolution, etc. For generalizing deleting strategy and further improving the efficiency of α-generalized resolution, in this paper we intend to discuss the properties of some deleting strategies in semantic way, restrict them to α-generalized lock resolution, and give their soundness and completeness.

2 Preliminaries

Among extensive research results on α-resolution automated reasoning in lattice-valued propositional logic LP(X) and first-order logic LF(X) based on LIA, we only outline elementary concepts of α-generalized lock resolution. For further details about the background and properties, we refer to the related references, e.g., [12–17].

Definition 1 [12, 15] Let $(L, \vee, \wedge, O, I)$ be a bounded lattice with an order-reversing involution "$'$", I and O the greatest and the smallest element of L, respectively, and $\rightarrow$: $L \times L \longrightarrow L$ a mapping. $\mathscr{L} = (L, \vee, \wedge, ', \rightarrow, O, I)$ is called a lattice implication algebra (LIA) if the following conditions hold for any $x, y, z \in L$:

(I_1) $\quad x \rightarrow (y \rightarrow z) = y \rightarrow (x \rightarrow z)$,

(I_2) $x \rightarrow x = I,$
(I_3) $x \rightarrow y = y' \rightarrow x',$
(I_4) $x \rightarrow y = y \rightarrow x = I$ implies $x = y,$
(I_5) $(x \rightarrow y) \rightarrow y = (y \rightarrow x) \rightarrow x,$
(L_1) $(x \vee y) \rightarrow z = (x \rightarrow z) \wedge (y \rightarrow z),$
(L_2) $(x \wedge y) \rightarrow z = (x \rightarrow z) \vee (y \rightarrow z).$

Definition 2 [15] (Łukasiewicz implication algebra on a finite chain L_n) Let L_n be a finite chain, $L_n = \{a_i | 1 \leq i \leq n\}$ and $a_1 < a_2 < \ldots < a_n$, define for any $a_j, a_k \in L_n$, $a_j \vee a_k = a_{max\{j,k\}}$, $a_j \wedge a_k = a_{min\{j,k\}}$, $(a_j)' = a_{n-j+1}$, $a_j \rightarrow a_k = a_{min\{n-j+k,n\}}$. Then $\mathscr{L}_n = (L_n, \vee, \wedge, ', \rightarrow, a_1, a_n)$ is an LIA.

Definition 3 [13] Let X be a set of propositional variables, $T = L \cup \{', \rightarrow\}$ be a type with ar($'$) = 1, ar($\rightarrow$) = 2 and ar(a) = 0 for every $a \in L$. The propositional algebra of the lattice-valued propositional calculus on the set X of propositional variables is the free T algebra on X and is denoted by LP(X).

Remark 1 Specially, when the field with valuation of LP(X) is an $\mathscr{L}_n$, this specific LP(X), i.e., $L_nP(X)$, is a linguistic truth-valued lattice-valued propositional logic system. Similarly, the truth-valued domain of $L_nP(X)$ is a Łukasiewicz implication algebra $\mathscr{L}_n$.

Definition 4 [13] A valuation of LP(X) is a propositional algebra homomorphism γ: LP(X) $\longrightarrow L$.

Definition 5 [13] Let F be a logical formula in LP(X), $\alpha \in L$. If there exists a valuation γ_0 of LP(X) such that $\gamma_0(F) \geq \alpha$, F is satisfiable by a truth-value level α, in short, α-satisfiable. If $\gamma(F) \geq \alpha$ for every valuation γ of LP(X), F is valid by the truth-value level α, in short, α-valid. If $\gamma(F) \leq \alpha$ for every valuation γ of LP(X), F is always false by the truth-value level α, in short, α-false.

Definition 6 [16] Let $g_1, g_2, \cdots, g_n$ be generalized literals in LP(X). A logical formula Φ is called a general generalized clause if these generalized literals are connected by $\wedge, \vee, \rightarrow, '$ and $\leftrightarrow$, denoted by $\Phi(g_1, g_2, \cdots, g_n)$.

The general generalized clause is the extension of generalized clause in LP(X).

Definition 7 [17] Let Φ be a general generalized clause in LP(X). A generalized literal g of Φ is called a local extremely complex form, if

(1) g cannot be expanded to a more complex generalized literal in Φ by adding $\rightarrow$ and $'$.
(2) If g is connected by $\leftrightarrow$, then g is the local extremely complex form as a whole.

All the generalized literals mentioned in this paper are the local extremely complex forms in their corresponding general generalized clauses.

Definition 8 [17] Let $g_1, g_2, \cdots, g_n$ be g-literals, a logical formula in LF(X) is called a general g-clause if these g-literals are connected by $\wedge, \vee, \rightarrow, '$ and $\leftrightarrow$, denoted by $\Phi(g_1, g_2, \cdots, g_n)$.

The general generalized clause in LP(X) is the ground form of general g-clause in LF(X).

Definition 9 [17] Let Φ be a general g-clause in LF(X). Φ is said to be locked if and only if for each g-literal g in Φ, there exists a positive integer i such that i is the index of g. This specific general g-clause Φ is called a locked general g-clause.

Definition 10 [17] Let $\Phi_1, \Phi_2, \cdots, \Phi_n$ be locked general g-clauses in LF(X), $\Phi_1^{\sigma_1}$ a factor of Φ_1 for g-literals $g_{11}, g_{12}, \cdots, g_{1r_1}$, $\Phi_2^{\sigma_2}$ a factor of Φ_2 for g-literals $g_{21}, g_{22}, \cdots, g_{2r_2}$, $\cdots$, and $\Phi_n^{\sigma_n}$ a factor of Φ_n for g-literals $g_{n1}, g_{n2}, \cdots, g_{nr_n}$, $\alpha \in L$. If there exists $g_{i1}^{\sigma_i}$ with the minimal index in $\Phi_i^{\sigma_i}(i = 1, 2, \cdots, n)$, such that $\wedge_{i=1}^{n} g_{i1}^{\sigma_i} \leq \alpha$, then

$$\Phi = \vee_{i=1}^{n} \Phi_i^{\sigma_i}(g_{i1}^{\sigma_i} = \alpha)$$

is called an α-generalized lock resolvent of Φ_1, Φ_2, $\cdots$, Φ_n, denoted by $\Phi = R_{\alpha-g-L}(\Phi_1(g_{11}), \Phi_2(g_{21}), \cdots, \Phi_n(g_{n1}))$.

3 α-Generalized Lock Resolution with Deleting Strategies in $L_nF(X)$

3.1 Compatibility with Strong Implication Deleting Strategies

Definition 11 Let g be a generalized literal in LP(X), then g is α-normal if there exists a valuation γ such that $\gamma(g) = \alpha$.

All the general generalized literals mentioned are α-normal as follows.

Definition 12 Let Φ, Ψ be general generalized clauses in LP(X). Φ is said to strongly imply Ψ if for any valuation γ, $\gamma(\Phi) \geq \alpha$ implies $\gamma(\Psi) \geq \alpha$, denoted by $\Phi \Rrightarrow_{\alpha} \Psi$.

Definition 13 Let Φ, Ψ be general generalized clauses in LP(X). Φ is said to weakly imply Ψ if for any valuation γ, $\gamma(\Phi \rightarrow \Psi) \geq \alpha$, denoted by $\Phi \Rightarrow_{\alpha} \Psi$.

Remark 2 Strong and weak implications are extensions of subsumption in form.

Proposition 1 *Let $\Phi_1, \Phi_2, \cdots, \Phi_n$ be general generalized clauses in LP(X), $R_{(g-\alpha)-g}(\Phi_1, \Phi_2, \cdots, \Phi_n)$ an α-generalized resolvent of Φ_1, Φ_2, $\cdots$, Φ_n.* If $\Phi_1 \geq \alpha$, then $R_{(g-\alpha)-g}(\Phi_1, \Phi_2, \cdots, \Phi_n) \geq \alpha$.

Proof Let $R_{(g-\alpha)-g}(\Phi_1, \Phi_2, \cdots, \Phi_n) = \vee_{i=1}^{n} \Phi_i^{\sigma_i}(g_{i1}^{\sigma_i} = \alpha) = \Phi_1^{\sigma_1}(g_{11}^{\sigma_1} = \alpha) \vee \Phi_2^{\sigma_2}(g_{21}^{\sigma_2} = \alpha) \vee \cdots \vee \Phi_n^{\sigma_n}(g_{n1}^{\sigma_n} = \alpha)$. Since all the general generalized literals in Φ_1 are α-normal, then $\Phi_1(g_{11} = \alpha) \geq \alpha$ by $\Phi_1 \geq \alpha$. Hence $\Phi_1^{\sigma_1}(g_{11}^{\sigma_1} = \alpha) \geq \Phi_1(g_{11} = \alpha) \geq \alpha$. Therefore, $R_{(g-\alpha)-g}(\Phi_1, \Phi_2, \cdots, \Phi_n) \geq \alpha$.

Proposition 2 *Let Φ_1^*, Φ_2, $\cdots$, Φ_n be general generalized clauses in LP(X), $R_{(g-\alpha)-g}(\Phi_1, \Phi_2, \cdots, \Phi_n)$ an α-generalized resolvent of Φ_1, Φ_2, $\cdots$, Φ_n. If $\Phi_1^* \Rrightarrow_\alpha \Phi_1$, then $R_{(g-\alpha)-g}(\Phi_1^*, \Phi_2, \cdots, \Phi_n) \Rrightarrow_\alpha R_{(g-\alpha)-g}(\Phi_1, \Phi_2, \cdots, \Phi_n)$, or $\Phi_1^* \Rrightarrow_\alpha R_{(g-\alpha)-g}(\Phi_1, \Phi_2, \cdots, \Phi_n)$.*

Proposition 3 *Let S be a set of general generalized clauses in $L_nP(X)$, $w = \{D_1, D_2, \cdots, D_m\}$ an α-generalized resolution deduction from S to the general generalized clause D_m. If $D_j \Rrightarrow_\alpha D_i (j < i)$, then there exists an α-generalized resolution deduction $w = \{D_1^*, D_2^*, \cdots, D_m^*\}$ of S, where $D_k^* \Rrightarrow_\alpha D_k (1 \leq k \leq n)$, and $D_j^* = D_i^* = D_j$.*

Definition 14 Let S be a set of locked general generalized clauses in L_nP(X), $w = \{D_1, D_2, \cdots, D_m\}$ an a_k-generalized lock resolution deduction from S to the general generalized clause D_m. If $D_j \Rrightarrow_\alpha D_i (j < i)$, then delete D_i, this process is called a_k-generalized lock resolution method with strong implication deleting.

The soundness of the hybrid resolution methods of a_k-generalized lock resolution hold obviously, hence we only discuss their completeness here.

Theorem 1 *Let S be a set of locked general generalized clauses in $L_nP(X)$, where the same generalized literals have the same indices. Then the strong implication deleting strategy is compatible with a_k-generalized lock resolution in $L_nP(X)$.*

Theorem 2 *Let S be a set of locked general g-clauses in $L_nF(X)$, where the same g-literals have the same indices. Then the strong implication deleting strategy is compatible with a_k-generalized lock resolution in $L_nF(X)$.*

Remark 3 Similar to Theorem and Theorem , the completeness also holds for α-generalized resolution with strong implication deleting strategy.

3.2 Compatibility with Weak Implication Deleting Strategies

Lemma 1 *Let Φ, Ψ be general generalized clauses in $L_nP(X)$. If $\Phi \Rightarrow_\alpha \Psi$, then $\Phi \Rrightarrow_\alpha \Psi$.*

Proposition 4 *Let Φ, Ψ be general g-clauses in $L_nF(X)$, Φ^*, Ψ^* ground instances of Φ and Ψ, respectively. If $\Phi \Rightarrow_\alpha \Psi$, then $\Phi^* \Rightarrow_\alpha \Psi^*$.*

Proof For any substitution λ and valuation γ, $\gamma(\Phi^\lambda) \geq \gamma(\Phi)$. Hence $\gamma((\Phi \to \Psi)^\lambda) \geq \gamma((\Phi \to \Psi))$, i.e., $\Phi^* \Rightarrow_\alpha \Psi^*$.

Definition 15 Let S be a set of locked general generalized clauses in L_nP(X), $w = \{D_1, D_2, \cdots, D_m\}$ an a_k-generalized lock resolution deduction from S to the general generalized clause D_m. If $D_j \Rightarrow_\alpha D_i (j < i)$, then delete D_i, this process is called a_k-generalized lock resolution method with weak implication deleting.

Theorem 3 *Let S be a set of locked general generalized clauses in $L_nP(X)$, where the same generalized literals have the same indices. Then the week implication deleting strategy is compatible with a_k-generalized lock resolution in $L_nP(X)$.*

Corollary 1 *Let S be a set of general generalized clauses in $L_nP(X)$, then the week implication deleting strategy is compatible with a_k-generalized resolution in $L_nP(X)$.*

Theorem 4 *Let S be a set of locked general g-clauses in $L_nF(X)$, where the same g-literals have the same indices. Then the week implication deleting strategy is compatible with a_k-generalized lock resolution in $L_nF(X)$.*

Corollary 2 *Let S be a set of general g-clauses in $L_nF(X)$, then the week implication deleting strategy is compatible with a_k-generalized resolution in $L_nF(X)$.*

4 An Algorithm for α-Generalized Lock Resolution with Deleting Strategies in L_nP(X)

In α-generalized resolution, the number of resolved literals is dynamic, hence it is vital to give a method to judge whether the given generalized literals are resolved or not.

Definition 16 Let $g_1, g_2, \cdots, g_m$ be generalized literals in LP(X). $g_1, g_2, \cdots, g_m$ are α-minimum resolved in α-generalized resolution if it satisfies

(1) $g_1 \wedge g_2 \wedge \cdots \wedge g_m \leq \alpha$.
(2) For any $\{g_{i1}, g_{i2}, \cdots, g_{ik}\} \subset \{g_1, g_2, \cdots, g_m\}$, $g_{i1} \wedge g_{i2} \wedge \cdots \wedge g_{ik} \nleq \alpha$.

The judgement for α-minimum resolved in α-generalized resolution can be seen as a function, i.e., Function $Res_m(g_1, g_2, \cdots, g_m, a_k)$. Let $g_1, g_2, \cdots, g_m$ be generalized literals in $L_nP(X)$. $g_1, g_2, \cdots, g_m$ are a_k-minimum resolved if $Res_m(g_1, g_2, \cdots, g_m, a_k) = 1$. Otherwise, it returns 0. Therefore, an algorithm for α-generalized lock resolution with deleting strategies follows.

Algorithm 1

Step 0. (Initiation) Let S be a set of locked general generalized clauses $\Phi_1, \Phi_2, \cdots, \Phi_n$. Assign to each occurrence of generalized literal a positive integer in Φ_i, the same generalized literals have the same indices, and all the indices of generalized literals in Φ_i are less than those of $\Phi_j (i < j)$. $c = 1, \alpha = a_k$. $S - \{\Phi_1\}$ be a_k-satisfiable. $\Phi = \Phi_1$.

Step 1. Let g be the generalized literal with the minimal index in Φ, H the set of generalized literals in $S - \{\Phi\}$.

Step 2. Let n_0 be the number of generalized literals in H. For $i_0 = 1$ to n_0, If there exist i_0 generalized literals $g_{j1}, g_{j2}, \cdots, g_{ji_0}$ ($g_{ji} \in \Phi_j$), such that $Res_m(g, g_{j1}, g_{j2}, \cdots, g_{ji_0}, a_k) = 1$, then $\Phi_m = R_{\alpha-g-L}(\Phi, \Phi_{j1}, \Phi_{j2}, \cdots, \Phi_{ji_0})$. If $\Phi_m = a_k$, then stop, $S \leq a_k$. Otherwise, stop, S is a_k-satisfiable.

Step 3. $S = S \cup \Phi_m$, $c = c + 1$. Set $\Phi = \Phi_m$.

Step 4. for any general generalized clauses Φ_i in S, if $\Phi_i \Rightarrow_\alpha \Phi_m$ (or $\Phi_i \Rrightarrow_\alpha \Phi_m$), then delete Φ_m. If $\Phi_m \Rightarrow_\alpha \Phi_i$ (or $\Phi_m \Rrightarrow_\alpha \Phi_i$), then delete Φ_i.

Step 5. If $c \leq c_0$, then go to Step 1. Otherwise, stop, S is a_k-satisfiable.

Remark 4 c_0 can be chosen according to the complexity of Algorithm 1 and numbers of generalized literals in S.

Theorem 5 *If Algorithm 1 terminates, then the a_k-unsatisfiability of S can be judged in $L_nP(X)$.*

Theorem 6 *If $S \leq a_k$ in $L_nP(X)$, then Algorithm 1 terminates in Step 2 or Step 5.*

5 Conclusion

Aiming to further improve the efficiency of α-generalized lock resolution, this paper presented α-generalized lock resolution with deleting strategies in Łukasiewicz implication first order logic $\mathscr{L}_n$F(X). The concepts of strong implication, weak implication, and α-generalized lock resolution with these two deleting strategies were given, as well as their completeness of α-generalized lock resolution and α-generalized resolution with these two deleting strategies were established in $\mathscr{L}_n$F(X). The further researches will be concentrated on discussing how to translate the semantic properties of deleting strategies into their syntactic properties, and further can be applicable to validate unsatisfiability of logical formulae in $\mathscr{L}_n$F(X).

Acknowledgments This work is partially supported by the National Natural Science Foundation of China (Grant No. 61175055, 61105059, 61100046) and Sichuan Key Technology Research and Development Program under Grant No. 2011FZ0051.

References

1. Boyer R (1971) Locking: a restriction of resolution. PhD thesis, University of Texas, Austin
2. Chen XC (1998) On compatibility of deletion strategy. Chin J Comput 21(2):176–182
3. Deng AS (1998) Deletion strategy in boolean operator fuzzy logic. J Northwest Normal Univ 1(1):1–2

4. He XX, Liu J, Xu Y, Martínez L, Ruan D (2012) On α-satisfiability and its α-lock resolution in a finite lattice-valued propositional logic. Logic J IGPL 20(3):579–588
5. He XX, Xu Y, Liu J, Ruan D (2011) α-lock resolution method for a lattice-valued first-order logic. Eng Appl Artif Intell 24(7):1274–1280
6. He XX, Xu Y, Liu J, Chen SW (2011) On compatibilities of α-lock resolution method in linguistic truth-valued lattice-valued logic. Soft Comput 16(4):699–709
7. Liu XH (1994) Automated reasoning based on resolution methods. Science Press, Beijing (In Chinese)
8. Murray NV (1982) Completely non-clausal theorem proving. Artif Intell 18:67–85
9. Robinson JA (1965) A machine-oriented logic based on the resolution principle. J ACM 12(1):23–41
10. Sun JG, Liu XH (1994) Deletion strategy using weak subsumption relation for modal resolution. Chin J Comput 17(5):321–329
11. Tang RK, Liu XH (1993) Deletion strategy in generalized λ-resolution. J JinLin Univ 2(2):37–41 (in Chinese)
12. Xu Y (1993) Lattice implication algebras. J Southwest Jiaotong Univ 89(1):20–27 (in Chinese)
13. Xu Y, Ruan D, Kerre EE, Liu J (2000) α-resolution principle based on lattice-valued propositional logic LP(X). Inf Sci 130:195–223
14. Xu Y, Ruan D, Kerre EE, Liu J (2001) α-resolution principle based on first-order lattice-valued logic LF(X). Inf Sci 132:221–239
15. Xu Y, Ruan D, Qin KY, Liu J (2003) Lattice-valued logic: an alternative approach to treat fuzziness and incomparability. Springer-Verlag, Berlin
16. Xu Y, Xu WT, Zhong XM, He XX (2010) α-generalized resolution principle based on lattice-valued propositional logic system LP(X). In: The 9th international FLINS conference on foundations and applications of computational intelligence (FLINS2010), Chengdu, 2–4 Aug, pp 66–71
17. Xu Y, He XX, Liu J, Chen SW (2011) A general form of α-generalized resolution based on lattice-valued logic (Submitted to Int J Computat Intell Syst)

Multivariate Time Series Prediction Based on Multi-Output Support Vector Regression

Yanning Cai, Hongqiao Wang, Xuemei Ye and Li An

Abstract An improved support vector regression (SVR) model is presented in this paper and the model can train and predict both the multiple input and output samples, which can avoid SVR's modeling for each output individually when predicting the multivariate time series. The proposed multi-output SVR (MOSVR) model can guarantee the regression ability of each output by choosing different kernel functions and model parameters for different outputs of one single optimization problem. On this basis, the norm summation of regression weight vector, the error summation of each output and the total error are minimized so as to be sure that the MOSVR model satisfies the structure risk minimization (SRM) principle. The experimental results based on several multivariable time series show that the MOSVR has better adaptability and gains less total regression error than the SVR.

Keywords Support vector regression · Multi-output SVR · Multivariate time series · Time series prediction

1 Introduction

Multivariate time series prediction is an important research focus in the time prediction field, which is popular in astronomy [1], finance [2], multimedia [3, 4], biomedicine [5–7] and control [8, 9]. As a good prediction tool for the single variable time series [10, 11], support vector regression (SVR) has shown nearly excellent performance. But for the multivariate time series, there still has not

Y. Cai (✉) · H. Wang · X. Ye
Xi'an Research Institute of Hi-Tech, Xi'an, Shaanxi 710025, China
e-mail: caiyanning666@yahoo.com.cn

L. An
Air Force Engineering University, Shaanxi, China

F. Sun et al. (eds.), *Knowledge Engineering and Management*,
Advances in Intelligent Systems and Computing 214,
DOI: 10.1007/978-3-642-37832-4_35, © Springer-Verlag Berlin Heidelberg 2014

effective solution using SVR. To predict multivariate time series, the support vector machine model is commonly established individually for each dimension time series respectively. This method can only guarantee better prediction precision of each dimension of the time series, which ignores the control of the total prediction precision in multivariate time series. Nowadays, there is less research on multi-output SVR model. Ref. [12] gives a basic idea about multi-output SVR's construction, which considers the regression estimation of all outputs; as a result the error summation can be minimized. In this paper, a multi-output SVR model is presented which can train and predict both the multiple input and output samples. The essence of the model is guaranteeing the minimization of both the total regression error and the model's constructural risk, together with the regression ability, namely the multi-output SVR model can choose different kernel function and model parameters for different output.

2 Support Vector Regression Machine

The form of samples is $(x_i, y_i), i = 1, 2, \cdots, l,$ let $x_i \in R^n, y_i \in R$, Through the function $\varphi(x)$, the samples are mapped to the high dimension space F where the linear regression function (1) is established.

$$f(x) = \omega^{\mathrm{T}}\varphi(x) + b \tag{1}$$

According to the Vapnik's structural risk minimization, f should be founded under minimizing the function (2),

$$\frac{1}{2}\|\omega\|^2 + C\sum_{i=1}^{l} L(f(x_i, y_i)) \tag{2}$$

In general, the loss function L can be defined as the ε insensitive loss function:

$$\begin{aligned} L(y, f(x, \alpha)) &= L(|y - f(x, \alpha)|_{\varepsilon}) \\ &= \max\{0, |y - f(x, \alpha)| - \varepsilon\} \end{aligned} \tag{3}$$

According to the Structural Risk Minimization (SRM) principle, the object function and the constraint conditions based on the ε insensitive loss function SVR are as follows:

$$\begin{aligned} &\min \frac{1}{2}(\omega \cdot \omega) + C\left(\sum_{i=1}^{l} \xi_i + \sum_{i=1}^{l} \xi_i^*\right) \\ s.t. \quad &\begin{cases} y_i - f(x_i) \le \varepsilon + \xi_i \\ f(x_i) - y_i \le \varepsilon + \xi_i^* \end{cases} \quad (i = 1, 2, \cdots l) \end{aligned} \tag{4}$$

where ε is the allowed error, $\varepsilon \ge 0$; ξ_i and ξ_i^* are slack factors, and $\xi_i, \xi_i^* \ge 0$; $C > 0$ is the penalty parameter, which controls the penalty degree to the samples exceeding the error ε. Using the Lagrange function, the duality principle and the quadratic programming, the regression function can be derived ultimately as

$$f(x,\beta)=\sum_{i=1}^{l}\beta_i K(x,x_i)+b \tag{5}$$

where $K(\cdot,\cdot)$ is the kernel function. Here the kernel function and the mapping $\varphi(x)$ satisfy

$$k(x_i,x_j)=\varphi(x_i)\cdot\varphi(x_j) \tag{6}$$

The coefficient $\beta_i=\alpha_i-\alpha_i^*$, $i=1,2,\cdots,l$. α_i and α_i^* are the optimal solutions maximizing the following functional,

$$\begin{aligned} W(\alpha,\alpha^*)= & \\ & -\varepsilon\sum_{i=1}^{l}(\alpha_i+\alpha_i^*)+\sum_{i=1}^{l}y_i(\alpha_i-\alpha_i^*) \\ & -\frac{1}{2}\sum_{i,j=1}^{l}(\alpha_i-\alpha_i^*)\left(\alpha_j-\alpha_j^*\right)K(x_i,x_j) \end{aligned} \tag{7}$$

$$\begin{aligned} \text{s.t.} \sum_{i=1}^{l}\alpha_i &= \sum_{i=1}^{l}\alpha_i^* \\ \alpha_i,\alpha_i^* &\in [0,C], \\ i &= 1,2,\cdots,l \end{aligned} \tag{8}$$

3 Multi-Output SVR Model

The essence of the multi-output SVR model has two aspects. Firstly, we consider the whole performance of all the outputs. The error summation from every output should be minimized, and the model must satisfy with the SRM principle. Secondly, both the whole performance of multi-output SVR model and the regression ability of the single output must be ensured, namely the multi-output SVR model can choose different kernel function and model parameters for different output. The training samples of the multi-output SVR model have the form as (x_i,y_i), $x_i\in R^n$, $y_i\in R^k$, the total number of the samples is l, for every n dimensional input x_i, the corresponding output y_i is a k dimensional vector. The objective function and the constraints are as follows,

$$\min \quad \Phi(w,\xi^*,\xi)=\frac{1}{2}\sum_{m=1}^{k}\|\omega_m\|^2+\sum_{m=1}^{k}C_m\left(\sum_{i=1}^{l}(\xi_{i,m}+\xi_{i,m}^*)\right)+C_0\sum_{i=1}^{l}(\eta_i+\eta_i^*) \tag{9}$$

$$\begin{aligned}&\sum_{m=1}^{k}\left(y_{i,m}-(\omega_m\cdot\varphi_m(x_i)-b_m)\right)\leq\varepsilon_0+\eta_i\\&\sum_{m=1}^{k}\left((\omega_m\cdot\varphi_m(x_i))+b_m-y_{i,m}\right)\leq\varepsilon_0+\eta_i^*\end{aligned}\tag{10}$$

$$\begin{aligned}&y_{i,m}-(\omega_m\cdot\varphi_m(x_i))-b_m\leq\varepsilon_m+\xi_{i,m}\\&(\omega_m\cdot\varphi_m(x_i))-b_m-y_{i,m}\leq\varepsilon_m+\xi_{i,m}^*\end{aligned}\tag{11}$$

$$\xi_{i,m},\xi_{i,m}^*,\eta_i,\eta_i^*\geq0\tag{12}$$

where $i=1,2,\cdots,l$, $m=1,2,\cdots,k$. $y_{i,m}$ is the mth output of the ith training sample. In Eq. (9), C_m is the penalty coefficient of the mth output, and C_0 is the penalty coefficient of all the outputs' error summation. In formula (10), ε_0 is the insensitive factor of all the outputs' error summation, correspondingly, ε_m is the insensitive factor of the mth dimension output error in formula (11). $\varphi_m(\cdot)$ is the mapping in high dimensional space of the mth dimension output. In the multi-output SVR algorithm, the flexibility of the model lies in its free selection of the insensitive coefficient, the penalty parameter and the kernel function for the outputs with different dimension. In this way, the multi-output SVR model's nonlinear mapping ability can be ensured even if that the outputs between different dimensions have different nonlinear degree, or even have big differences. From (9), we can see that the single output's regression ability of multi-output SVR model is guaranteed, on this basis, the error summation of all the outputs can be minimized and the model still satisfies with the SRM principle. The Lagrange function of the multi-output SVR model is

$$\begin{aligned}&\frac{1}{2}\sum_{m=1}^{k}\|\omega_m\|^2+\sum_{m=1}^{k}C_m\left(\sum_{i=1}^{l}(\xi_{i,m}+\xi_{i,m}^*)\right)+C_0\sum_{i=1}^{l}(\eta_i+\eta_i^*)\\&-\sum_{i=1}^{l}\sum_{m=1}^{k}\alpha_{i,m}(\varepsilon_m+\xi_{i,m}-y_{i,m}+\omega_m\cdot\varphi_m(x_i)+b_m)\\&-\sum_{i=1}^{l}\sum_{m=1}^{k}\alpha_{i,m}^*\left(\varepsilon_m+\xi_{i,m}^*+y_{i,m}-\omega_m\cdot\varphi_m(x_i)-b_m\right)\\&-\sum_{i=1}^{l}\gamma_i\left(\varepsilon_0+\eta_i-\sum_{m=1}^{k}(y_{i,m}-\omega_m\cdot\varphi_m(x_i)-b_m)\right)\\&-\sum_{i=1}^{l}\gamma_i^*\left(\varepsilon_0+\eta_i^*-\sum_{m=1}^{k}(\omega_m\cdot\varphi_m(x_i)+b_m-y_{i,m})\right)\\&-\sum_{i=1}^{l}\sum_{m=1}^{k}(\beta_{i,m}^*\xi_{i,m}^*+\beta_{i,m}\xi_{i,m})-\sum_{i=1}^{l}(\delta_i^*\eta_i^*+\delta_i\eta_i)\end{aligned}\tag{13}$$

Under KKT condition, we get the differential coefficients of ω_m, b_m, $\xi_{i,m}$, $\xi^*_{i,m}$, η_i, η^*_i,

$$\frac{\partial L}{\partial \omega_m} = \frac{\partial L}{\partial b_m} = \frac{\partial L}{\partial \xi_{i,m}} = \frac{\partial L}{\partial \xi^*_{i,m}} = \frac{\partial L}{\partial \eta_i} = \frac{\partial L}{\partial \eta^*_i} = 0 \tag{14}$$

Substituting the results from (14) into (9), the optimal functional and the constrains of the multi-output SVR model can be gained as Eqs. (15) and (16),

$$\begin{aligned} &\frac{1}{2}\sum_{m=1}^{k}\sum_{i,j=1}^{l}\{(\alpha_{i,m} - \alpha^*_{i,m} + \gamma_i - \gamma^*_i) \\ &\quad * (\alpha_{j,m} - \alpha^*_{j,m} + \gamma_j - \gamma^*_j)K_m(x_i \cdot x_j)\} \\ &+\sum_{m=1}^{k}\sum_{i=1}^{l}\varepsilon_m(\alpha_{i,m} + \alpha^*_{i,m}) - \sum_{m=1}^{k}\sum_{i=1}^{l} y_{i,m}(\alpha_{i,m} - \alpha^*_{i,m}) \\ &+\sum_{i=1}^{l}\varepsilon_0(\gamma_i + \gamma^*_i) - \sum_{m=1}^{k}\sum_{i=1}^{l} y_{i,m}(\gamma_i - \gamma^*_i) \end{aligned} \tag{15}$$

$$\begin{aligned} &\sum_{i=1}^{l}(\alpha^*_{i,m} + \gamma^*_i) = \sum_{i=1}^{l}(\alpha_{i,m} + \gamma_i), \\ &\alpha^*_{i,m}, \alpha_{i,m} \in [0, C_m], \\ &\gamma^*_i, \gamma_i \in [0, C_0], \\ &i = 1, 2, \cdots, l; m = 1, 2, \cdots, k. \end{aligned} \tag{16}$$

So, the regression estimation function of the mth output is

$$f_m(x) = \sum_{i=1}^{l}(\alpha_{i,m} - \alpha^*_{i,m} + \gamma_i - \gamma^*_i)K_m(x_i \cdot x) + b_m \tag{17}$$

where $\alpha_{i,m}$, $\alpha^*_{i,m}$, γ_i, γ^*_i is the optimal solution through minimizing functional (15), and b_m can be obtained under the KKT condition.

4 Experiments

In this section, several experiments are carried out to compare the estimation performance between the standard SVR and the multi-output SVR. The design idea of the experiments is as follows: firstly, get the SVR models for every single variable from the multivariable time series, then we can obtain the prediction results and note down the kernel function and model parameters. The parameters can be selected according to the Ref. [13], which is an evolutionary grid searching algorithm. The essence of the algorithm is as follows,

(1) Determine the searching boundary and steps of the parameters;
(2) Substitute the combinatorial parameters into the SVR and train the model;
(3) Select the combinatorial parameters with the best generalization performance;
(4) Multi-output SVR modeling for the multivariable time series.

In the model, the values of the variables are same with the corresponding SVR model's kernel function and the insensitive factors. The penalty coefficient is determined by the summation of C_m and C_0 from Eq. (9), which is equal to C in Eq. (4). To every single variable, the multi-output SVR model and the standard SVR have the same value ranges of the regression coefficients, which is helpful to compare the performance of the two models.

Let $x(t) = [x_1(t), \cdots, x_n(t)], t = 1, 2, \cdots, N$, which is the multivariable discrete time series to be studied. $x_1(t), \cdots, x_n(t)$ are correspondingly the n variables at time t. For each variable, select proper delays τ_i, embedded dimensions are m_i, $i = 1, 2, \cdots, n$, then the input samples of the multi-output SVR prediction model have the following form,

$$\begin{aligned} X(t) = \; & [x_1(t), x_1(t-\tau_1), \cdots, x_1(t-(m_1-1)\tau_1) \\ & \cdots \\ & x_i(t), x_i(t-\tau_i), \cdots, x_i(t-(m_i-1)\tau_i) \\ & \cdots \\ & x_n(t), x_n(t-\tau_n), \cdots, x_n(t-(m_n-1)\tau_n)]. \end{aligned} \tag{18}$$

The output samples are

$$Y(t) = [x_1(t+\tau_1), x_2(t+\tau_2), \cdots, x_n(t+\tau_n)], \tag{19}$$

so, the output sample corresponding to the ith single output SVR model is

$$Y_i(t) = x_i(t+\tau_1) \tag{20}$$

then predict the *Lorenz* and *Rossler* multivariable time series using the standard SVR model and the multi-output SVR model respectively. The time delay values of the two time series τ_i are 1, the embedded dimensions m_i are 2. The linear kernel function is used as the kernel function of the first variable, which has the form as (21)

$$K(x_i, x_j) = (x_i \cdot x_j) \tag{21}$$

The kernel functions correspond to the other variables are the polynomial kernels

$$K(x_i, x_j) = ((x_i \cdot x_j) + c)^d, c \geq 0 \tag{22}$$

The normalized mean square error (NMSE) is adopt as the test index

$$E_{NMSE} = \sqrt{\frac{1}{l}\sum_{i=1}^{l}(X_i - \hat{X}_i)^2} \tag{23}$$

4.1 Lorenz Time Series Prediction

Lorenz time series [14] has the form as

$$\frac{dx}{dt} = 10(y - x)$$
$$\frac{dy}{dt} = 28x - xz - yx$$
$$\frac{dz}{dt} = xy - \frac{8}{3}z$$

The initial values $x = 0.3$, $y = 0.2$, $z = 0.4$. Three time series with the time intervals 0–80 are generated using the ode45 function in Matlab7.0. Select 100 data behind the 1000th data from the three time series respectively, and then generate the training and test samples according to Eqs. (18) and (19). The prediction result is described in Fig. 1, where the solid line and dot are the actual value and the multi-output SVR model's prediction value.

From Fig. 1, we can see that the multi-output SVR model has good prediction performance for the *Lorenz* time series, but the result is restricted to the situation without noise. To evaluate the adaptability of the multi-output SVR model, some noises are added respectively to the training and test samples which obey normal distribution with parameters $N(0, 0.01)$, $N(0, 0.1)$, and $N(0, 1)$. The prediction results are shown in Table 1. In the table, the prediction accuracy of the standard SVR and the multi-output SVR (MOSVR) are indicated with different noise level. Compare to the SVR, the MOSVR has smaller summation of NMSE, in addition, it has good generalization.

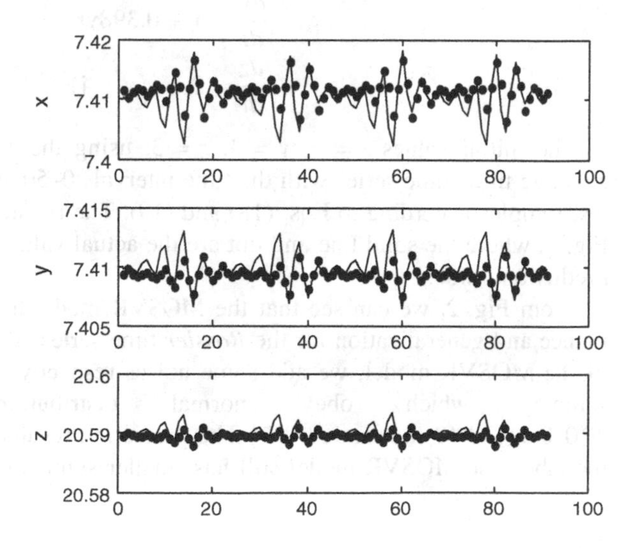

Fig. 1 Prediction results of Lorenz (without noise)

Table 1 Prediction results of Lorenz time series

Model	Noise	Parameters of models	NMSE
SVR	No noise	$C_x = C_y = C_z = 100;$ $\varepsilon_x = \varepsilon_y = \varepsilon_z = 0.001$	3.8600e-006
MOSVR		$C_0 = 15, C_x = C_y = C_z = 85;$ $\varepsilon_0 = 0.003, \varepsilon_x = \varepsilon_y = \varepsilon_z = 0.001$	3.8557e-006
SVR	0.01	$C_x = C_y = C_z = 100;$ $\varepsilon_x = \varepsilon_y = \varepsilon_z = 0.001$	2.4333e-004
MOSVR		$C_0 = 60, C_x = C_y = C_z = 40;$ $\varepsilon_0 = 0.003, \varepsilon_x = \varepsilon_y = \varepsilon_z = 0.001$	1.7252e-004
SVR	0.1	$C_x = C_y = C_z = 100;$ $\varepsilon_x = \varepsilon_y = \varepsilon_z = 0.001$	0.1479
MOSVR		$C_0 = 50, C_x = C_y = C_z = 50;$ $\varepsilon_0 = 0.003, \varepsilon_x = \varepsilon_y = \varepsilon_z = 0.001$	0.1174
SVR	1	$C_x = C_y = C_z = 100;$ $\varepsilon_x = \varepsilon_y = \varepsilon_z = 0.001$	8.6000
MOSVR		$C_0 = 60, C_x = C_y = C_z = 40;$ $\varepsilon_0 = 0.003, \varepsilon_x = \varepsilon_y = \varepsilon_z = 0.001$	6.9793

4.2 Rossler Time Series Prediction

Positioning Figures and Tables: *Place figures and tables Rossler* time series [14] has a form as

$$\frac{dx}{dt} = -y - z$$
$$\frac{dy}{dt} = x + 0.398yx$$
$$\frac{dz}{dt} = 2 + z(x - 4)$$

The initial values $x = 3$, $y = 3$, $z = 3$. using the same method in Sect. 4.1, generate three time series with the time intervals 0–50 and obtain the training and test samples according to Eqs. (18) and (19). The prediction result is described in Fig. 2, where the solid line and dot are the actual value and the MOSVR model's prediction value.

From Fig. 2, we can see that the MOSVR model has good prediction performance and generalization for the *Rossler* time series. To evaluate the adaptability of the MOSVR model, we add some noises respectively to the training and test samples which obey normal distribution with parameters $N(0, 0.01)$, $N(0, 0.1)$ and $N(0, 1)$ The prediction results are shown in Table 2. In the table, the MOSVR model still has smaller summation of NMSE than SVR.

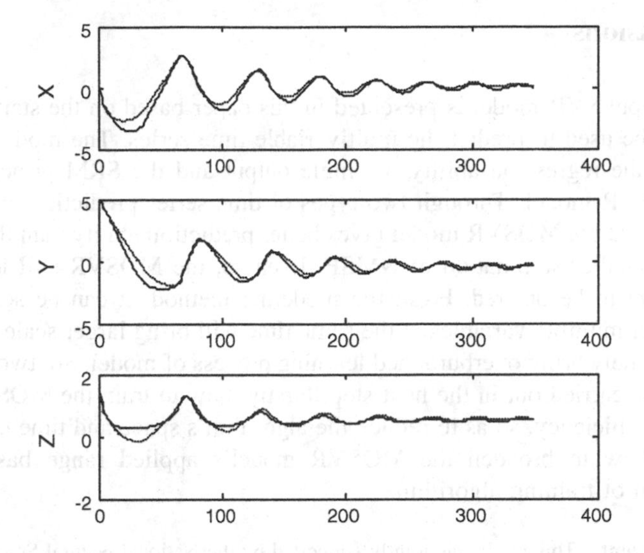

Fig. 2 Prediction results of Rossler (without noise)

Table 2 Prediction results of Rossler time series

Model	Noise	Parameters of models	NMSE
SVR	No noise	$C_x = 100, C_y = C_z = 80;$ $\varepsilon_x = \varepsilon_y = \varepsilon_z = 0.001$	0.2781
MOSVR		$C_0 = 5, C_x = 95, C_y = C_z = 75;$ $\varepsilon_0 = 0.003, \varepsilon_x = \varepsilon_y = \varepsilon_z = 0.001$	0.1763
SVR	0.001	$C_x = 100, C_y = C_z = 80;$ $\varepsilon_x = \varepsilon_y = \varepsilon_z = 0.001$	0.4590
MOSVR		$C_0 = 5, C_x = 95, C_y = C_z = 75;$ $\varepsilon_0 = 0.003, \varepsilon_x = \varepsilon_y = \varepsilon_z = 0.001$	0.3532
SVR	0.01	$C_x = 100, C_y = C_z = 80;$ $\varepsilon_x = \varepsilon_y = \varepsilon_z = 0.001$	2.3778
MOSVR		$C_0 = 15, C_x = 85, C_y = C_z = 65;$ $\varepsilon_0 = 0.003, \varepsilon_x = \varepsilon_y = \varepsilon_z = 0.001$	1.3593
SVR	0.1	$C_x = 100, C_y = C_z = 80;$ $\varepsilon_x = \varepsilon_y = \varepsilon_z = 0.001$	8.9540
MOSVR		$C_0 = 25, C_x = 75, C_y = C_z = 55;$ $\varepsilon_0 = 0.003, \varepsilon_x = \varepsilon_y = \varepsilon_z = 0.001$	5.2141

5 Conclusions

A multi-output SVR model is presented in this paper based on the standard SVR, which can be used to predict the multivariable time series. The model can guarantee both the regression ability of single output and the SRM principle of the whole MOSVR model. Through two types of time series prediction experiments, we can see that the MOSVR model gives better prediction ability than the standard SVR with smaller summation of NMSE. Even so, the MOSVR still leave much improvement to be desired. From the modeling method, it can be seen that the prediction of multiple variables at the same time will bring larger scale of training samples, namely bring overburdened learning process of model. So, two aspects of work will be carried out in the next step. Firstly, how to train the MOSVR model with higher efficiency, so as to reduce the algorithm's space and time complexity. Secondly, how to broaden the MOSVR model's applied range based on the optimization of training algorithm.

Acknowledgments This work was jointly supported by the National Natural Science Foundation for Young Scientists of China (Grant No: 61202332) and China Postdoctoral Science Foundation (Grant No: 2012M521905).

References

1. Yoon H, Yang K, Cyrus S (2005) Feature subset selection and feature ranking for multivariate time series. IEEE Trans Knowl Data Eng 17(9):1186–1198
2. Han M, Fan MM (2006) Application of neural on multivariate time series modeling and prediction. In: Proceedings of the 2006 American control conference, vol 7. Minneapolis, Minnesota, USA, pp 3698–3703
3. Tsang, S., Kao, B., Yip, K.Y.: Decision trees for uncertain data. In: Proceedings of the 2009 IEEE international conference on data engineering. Washington, DC: IEEE Computer Society, pp 441–444
4. Yang K, Shahabi C (2004) A PCA-based similarity measure for multivariate time series. In: Proceedings of the 2nd ACM international workshop on multimedia databases. New York, ACM Press, pp 65–74
5. Aboy M, Márquez OW, McNames J, Hornero R, Trong T, Goldstrein B (2005) Adaptive modeling and spectral estimation of nonstationary biomedical signals based on Kalman filtering. IEEE Trans Biomed Eng 52(8):1485–1489
6. Mukhopadhyay N, Chatterjee S (2007) Causality and pathway search in microarray time series experiment. Bioinformatics 23:442–449
7. Fieuws S, Verbeke G, Molenberghs G (2007) Random-effects models for multivariate repeated measures. Stat Methods Med Res 16(5):387–397
8. Yang K, Shahabi C (2007) An efficient k-nearest neighbor search for multivariate time series. Inf Comput 205(1):65–98
9. Guan HS, Jiang QS (2009) Pattern matching method based on point distribution for multivariate time series. J Softw 20(1): 67–79 (In Chinese)
10. Wu CH, Ho JM, Lee DT (2004) Travel-time prediction with support vector regression. IEEE Trans Intell Transp Syst 5(4):276–281

11. Cao LJ, Francis EHT (2003) Support vector machine with adaptive parameters in financial time series forecasting. IEEE Trans Neural Netw 14(6):1506–1518
12. Tang FM (2005) Research on support vector machine algorithms based on statistical learning theory. PhD Dissertation, Huazhong University of Science and Technology
13. Bi JB (2003) Support vector regression with application in automated drug discovery. Rensselaer Polytechnic Institute, Troy, New York, pp 87–92
14. Ren R, Xu J, Zhu SH (2006) Prediction of chaotic time sequence using least squares support vector domain. Acta Physica Sinica 55(2): 555–563 (In Chinese)

Classification Probability Estimation Based Multi-Class Image Retrieval

Hongqiao Wang, Yanning Cai, Shicheng Wang, Guangyuan Fu and Linlin Li

Abstract Aiming at multi-class large-scale image retrieval problem, a new image retrieval method based on classification probability estimation is proposed according to the thinking named "Classification First, Retrieval Later". According to the method, the image features are effectively fused using a composite kernel method first, and a composite kernel classifier with higher classification precision is designed. The optimal coefficients of the classifier are also obtained utilizing the classification result with small-amount image samples. Second, complete the classification probability estimation for the testing images using the composite machine. Third, realize the image retrieval based on the classification probability estimation values. In the experiments with multi-class large-scale image dataset, it is confirmed that the presented method can achieve better retrieval precision. Moreover, the generalization performance without prior knowledge is also studied.

Keywords Composite kernel · Multiple kernel learning · Support vector classifier · Content-based image retrieval

1 Introduction

Content-based image retrieval (CBIR) [1] is a new concept presented in 1990s, which gradually becomes a popular interdisciplinary. Centering on this theme, a lot of work [2] are studied such as the image vision information representation, image perceptual model, image perceptual organization from man, and high dimensional large-scale data processing and retrieval.

H. Wang (✉) · Y. Cai · S. Wang · G. Fu · L. Li
Xi'an Research Institute of Hi-Tech, Xi'an 710025 Shaanxi, China
e-mail: whq05@mails.tsinghua.edu.cn

F. Sun et al. (eds.), *Knowledge Engineering and Management*, Advances in Intelligent Systems and Computing 214,
DOI: 10.1007/978-3-642-37832-4_36,

The basic idea of image retrieval is mapping the image into a certain space by a mapping function, then measure the distance between images in the new space. The retrieval techniques are maturely developed at present, but multi-class large-scale image classification is still a problem. The main difficulty is that the object in reality can generate nearly infinite two-dimensional images with the difference of light, position, background, and perspective of observers [3]. The brain of man can recognize these images easily, but it is a very difficult task for a computer. So, it is also a research focus on how man recognizes object, which is studied by biology and neuroscience [4]. Wang [5] proposed a new method named "Classification First, Retrieval Later", which can improve the retrieval precision. But it becomes a key problem about how to construct a classifier with higher performance facing high dimensional, multi-class, and large-scale image retrieval background. Based on this thinking, a new image retrieval method is presented, which takes the classification probability estimation as a measure basis. According to the method, the image features are effectively fused using a composite kernel method first, and a composite kernel classifier with higher classification precision is designed. The optimal coefficients of the classifier are also obtained utilizing the classification result with small-amount image samples. Then the classification probability estimation for the testing images can be completed using the composite machine, and the image retrieval based on the classification probability estimation values can be realized.

In the remainder of this paper, we go along through different sections which are organized as follows: in Sect. 2 we provide a short introduction on multi-class image extraction, and then we introduce the principle and construction method of the composite kernel support vector classifier (SVC). In Sect. 3, the classification probability estimation-based image retrieval algorithm is detailed deduced. We present experiments in Sect. 4 on typical dataset, and finally, we conclude in Sect. 5.

2 Image Classification Based on Composite Kernel Support Vector Machine

To gain good classification precision, two works must be well treated. The first is image feature extraction, the features which can well characterize the multi-class images should be extracted. The second is construction of high-performance classifier.

2.1 Feature Extraction of Multi-class Images

The image class is usually large in the image retrieval problems, for example, the Wang dataset [6] is used in this paper, which possess large-scale images and the image class number is 10. So, to realize better classification, several feature extraction methods for images must be considered usually. According to the character of the images in Wang dataset, the color moment feature of image, the color distribution feature, the Fourier transform feature, and the fractal dimension feature [7] are extracted as the classification basic feature.

Color Moment Feature (CMF): the color moment is a simple and effective color-based image retrieval method. According to the principle that the probability distribution can uniquely be represented by its all order moments, the color histogram can be approximately represented by these moments, and the computational complexity can be reduced. As the color distribution information is mostly concentrated on the low-order moments, we only need the first-order moment (mean), the second-order moment (variance), and the third-order moment (skewness) to express the image's color distribution.

Color Distribution feature (CDF): To use the color distribution feature, the images should be transformed into the grayscale images first, and then the images can be discriminated based on distribution of the grayscale pixel values.

Fourier Transform Feature (FTF): images can be regarded as the signals distributed in two-dimensional space. So it is feasible that analyzing the images from both the space domain and the frequency domain transformations. Some rules and features can be found which cannot easily be found in the space domain through the Fourier transform.

Fractal Dimension Feature (FDF): in geometry, the fractal dimension is a very important feature, which reflects the effectiveness of complex object occupying the space, and it is the measurement of the object's irregularity. The box dimension of image is the most popular dimension, and the algorithm can be described as: (1) Transform the grayscale image into a curved surface, *x*, *y* axis is the position in the plane, the *z* axis is the grayscale value. (2) For a given scale L, it can divide the three-dimension space into several boxes with the size of $L \times L \times L'$, where L is the pixel number, L' is the grayscale value. Suppose the image size is $M \times M$, the grayscale level is G, then $L' = \lfloor L \times G/M \rfloor$. (3) Suppose the box scale which can occupy the image is $L_{\max}$, select $L = r \times L_{\max}$, where r is the scaling factor. (4) Let N_L is total of the boxes, so $N_L = 1/r^D = [L_{\max}/L]^D$, where D can be obtained by the slope from $\log N_L$ to $\log L$ according to the least square method.

2.2 *Construction of Composite Kernel Support Vector Classifier*

The basis of the kernel method is kernel function. There already exist some kernel functions at present, the kernel functions can be divide into two groups: the local kernel and the global kernel. The local kernel is that the kernel function value may be greatly affected only when the distance between input data is small or their features are similar. Instead, for a global kernel, when the input data is far from the other or their features have big difference, the kernel function value is also affected greatly. The typical local and global kernels are the Gaussian RBF kernel and the polynomial kernel, the calculation formulae are

Gaussian RBF Kernel:

$$k(x,x_0) = \exp(-\frac{\|x - x_0\|^2}{2\sigma^2}) \tag{1}$$

Polynomial Kernel:

$$k(x,x_0) = (x - x_0)^d (\text{Homogeneous}) \tag{2}$$

$$k(x,x_0) = (x - x_0 + C)^d (\text{Inhomogeneous}) \tag{3}$$

The kernels can be marked as RBF and Poly in short. Different kernel function reflects the diversity of mapping performance for the global data or the local data. In addition, for any local kernel, the kernel function presents different scale effect through the selection of coefficients.

So, if the kernel functions with different characters are combined, the advantages of multiple types of kernel functions can be gained, also the better mapping performance. Furthermore, the typical learning problems are usually concerned with the multi-class data of the heterogeneous data; the multiple kernel method can provide better flexicity. Moreover, it also can be regarded as a skillful method to explain the learning result, which can bring deeper comprehension for the application problem. This is the multiple kernel learning (MKL) method [8, 9]. The MKL method is a new focus in present kernel machine learning field, some applications [10, 11] testify that the composite kernel machine which is a typical MKL method can bring better classification result for most pattern recognition problems.

The general composite kernel can be composed with the linear combination of multiple kernels. Suppose $k(x,y)$ is a known kernel, $\hat{k}(x,y)$ is its normalization form. For example, kernel $k(x,y)$ can be normalized as $\sqrt{k(x,y)k(x,y)}$. Using the above-mentioned marks, some composite kernels can be defined:

(1) **Direct Summation Kernel**

$$k(x,y) = \sum_{j=1}^{M} \hat{k}_j(x,y) \tag{4}$$

(2) **Weighted Summation Kernel**

$$k(x,y) = \sum_{j=1}^{M} \beta_j \hat{k}_j(x,y), \text{ with } \beta_j \geq 0, \ \sum_{j=1}^{M} \beta_j = 1 \tag{5}$$

(3) **Weighted Polynomial Extended Kernel**

$$k(x,y) = \alpha\hat{k}_1^p(x,y) + (1-\alpha)\hat{k}_2^q(x,y) \tag{6}$$

where $k^p(x,y)$ is the polynomial extension of $k(x,y)$.

In order to get the coefficients of composite kernel, the kernels are commonly combined with the support vector machine (SVM), then the object function can be transformed into different optimization problems, such as the different regularization form or the restrains to the training samples, finally, we can obtain the solutions by different optimization method. On this basis, some new composite kernels are presented, for example, the non-stationary multiple kernel learning [12], the localized multiple kernel learning [13], the non-sparse Multiple Kernel Learning [14], the simple MKL [15], and the infinite Kernel Learning [16].

When all the inner product items are replaced by the kernel function

$$K_{i,j} = (\varphi(x_i)\cdot\varphi(x_j)) = \sum_{t=1}^{n}\beta_t k_t(x_{it}, x_{jt})$$

where $k_t(\cdot,\cdot)$ is the basic kernel function, β_t is the weight of kernel, $x_i = (x_{i1}, x_{i2}\ldots x_{in})$ $i,j = 1,2,\cdots,l$, then the single kernel SVC can be promoted to the composite kernel SVC. The ultimate classifier decision function can be gained as (7),

$$f(x) = \text{sgn}\left[\sum_{t=1}^{n}\sum_{i-1}^{l_t}\alpha_i y_i \beta_t k_t(x_t, x_{it}) + b\right] \tag{7}$$

3 Image Retrieval Method Based on Classification Probability Estimation

3.1 Classification Probability Estimation Based Measurement Method

In this paper, the basic features of images are not used directly to measure the distance of images. To realize the image retrieval, a classification probability estimation-based image retrieval method is studied in the paper. The main idea of the method is shown in Fig. 1. First, map the images in the original space $\mathbb{R}$ into the feature space $\mathbb{R}_f$ through the feature extraction function f. Second, mapping the classification probability estimation values from SVM into the classification probability space $\mathbb{R}_c$. Suppose $P_1, P_2\ldots, P_N$ are the probability estimation values,

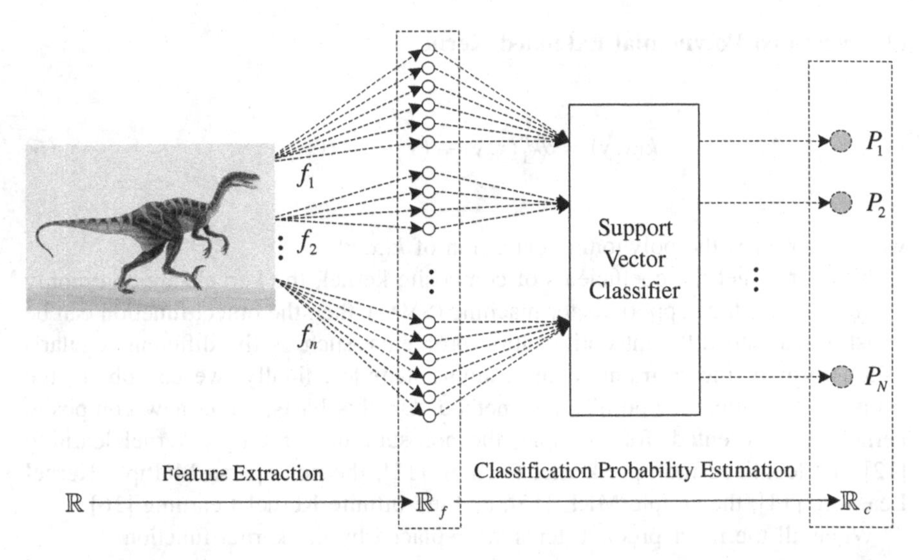

Fig. 1 Mapping relations of image space, feature space, and measurement space

which represent the probability of a sample that belongs to a certain class from $1, 2, \ldots, N$, respectively. Third, measure the distance between images and realize the image retrieval.

3.2 *Measurement and Evaluation Criterion for Image Retrieval Algorithms*

The key of image retrieval is determining the similarity between the image to be retrieved and the image in database, namely calculating the distance of feature vectors between the image to be retrieved and the object image in database. So, the similarity measurement method possesses an important position for image retrieval. The commonly used measurement criterions include

Minkowski Distance:

$$d_p(H, H') = \left(\sum_{m=1}^{M} (H_m - H'_m)^p \right)^{\frac{1}{p}} \tag{8}$$

Choosing different p, there are Euclidean distance d_2, the district distance d_1, and the max distance d_∞.

Histogram Crossing Distance:

The histogram crossing distance was first presented by Swain [17] in 1991. This method has a simple and fast calculation principle; moreover, it can well suppress the affections of background noise. The mathematical description is

$$d_{hi}(A,B) = 1 - \frac{\sum_{i=1}^{n} \min(a_i, b_i)}{\min(\sum_{i=1}^{n} a_i, \sum_{i=1}^{n} b_i)}. \quad (9)$$

Quadratic Form Distance:

For color-based image retrieval, it has been proved that the quadratic form distance is more effective than the Euclidean distance. The main reason is that the similarities of different colors are also considered in this method. The quadratic form distance can be expressed as

$$d_{qad}(A,B) = (A-B)^T M(A-B) \quad (10)$$

In the equation, $M = [m_{ij}]$, where m_{ij} is the similarity of two colors with subscripts i and j in the histogram. By introducing a color similarity matrix M, the method can examine the similarity factors for the similar but different colors.

The commonly used evaluation criterion of image retrieval algorithm includes the precision and the recall. Suppose Q is an image to be retrieved, the precision and the recall can be defined as

$$\text{Recall} = \frac{n}{N}, \ \text{Precision} = \frac{n}{T} \quad (11)$$

where N is the image number similar to Q in the database, n is the image number retrieved by the CBIR system. T is the image total number outputting from the retrieval system. The higher values the precision and the recall give, the better capacity the retrieval system owns. But the precision and the recall are a pair of contradiction. When the precision is required to be higher, the recall is usually lower and vice versa. So if we find an optimal balance point between the two criterions, a better retrieved performance will be gained.

4 Experiments

4.1 Image Classification Experiments and Result Analysis

The SVM coefficients optimizing tests are carried out first. As usual, the searching method is adopted, but we find a stepwise refinement method rather than searching in the full grid. The method shrinks the searching region step by step until find the optimal coefficient, which can reduce the searching computational complexity. Using this method, we can obtain the optimal coefficients with the best classification results.

Still using the four features, select different kernel for SVM, the optimal classification correct rates can be gained.

From the above tests, we can see that the image classification results are different when different features are used. For one type of feature, the classification precision is also various if different kernel is adopted. The **FD** feature and the sigmoid kernel have lower classification performance.

From Fig. 3, we can obviously see that the classification correct rates are various for each class of image with different features. Moreover, different features have different feature dimension, for example, the image FD feature has a dimension of 12, the CM feature's dimension is 21, the FT feature is 80, and the CD feature is 136, which can bring problems to the single kernel machine classification.

The upper tests also show that the classification will not be optimized if one single feature is used, so different features should be combined, also the kernels should be combined. According to the characters of features, the composite kernel-based image classification is carried out. Suppose the feature vector of image is $x = \{x_1, x_2 \ldots x_n\}$, $k(x, y)$ is the known kernel function, $\hat{k}(x, y)$ is the normalization form, and the composite kernel can be designed as (12)

$$k(x,y) = \sum_{j=1}^{M} \beta_j \hat{k}_j(x_j, y_j), \text{ with } \beta_j \geq 0, \tag{12}$$

where β_i is the composite weight of kernels, and we set β_i equals to the classification correct rate of K_i.

Using the combined feature vector, if only a single kernel is adopted for classification, the result is not so good. For example, the RBF kernel only has a classification correct rate of 61.23 % using the combined feature. But if we select the CM feature, the correct rate is 68.34 %. So, the simple composition of features cannot bring good classification effectiveness. In view of the above-mentioned facts, we must combine the kernel matrix in the kernel space from different feature mapping with different kernels.

The original weights of the kernel function are the classification correct rates, and the weights can be finely adjusted using a searching method. The RBF kernel and the Polynomial kernel are combined as the composite kernel. The ultimate weights of kernels are shown in Table 1. Using the composite kernel machine, the classification correct rate is greatly improved.

According to the composite kernel machine, the algorithm in this paper obtains the best classification correct rate of 80.57 %. In addition, the multi-scale RBF kernel is studied. From Fig. 2, we can see that the RBF kernel has better classification capacity. The multi-scale RBF kernels are also composed as a composite kernel, in this condition, the best classification correct rate is 80 %. In Ref. [18], an image classification algorithm named SMP is proposed, which gains the best classification correct rate of 75.32 %, we can see that both composite kernel machines have better performance than the SMP method. The precision comparison between the SMP method and the RBF and Polynomial composite kernel are shown in Fig. 3.

Table 1 The optimal classification weights of the composite kernel

Features	Dim	Kernels	Values in kernel space		Precision (%)	Weights
			Max	Min		
CD	72 + 64	RBF	1	0.089	74.5	0.645
CM	21	Poly	1187.33	875.86	67.4	1.04
FT	80	Poly	64.68	−52.77	64.2	0.301
FD	12	RBF	1	0.0047	50.1	0.574

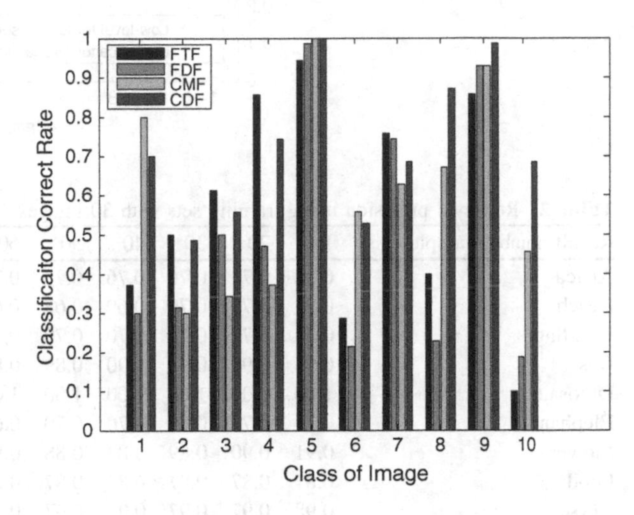

Fig. 2 Classification result for each class of image with different features

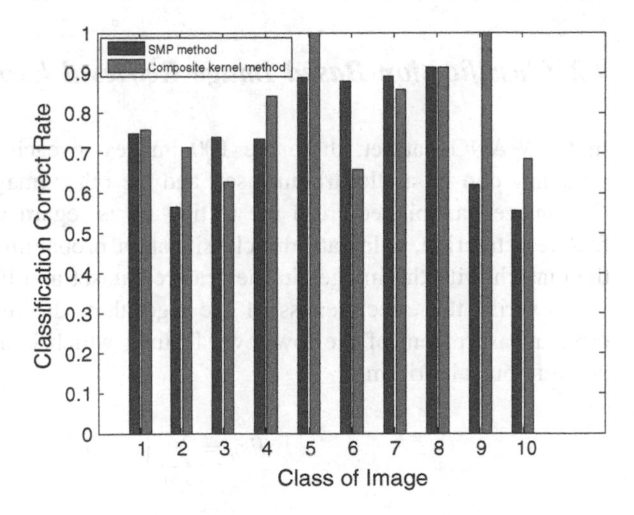

Fig. 3 Precision comparison between the SMP method and the composite kernel machine

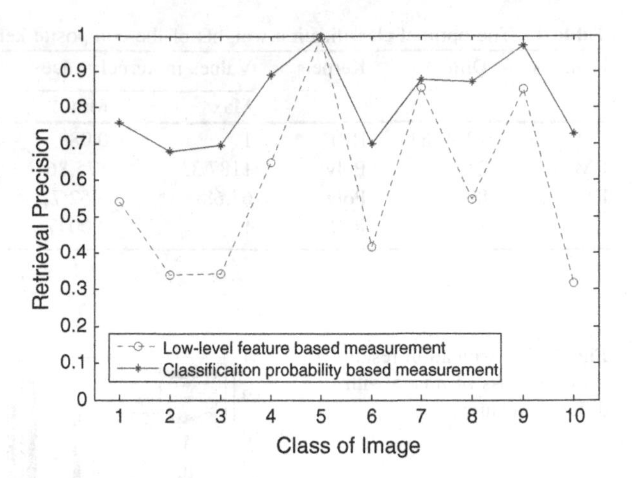

Fig. 4 Precisions of different measurement methods

Table 2 Retrieval precision using training sets with 30 images

Result number sample class	10	20	30	40	50	60	70	80	90	100
Africa	0.78	0.76	0.76	0.76	0.76	0.76	0.74	0.72	0.70	0.67
Beach	0.71	0.71	0.70	0.69	0.68	0.67	0.65	0.64	0.62	0.60
Building	0.73	0.72	0.71	0.70	0.70	0.68	0.65	0.63	0.59	0.55
Bus	0.94	0.92	0.91	0.90	0.89	0.88	0.88	0.86	0.85	0.81
Dinosaur	1.00	1.00	1.00	1.00	1.00	1.00	1.00	1.00	1.00	1.00
Elephant	0.77	0.73	0.71	0.70	0.70	0.69	0.68	0.65	0.61	0.57
Flower	0.91	0.90	0.89	0.88	0.88	0.87	0.87	0.84	0.84	0.81
Food	0.87	0.87	0.87	0.87	0.87	0.87	0.87	0.86	0.83	0.81
Horse	0.98	0.97	0.97	0.97	0.97	0.97	0.97	0.97	0.97	0.96
Mountain	0.75	0.74	0.73	0.73	0.73	0.72	0.69	0.65	0.64	0.62

4.2 Classification-Based Image Retrieval Experiments

In the WANG dataset, there are 100 images in each class. Thirty images are randomly chosen as the training set, and the other images are as the testing set. One image handpicked from the testing set is regard as the query image, after feature extraction, calculate the classification probability estimation using SVM, then match with the images in the feature dataset and find the similar image.

To verify the effectiveness of the algorithm, the retrieval method using the direct measurement of the low-level feature, which is described in (13), is compared to our algorithm,

$$dis_{ij} = \sum_{i=1}^{n} \left| x_i - x_i' \right| \tag{13}$$

where $x = (x_1, x_2, x_3, x_4)$, x_1 is the 12 dimensional FD feature, x_2 is the 21 dimensional CM feature, x_3 is the 64 + 72 dimensional CD feature, and x_4 is the 80 dimensional FT feature.

From Fig. 4, we can see that the precision is greatly improved using the algorithm presented in this paper than the distance measurement method directly using the low-level features. Table 2 is the retrieval precision and retrieval recall using training sets with 30 images.

5 Conclusions

The common method of image retrieval is mapping images into a space $\mathbb{R}$ through a certain function, and then the target of image retrieval will be achieved by the distance measurement between images in $\mathbb{R}$. The elements of space $\mathbb{R}$ are the image features, such as the color feature, the texture feature, the transformed feature and so on. A composite kernel support vector machine is applied for classification probability estimation of images in this paper, then the classification probability estimation values from SVM are mapped into the classification probability space $\mathbb{R}_c$. Finally, the distance between images are measured in $\mathbb{R}_c$, with this, we can realize the image retrieval method with higher performance.

Acknowledgments This work was jointly supported by the National Natural Science Foundation for Young Scientists of China (Grant No: 61202332) and China Postdoctoral Science Foundation (Grant No: 2012M521905).

References

1. Smeulders A, Worring M, Santini S, Gupta A, Jain R (2000) Content-based image retrieval: the end of the early years. IEEE Trans Pattern Anal Mach Intell 22(12):1349–1380
2. Datta R, Joshi D, Li J (2008) Image retrieval: ideas, influences, and trends of the new age. ACM Comput Surv 40:1–60
3. Ravela S, Manmatha R (1997) Image retrieval by appearance. ACM SIGIR Philadephia PA, USA, pp 278–285
4. Serre T, Wolf L, Bileschi S, Riesenhuber M, Poggio T (2007) Object recognition with cortex-like mechanisms. IEEE Trans Pattern Anal Mach Intell 9:411–426
5. Wang JZ, Li J, Wiederhold G (2001) SIMPLIcity: Semantics-Sensitive Integrated Matching for Picture LIbraries. IEEE Trans Pattern Anal Mach Intell 23(9):947–963
6. Content based image retrieval/image database search engine. http://wang.ist.psu.edu/docs/related/
7. Siggelkow S (2002) Feature histograms for content-based image retrieval. Ph.D.thesis, University of Freiburg, Institute for Computer Science, Freiburg, Germany
8. Bach FR (2008) Consistency of the group Lasso and multiple kernel learning. J Mach Learn Res 9:1179–1225
9. Gönen M, Alpaydin E (2011) Regularizing multiple kernel learning using response surface methodology. Pattern Recogn 44:159–171

10. Gönen M, Alpaydin E (2010) Cost-conscious multiple kernel learning. Pattern Recognit Lett 31:959–965
11. Wu ZP, Zhang XG (2011) Elastic multiple kernel learning. Acta Automatica Sinica 37(6):693–699
12. Lewis DP, Jebara T, Noble WS (2006) Nonstationary kernel combination. In: Proceedings of the 23rd international conference on machine learning. ACM, Pittsburgh, USA, pp 553–560
13. Gönen M, Alpaydin E (2008) Localized multiple kernel learning. In: Proceedings of ICML'08, pp. 352–359
14. Kloft M, Brefeld U, Laskov P, Sonnenburg S (2008) Non-sparse multiple kernel learning. In: Proceedings of the NIPS workshop on kernel learning: automatic selection of optimal kernels
15. Rakotomamonjy A, Bach FR, Canu S, Grandvalet Y (2008) SimpleMKL. J Mach Learn Res 9:2491–2521
16. Gehler PV, Nowozin S (2008) Infinite kernel learning. Technical report 178, Max Planck Institute for Biological Cybernetics, Germany
17. Myron F, Harpreet S (1995) Query by images and video content: the QBIR system. IEEE Comput 28(9):23–32
18. Paolo P, Sandrine A, Eric D, Michel B (2009) Sparse multiscale patches (SMP) for image categorization. In: Proceedings of the 15th international multimedia modeling conference on advances in multimedia modeling, Sophia Antipolis, France (2009)

A First Approach of a New Learning Strategy: Learning by Confirmation

Alejandro Carpio, Matilde Santos and José Antonio Martín

Abstract Learning by confirmation is a new learning approach, which combines two types of supervised learning strategies: reinforcement learning and learning by examples. In this paper, we show how this new strategy accelerates the learning process when some knowledge is introduced to the reinforcement algorithm. The learning proposal has been tested on a real-time device, a Lego Mindstorms NXT 2.0 robot that has been configured as an inverted pendulum. The methodology shows good performance and the results are quite promising.

Keywords Learning by Confirmation · Reinforcement Learning · Learning by examples · Lego NXT · Inverted Pendulum

1 Introduction

Learning by Confirmation is a new learning strategy, which combines two types of supervised learning: reinforcement learning and learning by examples [1–4]. Therefore learning by confirmation is a supervised learning mainly based on reinforcement learning but where in addition, an instructor gives the correct answers to some situations that the learner tries to repeat. The main idea is that when the learner repeats the answer, he introduces a variation as part of its exploration. For example, to find out the margin of error that can be accepted by

A. Carpio (✉) · M. Santos · J. A. Martín
Computer Architecture and Systems Engineering, Facultad de Informática,
Universidad Complutense de Madrid, 28040 Madrid, Spain
e-mail: acarsan2@gmail.com

M. Santos
e-mail: msantos@dacya.ucm.es

J. A. Martín
e-mail: jamartinh@fdi.ucm.es

F. Sun et al. (eds.), *Knowledge Engineering and Management*,
Advances in Intelligent Systems and Computing 214,
DOI: 10.1007/978-3-642-37832-4_37,

the instructor. In this way, it is possible to accelerate the learning process by introducing some knowledge within the reinforcement learning algorithm [5–8].

In order to validate this proposal, we have used a real-time device, the Lego Mindstorms NXT 2.0 robot configured as an inverted pendulum. It will be used as the agent of the learning by confirmation algorithm. The learning algorithm has been implemented and embedded into the robot.

Two different tests have been carried out in order to show the performance of the algorithm. It has been proved how we are able to influence the behavior of the agent when some extra knowledge about the environment is introduced.

The structure of the paper is as follows. First (Sect. 2), the scenario where the learning strategy is applied is described. In Sect. 3, the learning by confirmation algorithm is presented. Section 4 shows the results obtained when the learning proposal is applied to two different configurations. Conclusions end the paper.

2 Test Environment: Lego Mindstorms NXT 2.0

In order to implement and test the learning by confirmation strategy, we have used a real-time system: a Lego Mindstorms NXT 2.0 robot on inverted pendulum configuration. The learner must be able to learn the optimal route [9] to go from the starting point to a final point (previously fixed). This will show how the use of supervised learning plus reinforcement learning algorithm accelerates the learning process. Besides it is possible to influence the decision the robot makes, for example choosing a better path if there are several.

The inverted pendulum robot used in this work is the two-wheel robot Anyway, developed by Laurens Valk [10, 11]. This is able to move each of the wheels independently and therefore to turn and move forward. It is shown in Fig. 1.

Stabilization control is guaranteed by an embedded PID regulator. The PID controller is a feedback control whose purpose is to remove the error between the reference signal (desired) and the system output [12]. In addition, the controller includes a derivative action which gives a prediction on the error evolution. This controller is implemented in NXT-G language [13].

On the other hand, the robot will move in a well-defined environment. This layout has been divided into nine square tiles (3×3 grid), as shown in Fig. 2. In this experimental scenario, the robot starts at the lowest left corner (blue cross), facing the upper left side, and must reach the upper right corner (red cross). The tile in the middle is not allowed. The tile size is 30×30 cm.

3 Learning by Confirmation

Learning by Confirmation is a new type of supervised learning based on reinforcement learning; but when some knowledge is added by means of examples, the learner tries to repeat the correct actions given by the instructor (Fig. 3).

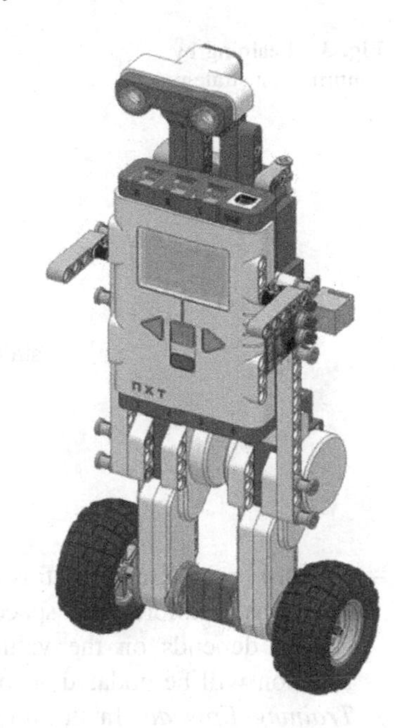

Fig. 1 Agent—anyway robot

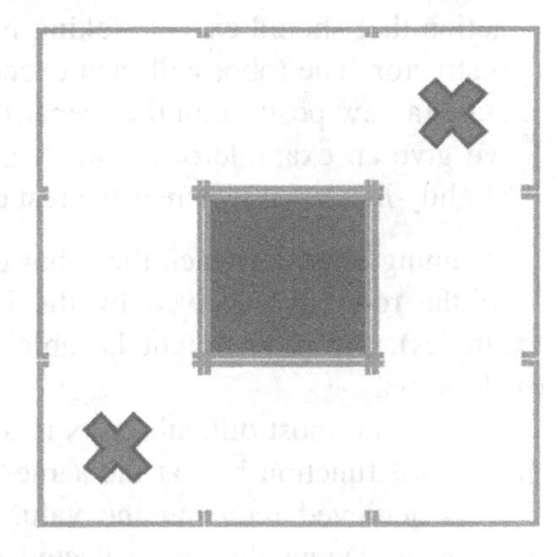

Fig. 2 Environment description

Some of the characteristics of learning by confirmation are the following: it can be applied online and it is an active learning as the agent can change the answers given by the expert. It takes the good points from the two learning strategies that it combines. For example, it allows generalization, speeds up the training, and it allows the instructor to specify some preferences (expert knowledge).

In order to introduce the expert knowledge, a series of "Training Episodes" are run. In those training steps, the right actions are shown to the agent. But these actions are introduced into the system as "opinions" because the robot must learn the optimal route by itself. If the instructor does not know what the right actions are, then the agent will explore the space using reinforcement learning.

Two types of different episodes have been defined and implemented (Fig. 4):

Fig. 3 Learning by confirmation strategy

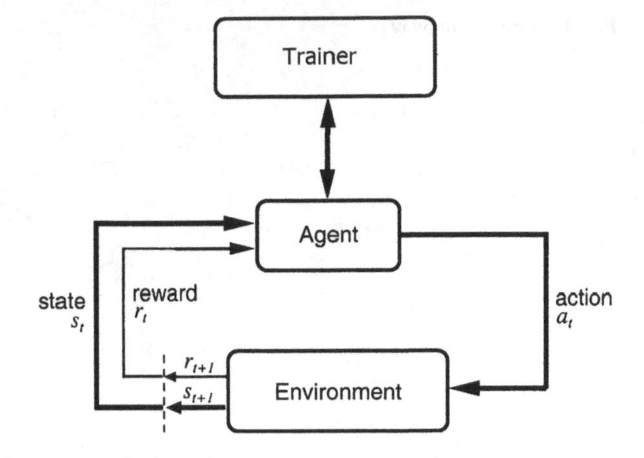

- *Exploratory Episode*: In this mode, the Q-Learning algorithm is applied [14]. The robot explores the space, choosing the next action based on its policy (which depends on the values of the reward function). The values of that function will be updated according to the exploration.
- *Training Episode*: In this mode, the instructor is asked by the robot for the action that should execute taking into account the privileged knowledge of the instructor. The robot will then execute the corresponding action. From this new state (a new position in the scenario), it will again asked for our opinion. When we give an example of a correct action for that state, the learning function is slightly increased, whereas the rest of possible actions will be penalized (Fig. 5).

Training episodes teach the robot through examples, as in supervised learning. If the robot is deceived by the information given by the instructor (wrong examples), the robot might be able to undo what it has been taught by using exploration.

One of the most difficult tasks is to influence the robot behavior by modifying the reward function but, at the same time, without spoiling the learning process. This is achieved reducing the values of the neighboring states before the last movement. Doing this, the selected path is strengthened without damaging the future learning of the reinforcement learning algorithm.

```
If it is an Exploratory Episode
     1) Select an action applying Q(S,A)
        (Reinforcement learning)
If it is a Training episode
     1) Ask the instructor (Learning by Confirmation)
     2) Execute the action
     3) Reduce the weight of the non-selection actions
     4) Increase the weight of the taken action
```

Fig. 4 Learning by confirmation algorithm (type of episodes)

```
    % Increase the weight of the right action and lower the
  rest.
    for t=1:(size(actionlist,2))
        Q(sp,t)=(Q(sp,t)-low);
        if(Q(sp,t)<=min) Q(sp,t)=min;end
    end
    Q(sp,ap)=Q(sp,ap)+hig;
    if(Q(sp,t)>=max) Q(sp,t)=max;end
```

Fig. 5 Agent policy based on the type of episode

4 Test and Results

In this section, we present the tests that have been carried out in order to validate the new learning algorithm and discuss the results.

The reward function used in each case is shown and the acceleration of the learning process is proved in terms of number of training episodes required.

Finally, we will show how to influence the final path chosen by the robot and how it reacts when it is deceived by the instructor.

4.1 Learning Speed up: Test 1

This section shows how the learning by confirmation strategy works. That is, adding knowledge of the environment facilitates the robot learning and reduces the number of steps and episodes of exploration required by the agent to find the solution.

In Fig. 6, the configuration of Test 1 is presented. The reward function has been set to 20 (upper right corner). The reward of each state has been fixed to different values, all of them negative ones. It means that we are giving some information to the robot on the best path.

Fig. 6 Reward associated to each state (Test 1)

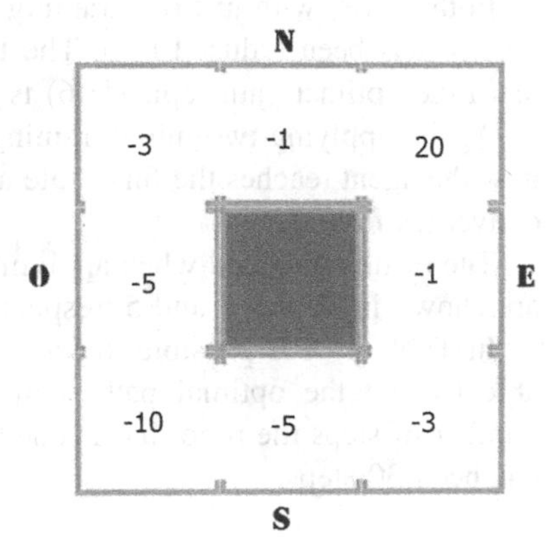

Table 1 Episodes to convergence (0 training episodes)

Episode	Number of steps	Reward
1	8	3
2	6	7
3	20	−70
4	14	−38
5	4	11
6	6	7
7	**4**	**11**

Table 2 Episodes to convergence (1 training episode)

Episode	Number of steps	Reward
1	4	11
2	6	7
3	20	−70
4	4	11
5	5	10
6	**4**	**11**
7	**4**	**11**

In the tables below, we have marked in bold the episodes in which the robot has learned the optimal route. The training episodes are light shaded.

In case the system does not need any training episode, we would obtain the same result as in the case of just reinforcement learning. The results obtained for test 1 (Fig. 6) when only reinforcement training is applied to the system (0 training episodes) can be seen in Table 1.

As it is possible to see, seven episodes of exploration have been required to find the optimal path. The total number of steps the robot has needed to finally learn the optimal solution (in episode 7) has been 62 steps (the sum of all the steps of every episode).

If we apply a single training episode (first row), the results are now shown in Table 2.

In this case, with just one training episode, the number of exploratory episodes needed has been reduced to 5. The total number of steps the robot has made to learn the optimal path (episode 6) is 43 steps.

Again, applying two initial training episodes, the results listed in Table 3 show how the agent reaches the final state at episode 5 with 38 steps. That is, the system converges faster.

The results obtained when applying 3 training episodes and 4 training episodes are shown in Tables 4 and 5, respectively.

In Table 4, it is possible to see how with three training episodes the robot is able to find the optimal path with only three exploratory episodes. The total number of steps the robot has needed to learn this optimal solution (at episode 6) has been 30 steps.

Table 3 Episodes to convergence (2 trainings episodes)

Episode	Number of steps	Reward
1	4	11
2	4	11
3	6	7
4	20	−70
5	**4**	**11**
6	**4**	**11**
7	**4**	**11**

Table 4 Episodes to convergence (3 trainings episodes)

Episode	Number of steps	Reward
1	4	11
2	4	11
3	4	11
4	6	7
5	8	−29
6	**4**	**11**
7	**4**	**11**

Table 5 Episodes to convergence (4 trainings episodes)

Episode	Number of steps	Reward
1	4	11
2	4	11
3	4	11
4	4	11
5	**4**	**11**
6	**4**	**11**
7	**4**	**11**

Table 5 shows that the number of required episodes of exploration when four initial training episodes have been applied is none. That is, the robot has found the optimal path at episode 5, with 20 steps.

For five or more training episodes, the robot does not learn faster, and therefore these episodes are unnecessary. As we can see the first training episode is the most helpful in the learning process, followed by the fourth which is the last episode of training that can be applied.

Figure 7 represents how, with each episode of training, the number of steps to find the correct solution decreases.

4.2 Learning Speed up: Test 2

We have carried out this experiment with another configuration of the reward function. In this case, we have normalized the values to -1 and 0 (final state), as it is shown in Fig. 8.

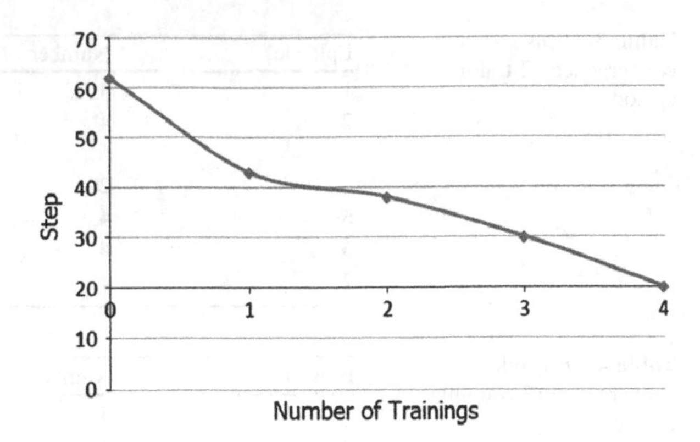

Fig. 7 Graph of learning acceleration (Test 1)

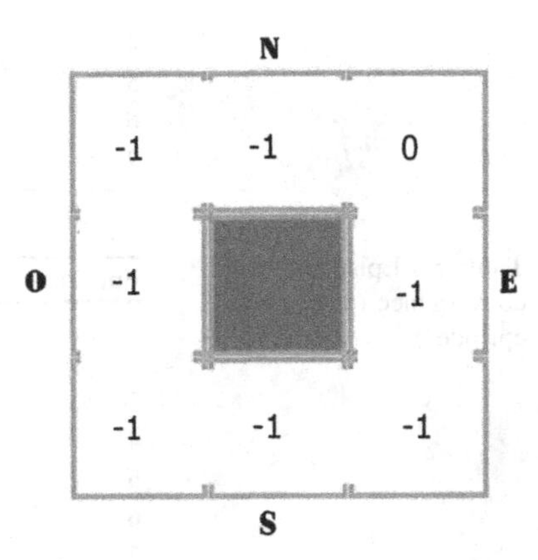

Fig. 8 Reward associated to each state (Test2)

In this example, we are not giving the agent any information about how useful a state is. All of them have the same value (-1). The robot learns the utility of each state by exploration. Obviously, in this case, the agent will initially require more steps to find the optimal solution than in the previous test.

After repeating the process made in Test 1, but with the new values of the reward function, the results are summarized in Fig. 9.

It is possible to see how the system converges faster as the number of training episodes increases. The initial number of steps is now 88 (reinforcement learning) instead of 62, and the final number when learning by confirmation is applied is 17. That is, it is even better than in the previous test.

4.3 Influencing the Final Path

We have also analyzed how many initial training episodes are necessary in order to influence the final route which the robot follows.

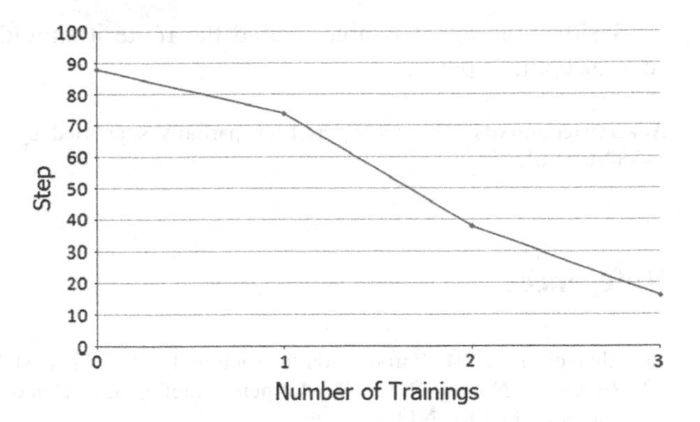

Fig. 9 Graph of learning acceleration (Test 2)

After several experiments, we have concluded that it is possible to modify the final path during the training after three episodes. It means that a route is set a very low reward value, and therefore that path is discarded by the robot. Then the agent will follow the route we wanted.

That is, if we want to change the route once it has been learnt by the robot, to teach it another path we need at least three episodes of training.

4.4 Deceiving the Agent

We can deceive the robot giving wrong responses when asked for the right actions. For example, if we suggest the action that leads to the same state again and again.

In this case, the robot will begin to explore the environment using the knowledge previously introduced. This contributes some information to the utility of the states regardless of what the instructor thought as correct.

But even misleading the robot, the agent needs fewer episodes of exploration (and therefore fewer steps) to find the final solution than without training episodes.

5 Conclusions

In this paper, we have shown how we can accelerate the learning process of an agent by introducing some examples as extra knowledge. We have called this combination of reinforcement learning and training by examples as Learning by confirmation.

We have applied this new strategy to a real-time system, a Lego NXT robot, in an inverted pendulum configuration.

We have proved how it is possible to help the robot providing some examples of correct actions based on the state where the robot is. Using these examples, we are able to introduce knowledge about the problem to the algorithm of reinforcement learning.

Besides, the agent is able to find the route it should follow even if there are several optimal paths.

Acknowledgments This work has been partially supported by the Spanish project DPI2009-14552-C02-01.

References

1. Alpaydin E (2004) Introduction to machine learning. The MIT Press, Cambridge, MA
2. Russel S, Norvig P (2004) Artificial intelligence: a modern approach. Prentice-Hall, Englewood Cliffs, NJ
3. Sutton S, Barto A (1998) Reinforcement learning: an introduction. The MIT Press, Cambridge, MA
4. Martín-H A, Santos M (2010) Aprendizaje por refuerzo. In: Aprendizaje automático, chapter 12, RA-MA, Madrid, Spain
5. Karamouzas I, Overmars MH (2008) Adding variation to path planning. Comp Anim Virtual Worlds 19: 283–293
6. Santos M, Martín-H JA, López V, Botella G (2012) Dyna-H: a heuristic planning reinforcement learning algorithm applied to role-playing-game strategy decision systems. Knowl-Based Syst 32:28–36
7. Garzés M, Kudenko D (2010) Online learning of shaping rewards in reinforcement learning. Neural Netw 23(4):541–550
8. Alvarez C, Santos M, López V (2010) Reinforcement learning vs. A* in a role playing game benchmark scenario. In: Ruan D, Li T, Xu Y, Chen G, Kerre E (eds) Computational intelligence: foundations and applications. World Scientific Proc. Series on Computer Engineering and Information Science Vol. 4. Computational Intelligente. Foundations and Applications. Proc. of the 9th Int. FLINS Conference, pp 644–650
9. Bertsekas DP (1995) Dynamic programming and optimal control. Athena Scientific, Belmont, Massachusetts
10. Anyway. http://robotsquare.com/2012/03/13/tutorial-segway-with-nxt-g/
11. Berthilsson S, Danmark A, Hammarqvist U, Nygren H, Savin V (2009) Embedded Control Systems LegoWay http://www.it.uu.se/edu/course/homepage/styrsystem/vt09/Nyheter/Grupper/g5_Final_Report.pdf
12. Astrom KJ, Hagglund T (2005) Advanced PID control. Research Triangle Park, NC: ISA–The Instrumentation, Systems, and Automation Society
13. Kelly JF (2007) Lego mindstorms 2.0 NXT-G programming guide. Apress, Berkeley, CA
14. Watkins C, Dayan P (1992) Q-learning. Mach Learn 8:279–292

Quantitative Evaluation of Interictal High Frequency Oscillations in Scalp EEGs for Epileptogenic Region Localization

Yaozhang Pan, Cuntai Guan, How-Lung Eng, Shuzhi Sam Ge, Yen ling Ng and Derrick Wei Shih Chan

Abstract Electroencephalography is a commonly used tool for presurgical evaluation of epilepsy patients. In this paper, we present a quantitative evaluation of interictal high frequency oscillations (HFOs) in scalp Electroencephalographies (EEGs) for epileptogenic region localization. We process multichannel EEGs using time-frequency spectral analysis in order to detect HFOs in each EEG channel. Comparison between the results of time-frequency analysis and visual assessment is performed to verify the reliability of time-frequency analysis. Later, *t*-test and Pearson correlation analysis are performed to analyze the relationships between ictal HFOs and interictal HFOs. The high correlations between interictal

Y. Pan (✉) · C. Guan · H.-L. Eng
Institute for Infocomm Research, Agency for Science, Technology and Research (ASTAR), 1 Fusionopolis Way, #21-01, Connexis 138632, Singapore
e-mail: yzpan@i2r.a-star.edu.sg

C. Guan
e-mail: ctguan@i2r.a-star.edu.sg

H.-L. Eng
e-mail: hleng@i2r.a-star.edu.sg

S. S. Ge
Department of Electrical and Computer Engineering, National University of Singapoire, Singapore 117576, Singapore
e-mail: samge@nus.edu.sg

S. S. Ge
Robotics Institute and School of Computer Science and Engineering, University of Electronic Science and Technology of China, Chengdu 11813, China

Y. l. Ng · D. W. S. Chan
KK Women's and Children's Hospital, 100 Bukit Timah Road, Singapore 229899, Singapore
e-mail: Ng.Yen.Ling@kkh.com.sg

D. W. S. Chan
e-mail: Derrick.Chan.WS@kkh.com.sg

F. Sun et al. (eds.), *Knowledge Engineering and Management*, Advances in Intelligent Systems and Computing 214,
DOI: 10.1007/978-3-642-37832-4_38,

and ictal HFOs imply that interictal HFOs, like ictal HFOs, are valuable in localizing the epileptogenic region. As a result, scalp interictal HFOs are valuable in epileptogenic region localization for presurgical evaluation of epilepsy patients. It holds great potential for reducing the long delay before patients can be referred for surgery.

Keywords High frequency oscillations · Scalp EEGs · Epileptogenic region localization

1 Introduction

Epilepsy is a common neurologic disorder affecting more than 1 % of the world population. Although antiepileptic drug (AED) is the most popular treatment of epilepsy, 40 % of epilepsy patients are pharmacoresistant. For these patients with pharmacoresistant seizures, surgery is a highly effective treatment instead of AED. However, a successful epilepsy surgery necessitates the accurate delineation of the epileptogenic region. Reliable biomarkers are needed to identify potential surgical candidates and localize the epileptogenic regions as early as possible after AED fails. A commonly used biomarker is the interictal EEG spike, which however is not reliable for localizing the extent of an epileptogenic region because it is difficult to differentiate two types of interictal epileptic discharges: those originating from the ictal onset zone; and those propagated from other regions or irritative regions [5, 11].

High frequency oscillations (HFOs) are oscillatory activities on EEG above 25 Hz. HFOs can be subgrouped into gamma waves, ripples, and fast ripples, etc. A gamma wave is a pattern of neural oscillation in humans with a frequency range between 25 and 100 Hz, though 40 Hz is prototypical. Ripple is HFO in the 80–200 Hz range, reflecting inhibitory field potentials, which synchronize neuronal activity [3]. HFOs of 250–600 Hz, which are referred to as fast ripples (FRs), are pathologic signals reflecting summated action potentials of spontaneously bursting neurons [3].

Clinical studies using direct brain recordings during presurgical evaluation have found ripples and FRs to be valuable in identifying the epileptogenic region [2, 3]. It has also been shown in [5, 11] that HFOs, with or without interictal spikes, are more reliable in identifying epileptogenic region than interictal spikes alone and even more reliable than ictal onset in determining the extent of brain tissue that must be resected.

HFOs in patients with focal epilepsies have been reported with ECoG recordings [9, 10]. In [6], HFOs' changes have been extracted based on a simple statistics of time-frequency representation. In [8], a reviewer-dependent approach for detection of HFOs has been proposed. In [4], HFOs have been detected using swarmed neural-network features.

Till recently, HFOs analysis has been confined to ECoG, which has limited its utility as a biomarker. Analysis of scalp EEG has demonstrated possibility to detect HFOs from scalp EEGs [1, 7]. In [1], Andrade-Valenca et al. have studied scalp EEGs recordings of 15 patients with focal epilepsy, and analyzed the rates of gamma waves (40–80 Hz) and ripples ($>$ 80 Hz) by visually rating the number of spikes, gamma waves, and ripples per minute for each channel of scalp EEGs. However, inspecting the bulky multi-channel EEG data visually is very tedious and time-consuming. In another work [7], HFOs in scalp EEGs recorded during slow-wave sleep (CSWS) have been investigated through visual analysis as well as time-frequency spectral analysis. However, the reliability of using the scalp EEG to identify the epileptogenic region is still an open question.

In this paper, we present a quantitative evaluation of HFOs in scalp EEGs in localizing epileptogenic region. Following [7], the quantitative analysis is based on a time-frequency spectral analysis of scalp EEGs recorded from 6 epilepsy patients. Comparison between the results of time-frequency analysis and visual assessment is performed to verify the reliability of time-frequency analysis. Furthermore, we use paired *t*-test and Pearson correlation analysis to look for relationships between the frequency-power characteristics of ictal and interictal HFOs. The results of *t*-test demonstrate significant differences between ictal and interictal HFOs in both values of peak frequency and corresponding power. This fact implies that the HFOs can be valuable for differentiating the ictal and interictal states. The Pearson correlation analysis shows high correlations between ictal and interictal HFOs. The high correlations imply that interictal HFOs, similar to ictal HFOs, hold potential in localizing the epileptogenic region.

Our results demonstrate the potential of using interictal HFOs of scalp EEGs as a biomarker of noninvasive localization of the epileptogenic region. The HFOs in interictal EEG are quite promising for detecting the epileptogenic region once it is proved to be reliable. The interictal EEGs have many advantages in localizing the epileptogenic region compare to ictal EEGs. They have less muscle artifacts. Moreover, they make it possible to diagnose epilepsy without waiting subsequent seizures and long-term presurgical evaluation. Thus, the findings in this paper would help to justify earlier surgical treatment and thereby reduce the long delay before patients are referred for surgery [2].

2 Methods

2.1 EEGs Recording

The EEGs recording was conducted at the KK Women's and Children's Hospital (KKH), Singapore. We recruited those pediatric epileptic patients with informed consent obtained from their guardians. The patients were subject to the standard video-EEG monitoring in the epilepsy monitoring unit (EMU) of KKH. Long-term

Table 1 Selected Patients and Patient Characteristics

Patient No.	Age[a]	N_s^b	Diagnosis
P003	12	3	Lennox Gastaut Syndrome
P004	9	13	Right Temporal Lobe Focal
P006	9	10	Right Frontal Lobe Focal
P023	8	2	Left Fronto-central Focal
P025	5	10	Right Fronto-central-vertex Focal
P027	11	6	Left Frontal Lobe Focal

[a] The age here is refer to age of screening.
[b] N_s is total number of seizures of one patient.

EEGs were recorded by 32 electrodes placed on scalp according to the international 10–20 system. All data were stored on a hard disk with 256 or 512 sampling rate.

The subjects recruited so far are 27 patients with diagnosis of different types of seizures. We focus only on focal seizure in this study. After careful assessment by the EEG technologist, 6 epileptic patients aged between 5 and 12 years (3 male, 3 female) are selected for this analytic work with 44 seizure events and frequently occurred HFOs in both ictal and interictal states. The related information of the patients are listed in Table 1.

2.2 Visual Assessment of EEGs

Each patient's EEGs are assessed visually. Seizure events are identified from the standard video-EEG system by a EEG technologist. EEGs are marked as ictal or interictal periods. In addition, some of the visible HFOs are also marked. The conventional EEG traces (10 s per page) are initially reviewed to identify the ictal events, as shown in Figs. 1a and 2a. Then, the traces are temporally expanded to 2 s per page in order to study the details of activity faster than 25 Hz (see Figs. 1b and 2b).

In interictal period, the HFOs are relatively clear, as shown in Fig. 2. But due to the large amount of artifacts during ictal state, the HFOs cannot be found easily by visual checking as in Fig. 1. We need to extract HFOs from EEG back ground noises in order to further study HFOs in epileptic EEGs.

2.3 Time-Frequency Analysis

We investigate the time evolution of the high-frequency power spectrum of the ictal and interictal activity by applying the Gabor transform, which is the Fourier transform with a sliding Gaussian window. For EEGs with 256 Hz sampling rate, we apply Fourier transform with a 500 ms wide Gaussian window (contains 128 data points, the frequency resolution is 2 Hz) and 10–100 Hz frequency range to

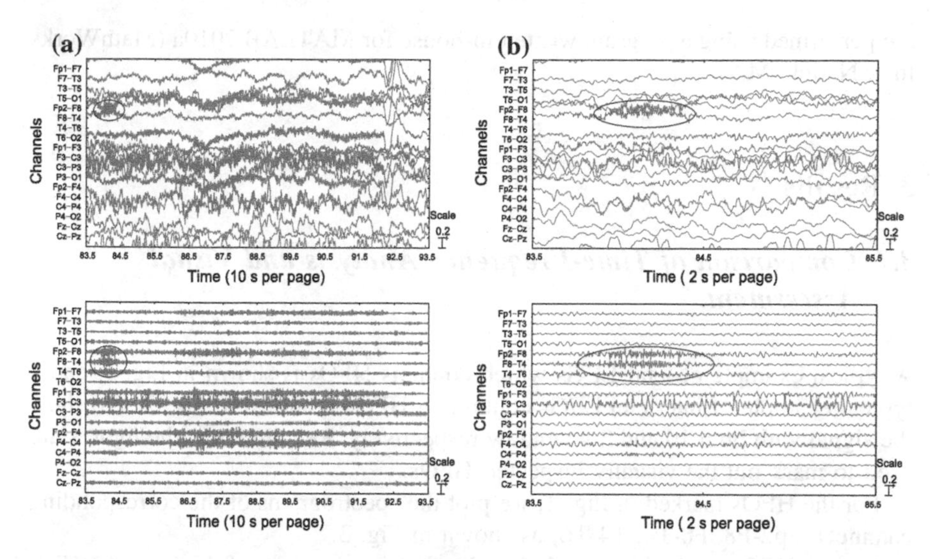

Fig. 1 Ictal EEGs of P025 (sampling rate 512 Hz): the above row is after 0.5–198 Hz bandpass filtering, the below row is after 20–198 Hz bandpass filtering. **a** P025 ictal EEGs with 10 s window. **b** P025 ictal EEGs with 2 s window

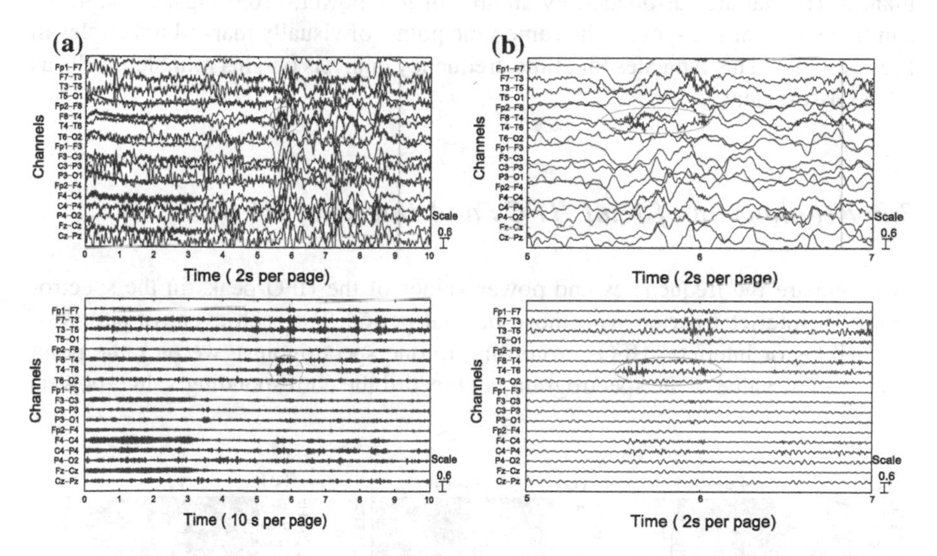

Fig. 2 Interictal EEGs of P006 (sampling rate 256 Hz): the above row is after 0.5–98 Hz bandpass filtering, the below row is after 20–98 Hz bandpass filtering. **a** P006 interictal EEGs with 10 s window. **b** P006 interictal EEGs with 2 s window

the raw EEG data. For EEGs with 512 Hz sampling rate, we apply Fourier transform with a 250 ms wide Gaussian window (contains 128 data points, the frequency resolution is 4 Hz) and 10–200 Hz frequency range. All computations are performed using a program written in-house for MATLAB 2010a (MathWorks Inc., Natick, MA).

3 Results

3.1 Comparison of Time-Frequency Analysis and Visual Assessment

A typical example of ictal EEGs which contains HFOs is shown in Fig. 1, and a typical example of interictal EEGs containing HFOs is shown in Fig. 2. We show the signal with both 10 and 2 s window respectively. Red circles in the figures are used to mark out the visually inspected HFOs.

For the HFOs marked in Fig. 1, we plot the spectrograms of the corresponding channels: Fp2-F8, F8-T4, T4-T6, as shown in Fig. 3.

For the HFOs marked in Fig. 2, we plot the spectrograms of the corresponding channels: Fp2-F8, F8-T4, T4-T6, as shown in Fig. 4.

HFOs are identified as clearly visible red spectral spots with frequencies faster than 20 Hz that are surrounded by an area of low power. From Figs. 3 and 4, we can find the clear red spots at the same time points of visually marked red circles in Figs. 1 and 2. This indicates that time-frequency analysis obtains reliable results as visual assessment.

3.2 Relationship of Ictal HFOs and Interictal HFOs

We compare the frequencies and power values of the HFO peaks in the spectrograms of interictal period with those of ictal period. For each event (a period of ictal HFOs or interictal HFOs, according to one spectrogram), we pick the point with highest value in the spectrogram, and record the corresponding value (power)

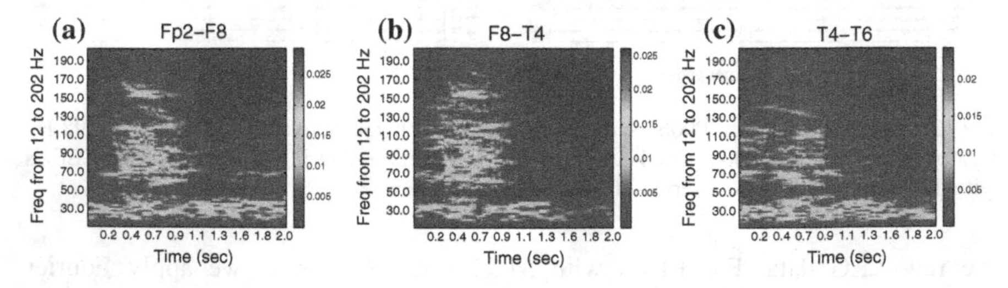

Fig. 3 Time-frequency spectrograms of EEGs in ictal state of P025, with sampling rate of 512 Hz, and bandpass filtering of 20–198 Hz. **a** Fp2-F8. **b** F8-T4. **c** T4-T6

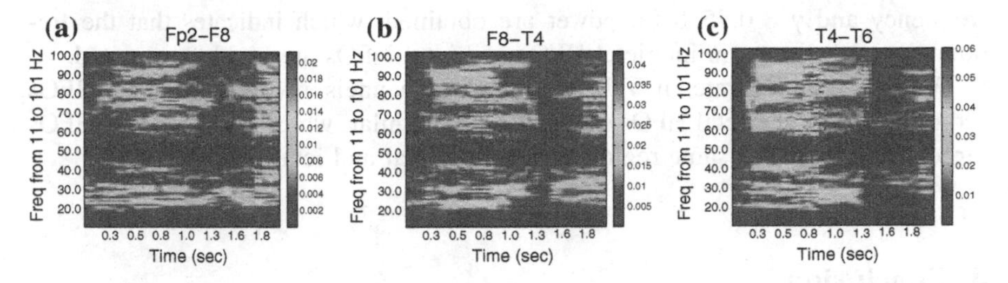

Fig. 4 Time-frequency spectrograms of EEGs in interictal state of P006, with sampling rate of 256 Hz, and bandpass filtering of 20–98 Hz. **a** Fp2-F8. **b** F8-T4. **c** T4-T6

Table 2 EEG observations and their associated HFOs: ictal and interictal

Patient id	N_i^b	N_{in}^b	Ictal channels involved	Interictal channels involved	Ictal (frequency (Hz)/power)	Interictal (frequency (Hz) /power)
P003	3	3	F3-C3, F4-C4	F3-C3, F4-C4	103.81/2.92 e-2	85.14/1.78 e-2
P004	5	3	T4-T6	T4-T6	55.95/2.15 e-2	43.35/1.42 e-2
P006	12	3	T4-T6, F8-T6	Fp2-F8, F8-T4	79.60/3.12 e-2	65.88/2.06 e-2
P023	2	3	Fp1-F7, Fp1-F3	F7-T3, F3-C3	76.86/2.14 e-2	47.91/1.70 e-2
P025	2	4	Fp2-F8, F8-T4	F4-C4, Fz-Cz	74.28/1.77 e-2	33.27/1.16 e-2
P027	4	9	Fp1-F3	Fp1-F7, Fp1-F3	92.47/2.78 e-2	58.08/2.59 e-2

[a] N_i is total number of ictal periods with HFOs analyzed for one patient.
[b] N_{in} is total number of interictal periods with HFOs analyzed for one patient.
Paired t-test between highest frequencies during ictal and interictal periods obtains p-value= 0.0328, and that for corresponding power values is p-value= 0.0423.
Pearson correlation coefficient for frequencies between ictal and interictal periods is $\gamma = 0.7808$ and that for power values is $\gamma = 0.7550$.

and frequency of that point. Then, for each patient with several events, we compute the average value of all events' power and frequency.

The average peak frequency and corresponding power value of ictal HFOs and interictal HFOs of each patient are shown in the 6th and 7th column of Table 2. The frequencies and powers of the total 6 patients are then analyzed using a paired t-test and Pearson correlation analysis between ictal HFOs and interictal HFOs.

For paired t-test, relationships are considered statistically significant if $p<0.05$. From the result of Table 2, HFOs detected during ictal period have significantly higher frequencies ($p = 0.0328$) and more power ($p = 0.0423$) than those detected during interictal period. This provides us evidence that HFOs can be used to differentiate interictal and ictal state for seizure detection.

For Pearson correlation analysis, larger Pearson correlation coefficient (range between [0,1]) implies higher correlation between two data sets. From the result of Table 2, the Pearson correlation coefficients are computed. Results $\gamma = 0.7808$ for frequency and $\gamma = 0.7550$ for power are obtained, which indicates that the frequencies and powers in interictal HFOs and ictal HFOs are highly correlated.

We also demonstrate in Table 2 that the channels involved in ictal HFOs (column 4) and interictal HFOs (column 5) are similar, which indicates that HFOs largely remain in the same region during interictal and ictal periods.

4 Conclusion

In this paper, we presented a quantitative evaluation of interictal scalp EEG for Epileptogenic Region Localization. The analysis was based on time-frequency spectral analysis of scalp EEG recorded from 6 epilepsy patients. The results of time-frequency analysis and visual assessment were compared to verify reliability of the time-frequency analysis. A paired *t*-test of peak frequencies and powers between ictal HFOs and interictal HFOs demonstrated that ictal HFOs had significantly higher frequencies ($p = 0.0328$) and more power ($p = 0.0423$) than interictal HFOs. Pearson correlation analysis demonstrated high correlations ($\gamma = 0.7808$ for frequency and $\gamma = 0.7550$ for power) between ictal and interictal HFOs. In conclusion, scalp interictal HFOs could be considered as an effective biomarker in epileptogenic region localization. In further study, investigations based on larger dataset using more advanced methods are worth performing.

References

1. Andrade-Valenca LP, Dubeau F, Mari F, Zelmann R, Gotman J (2011) Interictal scalp fast oscillations as a marker of the seizure onset zone. Neurology 77:524–531
2. Cendes F, Engel J (2011) Extending applications for high-frequency oscillations. Neurology 77:518–519
3. Engel JJ, Bragin A, Staba R, Mody I (2009) High-frequency oscillations: what is normal and what is not? Epilepsia 50:598–604
4. Firpi H, Smart O, Worreli G, Marsh E, Dlugos D, Litt B (2007) High-frequency oscillations detected in epileptic networks using swarmed neural-network features. Ann Biomed Eng 35(9):1573–1584
5. Jacobs J, LeVan P, Chander R, Hall J, Dubeau F, Gotman J (2008) Interictal high-frequency oscillations (80–500 Hz) are an indicator of seizure onset areas independent of spikes in the human epileptic brain. Epilepsia 49:1893–1907
6. Kobayashi K, Jacobs J, Gotman J (2009) Detection of changes of high-frequency activity by statistical time-frequency analysis in epileptic spikes. Clin Neurophysiol 120:1070–1077
7. Kobayashi K, Watanabe Y, Inoue T, Oka M, Yoshinaga H, Ohtsuka Y (2010) Scalp-recorded high-frequency oscillations in childhood sleep-induced electrical status epilepticus. Epilepsia 51:2190–2194
8. Le-Van-Quyen M (2007) bragin, A.: Analysis of dynamic brain oscillations: methodological advances. Trends Neurosci 30:365–373
9. Valderrama M, Quyen MV (2011) High-frequency oscillations and interictal spikes in partial epilepsy: joining the benefits. Clin Neurophysiol 122(1):3–4

10. Zijlmans M, Jacobs J, Kahn YU, Zelmann R, Dubeau F (2011) Ictal and interictal high frequency oscillations in patients with focal epilepsy. Clin Neurophysiol 122:664–671
11. Zijlmans M, Jacobs J, Kahn Y, Zelmann R, Dubeau F, Gotman J (2011) Ictal and interictal high frequency oscillations in patients with focal epilepsy. Clin Neurophysiol 122:664–671

Expert System of Ischemia Classification Based on Wavelet MLP

Javier F. Fornari and José I. Peláez

Abstract This paper proposes an expert system capable of identifying pathological ECG with signs of ischemia. The system design is based on the knowledge of a team of cardiologists who have been commissioned to identify ECG segments that contain information about the target disease, and subsequently validated the results of the system. The expert system comprises four modules, namely, a pre-processing module which is responsible for improving the SNR, a segmentation module, a DSP module which is responsible for applying the wavelet transform to improve the response of the last module, in charge of the classification. We used a database of about 800 ECG obtained in different clinical and extensively annotated by the team of cardiologists. The system achieves a sensitivity of 87.7 % and a specificity of 82.6 % with the set of ECG testing.

Keywords Classification · Expert system · ECG ischemia · MLP · DWT

1 Introduction

The Ischemic Heart Disease (IHD), also known as Coronary Artery Disease, is a condition that affects the supply of the blood to heart. The heart muscle depends on the coronary arteries for the supply of oxygen and nutrients, and only the inner layers (endocardium) profit from the oxygen rich blood that is being pumped.

J. F. Fornari (✉)
Rafaela Regional Faculty, Argentinian Technological University, 2300 Rafaela, Santa Fe, Argentina
e-mail: javier.fornari@frra.utn.edu.ar

J. I. Peláez
Department of Languages and Computer Sciences, Malaga University, Malaga, Spain
e-mail: jipelaez@uma.es

F. Sun et al. (eds.), *Knowledge Engineering and Management*, Advances in Intelligent Systems and Computing 214,
DOI: 10.1007/978-3-642-37832-4_39,

Treatment and complications of an inferior wall infarction is different than those of an anterior wall infarction. An inferior wall infarction may cause a decrease in heart rate because of involvement of the sinus node. Nevertheless, the anterior wall performs the main pump function, and alterations of the function of this wall will lead to decrease of blood pressure, increase of heart rate, shock, and on a longer-term heart failure. So, long-term effects of an anterior wall infarction are usually more severe than those of an inferior wall infarction.

The heart received oxygen and nutrients by the right and left coronary arteries. As shown in Fig. 1, the right coronary artery (RCA), which originates above the right cusp of the aortic valve, connects to the ramus descendens posterior (RDP). This artery supplies the right atrium, the inferior wall, the ventricular septum, and the posteromedial papillary muscle. This one usually also supplies both nodes (sinoatrial node and atrioventricular node).

The left coronary artery (LCA), which arises from the aorta above the left cusp of the aortic valve, divides itself in the left anterior descending artery (LAD) and the left circumflex artery (LCX, or ramus circumflexus—RCX). This artery supplies the left atrium, part of the right atrium, left third of the anterior wall of the right ventricle, left ventricle (except right half of lower side), and anterior two-thirds of the interventricular septum.

The early diagnosis of IHD is very important to minimize myocardial cell damage and initiate appropriate treatment, moreover, IHD is so dangerous than it is the most common cause of sudden death in different countries around the world. The occluded coronary can be identified with aid of ECG.

Electrocardiography expresses heart electrical activity, so diseases could be diagnosed by morphological study of recorded data. Cardiologist commonly used this technique since it consists of effective, non-invasive, and low-cost tool to the diagnosis of cardiovascular diseases. For this purpose, ECG record is made to examine and observe a patient. Early detection and treatment of the heart diseases can prevent permanent damages on tissues of the heart.

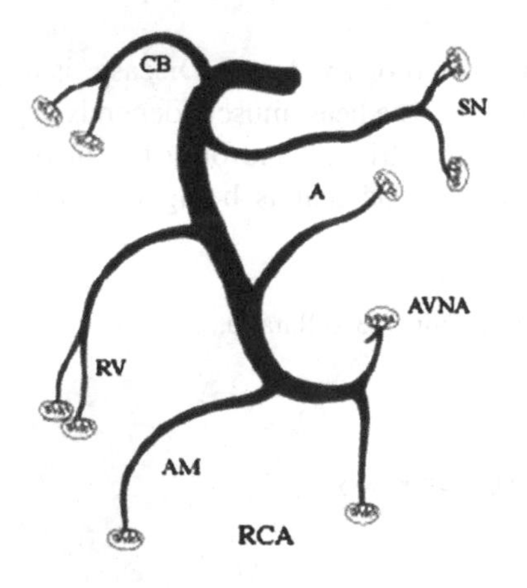

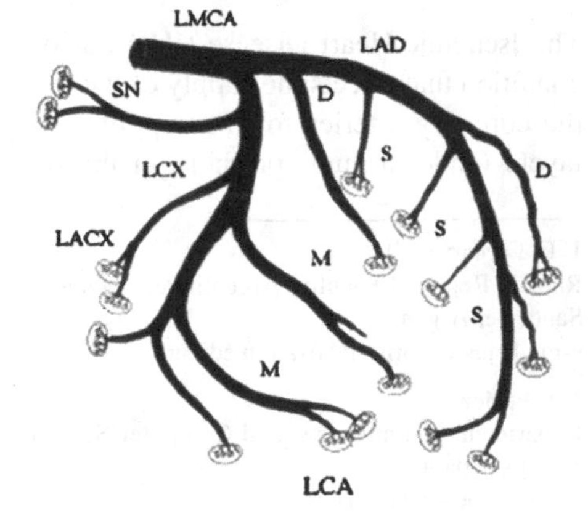

Fig. 1 Coronary arteries

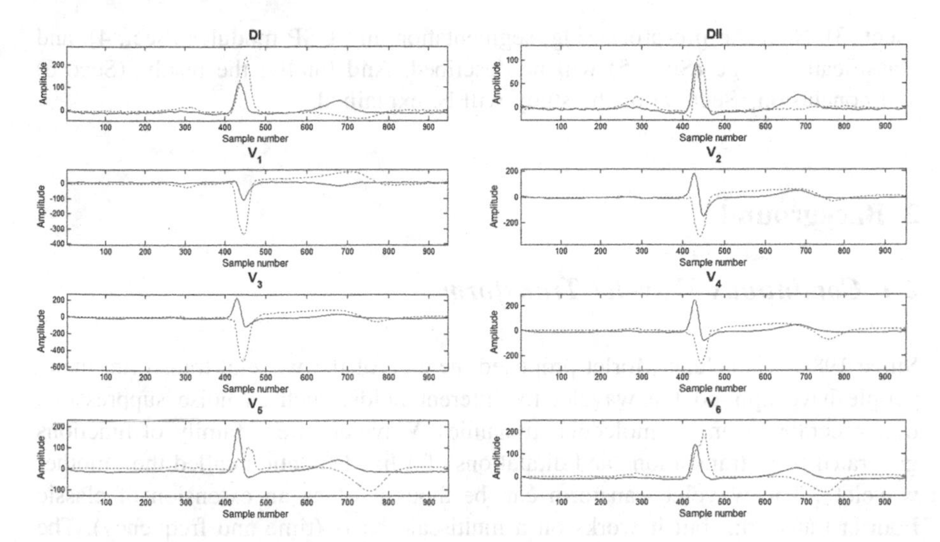

Fig. 2 Healthy patient (*solid line*) versus patient with extensive anterior ischemia (*dotted line*) [Gem-Med ECG DB]

In this study, the standard 12-lead ECG has been used in order to record the clinical information of each patient. The term 'lead' refers to the tracing of the voltage difference between two of the electrodes, namely, I, II, III, aVL, aVR, aVF, V1, V2, V3, V4, V5, and V6 Fig. 2. However, the first six leads are a linear combination, so only two of them are enough to gather all necessary information for analysis [1].

The most representative sign of myocardial ischemia is ST segment elevation or depression in two contiguous leads, in acute infarction, and pathological Q waves (high voltage), in chronic infarction [2]. There are many approaches for computer processing of ECG for diagnosing certain heart diseases. The two main used strategies are methods based in morphological analysis [3–5] and methods based in statistical models [6–8]. A third group, methods based in artificial neural networks (ANN) [9–11], is being developed lately focusing in ECG signal classification.

Several techniques have been applied to ECG for feature extraction, such as Discrete Cosine Transform (DCT) [12], Discrete Fourier Transform (DFT) [13], Continuous Wavelet Transform (CWT) [14], Wavelet Transform (DWT) [15], Principal Component Analysis (PCA) [16], and Multidimensional Component Analysis (MCA), among others.

In this study, we propose a two-stage system for ischemia detection. The first stage is responsible for signal enhancement and feature extraction using the signal averaging and wavelet transform. For the classification stage we opted for the neural network MLP.

This paper is divided into six sections, following this initial introductory section (Sect. 1), the basis of the employed signal processing will be commented (Sect. 2), and the most important details of the ECG database used will be discussed

(Sect. 3). Next the pre-processing, segmentation and DSP modules (Sect. 4), and classification stage (Sect. 5) will be described. And finally, the results (Sect. 6) and conclusion (Sect. 7) of the study will be explained.

2 Background

2.1 Continuous Wavelet Transform

Since 1982, when Jean Morlet proposed the idea of the wavelet transform, many people have applied the wavelet to different fields, such as noise suppression, image compression, or molecular dynamics. Wavelets are a family of functions generated from translations and dilatations of a fixed function called the "mother wavelet". The wavelet transform can be thought of as an extension of classic Fourier transform, but it works on a multiscale basis (time and frequency). The wavelet transform can be classified as continuous or discrete. Many researchers (Daubechies, Haar, Meyer, Mallat, etc.) enhanced and developed this signal-processing tool to make it more efficient [20].

The wavelet analysis has been introduced as a windowing technique with variable-sized regions. Wavelet transforms may be considered forms of time–frequency representation for continuous-time signals, and introduces the notion of scale as an alternative to frequency, mapping a signal into a time-scale plane. Mathematically, the continuous wavelet transform of a function is defined as the integral transform with a family of wavelet functions: in other words, the continuous wavelet transform (CWT) is defined as the sum of the signal multiplied by scaled and shifted versions of the wavelet function. A given signal of finite energy is projected on a continuous family of frequency bands of the form [f, $2f$] for all positive frequencies. This frequency bands are scaled versions of a subspace at scale 1. This subspace is in most situations generated by the shifts of the mother wavelet $\Psi(x)$. The projection of a function $f(t)$ onto the subspace of scale a has the form:

$$W_f(a,b) = \frac{1}{\sqrt{a}} + \int_{-\infty}^{\infty} f(t)\Psi\left(\frac{t-b}{a}\right)dt \quad (1)$$

For the continuous wavelet transform, the pair (a, b) varies over the full half-plane, while for the discrete wavelet transform this pair varies over a discrete subset of it.

2.2 Discrete Wavelet Transform

The discrete wavelet transform captures both frequency and time information as the continuous wavelet transform do, but using wavelets discretely sampled. The Fast Wavelet Transform (FWT) is an alternative to the conventional Fast Fourier Transform (FFT) due to it can be performed in O(n) operations and it captures as the notion of frequency content of the input (different scales—a) as the notion of temporal content (different positions—b).

The DWT can be implemented by a series of filters. First, the samples ($s[n]$) are passed through a low pass filter with impulse response $l[n]$ resulting in a convolution of the two, and simultaneously, the samples are passed through a high pass filter with impulse response $h[n]$ (Eq. 2).

$$\begin{aligned} Y_{\text{low}}[n] &= (s * l)[n] = \sum_{k=-\infty}^{\infty} s[k]l[[n-k] \\ Y_{\text{high}}[n] &= (s * h)[n] = \sum_{k=-\infty}^{\infty} s[k]h[[n-k] \end{aligned} \tag{2}$$

According to Nyquist's theorem, half of the frequencies of signal have now been removed, so half the samples can be discarded downsampling by 2 the outputs of the filters.

Calculating wavelet coefficients at every possible scale would waste computation time, and it generates a too large volume of data. However, only a subset of scales and positions are needed, scales and positions based on powers of two, so-called dyadic scales and positions. Then, the wavelet coefficients c_{jk} are given by (Eq. 3).

$$c_{jk} = W_f(2^{-j}, k2^{-j}) \tag{3}$$

where $a = 2^{-j}$, called the dyadic dilation, and $b = k2^{-j}$, is the dyadic position.

Such an analysis from the discrete wavelet transform (DWT) is obtained. DWT works like a band pass filter and DWT for a signal several levels can be calculated. Each level decomposes the input signal into approximations (low frequency part of initial signal) and details (high frequency part of initial signal). The next level of DWT is done upon approximations. y1 corresponds to the first level of DWT (approximations—low pass filter), y2 corresponds to the second level of DWT (approximations from details of first level), and so on. The penultimate level (ym) corresponds to the approximations of the last filter pair, and the last level (yn) corresponds to the details of the last filter pair [21].

3 Database

For this study we have used the ECG database of Gem-Med, S.L., which contains eight lead ECG signals of about 800 patients sampled with a frequency of 1 kHz. It is possible to recuperate all the 12 leads from the datasets recorded (I and II leads, and the six precordial leads V1 to V6) [1]. In order to suppress interferences (motion artefact, power line interference, etc.), a low pass filter with cutting frequency of 45 Hz has been used. The baseline wandering has been solved by using a high pass filter with cutting frequency of 0.5 Hz Among the ECG which composed the database, there are several diagnosis such as healthy patient, different types of bundle branch blocks (RBBB, LBBB, and hemiblocks), different types of ischemias (inferior, lateral, superior, anterior, septal, chronic, acute, etc.), and so on.

Four categories were used for diagnosis, namely, sure, discarding, probable, or negative. The team of cardiologists has marked as Sure those ECG showing unmistakable signs of the disease under study. In those cases in which there were not present all the signs, or any of them were not clear, ECG were marked as Probable. In cases in which only one sign was present, but cannot guarantee the diagnosis, was marked as Discarding. Finally, ECG which do not show any sign associated with the pathology under study were marked as Negative.

The original data is saved in SCP-ECG format, which stands for Standard Communications Protocol for Computer Assisted Electrocardiography. The ECG signals for training dataset contains eight lead of patients diagnosed with ischemia (Sure and Probable degrees) and those which were not diagnosed with this pathology. This dataset were divided into three groups of training, validation, and testing. Each ECG database record is composed with several beats and all the eight leads. After the pre-processing block, the mean beat has been calculated and the number of final variables has been reduced to about 20 samples per lead. Each subnet works with a different pre-processing, so each one manages a different number of samples as input. The input data storages all the required leads concatenated in a one-dimensional vector.

4 Signal Pre-Processing

The ECG signal pre-processing is performed in order to remove noise distortion (mean beat), to extract main features (wavelet transform) [22], and to data reduction (downsampling).

First, as the QRS complex is the most prominent wave in ECG, R wave is recognized in each beat with a peak-detection algorithm. A window of 950 samples centered in R wave is defined. Except in cases of severe bradycardia or tachycardia, complete heart cycle falls within that window and all the significant points are, approximately, at the same sample number. Locating R wave at the center of the exploration window is achieved that variations in heart rate increases or decreases number of isoelectric line samples at the edges of the exploration window.

As prognostic factors for ischemia are mainly located in the QRS complex and T wave [2], before the feature extraction stage proceeds to segment the ECG signal. As mentioned previously, except in cases of severe bradycardia or tachycardia, by placing the R wave at a specific position, the QRS complex and T wave maintain its lengths and can be extracted by windowing technique.

In this work, the Wavelet transform was selected in order to reduce the remaining number of samples. It is also essential to notice that determination of DWT level and the mother wavelet are very important in ECG feature extraction. Further we compared different six mother wavelets, namely, Haar, Daubechies 2, Daubechies 4, Coiflet 1, Symlet 2, Symlet 4, and a non-DWT process.

The Haar wavelet is the simplest possible wavelet, proposed in 1909 by Alfred Haar. Due to this wavelet is not continuous; it offers an advantage for detecting sudden transitions, such as the QRS complex in ECG.

The Daubechies wavelets named after her inventor Ingrid Daubechies, are a family of orthogonal wavelets characterized by a maximal number of vanishing moments. This type of wavelet is easy to put into practice using the fast wavelet transform but it is not possible to write down in closed form. Symlets are also known as the Daubechies least asymmetric wavelets and their construction is very similar to the Daubechies wavelets. Daubechies proposed modifications of her wavelets that increase their symmetry while retaining great simplicity. Coiflets are a family of orthogonal wavelets designed by Ingrid Daubechies to have better symmetry than the Daubechies wavelets.

Finally, the pre-processing section concludes reducing the number of variables remaining by the first level of wavelet transform. In the case of the subnet without integral transformation, simply the signal is downsampled.

Table 1 Results of number of neurons in hidden layer experiment

N°	Sens Max (%)	Sens Mean (%)	Sens Min (%)	Spec Max (%)	Spec Mean (%)	Spec Min (%)
14	94.23	89.56	82.69	97.18	94.71	90.40
15	91.35	87.64	83.65	96.05	94.39	90.96
16	93.27	89.70	86.54	96.05	93.99	90.40
17	95.19	89.29	81.73	98.31	94.71	91.53
18	94.23	89.84	85.58	98.31	94.27	89.27
19	93.16	88.87	83.76	98.12	94.23	90.40
20	92.94	89.24	84.23	97.87	94.12	91.53

5 MLP Implementation

In this study, we have designed two system focused in a different family of ischemia, the first family is composed by anteroposterior ischemia, and the second family is composed by inferolateral ischemia. For each system seven different MLP has been tested, one for each selected ECG signals pre-processing.

The basic MLP used in this study is formed by three layers: the input layer (with as many neurons as inputs have the net, generally about 20 neurons), one hidden layer (it has been estimated empirically the optimal number of neurons for each ANN—Table 1), and the output layer (which consists of a single neuron).

In Table 1 are shown the results obtained in the number of neurons in hidden layer experiment (only 14 to 20 neurons results are shown, but 1 to 50 neurons have been proved). The optimal number of neurons is not associated to the best value of sensitivity or specificity separately, but the compromise between the two factors. So, 17 neurons in the hidden layer were selected for our system.

Therefore, the training algorithm has to calculate the value of the weights which allow the network to correctly classify the ECG. The selected training algorithm was the scaled conjugate gradient back-propagation [23]. Since each ECG was previously diagnosed by a team of cardiologists, using a supervised method is a good choice.

6 Results and Discussion

6.1 Training Results

Training patterns were formed in mixed order from the ECG pre-processed. The size of the training patterns is 3 leads of about 20 samples for the first system (anteroposterior ischemia detector) and 2 leads of about 20 samples for the second system (inferolateral ischemia detector), i.e., 3 leads each including about 20

Table 2 Anteroposterior ischemia MLP quality parameters

Subnet	Training set			Test set		
	Sens (%)	Spec (%)	MC (%)	Sens (%)	Spec (%)	MC (%)
No WT	93.07	85.89	79.67	86.45	84.89	74.56
Haar	93.37	86.91	80.82	90.96	83.84	76.49
D2	94.28	82.82	77.65	90.66	81.19	74.04
D4	93.98	85.89	80.26	91.87	84.87	77.96
Coif1	86.45	91.00	79.92	81.93	81.80	69.52
Sym2	85.84	94.48	82.79	81.93	86.91	73.43
Sym4	85.24	95.09	82.95	82.23	84.87	72.00

Table 3 Inferolateral ischemia MLP quality parameters

Subnet	Training set			Test set		
	Sens (%)	Spec (%)	MC (%)	Sens (%)	Spec (%)	MC (%)
No WT	90.36	84.25	76.47	84.94	83.44	72.50
Haar	93.98	85.48	49.88	90.36	83.23	75.59
D2	94.88	82.21	77.47	91.57	78.73	72.52
D4	90.96	85.48	77.93	86.75	84.46	74.41
Coif1	84.94	89.57	77.61	79.82	79.96	67.07
Sym2	82.23	93.66	79.42	78.61	84.25	69.41
Sym4	83.73	93.46	80.30	80.42	83.64	70.01

samples, so, 60 samples are present. The combination of these training patterns was called as training set.

The optimum number of hidden nodes and learning rate were experimentally determined for each ANN structure. After the proposed structures were trained by the training set, they were tested for healthy ECG. For the stopping criterion of all the networks, maximum number of iterations was set to 10,000 and the desired error value (MSE) was set to 0.001. The test was implemented using ECG records taken from about 800 patients.

6.2 Test Results

The results of classification for the anteroposterior ischemia detector are resumed in Table 2, and those for the inferolateral ischemia detector are resumed in Table 3. Each row of the table represents the instances in a specific ANN.

The first group of columns represents the statistical measures of the performance of the ANN with the training set: Sensitivity (Sens), Specificity (Spec), and the Matthews correlation Coefficient (MC). The second group of columns represents the same quality parameters of the ANN with the test set. The statistical measures of the performance of the binary classification test used in this study are the following [23]:

Sens column represents sensitivity of the system, related to the system's ability to identify pathological patients. A high value means there are few pathological cases which are not detected by the system. This can also be written as Eq. 4.

$$\text{Sens} = \frac{TP}{TP + FN} \tag{4}$$

Spec column represents specificity of the system, related to the system's ability to identify healthy ECG. A high value means there are few healthy cases which are marked as pathological by the system. This can also be written as Eq. 5.

Table 4 Anteroposterior ischemia MLP accuracy

	No WT (%)	Haar (%)	D2 (%)	D4 (%)	Coif1 (%)	Sym2 (%)	Sym4 (%)
Sure	91.43	94.29	94.29	97.14	88.57	91.43	91.43
Probable	86.02	93.55	92.47	93.55	86.02	90.32	89.24
Discarding	83.33	88.24	90.19	81.86	73.53	71.08	74.51
Negative	83.89	83.84	81.19	84.87	81.80	86.91	84.87

Table 5 Inferolateral ischemia MLP accuracy

	No WT (%)	Haar (%)	D2 (%)	D4 (%)	Coif1 (%)	Sym2 (%)	Sym4 (%)
Sure	90.22	93.28	93.31	97.32	87.46	90.22	91.13
Probable	83.18	92.27	91.27	94.41	83.48	89.38	89.42
Discarding	81.45	86.12	89.26	82.86	72.45	72.32	74.40
Negative	83.84	83.23	78.73	84.46	79.96	84.25	83.64

$$\text{Spec} = \frac{TN}{TN + FN} \tag{5}$$

MC column represents the Matthews correlation Coefficient, which is used as a measure of the quality of binary classifications (two-class—pathological and non-pathological ECG). A high value means that both, sensitivity and specificity are high and the system is very reliable. This can also be written as Eq. 6.

$$MC = \frac{(TP * TN - FP * FN)}{\sqrt{(TP + FP) * (TP + FN) * (TN + FP) * (TN + FN)}} \tag{6}$$

Tables 4 and 5 show the results divided by degrees of certainty given by cardiologists. The first table shows the results of the detection system of antero-posterior ischemias, while the second shows the detection system inferolateral ischemia. As shown in the tables, the degree of success in the ECG in Sure category are clearly superior to those obtained by the other categories, with the Discarding category as the most complicated to classify.

7 Conclusion

An expert system consisting of four modules based on Wavelet MLP has been proposed in this paper. Following the instructions of the team of cardiologist the pathologic ECG have been grouped into two major families of ischemias, those occurring in the anteroposterior area of the heart and those presented in the inferolateral area of the heart. The first module is responsible for removing high frequency noise through a lowpass filter with cutoff frequency at 45 Hz and to correct the baseline wandering by a high pass filter with cutoff frequency of 0.5 Hz; segmentation module is responsible for extracting the segment of interest, namely, the QRS complex and the ST segment of the eight leads indicated by the cardiology team. The DSP module performs the wavelet transform of the selected segments, keeping the detail coefficients of the first level of decomposition. We tested several mother wavelet, namely, Coiflet 1, Daubechies 2 and 4, Haar and Simlet 2 and 4. Finally, for the classification module a MLP neural network was trained.

The optimum number of neurons in the hidden layer has been determined experimentally, concluding that 17 neurons is the best option. The best results among the networks of the anteroposterior ischemia detection belong to the DB4-MLP network, which has been able to differentiate healthy patients with a sensitivity of 91.87 % and a specificity of 84.87 %, with a associated Matthews coefficient of 77.96 %. The best results in the detection of inferolateral ischemia are provided by the Haar-MLP network, which has been able to differentiate healthy patients with a sensitivity of 90.36 %, a specificity of 83.23 % and a Matthews coefficient of 75.59 %.

Acknowledgments This investigation is allocated in the project TSI-020302-2010-136.

References

1. Klabunde RE (2011) Cardiovascular physiology concepts. Lippincott Williams & Wilkins, Philadelphia. ISBN 9781451113846
2. Bayès de Luna, Antoni. Bases de la electrocardiografía. De las variantes de la normalidad a los patrones diagnósticos (III): Isquemia, lesión y necrosis. Barcelona: Prous Science, 2007. B-1749-07
3. Taouli SA, Bereksi-Reguig F (2010) Noise and baseline wandering suppression of ECG signals by morphological filter. J Med Eng Technol 34:87–96
4. Christov I et al (2006) Comparative study of morphological and time-frequency ECG descriptors for heartbeat classification. Med Eng Phys 28:876–887
5. Sun Y, Chan KL, Krishnan SM (2002) ECG signal conditioning by morphological filtering. Comput Biol Med 32:465–479
6. Wiggins M et al (2008) Evolving a Bayesian classifier for ECG-based age classification in medical applications. Appl Soft Comput 8:599–608
7. Sharma LN, Dandapat S, Mahanta A (2010) ECG signal denoising using higher order statistics in wavelet subbands. Biomed Signal Process Control 5:214–222
8. Morise AP et al (1992) Comparison of logistic regression and Bayesian-based algorithms to estimate posttest probability in patients with suspected coronary artery disease undergoing exercise ECG. J Electrocardiol 25:89–99
9. Gholam H, Luo D, Reynolds, KJ (2006) The comparison of different feed forward neural network architectures for ECG signals diagnosis. Med Eng Phys 28:372–378
10. Sekkal M, Chick MA, Settouti N (2011) Evolving neural networks using a genetic algorithm for heartbeat classification. J Med Eng Technol 35:215–223
11. Moavenian M, Khorrami H (2010) A qualitative comparison of ANN and SVM in ECG arrhythmias classification. Expert Syst Appl 37:3088–3093
12. Ahmed SM et al (2009) ECG signal compression using combined modified discrete cosine and discrete wavelet transforms. J Med Eng Technol 33:1–8

13. Mitra S, Mitra M, Chaudhuri BB (2004) Generation of digital time database from paper ECG records and fourier transform-based analysis for disease identification. Comput Biol Med 34:551–560
14. Khorrami H, Moavenian M (2010) A comparative study of DWT, CWT, and DCT transformations in ECG arrhythmias classification. Expert Syst Appl 37:5151–5157
15. Ranjith P, Baby PC, Joseph P (2003) ECG analysis using wavelet transform: application to myocardial ischemia detection. ITBM RBM, 24:44–47
16. Ceylan R, Ozbay Y (2007) Comparison of FCM, PCA and WT techniques for classification ECG arrhythmias using artificial neural network. Expert Syst Appl 2007:286–295
17. Gaetano A et al (2009) A patient adaptable ECG beat classifier based on neural networks. Appl Math Comput 213:243–249
18. Ozbay Y, Tezel G (2010) A new method for classification for ECG arrhythmias using neural network with adaptative activation function. Digit Signal Process 20:1040–1049
19. Korurek M, Dogan B (2010) ECG beat classification using swarm optimization and radial basis function neural network. Expert Syst Appl 37:7563–7569
20. Daubechies I (1992) Ten lectures on wavelets. Society for Industrial and Applied Mathematics, Philadelphia. ISBN 0-89871-274-2
21. Vetterli M, Cormac H (1992) Wavelets and filter banks: theory and design. IEEE Trans Signal Process 40:2207–2232
22. Li C, Zheng C, Tai C (1995) Detection of ECG characteristic points using wavelet transform. IEEE Trans Biomed Eng 42:21–28
23. Baldi P et al (2000) Assessing the accuracy of prediction algorithms for classification: an overview. Bioinform Rev 16:412–424

Chinese Named Entity Recognition Using Improved Bi-gram Model Based on Dynamic Programming

Juan Le and ZhenDong Niu

Abstract This paper proposes a bi-gram model based on dynamic programming to Chinese person named entity recognition. By studying the previous work, we concluded that we can improve the precision of NER by improving the recall rate and narrowing the gap between the recall rate and the precision rate. The algorithm defines five recognition rules which ensure the names can be recognized and returned firstly to improve the recall rate. This paper's innovation is a filtering stage introduced to filter out the invalid names by combining the inverse-maximum-matching with bi-gram model. The bi-gram model takes four pairs of transition probability into consideration when segments the sentence which can effectively narrow the gap between precision rate and recall rate. We take the open test in different corpus and materials extracted from the Internet straightly, its precision rate achieves 83.53 %, recall rate achieves 91.43 % and its F-value achieves 87.3 %.

Keywords Named entity recognition · Chinese person named recognition · Bi-gram · Dynamic programming · Inverse-maximum-matching

1 Introduction

Named entity recognition (NER) is to recognize proper nouns in text and associating them with the predefined appropriate types including location, organization, person names, date, address, number, etc. NER is becoming important

J. Le (✉) · Z. Niu
College of Computer Science, Beijing Institute of Technology, Beijing, China
e-mail: 1659038116@qq.com

Z. Niu
e-mail: zniu@bit.edu.cn

F. Sun et al. (eds.), *Knowledge Engineering and Management*,
Advances in Intelligent Systems and Computing 214,
DOI: 10.1007/978-3-642-37832-4_40, © Springer-Verlag Berlin Heidelberg 2014

because of its significance in many areas referring to information extraction, machine learning, and question answering system. In order to mine semantic relations, NER is the first step:

- Nicolas Cage has played the role of gangster villain in *"The Green Hornet"*.
- The opera artist 梅兰芳 played the role of 杨贵妃 in 《贵妃醉酒》.
- Chairman of Microsoft Bill Gates praised Beijing 2008 Olympics.

Nicolas Cage, 梅兰芳, 杨贵妃 and Bill Gates are person names, *The Green Hornet* and 《贵妃醉酒》 are movie names and Microsoft is organization name. They are all defined as proper nouns and the task of NER is to recognize the names and classify them into proper types correctly.

More research institutes and researchers have proposed many methods to Chinese person name recognition which mainly including rule based [1–3], hidden Markov models (HMMs) [4–6], maximum entropy [7, 8], conditional Random Fields (CRFs) [9, 10], and hybrid method [11–13].

The rule-based method manually defined recognition rule according to the combination features of named entity. But rule based only is lacking of ability to tackle the problems of robustness and portability. The HMM-based tagger model appears in NER more and more which uses a segmentation method to a given sentence and segment it into a sequence of tagged tokens. Zhang et al. in [4] proposed a random role model to recognize Chinese NEs. They defined a set of roles about component tokens within a Chinese NE and the relevant contexts. Tokens after segmentation are tagged with different roles. Their experiments showed that the role-based model was effective but still some problems existing. In open test they got precision-rate 69.88 % and recall-rate 91.65 % separately. The big gap between them showed the algorithm returned too invalid person names. Fu et al. in [6] presented a lexicalized HMM-based approach to Chinese NEs. They unify unknown word identification and NER as a single tagging task on a sequence of known words. The experiment results of lexicalized HMMs using the PKU corpus are 91.72 % precision rate and 89.88 % recall rate. The obvious fact that precision rate is higher than the recall rate demonstrated a problem that is certain of amount valid person names were not recognized and returned correctly. The maximum entropy model extracted simple semantic features from a dictionary predefined to help improve the performance of the model NER. While people have made a big progress in previous work, it is still a big challenge to develop a high-performance NER system for Chinese person names because of the feature of Chinese.

The first difficulty is to determine the boundary of a person name in a sentence. e.g., in the sentence of " 《河间府》拿一撮毛侯七", "毛" and "候" are common surnames in a Chinese name, thus "毛候七" and "侯七" can be recognized as a lawful Chinese name. But in this sentence, "候七" should be correctly recognized rather than "毛候七" because "一撮毛" is a nickname of the name "候七".

The second difficulty is the ambiguity of named entities cased by segmentation. Take the sentence "南京市长江大桥" as an example: there are two segmentation results for the sentence namely '南京市长/江大桥' and '南京市/长江大桥'. In the first segmentation, if the phrase "南京市长" comprehended as mayor of Nanjing city, "江大桥" would be recognized as a person name. In the second, if "南京市" meaning a city of China, the phrase "长江大桥" would be recognized as a bridge name. Thus, the segmentation plays an important role to the final recognition result.

In this paper, we propose an improved bi-gram model-based dynamic programming and inverse-maximum-matching approach which consists of two stages separately solving the two difficulties: person name recognition stage which is to recognize the person name in the text by matching recognition rules and result filtering stage which is to filter out the invalid person names. The task of first stage is to segment a given sentence based on Chinese-name-terms-dictionary. The returned segmentation graph is to be matched against name-recognition-rules predefined according to the name features and context information. The matching will return a certain amount of results which meet up with the rules. But many of them are invalid person names which must be filtered out. The filtering stage loads all matching results to the duplicate of core dictionary prepared to the second segmentation and takes the second segmentation to the same sentence in the first stage using bi-gram model and inverse-maximum-matching duplicate of core and binary dictionaries. The bi-gram model improves segmentation accuracy by taking many considerations of transition probability between word–word, word-category, category-word and category–category.

This paper takes the open test in different corpus and materials extracted from the Internet straightly, its precision rate achieves 83.53 %, recall rate achieves 91.43 %, and its F-value achieves 87.3 %. In comparison to previous work, this paper solves two problems: (1) narrowing the big gap between recognition rate and recall rate and ensuring returning the valid person named entities; (2) ensuring the performance of recall rate, because the fact that recall rate lower than precision rate means the lawful person names cannot be recognized not to mentioning returned.

The rest of this paper is organized as follows: In Sect. 2, we discuss three key technologies: three dictionaries separately used for name recognition and segmentation; how to define Chinese person name recognition rules; and bi-gram model based on dynamic programming. In Sect. 3, we describe the implementation details of bi-gram model for Chinese NER. In Sect. 4, we give our experiment results. The conclusions work appears in Sect. 5.

2 Person Name Recognition Rules and BI-GRAM Model

2.1 Key Dictionaries

Chinese-person-name-terms-dictionary includes 7448 common Chinese characters, pre-context, and suffix-context. The characters are usually used to form a Chinese person name which including surname and other component. The pre context and suffix context usually appears before or after a Chinese name which can provide useful information to recognize the person name. e.g.:

Table 1 Chinese person names recognition rules

	Rules
1	preContext + surname + doubleNameFirst + doubleNameSecond e.g.:旦(preContext) + 梅(surname) + 兰(doubleNameFirst) + 芳(doubleNameSecond)
2	preContext + surname + singleName e.g.:院长(preContext) + 许(surName) + 翠(singleName)
3	preContext + surname + doubleNameFirst + doubleNameSecond + nextContext e.g.:院长(preContext) + 孙(surname) + 毓(doubleNameFirst) + 敏(doubleNameSecond) + 发言(nextContext)
4	surName + singleName + nextContext e.g.:许(surName) + 翠(singleName) + 教授(nextContext)
5	surName + doubleNameFirst + doubleNameSecond + nextContext e.g.:梅(surName) + 兰(doubleNameFirst) + 芳(doubleNameSecond) + 饰演(nextContext)

– 国家总理温家宝(Premier Wen J.B.), pre context: premier
– 胡锦涛主席(Hu J.T. President), suffix context: president

The dictionary divides all words into six categories including preContext, nextContext, surname, singleName, doubleNameFirst, and doubleNameSecond. The dictionary gives a term frequency to each word to demonstrate its probability as one of categories. The other two dictionaries are core and binary dictionaries which the filtering stage needs. The segmentation takes the inverse-maximum-matching on the duplicate of core dictionary and returns a sequence of tagged tokens. Bi-gram model gets the transition probability between tokens from the binary dictionary.

2.2 Defining Person Name Recognition Rules

This paper concludes four common Chinese person name form:

– One-character surname + one-character name, e.g. 许翠
– One-character surname + two-character name, e.g. 孙毓敏
– Two-character surname + one-character name, e.g. 令狐冲
– Two-character surname + two-character name, e.g. 上官婉容

By analyzing the above form and context information, we propose five rules to recognize Chinese person named entities in Table 1.

2.3 Bi-Gram Model Based on Dynamic Programming

The bi-gram model based on dynamic programming takes the input of segmentation as a character-string $C = C_1C_2\ldots\ldots C_n$ and the output as a word-string $S = W_1W_2\ldots\ldots W_m$ ($m <= n$). There is more than one possible S for a given character-string *C* and the task of dynamic programming aims to find the most appropriate segmentation *S*. We use *P(S)* to indicate the probability of *S* and the maximum *P(S)* is what we want. We divide dynamic programming into two subtasks:

(1) Find out all possible *S*s to the given *C*.
(2) Calculate the probability of all *S*s and pick out the maximum *P(S)*.

To the subtask one, we segment the input by inverse-maximum-matching duplicate of core dictionary to get all possible *S*s. To the subtask two, we use bi-gram model based on dynamic programming to calculate the probability of each possible *S* and get the maximum one. Each *S* is a sequence of tokens and bi-gram model get the transition probability between the two tokens from the binary dictionary. This paper takes four pairs of transition probability between the tokens into consideration which can effectively improve segmentation accuracy.

- word–word: indicate the relevancy between the two words. e.g. 'NBA@蓝球:10', the number '10' indicates the relevancy of two words which the larger number the more relevant.
- category–word: indicate the relevancy between a category and a word.
- wordcategory: indicate the relevancy between a and word a category.
- category–category: indicate the relevancy between two categories.

The equation of calculating the probability of *S* is:

$$P(S) = P(W_1 W_2 \ldots\ldots W_n) = P(W_1)P(W_2|W_1)\ldots\ldots P(W_i|W_{i-1}). \quad (1)$$

We take each token in the sequence as a node and transform (1) to (2):

$$P(S) = P(W_1 W_2 \ldots W_i) = P(\text{node}_i) = P(\text{StartNode}(W_k)) * P(W_k|\text{BestPrev}(\text{StartNode}(W_k))). \quad (2)$$

The term $P(\text{StartNode}(W_k))$ stands for the probability of the best precursor node of W_k and the term $P(W_k|\text{BestPrev}(\text{StartNode}(W_k)))$ stands for the conditional probability of W_k on condition that the occurrence of its best precursor node. This paper improves the calculation of term $P(W_k|\text{BestPrev}(\text{StartNode}(W_k)))$ namely

$$\begin{aligned} &p(w_k|Best\Pr ev(StartNode(w_k), w_k)) \\ &= \lambda_1 \times \frac{freq(w_k)}{TotalFreq} + \lambda_2 \times (\frac{freq(StartNode(w_k), w_k}{freq(StartNode(w_k))}) \\ &+ \frac{freq(p_k, w_k)}{TotalFreq} + \frac{freq(w_k, p_k)}{TotalFreq} + \frac{freq(p_k, p_k)}{TotalFreq} \end{aligned} \quad (3)$$

The illustration to each term of (3) is as Table 2.

Take the sentence "老生黄桂秋" as an example to illustrate dynamic programming. We can get three word strings after inverse-maximum-matching namely S1:老/生/黄桂秋/, S2:老生/黄桂秋/,and S3:老/生/黄/桂秋/. The calculation of P (S1), P (S2), and P (S3) is as follow:

Table 2 The illustration of term in (3)

Terms in equation	Demonstration
$\frac{freq(w_i)}{TotalFreq}$	term frequency of W_k
$\frac{freq(StartNode(w_k), w_k)}{freq(StartNode(w_k))}$	word–word
$\frac{freq(p_k, w_k)}{TotalFreq}$	category-word
$\frac{freq(w_k, p_k)}{TotalFreq}$	word-category
$\frac{freq(p_k, p_k)}{TotalFreq}$	category–category

P(S1) = P(老/生/黄桂秋/) = logP(老) × logP(生|老) × logP(黄桂秋|老生)
= (−9.3565) × (−10.717) × (−17.097) = −1714.3779
P(S2) = P(老生/黄桂秋/) = logP(老生) × logP(黄桂秋|老生)
=(−6.824) × (−17.097) = −107.4375
P(S3) = P(老/生/黄/桂秋/) = logP(老) × logP(生|老) × logP(黄|生) × logP(桂秋|黄)
= (−9.3565) × (−10.717) × (−11.6547) × (−26.277) = −30708.8485

We can find out P(S2) namely word-string '老生/黄桂秋' is the best segmentation to the given sentence '老生黄桂秋' because its probability is the least one.

3 Algorithm Implementation

3.1 Implementation Procedure

(1) Inputting the text.
(2) First segmentation: Segments the text based on Chinese-person-name-terms-dictionary and returns the segmentation graph in adjacency list.
(3) Rules matching: Matching the adjacency list to the person name recognition rules and loading the recognition results into the duplicate of core dictionary.
(4) Second segmentation: Segments the text secondly based on duplicate of core dictionary by inverse-maximum-matching using the bi-gram model which is based on dynamic programming.
(5) Part-of-Speech tagging: Using HMM model to tag the sequence of tokens
(6) Return the recognition result.

The above implementation procedure is as Fig. 1.

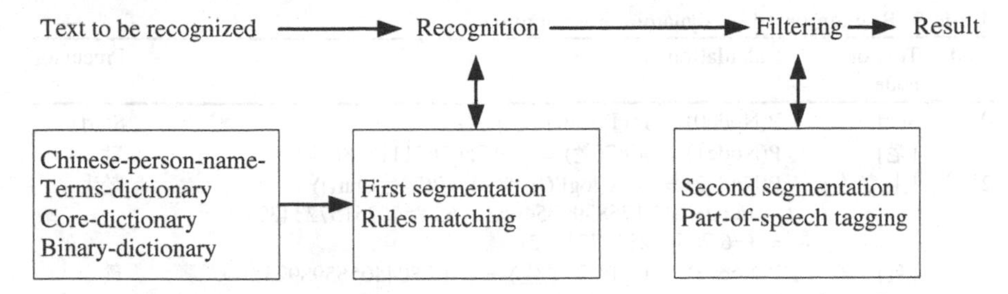

Fig. 1 Implementation procedure

3.2 Implementation Details

Take the sentence "老生黄桂秋" as an example to illustrate the details.

(a) Input the text.
(b) First segmentation (Fig. 2).
(c) Rules matching and loading the recognition results onto the duplicate of core dictionary (Fig. 3).
(d) Second segmentation (Table 3).
(e) Tag the Part-of-Speech and return the final results

0-2:老生:noun 2-5:黄桂秋:person-name

The final result is '黄桂秋' and the invalid names have been already filtered out. Figure 4 demonstrates the recognition results before and after filtering stage (Fig. 4).

node: 0:text:老 start:0 end:1--------data:surName:1 doubleName1:5 doubleName2:1 name:56 preContext:1 nextContext:4
text:老生 start:0 end:2--------data:preContext:5 nextContext:1 preContext:200
node: 1:text:生 start:1 end:2--------data:doubleName1:11 doubleName2:175 singleName:200
node: 2:text:黄 start:2 end:3--------data:surName:176 doubleName1:3
node:3: text:桂 start:3 end:4--------data:surName:7 doubleName1:31 doubleName2:3 singleName:4
node:4: text:秋 start:4 end:5--------data:surName:2 doubleName1:14 doubleName2:18 singleName:1

Fig. 2 Segmentation graphs in adjacency list

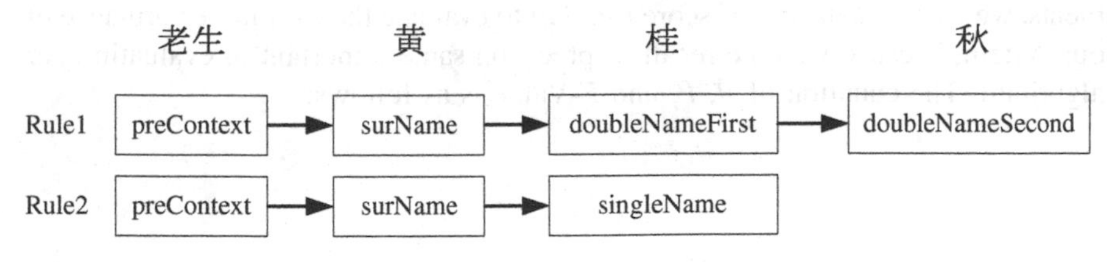

Fig. 3 Rules matching

Table 3 Bi-gram based on dynamic programming

<table>
<tr><th>Node</th><th>Text of node</th><th>Calculation</th><th>Precursor</th></tr>
<tr><td>0</td><td>Start</td><td>P(Node0) = logP(start) = 0</td><td>Start</td></tr>
<tr><td>1</td><td>{老}</td><td>P(Node1) = logP(老) = (−9.356562111818617)</td><td>老</td></tr>
<tr><td>2</td><td>{生,老生}</td><td>P(Node2) = max(logP(生|老),logP(老生|start)) = max(−10.71768390054663, −6.785602515723432) = (−6.785602515723432)</td><td>老生</td></tr>
<tr><td>3</td><td>{黄}</td><td>P(Node3) = logP(黄|老生) = (−17.80440565989232)</td><td>黄</td></tr>
<tr><td>4</td><td>{桂}</td><td>P(Node4) = logP(桂|黄) = (−28.478819058438223)</td><td>桂</td></tr>
<tr><td>5</td><td>{秋,桂秋,黄桂秋}</td><td>P(Node5) = max(logP(秋|桂), logP(桂秋|黄), logP(黄桂秋|老生)) = (−38.354373699017124,−26.174972876637284, −17.057998135067113) = (−17.057998135067113)</td><td>黄桂秋</td></tr>
</table>

Original Text：比如李多奎，先唱老生；夏月珊,先唱老生,后改丑角；赵如泉，先唱老生，后改武生，再改文武老生，兼演花脸、丑角；周信芳以文武老生为主,兼演过许多花脸、小生戏。在所演行当中兼演其他行当属常见，如京派演员黄桂秋以青衣出名，有时也演花旦；来自北方享名沪上的旦角李玉茹、童芷苓。

Recognition Result：[黄桂秋，李多，童芷，黄桂，言慧珠，夏月珊，老生为，周信，童芷苓，赵如泉，时也，李玉茹，武老生，周信芳，桂秋以，信芳以，李多奎，信芳，武生，许多，从北方，红生，老生，许多花，文武老，桂秋,文武]

Filtering Result：[李多奎，夏月珊，赵如泉，周信芳，黄桂秋，李玉茹，童芷苓]

Fig. 4 Recognition result and filtering result

4 Experiments

4.1 Experiment Measures

This section reports our experimental results. Researchers usually take close test and open test to evaluate NER system which close test uses subset of training corpus and open test uses the corpus different totally from training. We use open test to evaluate our system in terms of recall rate (*R*), precision rate (*P*), and F-measure (*F*). Here, recall (*R*) is defined as the number of correctly recognized NEs divided by the total number of NEs in the manually annotated corpus, and precision (*P*) is defined as the number of correctly recognized NEs divided by the total number of NEs recognized by the system. *F*-Value is a weighted harmonic mean of precision and recall where β is the weighing coefficient. In our experiments, we use the balanced *F*-score (viz.1F) to evaluate the overall performance of our system, because we take recall or precision same important in evaluating our algorithm. The equation of *R*, *P*, and *F*-Value is as follows:

Table 4 Demonstration of testing corpus

	Corpus	Size	Details
1	more person names	200 K	Raw materials form the Internet
2	less person names	200 K	Sogou corpus of text and news classification

Table 5 Experiment results of name entity recognition

Category	Corpus 1	Corpus 2
size (KB)	200	203
The total number of NEs	3375	991
The total number of NEs recognized	3522	1257
The number of correctly recognized NEs	3144	848
Precision-rate (%)	86.54	72.30
Recall-rate (%)	89.93	87.69

- Recall rate = the number of correctly recognized NEs/the total number of NEs in the corpus
- Precision rate = the number of correctly recognized NEs/the total number of NEs recognized
- $F_\beta = \frac{(\beta^2+1)\times P\times R}{\beta^2\times P+R}$

4.2 Experiment Data

We conducted a number of experiments to evaluate our algorithm using two kinds of corpus in Table 4 and the experiment results in Table 5.

We can concluded from Table 5 that thc recall rate and precision rate of corpus 1 are higher than corpus 2 which demonstrated that our algorithm is more efficient on recognizing the materials including more person names. The average recall rate and precision rate of the two corpuses are 83.53 and 91.43 % which is obvious higher than previous work. And our approach solves two main problems: (1) Narrow the big gap between recall rate and precision rate effectively to ensure the practicability of NER system. (2) Improved the precision rate not in sacrificing the recall rate. Low recall rate means the system improved the precision rate at the expense of losing a certain amount of valid person names. (3) The algorithm we proposed can recognize efficiently the materials including large amount of person names.

Table 6 Comparison after NER introduction

	Before NER applied	After NER applied
Segmentation accuracy (%)	95.21	99.1
Tagging accuracy (%)	94.22	98.2
Segmentation speed (words/second)	18.35	21.21

We take a test to our segmentation system after NER introduction using a text including 114869 characters. The segmentation needed 5414 s and got 9757 words. The segmentation accuracy is defined as the number of correctly segmented NEs divided by the total number of words in the text. The part-of-speech tagging accuracy is defined as the number of correctly tagged divided by the total number of words in the text. The segmentation speed is defined by the number of words segmented per second. Table 6 showed that we improved the segmentation accuracy and speed of our Chinese segmentation system after introducing the NER algorithm shown in Table 5.

5 Conclusions

In this paper, we have presented a two-stage NER system to Chinese NER system which consists of recognition and filtering. We defined person-name-recognition-rules which can decide the boundary of a Chinese person name in the context. And we proposed that using improved bi-gram model based on dynamic programming and inverse-maximum-matching duplicate of core dictionary to improve the NER recognition precision rate. In particular, our approach outperforms when used in recognizing the text including a large amount of person names. We have applied the approach in Beijing Opera Questioning and Answering System Based on Inference Rules and achieved good results. In the future work, we will mine the potential semantic relations among the named entities.

References

1. Kashif R (2010) Rule-based named entity recognition in Urdu. In: Proceedings of the 2010 named entities workshop, Curran Associates, Inc., Uppsala, Sweden, pp 126–135
2. Laura C, Rajasekar K, Yunyao Li Frederick R, Shivakumar V (2010) Domain adaptation of rule-based annotators for named-entity recognition tasks. In: Conference on empirical methods in natural language processing, Massachusetts, pp 1002–1012
3. Dilek K, Adnan Y (2009) Named entity recognition experiments on Turkish texts. In: 8th international conference flexible query answering systems, Springer, Denmark, pp 524–535
4. Zhang HuaPing, Liu Qun (2003) Chinese named entity recognition using role model. J Comput Linguist Chin Lang Process 8:29–60
5. GuoHong Fu, Jian Su (2002) Named entity recognition using an HMM-based tagger. In: 40th annual meeting of the Association for Computational Linguistics (ACL), Philadelphia, pp 473–480
6. GuoHong, Kang-Kwong Luke (2005) Chinese named entity recognition using lexicalized HMMs. J ACM SIGKDD Explor Newsl 7:19–25 (New York)
7. Hua Y, Tan Y, Hao W (2009) A method of Chinese named entity recognition based on maximum entropy model. In: ICMA international conference, IEEE Press, China, pp 2472–2477

8. Lufeng Z, Pascale F, Richard S, Marine C, Dekai W (2004) Using N-best lists for named entity recognition from Chinese speech. In: Proceedings of HLT-NAACL, short papers, IEEE Press, Boston, Massachusetts
9. FuChun Peng, FangFang Feng, McCallum A (2004) Chinese segmentation and new word detection using conditional random fields. In: *COLING*, Geneva, pp 562–568
10. HongPing Hu, HuiPing Zhou (2008) Chinese named entity recognition with CRFs. In: 2008 international conference on computational intelligence and security, NW Washington, pp 1–6
11. XiaoFeng Yu (2007) Chinese named entity recognition with cascaded hybrid model. In: Proceedings of the NAACL-Short 07' human language technologies 2007: the conference of the North American chapter of the association for computational linguistics, companion volume, short papers, New York, pp 197–200
12. Xiaoyan Zhang, TingWang, Tang J, Zhou H, HuoWang Chen (2005) Chinese named entity recognition with a hybrid-statistical model. In: Web technologies research and development —APWeb 2005, Lecture notes in computer science, Vol 3399/2005, pp 900–912
13. ZhuoYe Ding, DeGen Huang, Huiwei Zhou (2008) A hybrid model based on CRFs for Chinese named entity recognition. In: 2008 International conference on advanced language processing and web information technology, IEEE Press, China, pp 127–132

Improving Group Recommendations by Identifying Homogenous Subgroups

Maytiyanin Komkhao, Jie Lu, Zhong Li and Wolfgang A. Halang

Abstract Recommender systems have proven their effectiveness in supporting personalised purchasing decisions and e-service intelligence. In order to support members in user groups of recommender systems, recently designed group recommender systems search for data relevant to all group members and discover the agreements between members of online communities. This paper focuses on achieving common satisfaction for groups or communities by, e.g. finding a restaurant for a family or shoes for a group of cheerleaders. It establishes an algorithm, called I-GRS, to devise group recommender systems based on incremental model-based collaborative filtering and applying the Mahalanobis distance and fuzzy membership to create groups of users with similar interests. Finally, an algorithm and related design strategy to build group recommender systems is proposed. A set of experiments is set up to evaluate the performance of the I-GRS algorithm in group recommendations. The results show its effectiveness vis-à-vis the recommendations made by classical recommender systems to single or groups of individuals.

Keywords Group recommender systems · Subgroups · Rating confidence · Incremental group recommender systems · Model-based collaborative filtering

M. Komkhao (✉) · Z. Li · W. A. Halang
Chair of Computer Engineering, Fernuniversität in Hagen,
58084 Hagen, Germany
e-mail: maytiyanin.komkhao@fernuni-hagen.de

Z. Li
e-mail: zhong.li@fernuni-hagen.de

W. A. Halang
e-mail: wolfgang.halang@fernuni-hagen.de

J. Lu
Decision Systems and e-Service Intelligence Laboratory, School of Software,
Faculty of Engineering and IT, University of Technology Sydney, P.O. Box 123Broadway,
NSW 2007, Australia
e-mail: jie.lu@it.uts.edu.au

F. Sun et al. (eds.), *Knowledge Engineering and Management*,
Advances in Intelligent Systems and Computing 214,
DOI: 10.1007/978-3-642-37832-4_41,

1 Introduction

Recommender systems, as an effective personalisation approach, can suggest best suited items, such as products or services, to particular users based on their explicit and implicit preferences by applying information filtering technology [1]. The latter works by collecting user ratings for items in a given domain and computing the similarity between the profiles of several users in order to recommend items [1]. Recommender systems can be classified as making recommendations to individuals or to groups/communities. The former class of recommender systems makes recommendations of yet unrated items to an interested user based on his/her previous preferences [1]. The latter class, which is considered in this paper, makes recommendations to a group or cluster based on the common preferences of the members [2].

In the past decades, group recommender systems have been studied by several researchers [2–5]. PolyLens [5] reported that 77 % of their users were more satisfied by the results delivered by group recommender systems than those of individual recommender systems. So far, there are two methods to generate group recommendations, viz. recommending items to an interested group based on aggregated previous ratings by the group's members, and combining individual recommendations given to a group's members to produce a single one suiting all members [2, 3]. Chen et al. [2] indicate that either ratings from databases or explicit ratings given by the particular group members can be employed, and that a combination strategy is to be devised by domain experts. This study focuses on using ratings from databases. Amer-Yahia et al. [3] discussed how to make recommendations to a group of users who may not share similar tastes. They proposed to base group recommendations on consensus functions which are weighted combinations of group relevance and group disagreement.

In this paper, we propose an algorithm and a related design strategy to build a group recommender system based on incremental model-based collaborative filtering (I-GRS) algorithm. This is motivated by the approach's positive effect vis-à-vis the recommendations made by classical recommender systems to single or groups of individuals rendered by identifying the overlaps between the interests of groups members as expressed by similarity between them. In addition, the capability to handle users in online communities is increased enormously, which is very important for group recommender systems. The accuracy of individual recommendations to group members is considered as well.

Our framework employs the incremental model-based collaborative filtering algorithm InCF introduced in [6] and involving techniques from data mining and machine learning. To generate group recommendations, there are two main approaches, viz. just make an individual recommendation to each group member or to aggregate the individual recommendation values into a single one suitable for the whole group. For the former approach, first we shall be concerned with individual recommendations by creating models of users having the same interest based on the Mahalanobis distance instead of the Euclidean distance.

Furthermore, to handle confusion in decision making for overlapping groups, fuzzy sets are employed. The degrees of membership to them are expressed by the Mahalanobis radial basis function. Building genuine group recommender systems instead of somehow combining individual recommendations given to users is the objective of the second approach.

This paper is structured as follows. The pertaining group recommender systems in Sect. 2. The proposed algorithm is presented in detail in Sect. 3 including the previous work utilised. Employing the notion of a *rating confidence,* a strategy to build group recommender systems is designed in Sect. 4 as well as experimentally analysed and discussed there. A conclusion is given and directions for future work are indicated in Sect. 5.

2 Group Recommender Systems

Group recommender systems are being investigated for already two decades [2–5, 7]. There are various design strategies to build group recommender systems which we summarise briefly in the sequel.

In 2001, O'Connor et al. [5] proposed PolyLens: a collaborative filtering recommender system which recommends movies for groups of users. To make group recommendations, they use factors of social value such as aggregated group happiness and, with higher weights, opinions of expert members. As the easiest way to support group recommendations, some researchers create 'pseudo-users' to replace entire groups. One approach to create a pseudo-user is to manually determine an explicit consensus between the movie ratings given by the group members, which require obtaining additional information from them. The other one is automatic creation of pseudo-users by merging the rating profiles of the group members.

In designing PolyLens it was found out that automatic pseudo-user creation, which is based on nearest neighbour matching, has low efficiency, because not all group members can be satisfied. As a remedy, they developed a recommendation algorithm working with a social value function (group's happiness—the average value of the highest happiness score and the lowest one, called least misery principle). The results showed that 95 % of the users were satisfied with the group recommendations, and that 77 % were more satisfied with the group recommendations than with the individual ones.

Amer-Yahia [3] proposed consensus functions which combine two components, *relevance* and *disagreement.*

Group relevance is the relevance value (degree of ratings) of item i to a group G as calculated by one of two main score aggregation strategies: *average* and *least misery*. The former one aims to maximise the average of the group members' score for an item. The latter aims to maximise the lowest scores among all group members. Both aggregation strategies aim to minimise the number of group members disliking an item recommended.

Group disagreement is the disagreement within a group G over an item i as computed by two main methods: *average pair-wise disagreements and disagreement variance.* Average pair-wise disagreements are the differences between the relevance scores given by two group members. The variance of all group members' relevance scores is called disagreement variance. Both methods to compute a disagreement aim to propose an item with the lowest disagreement value for recommendation.

To generate individual recommendations, the collaborative filtering recommendation technique was utilised [3] with groups randomly composed from a pool of available users taking group size and group cohesiveness into account. Their results show that with their formalised notion of consensus functions, an increase of the effectiveness of group recommendation can be achieved.

3 A Group Recommender System Based on Incremental Model-based Collaborative filtering

The here proposed I-GRS algorithm is based on merging similar users taken from a large repository into clusters which form a behavioural model of the users. This model can then be employed to classify new (so-called interested) users, and to make individual recommendations to them, by measuring their similarity with the clusters. For model building, the algorithm InCF as introduced in [6] is applied.

In order to derive a group recommendation several approaches are possible. One is to obtain individual recommendations for each group member, form a common value in one way or the other, e.g. by averaging, and then use it as group recommendation. Or one may first determine, also in one way or the other, a centre point of the group, and then submit it to the model-based prediction process thus yielding a group recommendation.

Here, we propose to employ a different approach, which tries to capture certain structures existing in a given group. The basic idea is to recognise such a structure by forming relatively homogeneous subgroups. For the latter group recommendations will be derived, but also for the entire group in a further step.

Given a group of (interested) users, according to the above, we first employ algorithm InCF again, but extended by taking a so-called *rating confidence* into consideration, to construct a behavioural model of them. This model is called group model, and its clusters are called subgroups. Then, each subgroup's average point is compared with each cluster centroid of the repository model, and each subgroup is identified with its closest centroid's cluster.

Finally, in order to derive a group recommendation, i.e. an association to a cluster in the repository model common to all group members, a weighted average point is determined from all subgroups' average points and subjected to the prediction part of InCF.

3.1 Individual Recommendations with InCF

To create individual recommendations, the algorithm InCF introduced in [6] is employed, because by working incrementally it is well suited to cope with the tremendous growth of both items and interested users present in the web. Furthermore, it is able to generate overlapping clusters of users. A preprocessor removes redundant attributes in the user-item rating matrix and uses correlation-based feature subset selection (CFS) [8] to reduce the input datasets in size.

In InCF's learning phase, input vectors $\mathbf{L} = \{L_1, L_2,\ldots, L_l\}$ taken from a repository are considered one by one to create a behavioural model $\mathbf{W}$, to which new input vectors $\mathbf{U} = \{U_1, U_2,\ldots, U_h\}$ can be compared in the later prediction phase. The union of the sets of l attribute input vectors to the learning phase and of h attribute input vectors considered in the prediction phase is denoted by $\mathbf{P} = \{\mathbf{L}, \mathbf{U}\}$. The incremental algorithm carried out in the learning phase reads as follows [6].

Step 0: Apply CFS on the user-item rating.

Step 1: Let the first input vector $\mathbf{L}$ be the cluster forming the initial prototype of model $\mathbf{W}$. Set the initial variables as follows: a counter $C_{i,\,\text{initial}} = 1$, the distance threshold $(0 < d_{\text{th}} < 1)$ and the covariance matrix $(\mathbf{K}_{\text{initial}})$:

$$K_{\text{initial}} = \sigma \mathbf{I}_n \tag{1}$$

where $\mathbf{I}_n = \begin{bmatrix} 1 & 0 & \ldots & 0 \\ 0 & 1 & \ldots & 0 \\ \vdots & \vdots & \ddots & \vdots \\ 0 & 0 & \ldots & 1 \end{bmatrix}$, and

Step 2: Read the next $\mathbf{L} = \{\mathbf{L}_2,\ldots,\mathbf{L}_l\}$

Step 3: Calculate Mahalanobis distance $(\mathbf{dist}_M)$ values between $\mathbf{L}$ and the clusters in $\mathbf{W}$ using the non-singular covariance matrix.

$$\mathbf{dist}_M = \left[(\mathbf{P_L} - \mathbf{W})^{\mathrm{T}} \mathbf{K}^{-1} (\mathbf{P_L} - \mathbf{W}) \right]^{1/2} \tag{2}$$

Here, $\sigma \mathbf{I}$ is added to $\mathbf{K}$ in order to prevent the problem that the matrix becomes singular:

$$\mathbf{K}_{\text{new}} = \mathbf{K}_{\text{old}} + \sigma \mathbf{I}_n \tag{3}$$

Step 4: Calculate the membership value (find soft decision) for each cluster using the Mahalanobis radial basis function:

$$\mathbf{mem}(\mathbf{P}, \mathbf{W}) = \mathbf{exp}\left(-\frac{(\mathbf{P_L} - \mathbf{W})^{\mathrm{T}} \mathbf{K}^{-1} (\mathbf{P_L} - \mathbf{W})}{2} \right) \tag{4}$$

Step 5: Select the winner cluster as the one for which the maximum membership value identified by the fuzzy OR operator (i.e. max operator) is assumed:

$$\text{winner} = \arg\max_i(\mathbf{mem}_1 \vee \mathbf{mem}_2 \vee \ldots \vee \mathbf{mem}_n) \tag{5}$$

Step 6: Update the clusters of the model's prototypes, in particular (6) to increment the prototype count (C) as well as (7) to update the winner cluster ($\mathbf{w}$) and (8) the covariance matrix ($\mathbf{K}$) in case the input vector considered in the learning step does not give rise to the creation of a new cluster by applying the *learning rules* [9, 10]:

$$C_{i,\text{new}} = C_{i,\text{old}} + 1 \tag{6}$$

$$\mathbf{w}_{i,\text{new}} = (1 - \beta)\mathbf{w}_{i,\text{old}} + \beta\mathbf{P} \tag{7}$$

$$\mathbf{K}_{i,\text{new}} = (1 - \boldsymbol{\beta})\mathbf{K}_{i,\text{old}} + \boldsymbol{\beta}\left(1 - \beta^2\right)\left(\mathbf{P} - \mathbf{w}_{i,\text{new}}\right)\left(\mathbf{P} - \mathbf{w}_{i,\text{new}}\right)^{\mathrm{T}} \tag{8}$$

where the subscript 'old'/'new' refers to quantities before/after updating, C_i represents counting the number of attribute vectors at the ith cluster, $\mathbf{w}_i$ the vector being the centre of the ith cluster $\beta = 1/C_{i,\text{new}}$ and $\mathbf{K}_i$, represents the covariance matrix at the ith cluster.

Step 6a: If the distance between input vector and winner cluster does not exceed a given threshold, then update the cluster's variables, otherwise create a new cluster.

Step 7: If there is a further input vector in $\mathbf{L}$, go to step 2.

As result of running this algorithm, the prototype created last constitutes the model of the behaviour of the users characterised in the repository.

The part of InCF run in the prediction phase reads as follows [6].

Step 0: Read an interested user's data ($\mathbf{U}$).

Step 1: Calculate the distance between $\mathbf{U}$ and the model ($\mathbf{W}$) generated in the learning phase by using the Mahalanobis distance ($\mathbf{dist}_M$) as in (2), use $\mathbf{U}$ instead of $\mathbf{L}$.

Step 2: Calculate the value of the membership of $\mathbf{U}$ in each cluster using the Mahalanobis radial basis function:

$$\mathbf{mem}(\mathbf{P}, \mathbf{W}) = \mathbf{exp}\left(-\frac{(\mathbf{P}_{\mathrm{U}} - \mathbf{W})^{\mathbf{T}}\mathbf{K}^{-1}(\mathbf{P}_{\mathrm{U}} - \mathbf{W})}{2}\right) \tag{9}$$

Step 3: Find the winner cluster by identifying the highest degree of membership:

$$\text{winner} = \arg\ \max_i(\mathbf{mem}_1 \vee \mathbf{mem}_2 \vee \ldots \vee \mathbf{mem}_n) \tag{10}$$

Step 4: Associate $\mathbf{U}$ with the winner cluster and assign corresponding ratings and characteristics.

Step 5: If there is another interested user, go to step 0.

3.2 A Group Recommender System

To create group recommendations with the approach outlined above in this section, we first construct a group model and its subgroups of group members by utilising the learning phase of a slightly modified—in *Step* 6—algorithm InCF to be the group learning phase. To describe this modification, first the following definitions are required.

Definition 1 (*Rating Confidence* (*RC*)) With the factor *RC,* a group member's confidence in his/her item ratings is expressed. Consider group member *i*, then his/her rating confidence with respect to a set of items is defined as

$$RC_i = (\text{highest item rating given by } i)\left(\frac{\text{number of items rated by } i}{\text{total number of items}}\right) \quad (11)$$

The higher this factor is for a group member, the more items he/she rated in comparison with other group members.

Example 1 Table 1 presents the confidence values of four group members who rated five items.

Definition 2 (*Confidence Rating Threshold*) As the threshold mentioned in *Step 6a* of InCF, a value *0 < Confidence Rating Threshold < highest possible item rating* is used.

With this, the modified *Step 6a* reads as follows: If the distance between input vector and winner subgroup does not exceed a given threshold *and* if the *rating confidence* of the input vector is less than or equal to the *confidence rating threshold,* then update the subgroup's variables, otherwise create a new subgroup. The following example shows how the group model, its subgroups and their average points are constructed as results of the group learning phase.

Example 2 Incorporating the confidence rating values the group learning phase for the members mentioned in Table 1 yields the three subgroups $\{m_1\}$, $\{m_3\}$ and $\{m_2, m_4\}$ with the following average points: $g_1 = [0\ 4\ 3\ 0\ 0]$, $g_2 = [0\ 4\ 3\ 1\ 0]$ and $g_3 = [5\ 4\ 2.5\ 3.5\ 0]$.

Table 1 An artificial group member-item rating matrix

Group member	Item 1	Item 2	Item 3	Item 4	Item 5	*Rating confidence* (*RC*)
m_1	0	4	3	0	0	4(2/5) = 1.6
m_2	5	4	3	4	0	5(4/5) = 4
m_3	0	4	3	1	0	4(3/5) = 1.8
m_4	5	4	2	3	0	5(4/5) = 4

Each subgroup's average point may now be subjected to individual recommendation based on the behavioural model giving rise to a recommendation which is extended to the entire subgroup. From the individual group member's point of view, such a subgroup recommendation will usually fit better and be more specific than a recommendation to the entire group, since subgroups are more homogeneous. Many ways are imaginable to derive a group recommendation based on the ones for the subgroups. The way we propose here is to determine from the subgroups' average point a point to represent the whole group by weighted averaging.

Definition 3 (*Group Representation Point* φ) The point φ representing an entire group model is determined as

$$\varphi = \sum\nolimits_{\text{subgroups}} \text{weight} \times \text{average point} \tag{12}$$

with the weights summing up to 1.

Example 3 With the data of Example 2 we obtain $\varphi = [0\ 4\ 2.83\ 0\ 0]$. This representative point is compared to the behavioural model's clusters in the prediction phase to find the winner cluster, which we assume here to be $\mathbf{w} = [2.5\ 4\ 3\ 2\ 0.5]$. The ratings drawn from $\mathbf{w}$ are then applied to all items associated with the group members as group recommendation, which particularly holds for the ratings still missing in φ, and marked by 0, leading to the rating vector [2.5 4 2.83 2 0.5]. Finally, the item ranked highest ranking by the prediction is recommended to all group members, which is in this example the first one with the prediction value 2.5.

4 Experimental Analysis

In this section, we evaluate the proposed I-GRS algorithm for group recommender systems using a benchmark dataset. We investigate the effect the algorithm's use has on group recommendations in comparison to the performance of InCF in individual recommendations by measuring the average Normalised Mean Absolute Error (*NMAE*).

4.1 Experimental Set up

As a benchmark, the dataset 'MovieLens' is used containing real movie ratings gathered by the GroupLens research laboratory at the University of Minnesota, and available on-line at http://www.grouplens.org/node/73. It contains 100,000 ratings given by 943 users for 1,682 movies. Correlation-based feature subset selection is used to reduce the datasets in size and to remove irrelevant attributes. Since there is no standard to split the dataset into a testing and a training one, we apply five-fold cross-validation to divide the data randomly into five datasets. In addition, this prevents any bias caused by the particular sampling chosen for training and testing datasets.

4.2 Evaluation Matrices

In order to obtain evaluation results, we calculate predictive accuracy matrices using *NMAE*, which is determined from the Mean Absolute Error (*MAE*), i.e. the average of the absolute differences between actual and predicted ratings. Both *MAE* and *NMAE* are calculated as follows [11]: $MAE = \frac{\sum_{\{i,j\}} |p_{i,j} - r_{i,j}|}{n}$ where n is the total number of item ratings given by all members (which may be lower than the product of the number of items and the one of members), r_{ij} is the rating given by member i on item j and p_{ij} is the corresponding predicted rating. It can easily be seen that the lower *MAE* is, the better is the prediction [11]. As different recommender systems may use different numerical rating scales, a *Normalised Mean Absolute Error* is used to express errors as full-scale fractions [12]: $NMAE = \frac{MAE}{r_{\max} - r_{\min}}$ where $r_{\max}$ is the highest and is the lowest rating possible.

4.3 Effect of the I-GRS Algorithm Recommender Systems

A set of experiments is set up to compare, by average *NMAE*, the performance of the I-GRS algorithm on group recommendations vis-à-vis individual recommendations.

The results show that the proposed I-GRS algorithm applied in a group recommendation yields lower average *NMAE* values of recommendation quality than the individual recommender system considered. The average *NMAE* value for group recommendation gradually stabilises around 0.1433. On the other hand, for individual recommendation, this value gradually stabilises around 0.4154.

5 Conclusion and Future Work

We proposed an algorithm and a related strategy to design group recommendations in group recommender systems. This work does not only concern individual preferences of group members, but also their common satisfaction as a group. Beyond general ideas for group recommender systems found in the literature, we also take into account similarity in the behavioural classifications of group members by employing our previously devised InCF algorithm, which is effective by using ellipsoidal shapes to identify the boundaries of clusters and fuzzy sets.

The I-GRS algorithm's performance was shown to be higher when applied in group recommendations than in pure individual recommendations.

The objective of future work could be to increase the effectiveness of group recommender systems, e.g. by considering the interactions between group members, which could be represented by links. The effect, which the selection of weights in the determination of both the subgroups' representation points and the groups' representation points has on the overall quality of prediction, should be studied. In the process of forming subgroups, single members may be removed as outliers to yield more homogeneous groups and, thus, give rise to higher satisfaction of the members remaining.

References

1. Lu J, Shambour Q, Xu Y, Lin Q, Zhang G (2010) 'BizSeeker': A hybrid semantic recommendation system for personalized government-to-business e-services. Internet Res 20(3):342–365
2. Chen YL, Cheng LC, Chuang CN (2008) A group recommendation system with consideration of interactions among group members. Expert Syst Appl 34(3):2082–2090
3. Amer-Yahia S, Roy SB, Chawla A, Das G, Yu C (2009) Group recommendation: semantics and efficiency. In: Proceedings of the VLDB endowment, vol 2(1), pp 754–765
4. McCarthy JF., Anagnost TD (1998) MusicFX: An arbiter of group preferences for computer supported collaborative workouts. In: Proceedings of the 1998 ACM conference on computer supported cooperative work. ACM, New York, pp 363–372
5. O'Conner M, Cosley D, Konstan JA, Riedl J (2001) PolyLens: a recommender system for groups of users. In: Proceedings of the seventh conference on European conference on computer supported cooperative work. Kluwer Academic Publishers, Norwell, MA, pp 199–218
6. Komkhao M, Lu J, Li Z, Halang WA (2013) Incremental collaborative filtering based on mahalanobis distance and fuzzy membership for recommender systems. Int J Gen Syst 42(1):41–46
7. Gartrell M, Xing X, Lv Q, Beach A, Han R, Mishra S, Seada K (2010) Enhancing group recommendation by incorporating social relationship interaction. In: Proceedings of the 16th ACM international conference on supporting group work. ACM, New York, pp 97–106
8. Hall MA (2000) Correlation-based feature selection for discrete and numeric class machine learning. In: Proceedings of the 17th international conference on machine learning. Morgan Kaufmann Publishers Inc., San Francisco, CA, pp 359–366
9. Xu H, Vuskovic M (2004) Mahalanobis distance-based ARTMAP network. In: Proceedings of the IEEE international joint conference on neural networks. IEEE Press, Piscataway, NJ, vol 3, pp 2353–2359
10. Yen GG, Meesad P (2001) An effective neuro-fuzzy paradigm for machinery condition health monitoring. IEEE Trans Syst Man Cybern B Cybern 31(4):523–536
11. Su X, Khoshgoftaar TM (2009) A survey of collaborative filtering techniques. Adv Artif Intell 2009(4):1–20
12. Sarwar BM, Karypis G, Konstan JA, Riedl J (2002) Recommender systems for large-scale e-commerce: scalable neighbourhood formation using clustering. In: Proceedings of the 5th international conference on computer and information technology. San Jose State University, San Jose, CA, pp 158–167

Knowledge-Based Decision System for Sizing Grid-Connected Photovoltaic Arrays

Monir Mikati and Matilde Santos

Abstract In this paper we have developed a methodology and a numerical model for sizing photovoltaic (PV) arrays for grid-connected power consumers. The decision variables proposed are the PV contribution to the power demand and a transfer factor. The latter is a new index that we have defined, and it constitutes key knowledge for decision making when aiming to reduce power transfers between a local power system and the electric grid. The developed model generates days of solar irradiance and calculates PV power output for commercial PV modules. Sizing diagrams are presented that illustrate how the PV capacity, for a specific location and a local power demand, affects the PV contribution to the demand and the power trades with grid. For the particular simulation conditions studied, the sizing diagrams show PV contribution saturation levels and global transfer factor minimums.

Keywords Renewable energy · Photovoltaic solar power · Sizing · Optimization

1 Introduction

A frequent problem in the design of photovoltaic (PV) supply systems is that of sizing. Complexity is introduced by the intermediate and uncontrollable nature of the solar resource as well as by the particular characteristics of every power demand.

M. Mikati (✉)
ÅF Technology, Lindholmspiren 9 41756 Gothenburg, Sweden
e-mail: monir.mikati@afconsult.com

M. Santos
Department of Computer Architecture and Automatic Control,
Universidad Complutense de Madrid, 28040 Madrid, Spain
e-mail: msantos@dacya.ucm.es

F. Sun et al. (eds.), *Knowledge Engineering and Management*,
Advances in Intelligent Systems and Computing 214,
DOI: 10.1007/978-3-642-37832-4_42,

There are numerous simulation tools and methods for sizing grid-connected PV arrays. HOMER, a powerful simulation software has been especially developed for designing and analyzing hybrid power systems but may also be used for sizing single source power systems [1]. In [2], a review is given on the state of PV system sizing based on the application of artificial intelligence (AI) techniques. In [3] and [4], a particle swarm optimization (PSO) approach is applied in order to find optimized PV system dimensions. Papers [5–7] also present developed techniques for optimizing PV system characteristics, and include the correlation between the PV array and the inverter.

In the majority of the above-mentioned methods, the PV system size is optimized based on cost performance. More so, PV module characteristics are not modeled in detail and PV power generation is calculated on an hourly time basis.

In this study, we have implemented a knowledge-based approach. It is founded on a model of a renewable energy system where the solar resource and the PV modules have been modeled in detail [8]. We also put focus on the use of small time step calculations and on the site specific interaction between the solar resource and the power demand. The model may be used as a tool for decision making when analyzing how the size of the PV array affects the PV contribution to the power demand and the power trades with the electric grid.

In this way, we show how the representation of a renewable energy application can assist the decision making regarding its configuration, using modeling and generating knowledge by simulation about its behavior.

The structure of the paper is as follows. Section 2 presents the methodology based on several models of the whole system. It also includes explanations of the variables used for decision making regarding the size of the PV array. Section 3 describes the scenario simulated in order to obtain results that make validation of the method possible. In Sect. 4, the results are analyzed and discussed. The paper ends with the conclusions.

2 Methodology

The studied system contains the following main elements:

- The solar energy resource
- The PV arrays
- The electric power demand
- The electric grid connection.

The first two elements have been modeled in detail, as explained in Sect. 2.1 below. The power demand has simply been treated as a power requirement to be met by primarily the PV arrays and secondarily the electric grid. The presence of the grid connection has been modeled by registering the power transfers between the grid and the local power system (PV arrays and power demand).

Hence, the power distribution in the system is calculated in the model depending on the current relationship between the PV power generation and the power demand.

A complete simulation tool was developed in Matlab/Simulink. All simulations were carried out with a sampling period of 10 min.

Inputs for the simulation are (see Sect. 2.1):

- The characteristics of a specific location: latitude, altitude, and climate type.
- The PV module characteristics and the orientation of the PV array.
- The power demand.

For this study, the simulation outputs of interest are:

- The PV array power output.
- The PV contribution to the power demand.
- The transfer factor (see Sect. 2.2).

2.1 Modeling Daily Solar Irradiance and PV Power Output

The solar resource model calculates the daily solar irradiance (W/m^2 (squared meters)) for specific locations on clear days. Hence, the model does not account for cloud covered days, or days characterized by strong irradiance variations on the ground due to passing clouds. For a comprehensive description of the clear sky radiation model see [8–10]. A brief conceptual explanation is given below.

Solar irradiance on the ground can be modeled as the sum of two contributions: the direct and the diffuse radiation [9, 10]. The direct radiation represents the terrestrial solar radiation received by a plane without having been scattered by the atmosphere. The diffuse radiation represents the scattered solar radiation and is here modeled as isotropic.

Summing up the direct and the diffuse irradiance, daily irradiance curves are calculated for clear atmospheric conditions on a horizontal plane based on the zenith angle, the altitude of the location and correction factors which account for the type of climate. Four different climate types are used: tropical, midlatitude summer, subarctic summer, and midlatitude winter.

Having estimated the clear sky irradiance on the ground, it is necessary to calculate the irradiance incident on a tilted plane. A majority of all solar power installations are tilted (southwards on the northern hemisphere) to increase the collection of solar energy. During the summer, it is convenient to tilt the surface a little more toward the horizontal. During the winter, the surface should be tilted more to the vertical. Terrestrial irradiance on a titled plane is calculated at any instant by using the latitude, the orientation of the collector plane, the time of year, and the hour.

The model of the PV module was based on the work done by [11] and [12]. Mathematical expressions of a PV cell (or solar cell) are formulated based on a simplified equivalent circuit. The equivalent circuit consists of a photo current, a

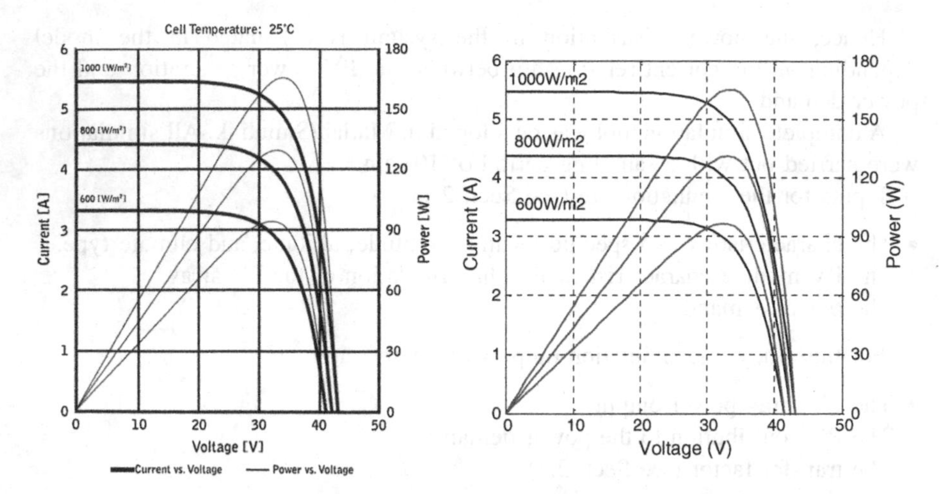

Fig. 1 Comparison between the current–voltage and power–voltage relationship specified by the PV module manufacturer [13] (*left*) and generated with the PV output model used in this work (*right*). All relationships are given for one PV cell at a fixed cell temperature of 25 °C

diode, a parallel resistor expressing a leakage current, and a series resistor describing an internal resistance to the current flow.

A PV module may be put together as the model allows for a number of cells in parallel and series to be defined. Also, parameters commonly known such as the open circuit voltage, the short circuit current, and the ideality factor are introduced.

In this way, commercial PV modules can be modeled using the manufacturer's data sheet. More so, the model may be validated since comparisons are easily made between the current–voltage and power–voltage relationships generated with the model and given by the manufacturer. Such a validation is shown in Fig. 1 for the particular PV module used in this study (see Sect. 3).

Thus, connecting the solar irradiance model and the model of the PV module it becomes possible to estimate the daily PV power output, on clear days, at any given location and any PV module orientation. PV module characteristics are defined using accessible information from the manufacturer's data sheets for specific commercial PV modules.

2.2 Decision Variables for Optimizing PV Capacity

It is difficult for any solar power system to guarantee a complete power supply at every instant. An auxiliary source is commonly required and in this case it is represented by the electric grid.

In this work, it is assumed that power can be traded with the grid, so that the PV array may deliver excess power to the grid. The PV capacity is analyzed in relation to the following two variables, which are explained below:

- The PV contribution to the power demand
- The transfer factor.

When the aim of the PV array is to supply power to a local demand, it is important to track how the installed PV power actually is contributing to the power demand, and how changes in PV capacity will affect the PV contribution.

The PV contribution to the power demand is calculated by firstly integrating the contributing PV power generation at every instant, and secondly dividing this value by the total energy demand for the time period between t_1 and t_2, that is (Eq. 1):

$$PV\ contribution = \frac{\int_{t_1}^{t_2} \left(contributing\ PV\ power\right)\, dt}{total\ energy\ demand} \tag{1}$$

Furthermore, a transfer factor is defined as (Eq. 2):

$$Transfer\ factor = \frac{total\ energy\ exchange\ with\ grid}{total\ energy\ demand} \tag{2}$$

where the numerator is the sum of the energy delivered and received from the electric grid during the analyzed time period.

The transfer factor can be seen as an indicator of the use of the electric grid. When the PV array delivers excess power to the grid, transformation and transmission losses reduce the efficiency of the supply chain. The power supplied to the local power demand by the grid also sums up to important losses on a global scale [14]. More so, trading power with the grid may be financially disadvantageous for the owner of the PV array. Hence, a low transfer factor is desirable from a PV system owner's point of view and to increase overall energy efficiency in the distribution system.

A transfer factor equal to 100 % implies that the local power system (power demand and PV array) is exchanging an amount of energy, with the grid, equal to the energy demand. There is, however, no upper limit. A transfer factor equal to zero suggests that the entire power demand is supplied by the PV array. In this case, no excess PV power is generated and no auxiliary power is required from the grid to supply the demand.

3 Simulation Conditions

Simulations were carried out for one characteristic day of every month. The characteristic day can be seen as representative for the respective month in terms of duration and intensity of the incoming solar irradiance.

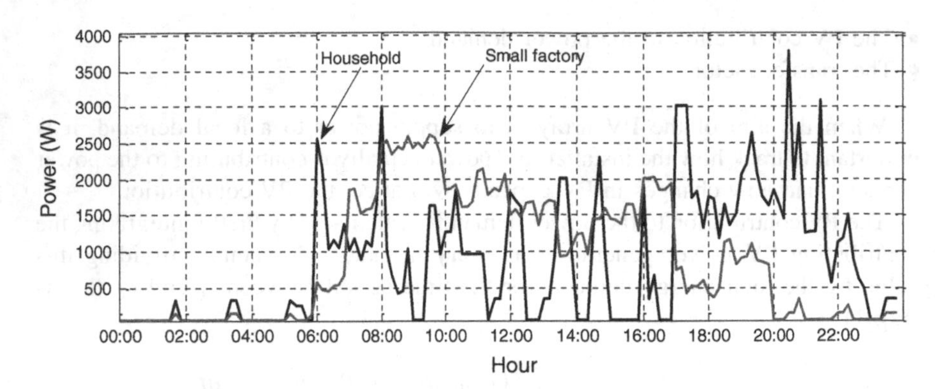

Fig. 2 The two power demands used during simulation. Both energy demands are 22 kWh/day

The latitude was set to 40°N in the solar resource model and the PV modules were oriented facing the equator with a plane tilt equal to the latitude. Correction factors corresponding to those of the climate type midlatitude summer were introduced and the altitude was set to 600 m above sea level.

The characteristics of a commercial 165 W PV module (Sharp NE-165U1) were introduced into the PV output model.

Two daily power demands were used, and can be seen in Fig. 2. They represent the electricity needed by a household and a very small factory or store. The electricity consumed at the end of the day is 22 kWh in both cases, but their power demand patterns are different.

Some additional elements were included in the model to account for losses in the power system. Their effect on the flow of power was calculated using their efficiencies. The following elements were included:

- Inverter to convert PV output from DC to AC, $\eta = f$(percent of full load) [15].
- Boost converter to adjust the PV output voltage, η = 80 % [16].
- Filter to smooth the inverter output, η = 97 %.
- Transformation between the inverter and the electric grid, η = 95 % [17].

In Fig. 3a–c, a series of graphs are shown for 1 day of simulation for the month of May. The apparent irradiance during the characteristic day is shown in Fig. 3a together with the dc power generation of 20 PV modules connected in series.

In Fig. 3b, the power demand of the household is shown together with the PV ac power generation. The actual PV power contribution is represented by the part of the power demand curve contained within the PV output curve.

Figure 3c shows the power distributed to the grid (positive values) and the power supplied to the demand by the grid (negative values).

These figures show that the PV power generation is completely dependent on the daily irradiance pattern while there is no particular relation between the PV power generation and the household electricity needs. Consequently, power exchange with the grid occurs throughout the day. More so, observing Fig. 3b it

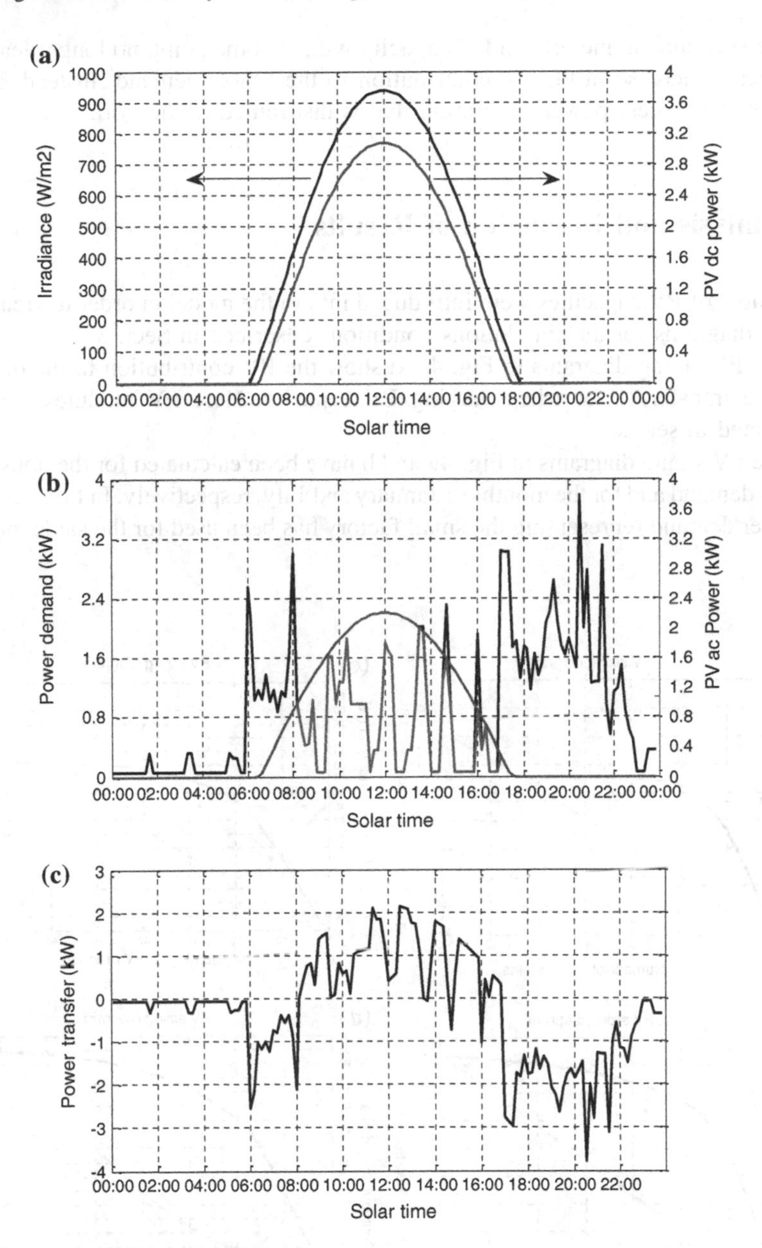

Fig. 3 **a** The modeled solar irradiance incident on the collector plane during 1 day in May and the power generation of 20 PV modules of 165 W connected in series, **b** The household power demand and the modeled ac power generation of the PV modules, **c** Modeled power transfers with the electric grid during the day. Negative values represent power supplied to the demand by the grid, and positive values indicate excess PV power distributed to the grid

can be seen that an increase in PV capacity will, at some point, no longer lead to a significant increase in the PV contribution to the power demand. Instead, it will increase the excess power generation that is distributed to the grid.

4 Analysis and Discussion of Results

A number of PV capacities were introduced into to the model in order to create PV sizing diagrams for the simulations conditions described in Sect. 3.

The PV sizing diagrams in Fig. 4a–d show the PV contribution to the demand and the transfer factor when varying PV capacity. Here PV modules are only connected in series.

The PV sizing diagrams in Fig. 4a and b have been calculated for the household power demand and for the months of January and July, respectively. In Fig. 4c and d, a power demand representing the small factory has been used for the same months.

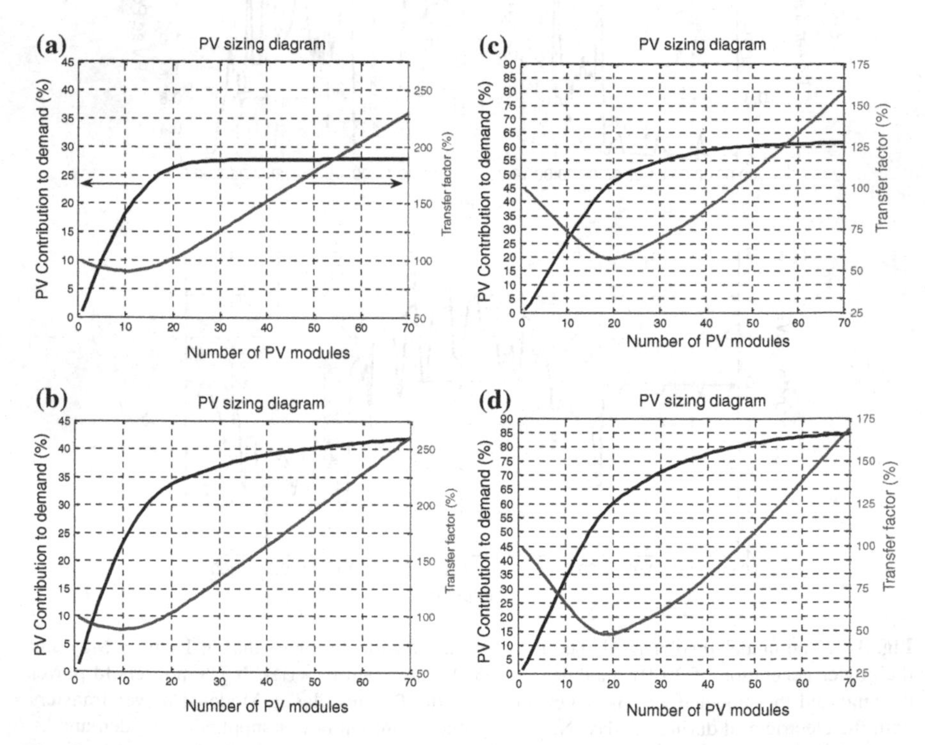

Fig. 4 **a** PV sizing diagram for the month of *January* and the *household* power demand, **b** PV sizing diagram for the month of *July* and the *household* power demand, **c** PV sizing diagram for the month of *January* and the *small factory* power demand, **d** PV sizing diagram for the month of *July* and the *small factory* power demand

As general results, it can be seen in Fig. 4a–d that the PV contribution increases rapidly at first, but after a certain point it saturates at a level less than 100 %. A further increase in capacity does not enhance PV contribution significantly. The transfer factor starts out at 100 % since without any PV modules the entire demand is supplied by the grid. As the PV capacity increases the transfer factor reaches a minimum. Hence, a global minimum is clearly present where the grid transfers are minimized.

It is observed that in July (Fig. 4b and d), when the hours of sunlight are abundant and the irradiance intensity is higher, the PV contribution saturates slower than in January (Fig. 4a and c). Furthermore, the PV contribution is significantly higher when the PV modules power the factory demand in comparison to the household demand. This can be explained by the better matching of the factory's electricity needs to the solar irradiance pattern throughout the day.

A lower transfer factor minimum is experienced when the PV modules power the small factory demand compared to the household power demand. Again, this can be explained by the better matching of the factory's power demand to the daily solar irradiance pattern. A stronger solar resource seems to slightly reduce the transfer factor minimum.

It is worth noting that optimum PV capacity for the household and the small factory is not the same, even though they have equal daily energy requirements.

5 Conclusions

A knowledge-based model for sizing grid-connected PV arrays has been presented. The model generates days of solar irradiance and calculates PV power output for commercial PV modules with a sampling period of 10 min.

The PV capacity for a specific location and local power demand is selected depending on the PV contribution to the demand and the power transfers with the electric grid, through the transfer factor.

The PV sizing diagrams shown in Fig. 4a–d provide a visual tool for decision making when sizing PV arrays. For example, a large community installation on a site known to have high power distribution losses might focus on a small transfer factor. On the other hand, a single private installation on a site characterized by low distribution losses and good financial trading terms might focus on a high PV contribution. Furthermore, this study indicates that the sizing of PV systems cannot be done in terms of daily energy requirements. It has been shown that two power demands with equal energy needs can have very different optimum PV capacities. Our results indicate that accurate modeling of solar power supply system requires a short simulation sampling period.

A future development of the presented model could include a sensitivity analysis on the results using measured solar irradiance data or simulated irradiance data from models that mimic the natural variations of terrestrial solar irradiance, see e.g. [8].

Acknowledgments. This work has been partially supported by Spanish project DPI2009-14552-C02-01.

References

1. Hybrid Optimization Model for Electric Renewables (HOMER). http://homerenergy.com/
2. Mellit A, Kalogirou SA, Hontoria L, Shaari S (2009) Artificial intelligence techniques for sizing photovoltaic systems: a review. Renew Sustain Energy Rev 13:406–419
3. Kornelakis A (2010) Multiobjective particle swarm optimization for the optimal design of photovoltaic grid-connected systems. Sol Energy 84:2022–2033
4. Kornelakis A, Marinakis Y (2010) Contribution for optimal sizing of grid-connected PV-systems using PSO. Renew Energy 35:1333–1341
5. Sulaiman SI, Rahman TKA, Musirin I, Shaari S, Sopian K (2012) An intelligent method for sizing optimization in grid-connected photovoltaic system. Sol Energy 86:2067–2082
6. Notton G, Lazarov V, Stoyanov L (2010) Optimal sizing of a grid-connected PV system for various PV module technologies and inclinations, inverter efficiency characteristics and locations. Renew Energy 35:541–554
7. Mondol JD, Yohanis YG, Norton B (2006) Optimal sizing of array and inverter for grid-connected photovoltaic systems. Sol Energy 80:1517–1539
8. Mikati M, Santos M, Armenta C (2012) Modeling and simulation of a hybrid wind and solar power system for the analysis of electricity grid dependency. Revista Iberoamericana de Automática e Informática Industrial 9:267–281
9. Duffie JA, Beckman WA (1991) Solar engineering of thermal processes, 2nd edn. 3–45, 73–75. Wiley Interscience, New York
10. Zekai S (2008) Solar energy fundamentals and modeling techniques: atmosphere, environment, climate change and renewable energy. Springer, London
11. Tsai HL, Tu CS, Su YJ (2008) Development of generalized photovoltaic model using MATLAB/SIMULINK. In: Proceedings of World congress on engineering and computer science, San Francisco, USA, 22–24 Oct 2008
12. Hansen AD, Sørensen P, Hansen LH, Bindner H (2000) Models for a stand-alone PV system. Risø National Laboratory, Roskilde
13. Data sheed from manufacturer of Sharp NE-165U1
14. Twidell J, Weir T (2006) Renewable energy resources, 2nd edn. Taylor & Francis, London and New York
15. Masters GM (2004) Renewable and efficient electric power systems. NJ Wiley & Sons, Hoboken
16. Power stream, DC/DC Converters. www.powerstream.com
17. Kubo T, Sachs H, Nadel S (2001) Opportunities for new appliance and equipment efficiency standards: energy and economics savings beyond current standard programs. Washington D.C., American Council for an Energy-Efficient Economy, Report no. A016

Feature-Based Matrix Factorization via Long- and Short-Term Interaction

Fei Ye and Jean Eskenazi

Abstract Recommender systems are designed to provide users with specific suggestions of products or services based on behavior history. Since both product popularity and customer taste constantly drift over time, temporal dynamics are of great importance when aiming at capturing user preferences over time with good accuracy. Traditional collaborative filtering is generally considered as a very promising technique regarding recommender systems and most recent studies focus mainly on long-term preferences. In this paper, we study how the interaction between short-term (session) behavior and long-term (historic) behavior can be integrated into a uniform framework which will use feature-based matrix factorization. We conduct experiments to test our algorithms on a movie-rating database, and we compare the actual performances of the different models tested. Results show that the approach we propose shows significant improvement over the baseline models.

1 Introduction

The large amount of information available on the Internet enriches people's life in a remarkable way. However, it is also very challenging for user-oriented Web services which would like to leverage this rich information in order to automatically provide users with satisfactory recommendations.

F. Ye (✉) · J. Eskenazi
Shanghai Jiaotong University, 800 Dongchuan Road, Shanghai, China
e-mail: sophyning@sjtu.edu.cn

J. Eskenazi
e-mail: jean.eskenazi@laposte.net

F. Sun et al. (eds.), *Knowledge Engineering and Management*,
Advances in Intelligent Systems and Computing 214,
DOI: 10.1007/978-3-642-37832-4_43,

Collaborative filtering (CF), a method based on the analysis of patterns among users and items without requiring the creation of exogenous information about user profile or item content, has been used very successfully by many Web services [1].

There are two fundamental approaches to model the relations between items and users in CF systems: the neighborhood models and latent factor models. Neighborhood models predict unknown ratings resorting to like-minded users or similar items previously rated by user. Meanwhile, latent factor models map users and items into the same low-dimensional space, and then uncover user-item interaction based on relations between factor vectors of users and items. These two classical models of CF consider users' preference as stable over time [2].

However, we can see in real-world applications that users' inclinations always drift over time. Being able to accurately and efficiently capture a time-evolving pattern which takes into account the multiple characteristics of the users and items is a big challenge to the traditional CF approach. Long-term trends can reflect the global evolution of specific data, thus indicating user's taste. Since shifts of user preferences are delicate, classical approaches, such as time-window or instance decay approaches, will lose a large proportion of usable information. Meanwhile in shorter session, users' choices could be largely influenced by any emergence of new items, while their characteristics are not yet fully uncovered because they are newly introduced into the market. In this case, users might probably be attracted to these new items because of their curiosity, and a general followership to popular trends. Moreover, external elements like seasonal changes, or specific holidays, can lead to a change of customers' focus, and then can influence the whole population via information diffusion. Although the trend of drift is general, a global concept drift [3] will not work well on these changes since they are driven by localized factors. Therefore a more informative approach is required. The main contributions of this paper are as follows:

- As drift of user preferences is driven by many different concept drifts, which act in different time frames and directions, we choose to use a feature-based matrix factorization which divides features into 3 types, thus allowing us to make a distinction between global factors and localized factors in our approach.
- Our goal is to extract user's long-term preference patterns, and to integrate it with short-session localized factors in our CF system.
- We build a uniform framework to make use of the efficiency of neighborhood models at detecting localized relationships, and of the superiority of latent factor models at capturing the overall structure. A combination of these two models has been studied previously in [2]. One of the improvements of this paper is to embed a similar combination into a feature-based matrix factorization, which could incorporate more efficiently other temporal and contextual information.

The rest of this paper is organized as follows. Section 2 presents classical models in recommender systems. Section 3 defines long- and short-term temporal factors. Section 4 integrates these two factors into a unified feature-based framework. We describe our experiment in Sect. 5. It includes data sets and comparisons of obtained results. We conclude in Sect. 6.

2 Preliminaries

2.1 Baseline Models

In this paper, special indexing letters are reserved to distinguish users from items: u, v are used to designate users, while i, j will refer to items. The user u attributes a rating r_{ui} to the item i. A baseline prediction for an unknown r_{ui} is denoted by b_{ui}:

$$b_{ui} = \mu + b_u + b_i \tag{1}$$

where μ is a constant indicating the global mean value of the rating, while b_u and b_i indicate the deviations observed for user u and item i. Baseline models aim at illustrating the effects of users and items separately.

2.2 Neighborhood Models

The most common approach used in CF systems is based on neighborhood models, which are divided into two primary types, corresponding to user-oriented and item-oriented methods. The user-oriented method [4] estimates unknown ratings based on existing ratings of users who share the same interest, and probably give the similar rating to a same item. We define a set of users $S^k(u;i)$ that tend to provide similar ratings to those of user u, and these "neighbors" already provided ratings on item i. The predicted value $\hat{r}_{ui}$ is taken as a weighted average of the neighbors' ratings:

$$\hat{r}_{ui} = b_{ui} + \frac{\sum_{v \in S^k(u;i)} s_{uv}(r_{vi} - b_{vi})}{\sum_{v \in S^k(u;i)} s_{uv}} \tag{2}$$

In the item-oriented method [2], unknown ratings are estimated based on historical ratings of the same user. Similarly to user-oriented approach, we identify a set of k neighboring items by $S^k(i;u)$, which are most similar to i and have already been rated by u. Then, the predicted value $\hat{r}_{ui}$ can be obtained as below:

$$\hat{r}_{ui} = b_{ui} + \frac{\sum_{j \in S^k(i;u)} s_{ij}(r_{uj} - b_{uj})}{\sum_{j \in S^k(i;u)} s_{ij}} \tag{3}$$

The similarity measures s are frequently based on the Pearson correlation coefficient.

2.3 Latent Factor Models

Latent factor models aim at capturing latent features of both users and items hidden under observed ratings. Matrix factorization models, which are well established for uncovering latent semantic factors, have recently become more and more popular as they are able to handle the scalability quite well. Each user u and item i is associated with a factor vector, we denote them $p_u, q_i \in \mathbb{R}^k$, respectively. Matrix factorization models then map both users and items to a joint latent factor space, such that user-item interactions are modeled as inner products in that space. A high correlation between item and user factors indicates a possible preference of a user to an item.

$$\hat{r}_{ui} = \mu + b_u + b_i + p_u^T q_i \tag{4}$$

By minimizing the regularized squared error, we can obtain all the model parameters.

$$\min_{b_*, q_*, p_*} \sum_{(u,i) \in K} \left(r_{ui} - \mu - b_u - b_i - p_u^T q_i\right)^2 + \lambda\left(b_u^2 + b_i^2 + \|p_u\|^2 + \|q_i\|^2\right) \tag{5}$$

Here the constant λ is a parameter controlling the extent of regularization. It is incorporated in order to avoid an over fitting in the learning process.

2.4 Temporal Dynamic Models

In certain literatures, studies have been conducted on the different approaches to model the temporal influence on the system. Xiong et al. computed the time weights for different items by decreasing weight far into time span [5]. Lu et al. proposed a spatiotemporal model for CF [6]. Li'ang et al. incorporated temporal factors in Euclidean embedding framework [7]. Yehuda built a variety of temporal factors of different aspects, and incorporated them into the latent factor models [3].

In our work, we define long- and short-term temporal factors in different ways in order to better indicate the different natures and influences of these two kinds of factors, thus interaction between long- and short-term temporal factors can be better integrated in a feature-based matrix factorization.

3 Long- and Short-Term Personalization

3.1 Long-Term User Preferences

Considering that popularity of items and the choices of users probably change over time, Yehuda suggested in [3] that a direct solution to track temporal dynamics in recommender systems is to integrate temporal variability into baseline predictors. By doing this we can consider the bias parameters of items b_i, and one of the users b_u, as functions of time.

$$b_{ui} = \mu + b_u(t_{ui}) + b_i(t_{ui}) \tag{6}$$

However, in our work we assume that we generally cannot easily separate changes of user bias over time from changes of item bias over time, because the drift of item bias over time is caused by an evolution of the user's taste. So regarding long-term time span, we only assign temporal dynamics to user bias. Thus we define a time interpolation model as follows:

$$b_{ui} = \mu + b_u^s \frac{e - t_{ui}}{e - s} + b_u^e \frac{t_{ui} - s}{e - s} + b_i \tag{7}$$

where s and e stand for the start and end of the time of all the ratings.

3.2 Short-Term User Preferences

In a view of short-term session, due to new launch, special event, or seasonal change, the influence from external elements is much larger and much more important than that in a long-term view. Our idea is to adjust neighborhood models and implicit information to capture this characteristic.

3.2.1 Shrunk Neighborhood Model

The neighborhood models introduced in the previous section are designed around user-specific similarity $s_{ij} \Big/ \sum_{j \in S^k(i;u)} s_{ij}$, relating item i to the items in a user-specific neighborhood $S^k(i;u)$. However, we would like to capture short-session information about a user by integrating its neighborhood relationships with other users. So here we adopt a modified neighborhood model which allows a global optimization by eliminating the user-specific similarity that we defined previously [2]. We introduce a new weight w_{ij} which is a global weight, independent of any specific user, and which will be obtained from the data through optimization.

$$\hat{r}_{ui} = b_{ui} + \sum\nolimits_{j \in R(u)} (r_{uj} - b_{uj}) w_{ij} \tag{8}$$

Here $R(u)$ is a set of all the items for which ratings by u are available. Weights w_{ij} represent offsets to baseline estimates. This way of seeing weights is different from the one adopted in classical neighborhood model. It is only when a user u rated item j higher than expected ($r_{uj} - b_{uj}$ is high), that the product $(r_{uj} - b_{uj})w_{ij}$ can have significant influence on the baseline estimate b_{ui} given the above relation between these observed rating r_{uj}, the unknown rating r_{ui}, and the global offset w_{ij}.

We suggest in this paper that neighborhood relationship impacts user preferences most largely during the current session. Although regarding long-term pattern of user behavior the preferences of a user varies from time to time, but as a whole part, it still strongly defines user's specific taste and personality. However in a short-term session when a user interacts in real time with recommender system, he/she is not only searching for items that suit perfectly to his/her taste but also searching for newly emerging items that will probably redefine his/her taste. For example, new movie launch, special event in cinema industry, or news reports on directors or actors of certain films might attract interest of users in a certain period. This influence could easily pass to every user since nowadays information is quickly and widely diffused via social networks, e-commerce, or other online activities. Thus through whole life time of an item, its popularity may not always be high, while during certain period, it could become a hot topic among the whole population.

To achieve the goal of recommending current popular items having a high potential to attract user's interest, we shrink the neighborhood model to short-term session by adding a temporal factor. Better results were obtained when we implemented this temporal factor into the global weight introduced previously.

$$\hat{r}_{ui}(t) = b_{ui} + \frac{\sum\nolimits_{j \in R(u,t)} \left[(r_{uj} - b_{uj}) w_{ij} \right]}{\sqrt{|R(u,t)|}} \tag{9}$$

Here $R(u,t)$ is the set of items rated by user u during the time interval t, which dates back from current moment when unknown rating needs to be predicted. Note that t here is different from t_{ui} in long-term modeling.

3.2.2 Shrunk Implicit Feedback

Different recommender systems provide various types of user inputs. The most common one is the explicit feedback which reflects directly users favoring products. However, users often forget to give this valuable explicit feedback, thus another implicit feedback is used which is more available and indirectly reflects preferences through other relative behavior [8]. The traditional matrix factorization model with implicit feedback, called SVD++ is as follows [9]:

$$\hat{r}_{ui} = b_{ui} + \left(p_u + \frac{\sum_{j \in N(u)} y_i}{\sqrt{|N(u)|}} \right)^T q_i \tag{10}$$

where $N(u)$ is the set of all the items for which u provided an implicit feedback. SVD++ can capture users' long-term preferences by incorporating users' whole rating history. Model parameters are obtained by minimizing the associated squared error function through gradient descent. Since short-term preferences depend more on the current situation, users' mood at the time, or recent activities, in [10], short-term property is extracted and integrated with the above SVD++ model characterized through users' implicit feedback.

$$\hat{r}_{ui}(t) = b_{ui} + \left(p_u + \frac{\sum_{j \in N(u)} y_i}{\sqrt{|N(u)|}} + \frac{\sum_{j \in N(u,t)} \varepsilon_i}{\sqrt{|N(u,t)|}} \right)^T q_i \tag{11}$$

where $N(u,t)$ is user u's short-term implicit feedback during time interval t. $\varepsilon_i (\in \mathbb{R}^k)$ is an item relevant vector.

4 Temporal Feature-Based Matrix Factorization

The contextual and dynamic information plays an important role in improving recommendation accuracy. If we can find a uniform model to implement all of this useful information, approach to recommendation will be more efficiently achieved.

4.1 Feature-Based Matrix Factorization

Tianqi et al. observes that some baseline CF algorithms are linear regression problems, thus he proposes a feature-based matrix factorization [11]. By adding a linear regression term to the traditional matrix factorization model, we are allowed to add more contextual bias and temporal dynamic information into the model.

$$\hat{r}_{ui} = \mu + \sum\nolimits_j w_j x_j + b_u + b_i + p_u^T q_i \tag{12}$$

A more flexible approach using features in factor is as follows:

$$\hat{r}_{ui} = \mu + \left(\sum\nolimits_j b_j^{(g)} \gamma_j + \sum\nolimits_j b_j^{(u)} \alpha_j + \sum\nolimits_j b_j^{(i)} \beta_j \right) + \left(\sum\nolimits_j p_j \alpha_j \right)^T \left(\sum\nolimits_j q_j \beta_j \right) \tag{13}$$

The input set $<\alpha, \beta, \gamma>$ denotes user, item, and global feature, respectively. With Eq. 1, we can transform basic MF model into the form below:

$$\gamma = \phi;\ \alpha_h = \begin{cases} 1 & h = u \\ 0 & h \neq u \end{cases};\ \beta_h = \begin{cases} 1 & h = i \\ 0 & h \neq i \end{cases} \tag{14}$$

Therefore, designing models with new features can be easily achieved by extending feature-based factorization models. For example, by incorporating contextual information of household membership as an addictive feature, the informative model presented in [12] successfully generates recommendations for households instead of individual users. Similar to baseline model, the long- and short-term indicators previously defined in Sect. 3 could both be completely integrated into a unified framework, which facilitates largely the optimization process.

4.2 Long-Term Information

The features of long-term information defined in Eq. 6 can also be transformed into the following forms:

$$\gamma = \phi;\ \alpha_h = \begin{cases} \frac{e-t}{e-s} & h = u \\ \frac{t-s}{e-s} & h = u+n; \\ 0 & \text{otherwise} \end{cases}\quad \beta_h = \begin{cases} 1 & h = i \\ 0 & h \neq i \end{cases} \tag{15}$$

4.3 Short-Term Neighboring Information

In order to integrate the above-mentioned temporal neighborhood information into the feature-based matrix factorization, we only need to implement it into the global feature. We did it as follows:

$$\hat{r}_{ui}(t) = b_{ui} + \frac{\sum_{j \in R(u,t)} \left[\left(r_{uj} - b_{uj} \right) w_{ij} \right]}{\sqrt{|R(u,t)|}} + p_u^T q_i \tag{16}$$

Similar to baseline model, we can get global feature as below:

$$\gamma_h = \begin{cases} \frac{r_{uj} - b_{uj}}{\sqrt{|R(u,t)|}} & h = \text{index}(i,j), j \in R(u,t) \\ 0 & \text{otherwise} \end{cases} \tag{17}$$

4.4 Short-Term Implicit Feedback Information

The features of short-term implicit feedback information defined in Eq. 11 can be translated into the following forms:

$$\alpha_h = \begin{cases} 1 & h = u \\ \frac{1}{\sqrt{|N(u)|}} & h = j + m, j \in N(u) \backslash N(u,t) \\ \frac{1}{\sqrt{|N(u,t)|}} & h = j + m, j \in N(u,t) \\ 0 & \text{otherwise} \end{cases} \quad (18)$$

where m means number of user.

5 Experiments

5.1 Data Set

In this paper, we test our model on Netflix movie-rating dataset. Over 480,000 customers have given ratings on over 17,770 movies. More than 100 million date-stamped ratings were recorded between 31 December 1999 and 31 December 2005. We use the official training and validation dataset to learn the parameters of our model. As is illustrated in Sect. 4, our approach integrates the interaction between long- and short- term temporal dynamics into an unified feature-based matrix factorization model. Rooted mean square error (RMSE) is used for accuracy comparison.

$$\text{RMSE} = \sqrt{\sum\nolimits_{(u,i) \in \text{TestSet}} (r_{ui} - \hat{r}_{\text{ui}})^2 / |\text{TestSet}|} \quad (19)$$

In the first step of our experiment, we use the factor b_{ui} as the long-term temporal factor and we follow its evolution through the whole rating history day after day. Time interval t in short-term session can be set to any value from years to a single minute depending on the amount of available data. Since long-term and short-term temporal factor are defined in two different ways, during training process we first use 5 kinds of short-term session: whole history, current season, current month, current week, and current day. These models are respectively called MF-Whole, MF-Season, MF-Month, MF-Week, and MF-Day. We compare them using RMSE averaged on all the users for a factorization dimension $f = 50$.

In the second step of our experiment, we compare the average RMSE (averaged on all users) of the basic, the neighboring, the long-term, the short-term, and our hybrid matrix factorization methods (respectively BMF, NMF, LMF, SMF, and HMF) over a range of factorization dimensions. The time interval t retained for this procedure is the minimum that was found during the first step.

5.2 Experimental Results

Results of the first step of our experiment are shown in Table 1. By comparing the performance regarding the average RMSE, we see that there is a minimum for t equal to 1 month. This result was expected since to an extreme point, a very tight time interval would only include very little data and make any prediction impossible, whereas on the other hand an extremely large interval would only be able to set boundaries for the prediction. If we have enough data to consider the RMSE being continuous with respect to the span of the time interval, this means it reaches a minimum for some t such that $0 < \text{span}(t) < +\infty$. We found the minimum to be reached for $t = 1$month which seems coherent with the pace of the film market.

Table 2 regroups the RMSE values for the different models and factorizations dimensions compared. The results show that the hybrid model introduced in this paper provides a better accuracy than the other models considered. It should be noted that the time interval that was fixed at this step was just a rough optimization of HMF for $f = 50$ so HMF was not optimized for other values of f. And the same optimization of time interval was applied for SMF.

In order to provide a better comparison of SMF and HMF we could improve the first step by finding a more precise value of the minimum span of t for HMF and also for SMF. Then we should do the same for different values of f. By doing this we would be able to compare the two methods optimized for average RMSE at each point. For even better results we could try to compare the results on some single users and see how it differs from the averaged RMSE.

Table 1 Average RMSE of different short-session models

Models (f = 50)	MF-whole	MF-season	MF-month	MF-week	MF-day
Average RMSE	0.9387	0.9236	0.9017	0.9115	0.9204

Table 2 RMSE for different models and factorization dimensions

Model	f = 10	f = 20	f = 50	f = 100
BMF	0.9201	0.9179	0.9153	0.9136
NMF	0.9178	0.9157	0.9122	0.9099
LMF	0.9141	0.9116	0.9089	0.9013
SMF	0.9138	0.9120	0.9092	0.9056
HMF	0.9041	0.9024	0.9017	0.9005

6 Conclusion

Temporal dynamics are of great importance in recommendation systems and it seems quite natural to think about it when dealing with real life experience. Unlike some other temporal approaches, the one we have introduced in this paper includes two types of time framework which gives us great flexibility when studying the impact of these two different kinds of time span. In order to make better use of the interaction between these two models, we unified them into a feature-based matrix factorization, which enables us to describe and incorporate different features more easily.

Temporal dynamics will mostly likely remain a hot topic as long as social networks continue to attract people and business. The flow of information on those networks is very dense and user preferences follow more obvious evolution patterns over time. However, these patterns can be easily influenced by external elements. The hybrid feature-based matrix factorization that we detailed in this paper could help provide recommendation which is more suited to capture relevant information in a fast changing society. Consequently, its range of applications also extends to marketing and customer targeting. We could imagine for example that some company would find interesting that we are able to determine the percentage of a given population for which recommendations are accurate only when considering a short time period, implying that certain part of population is influenced very easily by advertisement, media, or more generally any kind of communication, including social networks. This will be the direction that we will follow to further our research on recommender system and temporal dynamics.

References

1. Yehuda K, Robert B (2011) Advances in collaborative filtering. Recommender systems handbook, Part 1, pp 145–186
2. Yehuda K (2008) Factorization meets the neighborhood: a multifaceted collaborative filtering model. In: Proceeding of the 14th ACM SIGKDD international conference on knowledge discovery and data mining, KDD'08. New York, USA, pp 416–434
3. Yehuda K (2010) Collaborative filtering with temporal dynamics. Commun ACM 53(4):89–97
4. Paterek A (2007) Improving regularized singular value decomposition for collaborative filtering. In: Proceedings of KDD cup and workshop
5. Xiong L et al (2010) Temporal collaborative filtering with Bayesian probabilistic tenser factorization. In: Proceedings of SIAM data mining
6. Lu Z, Agalwal D, Dhillon IS (2009) A spatio-temporal approach to collaborative filtering. In: Proceedings of the 3rd ACM conference on recommender systems, pp 13–20
7. Li'ang Y, Yongqiang W, Yong Y (2012) Collaborative filtering via temporal Euclidean embedding. In: APWeb 2012, LNCS 7235, pp 513–520
8. Herlocker JL, Konstan JA, Borchers A et al (1999) An algorithmic framework for performing collaborative filtering. In: Proceedings of 22nd ACM SIGIR conference on information retrieval, pp 230–237

9. Salakhutdinov R, Mnih A (2008) Probabilistic matrix factorization. In: Advances in neural information processing systems 20 (NIPS'07), pp 1257–1264
10. Diyi Y et al (2012) Collaborative filtering with short term preferences mining. SIGIR'12, Portland, Oregon, USA
11. Tianqi C, Zhan Z et al (2011) Feature-based matrix factorization. CS, AI
12. Lu Q, Diyi Y, Tianqi C et al (2011) Informative household recommendation with feature-based matrix factorization. In: CAMra2011, Chicago, IL, USA

Lee's Theorem Extension for $\mathcal{IVFR}s$ Similarities

Garmendia Luis, Gonzalez del Campo Ramon and Lopez Victoria

Abstract In this paper an extension of Lee's theorem for interval-valued fuzzy relations is given. An algorithm to compute an $\mathcal{IV}$-similarity decomposition is given.

Keywords Interval-valued fuzzy relations · $\mathcal{IV}$-similarity · Bridge between two $\mathcal{IV}$-similarities

1 Introduction

Interval-valued fuzzy sets ($\mathcal{IVFS}s$) were introduced in the 1960s by Grattan-Guinness [12], Jahn [13], Sambuc [18], and Zadeh [21]. They are extensions of classical fuzzy sets where the membership value between 0 and 1 is replaced by an interval in $[0, 1] \times [0, 1]$. They easily allow to model uncertainty and vagueness: sometimes it is better to give a membership interval than a membership degree to describe a characteristic of objects on a universe. $\mathcal{IVFS}$ are a special case of type-2 fuzzy sets that simplifies the calculations while preserving their richness as well. $\mathcal{IVFS}$ have been widely applied [1–3, 5, 6, 8, 9, 11, 15–17, 19, 20].

Lee's theorem [14] was proposed for constructing the max–min equivalence closure of fuzzy similarity matrices.

This paper is organized as follows. In Sect. 2 some basic concepts about interval-valued fuzzy sets are recalled. Lee's theorem for fuzzy relations is

G. Luis (✉) · Gonzalez del Campo Ramon · L. Victoria
University Complutense of Madrid, Madrid, Spain
e-mail: lgarmend@fdi.ucm.es

Gonzalez del Campo Ramon
e-mail: rgonzale@estad.ucm.es

L. Victoria
e-mail: vlopez@fdi.ucm.es

F. Sun et al. (eds.), *Knowledge Engineering and Management*,
Advances in Intelligent Systems and Computing 214,
DOI: 10.1007/978-3-642-37832-4_44,

recalled as well. In Sect. 3 it is showed that the bridge between two $\mathcal{IV}$-similarities is a $\mathcal{IV}$-similarity. Moreover, for any $\mathcal{IV}$-similarity there exists at least a decomposition into two $\mathcal{IV}$-subsimilarities. In Sect. 3.1 an algorithm to compute the decomposition is given.

2 Preliminaries

Definition 1 [6] Let $(L, \leq_L)$ be a lattice such that:

1. $L = \{[x_1, x_2] \in [0,1]^2 \text{ with } x_1 \leq x_2\}$.
2. $[x_1, x_2] \leq_L [y_1, y_2]$ if and only if $x_1 \leq y_1$ and $x_2 \leq y_2$. In particular: $[x_1, x_2] =_L [y_1, y_2]$ if and only if $x_1 = y_1$ and $x_2 = y_2$.

$(L, \leq_L)$ is a complete lattice and the supremum and infimum are defined as follows.

Definition 2 [7] Let $\{[v_i, w_i]\}$ be a set of intervals on L. Then

1. $\wedge\{[v_i, w_i]\} \equiv [\mathit{infimun}\{v_i\}, \mathit{infimun}\{w_i\}]$
2. $\vee\{[v_i, w_i]\} \equiv [\mathit{supremun}\{v_i\}, \mathit{supremun}\{w_i\}]$

Definition 3 [6] An interval-valued fuzzy set A on a universe X can be represented by the mapping:

$$A = \{(a, [x_1, x_2]) \mid a \in X, [x_1, x_2] \in L\}$$

Definition 4 [6] Let A and B be two interval-valued fuzzy sets on a universe X. The equality between A and B is defined as follows:

$$A = B \text{ if and only if } A(a) =_L B(a) \text{ for all } a \in X.$$

Definition 5 [6] Let A and B be two interval-valued fuzzy sets on a universe X. The inclusion of A into B is defined as follows:

$$A \subseteq B \text{ if and only if } A(a) \leq_L B(a) \text{ for all } a \in X.$$

Definition 6 [6] Let A and B be two interval-valued fuzzy sets on a universe X. The intersection between A and B is defined as follows:

$$A \cap B(a) =_L \wedge(A(a), B(a)) \text{ for all } a \in X.$$

Definition 7 [6] A negation operator for interval-valued fuzzy sets $\mathcal{N}$ is a decreasing function $\mathcal{N} : L \to L$ that satisfies:

1. $\mathcal{N}(0_L) =_L 1_L$

2. $\mathcal{N}(1_L) =_L 0_L$

where $0_L = [0,0]$ and $1_L = [1,1]$.

If $\mathcal{N}(\mathcal{N}([x_1,x_2])) =_L [x_1,x_2]$ for all $[x_1,x_2]$ in L then $\mathcal{N}$ is called an involutive negation.

Definition 8 [6] A strong negation operator for interval-valued fuzzy sets $\mathcal{N}$ is a continuous, strictly decreasing, and involutive negation operator.

The t-norm operators are generalized to the lattice L in a straightforward way.

Definition 9 [6] A generalized t-norm function $\mathcal{T}$ is a monotone increasing, symmetric, and associative operator $\mathcal{T} : L^2 \to L$ that satisfies: $\mathcal{T}(1_L, [x_1,x_2]) =_L [x_1,x_2]$ for all $[x_1,x_2]$ in L.

Notation The lowest value of an interval $x = [x_1,x_2]$ is denoted $\underline{x}$ and the highest value of an interval $x = [x_1,x_2]$ is denoted $\overline{x}$, so $x = [x_1,x_2] = [\underline{x},\overline{x}]$.

Definition 10 [6] A generalized t-norm operator $\mathcal{T} : L^2 \to L$ is t-representable if there are two t-norms T_1 and T_2 that satisfy:

1. $\mathcal{T}([x_1,x_2],[y_1,y_2]) =_L [T_1(x_1,y_1), T_2(x_2,y_2)]$
2. $T_1(v,w) \leq T_2(v,w)$ for all $v,w \in [0,1]$

Definition 11 [4] Let X_1 and X_2 be two universes of discourse. An interval-valued fuzzy relation $R : X_1 \times X_2 \to L$ is a mapping:

$$R = \{((a,b),[x,y]) \mid a \in X_1, b \in X_2, [x,y] \in L\}$$

where $x = \underline{R(a,b)}$ and $y = \overline{R(a,b)}$.

In the rest of the paper, it is assumed that $X_1 = X_2 = X$ and that X is finite.

Definition 12 [10] An interval-valued fuzzy relation $R : X^2 \to [0,1]^2$ is reflexive if:

$$R(a,a) =_L 1_L \textit{ for all } a \in X$$

where $1_L = [1,1]$.

Definition 13 [10] An interval-valued fuzzy relation $R : X^2 \to [0,1]^2$ is $[\alpha_1,\alpha_2]$–reflexive if:

$$R(a,a) \geq_L [\alpha_1,\alpha_2] \textit{ for all } a \in X$$

where $\alpha_1, \alpha_2 \in [0,1]$ and $\alpha_1 \leq \alpha_2$.

Definition 14 [10] An interval-valued fuzzy relation $R : X^2 \to [0,1]^2$ is symmetric if:

$$R(a,b) =_L R(b,a) \textit{ for all } a,b \in X$$

Definition 15 [10] Let $\mathcal{T}$ be a generalized t-norm operator and let R be an interval-valued fuzzy relation on X. R is $\mathcal{T}$-transitive if:

$$\mathcal{T}(R(a,b),R(b,c)) \leq_L R(a,c) \text{ for all } a,b,c \in X$$

Definition 16 A fuzzy relation $R : X^2 \rightarrow [0,1]$ is a similarity if R is reflexive, symmetric, and min-transitive.

Definition 17 Let B be a similarity on X with cardinality n. Let $\pi = \{a_k\}$ be a permutation of the values 1...n. Then $P_\pi(B)$ is the similarity where the *kth* element of X is permuted by the element a_k of X.

Lemma 1 [14] Let B be a similarity. Let π be any permutation. Then $P_\pi(B)$ is a similarity.

Definition 18 [14] Let R and S be two similarities on universes X_1 and X_2 with cardinalities n_1 and n_2 respectively. Let t be an interval in $[0,1]$ such that $t \leq \wedge (\wedge(R(a_i,a_j)), \wedge(S(a_k,a_l)))$ for all i, j, k, l. Then, the t-bridge between R and S is a similarity defined as follows:

$$B_{t;R_{n_1 \times n_1},S_{n_2 \times n_2}} = \begin{pmatrix} R & (t)_{n_1 \times n_2} \\ (t)_{n_2 \times n_1} & S \end{pmatrix}$$

where $(t)_{n_2 \times n_1}$ and $(t)_{n_1 \times n_2}$ are a $n_2 \times n_1$ matrix and a $n_1 \times n_2$ respectively whose elements are t.

Sometimes it is written $B_{t;R,S}$ instead of $B_{t;R_{n_1 \times n_1},S_{n_2 \times n_2}}$.

Theorem 1 [14] Let C be a similarity on a universe with cardinality $n \geq 2$. Then, there exist two similarities (R, S) and a value t which verifies:

$$C = P_\pi(B_{t;R_{n_1 \times n_1},S_{n_2 \times n_2}})$$

for some permutation π.

Note that n_1 and n_2 are the cardinalities of the universes of R and S respectively with $n = n_1 + n_2$ and that t is the minimum value of R and S.

3 Decomposition of $\mathcal{IV}$-Similarities into $\mathcal{IV}$-Subsimilarities

Let $X = \{a_1, ..., a_n\}$ be a finite universe.

Definition 19 Let $\mathcal{M}([x_1,x_2],[y_1,y_2]) = [min(x_1,y_1), min(x_2,y_2)]$ be a t-generalized t-norm. Let R be an interval-valued fuzzy relation on a universe X. R is a $\mathcal{IV}$-similarity if it is reflexive, symmetric, and $\mathcal{M}$-transitive.

Definition 20 Let R and S be two $\mathcal{IV}$-similarities on universes X_1 and X_2 with cardinalities n_1 and n_2 respectively. Let $[t_1,t_2]$ be an interval in $[0,1]^2$ such that

$[t_1,t_2] \leq_L \wedge(\wedge(R(a_i,a_j)), \wedge(S(a_k,a_l)))$ for all i, j, k, l. Then the $[t_1,t_2]$-bridge between R and S is an interval-valued fuzzy relation on $X_1 \cup X_2$. B is defined as follows:

$$B_{[t_1,t_2];R,S} = \begin{pmatrix} R & ([t_1,t_2])_{n_1 \times n_2} \\ ([t_1,t_2])_{n_2 \times n_1} & S \end{pmatrix}$$

where $[t_1,t_2]$ is called the bridge interval.

Lemma 2 Let R and S be two reflexive $\mathcal{IVFR}s$ on universes X_1 and X_2. Then $B_{[t_1,t_2];R,S}$ is a reflexive interval-valued fuzzy relation on $X_1 \cup X_2$.

Proof Trivial because R and S are reflexive interval-valued fuzzy relations.

Lemma 3 Let R and S be two symmetric $\mathcal{IVFR}s$ on universes X_1 and X_2. Then $B_{[t_1,t_2];R,S}$ is a symmetric interval-valued fuzzy relation on $X_1 \cup X_2$.

Proof Cases:

1. If $i,j \leq n_1$ then $B_{[t_1,t_2];R,S}(a_i,a_j) = R(a_i,a_j) = R(a_j,a_i) = B_{[t_1,t_2];R,S}(a_j,a_i)$ because R is symmetric.
2. If $i,j > n_1$ then $B_{[t_1,t_2];R,S}(a_i,a_j) = S(a_i,a_j) = S(a_j,a_i) = B_{[t_1,t_2];R,S}(a_j,a_i)$ because S is symmetric.
3. Otherwise, $B_{[t_1,t_2];R,S}(a_i,a_j) = [t_1,t_2] = B_{[t_1,t_2];R,S}(a_j,a_i)$ by construction.

Lemma 4 Let R and S be two $(min - min)$-transitive interval-valued fuzzy relations on universes X_1 and X_2 with cardinalities n_1 and n_2 respectively. Let $B_{[t_1,t_2];R,S}$ be the $[t_1,t_2]$-bridge between R and S. Then, $B_{[t_1,t_2];R,S}$ is (min, min)-transitive interval-valued fuzzy relation.

Proof It will be proved that $\mathcal{M}(B_{[t_1,t_2];R,S}(a_i,a_k), B_{[t_1,t_2];R,S}(a_k,a_j)) \leq_L B_{[t_1,t_2];R,S}(a_i,a_j)$ for all i,j,k.

Cases:

1. For $i,j \leq n_1$:
   (a) For $k \leq n_1$:
   $\mathcal{M}(B_{[t_1,t_2];R,S}(a_i,a_k), B_{[t_1,t_2];R,S}(a_k,a_j)) =_L$
   $=_L \mathcal{M}(R(a_i,a_k), R(a_k,a_j))$
   $\leq_L R(a_i,a_j) =_L B_{[t_1,t_2];R,S}(a_i,a_j)$
   so $\mathcal{M}(B_{[t_1,t_2];R,S}(a_i,a_k), B_{[t_1,t_2];R,S}(a_k,a_j)) \leq_L B_{[t_1,t_2];R,S}(a_i,a_j)$ is verified due to the fact that R is a $\mathcal{IV}$-similarity.
   (b) $k > n_1$:
   $\mathcal{M}(B_{[t_1,t_2];R,S}(a_i,a_k), B_{[t_1,t_2];R,S}(a_k,a_j)) =_L \mathcal{M}([t_1,t_2],[t_1,t_2]) =_L [t_1,t_2] \leq_L B_{[t_1,t_2];R,S}(a_i,a_j)$.

2. For $i \leq n_1$ and $j > n_1$:
   (a) For $k \leq n_1$:
   $\mathcal{M}(B_{[t_1,t_2];R,S}(a_i,a_k), B_{[t_1,t_2];R,S}(a_k,a_j)) =_L \mathcal{M}(R(a_i,a_k),[t_1,t_2]) =_L$
   $[t_1,t_2] \leq_L B_{[t_1,t_2];R,S}(a_i,a_j)$
   (b) For $k > n_1$:
   $\mathcal{M}(B_{[t_1,t_2];R,S}(a_i,a_k), B_{[t_1,t_2];R,S}(a_k,a_j)) =_L \mathcal{M}([t_1,t_2],S(a_k,a_j)) =_L$
   $[t_1,t_2] \leq_L B_{[t_1,t_2];R,S}(a_i,a_j)$
   so $\mathcal{M}(B_{[t_1,t_2];R,S}(a_i,a_k), B_{[t_1,t_2];R,S}(a_k,a_j)) =_L [t_1,t_2] \leq_L B_{[t_1,t_2];R,S}(a_i,a_j)$
   for all k.
3. For $i,j > n_1$: similar to case 1 using S instead of R.
4. For $i > n_1$ and $j \leq n_1$: similar to case 2 using R instead of S.

Theorem 2 Let R and S be two $\mathcal{IV}$-similarities on universes X_1 and X_2. Let $B_{[t_1,t_2];R,S}$ be the $[t_1,t_2]$-bridge between R and S. Then, $B_{[t_1,t_2];R,S}$ is a $\mathcal{IV}$-similarity on $X_1 \cup X_2$.

Proof: Trivial according to the lemmas 2, 3, 4.

Lemma 5 Let R be an interval-valued fuzzy relation on a finite universe X. If R is a $\mathcal{IV}$-similarity then there exists an interval $R(a_p,a_q)$ such that $R(a_p,a_q) \leq_L R(a_i,a_j)$ for all i, j.

Proof It will be proved using mathematical induction:

1. Base case: Let X be a universe with cardinality 3. Let R be a $\mathcal{IV}$-similarity on X. Then R is as follows:

$$\begin{pmatrix} 1_L & [y_1,y_2] & [x_1,x_2] \\ [y_1,y_2] & 1_L & [x_1,x_2] \\ [x_1,x_2] & [x_1,x_2] & 1_L \end{pmatrix}$$

   where $[x_1,x_2] \leq_L [y_1,y_2]$, so there exists an interval $[y_1,y_2]$ which is minimum.
2. Inductive step: Let X be a universe with cardinality n. Let R be a $\mathcal{IV}$-similarity on X which has a minimum interval. Let's add an element a_{n+1} to the universe. Let S be a $\mathcal{IV}$-similarity on $X \cup \{a_{n+1}\}$ such that $S(a_i,a_j) = R(a_i,a_j)$ for all a_i, a_j in X. It is denoted $S(a_i,a_{n+1})$ as c_i for all $i : 1...n$. For any a_i, a_j in X, it is verified:
   $min(R(a_i,a_j),c_j) \leq_L c_i$
   $min(c_i,c_j) \leq_L R(a_i,a_j)$
   $min(R(a_i,a_j),c_i) \leq_L c_j$
   so $c_i \leq_L c_j$ or $c_j \leq_L c_i$. In similar way it is possible to prove that $R(a_i,a_j) \leq_L c_j$ or $c_j \leq_L R(a_i,a_j)$ and $c_i \leq_L R(a_i,a_j)$ or $R(a_i,a_j) \leq_L c_i$. Therefore, there exists a minimum interval in S.

Definition 21 Let R be a $\mathcal{IV}$-similarity on X. Let $I \subseteq X$ be a subset of X. A subsimilarity of R restricted to T $(R)_I$ is defined as $(R)_I : I \times I \longrightarrow [0,1]^2$ such that $(R)_I(a_r, a_s) = R(a_r, a_s)$ for all a_r and a_s are in I. $(R)_I$ is denoted I-subsimilarity of R.

Lemma 6 Let R be a $\mathcal{IV}$-similarity on X. Any I-subsimilarity of R on $I \subseteq X$ is a $\mathcal{IV}$-similarity.

Proof Trivial.

Lemma 7 Let C be a $\mathcal{IV}$-similarity and let $C(a_p, a_q)$ be the minimum value of C. Let $I = \{a_i \mid a_i \in X \text{ and } C(a_i, a_q) =_L C(a_p, a_q)\}$ and $I' = X \setminus I$. Then, $C(a_i, a_j) >_L C(a_p, a_q)$ for all a_i and a_j in I'.

Proof Trivial by construction of I'.

Lemma 8 Let C be a $\mathcal{IV}$-similarity and let $C(a_p, a_q)$ be the minimum value of C. Let $I = \{a_i \mid a_i \in X \text{ and } C(a_i, a_q) =_L C(a_p, a_q)\}$ and $I' = X \setminus I$. Then, $C(a_i, a_j) =_L C(a_p, a_q)$ for all a_i in I and for all a_j in I'.

Proof Let's suppose there exists a_r in I and a_s in I' such that $C(a_r, a_s) >_L C(a_p, a_q)$. Let $C(a_r, a'_s)$ be such that r is in I and s' in I'. By lemma 7 $C(a_s, a_{s'}) >_L C(a_p, a_q)$ for a_s and a'_s in I'. Then, it is verified $C(a_p, a_q) <_L min(C(a_r, a_s), C(a_s, a_{s'})) \leq_L C(a_r, a_{s'})$ so a_r and a'_s are in I' which contradicts the assumption.

Theorem 3 Let C be a $\mathcal{IV}$-similarity on a universe X with cardinality n. Then, there exist two similarities (R, S) and an interval $[t_1, t_2]$ which verifies:

$$C = P_\pi(B_{[t_1,t_2];R_{n_1 \times n_1}, S_{n_2 \times n_2}})$$

for some permutation π.

Note that n_1 and n_2 are the cardinalities of the universes of R and S respectively and $[t_1, t_2]$ is the minimum value of R and S.

Proof Let C be a $\mathcal{IV}$-similarity and let $C(a_p, a_q)$ be the minimum value of C according to lemma 5.

Let $I = \{a_i \mid a_i \in X \text{ and } C(a_i, a_q) = C(a_p, a_q)\}$ and $I' = X \setminus I$.

$(C)_I$ and $(C)_{I'}$ are $\mathcal{IV}$-similarities according to lemma 6

For all a_i in I and for all a_j in I' it is verified $C(a_i, a_j) = C(a_p, a_q)$ according to lemma 8.

Let $\pi = \{a_{i_1}, ..., a_{i_{n_1}}, a_{i'_1}, ..., a_{i'_{n_2}}\}$ be a permutation where $a_{i_1}, ..., a_{i_{n_1}}$ belong to I and $a_{i'_1}, ..., a_{i'_{n_2}}$ belong to I'.

Then: $C = P_\pi(B_{[t_1,t_2];R_{n_1 \times n_1}, S_{n_2 \times n_2}})$ where $[t_1, t_2] =_L C(a_p, a_q)$, $n_1 = card(I)$ and $n_2 = card(X \setminus I)$.

3.1 Algorithm to Decompose an $\mathcal{IV}$-Similarity into $\mathcal{IV}$-Subsimilarities

- *Input*: A $\mathcal{IV}$-similarity C on X with cardinality n such that $n \geq 2$.
- *Output*: Two $\mathcal{IV}$-subsimilarities R,S on X_1, X_2 such that $X = X_1 \cup X_2$ and a bridge interval t such that $C = P_\pi(B_{[t_1,t_2];R,S})$
- The algorithm runs as follows:
  - *Step 1*: Search for the minimum interval of C (it always exists by lemma 5): $t = [a_p, a_q]$.
  - *Step 2*: Choose X_1 such that $X_1 \subset X$. Compute I such that $I = \{a_i \mid a_i \in X_1$ and $C(a_i, a_q) =_L C(a_p, a_q)\}$
  - *Step 3*: Compute: $R \leftarrow C_I$ and $S \leftarrow C_{X \setminus I}$

Example 1 Let C be a $\mathcal{IV}$-similarity:

$$C = \begin{pmatrix} [1,1] & [0.1,0.2] & [0.8,1] \\ [0.1,0.2] & [1,1] & [0.1,0.2] \\ [0.8,1] & [0.1,0.2] & [1,1] \end{pmatrix}$$

Then: $R = \begin{pmatrix} [1,1] & [0.8,1] \\ [0.8,1] & [1,1] \end{pmatrix}$ $S = (\,[1,1]\,)$ and $t = [0.1, 0.2]$

$$B_{t;R,S} = \begin{pmatrix} [1,1] & [0.8,1] & [0.1,0.2] \\ [0.8,1] & [1,1] & [0.1,0.2] \\ [0.1,0.2] & [0.1,0.2] & [1,1] \end{pmatrix}$$

4 Conclusions

The decomposition of similarities into subsimilarities is a usefull result that can be applied, for example, in optimal computation of min-transitive closures [14]. In this paper those results are extended to decompose and $\mathcal{IV}$-similarity into $\mathcal{IV}$-subsimilarities, including an algorithm to do it.

Acknowledgments This research is partially supported by the Spanish Ministry of Science and Technology, grant number TIN2009-07901, the Research Group CAM GR35/10-A at Complutense University of Madrid.

References

1. Bilgic T (1998) Interval-valued preference structures. Eur J Oper Res 105(1):162–183
2. Bustince H, Barrenechea E, Pagola M, Orduna R (2007) Image thresholding computation using Atanassov's intuitionistic fuzzy sets. JACIII 11:187–194
3. Bustince H, Burillo P (2000) Mathematical analysis of interval-valued fuzzy relations: application to approximate reasoning. Fuzzy Sets Syst 113:205–219
4. Bustince H, Burillo P (2000) Mathematical analysis of interval-valued fuzzy relations: application to approximate reasoning. Fuzzy Sets Syst 113(2):205–219
5. Bustince H, Mohedano V, Barrenechea E, Pagola M (2006) An algorithm for calculating the threshold of an image representing uncertainty through a-ifss. In: Proceedings of 11th information processing and management of uncertainty in knowledge-based systems, pp 2383–2390
6. Cornelis C, Deschrijver G, Kerre E (2004) Implication in intuitionistic fuzzy and interval-valued fuzzy set theory: construction, classification, application. Int J Approx Reason 35(1):55–95
7. Cornelis C, Deschrijver G, Kerre E (2006) Advances and challenges in interval-valued fuzzy logic. Fuzzy Sets and Syst 157(5):622–627
8. Deschrijver G, Kerre EE (2003) On the composition of intuitionistic fuzzy relations. Fuzzy Sets Syst 136(3):333–361
9. Dziech A, Gorzalczany MB (August 1987) Decision making in signal transmission problems with interval-valued fuzzy sets. Fuzzy Sets Syst 23:191–203
10. González-del Campo R, Garmendia L, Recasens J (2009) Transitive closure of interval-valued relations. In: Proceedings IFSA-EUSFLAT'09, pp 837–842
11. Gorzalczany MB (1988) Interval-valued fuzzy controller based on verbal model of object. Fuzzy Sets Syst 28:45–53
12. Grattan-Guiness I (1975) Fuzzy membership mapped onto interval and many-valued quantities. Math Logik Grundladen Math 22:149–160
13. Jahn KU (1975) Intervall-wertige mengen. Math Nach 68:115–132
14. Lee H (2001) An optimal algorithm for computing the max-min transitive closure of a fuzzy similarity matrix. Fuzzy Sets and Syst 123(1):129–136
15. Pankowska A, Wygralak M (2004) On hesitation degrees in if-set theory. In: ICAISC, pp 338–343
16. Pankowska A, Wygralak M (2006) General if-sets with triangular norms and their applications to group decision making. Inf Sci 176(18):2713–2754
17. Roy MK, Biswas R (1992) I-v fuzzy relations and sanchez's approach for medical diagnosis. Fuzzy Sets Syst 47:35–38
18. Sanchez E, Sambuc R (1976) Fuzzy relationships. phi-fuzzy functions. application to diagnostic aid in thyroid pathology. In: Proceedings of an international symposium on medical data processing, pp 513–524
19. Szmidt E, Kacprzyk J (1998) Group decision making under intuitionistic fuzzy preference relations. In: Proceedings of information processing and management of uncertainty in knowledge-based systems, IPMU, pp 172–178
20. Tizhoosh HR (2005) Image thresholding using type-2 fuzzy sets. Pattern Recognit 38:2363–2372
21. Zadeh LA (1975) The concept of a linguistic variable and its application to approximate reasoning I. Inf Sci 8:199–249

Use of Idempotent Functions in the Aggregation of Different Filters for Noise Removal

Luis González-Jaime, Mike Nachtegael, Etienne Kerre and Humberto Bustince

Abstract The majority of existing denoising algorithms obtain good results for a specific noise model, and when it is known previously. Nonetheless, there is a lack in denoising algorithms that can deal with any unknown noisy images. Therefore, in this paper, we study the use of aggregation functions for denoising purposes, where the noise model is not necessarily known in advance; and how these functions affect the visual and quantitative results of the resultant images.

Keywords Denoising · Idempotent function · Aggregation function · OWA operator

1 Introduction

One of the most popular restoration techniques has been, and it is nowadays, the image denoising. No matter how good the capturing process is, an image improvement is always needed. The desired goals of any denoising algorithm are

L. González-Jaime (✉) · M. Nachtegael · E. Kerre
Applied Mathematics and Computer Science, Ghent University,
Krijgslaan 281 - S9 9000 Ghent, Belgium
e-mail: luis.gonzalez@ugent.be

M. Nachtegael
e-mail: mike.nachtegael@ugent.be

E. Kerre
e-mail: etienne.kerre@ugent.be

H. Bustince
Departamento de Automática y Computación, Universidad Pública de Navarra,
Campus Arrosadia 31006 Pamplona, Spain
e-mail: bustince@unavarra.es

F. Sun et al. (eds.), *Knowledge Engineering and Management*,
Advances in Intelligent Systems and Computing 214,
DOI: 10.1007/978-3-642-37832-4_45,

to completely remove noise, while effective information (edge, corner, texture, and contrast...) is preserved, at the same time that artifacts do not appear.

Along the years, many algorithms have been proposed by researchers, where the most popular noise assumption is the additive Gaussian noise. However, a Gaussian noise [1, 2] assumption is too simplistic for most applications, for instance, for medical or astronomical images [3]. The performance of the algorithms decays drastically when the images are contaminated with a noise distribution for which these algorithms are not reliable. It would be desirable to find a blind denoising algorithm being able to deal with any noise distribution, without any previous knowledge about the noise model. Therefore, we focus our work on the study of the aggregation functions for a set of filtered images previously filtered from a noisy image with unknown noise distribution. Specifically, filters for impulse, Poisson, Gaussian, and Rician noise are applied. Then, different aggregation functions are used to verify their behavior for the denoising task.

Figure 1 shows the proposed schema. We start from a noisy image I_0, the idea is to use multifuzzy sets to build a new set from the filtered images, so each pixel (i, j) is represented by several values. But, we need to get a single fused image, I_{result}. Thus, we use idempotent aggregation functions. In concrete, we select *min*, *max*, *arithmetic mean*, and three *OWA operators*. In particular, OWA operators are built from fuzzy quantifiers because they provide a more flexible knowledge representation than classic logic [4]. Our aim is to obtain consistent and stable results, regardless of the image nature (e.g., computer tomography (CT), magnetic resonance image (MRI), digital image). Although the main application of this work is with MRI, because they present a more sophisticated noise, it however can be applied to other images with different nature.

The paper is composed as follows: Sect. 2.1 introduces the different noise models and filters. In Sect. 2.2, multifuzzy sets are explained. Then, Sect. 3 presents the idempotent functions and a specific case: the OWAs operators, a family of idempotent averaging functions. Finally, in Sects. 4 and 5, specific results and a final conclusion are exposed.

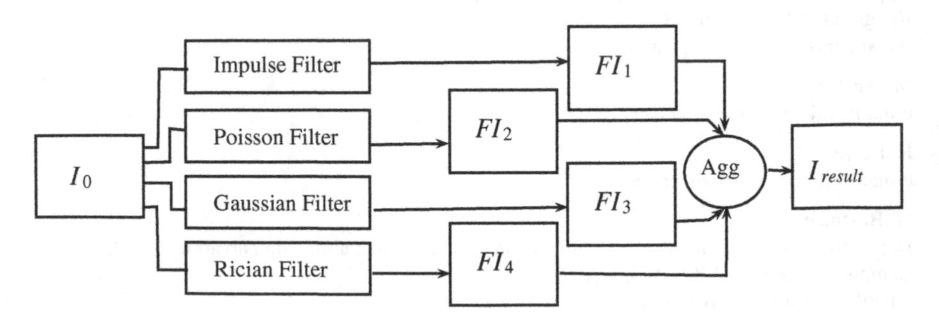

Fig. 1 Schema of the aggregation algorithm

2 Construction of Multifuzzy Sets from a Set of Filtered Images

Given an unknown noisy image, our first step consists in associating a multifuzzy set composed by several images. Each one of these images will be obtained by applying some filter optimized for a certain type of noise.

2.1 Noise Models and Filters

Several approaches exist that deal with Gaussian [1, 2] or impulse noise, although in some cases these are simple approximations compared to the real noise that is presented in the image. For instance, MRI, specifically MR magnitude image, are mainly characterized by Rician noise, although this noise is dependent on the number of coils or the reconstruction method [5]. Furthermore CT, single-photon emission computed tomography (SPECT), or positron emission tomography (PET) are identified by Poisson noise [6, 7]. The selected filters cover different approaches, as well as they perform better for a specific noise distribution. We give an overview of the characteristics of these filters. The first approach tackles the problem of impulse noise, and uses the DBAIN filter proposed by [8]. The considered filter to deal with white Gaussian noise has been the approach proposed by Goossens et al. [9]. This filter improves the non-local means (NLMeans) filter proposed by Buades et al. [2], dealing with noise in non-repetitive areas with a post-processing step and presenting a new acceleration technique. The approach used to estimate Rician noise is proposed by Aja-Fernandez et al. [10]. This filter adapts the linear minimum mean square error (LMMSE) to Rician distributed images. Finally, for Poisson noise, an extension of the NLMeans is proposed for images damaged by Poisson noise. Deledalle et al. [11] propose to adapt the similarity criteria of NLMeans algorithm to Poisson distribution data. For our experiments, the used parameters are those suggested in the original articles, as the algorithms are tuned to obtain good results.

2.2 Multifuzzy Sets

Once the set of filtered images is obtained, we represent them by means of multifuzzy sets, in which each element is given by a set of n memberships, taking n as the number of filters. A unique multifuzzy set is conformed with all the elements of the images.

Definition 1 A multifuzzy set of dimension $n \geq 2$ over a finite universe U is defined by a mapping $A : U \to [0,1]^n$ given by $A(u) = (A_1(u), \ldots, A_n(u))$ where each of the A_j for $j = 1, \ldots, n$ is a mapping $A_j : U \to [0,1]$.

We denote by $\mathcal{M}(U)$ the class of all multifuzzy sets on the referential set U.

Notice that the previous definition is equivalent to the following. Take a family of $n \geq 2$ fuzzy sets $Q_1, \ldots, Q_n$ on the same referential set U. Then a multifuzzy set on U is just the ordered combination of these n fuzzy sets as follows:

$$A = \{(u, A(u)) | u \in U\} \text{ given by } A(u) = (Q_1(u), \ldots, Q_n(u))$$

In this sense, the space of all multifuzzy sets inherits the order from the usual fuzzy sets, which endows it with a partial, bounded order.

In this work, we will deal with two finite referential sets $X = \{0, 1, \ldots, N-1\}$ and $Y = \{0, 1, \ldots, M-1\}$, where N and M are the number of rows and columns of the image, respectively. We will consider multifuzzy sets defined on the Cartesian product $X \times Y$.

Notice that a n dimensional multifuzzy set can also be understood as a type n fuzzy set, as well as an L-fuzzy set with $L = [0, 1]^n$ [12].

3 Idempotent Functions: Building a Fuzzy Set from Multifuzzy Sets

When the noisy image is filtered, we get a set of filtered images that composes the multifuzzy set. So, each pixel (i, j) is represented by n values, as many as filters used. This multifuzzy set needs to be fused in one single image, a fuzzy set. Therefore, we need functions that satisfy one condition: if all the values are the same, the value remains the same. For this reason, we decide to use idempotent functions.

Definition 2 An n-dimensional idempotent function is a mapping $\gamma : [0, 1]^n \rightarrow [0, 1]$ such that $\gamma(x, \ldots, x) = x$ for every $x \in [0, 1]$.

Example 1 An idempotent function is the *mode*, that is the value that occurs most frequently in a data set or a probability distribution.

Remark 1 Notice, that the *mode* from Example 1 is not monotone.

3.1 Construction of Idempotent Functions

In Proposition 1, we present a new method for constructing idempotent functions.

Proposition 1 *The mapping $\gamma : [0, 1]^n \rightarrow [0, 1]$ is an n-dimensional idempotent function if and only if there exist $f, g : [0, 1]^n \rightarrow [0, 1]$ such that*

(i) $g(x, \ldots, x) \neq 0$ *for every* $x \in [0, 1[$;
(ii) $f(x, \ldots, x) = \frac{x}{1-x} g(x, \ldots, x)$ for $x \in [0, 1[$, $f(1, \ldots, 1) = 1$, $g(1, \ldots, 1) = 0$;

(iii) $\gamma(x_1,\ldots,x_n) = \frac{f(x_1,\ldots,x_n)}{f(x_1,\ldots,x_n)+g(x_1,\ldots,x_n)}$.

Proof Assume that γ is an n-dimensional idempotent function. Take $f = \gamma$ and $g = 1 - \gamma$. Then

(i) $g(x,\ldots,x) = 1 - \gamma(x,\ldots,x) = 1 - x \neq 0$ for every $x \in [0,1[$.
(ii) $\frac{x}{1-x} g(x,\ldots,x) = \frac{x}{1-x}(1-x) = x = \gamma(x,\ldots,x) = f(x,\ldots,x)$ and $f(1,\ldots,1) = \gamma(1,\ldots,1) = 1$ and $g(1,\ldots,1) = 0$.
(iii) $\frac{f(x_1,\ldots,x_n)}{f(x_1,\ldots,x_n)+g(x_1,\ldots,x_n)} = \gamma(x_1,\ldots,x_n)$.

To see the converse, we only need to check the idempotency. But if γ is defined as in the statement of the proposition, we have that $\gamma(x,\ldots,x) = \frac{f(x,\ldots,x)}{f(x,\ldots,x)+g(x,\ldots,x)} = \frac{\frac{x}{1-x}g(x,\ldots,x)}{\frac{x}{1-x}g(x,\ldots,x)+g(x,\ldots,x)}$ which is equal to x for every $x \in [0,1[$. Finally, clearly $\gamma(1,\ldots,1) = 1$ □

Example 2

- Taking $f(x_1,\ldots,x_n) = \frac{1}{n}\sum_{i=1}^{n} x_i$ and $g(x_1,\ldots,x_n) = \frac{1}{n}\sum_{i=1}^{n}(1-x_i)$ we obtain as idempotent function the *arithmetic mean* (Eq. 1).

$$\gamma_{mean}(x_1,\ldots,x_n) = \frac{1}{n}\sum_{i=1}^{n} x_i \tag{1}$$

- Taking $f(x_1,\ldots,x_n) = \sqrt[n]{x_1 \cdot x_2 \cdot \ldots \cdot x_n}$ and $g(x_1,\ldots,x_n) = \max(1-x_1,\ldots, 1-x_n)$ we get Equation 2.

$$\gamma_{root}(x_1,\ldots,x_n) = \frac{\sqrt[n]{x_1 \cdot x_2 \cdot \ldots \cdot x_n}}{\sqrt[n]{x_1 \cdot \ldots \cdot x_n} + \max(1-x_1,\ldots,1-x_n)} \tag{2}$$

Regarding the structure of the space of n-dimensional idempotent functions, we also have the following:

Proposition 2 *Let* $\gamma_1, \gamma_2 : [0,1]^n \to [0,1]$ *be two n-dimensional idempotent functions. Then*:

1. $\frac{1}{2}(\gamma_1 + \gamma_2)$ *is also an n-dimensional idempotent function*;
2. $\sqrt{\gamma_1\gamma_2}$ *is also an n-dimensional idempotent function.*

3.2 Idempotent Aggregation Functions: Averaging Functions

Now we study *monotonic non-decreasing idempotent functions*, that are a special case of aggregation functions called *averaging functions*. With these functions we have not only idempotence, but also the result of the function will be bounded by the minimum and maximum of the arguments.

Definition 3 An aggregation function of dimension n (n-ary aggregation function) is a non-decreasing mapping $f : [0,1]^n \to [0,1]$ such that $f(0,\ldots,0) = 0$ and $f(1,\ldots,1) = 1$.

Definition 4 An aggregation function $f : [0,1]^n \to [0,1]$ is called averaging or a mean aggregation function if $\min(x_1,\ldots,x_n) \leq f(x_1,\ldots,x_n) \leq \max(x_1,\ldots,x_n)$.

Proposition 3 *Idempotent monotonic non-decreasing functions and idempotent averaging functions are the same.*

Example 3 Some averaging aggregation functions are the *arithmetic mean* (Eq. 1), *median* (Eq. 3) , *min*, or *max*.

$$\gamma_{med}(x_1,\ldots,x_n) = \begin{cases} \frac{1}{2}(x_k + x_{k+1}) & \text{if } n = 2k \\ x_k & \text{if } n = 2k-1 \end{cases} \quad (3)$$

3.3 Specific Case: OWA Operators and Fuzzy Quantifiers

Introduced by Yager [13], Ordered Weighted Averaging operators, commonly called OWA operators, are a parameterized family of idempotent averaging aggregation functions. They fill the gap between the operators *min* and *max*. The *min*, *max*, *arithmetic mean*, or *median* are particular cases of this family.

Definition 5 [13] A mapping $F : [0,1]^n \to [0,1]$ is called an OWA operator of dimension n if there exists a weighting vector W, $W = (w_1,\ldots,w_n) \in [0,1]^n$ with $\sum_{i=1}^n w_i = 1$ and such that $F(a_1,\ldots,a_n) = \sum_{j=1}^n w_j b_j$ with b_j the j-th largest of the a_i.

A natural question is how to obtain the associated weighting vector. Our idea is to calculate the weights for the aggregation operators using linguistic quantifiers, e.g., *about 5*, *a few*, *most*, and *nearly half*. The concept of fuzzy quantifiers was introduced by Zadeh [14], offering a more flexible tool for knowledge representation.

Yager suggested an interesting way to compute the weights of the OWA aggregation operator using fuzzy quantifiers [13], which, in the case of an increasing quantifier Q, is given by the expression 4.

$$w_i = Q(\frac{i}{n}) - Q(\frac{i-1}{n}) \qquad Q(r) = \begin{cases} 0 & \text{if } r<a \\ \frac{r-a}{b-a} & \text{if } a \leq r \leq b \\ 1 & \text{if } r>b \end{cases} \tag{4}$$

For the proportional increasing quantifiers, 'at least half' 'as many as possible' and 'most of them', the parameters (a, b) are $(0, 0.5)$, $(0.5, 1)$, and $(0.3, 0.8)$, respectively.

4 Results and Discussion

4.1 How the Algorithm Works: Visual Example

We start from a noisy image contaminated with Rician noise, Figure 2b. A multifuzzy set is composed with four filtered images (Figures 2c–f) optimized for a certain type of noise. We used the filters proposed in Sect. 2.1. Then, we need to build a fuzzy set from the multifuzzy set. For this, we use the idempotent functions. Those defined from the Eqs. 1, 2, and 3, over the multifuzzy set. Each obtained fuzzy set is presented as an image, shown in Figs. 3a–c, respectively.

4.2 Other Experiments

To be able to compare the results to a ground truth, we work with synthetic images artificially corrupted with noise. A magnitude MR volumen originally noise-free (from the BrainWeb database [15]) with 256 gray levels, is corrupted with Rician noise. The noisy images are processed using different filters (see Sect. 2.1). The aggregation functions used are *min*, *max*, *arithmetic mean*, and three OWA operators: *'at least half'*, *'as many as possible'*, and *'most of them'* (see Sect. 3.3).

To quantify the restoration performance of different methods, the PSNR is calculated. This is not bounded. A higher PSNR means better quality. However, it is not very well matched to perceived visual quality. This is our motivation to use also other quality indexes. In addition, the Mean Structural Similarity Index (MSSIM) [16] and the Quality Index based on Local Variance (QILV) [17] are used, giving a structural similarity measure. Nonetheless, the former is more sensitive to the level of noise and the latter to any possible blurring of the edges. Both indexes are bounded; the closer to one, the better the image. To avoid any bias due to background, the quality measures are only applied to those areas of the image that are relevant.

Two experiments were accomplished with noisy images corrupted with Rician noise, with $\sigma = 10$ and $\sigma = 20$. Table 1 shows that the Gaussian filter obtains the

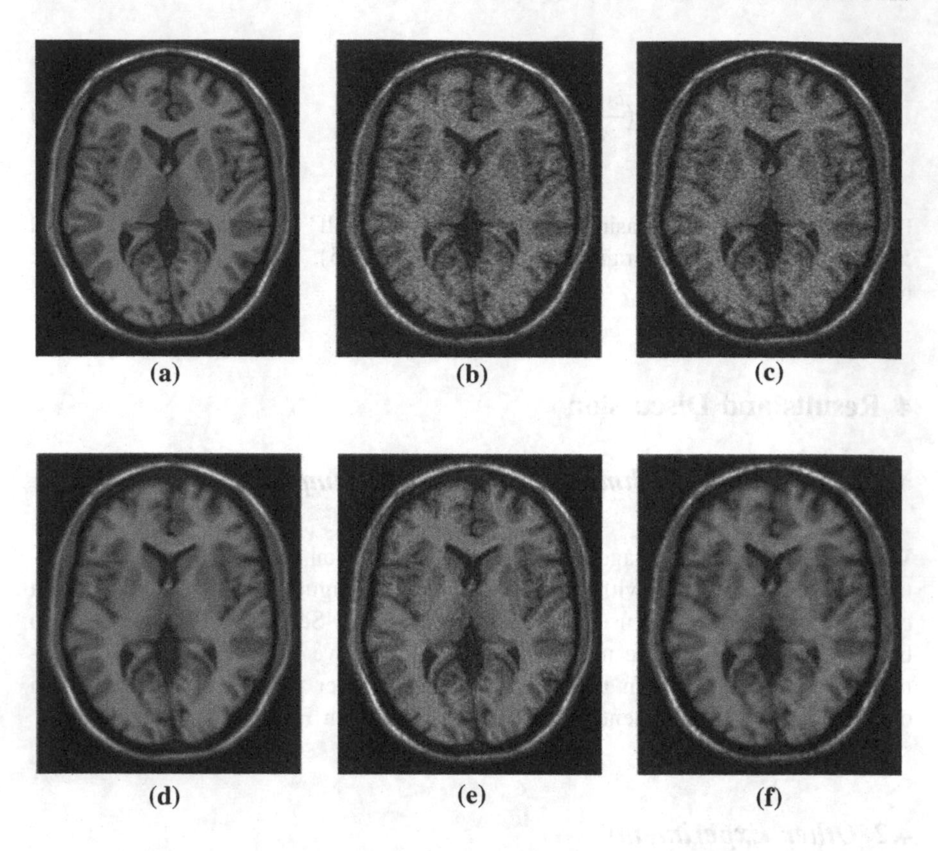

Fig. 2 Synthetic MR brain images (courtesy of Brainweb [15]). **a** Original MR image. **b** Noisy image ($\sigma = 20$). **c** Filtered image with an *impulse noise* filter. **d** Filtered image with a *Gaussian noise* filter. **e** Filtered image with a *Poisson noise* filter. **f** Filtered image with a *Rician noise* filter

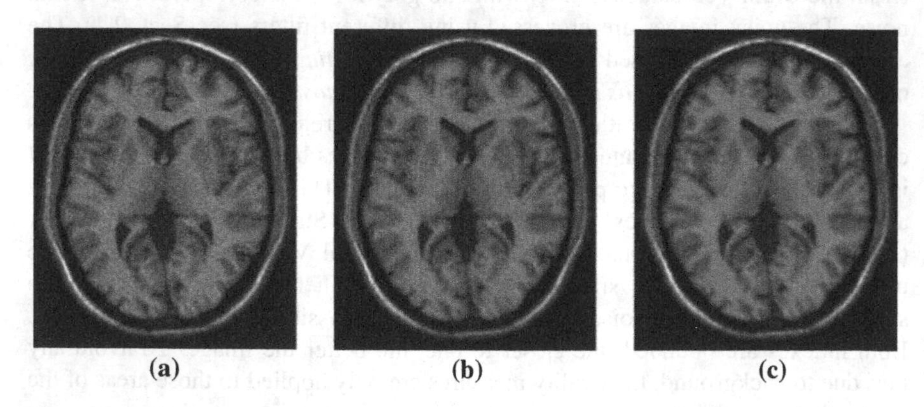

Fig. 3 Aggregated images using the Eq. 1 (γ_{mean}, **a**), Eq. 2 (γ_{root}, **b**), and Eq. 3 (γ_{median}, **c**)

Table 1 Results for the MR volumen, which contains 181 MR images contaminated with Rician noise with $\sigma = 10$ and $\sigma = 20$

	$\sigma = 10$						$\sigma = 20$					
Filter	PSNR		MSSIM		QILV		PSNR		MSSIM		QILV	
	mean	std	mean	std	mean	std	mean	std	mean	std	mean	std
Noisy	30.803	1.951	0.871	0.043	0.970	0.056	24.866	1.986	0.720	0.089	0.826	0.154
Impulse	30.803	1.960	0.872	0.972	0.043	0.053	24.906	2.005	0.720	0.089	0.835	0.150
Poisson	35.395	2.089	0.960	0.015	0.991	0.008	27.840	1.915	0.808	0.062	0.940	0.114
Gaussian	**36.966**	2.900	**0.970**	0.013	**0.994**	0.004	**32.629**	2.483	**0.927**	0.030	**0.970**	0.021
Rician	33.446	2.370	0.942	0.019	**0.994**	0.004	26.758	2.139	0.779	0.072	0.920	0.103
γ_{min}	33.128	2.273	0.930	0.026	**0.994**	0.006	27.736	2.347	0.823	0.063	0.965	0.048
γ_{max}	32.261	1.970	0.921	0.026	0.990	0.020	26.107	1.892	0.795	0.065	0.940	0.109
γ_{mean}	34.530	2.121	0.939	0.020	**0.994**	0.008	28.621	2.057	0.822	0.059	0.961	0.070
γ_{OWA_half}	32.830	1.992	0.928	0.023	0.991	0.017	26.915	1.911	0.802	0.063	0.944	0.100
γ_{OWA_many}	33.650	2.250	0.934	0.023	**0.994**	0.006	28.257	2.272	0.822	0.061	0.963	0.057
γ_{OWA_most}	34.324	2.114	0.940	0.019	**0.994**	0.008	28.291	2.064	0.815	0.061	0.954	0.079

Table 2 Results for the MR volumen, which contains 181 MR images contaminated with Poisson noise

Filter	PSNR		MSSIM		QILV		Filter	PSNR		MSSIM		QILV	
	mean	std	mean	std	mean	std		mean	std	mean	std	mean	std
(a)	31.314	2.661	0.883	0.051	0.977	0.023	(g)	32.501	2.887	0.925	0.035	0.993	0.005
(b)	31.290	2.650	0.883	0.052	0.978	0.024	(h)	35.293	2.836	0.941	0.028	**0.995**	0.002
(c)	**36.964**	3.496	**0.970**	0.015	0.993	0.004	(i)	32.993	2.816	0.927	0.034	0.993	0.005
(d)	34.166	3.021	0.941	0.028	0.967	0.022	(j)	33.129	2.737	0.929	0.033	0.991	0.004
(e)	31.314	2.661	0.883	0.052	0.977	0.024	(k)	34.807	2.797	0.937	0.030	**0.995**	0.002
(f)	32.488	2.660	0.928	0.032	0.985	0.010							

Legend: (a) Noisy; (b) Impulse; (c) Poisson; (d) Gaussian; (e) Rician; (f) γ_{min}; (g) γ_{max}; (h) γ_{mean}; (i) γ_{OWA_half}; (j) γ_{OWA_many}; (k) γ_{OWA_most}

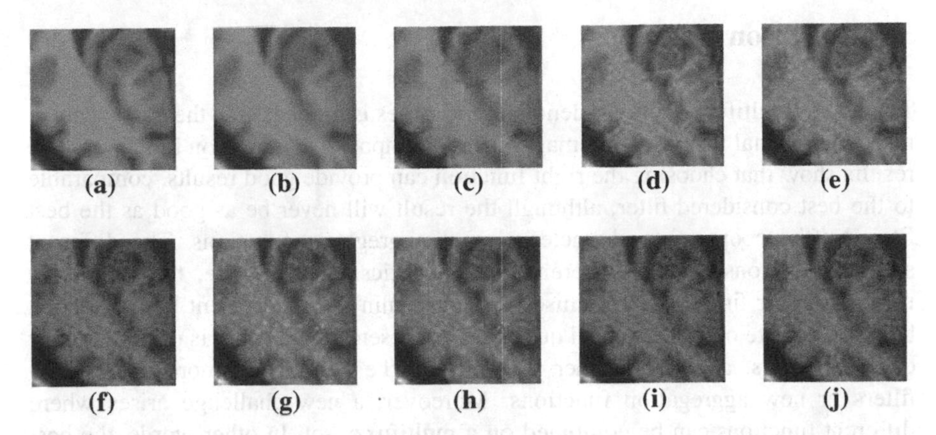

Fig. 4 Region cropped from a MR brain image. These are the results for the filtered and aggregated images from a noisy image contaminated with $\sigma = 20$. **a** Original. **b** Gaussian. **c** LMMSE. **d** Poisson. **e** γ_{mean}. **f** γ_{min}. **g** γ_{max}. **h** $\gamma_{OWAhalf}$. **i** $\gamma_{OWAmany}$. **j** $\gamma_{OWAmost}$

best results. However, the visual quality shows in Fig. 4b, that this filter over-filtered and blurries some regions, especially in the borders; consequently, it looses some important details (Fig. 4a). On the other side, Rician filter preserves more details (Fig. 4c), although visually it is less pleasant and almost does not filter close to the borders. Curiously, the Poisson filter obtains also good results, although the filter is not optimal for this type of noise; it mainly over-filters (Fig. 4d). It is also shown that results get affected when the noise level increases, since aggregation functions fuses the filtered images, that also get affected by noise. The statistics for the aggregation functions are quite similar, although their visual appearances are totally different. The min, max, or OWA 'at least half' present images that after being aggregated still look like are contaminated with impulse noise. These results are not interesting for denoising. However, some areas, close to the borders, are better defined for these functions (Figs. 4f, g), except for the presence of undesired noise. On the other side, the arithmetic mean or the OWA 'as many as possible' show a better compromise between the visual and quantitative quality (Figs. 4e, i).

We presented an approach that is noise type independent. For this reason, Table 2 also shows the use of the same algorithm for the same MR volumen contaminated with Poisson noise.[1] The results show that some aggregation functions, as the mean or 'OWA most of them', obtain comparable results to the Poisson filter.

[1] It is known that MR images are not contaminated with Poisson noise, but this was carried out for study purposes.

5 Conclusion

The use of multifuzzy sets for denoising purposes is probed, how these sets can be merged in a final fuzzy set (an image) using idempotent aggregation functions. The results show that choosing the right function can provide good results, comparable to the best considered filter, although the result will never be as good as the best filter by the cooperation characteristics of aggregation functions. The different studied functions present different characteristics. For instance, the arithmetic mean operator finds a compromise, while the min or max present better defined borders, despite of a poor global quality. The presented algorithm is used with four different filters, although further research can be done using more and/or new filters or new aggregation functions. Moreover, a new challenge arises where different functions can be combined on a multifuzzy set. In other words, the best aggregation function is chosen for each pixel, looking for a compromise and presenting a new tool for blind denoising.

Acknowledgments This work is supported by the European Commission under contract no. 238819 (MIBISOC Marie Curie ITN) and by the National Science Foundation of Spain, reference TIN2010-15055.

References

1. Rudin LI, Osher S, Fatemi E (1992) Nonlinear total variation based noise removal algorithms. Physica D 60(1–4):259–268
2. Buades A, Coll B, Morel JM (2005) A review of image denoising algorithms, with a new one. Multiscale Model Simul 4(2):490–530
3. Molina R, Nunez J, Cortijo FJ, Mateos J (2001) Image restoration in astronomy—a Bayesian perspective. IEEE Signal Process Mag 18(2):11–29
4. Chiclana F, Herrera F, Herrera-Viedma E (1998) Integrating three representation models in fuzzy multipurpose decision making based on fuzzy preference relations. Fuzzy Sets Syst 97(1):33–48
5. Aja-Fernandez S, Tristan-Vega A, Alberola-Lopez C (2009) Noise estimation in single- and multiple-coil magnetic resonance data based on statistical models. Magn Reson Imaging 27(10):1397–1409
6. Suzuki S (1985) A comparative-study on pre-smoothing techniques for projection data with poisson noise in computed-tomography. Opt Commun 55(4):253–258
7. Rosenthal MS, Cullom J, Hawkins W, Moore SC, Tsui BMW, Yester M (1995) Quantitative SPECT imaging—a review and recommendations by the focus committee of the society-of-nuclear-medicine computer and instrumentation council. J Nucl Med 36(8):1489–1513
8. Srinivasan KS, Ebenezer D (2007) A new fast and efficient decision-based algorithm for removal of high-density impulse noises. IEEE Signal Process Lett 14(3):189–192
9. Goossens B, Luong Q, Pizurica A, Philips W (2008) An improved non-local denoising algorithm, In: Proceedings of local and non-local approximation in image processing, international workshop, 2008, pp 143–156
10. Aja-Fernandez S, Alberola-Lopez C, Westin C-F (2008) Noise and signal estimation in magnitude MRI and Rician distributed images: a LMMSE approach. IEEE Trans Image Process 17(8):1383–1398

11. Deledalle C-A, Tupin F, Denis L (2010) Poisson NL means: unsupervised non local means for poisson noise. In: 2010 IEEE international conference on image processing
12. Klir GJ, Yuan B (1995) Fuzzy sets and fuzzy logic. Prentice Hall, Upper Saddle River
13. Yager RR (1988) On ordered weighted averaging aggregation operators in multicriteria decision-making. IEEE Trans Syst Man Cybern 18(1):183–190
14. Zadeh LA (1983) A computational approach to fuzzy quantifiers in natural languages. Comput Math Appl 9(1):149–184
15. Brain Web: Simulated Brain Database. URL http://mouldy.bic.mni.mcgill.ca/brainweb/
16. Wang Z, Bovik AC, Sheikh HR, Simoncelli EP (2004) Image quality assessment: from error visibility to structural similarity. IEEE Trans Image Proces 13(4):600–612
17. Aja-Fernandez S, Estepar RSJ, Alberola-Lopez C, Westin CF (2006) Image quality assessment based on local variance. In: Conference proceedings. Annual international conference of the IEEE engineering in medicine and biology society

Benchmarking for Stability Evaluation of Computer Systems

Victoria Lopez, Guadalupe Miñana, Juan Tejada and Raquel Caro

Abstract In this paper we propose a new benchmark to drive making decisions in maintenance of computer systems. This benchmark is made from load average sample data. The main goal is to improve reliability and performance of a set of devices or components. In particular, the stability of the system is measured in terms of variability of the load. A forecast of the behavior of this stability is also proposed as part of the reporting benchmark. At the final stage, a more stable system is obtained and its global reliability and performance can be then evaluated by means of appropriate specifications.

Keywords Reliability · Stability · Operating systems · Computer systems · Monitoring · Benchmarking

V. Lopez (✉) · G. Miñana
Mobile Technologies and Biotechnology Research Group (G-TeC),
Complutense University, Madrid, Spain
e-mail: vlopez@fdi.ucm.es

G. Miñana
e-mail: guamiro@fdi.ucm.es

J. Tejada
Mobile Technologies and Biotechnology Research Group (G-TeC),
Complutense University, Madrid, Spain
e-mail: jtejada@mat.ucm.es

R. Caro
Mobile Technologies and Biotechnology Research Group (G-TeC),
Industrial Organization Department, P. Comillas University, Madrid, Spain
e-mail: rcaro@upcomillas.es

F. Sun et al. (eds.), *Knowledge Engineering and Management*,
Advances in Intelligent Systems and Computing 214,
DOI: 10.1007/978-3-642-37832-4_46,

1 General Introduction

This paper presents a new indicator to measure the stability of a computer system with respect to its processors. Nowadays processors are becoming more vulnerable since hundreds of small applications are running on them. This situation makes it of particular interest not only the study of the stability of traditional systems but also in small mobile devices.

The great growth market of new technologies has meant that many of the applications that are installed daily in our devices have no real control of reliability. As a result, the systems often fail. Although mobile device manufacturers have included malfunction detectors in their operating systems, they need the help of the user to collect the data about fails of the system. In this sense, an index to report stability or instability of the system is always interesting if it can make predictions about the future behavior of the load and therefore failure prevention.

This paper is structured as follows. Sect. 2 serves as an introduction to performance evaluation of computer systems. In Sect. 3 we introduce the issue of reliable systems and their main characteristics. Section 4 shows how to deal with monitors and benchmarking. The mathematical model of the stability in computer systems is developed in Sect. 5. Then, in Sect. 6 a new benchmark for measuring stability is proposed as well as several examples. Finally, conclusions and future work is presented in Sect. 7.

2 Introduction to Performance Evaluation

Performance evaluation deals with studies of computers, computer communications, telecommunications, and distributed systems between others. That includes not only system reliability and risk studies but also resource allocation and control methods. Performance evaluation also includes design of algorithms as routing and flow control in networks, memory management, bandwidth allocation, processor scheduling, etc. System tuning, capacity planning, and others are typical performance evaluation applications. There are a lot of these studies on wireless sensor networks.

In this paper we are working on modeling and analysis of a monitor in which a data analysis is done about queuing and scheduling of the processor. The technique of measurement is by means of a Linux software monitor. The system architecture is taken into account and discussed by mean of numerical examples.

Finally, a benchmark about the stability of the system is developed as performance evaluation application.

3 Introduction to Reliability of Computer Systems

Computing systems are of growing importance because of their wide use in many areas including those in safety-critical systems. Computer system reliability deals with the basic models and it approaches to the reliability analysis of such systems [1, 2]. Improving system stability and reliability is very important. Improving application performance and usability are a good way to prevent system crashes that arise due to overloaded system processor and uncontrollable applications.

System stability problems arise due to undersized servers, improperly tuned services and upgrades, poor configurations, and hardware issues.

Mainly due to the desire for hardware cost reduction, server consolidation through performance optimization has been of primary interest. Fewer servers result in lower licensing, maintenance, electricity, and cooling costs. But the motivation behind improved performance is not solely about lowering costs. There is a perpetual balance being sought between system performance and ensuring end users receive a consistent experience.

System stability can be achieved through hardware resource optimization including CPU, physical and virtual memory, disk access, and application state. Whether the unstable system symptoms you experience are random segmentation faults, data corruption, hard locks, or lost data, hardware glitches can make your normally reliable operating system barely useable.

Under instability problems, it is a good idea to perform CPU and memory tests. In doing so, you may detect a hardware problem that could have bitten you at an inopportune time, something that could have caused data loss or hours of frustration as you engage in a frantic search for the source of the problem.

4 Network Monitoring and Benchmarking

4.1 Network Monitoring

This term deals with the use of a system that constantly monitors a computer network for events as slow or failing components and that notifies the network administrator (via email, SMS, or reporting files) in case of outages. It is part of the functions involved in network management. In this paper, network monitoring is used by means of Linux software monitors as $top or $uptime. Other Linux monitors (such us $time, $ps, and $free) also report information about performance and reliability of the system. More specifically we are interested in load average samples. These kinds of software monitors return the information with no significant overhead into the system.

Table 1 shows an example of execution of Top Linux command. The output includes information about users, processes, memory capacity, cpu performance, and tasks. Commonly measured metrics are response time, availability, and

Table 1 Example of top command Linux

Top	*$ top*
top—16:29:15 up 1:27, 2 users, load average: 0.18, 0.16, 0.09	
Tasks: 210 total, 1 running, 208 sleeping, 0 stopped, 1 zombie	
Cpu(s): 1.2 %us, 0.2 %sy, 0.0 %ni, 98.5 %id, 0.0 %wa, 0.0 %hi, 0.0 %si, 0.0 %st	
Mem: 6090036 k total, 1072944 k used, 5017092 k free, 13596 k buffers	
Swap: 10806264 k total, 0 k used, 10806264 k free, 644604 k cached	
2531 root 20 0 138 m 46 m 18 m S 5 0.8 1:58.54 Xorg	
3982 user 20 0 306 m 15 m 10 m S 3 0.3 0:09.04 gnome-terminal	
3509 user 20 0 758 m 174 m 92 m S 1 2.9 2:06.58 soffice.bin	
...	

uptime. However, both consistency and reliability metrics are starting to gain popularity since the widespread addition of mobile devices is having an adverse effect on most network monitoring tools.

Not only status request failures produce an action from the monitoring system. Other actions, such as a new task entering into the system or any kind of event, produce an action from the monitoring system as well. When a connection cannot be established, it times-out, or the document or message cannot be retrieved, for example, an alarm may be sent (via SMS, email, etc.) to the resident sysadmin to manage the system or remove the troubled server from duty until it can be repaired, etc.

Monitoring the performance of a network uplink is also known as network traffic measurement, and more software is listed there. In this sense, mostly operating systems store information from monitoring into files. For example, Linux stores information into/proc/cpuinfo folder.

4.2 Benchmarking

In computing, a benchmark is the act of running a computer program, a set of programs, or other operations, in order to assess the relative performance of an object, normally by monitoring the behavior of the device into the system or running a number of standard tests and trials against it. The benchmark can also work on the whole system as a set of devices within a net.
On the other hand, benchmarking is usually associated with assessing performance characteristics of computer hardware and software. For example, it is very popular to measure the floating point operation performance of a CPU. Benchmarks used to provide a method of comparing the performance of various subsystems across different chip/system architectures.

In this work, we consider the data from monitoring several processors (devices) and we propose a benchmark to study stability of the computers and to make comparisons between them. The benchmark process drives the expert to make a decision and help him/her as an aid support system in maintenance issues.

5 Stability Model in Computer Systems

In mathematics, stability theory addresses the stability of solutions of differential equations and of trajectories of dynamical systems under small perturbations of initial conditions. In computer systems, a configuration is stable if running process in the model lead to small variations in the performance. One must specify the metric used to measure the perturbations. In partial differential equations one may measure the distances between functions, while in differential geometry one may measure the distance between spaces. In computer systems one may measure the load average, for example, of the processors by time series techniques, more specifically by means of monitoring data and its softness with mobile moving average.

In dynamical systems, various criteria have been developed to prove stability or instability of a system. Under favorable circumstances, the question may be reduced to a well-studied problem involving eigenvalues of matrices. A more general method involves fixed points or equilibrium.

Although computer systems are dynamical systems, in this research we have had the decision to study stability in terms of no variability and then define a way based on statistics and time series analysis to analyze the behavior of the processor of a computer system as the main device and scheduler of task into the system.

5.1 Data Collection

Data is collected by means of monitoring. Linux commands can be used here to get all the information during a fixed period in the time. In particular, we are interested in load of the system as the measure of task that the system is processing at the moment of software monitor execution. This information can be got from the load average triplet (LA, [3]). This triplet consist of three real numbers that store the load average at the last minute, the load average during the last 5 min, and the load average during the last 15 min into the system. Figures 1, 2 and 3 show a sample of 30 LA data.

There are some studies in relation to the interpretation and analysis of the load average triplet [3], most of the times in order to get information about the future behavior of the system. Some of these studies drive to disagreements between the conclusions about increasing, decreasing, or stability of the load.

In this sense we define a new indicator as measure of the variability and therefore, the stability of the processor and system.

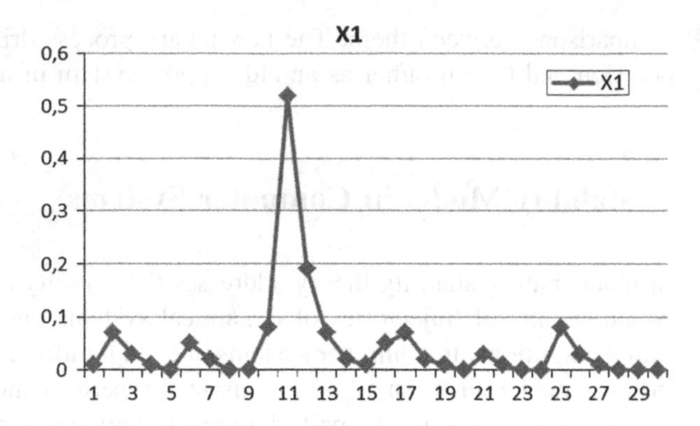

Fig. 1 30-sample LA last minute

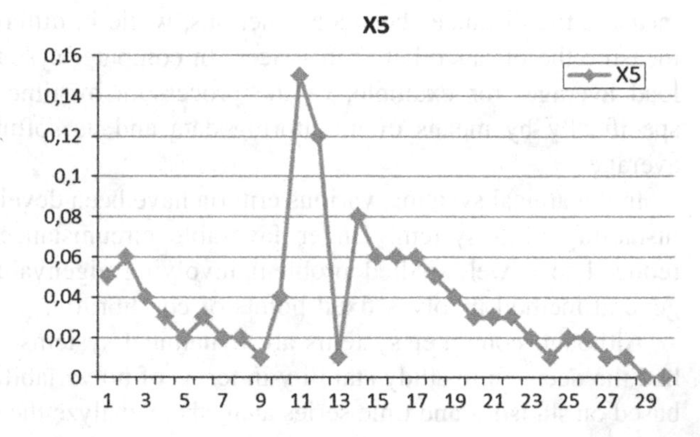

Fig. 2 30-sample LA last 5 min

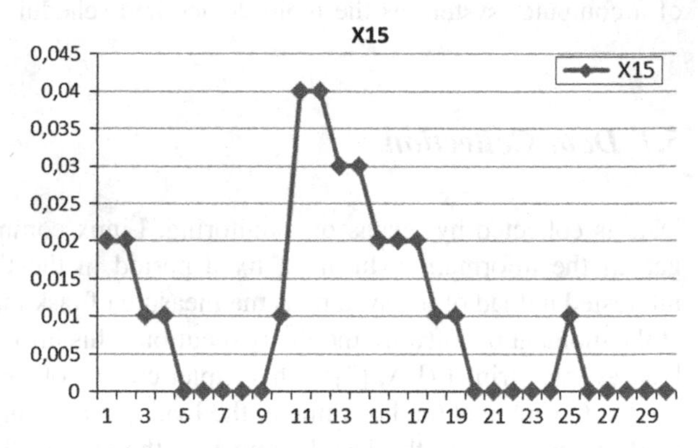

Fig. 3 30-sample LA last 15 min

5.2 Data Analysis

The Load Average [4] (LA) triplet is defined as follows

$$LA = (l_1, l_2, l_3) : l_i \geq 0 \forall i = \{1, 2, 3\} \tag{1}$$

Each number into the triplet is the solution of a system of equations as Eq. (2) shows. In the following x_i denotes the system load in the last ith minute.

$$\begin{aligned}&\exists(x_1, x_2, \ldots x_{15}) :\\&\forall i \in \{1\ldots 15\} : (x_i \geq 0)\Lambda\\&(l_1 = x_{15})\Lambda\left(5l_2 = \sum_{i=11}^{15} x_i\right)\Lambda\left(15l_3 = \sum_{i=1}^{15} x_i\right)\end{aligned} \tag{2}$$

It is possible to find the value of the 15 numbers (making an exhaustive monitoring), nevertheless it constitutes an overhead in monitoring and the information that they offer is not too much informative. The study of the most and the less stable samples is enough informative. These samples can be calculated without monitoring, and therefore without adding overhead to the system.

5.3 Mathematical Model

The model is a nonlinear optimization problem, under restrictions of (2) and the following cost functions.

$$\min_{x_i} \sum_{i=1}^{15} (x_i - l_3)^2 \ \max_{x_i} \sum_{i=1}^{15} (x_i - l_3)^2 \tag{3}$$

These models return the most and the less stable samples as solution, respectively.

The structure $(\alpha\,\alpha\,\alpha\,\alpha\,\alpha\,\alpha\,\alpha\,\alpha\,\alpha\,\alpha\,\beta\,\beta\,\beta\,\beta\,\delta)$ will be the solution of the first problem (most stable sample) and the structure $(\alpha'\ 0\ 0\ 0\ 0\ 0\ 0\ 0\ 0\ 0\ \beta'\ 0\ 0\ 0\ \delta)$ is the less stable. In both cases, the structure covers a set of solutions all of them with the same optimal value. It can be easily shown by the Karush–Kuhn–Tucker [5] (KKT) conditions which are first order necessary conditions for a solution in nonlinear programming to be optimal.

5.4 Some Properties and Solution

Thanks to the goodness of the relation between parameters [3] α, $\alpha'\beta$, β' and δ, the stability of the system can be defined in an easy way: the solution to the first problem (minimize) is:

$$\alpha = (3l_3 - l_2)/2,\ \beta = (5l_2 - l_1)/4 \text{ and } \delta = l_1.$$

and the solution to the second one (maximize) is:

$$\alpha' = (15l_3 - 5l_2),\ \beta' = (5l_2 - l_1) \text{ and } \delta = l_1.$$

Since the load measure is a non-negative number and there is not an only sample but several samples which also produce the same maximum (or the same minimum) value, under a discrete universe, the less stable sample has more probability than the more stable sample which only appears essentially once.

6 Benchmark Definition

Finally, we are ready to define and develop the benchmark. On one hand, we define the number

$$D = 1/14(90\alpha^2 + 12\beta^2) \quad (4)$$

D is a new index to measure stability of the system [3]. This number is the distance between the two variances of the samples solution, that is the difference between the cost function applied to their sample solutions.

On the other hand, we are ready to develop a new module into EMSI tool [6–8] to monitor, analyze, and manage the system at execution time.

7 Conclusions and Future Work

This work is a part of the modeling and software development of EMSI tool [6–8]. More specifically, in this article, we show a new benchmark to measure stability in a computer system and its mathematical model to be developed in the software tool as a new module. In the near future, this benchmark is going to be used to make predictions about the stability, then next goal will be to complete in this way the aid system and fit it into the EMSI tool (Performance and Reliability for Computer Systems) which is already available in [6].

References

1. López V (2007) Performance and reliability of computer systems. EME-Editorial, Madrid
2. Sahner RA, Trivedi KS, Puliafito A (2002) Performance and reliability of computer systems. Kluwer Academic, Amsterdam
3. López V, Miñana G (2012) Modeling the stability of a computer system. Int J Uncertain, Fuzziness Knowl Based Syst 20:81–90
4. Gunther NJ (2005) Analyzing computer system performance with perl: PDQ. Springer, New York
5. Kuhn HW, Tucker AW (1951) Nonlinear programming. In: Proceedings of 2nd Berkeley symposium. University of California Press, Berkeley, pp 481–492
6. http://www.tecnologiaucm.es

7. Lopez V, Santos M, Montero J (2009) Improving reliability and performance in computer systems by means of fuzzy specification. In: 4th international conference on intelligent systems and knowledge engineering, Hasselt
8. Lopez V, Santos M, Rodriguez T (2011) An approach to the analysis of performance, reliability and risk in computer systems. In: 6th international conference on intelligent systems and knowledge engineering, vol 2, pp 653–658, Shanghai, ISSN: 1867-5662

Relationships Between Average Depth and Number of Nodes for Decision Trees

Igor Chikalov, Shahid Hussain and Mikhail Moshkov

Abstract This paper presents a new tool for the study of relationships between total path length or average depth and number of nodes of decision trees. In addition to algorithm, the paper also presents the results of experiments with datasets from UCI ML Repository [1].

Keywords Decision trees · Number of nodes · Total path length · Average depth

1 Introduction

Decision trees are widely used as predictors, as a way of knowledge representation, and as algorithms for problem solving. These uses require optimizing decision trees for certain cost functions such as the number of misclassifications, depth/average depth, and number of nodes. That is, minimizing one of these cost functions yields more accurate, faster, or more understandable decision trees (respectively).

We have created a software system for decision trees (as well as decision rules) called DAGGER—a tool based on dynamic programming which allows us to optimize decision trees (and decision rules) relative to various cost functions such as depth (length), average depth (average length), total number of nodes, and number

---

I. Chikalov · S. Hussain (✉) · M. Moshkov
Computer, Electrical and Mathematical Sciences and Engineering, King Abdullah University of Science and Technology, Thuwal 23955-6900, Saudi Arabia
e-mail: shahid.hussain@kaust.edu.sa

I. Chikalov
e-mail: igor.chikalov@kaust.edu.sa

M. Moshkov
e-mail: mikhail.moshkov@kaust.edu.sa

F. Sun et al. (eds.), *Knowledge Engineering and Management*,
Advances in Intelligent Systems and Computing 214,
DOI: 10.1007/978-3-642-37832-4_47,

of misclassifications sequentially [2–5]. The aim of this paper is to study the relationships between total path length (average depth) and number of nodes of decision trees and present a new tool (an extension of our software) for computing such relationships. We also consider the work of this tool on decision tables from UCI ML Repository [1].

The presented algorithm and its implementation in the software tool DAGGER together with similar algorithms devised by the authors (see for example [6]) can be useful for investigations in Rough Sets [7, 8] where decision trees are used as classifiers [9].

This paper is divided into six sections including Introduction. Section 2 presents some basic notions related to decision tables, decision trees, and the two cost functions. Section 3 gives an algorithm to construct a directed acyclic graph (DAG) $\Delta(T)$ that captures all possible decision trees for a given decision table T. The main algorithm for computing relationships between total path length (average depth) and number of nodes is presented in Sect. 4. Section 5 shows some experimental results of work of the algorithm on data tables acquired from UCI ML Repository and Sect. 6 concludes the paper followed by References and an appendix for "transformation of functions" proposition used in Sect. 4.

2 Basic Notions

In the following section we define the main notions related to the study of decision trees and tables and two cost functions for decision trees.

2.1 Decision Tables and Decision Trees

In this paper, we consider only decision tables with discrete attributes. These tables do not contain missing values and equal rows. Consider a *decision table T* depicted in Fig. 1. Here $f_1, \ldots, f_m$ are the conditional attributes; $c_1, \ldots, c_N$ are nonnegative integers which can be interpreted as the decisions (values of the decision attribute d); b_{ij} are nonnegative integers which are interpreted as values of conditional attributes (we assume that the rows $(b_{11}, \ldots, b_{1m}), \ldots, (b_{N1}, \ldots, b_{Nm})$ are pairwise different). We denote by $E(T)$ the set of attributes (columns of the table T), each of

Fig. 1 Decision table

f_1	$\cdots$	f_m	d
b_{11}	$\cdots$	b_{1m}	c_1
	$\vdots$		$\vdots$
b_{N1}	$\cdots$	b_{Nm}	c_N

which contains different values. For $f_i \in E(T)$, let $E(T, f_i)$ be the set of values from the column f_i. We denote by $N(T)$ the number of rows in the decision table T.

Let $f_{i_1}, \ldots, f_{i_t} \in \{f_1, \ldots, f_m\}$ and $a_1, \ldots, a_t$ be nonnegative integers. We denote by $T(f_{i_1}, a_1)\ldots(f_{i_t}, a_t)$ the subtable of the table T, which consists of such and only such rows of T that at the intersection with columns $f_{i_1}, \ldots, f_{i_t}$ have numbers $a_1, \ldots, a_t$, respectively. Such nonempty tables (including the table T) will be called *separable subtables* of the table T.

For a subtable Θ of the table T we will denote by $R(\Theta)$ the number of unordered pairs of rows that are labeled with different decisions.

A *decision tree* Γ *over* the table T is a finite directed tree with a root in which each terminal node is labeled with a decision. Each nonterminal node is labeled with a conditional attribute, and for each nonterminal node, the outgoing edges are labeled with pairwise different nonnegative integers. Let v be an arbitrary node of Γ. We now define a subtable $T(v)$ of the table T. If v is the root then $T(v) = T$. Let v be a node of Γ that is not the root, nodes in the path from the root to v be labeled with attributes $f_{i_1}, \ldots, f_{i_t}$, and edges in this path be labeled with values $a_1, \ldots, a_t$, respectively. Then $T(v) = T(f_{i_1}, a_1)\ldots(f_{i_t}, a_t)$.

Let Γ be a decision tree. We say that Γ is a *decision tree for* T if any node v of Γ satisfies the following conditions:

- If $R(T(v)) = 0$ then v is a terminal node labeled with the common decision for $T(v)$.
- Otherwise, v is labeled with an attribute $f_i \in E(T(v))$ and, if $E(T(v), f_i) = \{a_1, \ldots, a_t\}$, then t edges leave node v, and these edges are labeled with $a_1, \ldots, a_t$ respectively.

Let Γ be a decision tree for T. For any row r of T, there exists exactly one terminal node v of Γ such that r belongs to the table $T(v)$. Let v be labeled with the decision b. We will say about b as the *result of the work of decision tree* Γ *on* r.

For an arbitrary row r of the decision table T, we denote by $l(r)$ *the length of path* from the root to the terminal node v of T such that r is in $T(v)$. We say that *the total path length*, represented as Λ, is the sum of path lengths for all rows in T. That is

$$\Lambda(T, \Gamma) = \sum_r l(r),$$

where we take the sum on all rows r of the table T. Note that *average depth*, represented as h_{avg} of Γ is equal to the total path length divided by the total number of rows in T i.e.,

$$h_{\mathrm{avg}}(T, \Gamma) = \frac{\Lambda(T, \Gamma)}{N(T)}.$$

We will drop T when it is obvious from the context. That is, we will write $\Lambda(\Gamma)$ instead of $\Lambda(T, \Gamma)$ if T is known.

For a decision tree Γ of a decision table T, we represent the total number of nodes of Γ by $L(\Gamma)$. It is interesting to note that the cost functions Λ and L are bounded above by values depending upon the size of the table. That is, mN and $2N-1$ are the upper bounds for Λ and L for a decision table with m conditional attributes and N rows.

3 Representation of Sets of Decision Trees

Consider an algorithm for construction of a graph $\Delta(T)$, which represents the set of all decision trees for the table T. Nodes of this graph are some separable subtables of the table T. During each step we process one node and mark it with the symbol *. We start with the graph that consists of one node T and finish when all nodes of the graph are processed.

Assume the algorithm has already performed p steps. We now describe the step number $(p+1)$. If all nodes are processed then the work of the algorithm is finished, and the resulting graph is $\Delta(T)$. Otherwise, choose a node (table) Θ that has not been processed yet. If $R(\Theta)=0$, label the considered node with the *common decision b* for Θ, mark it with symbol * and proceed to the step number $(p+2)$. If $R(\Theta)>0$, then for each $f_i \in E(\Theta)$ draw a bundle of edges from the node Θ (this bundle of edges will be called *f_i-bundle*). Let $E(\Theta,f_i)=\{a_1,\ldots,a_t\}$. Then draw t edges from Θ and label these edges with pairs $(f_i,a_1),\ldots,(f_i,a_t)$ respectively. These edges enter into nodes $\Theta(f_i,a_1),\ldots,\Theta(f_i,a_t)$. If some of the nodes $\Theta(f_i,a_1),\ldots,\Theta(f_i,a_t)$ are not present in the graph then add these nodes to the graph. Mark the node Θ with the symbol * and proceed to the step number $(p+2)$. Now for each node Θ of the graph $\Delta(T)$, we describe the set of decision trees corresponding to the node Θ. We will move from terminal nodes, which are labeled with numbers, to the node T. Let Θ be a node, which is labeled with a number b. Then the only trivial decision tree depicted in Fig. 2 corresponds to the node Θ.

Let Θ be a nonterminal node (table) then there is a number of bundles of edges starting in Θ. We consider an arbitrary bundle and describe the set of decision trees corresponding to this bundle. Let the considered bundle be an f_i-bundle where $f_i \in (\Theta)$ and $E(\Theta,f_i)=\{a_1,\ldots,a_t\}$. Let $\Gamma_1,\ldots,\Gamma_t$ be decision trees from sets corresponding to the nodes $\Theta(f_i,a_1),\ldots,\Theta(f_i,a_t)$. Then the decision tree depicted in Fig. 3 belongs to the set of decision trees, which correspond to this bundle. All such decision trees belong to the considered set, and this set does not contain any other decision trees. Then the set of decision trees corresponding to the node Θ coincides with the union of sets of decision trees corresponding to the bundles starting in Θ. We denote by $D(\Theta)$ the set of decision trees corresponding to the node Θ.

The following proposition shows that the graph $\Delta(T)$ can represent all decision trees for the table T.

Proposition 1 *Let T be a decision table and Θ a node in the graph $\Delta(T)$. Then the set $D(\Theta)$ coincides with the set of all decision trees for the table Θ.*

Fig. 2 Trivial DT

b

Fig. 3 Aggregated DT

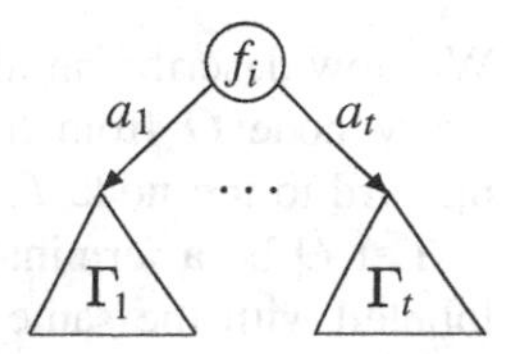

Proof We prove this proposition by induction on nodes in the graph $\Delta(T)$. For each terminal node Θ, only one decision tree exists as depicted in Fig. 2, and the set $D(T)$ contains only this tree. Let Θ be a nonterminal node and the statement of proposition hold for all its descendants.

Consider an arbitrary decision tree $\Gamma \in D(\Theta)$. Obviously, Γ contains more than one node. Let the root of Γ be labeled with an attribute f_i and the edges leaving root be labeled with the numbers $a_1, \ldots, a_t$. For $j = 1, \ldots, t$, denote by Γ_j the decision tree connected to the root with the edge labeled with the number a_j. From the definition of the set $D(\Theta)$ it follows that f_i is contained in the set $E(\Theta)$, $E(\Theta, f_i) = \{a_1, \ldots, a_t\}$ and for $j = 1, \ldots, t$, the decision tree Γ_j belongs to the set $D(\Theta(f_i, a_j))$. According to the inductive hypothesis, the tree Γ_j is a decision tree for the table $\Theta(f_i, a_j)$. Then the tree Γ is a decision tree for the table Θ.

Now we consider an arbitrary decision tree Γ for the table Θ. According to the definition, the root of Γ is labeled with an attribute f_i from the set $E(\Theta)$, edges leaving the root are labeled with numbers from the set $E(\Theta, f_i)$ and the subtrees whose roots are nodes, to which these edges enter, are decision trees for corresponding descendants of the node Θ. Then, according to the definition of the set $D(\Theta)$ and to inductive hypothesis, the tree Γ belongs to the set $D(\Theta)$.

4 Relationships

In the following we consider relationships between average depth (total path length) and number of nodes for decision trees and give an algorithm to compute the relationships. We also provide an illustration of working of the algorithm on an example decision table.

Let T be a decision table with N rows and m columns labeled with $f_1, \ldots, f_m$, and $D(T)$ be the set of all decision trees for T (as discussed in Sects. 2 and 3). We will use the notion of total path length instead of average depth for clarity and ease of implementation.

We denote $B_{\Lambda,T} = \{\beta, \beta + 1, \ldots, mN\}$ and $B_{L,T} = \{\alpha, \alpha + 1, \ldots, 2N - 1\}$, here $\beta = \beta(T)$ and $\alpha = \alpha(T)$ are minimum total path length and minimum number of nodes, respectively, of some decision tree in $D(T)$ (not necessarily the same tree). We define two functions $\mathcal{G}_T : B_{\Lambda,T} \to B_{L,T}$ and $\mathcal{F}_T : B_{L,T} \to B_{\Lambda,T}$ as follows:

$$\mathcal{F}_T(n) = \min\{\Lambda(\Gamma) : \Gamma \in D(T) : L(\Gamma) \leq n\}, \quad n \in B_{L,T}$$
$$\mathcal{G}_T(n) = \min\{L(\Gamma) : \Gamma \in D(T) : \Lambda(\Gamma) \leq n\}, \quad n \in B_{\Lambda,T}.$$

We now describe an algorithm which allows us to construct the function $\mathcal{F}_\Theta$ for every node Θ from the graph $\Delta(T)$. We begin from terminal nodes and move upward to the node T.

Let Θ be a terminal node. It means that all the rows of decision table Θ are labeled with the same decision b and the decision tree Γ_b as depicted in Fig. 2 belongs to $D(\Theta)$. It is clear that $\Lambda(\Gamma_b) = 0$ and $L(\Gamma_b) = 1$ for the table Θ as well as $\alpha(\Theta) = 1$, therefore, $\mathcal{F}_\Theta(n) = 0$ for any $n \in B_{L,\Theta}$.

Let us consider a nonterminal node Θ and a bundle of edges, which start from this node. Let these nodes be labeled with the pairs $(f_i, a_1), \ldots, (f_i, a_t)$ and enter into the nodes $\Theta(f_i, a_1), \ldots, \Theta(f_i, a_t)$, respectively, to which the functions $\mathcal{F}_{\Theta(f_i,a_1)}, \ldots, \mathcal{F}_{\Theta(f_i,a_t)}$ are already attached.

Let $v_1, \ldots, v_t$ be the minimum values from $B_{L,\Theta(f_i,a_1)}, \ldots, B_{L,\Theta(f_i,a_t)}$, respectively. Let

$$B_{L,\Theta,f_i} = \{\alpha_i, \alpha_i + 1, \ldots, 2N - 1\}, \text{ where } \alpha_i = 1 + \sum_{j=1}^{t} v_j.$$

One can show that α_i is the minimum number of nodes of a decision tree from $D(\Theta)$ for which f_i is attached to the root and $\alpha(\Theta) = \min\{\alpha_i : f_i \in E(\Theta)\}$, where $\alpha(\Theta)$ is the minimum value from $B_{L,\Theta}$.

We correspond to the bundle (f_i-bundle) the function $\mathcal{F}^{f_i}_\Theta$: for any $n \in B_{L,\Theta,f_i}$,

$$\mathcal{F}^{f_i}_\Theta(n) = \min \sum_{j=1}^{t} \mathcal{F}_{\Theta(f_i,a_j)}(n_j) + N(\Theta),$$

where the minimum is taken over all $n_1, \ldots, n_t$ such that $n_j \in B_{L,\Theta(f_i,a_j)}$ for $j = 1, \ldots, t$ and $n_1 + \cdots + n_t + 1 \leq n$. [It should be noted that computing $\mathcal{F}^{f_i}_\Theta$ is a nontrivial task. We describe the method in detail in the following subsection.] It is not difficult to show that for all $n \in B_{L,\Theta}$,

$$\mathcal{F}_\Theta(n) = \min\{\mathcal{F}^{f_i}_\Theta(n) : f_i \in E(\Theta), n \in B_{L,\Theta,f_i}\}.$$

We can use the following proposition to construct the function $\mathcal{G}_T$ (using the method of transformation of functions described in the Appendix).

Proposition 2 *For any $n \in B_{\Lambda,T}$, $\mathcal{G}_T(n) = \min\{p \in B_{L,T} : \mathcal{F}_T(p) \leq n\}$.*

Note that to find the value $\mathcal{G}_T(n)$ for some $n \in B_{\Lambda,T}$ it is enough to make $O(\log |B_{\Lambda,T}|) = O(\log(mN))$ operations of comparisons.

4.1 Computing $F^{f_i}_\Theta$

Let Θ be a nonterminal node in $\Delta(T)$, $f_i \in E(\Theta)$ and $E(\Theta, f_i) = \{a_1, \ldots, a_t\}$. Furthermore, we assume the functions $\mathcal{F}_{\Theta(f_i,a_j)}$ for $j = 1, \ldots, t$, have already been computed. Let the values of $\mathcal{F}_{\Theta(f_i,a_j)}$ be given by the tuple of pairs, $\left((\gamma_j, \lambda^j_{\gamma_j}), (\gamma_j+1, \lambda^j_{\gamma_j+1}), \ldots, (2N-1, \lambda^j_{2N-1})\right)$, where $\gamma_j = \alpha(\Theta(f_i, a_j))$ and $\lambda^k_j = \mathcal{F}_{\Theta(f_i,a_j)}(k)$. We need to compute $\mathcal{F}^{f_i}_{\Theta}(n)$ for all $n \in B_{L,\Theta,f_j}$;

$$\mathcal{F}^{f_i}_{\Theta}(n) = \min \sum_{j=1}^{t} \mathcal{F}_{\Theta(f_i,a_j)}(n_j) + N(\Theta),$$

for $n_j \in B_{L,\Theta(f_i,a_j)}$, such that $n_1 + \cdots + n_t + 1 \leq n$.

We construct a layered directed acyclic graph (DAG) $\delta(\Theta, f_i)$ to compute $\mathcal{F}^{f_i}_{\Theta}$ as following. The DAG $\delta(\Theta, f_i)$ contains nodes arranged in $t+1$ layers $(l_0, l_1, \ldots, l_t)$. Each node has a pair of labels and each layer $l_j (1 \leq j \leq t)$ contains at most $j(2N-1)$ nodes. The first entry of labels for nodes in a layer l_j is an integer from $\{1, 2, \ldots, j(2N-1)\}$. The layer l_0 contains only one node labeled with $(0, 0)$.

Each node in a layer l_j $(0 \leq jt)$ has at most $2N-1$ outgoing edges to nodes in layer l_{j+1}. These edges are labeled with the corresponding pairs in $\mathcal{F}_{\Theta(f_i,a_{j+1})}$. A node with label x as a first entry in its label-pair in a layer l_j connects to nodes with labels $x + \gamma_j$ to $x + 2N - 1$ (as a first entry in their label-pairs) in layer l_{j+1}, with edges labeled as $(\gamma_{j+1}, \lambda^{j+1}_{\gamma_{j+1}}), (\gamma_{j+1}+1, \lambda^{j+1}_{\gamma_{j+1}+1}), \ldots, (2N-1, \lambda^{j+1}_{2N-1})$, respectively.

The function $\mathcal{F}^{f_i}_{\Theta}(n)$ for $n \in B_L$ can be easily computed using the DAG $\delta(\Theta, f_i)$ for $\Theta \in \Delta(T)$ and for the considered bundle of edges for the attribute $f_i \in E(\Theta)$ as follows:

Each node in layer l_1 gets its second value copied from the corresponding second value in incoming edge label to the node (since there is only one incoming edge for each node in layer l_1). Let (k, λ) be a node in layer l_j, $2 \leq j \leq t$. Let $E = \{(v_1, \lambda_1), (v_2, \lambda_2), \ldots, (v_r, \lambda_r)\}$ be the set of incoming nodes to (k, λ) such that $(\alpha_1, \beta_1), (\alpha_2, \beta_2), \ldots, (\alpha_r, \beta_r)$ are the labels of these edges between the nodes in E and (k, λ), respectively. It is clear that $k = v_i + \alpha_i$, $1 \leq i \leq r$. Then $\lambda = \min_{1 \leq i \leq r}\{\lambda_i + \beta_i\}$. We do this for every node layer-by-layer till all nodes in $\delta(\Theta, f_i)$ have received their second label.

Once we finish computing the second value of label pairs for the nodes in layer l_t, we can use these labels to compute $\mathcal{F}^{f_i}_{\Theta}(n)$. Let $(k_1, \lambda_1), \ldots, (k_s, \lambda_s)$ be all label-pairs attached to the nodes in l_t. One can show that

$$\mathcal{F}^{f_i}_{\Theta}(n) = \min\{\lambda_q : q \in \{1, \ldots, s\}, k_q \leq n-1\} + N(\Theta).$$

An example of working of the algorithm can be found in Fig. 4.

Let us evaluate the number of arithmetic operations of the considered algorithm. The DAG $\delta = \delta(\Theta, f_i)$ has $t+1$ layers and each layer l_j has at most $j(2N-1)$ nodes. Therefore, the total number of nodes in δ is $O(t^2 N)$. Since every node has at most $2N-1$ outgoing edges (except the nodes in layer l_t), the number of edges in δ is $O(t^2 N^2)$. Hence, to build the graph δ, we need $O(t^2 N^2)$ operations. To find the second labels we need a number of additions and comparisons bounded from

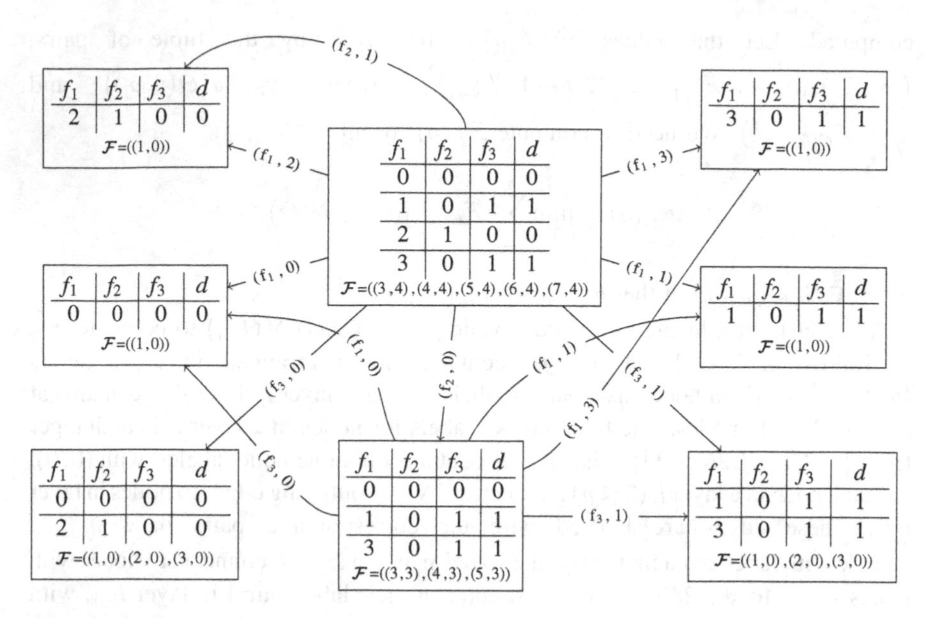

Fig. 4 Example illustrating the working of the algorithm

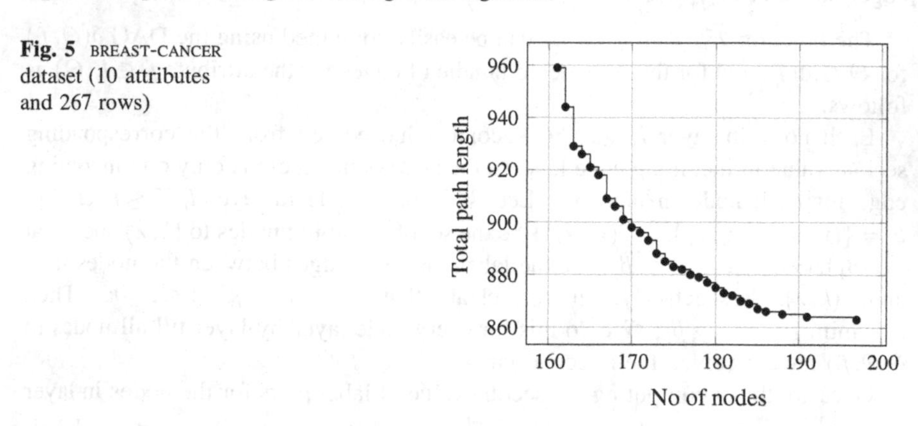

Fig. 5 BREAST-CANCER dataset (10 attributes and 267 rows)

above by the number of edges, i.e, $O(t^2N^2)$. Similarly, to find values of $\mathcal{F}^{f_i}_{\Theta}$ we need $O(tN)$ comparisons. Therefore, the total number of additions and comparisons is $O(t^2N^2)$.

5 Experimental Results

We performed several experiments on datasets (decision tables) acquired from UCL ML Repository [1]. The resulting plots are depicted in Figs. 5, 6, 7, and 8. These plots show the relationship between two cost functions, take for example the

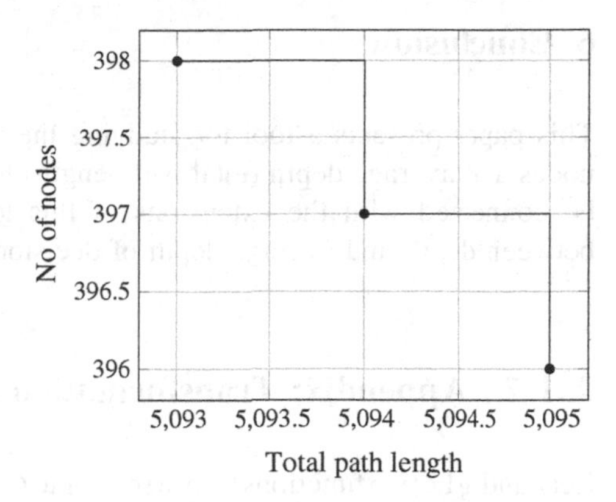

Fig. 6 CARS dataset (6 attributes and 1729 rows)

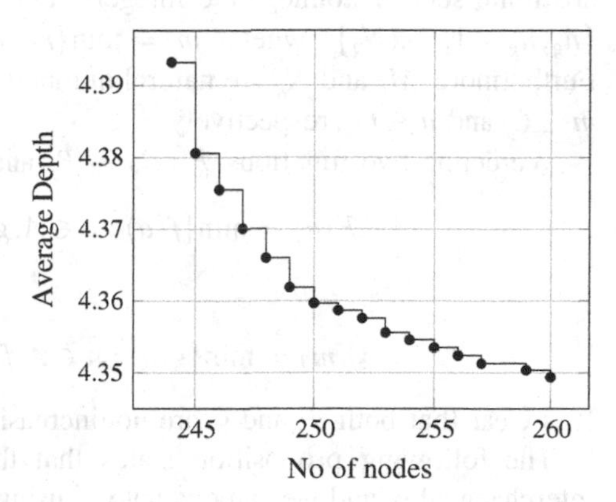

Fig. 7 TIC-TAC-TOE dataset (9 attributes and 959 rows)

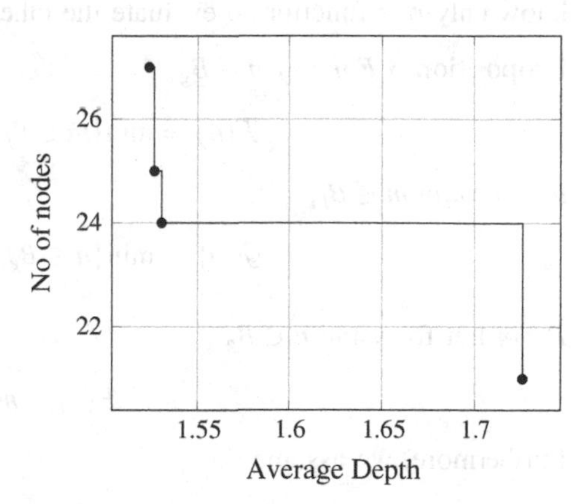

Fig. 8 MUSHROOM dataset (22 attributes and 8125 rows)

plot in Fig. 7. It shows that when the number of nodes is 250 the average depth of the tree is 4.36. These plots help us understand the nature of trees for the particular dataset. The plots in Figs. 7 and 8 use average depth instead of total path length.

6 Conclusion

This paper presents a tool for studying the relationships between the number of nodes and average depth (total path length) for decision trees. Further studies will be connected with the extensions of this tool including studying relationships between depth and average depth of decision trees.

A.1 7 Appendix: Transformation of Functions

Let f and g be two functions from a set A onto C_f and C_g respectively, where C_f and C_g are finite sets of nonnegative integers. Let $B_f = \{m_f, m_f + 1, \ldots, M_f\}$ and $B_g = \{n_g, n_g + 1, \ldots, N_g\}$ where $m_f = \min\{m : m \in C_f\}$ and $n_g = \min\{n : n \in C_g\}$. Furthermore, M_f and N_g are natural numbers such that $m \leq M_f$ and $n \leq N_g$ for any $m \in C_f$ and $n \in C_g$, respectively.

We define two functions $\mathcal{F} : B_g \rightarrow B_f$ and $\mathcal{G} : B_f \rightarrow B_g$ as follows:

$$\mathcal{F}(n) = \min\{f(a) : a \in A, g(a) \leq n\}, \ \forall n \in B_g, \tag{1}$$

$$\mathcal{G}(m) = \min\{g(a) : a \in A, f(a) \leq m\}, \ \forall m \in B_f. \tag{2}$$

It is clear that both $\mathcal{F}$ and $\mathcal{G}$ are nonincreasing functions.

The following proposition states that the functions $\mathcal{F}$ and $\mathcal{G}$ can be used interchangeably and we can evaluate $\mathcal{F}$ using $\mathcal{G}$ and vice versa, i.e., it is enough to know only one function to evaluate the other.

Proposition 3 *For any $n \in B_g$,*

$$\mathcal{F}(n) = \min\{m \in B_f : \mathcal{G}(m) \leq n\},$$

and for any $m \in B_f$,

$$\mathcal{G}(m) = \min\{n \in B_g : \mathcal{F}(n) \leq m\}.$$

Proof Let for some $n \in B_g$

$$\mathcal{F}(n) = m_0. \tag{3}$$

Furthermore, we assume that

$$\min\{m \in B_f : \mathcal{G}(m) \leq n\} = t. \tag{4}$$

From (3) it follows that

(i) there exists $b \in A$ such that $g(b) \leq n$ and $f(b) = m_0$;

(ii) for any $a \in A$ if $g(a) \leq n$ then $f(a) \geq m_0$.From (i) it follows that $\mathcal{G}(m_0) \leq n$. This implies $t \leq m_0$. Let us assume that t. In this case, there exits m_1 for which $\mathcal{G}(m_1) \leq n$. Therefore, there exists $a \in A$ such that $f(a) \leq m_1$ and $g(a) \leq n$, but from (ii) it follows that $f(a) \geq m_0$, which is impossible. So $t = m_0$.

Similarly, we can prove the second part of the statement.

Proposition 3 allows us to transform the function $\mathcal{G}$ given by a tuple $(\mathcal{G}(m_f), \mathcal{G}(m_f+1), \ldots, \mathcal{G}(M_f))$ into the function $\mathcal{F}$ and vice versa. We know that $\mathcal{G}(m_f) \geq \mathcal{G}(m_f+1) \geq \cdots \geq \mathcal{G}(M_f)$, to find the minimum $m \in B_f$ such that $\mathcal{G}(m) \leq m$ we can use binary search which requires $O(\log |B_f|)$ comparisons of numbers. So to find the value $\mathcal{F}(n)$ for $n \in B_g$ it is enough to make $O(\log |B_f|)$ operations of comparison.

References

1. Frank A, Asuncion A (2010) UCI Machine Learning Repository
2. Alkhalid A, Chikalov I, Moshkov M (2010) On algorithm for building of optimal α-decision trees. In: Szczuka MS, Kryszkiewicz M, Ramanna S, Jensen R, Hu Q (eds) RSCTC. Springer, Heidelberg, pp 438–445
3. Alkhalid A, Chikalov I, Moshkov M (2010) A tool for study of optimal decision trees. In: Yu J, Greco S, Lingras P, Wang G, Skowron A (eds) RSKT. LNCS, vol 6401. Springer, Heidelberg, pp 353–360
4. Alkhalid A, Chikalov I, Hussain S, Moshkov M (2012) In: Extensions of dynamic programming as a new tool for decision tree optimization. SIST, vol 13. Springer, Heidelberg, pp 16–36
5. Alkhalid A, Amin T, Chikalov I, Hussain S, Moshkov M, Zielosko B (2011) Dagger: a tool for analysis and optimization of decision trees and rules. In: Francisco V. C. Ficarra (ed) Computational informatics, social factors and new information technologies: hypermedia perspectives and avant-garde experiencies in the Era of communicability expansion. Blue Herons, Bergamo, pp 29–39
6. Chikalov I, Hussain S, Moshkov M (2011) Relationships between depth and number of missclassifications for decision trees. In: Kuznetsov SO, Slezak D, Hepting DH, Mirkin B (eds) Thirteenth international conference on rough sets, fuzzy sets, data mining and granualr computing (RSFDGrC 2011). LNCS, vol 6743. Springer, Heidelberg, pp 286–292
7. Pawlak Z (1991) Theoretical aspects of reasoning about data. Kluwer Academic Publishers, Dordrecht
8. Skowron A, Rauszer C (1992) The discernibility matrices and functions in information systems. In: Slowinski R (ed) Intelligent decision support. Handbook of applications and advances of the rough set theory. Kluwer Academic Publishers, Dordrecht, pp 331–362
9. Nguyen HS (1998) From optimal hyperplanes to optimal decision trees. Fundam Inf 34(1–2): 145–174

On The Use of LTSs to Analyze Software Product Line Products Composed of Features

Jyrki Nummenmaa, Timo Nummenmaa and Zheying Zhang

Abstract In product line engineering, it is common to define the products as sets of features, where each feature has a related set of requirements. Typically, there is a common set of features/requirements, and some variable features/requirements for building different products. In an earlier proposal to use labeled transition systems (LTSs) to model and check the products, the products were composed using the feature-oriented approach and LTS models were analyzed using a related LTS analyzer tool. However, no further details or analysis about the models and possible conflicts were given. We investigate in more detail the types of conflicts that may arise and discuss the integration strategies for building an integrated LTS for the product composed of features.

Keywords Software product line · Functional requirement · Feature model · Labeled transition system

1 Introduction

A software product line (SPL) is expected to provide customized products at reasonable costs. Unlike the conventional software development, software artifacts change continually throughout the SPL life cycle. It is a challenge to suit the customer's requirements with the right combination of these software artifacts.

J. Nummenmaa (✉) · T. Nummenmaa · Z. Zhang
School of Information Sciences, University of Tampere, Tampere, Finland
e-mail: jyrki.nummenmaa@uta.fi

T. Nummenmaa
e-mail: timo.nummenmaa@uta.fi

Z. Zhang
e-mail: zheying.zhang@uta.fi

F. Sun et al. (eds.), *Knowledge Engineering and Management*,
Advances in Intelligent Systems and Computing 214,
DOI: 10.1007/978-3-642-37832-4_48,

Application engineering is the process of SPL engineering in which the applications of the product line are built by reusing domain artifacts and exploiting the product line variability [15]. It starts by reusing the domain requirements artifacts and identifying new ones for a particular application. Different from a single software development process, it reuses as many domain artifacts as possible by binding the variants from requirements to the architecture, to the components, and to the tests cases. Such a process typically involves trade-off decisions with regard to the application–specific requirements proposed by customers. Adding requirements may introduce conflicts among reusable artifacts, which may lead to significant modification of the architecture and the reusable components. It consequently impacts the development cost and the quality of the application. Therefore, an effort on validating the requirements and feasibility of applications is of importance to reveal the problems introduced by adding new requirements or changing the existing ones early enough, which effectively save the cost of the application development.

In application requirements engineering, an application is declaratively specified by selecting or deselecting features according to customers' preferences. The features and their dependencies are commonly specified in a feature model [8, 9, 15], which expresses the high-level requirements of a product line architecture. The process of specifying application specific requirements can result in changes of the dependency among features, and consequently the relationships between reusable components in application design and implementation. As a feature model itself lacks a formal mechanism to reason the validity of the model, many researchers analyze the validity of feature models, with a focus on the consistency [2, 9, 17], completeness [3, 14], redundancy [19], or omissions [7, 18] of a model. Benavides et al. [3] had a comprehensive literature review on the automated analysis of feature models and addressed a number of analysis operations and approaches providing automated support. The approaches include the use of propositional logic [2, 12], constraint programming [3, 6], description logic [20], etc. They detect the dead features, deadlocks, false optional features, and verify the constraint dependencies for applications.

As a feature is an end-user visible characteristics of an application, it can be elaborated into a set of textual software requirements describing application services or functions, as well as constrains on the functions [16]. When a feature is selected for an application of the product line, adaptation may be done according to the customers' needs. This sometimes means changes insides a feature, that is, changes to individual requirements within the feature. In such a situation, the above-mentioned analysis approaches for feature models can only play a very limited role in validating the feasibility and impact of change requests, as they cannot validate the changes embedded in a feature. Meanwhile, the features and the associated requirements are not necessarily analyzed and validated in one "big bang". Validation of the combinations of domain requirements with new ones or changes of an existing feature can be done incrementally, which makes it easy to identify the potential problems or conflicts within a set of requirements. Following the results of prior work on automated analysis of feature models [3], a formalized approach seems an effective option to specifying and validating the behavior of an application according to customers' requirements [13].

Formal specifications are a powerful method for defining system behavior. Moreover, composed specifications can be taken as models of software, which introduce further options. A formal specification can be written before implementing a piece of software [5]. That specification can then be used to prove that the written software meets its specification. Moreover, simulations can be used to validate and test such specifications [1]. Despite the benefits, such specifications do not seem to be used in software development. On the contrary, even in the field of software development, formal specifications do not enjoy a lot of popularity, as formal methods alienate the users and are not widely used by commercial software companies [10]. Even though formal specifications are not so popular amongst end users, with some expert help they can be used to provide important insights and analysis of the dynamic behavior of the system.

Yang et al. [21] proposed an approach to feature-oriented customization on the component level. This approach integrates a Labeled Transition System Analyzer (LTSA) tool [11] to validate the component behavior. The Labeled Transition Systems (LTSs) are high-level abstract specifications of concurrent systems, and once they are specified, then LTSA tool can be used to check various properties related to fairness, correctness, and deadlocking. The work of Yang et al. [21] contains a higher level process view that positions the use of LTSs in feature-driven product composition. However, their work does not go into the details of the use of LTSs, and it lacks in-depth discussion of various properties like correctness that may be identified and verified in the LTSA tool. In this work, we study the details of using LTSs in feature-driven product composition. We analyze the potential problems in integrating the specifications, exemplifying our findings with a feature-driven example SPL and related specifications.

In Section 2, we discuss how LTSs can be formed to specify the features and related functional requirements. Section 3 contains a discussion on how these specifications can be integrated, and how certain types of problems may be created in the integrated model, along with a discussion how a related toolset can be used to identify these problems. Section 4 contains our conclusions.

2 FRs and LTSs for Product Lines

We adopt the use of a feature model [8, 9, 15] as discussed in the Introduction, for the management of product lines. Our running example is a home security system (HSS) SPL. Typical components of a HSS are identification mechanisms (e.g., password authentication), surveillance equipment (e.g., infrared motion sensors, outdoor cameras), and notification mechanisms (e.g., immediate reaction in the case of an attempted burglary). To exemplify our work, we use the HSS SPL features and requirements specifications presented in [15] for illustrative purposes. Figure 1 shows the example features and requirements specification that we use.

Product_1	Product_2	Product_3
Goals G1: Protection against burglary. G2: Catching the thief. G...: ... G5: Video surveillance of the house. **Desired Features** F1: Video surveillance. F...: ... F3: Generate reports. **Requirements** ... R2: The alarm signal shall be deactivated by the police, by the owner, or automatically after 20 minutes. R2.1: The alarm signal shall start immediately after the detection of the open window or door. R11: The camera shall have enough storage for 5 minutes' video stream to be stored as alarm buffer. R11.1: The recording is only initiated if motion is detected. R28: The system shall be able to generate user-specific reports that document the system events. ...	**Goals** G1: Protection against burglary in the office. G...: ... G5: Determent of thieves through alarm signal. **Desired Features** F1: Video surveillance. F1.1: Alarm activation. F...: ... **Requirements** R1: The recording of the video surveillance system shall start after the detection of the open door or window. R2: The police shall be informed about the burglary via radio transmission. R3: The alarm shall be activated in case of burglary. R3.1: The alarm signal shall only start if the police do not arrive within 5 minutes of alarm detection. R3.2: The alarm (signal) shall only be deactivated by the police or by the owner, via password and transaction number. R13: The camera shall have enough storage for 5 minutes' video stream to be stored as alarm buffer. ...	**Goals** G1: Safety against thieves. G...: ... G5: Video surveillance of the flat. **Desired Features** F1: Video surveillance. F...: ... F2.3: Manual deactivation of alarm. **Requirements** R1: The police shall be informed immediately after the detection of an open window or door. R1.1: The police shall be informed via internet message or SMS. R2: The alarm signal shall start immediately after the detection of the open window or door. R2.1: The alarm signal shall be deactivated by the police, by the owner, or automatically after 20 minutes. R9: The video surveillance shall be active as soon as activated by the user. R9.1: A recording is only initiated if motion is detected. R9.2: The camera shall have enough storage for 5 minutes' video stream to be stored as alarm buffer. ...

Fig. 1 Example feature model for a HSS software product line

The figure should be relatively self-explanatory on its abstraction level. In this section, we discuss the representation of the SPL features and requirements formally, to enable their analysis.

In LTSs, the basic building block of a specification is finite-state processes (FSPs), composed of actions by using action prefix (sequentiality) ("->"), choice ("|"), guards "`when` (boolean_cond)", parallel composition ("||"), and other operations, not used here. The process names start with uppercase and the action names with lowercase. This way, we can define e.g., the process `CAMERA = (camera_activates -> camera_deactivates -> CAMERA)` to say that the camera process performs alternately actions to activate and deactivate the camera. A process `CAMERA_CONTROL = (motion_detected -> camera_activates -> CAMERA_CONTROL | motion_not_detected -> CAMERA_CONTROL)` can be parallel composed with `CAMERA` as follows `||CAMERA_AND_CONTROL = (CAMERA || CAMERA_CONTROL)`. Then, the processes are synchronized using the names of actions. So, e.g., in `||CAMERA_AND_CONTROL` the action camera_activates can only be executed in `CAMERA` when it is at the same time executed in `CAMERA_CONTROL`. This, in effect, means that if motion_detected action is executed, then the `CAMERA_CONTROL` process is ready to execute the camera_activates action, and it can be executed, if the `CAMERA` process is ready for that (either initially or after the camera has been de-activated after the previous activation. Due to the complex nature of concurrency, the collective behavior of the SPL product should be examined when it is composed.

In addition to the actual processes, the LTSs may contain formal properties, which can be either safety or progress properties. The safety properties define correct ways in which the actions may progress. For instance, we may define `property Mot_Cam = motion_detected -> camera_activates -> motion_not_detected -> camera_deactivates -> Mot_Cam.`

The property `Mot_Cam` is a safety property and it defines how the events mentioned in it are allowed to interleave. Once the property is included in the model, the LTSA can be used to examine if the specification allows any incorrect interleaving of the actions in `Mot_Cam`, which will be reported as an error. Progress properties can be used to check if all or a desired subset of the actions specified will always be reachable in the future. For this testing, it is also possible to prioritize the actions.

Even though there may be some variance between how FRs are represented in practice, in general a FR represents and describes a functionality. In this section, we want to study how we can represent FRs using LTSAs in the context of product lines. Ideally, we might hope to have a one-to-one correspondence between a requirement that belongs to a product and an element of LTSAs. LTSAs are process based, so we might hope for a one to one between FRs and LTSAs in product lines. However, some of the FRs may only represent a part of a process, possibly only a condition such as: "The recording of the video surveillance system shall start after the detection of the open door or window". However, when a feature model is used, single features are much more likely to contain suitable "chunks" of requirements to be implemented as one or several LTSA processes. The LTSA processes are fairly succinct, so in Listing 1 we give the code for a major part of the features presented in Fig. 1. For clarity, we have added also some comments in the specification. The comment lines start with "`//`".

Listing 1. Process definitions for the safe home features

```
// Detects motion when a window or a door is opened:
MOTION_DETECTOR = (door_opens->motion_detected->
   detector_reset-> MOTION_DETECTOR
  | window_opens->motion_detected -> detector_reset->MOTION_DETECTOR
  | motion_not_detected -> MOTION_DETECTOR).
// When motion is detected, an alarm is activated:
ALARM_CONTROL = (motion_detected -> alarm_activates
   -> alarm_deactivates -> ALARM_CONTROL
  | motion_not_detected -> ALARM_CONTROL).
// Alarm MUST activate directly after motion detection:
ALARM_CONTROL_PROP = (motion_detected->alarm_activates
   ->detector_reset->ALARM_CONTROL_PROP).
// Alarm activation is followed by detector reset:
MOTION_DETECTOR_ALARM_MOD = (alarm_activates->detector_reset->
    MOTION_DETECTOR_ALARM_MOD).
// Messaging takes place after motion detection:
MSG_SYSTEM =  (motion_detected->internet_message_to_police->MSG_SYSTEM).
// Messaging MUST happen directly after motion detection:
MSG_SYSTEM_PROP = (motion_detected->internet_message_to_police
  ->alarm_activates->detector_reset->MSG_SYSTEM_PROP).
// After message to police, the alarm activates:
MSG_SYSTEM_ALARM_CONTROL_MOD = (internet_message_to_police -> alarm_activates
  -> MSG_SYSTEM_ALARM_CONTROL_MOD
  | motion_not_detected -> MSG_SYSTEM_ALARM_CONTROL_MOD).
// After message to police detector reset takes place:
MSG_SYSTEM_MOTION_DETECTOR_MOD =  (internet_message_to_police->detector_reset
                          ->MSG_SYSTEM_MOTION_DETECTOR_MOD).
// Defines the right order how these events cycle:
property ALARM_CONTROL_PROP_WITH_MSG = (motion_detected->alarm_activates->
   (detector_reset->ALARM_CONTROL_PROP_WITH_MSG
   |internet_message_to_police->ALARM_CONTROL_PROP_WITH_MSG).
// alarm is deactivated by police or automatically:
property ALARM_DEACTIVATION_SYSTEM_PROPERTY = ( alarm_activates -> (
    police_alarm_deactivation_signal-> alarm_deactivates->
    ALARM_DEACTIVATION_SYSTEM_PROPERTY)
  | automatic_alarm_deactivation_signal->alarm_deactivates ->
      ALARM_DEACTIVATION_SYSTEM_PROPERTY)).
// Alarm activation and deactivation alternate:
AUTOMATIC_ALARM_DEACTIVATION_SYSTEM = (alarm_activates
  -> automatic_alarm_deactivation_signal ->
      AUTOMATIC_ALARM_DEACTIVATION_SYSTEM).
// When activated alarm is resolved, alarm is deactivated by signal:
POLICE_ALARM_DEACTIVATION_SYSTEM = (alarm_activates ->alarm_resolved
   -> police_alarm_deactivation_signal -> POLICE_ALARM_DEACTIVATION_SYSTEM).
```

3 Testing the Feasibility of a Product Line Product

On a high level, the process of using the LTSs for checking the composition of a product line product goes as follows:

1. Merge the specifications (processes and properties) of the components related to the set of required features into the same file.
2. Compile the specifications. (This step may require some manual assistance for compatibility, but ideally none.)
3. Test the specifications against deadlock, progress, and safety properties.

However, this is an oversimplification, particularly if some problems arise. We may consider an alternative strategy, where we add incrementally the features to the product. This may be insightful, if there are some problems or conflicts when merging a certain set of features into a product. Integrating feature by feature may explain why a certain composition of features is not successful.

Another factor in integration is whether the common parts of the SPL produce a product or not. Consider, for instance, a mobile phone software that as a common part contains the basic functionalities, but no communication method (such as WLAN, GSM, and UMTS), then we can claim that the common parts do not comprise a meaningful product. If, on the other hand, for some simple phone only one communication method, such as GSM, is available and included in the common part, then in that case it may be the case that the common part produces a product. Clearly, if the communal part does not produce a product, then the incremental integration may not provide positive results until some necessary amount of variable features are added.

Even though the process is of some interest, our main interest is in the potential problems when the specifications are integrated. In addition to compiling the specifications in the same compilation, the processes must be combined. By default, this means that different processes execute in the same system concurrently, synchronized by their common actions. We assume here that the actions for synchronization are named the same. This is not a necessary requirement when using the FSPs. In the parallel composition, it is also possible to rename the action names for a process to make them match some differently named actions in another process.

In our example, we give just some of the more interesting parallel compositions, and discuss their behavior.

Listing 2. Parallel compositions of the processes

```
// Combining motion detection and alarm control,
// the ALARM_CONTROL_PROP is for checking correctness of executions:
||SAFEHOME_ALARM_FAIL = (MOTION_DETECTOR||ALARM_CONTROL||ALARM_CONTROL_PROP).
||SAFEHOME_ALARM_FIX = (MOTION_DETECTOR||ALARM_CONTROL||ALARM_CONTROL_PROP||
  MOTION_DETECTOR_ALARM_MOD).
// Here we add motion detection alarm module and a messaging system,
// the messaging system comes with its property for correctness:
||SAFEHOME_MSG_FAIL (MOTION_DETECTOR||ALARM_CONTROL||ALARM_CONTROL_PROP
  ||MOTION_DETECTOR_ALARM_MOD||MSG_SYSTEM||MSG_SYSTEM_PROP).
// This violates the requirement that also the alarm must go on
// directly after the detection of movement:
||SAFEHOME_MSG_FIX = (MOTION_DETECTOR||ALARM_CONTROL||ALARM_CONTROL_PROP
  ||MOTION_DETECTOR_ALARM_MOD||MSG_SYSTEM||MSG_SYSTEM_MOTION_DETECTOR_MOD
  ||MSG_SYSTEM_ALARM_CONTROL_MOD||MSG_SYSTEM_PROP).
// But the requirements say that also the alarm
// must happen before everything..
||SAFEHOME_MSG_FIX_MSG_EQUAL_ALARM = (MOTION_DETECTOR||ALARM_CONTROL
      ||ALARM_CONTROL_PROP_WITH_MSG||MOTION_DETECTOR_ALARM_MOD||MSG_SYSTEM
      ||MSG_SYSTEM_MOTION_DETECTOR_MOD||MSG_SYSTEM_ALARM_CONTROL_MOD||
         MSG_SYSTEM_PROP).
// Alarm is deactivated by signal from police or automatically.
// Now both deactivation signals are set because
// it is not detected if the alarm is off or not:
||SAFEHOME_MSG_FIX_WITH_DEACTIVATION_FAIL = (MOTION_DETECTOR||ALARM_CONTROL||
   ALARM_CONTROL_PROP
 ||MOTION_DETECTOR_ALARM_MOD||MSG_SYSTEM||MSG_SYSTEM_MOTION_DETECTOR_MOD
 ||MSG_SYSTEM_ALARM_CONTROL_MOD||MSG_SYSTEM_PROP||
   ALARM_DEACTIVATION_SYSTEM_PROPERTY
 ||AUTOMATIC_ALARM_DEACTIVATION_SYSTEM_CHECK_ALARM||
    POLICE_ALARM_DEACTIVATION_SYSTEM_CHECK_ALARM).
```

In Listing 2, we present a number of parallel compositions that can be used to study the compatibility of different processes. The LTSA tool provides the analyses automatically, and all the specification code that is necessary to do the checking, is included in this article. When testing the parallel process ||SAFEHOME_ALARM_FAIL with the LTSA tool, the tool claims that there is a safety violation in ALARM_CONTROL_PROP and the trace leading to the violation is door_opens -> motion_detected -> detector_reset. Checking back, ALARM_CONTROL_PROP = (motion_detected -> alarm_activates -> detector_reset -> ALARM_CONTROL_PROP) means, in addition to other things, that alarm_activates needs to be executed after motion_detected before detector_reset can take place. It is, thus, easy to see that the trace contains the property violation. To get more insight into this situation, we may be interested in checking what allows this trace of actions. Apart from the property that was violated, ||SAFEHOME_ALARM_FAIL only contains the processes MOTION_DETECTOR and ALARM_CONTROL, which need to synchronize for the common action motion_detected to be executed at the same time. However, after motion_detected the process MOTION_DETECTOR can perform detector_reset without a need to synchronize just as ALARM_CONTROL can execute the action alarm_activates independently of MOTION_DETECTOR. This way, the specification allows x and y to be executed in whatever global order, which allows the property violation to take place.

Having identified the problem, a natural question is "How to fix it?" In principle, there are several optional ways to go:

1. It may be that adding some further components will fix the problem by forcing the actions x and y to be executed in the right order. Typically, this would be some extra control software that does not allow the incorrect global ordering of the events.
2. In general, some features may be incompatible. Even though in this case it does not seem likely that this should be the case, there may also be different types of control systems and some may allow only correct global orders, while others may allow also allow incorrect global orders.
3. It may be that the SPL is lacking some components for the successful integration of the components. In this case, new components need to be modelled and implemented.
4. It may be feasible to fix one of the existing components in the software product line. Just from the point of view of getting the property violation solved this seems like a straightforward case. Even though in this case it would seem a straightforward solution from the specification point of view, it may not be feasible from the component point of view (considering the different environments where the component needs to function).

The above example is, of course, extremely simple, which is good to be able to manually trace its behavior. The LTSA tool gives also facilities for running the action system and re-running the traces of the analyses. In this first example, intuitively it feels that the safety property is not achieved because the system lacks some component to coordinate the concurrent execution, which means that Option 1 for rectifying the problem seems like a proper one.

Next, let us consider the simple additions to our specification given in Listing 3. We have added a logging facility which, at a first glance may seem harmless. It performs an action for writing the detector log as a part of the activation and reset cycles. As the reader might have observed by now, this new composed process will, however, deadlock, as the process MOTION_DETECTOR_LOGGER requires the reset to take place before activation and the process MOTION_DETECTOR_ALARM_MOD requires the activation to take place before reset. The ||SAFEHOME_MSG_FIX_MSG_EQUAL_ALARM process also has a similar deadlock. These are extremely simple cases of deadlock. However, the LTSA tool requires deadlock even in the more general case, where there is shared resource contention and a wait-for cycle.

Listing 3. Deadlocking addition

```
// Seemingly innocent process:
MOTION_DETECTOR_LOGGER = (detector_reset -> alarm_activates ->
    write_detector_log -> MOTION_DETECTOR_LOGGER).

// But when combined with others, a deadlock occurs:
||SAFEHOME_ALARM_LOGGED = (MOTION_DETECTOR || MOTION_DETECTOR_LOGGER ||
    MOTION_DETECTOR_ALARM_MOD || ALARM_CONTROL || ALARM_CONTROL_PROP).
```

In addition to deadlocks, error states, and safety violations, the LTSA can also be used to check for progress violations. In progress violations, some set of actions will never become available for execution in the future, and the execution will only allow a proper subset of the actions (called terminal set) to be executed. In some cases, it may be natural that there are some actions only used at something like an initialization stage, which may not be a problem at all. In some other cases, there may be a design flaw that never allows us to reach some actions—this in fact is also the case with deadlock. The "trickier" cases are such that if some events take place too quickly, then some others may never get the chance to get executed. This can be modeled for the LTSA by giving some actions higher or lower priority. In our example case consider, for example, the possibility that the motion detector very quickly turns on and off . We add the priorities in the model with the << operator to show high priority or the>> operator to show low priority, the syntax should be clear from the following listing where we have prioritized the set {motion_detected, motion_not_detected} with a higher priority. Now, the LTSA tool will reveal that the composition ||SAFEHOME_MSG_FIX suffers from a progress violation which e.g., leads to the fact that the police will never be alarmed.

Listing 4. Fairness problems introduced

```
// Using priorities to examine fairness in extreme conditions:
||SAFEHOME_MSG_FIX = (MOTION_DETECTOR || ALARM_CONTROL || ALARM_CONTROL_PROP
  || MOTION_DETECTOR_ALARM_MOD || MSG_SYSTEM ||
     MSG_SYSTEM_MOTION_DETECTOR_MOD
  || MSG_SYSTEM_ALARM_CONTROL_MOD || MSG_SYSTEM_PROP)<<{motion_detected,
     motion_not_detected}.
```

In this section, we have discussed the ways in which the feature-driven software composition may introduce problematic behavior in the SPL product to be composed, and how the LTSA tool can be used to check for these problems. A further problem may simply be that certain features have overlapping specifications, i. e., they specify same actions and processes. It would probably be advisable to check the models always to be aware whether this is the case or not, because additional problems, like deadlocks, may easily be introduced because of this.

Magee and Kramer [11] do not give explicit analysis of the efficiency of model checking with the LTSA toolset. In practice, the tool seems to be efficient as long as the models keep to a reasonable size. In our example case, the models (states and transitions) have very reasonable sizes, for instance the model of Listing 1 gives a state model with 2^{20}(a little over one million) states and the safety and progress checks run typically in 1 ms on a 2.67 GHz Intel(R) Core(TM) i7 CPU. Already in 1997, Cheung, Giannakopoulou, and Kramer demonstrated the use of model checking of a system with 96, 000 states and 660,000 transitions [4]. However, if the FSPs contain indexed processes (processes that have "indexed copies" in the state system), the state space may easily explode if one is not careful with the index values. Also, the LTSA toolset's graphical interface has limited capacity, but models can be checked without it as well.

4 Conclusions

Previous work in utilization of LTS specifications to manage feature-driven software process lines has left open some technicalities on how the specifications can be used to check various properties of a product from a product line. In this work, we have studied some of those technicalities. We have identified certain potential problems that may arise when the specifications representing the features of the product line are integrated. The problems in the integration may be caused by design issues related to the features; however, they may also be caused by some features just not being compatible with others or some features expecting some further features to be added in the model to behave appropriately.

Based on analysis of the categories of integration problems, it may be possible to design more carefully the components to avoid problems in composing products using the feature-oriented approach. However, it is also important to understand what the potential problems in the composition are, and how it may be possible to resolve them.

References

1. Aaltonen T, Katara M, Pitkänen R (2001) Disco toolset—the new generation. J Univ Comput Sci 7(1):3–18
2. Batory D (2005) Feature models, grammars, and propositional formulas. In: Proceedings of the 9th international conference on Software Product Lines, SPLC'05, 2005. Springer, Berlin, pp 7–20

3. Benavides D, Segura S, Ruiz-Cortés A (2010) Automated analysis of feature models 20 years later: a literature review. Inf Syst 35(6):615–636
4. Cheung SC, Giannakopoulou D, Kramer J (1997) Verification of liveness properties using compositional reachability analysis. In: Proceedings of the 6th European software engineering conference held jointly with the 5th ACM SIGSOFT international symposium on Foundations of software engineering, ESEC '97/FSE-5. Springer, New York, pp 227–243
5. Diller A (1990) Z: an introduction to formal methods. Wiley, Chichester
6. Djebbi O, Salinesi C, Diaz D (2007) Deriving product line requirements: the red-pl guidance approach. In: Proceedings of the 14th Asia-Pacific software engineering conference, APSEC '07. IEEE Computer Society, Washington, DC, USA, pp 494–501
7. Hemakumar A (2008) Finding contradictions in feature models. In: First international workshop on analyses of software product lines (ASPL'08), pp 183–190, 2008
8. Hess J, Novak W, Kang K, Cohen S, Peterson A (1990) Feature-oriented domain analysis (FODA) feasibility study (CMU/SEI-90-TR-021). Software Engineering Institute, Carnegie Mellon University, Pittsburgh, Pennsylvania 15213
9. Kang KC, Kim S, Lee J, Kim K, Shin E, Huh M (1998) Form: a feature-oriented reuse method with domain-specific reference architectures. Ann Softw Eng 5:143–168
10. Knight JC, DeJong CL, Gibble MS, Nakano LG (1997) Why are formal methods not used more widely?. In: Fourth NASA formal methods workshop, pp 1–12
11. Magee J, Kramer J (2006) Concurrency: state models & Java Programs. Wiley, New York
12. Mannion M (2002) Using first-order logic for product line model validation. In: Proceedings of the second international conference on software product lines, SPLC 2. Springer, London, UK, pp 176–187
13. Nummenmaa J, Zhang Z, Nummenmaa T, Berki E, Guo J, Wang Y (2010) On the generation of disCo specifications from functional requirements, D-2010-13. University of Tampere, Tampere
14. Peña J, Hinchey MG, Ruiz-Cortés A, Trinidad P (2007) Building the core architecture of a nasa multiagent system product line. In: Proceedings of the 7th international conference on Agent-oriented software engineering VII, AOSE'06. Springer, Berlin, pp 208–224
15. Pohl K, Böckle G, van der Linden FJ (2005) Software product line engineering: foundations, principles and techniques. Springer, New York
16. Sommerville I, Kotonya G (1998) Requirements engineering: processes and techniques. Wiley, New York
17. Trinidad P, Benavides D, Durán A, Ruiz-Cortés A, Toro M (2008) Automated error analysis for the agilization of feature modeling. J Syst Softw 81(6):883–896
18. Trinidad P, Ruiz Cortés A (2009) Abductive reasoning and automated analysis of feature models: how are they connected?. In: VaMoS'09, pp 145–153, 2009
19. von der Massen T, Lichter HH (2004) Deficiencies in feature models. In: Mannisto T, Bosch J (ed) Workshop on software variability management for product derivation—towards tool support. LNCS Vol. 3154, Springer, 2004
20. Wang H, Li YF, Sun J, Zhang H, Pan J (2005) A semantic web approach to feature modeling and verification. In: Workshop on semantic web enabled software engineering, SWESE'05, 2005
21. Yang Y, Peng X, Zhao W (2008) Feature-oriented software product line design and implementation based on adaptive component model. J Front Comput Sci Technol 2(3):274

A Human Perception-Based Fashion Design Support System for Mass Customization

Lichuan Wang, Xianyi Zeng, Ludovic Koehl and Yan Chen

Abstract When developing mass customized products, human perception on products, including consumer's and design expert's perception, should be integrated into the process of design. So in this paper, we originally propose a fashion decision support system for supporting designer's work. In this system, we first characterize and acquire fashion expert perception and consumer perception on human body shapes. Next, these perceptual data are formalized and analyzed using the intelligent techniques, such as fuzzy set theory, decision tree, and fuzzy cognitive map. The complex relations between these perceptions as well as the physical measurements of body shapes are modeled, leading to the criteria which will permit to determine if a new design style is feasible or not for a given fashion theme. The proposed system is aimed to support customized design and mass market selection in practice.

Keywords Human perception · Intelligent system · Fuzzy sets · Decision tree · Fashion design · Mass customization

1 Introduction

In modern apparel industry, faced to serious competitions due to globalization, diversification of markets and digital revolution, product designers and managers need to pay much more attention to consumer's personalized needs in order to

L. Wang (✉) · Y. Chen
National Engineering Laboratory for Modern Silk, College of Textile and Clothing Engineering, Soochow University, Suzhou 215021, People's Republic of China
e-mail: sebastian.wlc@gmail.com

L. Wang · X. Zeng · L. Koehl
GEMTEX Laboratory, Roubaix 59100, France

L. Wang · X. Zeng · L. Koehl
University of Lille, Nord de France, Lille 59000, France

F. Sun et al. (eds.), *Knowledge Engineering and Management*,
Advances in Intelligent Systems and Computing 214,
DOI: 10.1007/978-3-642-37832-4_49,

quickly submit new products to the markets [1]. Even in mass market, consumers lose their tolerance for regular products and become more and more demanding for customized products meeting their individual expectations. They desire to personalize the style, fit, and other characters of fashion products. For copying with this new challenge from consumers, fashion companies are more and more interested by the concept of mass customization, i.e., producing goods and services catering to individual customers' needs with near mass production efficiency [2, 3].

In this paper, we propose a new mass customization-oriented and human perception-based decision support system for supporting designer's work. In this system, both the fashion expert's perception and consumer's perception are important for fashion mass customization. The techniques of sensory evaluation are used to extract data from design experts on the basic sensory attributes of human body shapes. These data are independent of any socio-cultural context. In the same time, consumer's perception on relations between the basic sensory attributes of body shapes and the concerned fashion theme are also evaluated. These consumer's data are strongly related to a socio-cultural context.

In our proposed support system, the techniques of artificial intelligence constitute the main computational tools for formalization and modeling of perceptual data. Especially, fuzzy techniques play an important role in the modeling of this system. The perceptual data of both consumers and designers are formalized mathematically using fuzzy sets and fuzzy relations. Moreover, the complex relation between human body measurements and basic sensory attributes, provided by designers, is modeled using fuzzy decision trees. And the complex relation between basic sensory attributes and fashion themes, given by consumers, is modeled using fuzzy cognitive maps. The former is an empirical model based on learning data measured and evaluated on a set of representative samples. The latter is a conceptual model obtained from cognition of consumers. The combination of the two previous models can provide more complete information to the fashion design support system, permitting to evaluate a specific body shape related to a desired fashion theme and obtain the design orientation in order to improve the image of the body shape.

2 Perceptual Data Acquisition and Formalization

In practice, trained fashion professionals should play an important role in garment mass customization in order to recommend or create personalized fashion designs. For this purpose, our study will focus on development of a fashion decision support system based on the perceptions of consumers and fashion experts.

In our system, the expert perception is involved in the evaluation of basic and concrete sensory descriptors describing the basic nature of male body shapes (naked bodies and dressed bodies) to be evaluated [4]. A regular sensory evaluation procedure is adopted to perform this task.

In the evaluation, sensory descriptors are more concrete and more related to the basic nature of body shapes, independent of socio-cultural context of wearers [5]. From these concrete parameters, design of garment becomes easier. Moreover, these sensory descriptors, such as "bulgy-slim", "swollen-dented", and "forceful-atrophic" are provided by fashion experts, because they are more sensible to evaluate a body shape by using their professional experience.

Unlike the expert perception, the consumer perception is involved in the evaluation of the relationship between fashion themes such as "sporty", "nature", and "attractive" and the previously determined sensory descriptors. These fashion themes are abstract and complex concepts, strongly related to the socio-cultural context of the description on male body shape [4]. They represent the general ambiance to consumers provided by the garment or human body shape. In fact, this consumer perception procedure can be considered as a cognition procedure, i.e., the last stage of perception. It is obtained from evaluators' common knowledge and not related to specific human bodies.

Three experiments are carried out for the sensory evaluation, which are purposed to obtain the relation between naked visual body shapes and sensory descriptors describing body shapes in Experiment I, the relation between visual body shapes with a garment style and these sensory descriptors in Experiment II, and the relation between fashion themes and sensory descriptors in Experiment III. The first two are established to acquire expert perception, and the last one is for consumer perception.

The mathematical formalization of the perceptual data will be as in the followings:

Let $T = \{t_1, t_2, \ldots, t_n\}$ be a set of n fashion themes characterizing the socio-cultural categories of body shapes. In my project, we have identified three fashion themes ($n = 3$) in sensory evaluation, which are "*sporty*", "*nature*" and "*attractive*".

Let $D = \{d_1, d_2, \ldots, d_m\}$ be a set of m basic sensory descriptors extracted by the fashion experts for describing the fashion themes in T. In previous sensory evaluation, 22 sensory descriptors have been extracted for describing body shapes.

Let $W = \{w_1, w_2, \ldots, w_p\}$ be a set of p virtual body shapes generated by using the software LECTRA-Modaris 3D fit [6].

Let $BM = \{bm_1, bm_2, \ldots, bm_h\}$ be a set of h body measurement features characterizing human body shapes. The body measurements of all the p virtual body shapes in W constitutes a matrix, denoted as $(bm_{ij})h \times p$ with $i = 1, \ldots, h, j = 1, \ldots, p$.

Let $BR = \{br_1, br_2, \ldots, br_g\}$ be a set of g body ratio indexes calculated from the body measurements of the set B. The body ratios of all the p virtual body shapes in W also constitutes a matrix, denoted as $(br_{ij})g \times p$ with $i = 1, \ldots, g, j = 1, \ldots, p$.

Let $EX = \{ex_1, ex_2, \ldots, ex_r\}$ be a set of r evaluators (design experts) evaluating the relevancy of sensory descriptors to virtual body shapes (naked or with garment). For any given sensory descriptor, each evaluator compares each virtual body shape with the standard body shape CA170 by selecting a linguistic score from {C_1(*very inferior*), C_2(*inferior*), C_3(*fairly inferior*), C_4(*a little inferior*), C_5(*neutral*), C_6(*a little superior*), C_7(*fairly superior*), C_8(*superior*), C_9(*very superior*)}.

Let $EC = \{ec_1, ec_2, \ldots, ec_z\}$ be a set of z evaluators (consumers) evaluating the relevancy between fashion themes and sensory descriptors. For any specific sensory descriptor and fashion theme, each evaluator gives a degree of their relevancy by selecting a linguistic score from the set (R_1 *very irrelevant*), R_2(*fairly irrelevant*), R_3(*neutral*), R_4(*fairly relevant*), R_5(*very relevant*)}.

Let $S = \{s_1, s_2, \ldots, s_\xi\}$ be a set of ξ existing garment styles which will be used as references in the design process. In our experiment, we give five reference styles for men's overcoat, which are "*Chester*", "*Ulster*", "*Balmacaan*", "*Trench*", and "*Duffle*".

Let $DE = \{de_1, de_2, \ldots, de_\lambda\}$ be a set of λ new garment styles generated for a special body shape. These new styles are generated by making combinations of the reference styles. It can be given in the form of photos or sketches by fashion designers.

3 Constitution of the Proposed System

The perceptual data from both design experts and consumers, obtained from the experiments, will be taken as learning data of the decision support unit inside the system. This decision support unit permits to determine from a series of computations whether a specific body shape is conform to a given fashion theme. The three perception databases are connected to the different stages of this decision support unit (see Fig. 1). Furthermore, the special body shape, represented by its body ratios BR^Y, the required fashion theme t_i, and the new garment design de_υ, obtained by combining different reference styles, are the three input components of the decision support unit.

The first stage of the decision support unit is Model I, whose aim is to evaluate the relevancy of a specific naked body shape Y, expressed by body ratios BR^Y, related to a required fashion theme t_i. The second stage of the decision support unit is Model II, whose aim is to evaluate the relevancy of a specific body Y with a garment design style de_υ related to t_i. In the last stage of the decision support unit, the previous two relevancy results are compared by using a gravity center-based criterion in order to evaluate whether the new garment design de_υ is suitable for the body shape Y in terms of image improvement related to the fashion theme t_i.

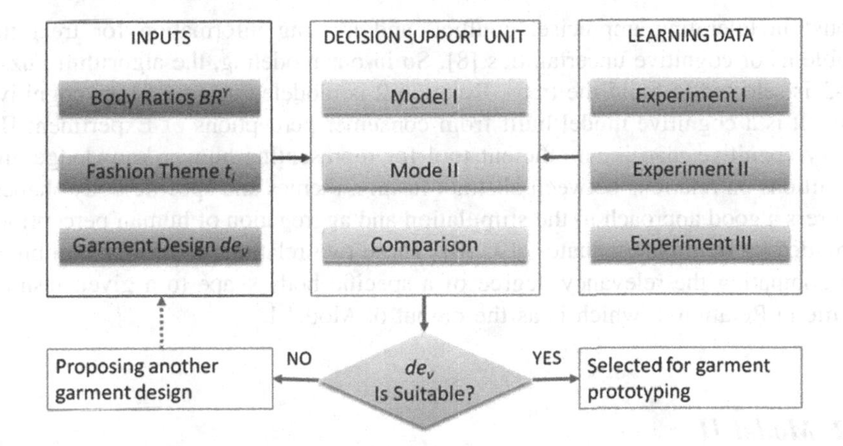

Fig. 1 The proposed fashion design support system

3.1 Model I

Model I deals with three relations, i.e., the relation between sensory descriptors and naked body ratios (**Relation 1**: **REL**(D, BR^Y)), the relation between fashion themes and sensory descriptors (**Relation 2**: **REL**(t_i, D)), and the relation between naked body ratios and a specific fashion theme (**Relation 3**: **REL**(t_i, BR^Y)). The data flow of Model I is presented by its functional structure in Fig. 2.

Relation 1 is modeled using Fuzzy decision trees. It is an empirical model built by learning from the perceptual data in Experiment I. Fuzzy decision trees are more efficient for treating learning data of mixed type, including both numerical and categorical data [7]. Furthermore, fuzzy decision trees can represent classification knowledge more naturally to the way of human thinking. They are more

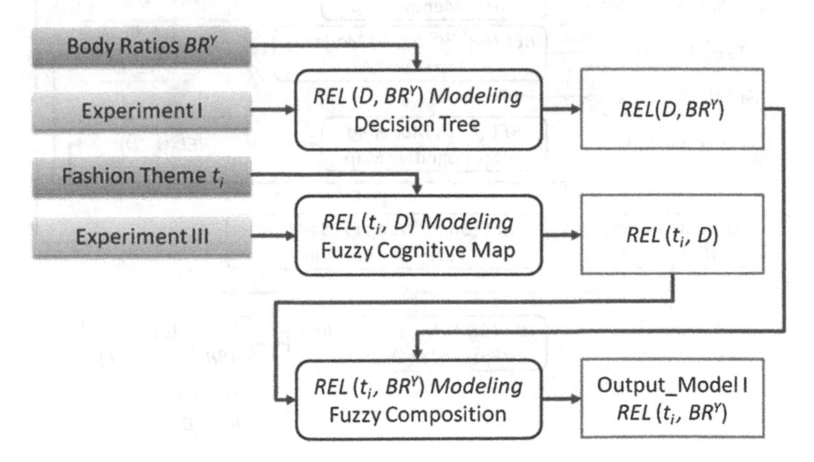

Fig. 2 Functional structure of Model I

robust in tolerating imprecise, conflict, and missing information for treat the problems of cognitive uncertainties [8]. So in our modeling, the algorithm Fuzzy ID-3 is selected to build the trees. Relation 2 is modeled using a fuzzy cognitive map. It is a cognitive model built from consumer perceptions of Experiment III. Fuzzy cognitive map is an efficient tool for representing human knowledge and cognitions on relations between abstract fashion themes and specific body shapes. It offers a good approach to the stimulation and aggregation of human perceptions provided by multiple evaluators [9, 10]. These two relations were then combined for computing the relevancy degree of a specific body shape to a given fashion theme in Relation 3, which is as the output of Model I.

3.2 Model II

The data flow of Model II is presented by its functional structure in Fig. 3. Model II first treats the fuzzy relation between a new garment design style and a number of reference styles (***Relation*** **4**: $\mathbf{REL}(de_\upsilon, S)$) in order to express this new design using its membership degrees to the existing reference styles. Next, we extract the fuzzy relation between the standard body *CA170* with each reference style and the identified sensory descriptors from the expert perceptual data in Experiment II (***Relation*** **5**: $\mathbf{REL}(D, s_\lambda)$). By combining the two previous fuzzy relations and Relation 2 calculated in Model I, we obtain the relevancy of the standard body shape with a new garment design to a given fashion theme for ***Relation*** **6** ($\mathbf{REL}(t_i, BR^{CA170} \bigvee de_\upsilon)$). Next, by using the union operation between this relevancy and the output of Model I (relevancy of a specific naked body shape to a given fashion theme), we obtain the relevancy of a specific body shape with a

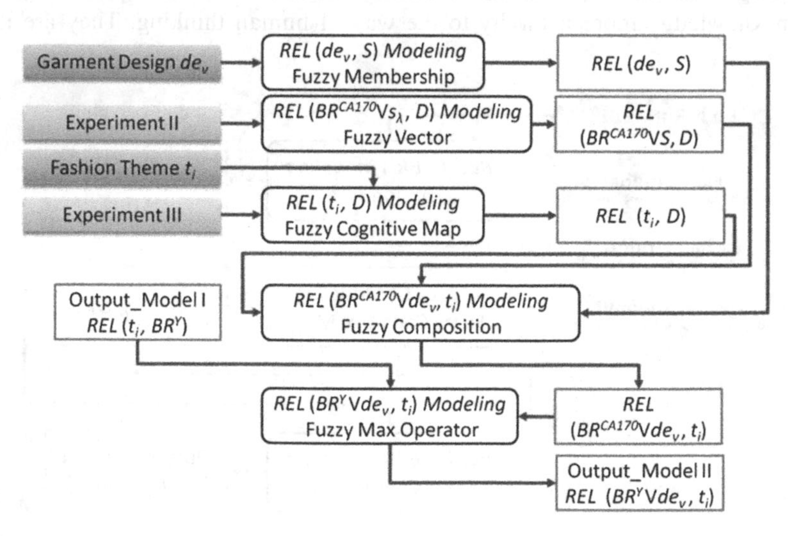

Fig. 3 Functional structure of Model II

specific garment design to a given fashion theme in ***Relation* 7** ($\mathbf{REL}(t_i, BR^Y \bigvee de_\upsilon)$), which constitutes the output of Model II.

3.3 Comparison

The outputs of the above two models can be represented by a distribution fuzzy matrix as the example showed in Fig. 4a. They are compared in the stage of Comparison to determine whether the garment design de_υ is feasible for the special body ratios for promoting the relevancy of the fashion theme t_i.

For this propose, we propose a criterion based on the principle of gravity center.

For computing the gravity center of the fuzzy matrix $\mathrm{REL}(t_i, BR^Y)$, we first transform the linguistic values $\{R_1, R_2, R_3, R_4, R_5\}$ into the equivalent numeric values $\{0, 0.25, 0.5, 0.75, 1\}$ and $\{C_1, C_2, C_3, C_4, C_5, C_6, C_7, C_8, C_9\}$ into $\{-1, -0.75, -0.5, -0.25, 0, 0.25, 0.5, 0.75, 1\}$.

The gravity of the fuzzy relation matrix (see example in Fig. 4b

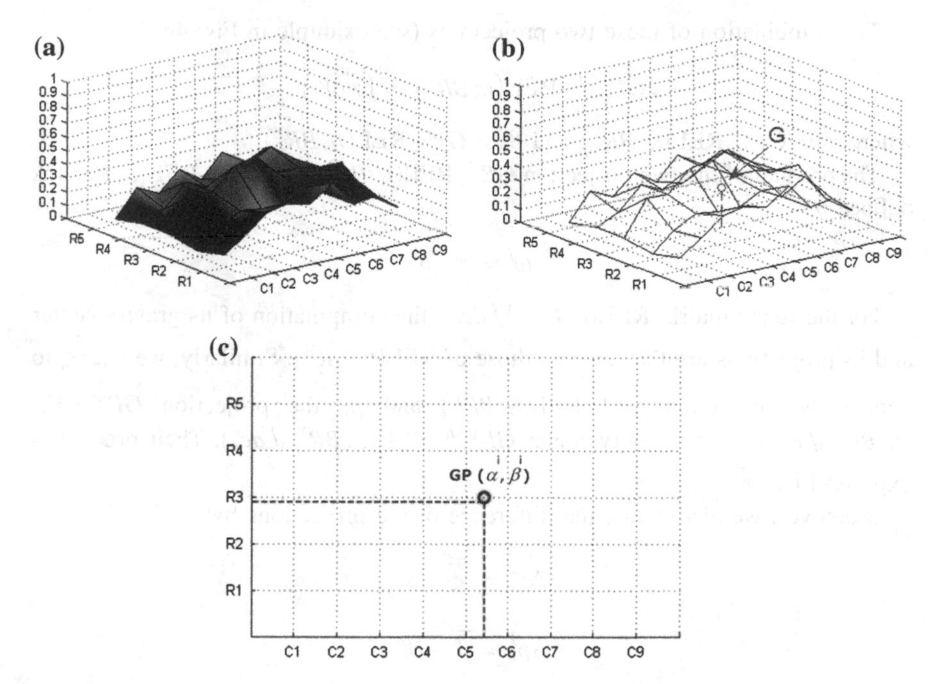

Fig. 4 Distribution, gravity center and projection of a fuzzy relation matrix. **a** Distribution of a fuzzy relation matrix, **b** Gravity center of the distribution, **c** Projection of the gravity center on the $(Y \times t_i)$ plan

$$\mathrm{REL}(t_i, BR^Y) = \begin{bmatrix} \mu_i^Y(R_1, C_1) & \cdots & \mu_i^Y(R_1, C_9) \\ \vdots & \ddots & \vdots \\ \mu_i^Y(R_5, C_1) & \cdots & \mu_i^Y(R_5, C_9) \end{bmatrix}$$

is calculated by

$$G\_\mathrm{REL}(t_i, BR^Y) = \sum_{j=1}^{5}\sum_{k=1}^{9} \mathrm{eq}(R_j)\mathrm{eq}(C_k)$$

where $\mathrm{eq}(R_j)$ and $\mathrm{eq}(C_k)$ are the equivalence values of R_j and C_k respectively.

The projection of the gravity center on the axis Y is calculated by

$$GP^Y_{\mathrm{REL}(t_i, BR^Y)} = G\_\mathrm{REL}(t_i, BR^Y) / \sum_{j=1}^{5}\sum_{k=1}^{9} \mu_{jk}\mathrm{eq}(R_j)$$

Similarly, The projection of the gravity center on the axis t_i is calculated by

$$GP^{t_i}\_\mathrm{REL}(t_i, BR^Y) = G\_\mathrm{REL}(t_i, BR^Y) / \sum_{j=1}^{5}\sum_{k=1}^{9} \mu_{jk}\mathrm{eq}(C_k)$$

The combination of these two projections (see example in Fig. 4c) is:

$$GP^{Y \times t_i}\_\mathrm{REL}(t_i, BR^Y) = (\alpha^i, \beta^i)$$

where $\alpha^i = GP^Y\_\mathrm{REL}(t_i, BR^Y)$ and $\beta^i = GP^{t_i}\_\mathrm{REL}(t_i, BR^Y)$.

The product of the two projection $GP^Y\_\mathrm{REL}(t_i, BR^Y)$ and $GP^{t_i}\_\mathrm{REL}(t_i, BR^Y)$ is defined by

$$\omega^i = \alpha^i \cdot \beta^i$$

For the fuzzy matrix $\mathrm{REL}(t_i, BR^Y \bigvee de_\upsilon)$, the computation of its gravity center and its projections are the same as those of $\mathrm{REL}(t_i, BR^Y)$. Similarly, we use $\overline{\overline{\alpha^i_\upsilon}}$ to denote the projection $GP^Y\_\mathrm{REL}(t_i, BR^Y)$ and $\overline{\overline{\beta^i_\upsilon}}$ the projection $GP^{t_i}\_\mathrm{REL}(t_i, BR^Y \bigvee de_\upsilon)$ of the gravity center $GP^{Y \times t_i}\_\mathrm{REL}(t_i, BR^Y \bigvee de_\upsilon)$. Their product is expressed by $\overline{\overline{\omega^i_\upsilon}}$.

Moreover, we also define the difference of the projections by

$$\Delta\alpha^i_\upsilon = \overline{\overline{\alpha^i_\upsilon}} - \alpha^i$$

$$\Delta\beta^i_\upsilon = \overline{\overline{\beta^i_\upsilon}} - \beta^i$$

And the difference of the products is

$$\Delta\omega^i_\upsilon = \overline{\overline{\omega^i_\upsilon}} - \omega^i$$

For comparison and analysis of the two relevancy degrees obtained from Model 1 and Model 2, we define the following rules:

L-1). For the value of α (α^i or $\overline{\overline{\alpha_\upsilon^i}}$):

Bigger is, higher is the relevancy degree of the special body ratios (naked or with garment design) to the fashion theme t_i for the design expert perception. Otherwise, the relevancy degree is lower.

L-2). For the value of β (β^i or $\overline{\overline{\beta_\upsilon^i}}$):

Bigger is, higher is the relevancy degree of the special body ratios (naked or with garment design) to the fashion theme t_i for the consumer perception. Otherwise, the relevancy degree is lower.

L-3). For the value of ω (ω^i or $\overline{\overline{\omega_\upsilon^i}}$):

Bigger is, higher is the relevancy degree of the special body ratios (naked or with garment design) to the fashion theme t_i for the compromise between design expert perception and consumer perception. Otherwise, the relevancy degree is lower.

Therefore, if $\Delta\omega_\upsilon^i$ is a positive value, the relevancy degree of the body shape Y with the garment design de_υ to the fashion theme t_i is higher than that of the naked body shape Y for the compromise between the expert perception and the consumer perception;

If $\Delta\omega_\upsilon^i$ is 0, there is no difference;

If $\Delta\omega_\upsilon^i$ is a negative value, the relevancy degree of the body shape Y with the garment design de_υ to the fashion theme t_i is smaller than that of the naked body shape Y for the compromise between the expert perception and the consumer perception.

According to the previous rules of comparison and analysis, we define the evaluation criterion of new garment design styles as follows:

If the value of $\Delta\omega_\upsilon^i$ is positive, the garment design style de_υ will be selected as a feasible design style for promoting the relevancy of the body shape Y to the fashion theme t_i; otherwise, design style de_υ will be rejected.

4 Application and Analysis

Two real case studies, customized design and mass market selection are proposed for showing the effectiveness of the design support system. In the related experiments, the special body ratios are obtained from the body measurements taken on real consumers. The fashion themes are those provided by the marketing experts, i.e., "*Sporty*", "*Nature*", and "*Attractive*". In order to minimize the development

Table 1 Body measurements and ratios in Case Study I

Body Measurements (Unit: cm)	*Special Body Shapes*		
	*Y*01	*Y*02	*Y*03
b_1 (Stature):	169	171.5	162.5
b_2 (Arm length):	53.5	54.5	49
b_3 (Chest circumstance):	89.5	80	84.6
b_4 (Neck circumstance):	37	36	36.5
b_5 (Waist Circumstance):	76.5	65	68.8
b_6 (Hip circumstance):	94	87	86.1

cost, our design strategy is to create a series of new design styles by copying, modifying and combining five classical references of styles: "*Chester*", "*Ulster*", "*Balmacaan*", "*Trench*", and "*Duffle*".

Case Study I: *Customized Design*

(1) ***Input:***

(a) **Target market**: Three different body shapes are selected from a database of body measurements, taken from young men of 25–35 years old in Eastern China. The corresponding data are given in Table 1.

(b) **Design objective**: promotion of the fashion theme "*Attractive*" for male overcoats.

(c) **New garment styles**: These new styles are described using their linguistic membership degrees related to five existing reference styles, "*Chester*", "*Ulster*", "*Balmacaan*", "*Trench*", and "*Duffle*". We have

de_1 : Totally belong to *Chester*
de_2 :Totally belong to *Duffle*
de_3 :Fairly belong to *Chester* AND a little belong to *Ulster*

(2) ***Output:***

The results in Table 2 show that given a specific design style and a fashion theme, the perceptual effects are different for different body shapes. It means that design strategy should be personalized. As shown in Table 2, two design styles, de_1 and de_3, are feasible for *Y01*. For *Y*02, all the three design styles are feasible. For *Y*03, only de_1 is feasible.

For a specific body shape, there exists some difference even for various feasible styles. For example, de_1 and de_3 are both feasible styles for *Y*01 but their relevancy degrees are different (0.5206 for de_1 and 0.1489 for de_3). The style de_1 can be considered as the best recommendation. In practice, we just need to deliver both solutions to the designer who will make the final choice according to his/her personal experience.

Case Study II: *Mass Market Selection*

The purpose of this case is to apply our design support system to mass fashion market in order to select a set of feasible design styles meeting the needs of the market. Different from Case Study I, the personalized body shapes should be taken

Table 2 Results in Case Study I for *Y*01, *Y*02, *Y*03 with the garment design styles (de_1, de_2, de_3) relevant to the fashion theme "*Attractive*"

Attractive (Fuzzy)								
de_1			de_2			de_3		
	$\Delta\omega_v^i$	Feasible?		$\Delta\omega_v^i$	Feasible?		$\Delta\omega_v^i$	Feasible?
*Y*01	0.5206	YES	*Y*01	−0.2754	NO	*Y*01	0.1489	YES
*Y*02	2.5990	YES	*Y*02	0.3411	YES	*Y*02	1.8507	YES
*Y*03	0.7446	YES	*Y*03	−0.3937	NO	*Y*03	−0.0902	NO

into account in the context of one population. This application is significant for developing new garment products with low costs.

(1) ***Input:***

(a) **Target market:** 60 special body shapes of male consumers whose ages vary from 25 to 35. Their body measurements are also taken from the database of the population in Eastern China.

(b) **Design objective:** Promotion by combining the previous three fashion themes for the following objectives.
Objective 1: "*Sporty*" AND "*Nature*"
Objective 2: "*Sporty*" AND "*Attractive*"
Objective 3: "*Nature*" AND "*Attractive*"

(c) **New design styles:** Six new garment design styles are described using their linguistic membership degrees related to five existing reference styles, "*Chester*", "*Ulster*", "*Balmacaan*", "*Trench*", and "*Duffle*":
de_1: Totally belong to *Chester*
de_2: Totally belong to *Duffle*
de_3: Fairly belong to *Chester* AND a little belong to *Ulster*
de_4: A little belong to *Chester* AND fairly belong to *Ulster*
de_5: Fairly belong to *Balmacaan* AND a little belong to *Trench*
de_6: A little belong to *Balmacaan* AND totally belong to *Trench*.

2) ***Output***

Table 3 gives the rates of feasible responses of all the body shapes for each design style and each objective. A feasible response is "YES" if and only if the corresponding evaluation criterion, defined previously, is positive (comparison between the naked body shape and the body shape with the given design style for a

Table 3 Results in Case Study II for selection

	de_1 (%)	de_2 (%)	de_3 (%)	de_4 (%)	de_5 (%)	de_6 (%)
Objective 1	53.33	16.67	53.33	43.33	40.00	65.00
Objective 2	68.33	16.67	63.33	43.33	31.67	60.00
Objective 3	53.33	21.67	53.33	55.00	31.67	58.33

Table 4 Results of selection in Case Study II

	Feasible selection	*Best selection*
Objective 1	de_1, de_3, de_6	de_6
Objective 2	de_1, de_3, de_6	de_1
Objective 3	de_1, de_3, de_4, de_6	de_6

specific fashion theme). As each objective is composed of two fashion themes, a feasible response is "YES" if and only if the evaluation criteria are positive for both fashion themes. In any cases, higher is the evaluation criterion, more feasible the corresponding design style is to the design objective.

3) ***Selection***

Based on the statistics of all the individual feasible responses, we define the following rules for selecting design styles for the whole population:

(a) If the rate of feasible responses is equal to or more than 50 %, the corresponding design style will be selected for the population.
(b) The design style having the highest rate of feasible responses will be regarded as the best solution.

According to these rules, the final result of selection is showed in Table 4 for each design objective.

5 Conclusion

In this paper, we offer a quantitative evaluation criterion for selecting feasible fashion products in a mass market. Based on this work, a new classification of human body shapes using human perceptions is developed. It should be more significant than the existing physical measurement-based body shape classification methods, especially in garment design. The classified results are capable of taking into account both aesthetics and physical measurements of human bodies.

The proposed fashion design support system can be further developed toward the concept of mass customization. Using the previous evaluation criterion, we can classify the whole set of design styles into different classes, each corresponding to a specific class of body shapes.

The human perception-based classification of body shapes can be further improved and extended by integrating other physical features of consumers, like skin color and face, and consumer's socio-cultural features such as age, profession, and wearing occasion. The new classification will permit to realize market segmentation for a specific population. For each segmented market, consumers will have similar requirements on fashion products. In the same time, we will also extend the perception-based classification of fashion styles by integrating other design elements, like colors and material textures. The final objective is to find the most appropriate class of design elements for each segmented market. In this way, the customization of fashion products can be realized with high-profit and low-cost in mass market.

References

1. Lim H et al (2009) Advance mass customization in apparel. J Text Apparel: Technol Manage 6(1):1–16
2. Tseng M, Jiao J (2001) Mass customization. In: Salvendy G (ed) Handbook of industrial engineering, 3rd edn. Wiley Press, New York, pp 684–709
3. Wilson JO, Rosen D (2005) Selection for rapid manufacturing under epistemic uncertainty. In: Design for manufacturing and the life cycle conference, vol 4b. ASME, Long Beach, CA, pp 451–460, 24–28 Sep 2005
4. Brangier E, Barcenilla J (2003) Concevoir un Produit Facile à Utiliser. Editions d'Organisation, Paris, pp 662
5. Zhu Y et al (2010) A general methodology for analyzing fashion oriented textile products using sensory evaluation. Food Qual Prefer 21(8):1068–1076
6. Cichocka A et al (2007) Introduction to modeling of virtual garment design in 3D, ISC'2007. Delft University of Technology, Netherlands, 11–13 June 2007
7. Janikow CZ (1996) Exemplar learning in fuzzy decision trees. In: Proceedings of FUZZ-IEEE, pp 1500–1505
8. Yuan YF, Shaw MJ (1995) Introduction of fuzzy decision trees. Fuzzy Sets Syst 69(2):125–139
9. Khan MS, Quaddus M (2004) Group decision support using fuzzy cognitive maps for causal reasoning. Group Decis Negot 13:463–480
10. Papageorigiou EI et al (2008) Fuzzy cognitive map based decision support system for thyroid diagnosis management. In: FUZZ-IEEE 2008, Fuzzy Systems, pp 1204–1211, 1–6 June 2008

An Approach of Knowledge Representation Based on LTV-LIA

Yunguo Hong and Li Zou

Abstract In order to represent the truth degree or credibility degree, people often use some linguistic hedges in knowledge representation. This paper represents the credibility of uncertain knowledge using linguistic values factors. Based on lattice implication algebra, we construct a kind of model for knowledge representation with ten linguistic-valued credibility factors. This approach can better express both comparable and incomparable linguistic information in knowledge representation. And it can also fit to knowledge reasoning under a fuzzy environment with both comparable and incomparable linguistic credibilities. Examples illustrate that the proposed approach can retain the original information as much as possible. It can also simulate people's thinking model with linguistic values so as to deal with the knowledge more intelligence.

Keywords Knowledge representation · Linguistic-valued credibility factor · Lattice implication algebra

1 Introduction

Most problems in real world are inaccurate and imperfect. Uncertainty of knowledge is the essential feature of intelligence problems. In reality, people often express uncertain knowledge with linguistic values and reason based on it [1–3]. Now there are many approaches [4–7] for the representation and reasoning of

Y. Hong
Dalian Vocational and Technical College, Dalian 116035, China

L. Zou (✉)
School of Computer and Information Technology, Liaoning Normal University,
Dalian 116081, China
e-mail: zoulicn@163.com

F. Sun et al. (eds.), *Knowledge Engineering and Management*,
Advances in Intelligent Systems and Computing 214,
DOI: 10.1007/978-3-642-37832-4_50,

uncertain knowledge. Among all these, when expressing the credibility of uncertain knowledge in traditional approach, linguistic values will be changed into values factors, and determine the scope of value factors and factual meaning. Many scholars also have proposed some methods [8–10] to deal with linguistic value such as 2-tuple.

Lattice implication algebra can deal with both comparable and incomparable linguistic values, and also can process a lot of lattice implication operations of uncertain knowledge for knowledge reasoning [11, 12]. This paper presents a new kind of approach of uncertain knowledge, i.e., representing knowledge credibility directly with linguistic value factors instead of changing it into values when processing knowledge from facts or rules. Based on lattice implication algebra, we define lattice implication algebra with ten linguistic-valued credibility factors. We construct a kind of model for knowledge representation with ten linguistic-valued credibility factors to deal with linguistic-valued credibility about uncertain knowledge. Credibility factors can be used in operations directly, thus knowledge reasoning of uncertain knowledge can be achieved.

2 Representation Method of Linguistic-valued Credibility Factors

Typical knowledge representation method is production representation. It is easy to describe facts and rules, and production representation can not only express certain knowledge but also uncertain knowledge.

There is a fact: representation of certain knowledge is a triple (objects, attributes, and values). Representation of uncertain knowledge is a 4-tuple (objects, attributes, values, and credibility factors). For example, the sentence "John has successful academic achievements." can be expressed as (John, academic achievements, and successful). The sentence "John has very successful academic achievements." can be expressed as (John, academic achievements, successful, and 0.8). Here, 0.8 is a credibility factor, which refers to the degree of which John's academic achievements is high. It means the information "John has successful academic achievements." is highly credible. When credibility factor is replaced by 0.4, it means the degree of which John's academic achievements is low, and even it is not true.

As for rules, the basic form is:

$$\text{IF } P \text{ THEN } Q,$$

where P is a condition, Q is a conclusion. It means that if the condition is satisfied, then the conclusion Q can be referred. For example, the sentence, IF power on THEN the light will be on, is certain knowledge. While the rule:

$$\text{IF windly THEN rain, } (0.7)$$

is uncertain knowledge. 0.7 means that this information is not absolutely credible, i.e., there will probably be rain if it is windy.

Credibility factors can be used in knowledge representation so as to better express the meaning of uncertain knowledge, solve the problems more efficiently and perform uncertain knowledge reasoning. So a new description of credibility is represented here.

Definition 1 Knowledge credibility can be divided into ten levels, i.e., credibility factors set $C = (\{$absolutely credible, very credible, exactly credible, somewhat credible, slight credible, slight incredible, somewhat incredible, exactly incredible, very incredible, and absolutely incredible$\})$, knowledge representation model about ten linguistic-valued credibility factors is the following 4-tuple:

(objects, attributes, values, ten linguistic-valued credibility factors),

For example, the sentence "John has successful academic achievements." can be expressed as (John, academic achievements, successful, and very credible). Here, "very credible" is a linguistic-valued credibility factor.

To better express knowledge, in the model for knowledge representation, the former three in the 4-tuple can be represented descriptive sentences, i.e., (John has successful academic achievements, great credible).

Definition 2 If–Then Rule is represented as

IF (P, Credibility_1) THEN (Q, Credibility_2) (Credibility_3),

Here, Credibility_1, Credibility_2, and Credibility_3 are the credibility of Evidence P, Conclusion Q and this rule respectively. The credibility can be applied into knowledge in accordance with the conditions.

For example,

(1) IF strongly windy, THEN rain, (very credible),
(2) IF (strongly windy, very credible) THEN (rain, very credible),
(3) IF (strongly windy, very credible) THEN (rain, somewhat credible) (very credible).

The more often used way is to deal with and express the credibility of the original knowledge in the form of number values factors. However, some linguistic values are incomparable, so linguistic values which describe the original knowledge can be firstly extracted and then shown as knowledge credibility. Appropriate reasoning result can be obtained from the operation of linguistic truth-valued lattice implication algebra.

There are two ways of extracting linguistic values are as follows:

(1) Add credibility factors directly to the certain knowledge.
(2) Extract credibility factors from uncertain knowledge and the former three in the 4-tuple represent certain knowledge.

As for one fact, descriptions are not the same when perspectives are different. Suppose "high, strong, good and so on" are positive value, and then the opposites "low, weak, bad and so on" are negative value. In reality, many descriptions are not positive but negative. For example, "John has successful academic achievements", it describes in a positive way. "John has unsuccessful academic achievements", it describes in a negative way. It can be easy to deal with it in a positive form in real life and it can also be fit to knowledge reasoning. So it is necessary to deal with the original knowledge for the sake of knowledge reasoning when the knowledge is described in a negative form. For example, "John has unsuccessful academic achievements". Without dealing with the original knowledge, the sentence can be expressed as:

(John has unsuccessful academic achievements, very credible),

To have better knowledge representation and knowledge reasoning, now deal with the original knowledge and make the attribute and the linguistic values factors opposite. For the sake of knowledge reasoning when the knowledge is described in a negative form, the knowledge is as follows:

(John has successful academic achievements, very incredible),

So clearly the second form can be more easily accepted. Knowledge reasoning is easy to perform when the attribute values of knowledge are classified.

Definition 3 Let (L, $\vee$, $\wedge$, O, I) be a bounded lattice with universal boundaries O (the least element) and I (the greatest element) respectively, and "$'$" be an order-reversing involution [8]. For any $x, y, z \in L$, if mapping $\rightarrow: L \times L \rightarrow L$ satisfies:

$$(I1): x \rightarrow (y \rightarrow z) = y \rightarrow (x \rightarrow z);$$
$$(I2): x \rightarrow x = I;$$
$$(I3): x \rightarrow y = y' \rightarrow x';$$
$$(I4): x \rightarrow y = y \rightarrow x \text{ implies } x = y;$$
$$(I5): (x \rightarrow y) \rightarrow y = (y \rightarrow x) \rightarrow x;$$
$$(I6): (x \vee y) \rightarrow z = (x \rightarrow z) \wedge (y \rightarrow z);$$
$$(I7): (x \wedge y) \rightarrow z = (x \rightarrow z) \vee (y \rightarrow z).$$

Then (L, $\vee$, $\wedge$, ', $\rightarrow$, O, I) is a lattice implication algebra (LIA for short).

Based on LIA with ten linguistic valued, we discuss lattice implication algebra with ten linguistic-valued credibilities.

Definition 4 In linguistic set $H = \{h_i | i = 0, 1, 2, 3, 4\}$, linguistic hedges h_0 means slightly, h_1 means somewhat, h_2 means exactly, h_3 means very and h_4 means absolutely. Evaluating values set is $\{L_1, L_2\}$, where L_1 means credibility and L_2 means incredibility. As for 10-element lattice implication algebra $(V, \vee, \wedge, \rightarrow)$, the operation "$\vee$" and "$\wedge$" are shown in Fig. 1 and Table 1. And the operation "$'$" is $(h_i, L_1)' = (h_i, L_2), (h_i, L_2)' = (h_i, L_1)$.

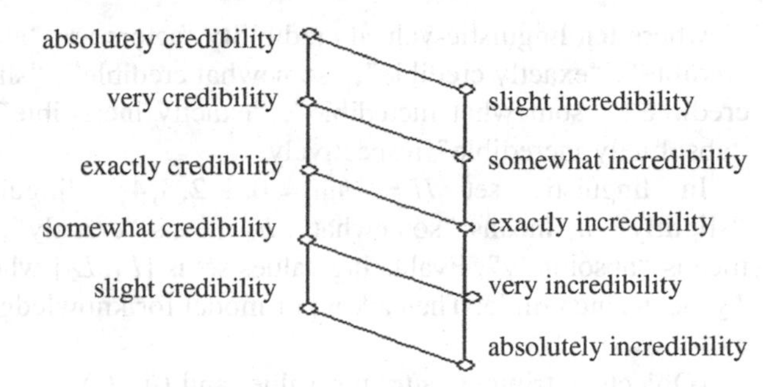

Fig. 1 Hasse diagram of L_{10}

Table 1 Implication operation of ten linguistic-valued credibility factors LIA

$\to$	(h_4, L_1)	(h_3, L_1)	(h_2, L_1)	(h_1, L_1)	(h_0, L_1)	(h_0, L_2)	(h_1, L_2)	(h_2, L_2)	(h_3, L_2)	(h_4, L_2)
(h_4, L_1)	(h_4, L_1)	(h_3, L_1)	(h_2, L_1)	(h_1, L_1)	(h_0, L_1)	(h_0, L_2)	(h_1, L_2)	(h_2, L_2)	(h_3, L_2)	(h_4, L_2)
(h_3, L_1)	(h_4, L_1)	(h_4, L_1)	(h_3, L_1)	(h_2, L_1)	(h_1, L_1)	(h_0, L_2)	(h_0, L_2)	(h_1, L_2)	(h_2, L_2)	(h_3, L_2)
(h_2, L_1)	(h_4, L_1)	(h_4, L_1)	(h_4, L_1)	(h_3, L_1)	(h_2, L_1)	(h_0, L_2)	(h_0, L_2)	(h_0, L_2)	(h_1, L_2)	(h_2, L_2)
(h_1, L_1)	(h_4, L_1)	(h_4, L_1)	(h_4, L_1)	(h_4, L_1)	(h_3, L_1)	(h_0, L_2)	(h_0, L_2)	(h_0, L_2)	(h_0, L_2)	(h_1, L_2)
(h_0, L_1)	(h_4, L_1)	(h_4, L_1)	(h_4, L_1)	(h_4, L_1)	(h_4, L_1)	(h_0, L_2)	(h_0, L_2)	(h_0, L_2)	(h_0, L_2)	(h_0, L_2)
(h_0, L_2)	(h_4, L_1)	(h_3, L_1)	(h_2, L_1)	(h_1, L_1)	(h_0, L_1)	(h_4, L_1)	(h_3, L_1)	(h_2, L_1)	(h_1, L_1)	(h_0, L_1)
(h_1, L_2)	(h_4, L_1)	(h_4, L_1)	(h_3, L_1)	(h_2, L_1)	(h_1, L_1)	(h_4, L_1)	(h_4, L_1)	(h_3, L_1)	(h_2, L_1)	(h_1, L_1)
(h_2, L_2)	(h_4, L_1)	(h_4, L_1)	(h_4, L_1)	(h_3, L_1)	(h_2, L_1)	(h_4, L_1)	(h_4, L_1)	(h_4, L_1)	(h_3, L_1)	(h_2, L_1)
(h_3, L_2)	(h_4, L_1)	(h_4, L_1)	(h_4, L_1)	(h_4, L_1)	(h_3, L_1)	(h_4, L_1)	(h_4, L_1)	(h_4, L_1)	(h_4, L_1)	(h_3, L_1)
(h_4, L_2)	(h_4, L_1)	(h_4, L_1)	(h_4, L_1)	(h_4, L_1)	(h_4, L_1)	(h_4, L_1)	(h_4, L_1)	(h_4, L_1)	(h_4, L_1)	(h_4, L_1)

The operation "$\to$" is

$$\begin{cases} (h_i, L_1) \to (h_j, L_2) = (h_{\max\{0,\, i+j-4\}}, L_2) \\ (h_i, L_2) \to (h_j, L_1) = (h_{\min\{4,\, i+j\}}, L_1) \\ (h_i, L_1) \to (h_j, L_2) = (h_{\min\{4, 4-i+j\}}, L_1) \\ (h_i, L_2) \to (h_j, L_2) = (h_{\min\{4, 4-j+i\}}, L_1) \end{cases},$$

Then $(V, \vee, \wedge, \to)$ is ten linguistic-valued credibility factors LIA.

Table 1 is obtained based on operation "$\to$" of ten linguistic-valued credibility factors LIA.

3 Knowledge Representation and Knowledge Reasoning Based on Ten Linguistic-valued Credibility Factors LIA

The model for knowledge representation with ten linguistic-valued credibility factors LIA is 4-tuple set:

(Objects, attributes, attribute values, and ten linguistic-valued credibility factors);

where ten linguistic-valued credibility factors are “absolutely credible”, “very credible”, “exactly credible”, “somewhat credible”, “slightly credible”, “slightly credible”, “somewhat incredible”, “exactly incredible”, “very incredible”, and “absolutely incredible” respectively.

In linguistic set $H = \{h_i | i = 0, 1, 2, 3, 4\}$, linguistic hedges h_0 means “slightly”, h_1 means “somewhat”, h_2 means “exactly”, h_3 means “very” and h_4 means “absolutely”. Evaluating values set is $\{L_1, L_2\}$ where L_1 means credible and L_2 means incredible. Then a kind of model for knowledge representation is shown as follows:

(Objects, attributes, attribute value, and (h_i, L_j));

It is easy and feasible to show in this way and it is helpful for the operation of knowledge reasoning.

While knowledge reasoning, the values between [−1, 1] are used as credibility factors and uncertain linguistic values are changed into accurate number values in traditional approach, then the number values should be changed into linguistic values again in the result. But sometimes it is difficult to find the appropriate linguistic values corresponding to them. The characteristic of human thinking is to describe and reason directly with linguistic values, so here we have the rules for linguistic-valued credibility (in short LCF):

$$\text{IF } P\ (h_i, L_m)(AND/OR\ \ Q\ (h_j, L_n))\ \text{THEN}\ \ R\ (h_s, L_t),\ (h_u, L_v).$$

The following methods are obtained based on ten linguistic-valued credibility factors LIA:

(1) $P(h_i, L_m) AND\ \ Q\ (h_j, L_n)\ = P\, AND\, Q(h_p, L_q),$ where (h_p, L_q)
$= (h_i, L_m) \wedge (h_j, L_n);$

(2) $P(h_i, L_m)\ OR\ \ Q\ (h_j, L_n) = (P\, OR\, Q)\ (h_p, L_q),$ where (h_p, L_q)
$= (h_i, L_m) \vee (h_j, L_n);$

(3) $P(h_p, L_q)\ \rightarrow R(h_i, L_m)\ = (P \rightarrow R)\ (h_j, L_n)\ ,$ where (h_j, L_n)
$= \ (h_p, L_q)\ \rightarrow\ (h_i, L_m);$

(4) $(E(h_a, L_b),\ (h_u, L_v)) = E(h_x, L_y),$ where$(h_x, L_y)\ =\ (h_a, L_b)\ \rightarrow\ (h_u, L_v).$
This case is for describing an uncertain fact with credibility;

(5) $E((h_a, L_b),\ (h_u, L_v)) = E(h_x, L_y),$ where $(h_x, L_y) = (h_{[(a+u)/2]}, L_1)$. This case is for describing several uncertain facts with multiple credibilities.

Example 1 We can do the uncertain knowledge reasoning under the knowledge representation with the linguistic-valued credibility factors. The knowledge is as follows:

r_1: IF E_1 THEN H (absolutely credible)
r_2: IF (E_2, very credible) THEN H (somewhat credible)
r_3: IF (E_3, very credible) *AND* ((E_4, somewhat credible) *OR* (E_5, somewhat incredible)) then E_1 (very credible)

Resolve the linguistic-valued credibility of the conclusion H.

First, resolve the linguistic-valued credibility of evidence E_1 in accordance to r_3. Next resolve the two linguistic-valued credibilities of two supporting evidences in accordance to r_1 and r_2 respectively, and then resolve the final result by aggregation operation.

Resolve: In r_3, there is certain aggregation relation in evidences, so do the aggregation operation first:

$$(h_3, L_1) \wedge (h_1, L_1) \vee (h_1, L_2),$$

We can obtain the linguistic-valued credibility of these evidences according to the linguistic-valued credibility factor LIA: (h_3, L_1);

Suppose the value be X, and thus we obtain a reasoning solution:

$$(h_3, L_1) \to X = (h_3, L_1),$$

Perform the implication operation, we get $LCF(e_1) = (h_4, L_1)$;

We can obtain the following solution similar,

$$r_1 : (h_4, L_1) \to Y = (h_4, L_1);$$

$$r_2 : (h_3, L_1) \to Z = (h_1, L_1);$$

Then we get $LCF(H) = (h_4, L_1)$ according to r_1 and $LCF(H) = (h_2, L_1)$ according to r_2;

At last, operate the two credibilities in accordance with the method (5), and obtain the final result: (h_3, L_1), i. e., the conclusion H is very credible.

4 Conclusions

This paper is based on 10-element lattice implication algebra and represents knowledge representation and knowledge reasoning with linguistic-valued credibility factors. Examples illustrate that the proposed approach can make it not only easy to understand the knowledge with the model for knowledge representation but also effective to perform reasoning operations. Based on ten linguistic-valued credibility factors lattice implication algebra, the reasoning of different kinds of credibility can be solved in knowledge representation by using different operations of linguistic-valued credibility factors. Therefore, the more complicated knowledge representation and knowledge reasoning will be further studied and applied to intelligence information field.

Acknowledgments This work is partly supported by National Nature Science Foundation of China (Grant No. 61105059, 61175055, 61173100), International Cooperation and Exchange of the National Natural Science Foundation of China (Grant No. 61210306079), China Postdoctoral Science Foundation (2012M510815), Liaoning Excellent Talents in University (LJQ2011116), Sichuan Key Technology Research and Development Program under Grant No.2011FZ0051,

Sichuan Key Laboratory of Intelligent Network Information Processing (SGXZD1002-10), and Key Laboratory of Radio Signals Intelligent Processing (Xihua University) (XZD0818-09).

References

1. Pei Z, Ruan D, Liu J, Xu Y (2009) Linguistic values based intelligent information processing: theory, methods, and application. Atlantis Computational Intelligence Systems. Atlantis press/World Scientific, Paris/Singapore
2. Pei Z, Xu Y, Ruan D, Qin K (2009) Extracting complex linguistic data summaries from personnel database via simple linguistic aggregations. Inf Sci 179:2325–2332
3. Deschrijver G, Kerre EE (2003) On the composition of intuitionistic fuzzy relations. Fuzzy Sets Syst 136:333–361
4. Van Gasse B, Cornelis C, Deschrijver G, Kerre EE (2008) A characterization of interval-valued residuated lattices. Int J Approximate Reasoning 49(2):478–487
5. Zou L, Liu X, Wu Z, Xu Y (2008) A uniform approach of linguistic truth values in sensor evaluation. Int J Fuzzy Optim Decis Making 7(4):387–397
6. Xu Y, Ruan D, Qin KY, Liu J (2004) Lattice-valued logic. Springer, Heidelberg
7. Xu Y, Liu J, Ruan D, Lee TT (2006) On the consistency of rule bases based on lattice-valued first-order logic LF(X). Int J Intell Syst 21:399–424
8. Martinez L, Ruan D, Herrera F (2010) Computing with words in decision support systems: an overview on models and applications. Int J Comput Intell Syst 3(4):382–395
9. Herrera F, Martinez L (2001) The 2-tuple fuzzy linguistic computational model advantages of its linguistic description, accuracy and consistency. Int J Uncertainty Fuzziness Knowl Based Syst 9:33–48
10. Herrera F, Alonso S, Chiclana F, Herrera-Viedma E (2009) Computing with words in decision making: foundations, trends and prospects. Int J Fuzzy Optim Decis Making 8(4):337–364
11. Zou L, Ruan D, Pei Z, Xu Y (2008) A linguistic truth-valued reasoning approach in decision making with incomparable information. J Intell Fuzzy Syst 19(4–5):335–343
12. Zou L, Ruan D, Pei Z, Xu Y (2011) A linguistic-valued lattice implication algebra approach for risk analysis. J Multi-Valued Logic Soft Comput 17(4):293–303

Designing a Client Application with Information Extraction for Mobile Phone Users

Luke Chen, Peiqiang Chen, Chu Zhao and Jianming Ji

Abstract Pervasive diffusion of mobile phones nowadays attracts numerous research attention of scholars and engineers to providing client applications. This paper is aimed at providing a conceptual design for a client application as a service from home web platforms in consideration of limited screen size and navigability of mobile device. In doing so, the work is centered on the design of the information extraction module by incorporating various ways of extracting entries of interest from the perspectives of relevance, coverage, and redundancy, as well as introducing a combined measure. Moreover, a preliminary prototype is developed to show its applicability in an Android environment.

Keywords Client application · Mobile phone · Information extraction · Mobile search

1 Introduction

Rapid advances in information and telecommunication technologies and applications have enabled a pervasive diffusion of mobile phone use in our lives nowadays. Surfing and browsing with mobile phones has become a major way for users to search for information in business and leisure. Recent reports have revealed that the world mobile users amounted to 5.9 billion, of which about one-sixth are Chinese users, resulting in twice as many mobile-broadband as fixed broadband subscriptions [1, 2]. Moreover, looking at mobile search in various services as an example, it was reported to account for about 15 % of mobile advertising market value in 2011 [3, 4].

L. Chen (✉) · P. Chen · C. Zhao · J. Ji
Department of Computer Science at Century College, Beijing University of Posts and Telecommunications, Beijing 102000, People's Republic of China
e-mail: chenluke@outlook.com

F. Sun et al. (eds.), *Knowledge Engineering and Management*, Advances in Intelligent Systems and Computing 214,
DOI: 10.1007/978-3-642-37832-4_51, © Springer-Verlag Berlin Heidelberg 2014

Usually, a client service for mobile devices is an application mapping from wired web platforms to wireless service platforms, with newly designed client services software and functionalities. Due to the limitation of the screen size and navigability of mobile devices such as smart/cell handsets compared with those of PCs, many client applications call for features that are compact in content and effective in search.

First of all, mobile search is a key element of a mobile client service as it is a basis for information retrieval and knowledge engineering, on which users browsing is based directly or indirectly via intermediate software. Examples of related services/applications include optimized engines, Q&A, directory, navigation, dynamic selection interface, etc. [5].

Next, when designing a mobile client from a home platform, we are also concerned with the problem of how to extract information that is of high interest and rich content [6–14]. Often, the volume of information in different webpage channels (including search portals) may be huge and can hardly be uploaded or transplanted entirely to the client. Hence, information selection is a crucial issue of content extraction.

Furthermore, as a matter of fact, to design a mobile client is, to a large extent, to extract information from various channels/sources of massive data on the home web platforms in terms of keywords, records, and/or texts. The selection metrics to use reflect the quality of the extraction process. As will be seen in later sections, the metrics of concern include relevance, coverage, and redundancy.

Accordingly, in designing a client application for mobile phone users, primary attention of our work is paid to incorporation of information extraction methods into the application. The rest of the paper is organized as follows. Section 2 presents a conceptual design framework, Sect. 3 describes the metrics of consideration in application of the methods, Sect. 4 provides a prototype implemented preliminarily in an Android environment, and Sect. 5 concludes the work and highlights certain possible areas of future work.

2 The Design Framework

As mentioned in the prior section, the mobile client design is a mapping from a home web platform, which can usually be characterized in two respects. One respect is that it inherits the identity features of its home platform with major content aspects and brand image; the other respect is that it is not meant to be a simplified version with proportionally shortened contents, but a new service with more focused aspects and/or richer content.

For example, with a search engine such as Google (www.google.com) or Baidu (www.baidu.com), a keyword search may result in a large number of entries or web pages, displayed in ranking of relevance degrees (e.g., PageRank values in Google [15, 16]). Since these entries of search results are huge in size, even PC users will often only look at the first several entries/pages and neglect the rest,

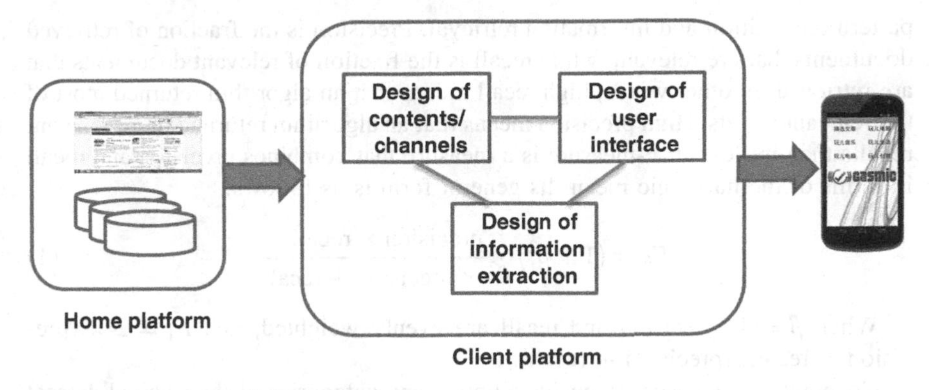

Fig. 1 The conceptual design framework

let along mobile users. However, the screen size and large-scale browsing incapability of mobile phones further attach the importance of the search quality in both content and speed. This means that separate techniques and algorithms are often needed in the mobile context.

In design of a mobile client, the mapping process is a transformation, which is composed of three major components in the client platform, namely, design of contents (e.g., service blocks/channels), design of user interface (UI), and design of information extraction. Figure 1 depicts the corresponding framework for the conceptual design.

3 Metrics and Methods for Information Extraction

In a client service, information extraction is to fetch related data/information selectively and process it in a manner that meets the requirements of mobile client applications concerned. There are a variety of metrics in assessment of the information extraction quality, usually coupled with different methods in favor of various metrics. These methods can be invoked directly such that the extraction outcomes are displayed as output of the client platform, or can be embedded into other modules to formulate integrated functions. For example, the ranking (rank listing) of the extracted entries/records is a direct procedural call, while extracted blog articles could be further used to in a module for generating summaries and statistics and visualizing the output.

Unlike commonly used search engines where relevance is a major concern, our design will take into account other metrics, in addition to relevance, so as to provide more perspectives of the quality. Concretely, three types of metrics are considered, i.e., relevance, coverage, and redundancy, each having different variants and being desirable in different application contexts.

For relevance, nowadays, the well-known measures are precision, recall, and F-measure (or their variants) [17]. Precision and recall are two important notions of pattern recognition and information retrieval. Precision is the fraction of retrieved documents that are relevant, while recall is the fraction of relevant documents that are retrieved. In other words, high recall means that an algorithm returned most of the relevant results. High precision means that an algorithm returned more relevant results than irrelevant. F-measure is a measure that combines precision and recall in terms of the harmonic mean. Its general form is as follows:

$$F_{\beta} = (1 + \beta^2)\frac{\text{precision} \times \text{recall}}{\beta^2 \times \text{precision} + \text{recall}} \tag{1}$$

When $\beta = 1$, precision and recall are evenly weighted, i.e., $F_1 = 2 \times$ (precision $\times$ recall)/(precision + recall).

Coverage is a measure to evaluate how many categories of the original dataset are contained in the extracted set. In other words, it is an assessment of the extent that the content of the original data is covered in the extraction result. Though with many ways, a common one is the following [18]:

$$\text{coverage} = \frac{C(E)}{C} \tag{2}$$

where C is the total number of categories and $C(E)$ is the number of different categories in the extraction set E. In a sense, coverage reflects the degree of diversity, and the higher coverage the richer the content.

Redundancy has been studied intensively in many fields and perspectives. It has also been investigated in terms of inequality with numerous notions proposed [19]. Here we adopt a similarity-based perspective in web search context where entries are generally treated to be similar to each other in various degrees. Concretely, the degree of redundancy for an extracted set E is based on an average of pair-wise similarities [20]:

$$\text{redundancy} = \sum\nolimits_{x \in E}\left(1 - \frac{1}{\sum\nolimits_{y \in E} \text{similarity}(x,y)}\right) \Big/ |E| \tag{3}$$

It is worth mentioning that the calculation of all these metrics relies on the similarity measures used. For instance, a widely used way to calculate the degree of similarity between two documents is to calculate the Cosine similarity according to keyword matching and corresponding frequencies [21]. Specifically, let $D = \{x_1, x_2, \ldots, x_n\}$ be a set of documents, $K = \{k_1, k_2, \ldots, k_m\}$ be a set of keywords, q_{ij} be the number of times that keyword k_i appears in document x_j, and n_i be the number of documents containing k_i ($1 \le i \le m$, $1 \le j \le n$). Then, the importance (weight) of keyword k_i in document x_j is defined as w_{ij}:

$$w_{ij} = \left(\frac{q_{ij}}{\max\limits_{1 \le i \le m}\{q_{ij}\}}\right) \times \log\left(\frac{n}{n_i}\right) \tag{4}$$

where $\max_{1 \le i \le m} \{q_{ij}\}$ is the maximum frequency of keywords in document x_j. Furthermore, the similarity between documents x_p and x_q $(1 \le p,\ q \le n)$ is measured as:

$$\text{similarity}(x_p, x_q) = \frac{\sum_{i=1}^{m} w_{ip} \times w_{iq}}{\sqrt{\left(\sum_{i=1}^{m} w_{ip}^2\right) \times \left(\sum_{i=1}^{m} w_{iq}^2\right)}} \quad (5)$$

In light of these metrics, different methods have been developed to take consideration of one or more of the related measures, each having respective advantages. In design of our client service, three methods are chosen in the module of information extraction in accordance with the three metrics and will be applied upon the needs of the client functions, namely, the top-k extraction, heuristic extraction, and random extraction. The Top-k extraction method (briefly, Top-k) belongs to the relevance-oriented type that usually extracts the first k results sequentially according to a ranking based on a certain evaluation function, which is commonly used in today's databases and web search engines [8–14]. The heuristic extraction method (briefly, Heuristic) is designed with a heuristic means to seek information with high coverage degree [18]. The random extraction method (briefly, Random) selects k results randomly from the original dataset, which may help reduce redundancy to certain extent (also helping increase coverage sometimes). Notably, the degrees of coverage and redundancy may or may not be mutually affected depending upon the specific measures used in applications and implementation strategies developed in algorithms.

Furthermore, we could also propose a combined way to integrate the Top-k and Heuristic methods in a manner that takes relevance and coverage into consideration. The idea is to use F-measure to rearrange the results of the two methods. Since each result (extracted set) will be sorted in display (e.g., according to the degrees of the measure concerned), the entries on top will be selected depending on β for the importance attached to the measure. For instance, when relevance is placed heavier than coverage, we will have β attached as a coefficient of relevance, leading to the weight of relevance from 50 % ($\beta = 1$) to 100 % ($\beta = 0$).

Then, due to the nature of Top-k and Heuristic in favor of relevance and coverage, an overall degree of p in [0, 1] will be used as a percentage at which the corresponding top potion of Top-k result will be taken, and the rest portion will be taken from the Heuristic result. Concretely, let k be the number of entries to extract, and R and H be the two sets extracted by Top-k and Heuristic with their relevance and coverage degrees being $F_\beta(R)$ and coverage(H), respectively. Thus, the percentage for R is:

$$p = \frac{F_\beta(R) + \alpha \times \text{coverage}(H)}{F_\beta(R) + \text{coverage}(H)} \quad (6)$$

```
Input: source[n];  k; metric; α;  /* source[n] =
source data of size n on the home platform; k = the
number of extracted entries; metric = {relevance,
coverage, redundancy, combined};   α = the extra. */
 Output:  outcome[k];  /* extracted set of size k*/
 Begin
 if metric   = Relevance
 outcome[k] = extracted entries using Top-k; else
 if metric   = Coverage
 outcome[k] = extracted entries using Heuristic; else
 if metric   = Redundancy
 outcome[k] = extracted entries using Random; else
 if metric   = Combined
 R[k] = extracted entries using Top-k;
 Compute Fβ(R[k]);
 H[k] = extracted entries using Heuristic;
 Compute Fβ(H[k]);
 p = (Fβ(R[k])+α×coverag(H[k])) / (Fβ(R[k])+coverag(H[k])) ;
 outcome[k] = selected entries from p of R[k] and (1-
p)of H[k];
 /* generate extracted set outcome[k]. */
 End
```

Fig. 2 Pseudo-codes of the module for information extraction

where α is a designer-specified parameter expressing the extra taken from coverage to relevance, $0 \leq \alpha \leq 1$. Then, the percentage for H is $q = 1 - p$. In this way, when $\alpha = 0$, R takes its own proportion; when $\alpha = 1$, R takes all extra proportion from H amounting to an entire consideration ($p = 1$). For example, suppose $F_\beta(R) = 0.92$, and coverage(H) = 0.86, then for $\alpha = 0$, $p = 0.52$, meaning that 52 % of the entries are selected from R; and for $\alpha = 0.5$, $p = 0.76$, which takes extra 24 % of the entries from H than that of its own proportion. Note that q could be formulated in the same manner if coverage is in more favor. For illustrative purposes, the pseudo-codes of the module are provided in Fig. 2.

4 A Prototype Application

This section describes a prototype for the design of the mobile client in an Android environment. The home web platform is CASMIC.cn, which is a website (in Chinese) that the authors developed mainly for college students. Figure 3 shows its homepage.

For the contents of the mobile client named CASMIC Client, five channels were included, such as movie, music, life, IT, and DIY, along with a listing of selected articles, which are main and popular functions. The design of the user interface of the client kept the main image of the home platform, such as logo and major color (red), which was styled in a new look (see Fig. 4).

Fig. 3 Home platform CASMIC.cn

Fig. 4 Contents of the mobile client

The mobile client was implemented using Android SDK 7. The application supports and is compatible with Android version 2.1 (or higher).

There were a few functions of the client, in that the information extraction module discussed in the previous section could be applied, such as for selected

Fig. 5 The listings of selected articles

articles and movie reviews. For instance, the entries of selected articles could result from the extraction of the source articles in forms of different listings.

As can be seen in Fig. 5, five choices of the listings were designed to support various extraction preferences, with radio buttons such as Hotness, Recent, Sequential, Diversity, and Combined. The Hotness listing was designed to provide the articles that were read in most frequency; the Recent listing was designed to provide the articles that were released most recently; the Sequential listing was designed to provide the articles that were selected on relevance to interested keywords (w.r.t. (1)) using the Top-k method; the Diversity listing was designed to provide the articles that were selected on coverage (w.r.t. (2)) using the Heuristic method; and the Combined listing was designed to provide the articles that were selected on both relevance to interested keywords and coverage (w.r.t. (6)) using the combined way. Figure 5 shows the corresponding screen in the client. Noteworthy is that, as above-mentioned, the listings of Diversity and Combined via the Heuristic and Combined methods are in favor of coverage and meanwhile could help reduce the redundancy (w.r.t. (3)).

Moreover, a preliminary testing for the applicability of the mobile client was made with HTC A810e smart phones on browsing the articles of the first pages. CASMIC Client was compared with two other browsers (i.e., HTC browser, and UC browser which is a well-known browser in China for its data traffic saving features (http://www.uc.cn)), revealing its advantage in data traffic saving. with CASMIC Client. Figure 6 shows the data traffics of the methods briefly named as CASMIC, HTC, and UC, respectively.

Fig. 6 Data traffics

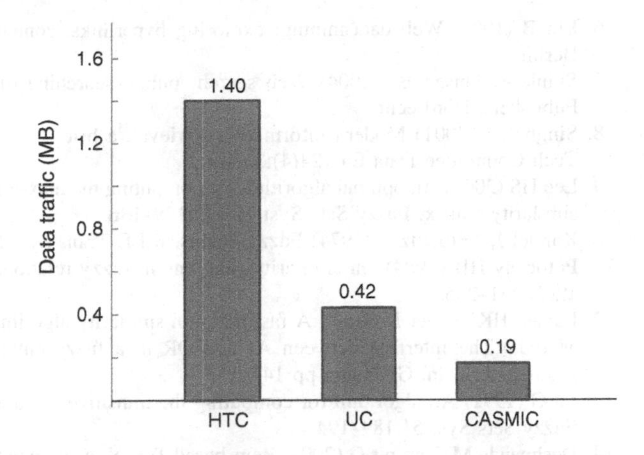

5 Conclusion and Future Work

Mobile applications are widely seen in social and business lives. This paper has presented a conceptual design of a mobile client application as a service provided to smart/cell phone users. With a design framework, primary attention of the work has been paid to the module for information extraction, aimed at fetching and selecting information from the home platforms in light of metrics on not only relevance, but also coverage and redundancy. The module has embraced the methods, i.e., Top-k, Heuristic, and Random, in accordance with the metrics of concern. In particular, a combined measure has been proposed for relevance and coverage, coupled with a combined way to integrate the extracted results. Finally, a prototype application has been designed and partly realized in an Android environment to show the applicability.

Future work can be carried out in a few directions, such as detailed treatments of the client functions, optimization of the extraction module and related methods, and a further testing for the in-depth scalability and use on large platforms. Moreover, the design of user interface could be enhanced with certain affective technologies such as eye-tracking equipment so as to further test the users' experiences and optimize the screen layout.

References

1. ITU (2011) The world in 2011: ICT facts and figures. ITU Telecom 2011, Geneva, 25–27 October (2011). (http://www.itu.int/)
2. CCID reports. http://miit.ccidnet.com/20120321
3. Mobile Search in the US, eMarketer report. http://www.emarketer.com. Accessed Jan 2010
4. The search wars are going mobile, eMarketer report. http://www.emarketer.com. Accessed July 2007
5. Mobile Search. www.wikipedia.org

6. Liu B (1998) Web data mining: exploring hyperlinks, contents, and usage data. Springer, Berlin
7. Spink A, Jansen BJ (2004) Web search: public searching of the web. Kluwer Academic Publishers, Dordrecht
8. Singhal A (2001) Modern information retrieval: a brief overview. Bull IEEE Comput Soc Tech Committee Data Eng 24(4):35–43
9. Lee HS (2001) An optimal algorithm for computing the max-min transitive closure of a fuzzy similarity matrix. Fuzzy Sets Syst 123(1):129–136
10. Kandel J, Yelowitz L (1974) Fuzzy chains. IEEE Trans Syst Man Cybern 4:472–475
11. Potoczny HB (1984) On similarity relations in fuzzy relational databases. Fuzzy Sets Syst 12(3):231–235
12. Larsen HK, Yager R (1989) A fast maxmin similarity algorithm. In: Verdegay JC, Delgado M (eds) The interface between AI and OR in a fuzzy environment, IS 95. Verlag TUV Rheinland, Koln, Germany, pp 147–155
13. Fu G (1992) An algorithm for computing the transitive closure of a fuzzy similarity matrix. Fuzzy Sets Syst 51:189–194
14. Deshpande M, Karypis G (2004) Item-based Top-N recommendation algorithms. ACM Trans Inf Syst 22(1):143–177
15. Langville AN, Meyer CD (2006) Google's page rank and beyond: the science of search engine rankings. Princeton University Press, New Jersey
16. Page L, Brin S, Motwani R, Winograd T (1999) The pagerank citation ranking: bringing order to the web
17. Tan P-N, Steinbach M, Kumar V (2005) Introduction to data mining, Addison-Wesley, Upper Saddle River
18. Pan F, Wang W, Anthony KHT, Yang J (2005) Finding representative set from massive data. In: The fifth international conference on data mining, pp 338–345
19. Coulter PB (1989) Measuring Inequality. Westview Press, Boulder
20. Ma BJ, Wei Q, Chen GQ (2010) A combined measure for representativeness on information retrieval in web search. In: Proceedings of FLINS2010, Chengdu, August 2010
21. Salton G (1983) Extended boolean information retrieval. Commun ACM 26:1022–1036

Real-Time Road Detection Using Lidar Data

Chunjia Zhang, Jianru Xue, Shaoyi Du, Xiaolin Qi and Ye Song

Abstract Real-Time road detection is a demanding task in active safety and auto-driving of vehicle. Vision is the most popular sensing method for road detection, but it is easier to be influenced by illumination, shadows, shield, etc. To overcome those difficulties, the light detection and ranging (Lidar) sensor is a good choice for road detection. For either the urban environment or the rural areas, the important feature of the road is that the road surface could be approximately represented by some planes. Hence, the Lidar's scanning plane and the road surface intersect at a set of line segments, and a line segment means a road plane. To extract the line segments from a scan, a least mean square problem is proposed, which is solved by a distance segment approach and an iterative line fitting approach. To eliminate the perception dead zone, some suitable historical data are adopted combining with the fresh data. In order to initial the road search range, a hypothesis is given that the area under the vehicle is road. The road detection is achieved for the scans from the close to the distance and in the meanwhile the search range of the next scan is updated. A lot of experiment results demonstrate the robustness and efficiency of the proposed approach for real-time auto-driving of the intelligent vehicles.

C. Zhang (✉) · J. Xue · S. Du · X. Qi · Y. Song
Institute of Artificial Intelligence and Robotics, Xi'an Jiaotong University, Xi'an, China
e-mail: cjzhang.china@gmail.com

J. Xue
e-mail: jrxue@gmail.com

S. Du
e-mail: dushaoyi@gmail.com

X. Qi
e-mail: qixiaolin@gmail.com

Y. Song
e-mail: songye@gmail.com

F. Sun et al. (eds.), *Knowledge Engineering and Management*, Advances in Intelligent Systems and Computing 214,
DOI: 10.1007/978-3-642-37832-4_52,

Keywords Road detection · Lidar data · Line fit

1 Introduction

Real-Time road detection is one of the most important issues in intelligent vehicle due to its essentiality and necessary in active safety, auto-driving and navigation. To solve this problem a lot of research have been done using visible image [1–3], infrared image [4, 5] and point set of light detection and ranging (Lidar) [6, 7]. Visible image is the most common perception data. A lot of research based on Visible image have done to deal with this problem. To detect road from a single color image, Hui et al. [1] proposed a vanishing point and Foedisch et al. [2] introduced a neural network. For stereo visible image, Son et al. [3] combined a posteriori probability and visual information for image segmentation. Visible image is widely applied for its high resolution, abundant texture information, and lower cost, but it is easier to be influenced by illumination, shadows, shield, etc. Infrared image is mainly used to detect pedestrian [4], but Fardi et al. [5] proposed a Hough transformation-based approach to detect road. The weakness of infrared type sensors is too sensitive to the temperature. To overcome those difficulties, Lidar sensor is a good choice for it stability of illumination, shadows, and temperature. Hence, some research has been done using this type of active sensor. For example, Kirchner and Heinrich [6] proposed a model-based approach and Wijesoma et al. [7] introduced a extend kalman filter. However, the computational complexities of those methods are high and cannot be used to achieve real-time road detection.

To solve the aforementioned problem, a new method based on line segments extraction is approached. For either the urban environment or the rural areas, the important feature of the road is that the road surface could be approximately represented by some planes. Hence, the Lidar's scanning plane and the road surface intersect at a set of line segments, and a line segment means a road plane. To extract the line segments from these segments, a least mean square problem is proposed, which can be solved by an iterative line fitting approach. To eliminate the perception dead zone, some suitable historical data are adopted combining with the fresh data. To extract road from those data, an initial road range is given based on the hypothesis that the area under the vehicle is road. A lot of experiment results demonstrate the robustness and efficiency of the proposed approach for real-time auto-driving of the intelligent vehicles.

The rest of this paper is organized as follows. In Sect. 2, the model of road is built, and the point set segment method and the iterative line fitting approach are proposed. In Sect. 3 the perception dead zone elimination and the road detection are presented. In Sect. 4, experiments in the urban environment and campus are carried out to demonstrate the robustness and efficiency of the proposed approach for real-time auto-driving of the intelligent vehicles.

Fig. 1 The actual road. a Urban environment. b Highway. c Rural environment

2 Road Model and Line Segment Extraction

2.1 Road Model

The most important feature of the road in the urban and highway environment is the well-paved surface and the distinct edge such as isolation belt and curb. For the rural areas in Fig. 1, the road is not so paved and the boundary is ambiguous too.

Hence, a unified model is needed to present the road structure for all the above-mentioned environments. For either the urban environment or the rural areas, the important feature of the road is that the road surface could be approximately represented by some planes. The isolation belt and curb could also be treated as a plane perpendicular to the road surface. Hence, the Lidar's scanning plane intersects with the road surface or edge at a set of line segments, and a line segment means a plane.

2.2 Point Set Segment

If the Lidar beams fire at the same object, the distance of the adjacent points is small, and it is clearly shown in Fig. 2. Otherwise a large distance means the adjacent points are belong to the different objects. Hence, the point set is segment according to the distances of each pair's adjacent points, the unified road model is shown in Fig. 3 and Table 1 present the process of segmentation.

In the implementation, an average filtering is used after the segmentation to smooth the data in the same segment. The segments that have only one point are treated as the isolated point and those segments are removed, and the result of segmentation is shown in Fig. 4.

2.3 Line Segment Extraction

The point set is segmented in the previous step, but it is not mean that one segment contain only one line segment. Hence, the problem how many line segments are

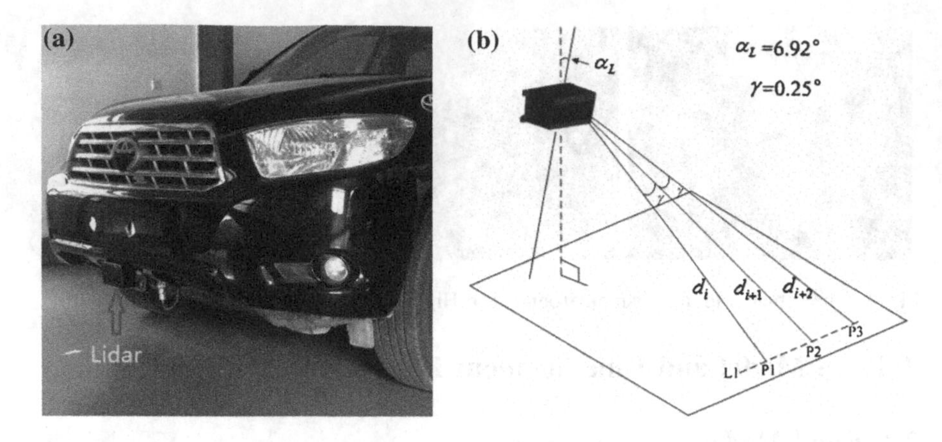

Fig. 2 Configuration of the Lidar. **a** Location of Lidar on the intelligent vehicle. **b** Title angle and angle resolution of the Lidar

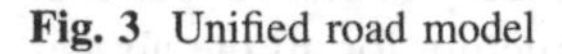

Fig. 3 Unified road model

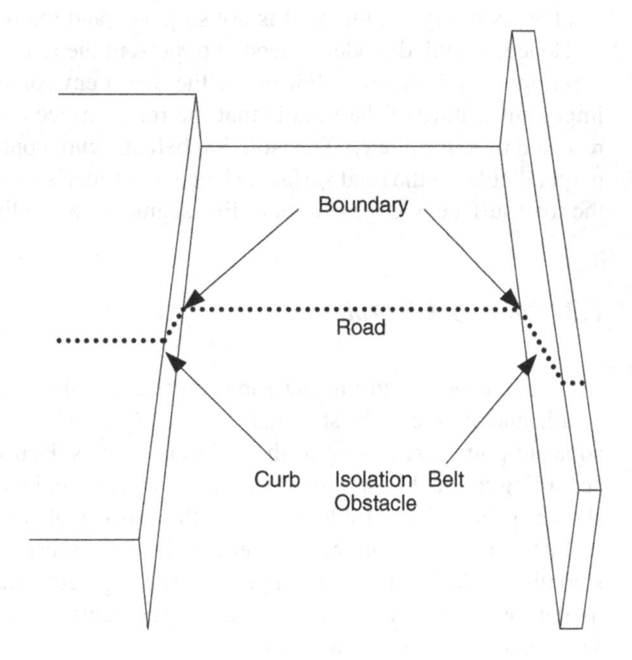

contained and how to extract the line segments in each segment is proposed. To solve this problem, an iterative line fitting approach is adopted [8], and Table 2 presents the process of line fitting.

This iterative approach for extracting the line segments from the segments is terminated until all the segments are fitted to the proper line, and the computation complexity of this algorithm is O(log2(N)).

Table 1 The process of point set segment

Point set segment
1. Save all the N points along the scanning sequence
2. Create a current segment S_j contain the first point P_1
3. Compute the distance between point P_i and its adjacent point P_{i-1}
4. If the distance is smaller than a given threshold T_s, save point P_i to the current Segment S_j. (go to 3)
5. Otherwise, a new segment S_{j+1} with the point P_i is created, and it is treated as the current segment (go to 3)
6. When all the points are computed, save the segments as its subscript

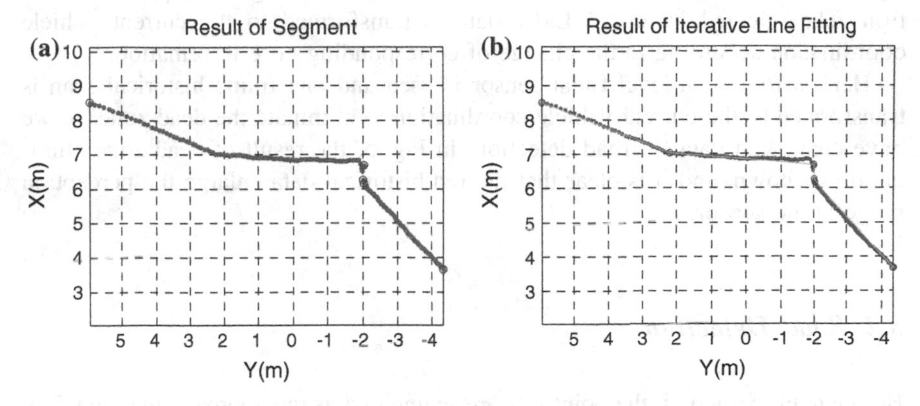

Fig. 4 Result of segment and line segment extraction. **a** Result of segment. **b** Result of line segment extraction based on the result of segment

Table 2 The process of line segment extraction

Line segment extraction
1. Get the segment $S_j = \{P_1 \ldots P_M\}$
2. Fit the segment S_j to a line using least mean square algorithm
3. If the least mean square is smaller than a given threshold T_{lms}, save the line segment (go to 1)
4. Otherwise, a new line segment is created by the first point and the last point of the segment
5. Find the point P_i with the maximum distance to the new line
6. Divide S_j into two new segments, one is $S_j' = \{P_1 \ldots P_i\}$ and the other is $S_{j+1}' = \{P_{i+1} \ldots P_M\}$.
7. Replace S_j by S_j', and insert S_{j+1}' into the storage sequence behind S_j' (go to 1)

3 Road Detection

3.1 Eliminate Perception Dead Zone

A small vertical perception angle is the main disadvantage of the Lidar sensor, consequently, it is important to enlarge the Lidar's scanning rang to eliminate the perception dead zone in front of the intelligent vehicle. Here we suppose that the vehicle is driving forward and its pose information is available at each moment. Therefore, the suitable historical data can be adopted to eliminate the perception dead zone.

The scanning data of Lidar could be easily corresponding with the pose information. To eliminate perception dead zone, the suitable historical Lidar data right corresponding to the dead zone are selected according to the pose information. The selected historical Lidar data is transformed to the current vehicle coordination according to the change of corresponding pose information.

Here, a four scan-level Lidar sensor is used and one frame historical scan is transformed to the current vehicle coordination to eliminate the dead zone, so we have eight scan data for road detection. In Fig. 5, the result of dead zone elimination is shown, and it is clear that the red historical data enlarge the perception range of the sensor.

3.2 Road Detection

Before road extraction, the point sets are segmented as the approach mentioned in part B of Sect. 2, and then the line segments are extracted from those segments using the method in part C of Sect. 2. To extract road from those line segments, we

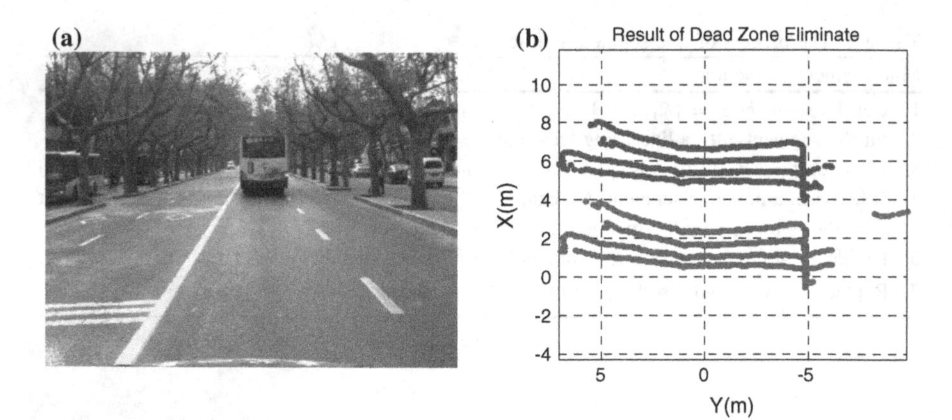

Fig. 5 Result of perception dead zone elimination. **a** Image of the real road environment. **b** Result of perception dead zone elimination, the *blue* data is the fresh data and the *red* data is the historical data

Table 3 The process of road extraction

Road extraction
1. Initialize road range value with the areas occupied by the car
2. Get the line segments set $L_m = \{l_1\ l_k\}$ of next scan
3. As the sequence of subscript get the line segment l_i
4. Compute overlap ratio of line segment l_i and the road range
5. If the overlap ratio is smaller than a given threshold T_s, save the line segment as road (go to 3)
6. Otherwise, (go to 3)
7. When all line segments in L_m is computed, updated the road range with the saved line segments

hypothesize that the area under the vehicle is road. Hence, the initial road range is given as the areas occupied by the car. Then the road detection is achieved for the scans from the close to the distance with a road range value that is updated by the previous scan. The process of road extraction is presented in Table 3.

Here we have eight scan data to detection road. To ensure the accuracy, all the eight scan data are computed as the above-mentioned steps. The line segment extraction based on least mean square algorithm is the most time consuming step. The computation complexity of each scan data is O(log2(N)), so the total computation complexity of the road detection is O(8log2(N)). It is not a calculation burden for the current computer while N is 220 in this paper. The result of road extraction is shown in Fig. 6.

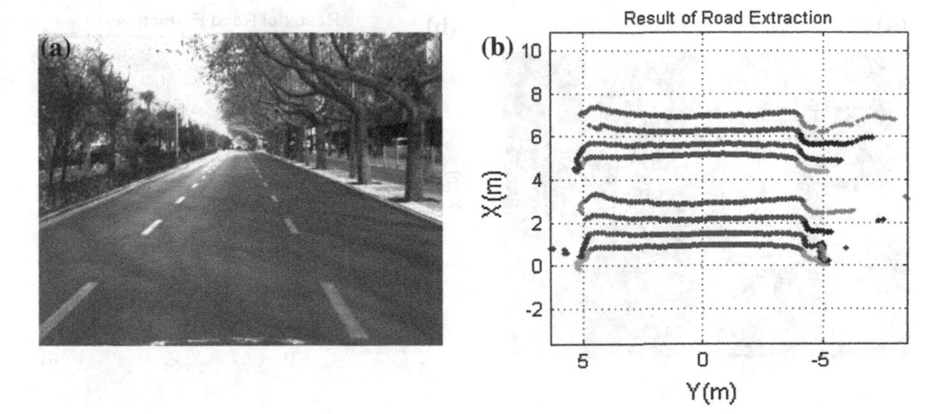

Fig. 6 Road extraction result of the urban environment without obstacle. **a** Image the real road environment. **b** Road extraction result, *green* is the road area

4 Experiment Results

To demonstrate the robustness and efficiency of the proposed approach for real-time auto-driving of the intelligent vehicles, a lot of experiments in the real urban traffic environment and campus have been done. Here some experiment results of different traffic scenes are carried out.

The most common case in the real traffic scene is the obstacle which mainly is vehicle and pedestrian on the adjacent lane or in front of the vehicle, so it is significant to detect the road areas in those situations robustly. The road extraction results of those environments are given in Figs. 7 and 8, and it is clear that our approach works well in this kind of scene.

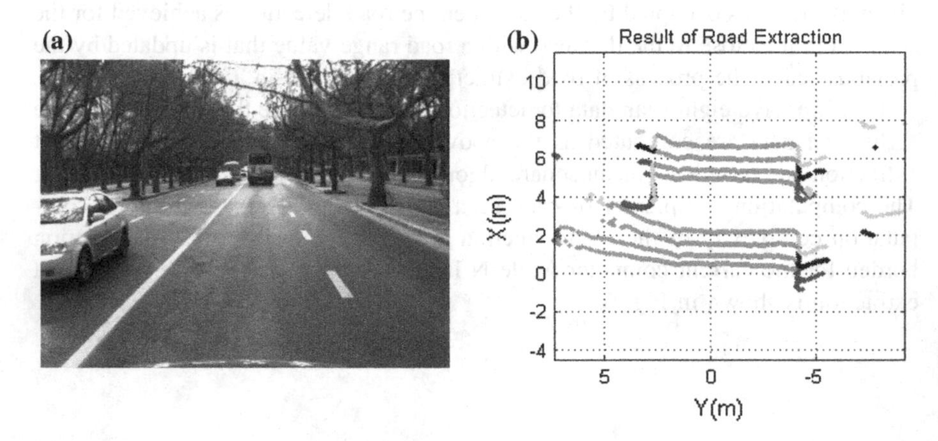

Fig. 7 Road extraction result of the urban environment with vehicles. **a** Image of the real road environment. **b** Road extraction result, *green* is the road area

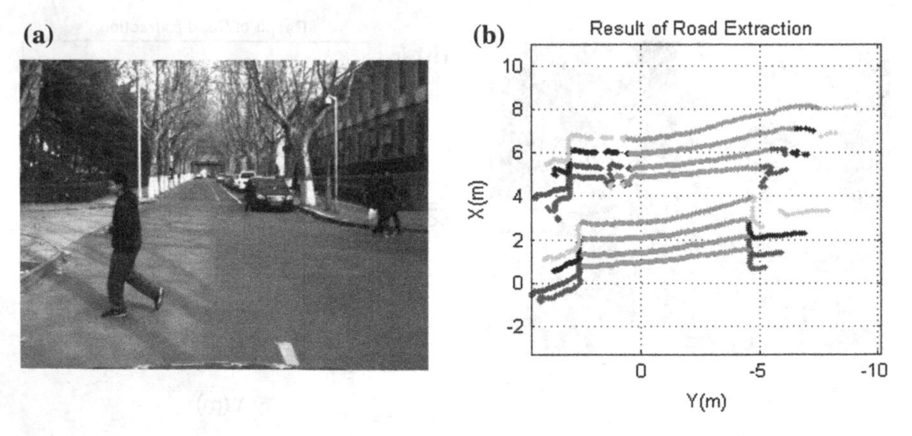

Fig. 8 Road extraction result of the urban environment with pedestrian. **a** Image of the real road environment. **b** Road extraction result, *green* is the road area

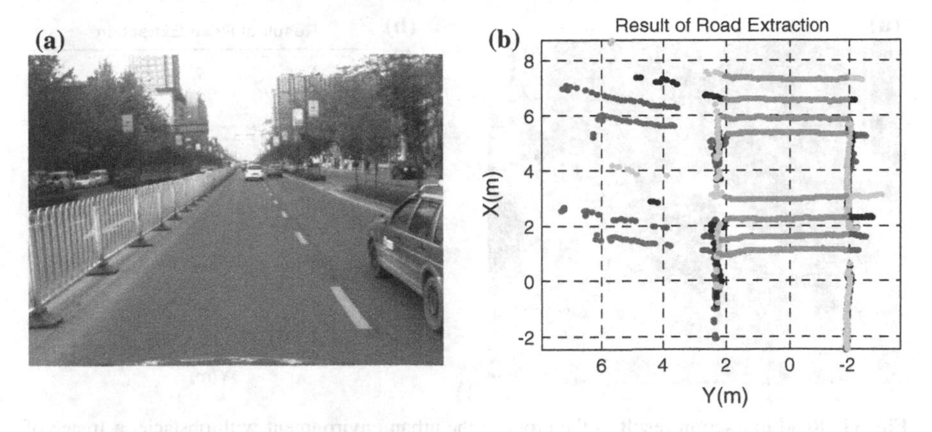

Fig. 9 Road extraction result of the urban environment with barrier and obstacle. **a** Image of the real road environment. **b** Road extraction result, *green* is the road area

The barrier is an important instrument on the road for it is essential to defend the vehicles turning back arbitrarily and the humans across the road. Consequently, it is important to detect road areas in this kind of situation and Fig. 9 illuminate that our approach is efficient.

Cross is a complex traffic scene, it is essential to deal with this kind of scene for auto-driving of intelligent vehicle. The result is shown in Figs. 10 and 11, it is obvious that our approach can give out right road areas for vehicle driving through the cross.

In some particular situation such as the traffic accident and road works, the road is bounded by the traffic cones. Hence, it is necessary to detect the safe areas of the road. Figure 12 demonstrates that our approach is valid for this situation.

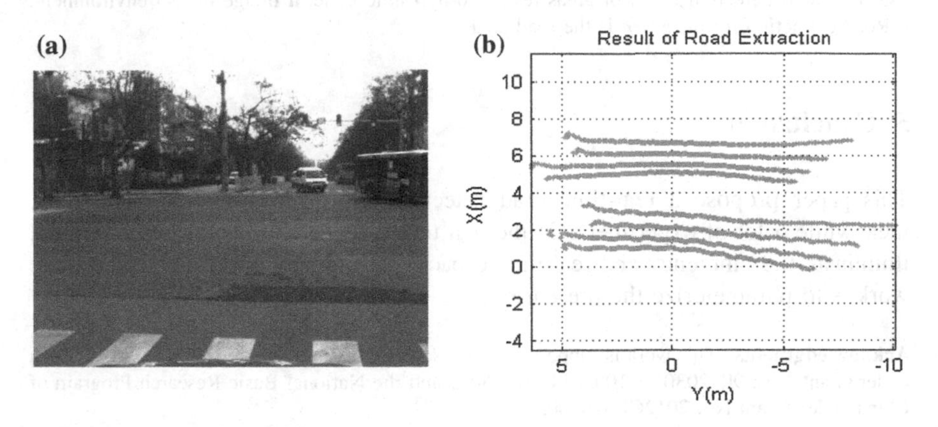

Fig. 10 Road extraction result in the cross of the urban environment without obstacle. **a** Image of the real road environment. **b** Road extraction result without obstacle, *green* is the road area

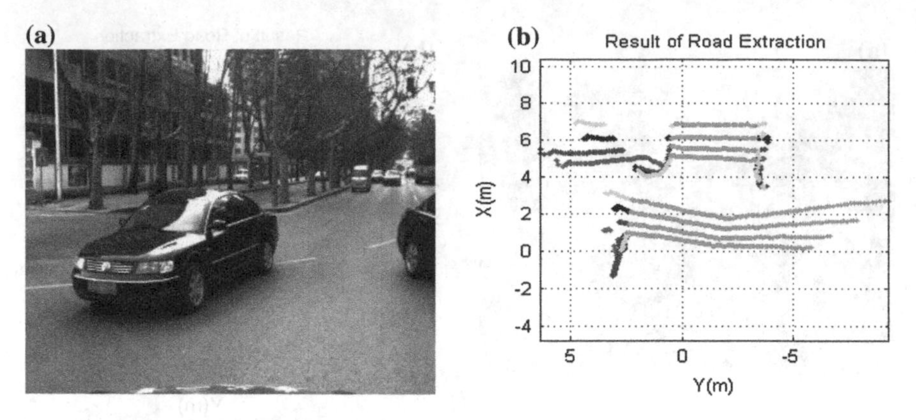

Fig. 11 Road extraction result in the cross of the urban environment with obstacle. **a** Image of the real road environment. **b** Road extraction result with obstacle, *green* is the road area

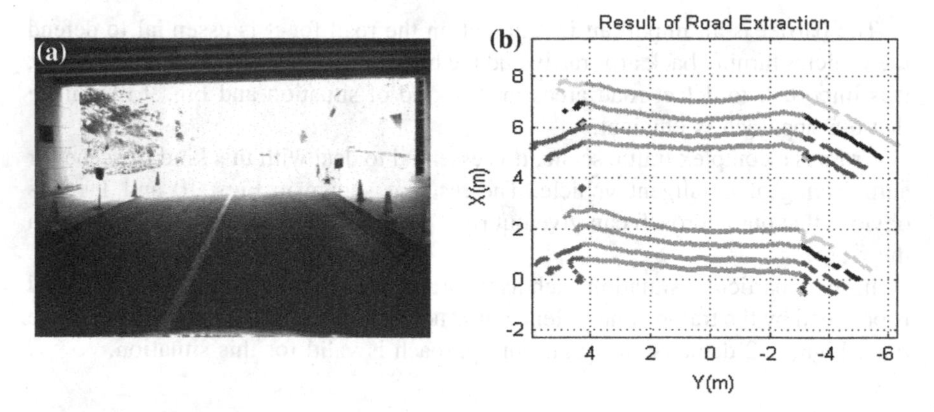

Fig. 12 Road detection result of areas restricted by traffic cone. **a** Image of real environment. **b** Road extraction result, *green* is the road area

5 Conclusion

This paper propose a real-time road detection approach using the Lidar data, meanwhile a lot of experiments in the real traffic scene and campus environment illuminate the efficiency and real-time capacity of the new approach. The feature work is to parameterize the present road.

Acknowledgments This work is supported by the National Natural Science Foundation of China under Grant Nos. 90920301, 61005014, 61005002 and the National Basic Research Program of China under Grant No. 2012CB316400.

References

1. Hui K, Audibert JY, Ponce J (2009) Vanishing point detection for road detection. In: IEEE conference on computer vision and pattern recognition 2009, CVPR 2009, pp 96–103
2. Foedisch M, Takeuchi A (2004) Adaptive real-time road detection using neural networks. In: Proceedings of the 7th international IEEE conference on intelligent transportation systems, 2004, pp 167–172
3. Son TT, Mita S, Takeuchi A (2008) Road detection using segmentation by weighted aggregation based on visual information and a posteriori probability of road regions. In: IEEE international conference on systems, man and cybernetics, 2008, SMC 2008, pp 3018–3025
4. Suard F, Rakotomamonjy A, Bensrhair A, Broggi A (2006) Pedestrian detection using infrared images and histograms of oriented gradients. In: IEEE intelligent vehicles symposium, 2006, pp 206–212
5. Fardi B, Wanielik G (2004) Hough transformation based approach for road border detection in infrared images. In: IEEE intelligent vehicles symposium, 2004, pp 549–554
6. Kirchner A, Heinrich T (1998) model based detection of road boundaries with a laser scanner. In: Proceedings of the 1998 IEEE international conference on intelligent vehicles, vol 1, pp 93–98
7. Wijesoma WS, Kodagoda KRS, Balasuriya AP (2004) Road-boundary detection and tracking using ladar sensing. IEEE Trans Robotics Autom 20:456–464
8. Nguyen V, Martinelli A, Tomatis N, Siegwart R (2005) A comparison of line extraction algorithms using 2D laser rangefinder for indoor mobile robotics. In: IEEE/RSJ international conference on intelligent robots and systems, 2005 (IROS 2005), pp 1929–1934

Design and Analysis of a Two-Axis-Magnetic Bearing with Permanent Magnet Bias for Magnetically Suspended Reaction Wheel

Bangcheng Han, Shiqiang Zheng and Haitao Li

Abstract A magnetically suspended reaction wheel (MSRW) for high-precise stabilization of spacecraft attitude is described in this paper. The speed range of the MSFW is from −6000 r/min to 6000 r/min, and angular momentum at normal speed is 30 Nms. A two-axis-magnetic bearing with permanent magnet bias is designed for reducing the power and minimizing the size, etc. The magnetic force, current stiffness, and negative position stiffness are derived by the equivalent magnetic circuit. The nonlinearity of this radial magnetic bearing is shown by using the characteristic curves of force-current-position, current stiffness, and position stiffness considering all the positions of the rotor within the clearance space and the control current. The maximum bearing capacity is given in this paper. The analysis and design method in this paper that can predict all the performance of the radial magnetic bearing with permanent magnet bias can be supplied for the design and analysis of the magnetic bearing wheel system. The errors of current stiffness and position stiffness for the two-axis-active magnetic bearing calculated by the linearized model compared with the FEM method are 5.9 % and 4.5 % respectively. The linearized model is validated by the FEM.

This work was supported in part by the National Nature Science Funds of China under Grant 61203203, and in part by the Aviation Science Fund of China under Grant 2012ZB51019, and in part by the National Major Project for the Development and Application of Scientific Instrument Equipment of China under Grant 2012YQ040235.

B. Han · S. Zheng (✉) · H. Li
Science and Technology on Inertial Laboratory, Beijing 100191, China
e-mail: zhengshiqiang@buaa.edu.cn

B. Han · S. Zheng · H. Li
Key Laboratory of Fundamental Science for National Defense
of Novel Inertial Instrument and Navigation System Technology,
Beijing 100191, China

B. Han · S. Zheng · H. Li
School of Instrument Science and Opto-Electronics Engineering,
Beijing University of Aeronautics and Astronautics, Beijing 100191, China

F. Sun et al. (eds.), *Knowledge Engineering and Management*,
Advances in Intelligent Systems and Computing 214,
DOI: 10.1007/978-3-642-37832-4_53,

Keywords Magnetically suspended wheel · Two-axis-active magnetic bearing · Permanent magnet bias · Equivalent magnetic circuit

1 Introduction

The three-axis-stabilized spacecraft is equipped with reaction or momentum wheels for its attitude control. The precision and life are affected by the bearing to support wheel rotor. The angular contact ball bearings are applied to the conventional ball bearing wheel, but the ball-bearing cannot absorb the centrifugal force by imbalance of the wheel rotor. Accordingly, the ball-bearing wheel can be one of the most harmful disturbance sources for the spacecraft attitude control, and the lubrication remains the principal life-limiting factor for momentum and reaction wheels [1, 2]. At zero speed for ball-bearing wheel, there will be a torque spike. To avoid some of these problems, wheels supported on magnetic bearings have been the subject of intense development. Compared with traditional ball-bearing wheel, to support a wheel magnetically in five degrees of freedom (DOF), no contact and lubrication are the remarkable features of the magnetic bearings, which solve the long-life and high precision problems of momentum or reaction wheel through the active vibration suppression methods for disturbance suppression. The advanced magnetic bearing wheel has been flight-proven on the SPOT satellite [3].

The magnetic bearing wheel is supported by radial magnetic bearings and thrust magnetic bearings in five degrees of freedom (DOF). The radial magnetic bearing will affect the stability, precision, and dimension, etc. of wheel. The first to fly in Europe was a French 1 DOF wheel in SPOT 1, 2 and 3 and in ERS 1 and 2. And then, the 2 DOF wheels were developed to suppress the axial dimension of the wheel [4].

A new MSRW (rating angular momentum is 30 Nms @ 6,000 r/min) is described in this paper based on other MSRWs [5–7]. This magnetically suspended wheel is radially active and axially passive type. With the aim of decreasing the nonlinearity of magnetic force and axial dimension, save energy, and minimize the size of the bearing [8], a two-axis-active magnetic bearing with a permanent magnet bias (referred to as hybrid magnetic bearing) is designed. The permanent magnet is used to provide a steady-state bias flux for suspension and the electromagnets are used to provide the necessary stability and control for the radial hybrid magnetic bearing. The performance of the radial hybrid magnetic bearing is calculated and analyzed. Its magnetic force, current stiffness, and negative stiffness are derived by using equivalent magnetic circuit and their nonlinearity is shown by the curve of force-current-position characteristic, the curve of current stiffness, and the curve of position stiffness considering all the positions of the rotor within the clearance space and the control current. The maximum bearing capacity is also given in this paper. The results of design and analysis are validated by FEM.

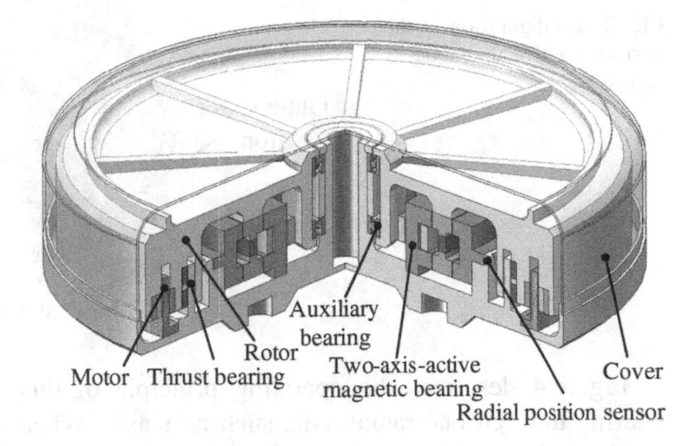

Fig. 1 Configuration of the MSRW

2 Physical Descriptions of MSRW and RHMB

Figure 1 illustrates the configuration of the MSRW, the rotation speed of magnetically suspended rotor is from −6000 r/min to 6000 r/min (rating angular momentum is 30 Nms), the maximum rotation speed is 8,000 r/min, and is supported by radial hybrid magnetic bearing and passive thrust magnetic bearing.

The wheel in this paper has advantages of high precision, low power consumption, and flatter geometry compared to the other magnetic bearing reaction or momentum wheel [9–11].

The two-axis-active magnetic bearing is shown in Fig. 2. The two opposing magnetic poles allow control forces on the axes perpendicular to the rotation axis, respectively, taking account of linearization effect. This is beneficial for the position control. This kind of two-axis-active magnetic bearing consists of rotor iron, magnet, coils, inner stator iron, and outer stator iron. This feature leads to a very compact design. Four coils are distributed at intervals of 90° around the inner stator iron. The bias flux originates from a permanent magnet, and the magnetization direction of the permanent magnet is in axial direction. The control flux originates from the control coil. It allows adding or subtracting flux into the air gaps according to the sign of the current.

3 Magnetic Circuit Model

According to the structure of two-axis-active magnetic bearing (Fig. 2), the magnetic resistance of the iron path and flux leakage is neglected, the permeability of the iron is assumed to be infinite, and their equivalent magnetic circuits can be gained and are shown in Fig. 3. Figure 3a is the equivalent permanent bias circuit, and Fig. 3b is the equivalent control circuit of two magnetic poles on one radial axis (referred to as x-axis) perpendicular to the rotation axis. The control flux and the bias flux pass through the different magnetic circuit, which are decoupled.

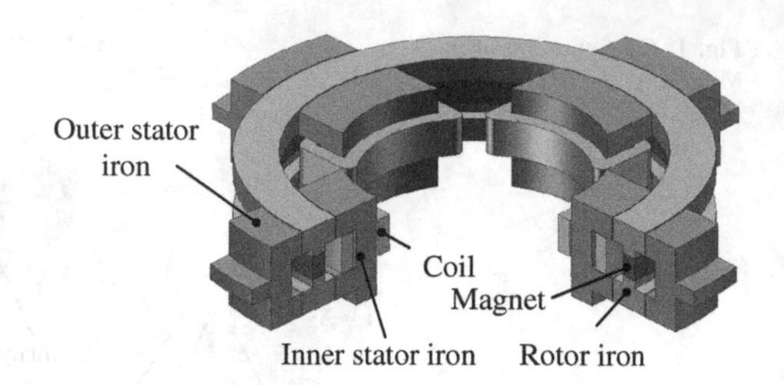

Fig. 2 Configuration of the two-axis-active magnetic bearing

Figure 4 describes the operating principle of this two-axis-active magnetic bearing through one radial axis, such as x-axis. When the rotor departs from its centre position with disturbance force at negative x-axis direction, the control flux on the left side of the rotor will add to the bias flux in outer air gaps and subtract from the bias flux in inner air gaps, and the control flux on the right side of the rotor will subtract from the bias flux in outer air gaps and add to the bias flux in inner air gaps. The radial hybrid magnetic bearing will produce a net restoring force on the rotor at positive x-axis direction.

According to the structure of two-axis-active magnetic bearing (Figs. 2 and 4) and its equivalent magnetic circuits (Fig. 3), the net force at the x-axis, f_x, is,

$$f_x = \left[\frac{(\phi_{px3} + \phi_{cx1})^2}{\mu_0 A_2} - \frac{(\phi_{px1} - \phi_{cx1})^2}{\mu_0 A_1}\right] + \left[\frac{(\phi_{px2} + \phi_{cx2})^2}{\mu_0 A_1} - \frac{(\phi_{px4} - \phi_{cx2})^2}{\mu_0 A_2}\right] \quad (1)$$

where the constant μ_0 is the permeability of free space and is equal to $4\pi \times 10^{-7}$ H/m, the A_1 and A_2 are the areas of the inner pole face and the outer pole face, respectively, the ϕ_{px1}, ϕ_{px2}, ϕ_{px3} and ϕ_{px4} are the bias flux of the permanent bias circuit on x-axis (Figs. 3 and 4), the ϕ_{cx1} and ϕ_{cx2} are the control flux of magnetic circuits (Figs. 3 and 4).

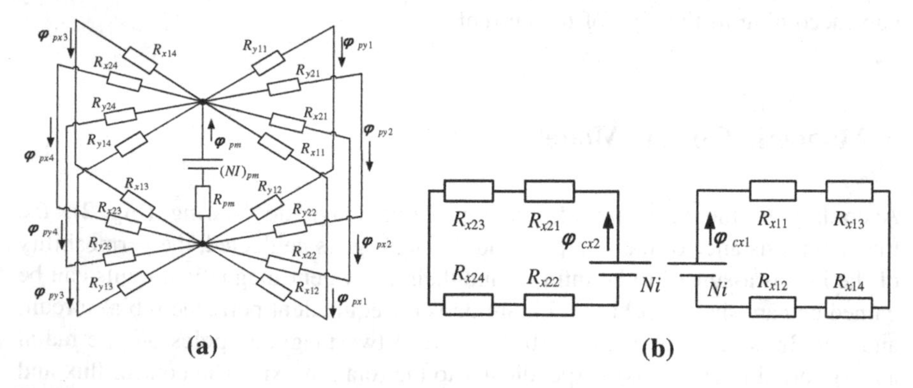

Fig. 3 Equivalent magnetic circuits of the two-axis-active magnetic bearing. **a** Equivalent permanent bias circuit. **b** Equivalent control circuit

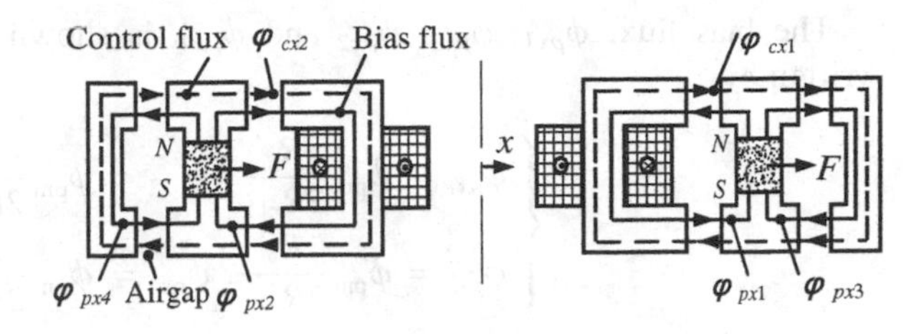

Fig. 4 The working principle of two-axis-active magnetic bearing

The relationship of air gap resistance is shown in Fig. 3a and is: $R_{x11} = R_{x12}$, $R_{x21} = R_{x22}$, $R_{x13} = R_{x14}$, $R_{x23} = R_{x24}$, $R_{y11} = R_{y12}$, $R_{y21} = R_{y22}$, $R_{y13} = R_{y14}$, $R_{y23} = R_{y24}$.

The total magnetic resistance, R_{pmsum}, of the permanent bias circuit in Fig. 3a is

$$R_{\text{pmsum}} = R_{\text{pm}} + R_g \tag{2}$$

where R_g is the total air gap resistance, R_{pm} is the magnetic resistance of permanent magnet, and they can be written as

$$\begin{cases} R_g = \dfrac{1}{\frac{1}{2R_{x11}} + \frac{1}{2R_{x21}} + \frac{1}{2R_{x13}} + \frac{1}{2R_{x23}} + \frac{1}{2R_{y11}} + \frac{1}{2R_{y21}} + \frac{1}{2R_{y13}} + \frac{1}{2R_{y23}}} \\ R_{\text{pm}} = \dfrac{h_{\text{pm}}}{\mu_0 \mu_r A_{\text{pm}}} \end{cases} \tag{3}$$

where h_{pm} is magnet thickness, the constant μ_r is the relative permeability of permanent magnet, and A_{pm} is the cross-section area of the ring permanent magnet. The air gap, s_O, is the length between the pole surface of rotor and the pole surface of the stator when the rotor is in the center position, the displacement, x, is the distance of rotor depart from its center position on x-axis direction, and the displacement, y, is the distance of rotor depart from its center position on y-axis direction.

$$\begin{cases} R_{x11} = \dfrac{s_0 - x}{\mu_0 A_1}, R_{x21} = \dfrac{s_0 + x}{\mu_0 A_1}, \ R_{x13} = \dfrac{s_0 + x}{\mu_0 A_2}, \ R_{x23} = \dfrac{s_0 - x}{\mu_0 A_2} \\ R_{y11} = \dfrac{s_0 - y}{\mu_0 A_1}, R_{y21} = \dfrac{s_0 + y}{\mu_0 A_1}, \ R_{y13} = \dfrac{s_0 + y}{\mu_0 A_2}, \ R_{y23} = \dfrac{s_0 - y}{\mu_0 A_2} \end{cases} \tag{4}$$

The magnetic flux, ϕ_{pm}, in permanent bias circuit can be written by

$$\phi_{\text{pm}} = \frac{(NI)_{\text{pm}}}{R_{\text{pmsum}}} = \frac{H_c h_{\text{pm}}}{R_{\text{pmsum}}} \tag{5}$$

where permanent magnet is simply described as magnetomotive force $(NI)_{\text{pm}}$, which depends on its thickness and material properties, and a resistance R_{pm}, H_c is the coercive force of the permanent magnet.

The bias flux, ϕ_{px1}, ϕ_{px3}, ϕ_{px2} and ϕ_{px4}, is shown in Figs. 3 and 4 can be written by

$$\begin{cases} \phi_{px1} = \phi_{\text{pm}} \dfrac{R_g}{2R_{x11}}, \ \phi_{px3} = \phi_{\text{pm}} \dfrac{R_g}{2R_{x13}} \\ \phi_{px2} = \phi_{\text{pm}} \dfrac{R_g}{2R_{x21}}, \ \phi_{px4} = \phi_{\text{pm}} \dfrac{R_g}{2R_{x23}} \end{cases} \tag{6}$$

The control flux, ϕ_{cx1} and ϕ_{cx2}, are shown in Figs. 3 and 4 can be written by

$$\begin{cases} \phi_{cx1} = \dfrac{Ni}{R_{x11} + R_{x12} + R_{x13} + R_{x14}} = \dfrac{Ni}{2R_{x11} + 2R_{x13}} \\ \phi_{cx2} = \dfrac{Ni}{R_{x21} + R_{x22} + R_{x23} + R_{x24}} = \dfrac{Ni}{2R_{x21} + 2R_{x23}} \end{cases} \tag{7}$$

where N is the number of winding turns and i is the control current.

Substituting Eqs. (6) and (7) into the right side of Eq. (1), results in the net force at x-axis direction. The force generated by a magnetic bearing is generally described by an equation which is linear in the control current but is most likely nonlinear in the displacement of the supported object. For the purpose of studying the stability and dynamic performance of the closed-loop system, a linearized actuator mathematical model of the magnetic bearing is required. Using the Talyor series expansion for small values of x and i, we can get the following attractive force with linear terms at x-axis direction:

$$f_x \cong f_x|_{\substack{x=0 \\ i=0}} + k_i i + k_x x \tag{8}$$

Since the net force is zero at the center position, Eq. (8) can be written as

$$f_x(x, i) = k_i i + k_x x \tag{9}$$

where k_i is the bearing positive current stiffness and k_x is the bearing negative position stiffness caused by the increasing attractive force as the air gap is reduced. This must be overcome for stable suspension since the homogeneous solution to magnetic bearing force is in the form of hyperbolic functions indicating that x grows with time. They can be calculated by Eq. (10).

$$\begin{aligned} k_i &= \left.\frac{\partial f_x}{\partial i}\right|_{\substack{i=0 \\ x=0}} = \frac{2\mu_0 N A_1 A_2 H_c h_{\text{pm}}}{s_0 (A_1 + A_2) \left[\frac{2(A_1 + A_2) h_{\text{pm}}}{\mu_r A_{\text{pm}}} + s_0 \right]}, \ k_x = \left.\frac{\partial f_x}{\partial x}\right|_{\substack{i=0 \\ x=0}} \\ &= -\frac{(H_c h_{\text{pm}})^2 \mu_0 (A_1 + A_2)}{\left[2(A_1 + A_2) \frac{h_{\text{pm}}}{\mu_r A_{\text{pm}}} + s_0 \right]^2 s_0} \end{aligned} \tag{10}$$

4 Case Study

4.1 The Parameters Design

The main parameters of the two-axis-active magnetic bearing are the maximum magnetic force, the current stiffness, the position stiffness, inductance, etc. The performance and application area are decided by these parameters. The two-axis-active magnetic bearing is designed according to the assumed conditions and a 30 Nms magnetic bearing wheel (the rotor mass is 4.4 kg) to control the attitude of spacecraft. The main design parameters, which are chosen by calculation, are shown in Table 1. The value of the current stiffness, k_i, and the position stiffness, k_x, are calculated by Eq. (10) for small values of x and i in Table 1.

4.2 The Finite Element Model

In this paper, the finite element method is used to calculate the parameters of the two-axis-active magnetic bearing, and validate the linearized model.

The finite element model is shown in Fig. 5 except the air.

4.3 Analysis Results by the Linearized Model and the FEM

The performance of the radial hybrid magnetic bearing two-axis-active magnetic bearing can be calculated according to design parameters and its equivalent magnetic circuits (Fig. 3). The relationship between magnetic force and the control current is calculated, respectively, and is shown in Fig. 6 by the linearized model and the FEM. The magnetic force is near-linear relationship to the control current, which helps to improve the control behavior of such bearings. The force calculated by the linearized model is greater than the value calculated by the FEM for the

Table 1 The design parameters of two-axis-active magnetic bearing

Symbol	Parameter	Value
s_0	Air gap	0.3 mm
A_1	The area of the inner magnetic pole face	251.9 mm^2
A_2	The area of the outer magnetic pole face	312.7 mm^2
B_{bias}	The bias flux density of air gap	0.445 T
N	The number of winding turns	350
h_m	The magnet thickness	6 mm
A_{pm}	The magnet cross-section area	1001.7 mm^2
$\text{I}_{\max}$	The maximum control current	1 A

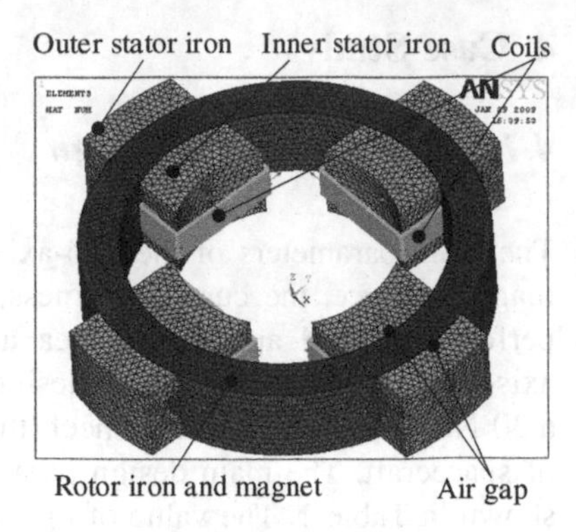

Fig. 5 The finite element model

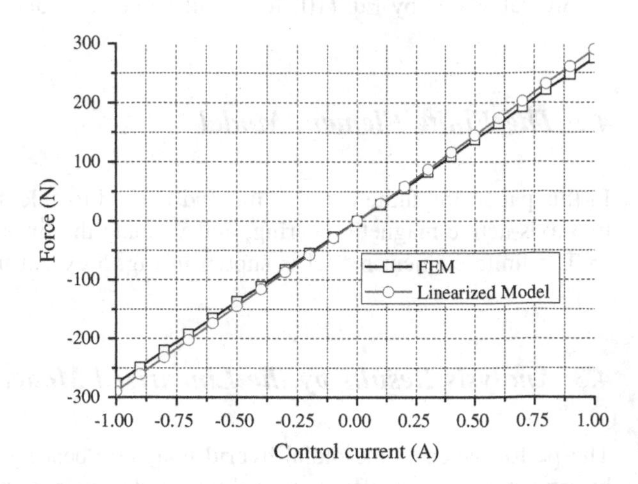

Fig. 6 Calculated magnetic force versus control current by the linearized model and the FEM

magnetic resistance of the iron path and flux leakage is neglected and the permeability of the iron is assumed to be finite in the linearized model, but it is almost equal for small values of control current.

The current stiffnesses are 289.7 N/A and 273.45 N/A calculated by the linearized model and the FEM (Fig. 7), respectively. The calculated error is 5.9 % between the linearized model and the FEM.

The relationship curves between magnetic force and the rotor position are shown in Fig. 8, which are calculated by the linearized model and the FEM. It is obvious that the magnetic force is nonlinear. But the magnetic force is near-linear relationship for small values of rotor position. The force calculated by the linearized model is greater than the value calculated by the FEM, but it is almost equal for small values of rotor position.

The position stiffnesses are −1186 N/mm and −1134.5 N/mm calculated by the linearized model and the FEM (Fig. 9), respectively. The calculated error is 4.5 % between the linearized model and the FEM. And the position stiffness is

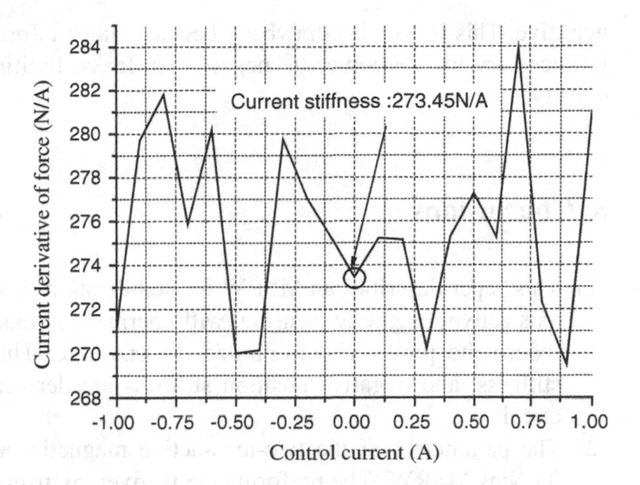

Fig. 7 Calculated current stiffness by FEM

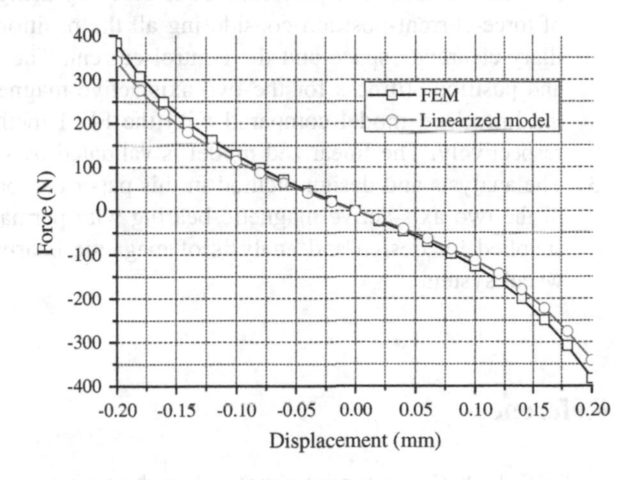

Fig. 8 Calculated force versus displacement by the linearized model and the FEM

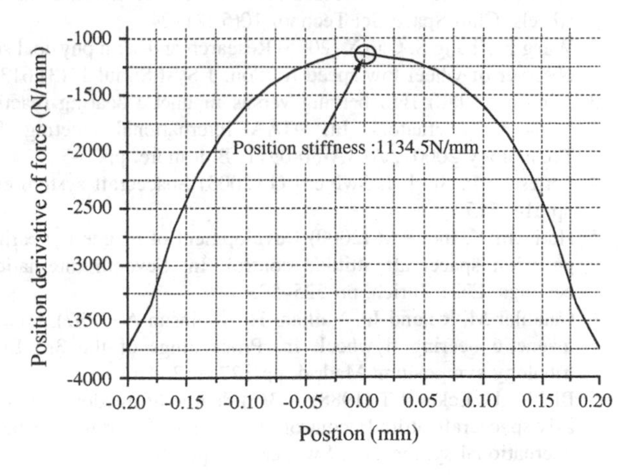

Fig. 9 Calculated position stiffness by FEM

negative. This is a major drawback because much effort of the position controller is necessary to overcome the negative stiffness, limiting the achievable control quality.

5 Conclusions

(1) This paper describes an MSRW for stabilization of spacecraft attitude. A two-axis-active magnetic bearing with permanent magnet bias is designed to reduce the power and minimize the size, etc. The magnetic force, current stiffness, and negative position stiffness are derived by equivalent magnetic circuits.
(2) The parameters of the two-axis-active magnetic bearing are designed for a 30 Nms MSRW. The performance is given by using the characteristic curves of force-current-position considering all the position of rotor within the auxiliary clearance space and the control current. The errors of current stiffness and position stiffness for the two-axis-active magnetic bearing calculated by the linearized model compared with the FEM method are 5.9 % and 4.5 % respectively. The linearized model is validated by the FEM.
(3) The analysis and design method in this paper can predict all the performance of the two-axis-active magnetic bearing with permanent magnet bias, can be supplied for design and analysis of magnetic bearings and magnetic bearing wheel system.

References

1. Fei C, Li Y (2001) Nonlinear control law of satellites attitude at zero-speed of reaction wheels. Chin Space Sci Technol 10(5):21–24
2. Wang F, Zhang S, Cao X (2005) Research and semi physical simulation for the compensation observer of wheel low speed friction. J Syst Simul 17(3):613–616 (in Chinese)
3. Auer W (1980) Ball bearing versus magnetic bearing reaction and momentum wheels as momentum actuators. In: AIAA international meeting & technical display "Global Technology 2000", AIAA-80-0911, Baltimore, pp 1–5
4. Fortescue P, Stark J, Swinerd G (2003) Spacecraft system engineering. Wiley, Chichester, pp 514–515
5. Horiuchi Y, Inoue M (2000) Development of magnetic bearing momentum wheel for ultra-precision spacecraft attitude control. In: Seventh international symposium on magnetic bearings, ETH Zurich, pp 525–503
6. Gauthler M, Roland JP, Vaillant H, Robinson A (1987) An advanced low-cost 2-axis active magnetic bearing flywheel. In: Proceedings of the 3rd European space mechanisms & tribology symposium, Madrid, pp 177–182
7. Bichler U, Eckardt T (1988) A 3(5) degree of freedom electrodynamic-bearing wheel for 3-axis spacecraft attitude control application. In: Magnetic bearings proceedings of the first international symposium, Switzerland, pp 15–22

8. Ehmann C, Sielaff T, Nordmann R (2004) Comparison of active magnetic bearings with and without permanent magnet bias. In: Ninth international symposium on magnetic bearings, USA
9. Han B, Hu G, Fang J (2006) Optimization design of magnetic bearing reaction wheel rotor. J Syst Simul, J Astronaut 27(3):272–276
10. Horiuchi Y, Inoue M, Sato N (2000) Development of magnetic bearing momentum wheel for ultra-precision spacecraft attitude control. In: The seventh international symposium on magnetic bearings, Switzerland, pp 525–530
11. Nakajima A (1988) Research and development of magnetic bearing flywheels for attitude control of spacecraft. In: Magnetic bearings proceedings of the first international symposium, Switzerland, pp 3–12
12. Hagiwara S (1981) Rotational tests of magnetically suspended flywheel for spacecraft (2nd report). JSME No. 810-4:97-69

Vibration Control of a High-Speed Manipulator Using Input Shaper and Positive Position Feedback

Zhongyi Chu, Jing Cui and Fuchun Sun

Abstract This paper presents an experimental study on the dynamics and vibration control of a high-speed manipulator. Light weight and high-speed manipulators are flexible structures, vibration will be unavoidable due to motion of inertial components or uncertainty disturbance excitation. To solve this problem, input shaping feed forward controller is adopted to suppress vibration of a flexible smart manipulator. Also, multi-mode positive position feedback (PPF) controller is designed with piezoelectric actuator, for suppressing the lower amplitude vibration near the equilibrium point significantly. Especially, the experiment setup that includes the test-bed mechanism of a flexible planar parallel smart manipulator and the hardware and software structures of the control system are then developed. Experimental research is conducted to show that the adopted input shaping algorithm can substantially suppress the larger amplitude vibration, and the PPF controller can also damp out the lower amplitude vibration significantly. The experimental results demonstrate that the proposed controllers can suppress vibration effectively.

Keywords High-speed manipulator · Vibration control · Input shaping · Positive position feedback

---

Z. Chu (✉)
School of Instrument and Opto-Electronics, Beihang University,
Beijing, China
e-mail: chuzystar@gmail.com

Z. Chu
Science and Technology, Interial Lab, Beijing, China

J. Cui
College of Mechanical Engineering and Applied Electrical Technology,
Beijing University of Technology, Beijing, China

F. Sun
Department of Computer Science and Technology, Tsinghua University, Beijing, China

F. Sun et al. (eds.), *Knowledge Engineering and Management*,
Advances in Intelligent Systems and Computing 214,
DOI: 10.1007/978-3-642-37832-4_54,

1 Introduction

Compared with rigid manipulator systems, flexible counterparts are provided with a series of advantages in terms of higher load/mass ratio, higher velocity/acceleration, lower power, inexpensive and smaller actuators, etc [1]. Unfortunately, taking into account the flexibility of the arm leads to the appearance of oscillations during the motion, it is a challenging task to achieve high accuracy end-effector motion for lightweight mechanisms and manipulators, which forces the controller to carry out two tasks: rigid-body positioning and the suppression of tip oscillations [2]. The issue of active vibration control is an important concern, many strategies have been used in the control of flexible structures.

The usual approach was to control the link vibrations through joint actuators only. Recently [3–5]. However, the feedback control of the end-effector position using joint actuators is a non-collocated control problem and difficult to stabilize, with an undesirable non-minimum phase behavior. An alternative control strategy advocates reduction of structural vibrations through controlling joint motions or torques based on input shaping [6]. However, if the system has a large range of unknown or varying frequencies, the approach is needed to make the control method robust, one important technique is model reference input shaper design [7], where an adaptive scheme to update shaper parameters is presented. Also, the adaptive tunning scheme usually requires very tedious analysis to implement, these become significant drawbacks in manipulator application. A rather different approach to closed loop control of flexible link robots was developed in recent years, relying on the use of what have become known as 'smart structures' that actively enhance the links vibration damping property, and positive position feedback (PPF) has been shown to be a solid vibration control strategy for flexible systems with smart materials [8], particularly with the PZT (lead zirconium titanate) type of piezoelectric material. However, the effectiveness of PPF will deteriorate when the natural frequencies are poorly known or have changed due to, for example, the presence of a tip mass. These problems may be even more critical in the case of a time varying payload during a motion execution [9]. Although great progress has been achieved in this field, the issue is still far from completely solved. Especially, dynamics model and parameters are highly nonlinear and the induced frequency uncertainty is hard to quantify exactly. To tackle this problem, the combination of optimal input shaper with closed loop PPF design raises the issue of the roles and the contribution of each actuator type to the overall performance of the maneuver [10, 11]. However, most of these works investigate the vibration control of manipulators and mechanisms with a single flexible link [12] and bonded with a single actuator and sensor [13], moreover, very few attempts have been made toward experimental investigations compared with numerical simulations, especially for high-speed manipulators with multiple flexible links.

This paper presents a new approach to overcome the problem of large parameter uncertainties. Not limited to using input shaping techniques, a control strategy is proposed here to combine the input shaping with a multimode PPF in order to suppress the flexible vibration of a planar parallel manipulator. The organization of the paper is as follows. Section II describes the dynamic modeling of a flexible manipulator with bonded piezoelectric actuators using the finite element method. In Sect. 3, an active vibration control system is designed by combining the input shaping technique and multimode PPF. In Sect. 4, the results of experiments and their discussion are given. Section 5 presents the conclusions.

2 Modeling

In this study, a single flexible manipulator (see Fig. 1) will be modeled first and the system dynamics equation will be augmented later to consider the effects of the piezoelectric actuator using finite element method. The energy of the single manipulator assuming an Euler-Bernoulli beam is

$$\mathbf{T} = \frac{1}{2} I_h{}^2 + \frac{1}{2} \int_0^L \rho A [\dot{w} + (b + x)\dot{\theta}]^2 dx \tag{1}$$

where the first term is due to rigid motion of base, the second term is due to the energy of the beam with respect to its velocity normal to x.

It can be seen that after discretizing the body into elements, kinetic energy expression of ith element is given by

$$\mathbf{T}_i = \frac{1}{2} \int_0^l \rho A [\dot{w} + (b + x + x_i)\dot{\theta}]^2 dx \tag{2}$$

where l is the length of the element, x_i is the distance from the root of the beam to the closest side of the ith beam element, this naturally leads to an augmentation of the traditional Euler-Bernoulli finite element shape functions to

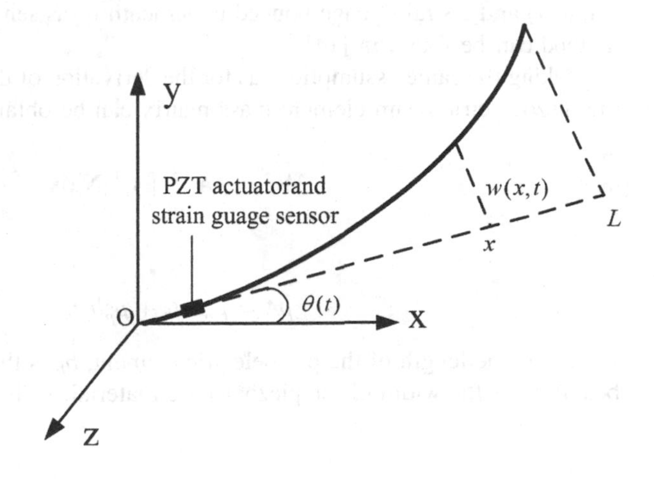

Fig. 1 diagram of single flexible manipulator

$$[N] = [N_0 \quad N_1 \quad N_2 \quad N_3 \quad N_4] \tag{3}$$

where $N_0 = b + x_i + x, N_1 = 1 - 3\zeta^2 + 2\zeta^3, N_2 = x(1 - 2\zeta^2 + \zeta^3), N_3 = 3\zeta^2 - 2\zeta^3, N_4 = x(-\zeta + \zeta^2), \zeta = x/l$, the mass matrix of each element can be seen as

$$\mathbf{M^e} = \rho \mathbf{A} \int_0^{\mathbf{l}} [\mathbf{N}]^{\mathbf{T}}[\mathbf{N}]\mathbf{dx} = \begin{bmatrix} \mathbf{M}^{\mathbf{e}}_{\theta\theta} & \mathbf{M}^{\mathbf{e}}_{\theta\mathbf{w}} \\ \mathbf{M}^{\mathbf{eT}}_{\theta\mathbf{w}} & \mathbf{M}^{\mathbf{e}}_{\mathbf{ww}} \end{bmatrix} \tag{4}$$

the corresponding element stiffness matrix will be

$$\mathbf{K^e} = \frac{\mathbf{EI}}{\mathbf{l^3}} \begin{bmatrix} \mathbf{0} & \mathbf{0} & \mathbf{0} & \mathbf{0} & \mathbf{0} \\ \mathbf{0} & \mathbf{12} & \mathbf{6l} & -\mathbf{12} & \mathbf{6l} \\ \mathbf{0} & \mathbf{6l} & \mathbf{4l^2} & -\mathbf{6l} & \mathbf{2l^2} \\ \mathbf{0} & -\mathbf{12} & -\mathbf{6l} & \mathbf{12} & -\mathbf{6l} \\ \mathbf{0} & \mathbf{6l} & \mathbf{2l^2} & -\mathbf{6l} & \mathbf{4l^2} \end{bmatrix} \tag{5}$$

then, the systematic mass and stiffness matrices can be assembled by finite element method [10].

2.1 Piezoelectric Elements in Full System Model

The piezoelectric elements are bonded to the displacement of the beam where maximum strain happened, and are assumed to have two structural degrees of freedom like the regular beam element. In addition, the piezoelectric element has one electrical degree of freedom which is the voltage. Voltage can be used as a control input to the actuator, which will cause the actuator to apply moments that are equal and opposite to each other at either end of the element. Generally, it is supposed that the PZT beam element can be seen as a sandwich with the regular beam element in the middle with a piezoelectric element bonded on top of it (as actuator) and a strain guage bonded underneath it (as sensor). More details of this method can be found in [14].

Making the same assumptions as for the derivation of the regular beam element, the piezoelectric beam element mass matrix can be obtained as

$$\mathbf{M}^{pe} = \rho \mathbf{A} \int_0^{l} [\mathbf{N}]^{\mathbf{T}}[\mathbf{N}]\mathbf{dx} \tag{6}$$

$$\rho A = \rho_b b_b t_b + \rho_p b_p t_p \tag{7}$$

where l is the length of the piezoelectric element, b_b is the height (or width of the beam), b_p is the width of the piezoelectric material, ρ_b is the density of the beam,

ρ_p is the density of the piezoelectric material, t_b is the thickness of the beam, and t_p is the thickness of the piezoelectric patch. It can be clearly seen that the mass matrix of the piezoelectric element can be written as

$$\mathbf{M^{pe}} = \mathbf{M^e} + \mathbf{M^p} \tag{8}$$

similarly, the piezoelectric beam element stiffness matrix can also be written as

$$\mathbf{K^{pe}} = \mathbf{K^e} + \mathbf{K^p} \tag{9}$$

2.2 System Assembly Equations

The entire second-order differential equation can be assembled using the standard finite element technique with Rayleigh damping included as [10]

$$\mathbf{M\ddot{q}} + \mathbf{D\dot{q}} + \mathbf{Kq} = \mathbf{F} \tag{10}$$

where $\mathbf{M}, \mathbf{D}, \mathbf{K}$ are the global mass, damping, and stiffness matrices respectively.

Modal decoupling can be performed to obtain the normal mode system through similarity transformations

$$\ddot{x} + Z\dot{x} + \Omega x = S_m^T F \tag{11}$$

3 Controller Design

Practically, the joint motion of the manipulator is controlled using a simple proportional-derivative (PD) feedback law to reduce computation complexity, vibrations during the motion should be controlled using other control laws. To do this, the PD controller will be first combined with input shaper.

3.1 Optimal Input Shaper

Basically, any type of input shaper can be combined with the proposed multimode adaptive PPF control for vibration suppression. Here, for simplicity, optimal arbitrary time-delay filter (OATF) [15] as a particular three-term command shaper will be used. The amplitudes of three impulses are as followings:

$$A_1 = \frac{1}{Q_0}, \quad A_2 = \frac{2\,cos(\omega_d T_1)}{e^{-\xi\omega_n T_1}}, \quad A_3 = \frac{e^{-2\xi\omega_n T_1}}{Q_0} \tag{12}$$

where ξ, ω_d represent the corresponding damping ratio, damped natural frequency respectively, and $Q_0 = 1 - 2\,cos(\omega_d T_1)e^{-\xi\omega_n T_1} + e^{-2\xi\omega_n T_1}$. It has been shown that if the impulse magnitudes of the OATF are properly decided, we can choose any location of the impulses to reduce the residual vibration.

3.2 Positive Position Feedback

In theory, flexible manipulator has infinite vibrational modes, and sometimes there is more than one dominant mode. To damp the dominant modes, multiple PPF controllers are required in parallel, where each controller is tuned to the natural frequency of the mode it is to damp. In our system, the first two vibration modes are dominant and need to be suppressed.

For the general multivariable case, PPF can be described by two coupled differential equations

$$\ddot{\xi} + D\dot{\xi} + \Omega\xi = C^T G\eta, \quad \ddot{\eta} + D_f\dot{\eta} + \Omega_f\eta = \Omega_f C\xi \tag{13}$$

where the first equation describes the structure, and the second describes the compensator [10], G is the diagonal gain matrix, C is the participation matrix, Ω and Ω_f are the diagonal modal and filter frequency matrices, and D and D_f are the diagonal modal and filter damping matrices. In this case, stability can be guaranteed if and only if

$$\Omega - C^T GC > 0 \tag{14}$$

Considering for a moment, only the flexible modes of the structure, a PPF controller is developed in this section for the beam using a single collocated PZT actuator pair. The dynamic equation of the structure in modal coordinates is

$$\ddot{x} + Z_s\dot{x} + \Omega_s x = S_m^T pu \tag{15}$$

where x is the vector of modal coordinates, Z_s is the damping matrix, Ω_s is the frequency matrix, S_m^T is the matrix of mass normalized eigenvectors of the system, p is the actuator influence matrix, and u is the input to the actuator (voltage in this case).

The sensor (or output) equation can be seen as

$$y = p^T S_m x \tag{16}$$

where p is the sensor influence matrix.

The equation describing the controller is given as

$$\ddot{\eta} + Z_f\dot{\eta} + \Omega_f\eta = \Omega_f Ey \tag{17}$$

where η is the vector of controller coordinates, Z_f is the controller damping matrix, Ω_f is the controller frequency matrix, and E is the modal participation factor matrix, which will be defined shortly. The actuator input equation is given as

$$u = E^T G\eta \tag{18}$$

Combining Eqs. (15, 16, 17, 18) will generate two second-order differential equations to describe the system, the closed loop system was proven to be stable.

$$\ddot{x} + Z_s\dot{x} + \Omega_s x = S_m^T p E^T G\eta \tag{19}$$

$$\ddot{\eta} + Z_f\dot{\eta} + \Omega_f\eta = \Omega_f E p^T S_m x \tag{20}$$

Similarly in this case, the Schur complement of the closed loop stiffness matrix will give the condition that guarantee stability for the system with feedthrough as

$$\Omega_s - S_m^T p E^T G E p^T S_m > 0 \tag{21}$$

4 Experimental Results

The experimental setup illustrated in Fig. 2 consists of the following components. First, two Parker linear brushless DC server motors (404LXRxx) are attached to the base to driving a flexible planar parallel mechanism (Fig. 3), every flexible link is instrumented with a pair of monolithic piezoelectric actuator plates, and a pair of strain gauges assembled in full-bridge configuration, with the adopted instrumentation. The system repeatability is measured by OFV3001 type Fiber Vibrometer made in Polytec Co. Ltd, whose resolution is 0.32 μm, and the system settling time is measured by capaNCDT 620 type Non-contact capacitive displacement measurement made in Micro Epsilon Co. Ltd, whose resolution is 0.04 μm.

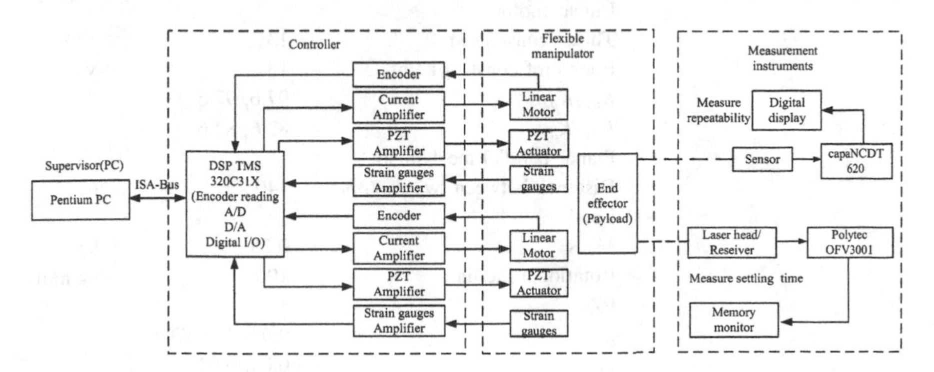

Fig. 2 Experimental and test system setup

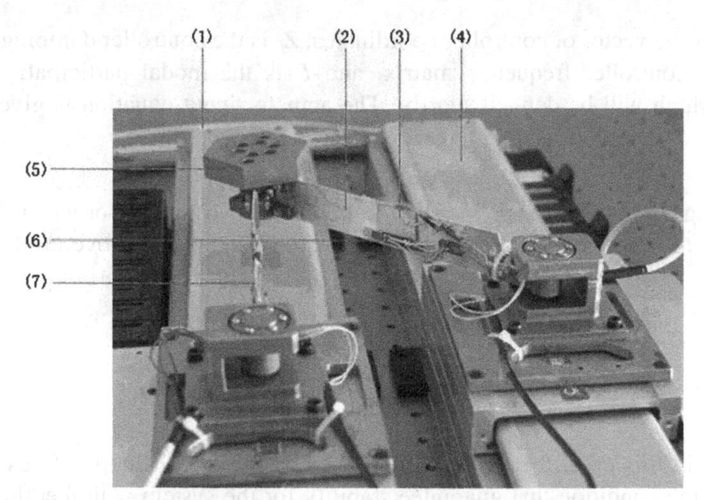

(1) Linear motor A; (2) Flexible link B; (3) PZT; (4) Linear motor B;
(5) Payload; (6) Strain gauge; (7) Flexible link A

Fig. 3 Schematic of planar parallel mechanism

4.1 Results

The physical properties of the system and the controller parameters are listed in Table 1. Finite element analysis shows that the actual frequencies of the first two vibration modes are 23.333 and 28.1 Hz. It is desired to slew the linear motor 20 mm, the corresponding maximum velocity and acceleration are 0.5 m/s, 19.6 m/s^2. In this study, two scenarios are considered: (1) same direction motion of two linear motors (constant payload situation); (2) reverse direction motion of two linear motors (variable payload situation).

Table 1 Experimental system specifications

Item	Value	Unit
Linear motor		
Thrust constant kf	13	N/A
Back emf constant kemf	13	Vs/m
K_{p1}, K_{p2}	97.6, 97.6	
K_{d1}, K_{d2}	82.6, 82.6	
Planar parallel mechanism		
Distance between two motors	140	mm
Load		
Mass	0.25	kg
Rotational inertia	100	kg.mm^2
PZT		
g	0.9012, 0.3708	
ξ_f	0.04, 0.04	
Ω_f	23.333, 28.1	

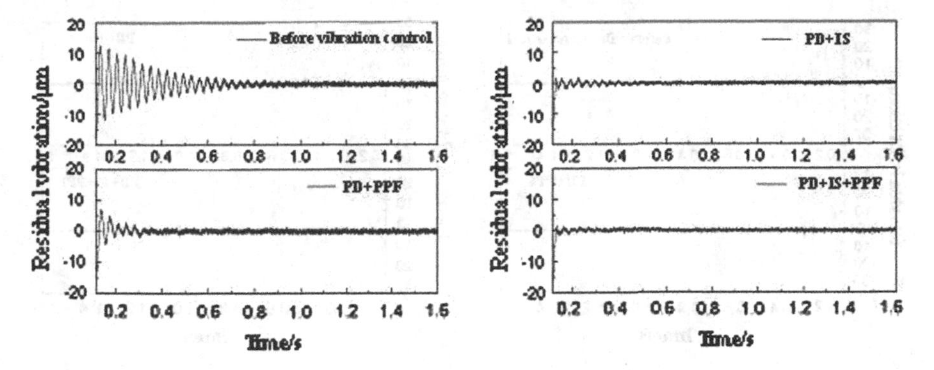

Fig. 4 Experimental results when linear motor moving in same direction

In the first scenario, First, PD feedback (PD) is applied to control linear motor, however, neither input shaper nor PPF is used. Since no active vibration control is used, the residual vibration of the flexible beam is big, as seen in Fig. 4. To suppress flexible vibration, in the next experiment, the proposed multimode PPF controller (PD+PPF) is employed to suppress the vibration of the first two modes. The experimental results are shown in Fig. 4 for comparison. It can be seen that the flexible vibration has been suppressed to a very low level after 100 s.

Finally, the input shaper and PPF (PD+IS+PPF) are combined to suppress vibration. Input shaper is designed to suppress the first flexible vibration mode during motion. Because of the nonlinear uncertainty, however, there will be residual vibration, which will be suppressed through the PZT actuator with multimode PPF. The results have been included in Fig. 4, it can be seen that the vibration has been suppressed very well, the combined law has no residual vibration after roughly 40 s.

In the next case, the induced variable payload of joints become violent, the corresponding experimental results (PD, PD+PPF, and PD+IS+PPF) are shown in Fig. 5 for comparison. It can be seen that the flexible vibration has been suppressed to a very low level after 210 s using the proposed adaptive PPF controller, the combined law has no residual vibration after roughly 115 s.

The results indicate that better vibration suppression performance can be obtained by combining input shaper with the proposed multimode adaptive PPF controller.

5 Conclusion

In this paper, the dynamic modeling of flexible manipulator was performed using the finite element method through Euler-Bernoulli beam theory. Optimal input shaping was used with a proportional-derivative controller to slew the flexible beam in order to minimize the induced vibration during the maneuver. The

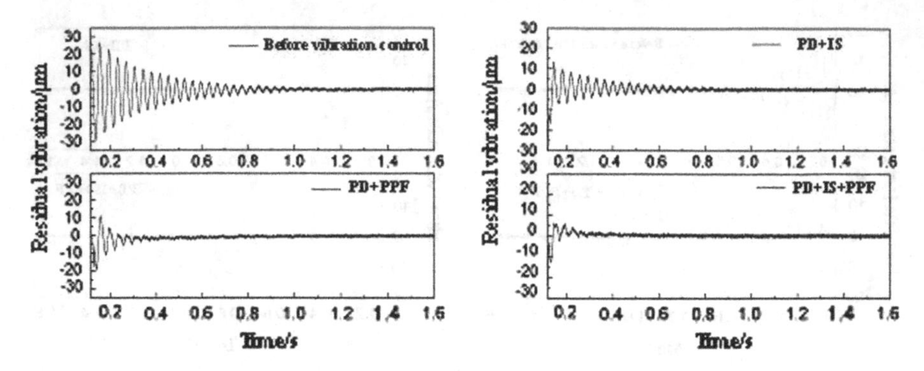

Fig. 5 Experimental results when linear motor moving in reverse direction

residual vibration due to parameter uncertainty was suppressed by the PZT actuator and the proposed multimode adaptive PPF. Experiment were run through which the combination of input shaping with the multimode adaptive PPF control clearly offers advantages in the vibration suppression over either of these control methods alone, particularly for systems where there are frequency Euler-CBernoulli beam theory. Input shaping was used with a proportional-derivative controller to slew the flexible beam in order to minimize the induced vibration during the maneuver. The residual vibration due to parameter uncertainty was suppressed by the PZT actuator and the proposed multimode adaptive PPF. An adaptation law based on the recursive least-squares method was developed to update the systems of first two natural frequencies. These frequencies were then used by the multimode PPF controller to suppress the residual vibrations. Simulations were run through which the combination of input shaping with the multimode adaptive PPF control clearly offers advantages in the vibration suppression over either of these control methods alone, particularly for systems where there are frequency variety.

Acknowledgments This work was jointly supported by the National Natural Science Foundation of China (Grant No. 50905006, 61005066) and the Research Fund for the Doctoral Program of Higher Education (Grant No. 20091102120027).

References

1. Yu Y, Du Z, Yang J et al (2011) An experimental study on dynamics of a 3-RRR flxible parallel robot. IEEE Trans Robot 27(5):992–997
2. Zheng J, Salton A, Fu M (2011) Design and control of a rotary dual-stage actuator positioning system. Mechatronics 21:1003–1021
3. Ozer A, Semercigil E (2008) An event-based vibration control for a two-link flexible robotic arm: numerical and experimental observations. J Sound Vibr 313:375–394
4. Morales R, Feliu V, Jaramillo V (2012) Position control of very lightweight single-link flexible arms with payload variations by using disturbance observers. Robot Auton Syst 60:532–547

5. Boscariol P, Gasparetto A, Zanotto V (2011) Simultaneous position and vibration control system for flexible link mechanisms. Meccanica 46:723–737
6. Singhose W (2009) Command shaping for flexible systems: a review of the first 50 years. Int J Precis Eng Manufact 10(4):153–168
7. Yuan D, Chang T (2008) Model reference input shaper design with application to a high-speed robotic workcell with variable loads. IEEE Trans Ind Electron 55(2):842–851
8. Orszulik RR, Shan J (2012) Active vibration control using genetic algorithm-based system identification and positive position feedback. Smart Mater Struct 21:055002(1–10)
9. Feliu V, Castillo FJ, Ramos F et al (2012) Robust tip trajectory tracking of a very lightweight single-link flexible arme in presence of large payload changes. Mechatronics 22:594–613
10. Orszulik RR, Shan J (2011) Vibration control using input shaping and adaptive positive position feedback. J Guid Control Dyn 34(4):1031–1044
11. Reis JCP, Costa JSD (2012) Motion planning and actuator specialization in the control of active-flexible link robots. J Sound Vibr 331:3255–3270
12. Diaz IM, Pereira E, Feliu V et al (2010) Concurrent design of multimode input shapers and link dynamics for flexible manipulators. IEEE Trans Mechatron 15(4):646–651
13. Gurses K, Buckham BJ, Park EJ (2009) Vibration control of a single-link flexible manipulator using an array of fiber optic curvature sensors and PZT actuators. Mechatronics 19:167–177
14. Zhang X, Mills JK, Cleghorn WL (2009) Flexible linkage structural vibration control on a 3-PRR planar parallel manipulator: experiments results. Proc IMechE Part I J Syst Control Eng 223:71–84
15. Rhim S, Book WJ (1999) Adaptive command shaping using adaptive filter approach in time domain. In: Proceeding American Control Conference, vol 6, issue 1. San Digeo, pp 81–85

Recognition of Human Head Movement Trajectory Based on Three-Dimensional Somatosensory Technology

Zheng Chang, Xiaojuan Ban and Xu Liu

Abstract Studying the traditional robot visual tracking technology on human head movement trajectory based on two-dimensional image, this paper summarizes the shortcomings of the traditional robot visual tracking technology. By combining three-dimensional somatosensory information technology and the Hidden Markov Model, this paper applies the technology of robot visual tracking to human head movement trajectory. In detail, this paper describes the algorithm and the realization scheme of the human head movement trajectory identification based on three-dimensional somatosensory information technology. Finally, through comparing with the results of the recognition experiment, we verify that the method of human head movement trajectory identification based on three-dimensional somatosensory information technology has more accurate recognition rate and better robustness.

Keywords Robot visual tracking · Hidden markov model · Human head movement trajectory recognition

1 Introduction

With the rapid development of robot technology, robot visual tracking technology and human–computer interaction technology have also become important research directions in the field of robot technology. As the human head movement is a

Z. Chang · X. Ban (✉) · X. Liu
School of Computer and Communication Engineering, University of Science and Technology Beijing, Beijing, Haidian, China
e-mail: banxj@ies.ustb.edu.cn

Z. Chang
e-mail: elephant.cz@gmail.com

X. Liu
e-mail: liuxu.ustb@gmail.com

F. Sun et al. (eds.), *Knowledge Engineering and Management*, Advances in Intelligent Systems and Computing 214,
DOI: 10.1007/978-3-642-37832-4_55,

natural and intuitive communication mode, beyond all doubt, the human head recognition technology has become an indispensable technology to the new generation of human–computer interaction interface; especially for the disabled, just a swinging of his head can give the robot orders, which will bring him more convenience [1].

Previous human head movement trajectory recognition researches in human–computer interaction mainly focus on human skin color modeling and extraction of dynamic head movements based on image attributes of the robust feature; however, due to the diversity, ambiguity, and disparity in time and space of human head movement, the traditional head movement trajectory recognition researches have significant limitations [2]. This paper attempts to introduce three-dimensional somatosensory technology into the robot recognition of human head movement trajectory, which will make robot identification more accurate and possess more robust features.

2 Problem Description

Human head trajectory recognition is the matching between the human head movement trajectory captured by the sensor and the predefined sample movement trajectory [3]. In this paper, we use the Hidden Markov Model to solve the human head trajectory matching problem. Human head movement trajectory data $\mathcal{M}$ can be regarded as the state series $\mathcal{M} = (\mathcal{F}_1, \mathcal{F}_2, \mathcal{F}_3 \ldots \mathcal{F}_T)$ obtained from sampling frames. In one frame, the state of the human head is represented as $\mathcal{F} = (r_1, r_2, r_3 \ldots r_n)$, wherein r_n represents the value of the nth characteristic values in the current frame. The movement trajectory recognition is that we indicate the real-time trajectory captured by sensor with the state series $\mathcal{M}$, and then $\mathcal{M}$ is matched with the state series of the predefined sample trajectory, $\mathcal{M}^s$ in order to identify the meaning of the current head movement trajectory.

As shown in Fig. 1, based on the two-dimensional image, Hidden Markov Model names the center of the human head image as the origin $\mathcal{N}_0$, and two eyes, nose, two ears, lips, head, and chin as the key points for identification $\{\mathcal{N}_1, \mathcal{N}_2, \mathcal{N}_3, \mathcal{N}_4, \mathcal{N}_5, \mathcal{N}_6, \mathcal{N}_7, \mathcal{N}_8\}$. In each frame, the relative position of these key points and the origin point are called trajectory eigenvectors $\mathcal{D} = [d_1, d_2, d_3, d_4, d_5, d_6, d_7, d_8]$. Finally, we compare the trajectory eigenvectors between the real-time trajectory and the predefined sample trajectory. The process is shown in Fig. 2. However, the recognition based on two dimensions still has several difficulties, as follows:

(a) *Light* when the light condition is changed, the luminance information of the human head is changed since the images captured by the sensor are easily affected by natural and artificial light [4].

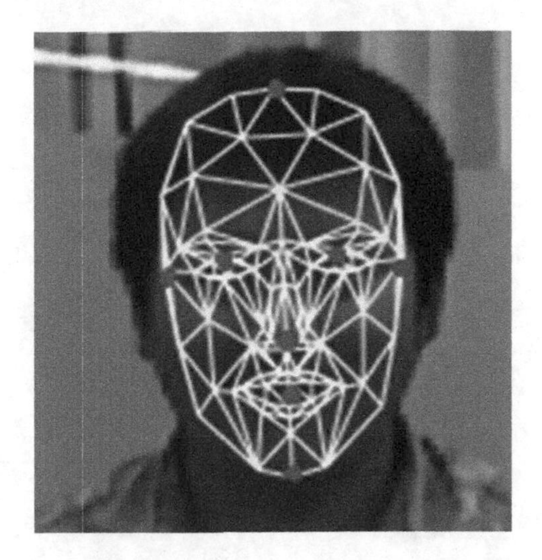

Fig.1 Two-dimensional image of the human head

(b) *Obstruction* In the recognition process, the trajectory of the human head may be obscured by the stationary background; glasses, hats, or other objects. While the obstruction can cause the loss of the identification information, it will affect the reliability of the recognition greatly.

(c) *Background* In the actual recognition process, if the factors (color, texture, shape, and so on) of the head and background areas are similar, it will also increase the difficulty of the recognition.

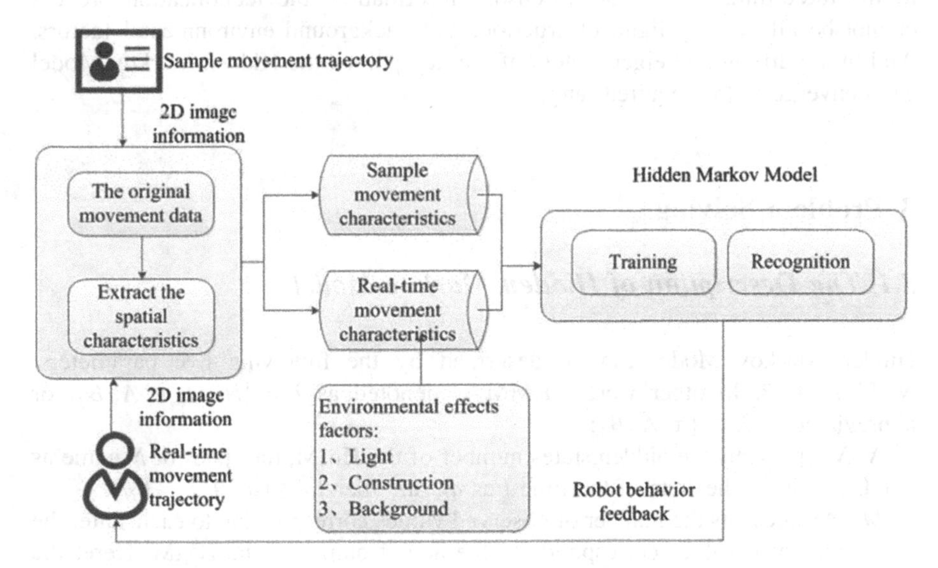

Fig. 2 HMM based on two-dimensional image

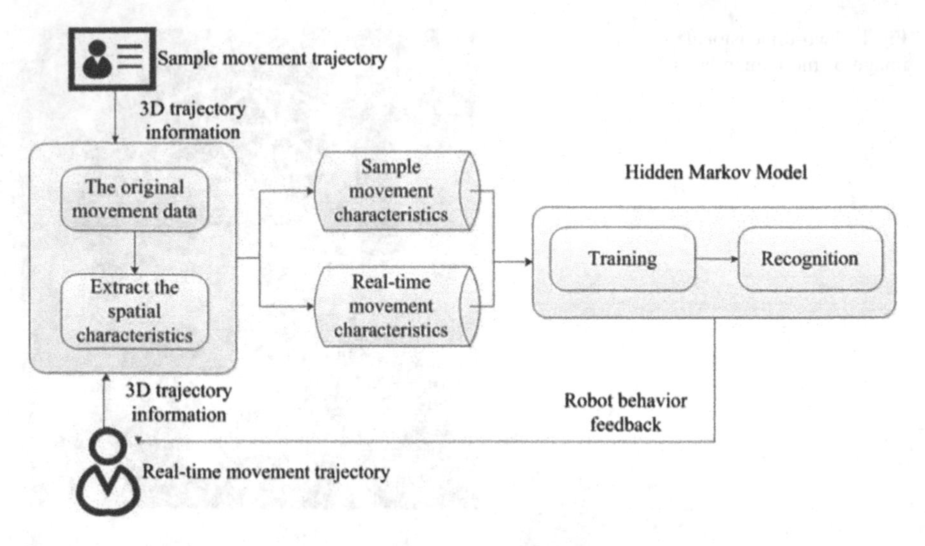

Fig. 3 HMM based on three-dimensional somatosensory information

Based on the three-dimensional somatosensory information, Hidden Markov Model names two eyes, nose, two ears, lips, head, and chin as the key points $\{\mathcal{N}_1, \mathcal{N}_2, \mathcal{N}_3, \mathcal{N}_4, \mathcal{N}_5, \mathcal{N}_6, \mathcal{N}_7, \mathcal{N}_8\}$ for identification, and refers to the three-dimensional somatosensory information; hence it adopts the three-dimensional angle of these key points and the origin points as the trajectory eigenvectors $\mathcal{A} = [a_1, a_2, a_3, a_4, a_5, a_6, a_7, a_8]$. The matching process is shown in Fig. 3. Owing to the three-dimensional somatosensory information, the identification process cannot be affected by light, obstruction, and background environmental factors. And by modifying the eigenvectors of the key points, the Hidden Markov Model can converge to the required range.

3 Problem Solving

3.1 The Description of Hidden Markov Model

Hidden Markov Model can be described by the following five parameters: N, M, π, A, B. In other words, HMM is denoted as $\lambda = \{N, M, \pi, A, B\}$, or abbreviated as $\lambda = \{\pi, A, B\}$:

N: N represents the hidden states number of the HMM, the states of N name as $(\theta_1, \theta_2, \ldots, \theta_N)$, the state of the time t as q_t, and then $q_t \in (\theta_1, \theta_2, \ldots, \theta_N)$.

M: M represents the number of observed values corresponding to each state; the set of observed values corresponds to the actual output of the HMM. Here, the observed values can be observed directly. These observed values are denoted as $(V_1, V_2, \ldots, V_M)$. If we name the observed values of the model at time t as O_t, then we have $O_t \in (V_1, V_2, \ldots, V_M)$.

A: A represents the state transition probability matrix $A = (a_{ij})_{N\times N}$, in the formula,

$$a_{ij} = P(q_{t+1} = \theta_j / q_t = \theta_i), \quad 1 \le i, j \le N$$

a_{ij} represents the probability of transferring from state i to state j, and $a_{ij} \ge 0, \forall i, j \bot \sum_j a_{ij} = 1, \forall i$.

B: B represents observed values probability matrix $B = (b_{jk})_{N\times M}$, in the formula,

$$b_{jk} = P(O_t = V_k / q_t = \theta_j), \ 1 \le i \le N, \quad 1 \le k \le N$$

b_{jk} represents the probability of the HMM outputting the corresponding observed values k in the HMM state j.

Π: Π represents the initial state probability matrix, $\pi = [\pi_1, \pi_2, \pi_3 \ldots, \pi_n]$, in the formula

$$\pi_i = P(q_1 = \theta_i), \quad i = 1, 2, \ldots N$$

In other words, at time $t = 1$, $\sum_{i=1}^{N} \pi_i = 1$, the π_i represents the probability of the HMM in the state of θ_i.

3.2 *Vector Selection of the Characteristics of Head Movement Trajectory*

The basic characteristics of the head trajectory are position, velocity, and angle. In terms of speed, the meaning of different head movement speeds trajectory has almost no difference. In terms of position, the same head trajectory will be affected by the head size, sensor distance, and sensor angle. Hence, the paper uses the angle between the key parts of the head and the head space centroid as the critical head trajectory eigenvectors which can be used to identify the head movement.

3.3 *HMM Training and Classification*

The use of HMM involves two stages: training and classification. The training puts the predefined trajectory template into the model for calculating, to determine the parameters the model contains. After the parameters of the model are determined, the classification involves the matching and classification process between the predefined trajectory template and the real-time head trajectory captured by the Kinect sensor. Finally, we achieve the goal of recognition of the real-time head trajectory.

(1) *HMM Training* In this paper, we use the Baum-Welch algorithm to train HMM. The algorithm is an iterative algorithm based on the expected regulation concept, in the initial case. We first give empirical estimates, and then through the repeated iterative, we finally choose a more reasonable optimal value among all the parameters.

The definition of the front variable $\alpha_t(i)$ is the probability when the model stays in the state θ_i at time t, and has outputted the visible result sequence up to time t. The definition of the backward variable $\beta_t(i)$ is the probability when the model stays in the state θ_i at time t, and will output the visible result sequence $O_1, O_2, \ldots, O_T$ after time t.

$$P(O/\lambda) = \sum_{i=1}^{N}\sum_{j=1}^{N} a_t(i)a_{ij}b_j(O_{t+1})\beta_{t+1}(j),\ 1 \leq t \leq T-1$$

The definition of variable $\xi_t(i,j)$ is the expected probability when the model stays in state i at time t, and stays in state j at time $t+1$, in a giving model λ and the observation sequence O. Hence we can denote $\xi_t(i,j)$ as the following formula:

$$\xi_t(i,j) = P(q_t = i, q_{t+1} = j|O,\ \lambda)$$

Assumed that in case of the given model λ and the observation sequence O, at time t, and the model stays in state i, the probability of this case is as shown in the following formula:

$$\gamma_t(i) = \sum_{j=1}^{N} \xi_t(i,j)$$

The core revaluation formula of the Baum-Welch algorithm is as shown in the following formula:

$$\overline{\pi} = \gamma_1(i)$$

$\gamma_1(i)$ represents the expectation value when the model stays in state θ_i at time t.

$$\overline{a}_{ij} = \frac{\sum_{t=1}^{T-1} \xi_t(i,j)}{\sum_{t=1}^{T-1} \xi_t(i)}$$

$\sum_{t=1}^{T-1} \xi_t(i,j)$ represents the expectation probability of transferring from state i to state j. $\sum_{t=1}^{T-1} \xi_t(i)$ represents the expectation probability from state i.

$$\overline{b}_{jk} = \frac{\sum_{t=1 \text{ and } O_t = V_k}^{T} \xi_t(j)}{\sum_{t=1}^{T} \xi_t(j)}$$

$\sum_{t=1 \text{ and } O_t = V_k}^{T} \xi_t(j)$ represents the probability when the model stays in state k and the observed value is V_k. $\sum_{t=1}^{T} \xi_t(j)$ represents the probability when the model stays in state k.

Use of this revaluation formula can train the model iteratively. The specific process is as follows:

(a) Select the initial parameters of the model $\lambda = \{\pi, A, B\}$.
(b) Based on the initial parameters of the model λ and the observed values sequence o, dealing with a new set of the parameters by revaluation formula, we can get the new model $\overline{\lambda}$.
(c) If $\left|\log P(O/\overline{\lambda}) - \log P(O/\lambda)\right| < \varepsilon$, ε is the predefined threshold value, this training has met the error requirement, and the algorithm ends.
(d) Otherwise, let $\lambda = \overline{\lambda}$, repeat step b.
Repeat this process and gradually adjust the model parameters, until $P(O/\lambda)$ converges to an ideal error range; finally get the trained HMM model; here the selected error range is 0.00001.

(2) *The Recognition of the HMM Model* HMM recognition is the match of likelihood between the predefined sample trajectory and the real-time head movement trajectory, therefore the process involves the maximum likelihood calculation. This paper applies the Viterbi algorithm to calculate the maximum likelihood in order to identify the head movement trajectory. The specific process of Viterbi algorithm is as follows:

(a) Initialization

$$\delta_1(i) = \pi_i b_i(o_1), \ 1 \leq i \leq N \quad \psi_1(i) = 0$$

(b) Iterative computation

$$\delta_t(j) = \max_{1 \leq i \leq N} [\delta_{t-1}(i) a_{ij}] b_j(o_t), \ 1 \leq j \leq N, \quad 2 \leq t \leq T$$

$$\psi_t(j) = \max_{1 \leq i \leq N} [\delta_{t-1}(i) a_{ij}], \ 1 \leq j \leq N, \quad 2 \leq t \leq T$$

(c) The termination of the computation

$$P^* = \max_{1 \leq i \leq N} \delta_T(i) \quad q_T^* = \arg \max_{1 \leq i \leq N} \delta_T(i)$$

The best path obtained is shown in the following:

$$q_t^* = \psi_{t+i}(q_{t+1}^*), \quad 1 \leq t \leq T-1$$

In the iterative process, $\delta_t(i)$ represents the cumulative probability in state i at time t. $\psi_t(i)$ represents the preorder state in state i at time t. q_t^* represents the model staying in the state at time t. P^* represents the final output probability. The final classification results are the q_t^* sequence which maximum P^* corresponds to.

4 Results of the Experiment

4.1 Data Pre-Processing

We are in a common laboratory environment for human head movement trajectory recognition experiment. In the experiment, the experimenter keeps the head face forward, keeps the head perpendicular to the horizontal plane, and keeps the head and the sensor at a distance of 0.8–1 m. Head movement speed cannot be too fast or too slow, and also requires a certain amount of movement amplitude so as to be detected by the sensor. After being detected, we begin to identify head motion trajectory. In the experiment, we pre-process the head movement trajectory in order to remove the jitter factor of the trajectory. First, we record the origin position data in the previous frame, and compare with the origin position in the current frame; if the deviation between two position data is in the threshold range, we can regard the movement in the current frame as a small amplitude jitter, and thus we can ignore the movement in the current frame.

In the experiment, we define four different head movement trajectories, including head to the left (as shown in Fig. 4), head to the right (as shown in Fig. 5), head up (as shown in Fig. 6, and head down (as shown in Fig. 7).

4.2 The Analysis of the Data Results

In order to verify the HMM robustness based on three-dimensional somatosensory information technology, we performed 50 experiments, under different light conditions. Table 1 records the number and rate of each kind of action identified correctly under sunlight lighting environment and dark environment. Figures 8 and 9 show the comparison results under two different light conditions. The experiment proved that HMM based on three-dimensional somatosensory information technology has better robustness.

Fig. 4 Head to the *left*

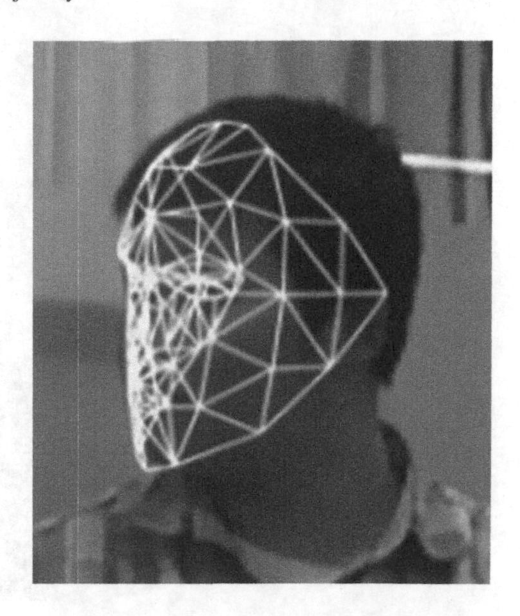

Fig. 5 Head to the *right*

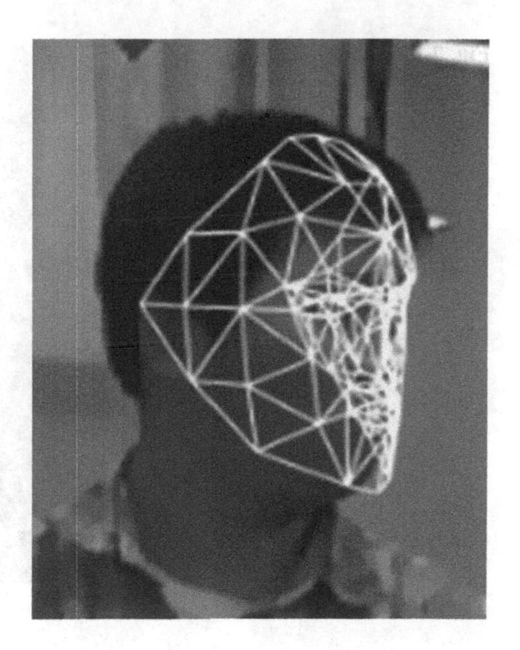

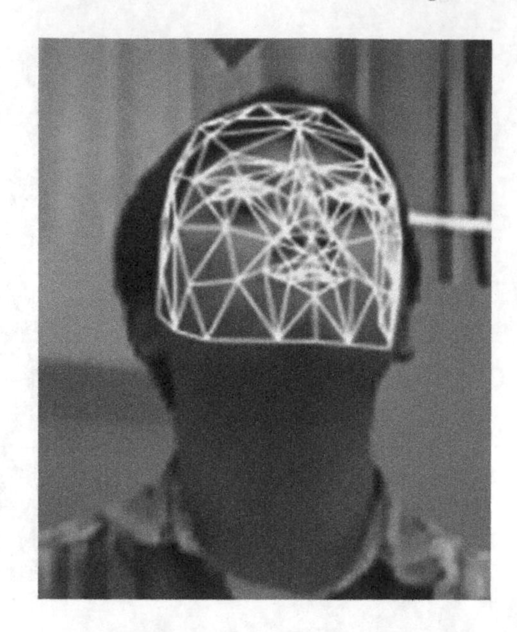

Fig. 6 Head up

Fig. 7 Head down

Table 1 Action recognition rate under different light conditions

Action	Light		Dark	
	Correct times	Correct rate	Correct times	Correct rate
Head left	48	0.96	47	0.94
Head right	49	0.98	48	0.96
Head up	47	0.94	47	0.94
Head down	48	0.96	48	0.96

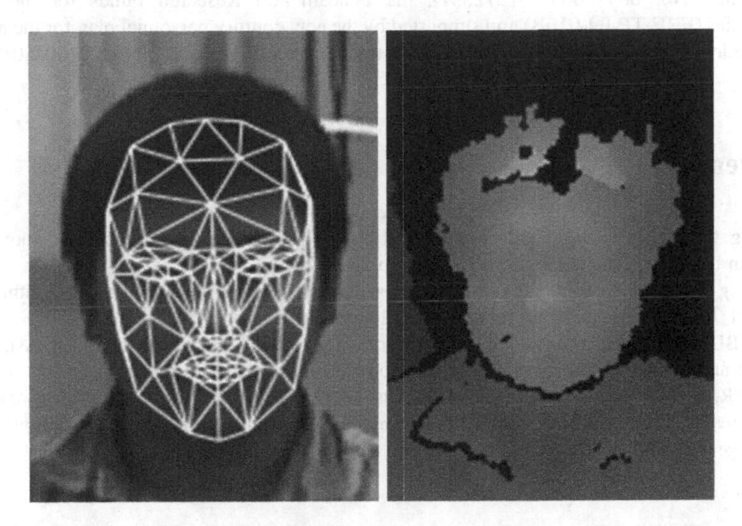

Fig. 8 Human head color and depth image under fluorescent lamp environment

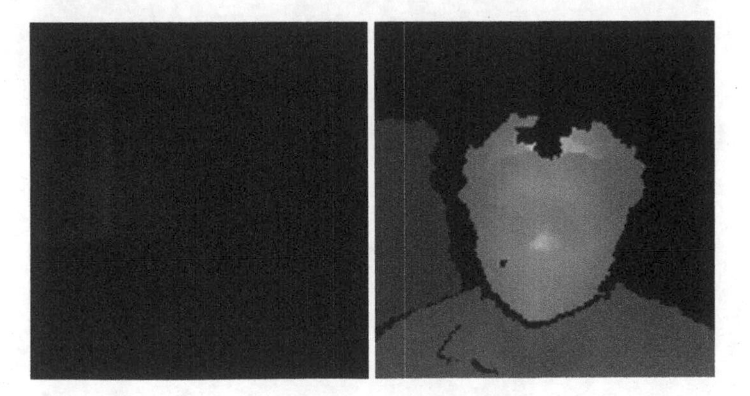

Fig. 9 Human head color and depth image under dark environment

5 Conclusion

Combined with three-dimensional somatosensory information technology and Hidden Markov Model to realize the recognition of human head movement trajectory, the paper overcomes the shortcomings of the traditional recognition method based on two-dimensional image, and finally by experiments, proves that the method in the paper has more accurate recognition rate and better robustness.

Acknowledgments This work was supported by National Nature Science Foundation of P.R. China (No. 60973063, 61272357), the Fundamental Research Funds for the Central Universities (FRF-TP-09-016B) and supported by the new century personnel plan for the Ministry of Education (NCET-10-0221), China Post-doctoral Science Foundation (No. 20100480199).

References

1. Liying C (2012) Design and implementation of Human-Robot interactive demonstration system based on Kinect. In: 24th CCDC, pp 971–975
2. Yiqiang Q, Kazumasa S. (2011) EK-means tracker: a pixel-wise tracking algorithm using Kinect. In: 3rd Chinese conference on intelligent visual surveillance, pp 277–80
3. Sang BL (2011) Real-time stereo view generation using Kinect depth camera. In: Asia-pacific signal and information processing association annual summit and conference
4. Omer R, Ayoub A-H, Bernd M (2011) Robust hand posture recognition with micro and macro level features using Kinect. In: IEEE international conference on intelligent computing and intelligent systems

Study on Wideband Array Processing Algorithms Based on Single Vector Hydrophone

Wei Zhang, Jin-fang Cheng and Jie Xu

Abstract This paper presents a method of wideband coherent signal-subspace estimate by single vector hydrophone (*SVH*). The wideband array signal model of *SVH* is constructed, and the wideband focusing matrixes is first introduced into wideband processing of *SVH*. The relationship between focusing matrixes of *SVH* and acoustic pressure-sensor array is derived, thus it provides a method for the construction of wideband focusing matrixes of *SVH*. The wideband spatial spectrum of *SVH* is derived; the wideband focusing data of *SVH* in this paper is achieved by two-sided correlation transformation (*TCT*) technology without preliminary estimates or a priori knowledge of the spatial distribution of the sources. The simulation results demonstrate that the single vector hydrophone wideband *TCT* focusing algorithm has more superior performances, it can resolve wideband coherent source problems; realize unambiguous direction finding in the whole space; improve the signal-to-noise ratio (*SNR*); and lower resolution threshold.

Keywords Single vector hydrophone · Wideband focusing algorithm · Two-sided correlation transformation · DOA

1 Introduction

Wideband signal is helpful for target detection, parameter estimation, and target feature extraction with the characteristics of the target echo carrying large amount of information, and reverberation weak. So it is an important means in passive

W. Zhang (✉) · J. Cheng
Department of Weaponry Engineering, Naval University of Engineering,
Wuhan 430033, China
e-mail: tgzy_2010@163.com

J. Xu
Department of Weaponry System, Naval Academy of Armament, Beijing 100073, China

F. Sun et al. (eds.), *Knowledge Engineering and Management*,
Advances in Intelligent Systems and Computing 214,
DOI: 10.1007/978-3-642-37832-4_56, © Springer-Verlag Berlin Heidelberg 2014

detection system, and target radiation wideband continuous spectrum for target detection is effective to detect the target. Underwater signal processing forward wideband direction [1].

Wideband processing arises in many applications such as audio conferencing, spread spectrum transmission, and passive sonar. In the so-called incoherent signal-subspace method [2] (*ISS*), the narrow-band signals are processed as a vector to estimate the *DOA*'s. Then, these results are combined to obtain the final solution. Perfectly correlated (coherent) sources cannot be handled by this approach. Furthermore, the efficiency of this method deteriorates for closely separated sources and low *SNR*. The coherent signal-subspace method [3] (*CSS*) is an alternative to *ISM* that improves the efficiency of the estimation by condensing the energy of narrowband signals in a predefined subspace. This process is called focusing. A high-resolution method such as *MUSIC* is then used to find the *DOA*'s. The *DOA*'s are estimated by determining the angular location of peaks in the spatial spectrum of the MUSIC algorithm. It has been shown that *CSM* improves the resolution threshold and resolves coherent sources. However, the *CSS* class methods need a priori information of *DOA* to form focusing matrix, and increase the computation quantity of the algorithm.

The above wideband array algorithms are based on sound pressure hydrophone array, the new underwater acoustic transducer-vector hydrophone can measure both pressure and particle velocity of acoustic field at a point in space, whereas a traditional pressure sensor can only extract the pressure information. This would not only help to improve the performance of the underwater acoustic system, and also widened the signal processing space. And a single vector hydrophone array itself flow pattern can obtain array signal processing to get high resolution.

So in this paper, it combines with the broadband array algorithm and the single vector hydrophone, proposed two-sided correlation transformation (*TCT*) focusing algorithm based on single vector hydrophone array broadband. Simulation results show that it is better than the three element acoustic pressure line array.

2 Single Vector Hydrophone Wideband Mathematical Model

Single vector hydrophone measurement model scheme is shown in Fig. 1. Point is hydrophone k position, three of the velocity vector coincides with the x, y, z axis, $\theta \in [0, 2\pi)$ is the azimuth of incident sound waves. The azimuth of vector hydrophone measures three quadrate component $v_x(r,t)$, $v_y(r,t)$ of acoustic particle velocity vector v, as well as the acoustic pressure $p(r,t)$ at one location, then, they respectively can be expressed as:

Fig. 1 Coordinate system

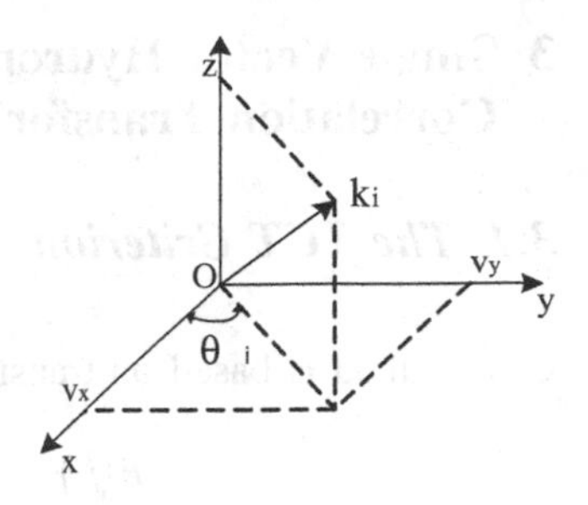

$$\begin{cases} p(r,t) = A(r)\exp\{-i[\omega t - \varphi(r)]\} \\ v_x(r,t) = \dfrac{1}{\rho_0(r)c}\cos\theta \cdot p(r,t) \\ v_y(r,t) = \dfrac{1}{\rho_0(r)c}\sin\theta \cdot p(r,t) \end{cases} \tag{1}$$

As an example of the two-dimensional vector sensor, it can measure both pressure and particle velocity of acoustic field at a point in space. There are two points of source k radiation wideband acoustic signal, denoted as No. 1 and 2, respectively, horizontal azimuth θ_1 and θ_2, Fig. 1. The two targets' sound pressure and vibration velocity signal of the horizontal x-direction and the y-direction received by single vector hydrophone are respectively: $p_1(t)$, $v_{x1}(t)$, $v_{y1}(t)$, $p_2(t)$, $v_{x2}(t)$, $v_{y2}(t)$.

Generally speaking, the far field condition pressure and horizontal vibration speed has good correlation[4], $v_{xi}(t) \approx p_i(t)\cos\theta_i$ and $v_{yi}(t) \approx p_i(t)\sin\theta_i$ Simple as writing, the vibration speed signal passes through acoustic impedance compensation, so the sound pressure and horizontal vibration speed constitute signal space, For No. i target ($i = 1, 2$), the direction vector of the reception signal can be expressed as:

$$a^{(2)}(\Theta_k) = \begin{bmatrix} 1 \\ a_x(\theta_k) \\ a_y(\theta_k) \end{bmatrix} = \begin{bmatrix} 1 \\ \cos\theta_k \\ \sin\theta_k \end{bmatrix} \tag{2}$$

Consider the kth wideband signal, the frequency band range of (f_l, f_h). The mathematical model of certain position in the space of a single two-dimensional vector hydrophone that received kth source can be expressed as:

$$\begin{aligned} X^{(2)}(f_i) &= \begin{bmatrix} X_p(f_i) \\ X_{vx}(f_i) \\ X_{vy}(f_i) \end{bmatrix} \\ &= \sum_{k=1}^{K} a^{(2)}(\theta_k)\exp(-j2\pi f_i \tau_k) S_k(f_i) + N^{(2)}(f_i) \end{aligned} \tag{3}$$

3 Single Vector Hydrophone Wideband Two-Sided Correlation Transformation Focusing Algorithm

3.1 The TCT Criterion

Our method is based on transformation of the matrixes:

$$P(f_j) = A(f_j)S(f_j)A(f_j)^H, \quad j = 1\ldots J \tag{4}$$

where $P(f_j)$ is the correlation matrix of the sensor output at the jth frequency bin in a noise-free environment. Let $P(f_0)$ be the focusing noise-free correlation matrix. The TCT focusing matrixes are found by minimizing:

$$\min_{U(f_j)} \left\| P(f_0) - U(f_j)P(f_j)U(f_j)^H \right\| \tag{5}$$

$$s.t. U(f_j)^H U(f_j) = I$$

For $j = 1, \ldots, J$, the solution of (4) is obtained as

$$U(f_j) = Q_0 Q_j^H \tag{6}$$

where Q_0 and Q_j are the eigenvector matrices of $P(f_0)$ and $P(f_j)$, respectively. The matrix $U(f_j)$ can be used to transfer the observation vector $x(f_j)$ into $y(f_j)$ through:

$$y(f_j) = U(f_j)x(f_j) \tag{7}$$

The observation vectors $y(f_j), \quad j = 1, \ldots, J$ are in the focusing subspace. Their correlation matrices can be averaged to find the universal focused sample correlation matrix.

3.2 The Focusing Frequency and Subspace

Based on principle of the narrow-band algorithm, the direction vector of signal-subspace constitutes a set of orthogonal basis. By the definition of the direction vector, it is not only a function of the azimuth, but also a function of frequency. So it can use $A(f, \theta)$ to express. For narrow-band signals, the direction of the matrix A is a Vander monde matrix. In the case of $\theta_i \neq \theta_j, i \neq j$, it is mutually independent of each column. In the treatment of the broadband signal, because of A in the changes with frequency, when $f_k \sin\theta_m = f_j \sin\theta_n$, it will inevitably destroy the square column independence; thus there is no longer $\mathrm{rank}(AR_sA^H) = P$ established. Due to the wide frequency of the signal factor the direction matrix phase fuzzy make high-resolution algorithm which is suitable for narrow-band signals failure.

Therefore in computing $U(f_j)$, the matrices $A(f_j)$ and $S(f_j)$ are assumed to be known. In practice, a preprocessing step is required to estimate these matrices. A low-resolution beam former is applied to estimate the number and the *DOA*'s of the sources. For instance, if the *i*th *DOA* is found at θ_i by the preprocessing, the focusing angles are chosen at $(\theta_i - 0.25B_W,\ \theta_i,\ \theta_i + 0.25B_W)$.

From formula (2) can be obtained the direction vector of the vector hydrophone that does not vary with frequency, it is only concerned with the signal azimuth. Therefore, in constructing the direction vector, it does not need to estimate azimuth angle, but just to estimate the number of sources. This simplifies the calculation and avoids the estimated deviation of the orientation of the target due to the deviation of the estimated angle.

The source correlation matrix is then found from:

$$S(f_j) = (A^H A)^{-1} A^H [R(f_j) - \sigma^2(f_j) I] A (A^H A)^{-1} \tag{8}$$

Since $P(f_j)$'s are independent of the focusing frequency, f_0 can be determined from:

$$\min_{f_0} \sum_{i=1}^{q} \left| \sigma_i(P_0) - \frac{\mu_i}{J} \right|^2 \tag{9}$$

where, $\mu_i = \sum_{j=1}^{J} \sigma(P(f_j))$, $\sigma(P(f_0)) = \frac{\mu_i}{J}$, $\quad i = 1, \ldots, q$.

This is a one-variable optimization problem, and a search procedure can be applied to find the minimum point. In practice, it is sometimes convenient to choose a predefined frequency such as the center frequency of the spectrum for focusing. However, to improve the performance, a focusing frequency that produces the smallest error should be selected.

3.3 The TCT Algorithm

The TCT algorithm is summarized as follows:[7]

(1) Use *AIC*, *MDL*, or *PSC* to find the true number of sources.
(2) Apply a *DFT* to the array output to sample the spectrum of data.
(3) Form A and $S(f_j)$ using the results of the preprocessing step and formula (8).
(4) Find the Focusing frequency and Subspace, formula (8) and (9).
(5) Find $P(f_0) = AS(f_0)A^H$ and the $P(f_j)$'s using: $P(f_j) = Rf_j - \sigma^2(f_j)I$.
(6) Determine the unitary transformation matrices formula (6).
(7) Multiply these matrices by the sample correlation matrices, and average the results.
(8) Apply *MUSIC* [5, 6] or any other high-resolution spectral estimation method to find the *DOA*'s.

4 Simulation and Verification

Simulation experiments using the same type acoustic pressure arrays to comparison to assess single vector hydrophone target azimuth estimation performance. All these results given in the paper are 50 times *Monte Carlo* simulation results.

(1) Orientation estimates of two space coherent source in different *SNR*

The simulation conditions: sound pressure hydrophone arrays' element $M = 3$, the array element spacing d corresponding to half the wavelength of the signal center frequency, signal is of two coherent gaussian white noise, Bandwidth of 200–300 Hz, Spatial azimuth angle of 10° and 20°, the sampling frequency $fs = 1200$ Hz, signal bandwidth in the frequency domain divided into five sub-bands. The simulation results of different *SNR* under the power spectrum, adaptive linear spectrum enhancement and pressure.

As can be seen from Fig. 2, the single vector hydrophone compared with acoustic pressure line arrays in the same coherent signals simulation conditions, has a better noise suppression performance and no starboard fuzzy. The new algorithm estimate direction of beam side lobe and beam width is significantly less than the original algorithm. In Fig. 3, because the *SNR* decreased, acoustic pressure line arrays can not be given to the exact location of the two targets, but using vector hydrophone can still accurately estimate the orientation of the two targets.

(2) SNR = −20 dB, two different angles interval coherent signal source azimuth estimation, respectively (10, 15°) in Fig. 4 and (10°, 13°) in Fig. 5.

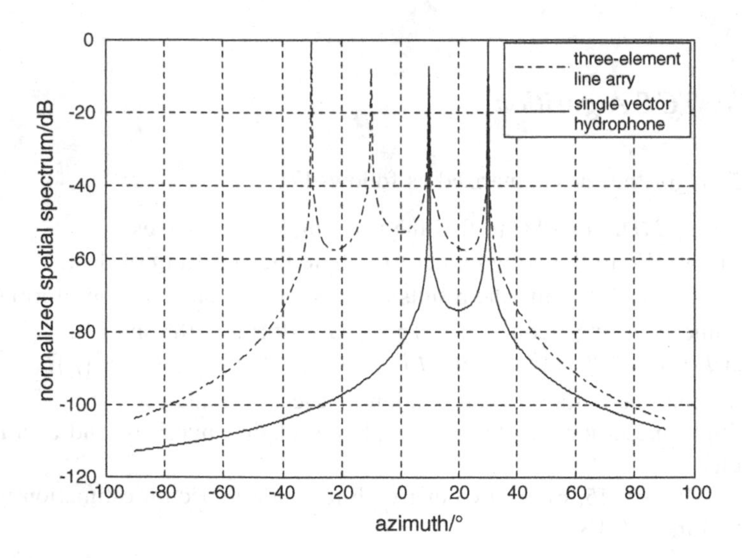

Fig. 2 $SNR = -20\,\text{dB}$ the spatial spectrum

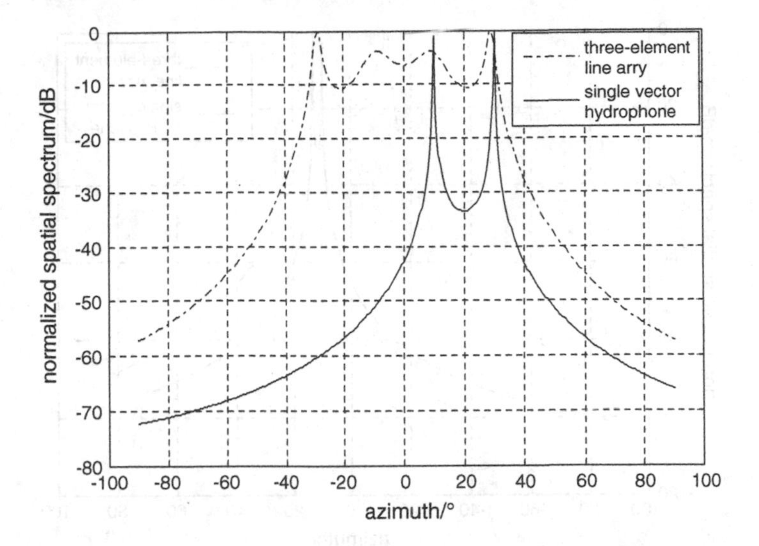

Fig. 3 *SNR* = −33 dB the spatial spectrum

Acoustic pressure line arrays' azimuth estimation performance decreased obviously when the target bearing is 5° and it can not distinguish the two goals between 3°. But the single vector hydrophone is still estimate of the exact location of the two goals when azimuth difference is 3°. Thus the use of a single vector hydrophone is to have a better estimate performance of a number of similar orientation signals.

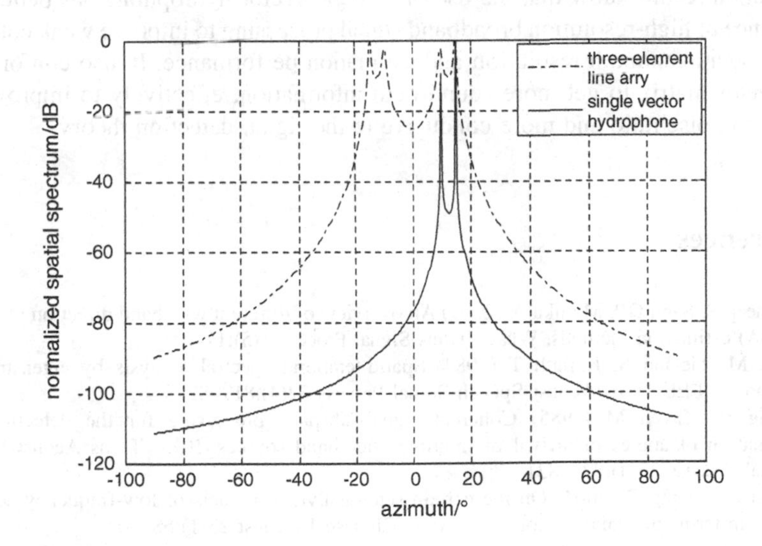

Fig. 4 The spatial spectrum between 5°

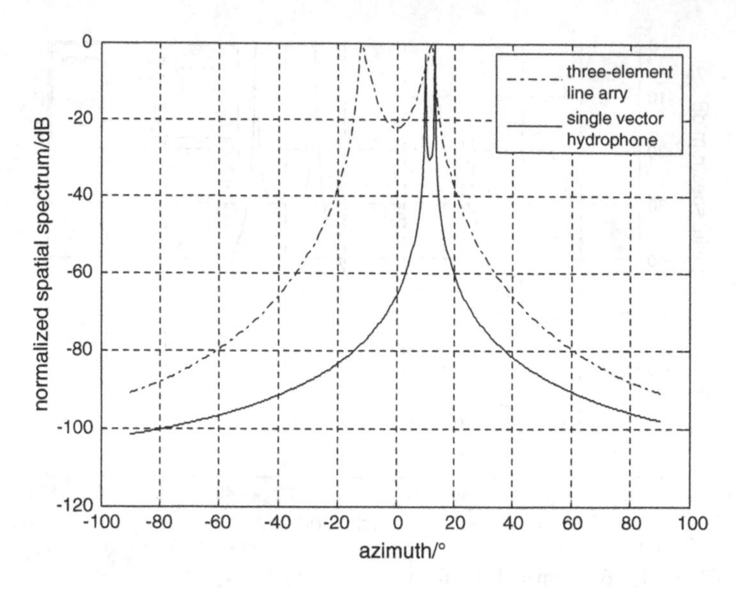

Fig. 5 The spatial spectrum between 3°

5 Conclusions

In this paper, a single vector hydrophone wideband signal *DOA* estimation of coherent sources is studied. The broadband single vector hydrophone array model is derived, and using the same type acoustic pressure arrays to comparison, use *MUSIC* azimuth estimation algorithm makes the simulation experiment. The simulation results show that the use of a single vector hydrophone has better performance of high-resolution broadband signal processing to improve weak coherent signal source azimuth resolution and estimation performance. It also conforms to the vector matrix to get more sound field information, effectively to improve the signal-to-noise ratio and more conducive to the signal detection theory.

References

1. Sandeep S, Sahu OP, Monika A (2009) An overview of different wideband direction of arrival (DOA) estimation methods. Wsaes Trans Signal Process 1(5):11–21
2. Wax M, Tie-Jun S, Kailath T (1984) Spatio-temporal spectral analysis by eigenstructure methods. IEEE Trans Acoust Speech Signal Process 32(4):817–827
3. Wang H, Kaveh M (1985) Coherent signal-subspace processing for the detection and estimation of angles of arrival of multiple wide-band sources. IEEE Trans Acoust Speech Signal Process 33(4):823–831
4. Gulin OE, Yang D (2004) On the certain semi-analytical models of low-frequency acoustic fields in terms of scalar vector description. Chinese J Acoust 23(1):58–70

5. Chen H, Zhao J (2005) Coherent signal subspace wideband optimal beam forming for acoustic vector-sensor arry. Acta Acustica 30(1):277–282
6. Xu H, Liang G, Hui J (2005) Acoustic vector array beam-space broadband focused music algorithm. J Harbin Eng Univ 26(3):349–354
7. Wang Y, Chen H, Peng Y (2004) Spatial spectrum estimation methods and algorithms. Tsinghua University Press, Beijing

Kernel Principal Component Analysis-Based Method for Fault Diagnosis of SINS

Zhiguo Liu, Shicheng Wang and Lihua Chen

Abstract Fault diagnosis is necessary in inertial navigation system to ensure navigation success. The kernel principal component analysis (KPCA) method is applied to fault diagnosis of inertial navigation system. At first, the square prediction error is used as the fault monitoring indictor. In addition, this paper locates the faults by using sensor variables changes. In order to reduce the dependence to the prior knowledge in selecting kernel function parameter, genetic algorithm is introduced into the Gauss RBF kernel function parameter optimization. It can improve the scientific nature of parameter choice. Experimental results show that the method based on KPCA of the inertial navigation system has good fault monitoring and recognition ability.

Keywords Strap down inertial navigation system · Sensor · Fault diagnosis · Kernel principal component analysis · Genetic algorithm

1 Introduction

With the development of navigation technique and maturity, strap down inertial navigation system (SINS) in aviation, aerospace, and other fields has been widely applied in. SINS (mainly by gyro angular rate sensor and accelerometer) measurement information is transferred to the host computer to be carrier positioning calculation basis. The quality of gyro and accelerometer data provided is related to the success of the solution. Therefore, we first need to develop fault detection and diagnosis to SINS sensor output, find the fault timely and effectively and carry out further processing to ensure the measurement parameters reliability. At present,

Z. Liu (✉) · S. Wang · L. Chen
The 301 Teaching and Research Section, The Second Artillery Engineering University, Xi'an, People's Republic of China
e-mail: lzgc@163.com

F. Sun et al. (eds.), *Knowledge Engineering and Management*, Advances in Intelligent Systems and Computing 214,
DOI: 10.1007/978-3-642-37832-4_57,

the sensor fault diagnosis is mainly based on model and knowledge. These methods have certain limitations such as the first method needs to establish the precise mathematical model and the second method relies on expert knowledge. Aiming at the insufficiency of these two approaches, in recent years, the principal component analysis (PCA) method in fault diagnosis of sensors has been widely applied and also got a good effect [1, 2]. But theoretically, PCA is a linear method. In order to apply to nonlinear systems, a nonlinear method—kernel principal component analysis (KPCA) is applied to the fault diagnosis of SINS sensors. At the same time this paper put forward a genetic algorithm for parameter optimization of kernel function to get the optimized KPCA model.

2 Common Types of SINS Sensor Fault

The sensor failures are divided into four types which are bias, drift, precision drops, and complete failure. The former three kinds of fault is also called soft fault, the latter fault is called a hard fault. In SINS, although soft fault accounts for a small part of, but it is often difficult to be found, which will produce great harm, affect the calculation accuracy of navigation and positioning.

A measurement commonly includes three parts, i.e.

$$x_m = x_r + f + u \tag{1}$$

Among them, x_m is the measure value, x_r is the real value, f is measuring systematic error, u is measuring random error. Usually random error obeys normal distributions with zero mean, i.e. $u \sim N(0, \sigma^2)$; and the system error of f is mainly caused by a glitch, corresponding to different fault, there are different expressions.

(1) Bias fault
Bias fault mainly refers to the fault measurement value and correct measurement has a constant deviation, i.e.

$$f = b \tag{2}$$

where, b is a constant.

(2) Drift failure
Drift failure refers to the magnitude of fault changes linearly with the time, i.e.

$$f = d(t - t_s) \tag{3}$$

where, d is a constant, t_s is the fault starting time, t is the time after which faults occur.

(3) Precision grade decline fault
Bias, drift fault performs measured average value's deviation. If precision

grade declines, the measured average value does not change, but the measured variance will change i.e.

$$f = N(0,\ \sigma_1^2) \tag{4}$$

where σ_1^2 is the change of variance. Put Eq. (4) into Eq. (1), available:

$$x_m = x_r + N(0,\ \sigma_1^2) + u \tag{5}$$

As $u \sim N(0, \sigma^2)$, by probability theory, the sum of two normal distributions is still normal distribution, i.e.

$$x_m = x_r + N(0,\ \sigma^2 + \sigma_1^2) \tag{6}$$

Therefore, this kind of fault is similar to the variances of random errors increase.

(4) Complete failure

The measured values always keep a constant when sensors are failure. But these constant values are typically zero or maximum. This kind of fault can be expressed as

$$f = c - x_m - u \tag{7}$$

3 Sensor Fault Diagnosis Based on GA-KPCA

3.1 Fault detection

The monitoring method based on KPCA is similar to PCA, in feature space Q still used in statistics [3]. It is also named as the square prediction error. SPE is an important performance monitoring indicator, which expresses the change tendency of each sampling and statistical error between the models. It is a measure when the model's external data changes.

SPE is defined as follows:

$$\text{SPE} = \left\| \varphi(x) - \varphi_p(x) \right\|^2 \tag{8}$$

Herein, $\varphi(x)$ is the sum of not equal to zero for all eigenvalues of the corresponding vector and feature vector product. Then we have

$$\text{SPE} = \sum_{j=1}^{n} t_j^2 - \sum_{j=1}^{p} t_j^2 \tag{9}$$

where, n is nonzero eigenvalue of a number, p is the main element, in general can be selected by the following equation:

$$\frac{\sum_{k=1}^{p} \lambda_k}{\sum_{k=1}^{m} \lambda_k} \geq cn \tag{10}$$

where cn is a constant, this paper take $cn = 0.90$.

SPE's confidence limit can be calculated through its approximate distribution:

$$\mathrm{SPE}_\alpha \sim g\chi_h^2 \tag{11}$$

where α is the significant level, g and h, respectively denotes right parameters and the degree of freedom of SPE; if a and b are SPE's mean and variance of estimate, then g and h can be approximated as $g = b/2a$, $h = 2a^2/b$.

3.2 Fault diagnosis

It is needed to further find fault variables if the fault is detected by monitoring. The SPE's indictors can only monitor the fault and cannot identify the fault effectively. In PCA method, fault variables and the monitored variables has a linear relationship. So it is easy to calculate variable contribution for drawing contribution chart. While the KPCA nonlinear transformation process does not use an explicit nonlinear transfer function, and the kernel function method cannot provide the relation between original measured variables and monitored variables. So the contribution graph like PCA cannot be applied in KPCA fault variable identification directly.

In order to complete the calculation of the original measured variables to fault contribution and realize fault location, we define the contribution amount of the original measured variables j as follows:

$$\mathrm{cntr}_{j,\,i} = \sum_{i=1}^{p} \left| t_i^T x_j / \lambda_i \right| \tag{12}$$

Contribution percentage change

$$\Delta \mathrm{cper}_j = \mathrm{cper}_{t1,\,j} - \mathrm{cper}_{t0,\,j}, \tag{13}$$

where, $\mathrm{cper}_{t0/1,\,j} = \dfrac{\mathrm{cntr}_{j,\,i}}{\sum_{j=1}^{n} \mathrm{cntr}_{j,\,i}}$

Among them, p is the chosen principal component, $\boldsymbol{t}_i$ and $\boldsymbol{x}_j$ denote the ith nonlinear principal component and the jth sensor measured variables, λ_i is the ith characteristic value, $t1$ and $t0$ respectively denotes moment of sensors with failure and without failure, sn is the number of sensors. By comparing amount of contribution percentage changes before and after fault occurrence, the fault variables can be isolated.

3.3 Kernel parameter optimization

In the selection of Gauss radial basis function as the KPCA kernel function of the model, kernel width selection is an important content. It is related to the model precision and generalization ability. The usual approach is based on the experience and repeated experiments. This method is not scientific, also the workload is hard and blindness. Aiming at this problem, this paper uses a genetic algorithm to optimize the choice of unclear parameter σ [4–6] (Fig. 1).

4 Simulations

In order to test the effectiveness of the method, 400 SINS output data into the KPCA optimization model are analyzed. The former 200 groups of data are the normal output data, the latter 200 groups of data are simulated on the first gyro output artificially introduced into the previously mentioned four kinds of fault. Fault monitored results are shown in Fig. 2a–d. Through the graph, four kinds of failure can be obviously monitoring. Wherein the solid line shows the actual SPE value, dotted lines indicate the detection limit. Significance level $\alpha = 0.99$.

The percentage changes of the fault occurred before and after each sensor variable contribution are calculated. The results are shown in Fig. 2. Seen from the graph, fault occurred after the first sensor variable contribution percentage change in maximum, indicating failure occurs at the first sensor, which is consistent with the actual. This also explains the validity of the method.

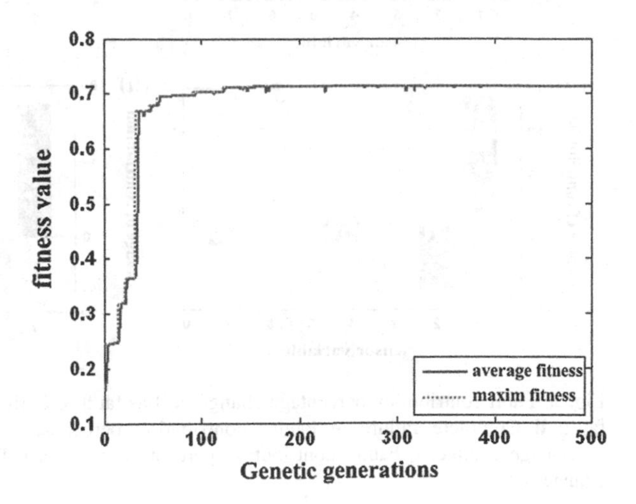

Fig. 1 Process of fitness change for resolving

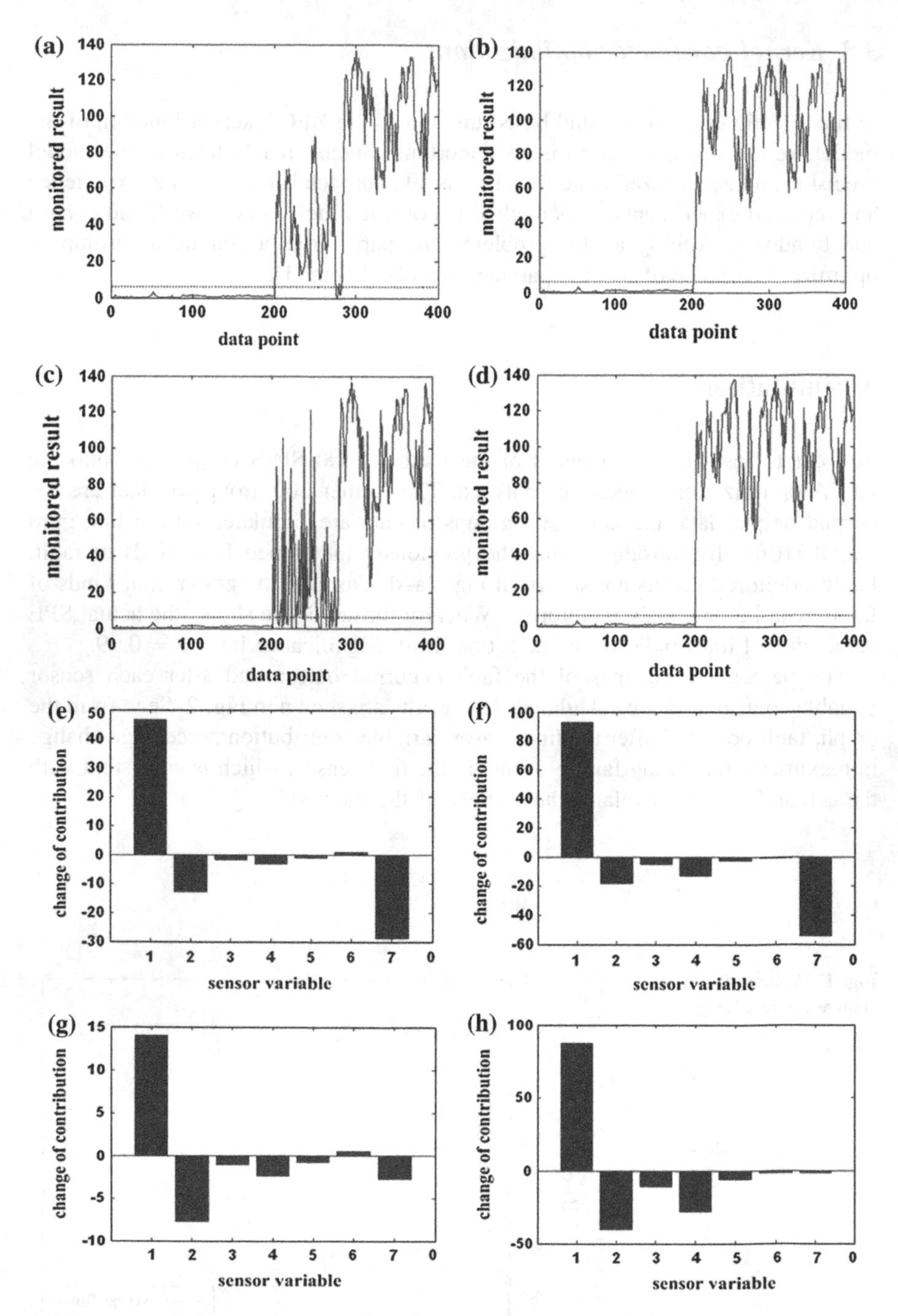

Fig. 2 Fault contribution percentage change. **a** Bias fault. **b** Drift fault. **c** Precision grade reduce fault. **d** Complete failure. **e** Fault1 contribution percentage change. **f** Fault2 contribution percentage change. **g** Fault3 contribution percentage change. **h** Fault4 contribution percentage change

5 Conclusion

Sensor fault diagnosis based on PCA method does not need to establish the precise mathematical model, also do not need to be overly dependent on prior knowledge, has been widely used. But PCA is just a linear model. In order to adapt to the actual nonlinear system, the application of KPCA to the SINS sensor fault diagnosis using SPE as fault monitoring indictors and the fault positioning method are studied. In order to establish the optimized KPCA model and improve the choice of kernel function parameter scientifically, genetic algorithm is introduced to Gauss RBF kernel function parameter automatic optimization. The experimental results show that the proposed GA-KPCA method can monitor and diagnose SINS common sensor fault effectively.

The Lecture Notes in Computer Science volumes are sent to ISI for inclusion in their Science Citation Index Expanded.

References

1. Wang SW, Cui JT (2006) A Robust fault detection and diagnosis strategy for centrifugal chillers. HVAC&R Res 12(3):407–428
2. Gao D, Wu C-g, Zhang B-k (2011) Method combining PCA and SDG for fault diagnosis of sensors and its application. J Syst Simul 23(3):567–573
3. Lee JM, Yoo CK, Choi SW et al (2004) Nonlinear process monitoring using kernel principal component analysis. Chem Eng Sci 59(1):223–234
4. Chi HM, Moskowitz H, Ersoy OK (2009) Machine learning and genetic algorithms in pharmaceutical development and manufacturing processes. Decis Support Syst 48(1):69–80
5. Gao Y-g, Li Z Shi D-q (2008) A hybrid learning method for RBFNN based on AGA. Radio Eng China 38(11):25–28
6. Deng X-g, Tian X-m (2005) Nonlinear process fault diagnosis method using kernel principal component analysis. J Shandong Univ (Eng Sci) 35(3):103–106

Design of Insulator Resistance Tester for Multi-Core Cables

Dacheng Luo, Yan Liu, Zhiguo Liu, Dong Chen and Qiuyan Wang

Abstract By the long-time of the insulator resistance testing of multi-core cables, a new insulator resistance tester based on an intelligent test system consisting of an embedded controller, two relay matrixes, and so on is presented. To meet the demands of the cables with kinds of BNC connector plugs, the connecting cables, one port of which is with the standard connecter to plug in the design tester and the other port of which is with the special connecter to interconnect testing cables, are also designed. With the using of the tester, the insulator resistance testing time is shorted much.

Keywords Embedded controller · Insulation résistance · Relay matrix

1 Introduction

Multi-core cables are widely used in connecting kinds of measurement devices. For example, in the ordinary maintenance of airplanes, the multi-core cables are used to connect airplanes and measurements. Before the testing, the first thing to be done is to confirm the validation of multi-core cables by insulator resistance testing. As we all know that multi-core cables are with kinds of BNC connector plugs and the contents of insulator resistance testing are complex, the testing of insulator resistance is commonly based on Meg-ohmmeter. There are some shortages in this testing. First of all, the time of testing is too much because of only the insulator resistance between two pins of multi-core cables are tested one time, second, the testing results may include mistake records by the whole testing is operated by man. So the aims of the design of insulator resistance tester are to short the testing time and to decrease the mistake records.

D. Luo (✉) · Y. Liu · Z. Liu · D. Chen · Q. Wang
Xi'an Research Institute of High-Tech, Xi'an, People's Republic of China
e-mail: luodcheng@163.com

F. Sun et al. (eds.), *Knowledge Engineering and Management*,
Advances in Intelligent Systems and Computing 214,
DOI: 10.1007/978-3-642-37832-4_58,

2 The Principle of Insulator Resistance

(A.) *Analysis of Insulator Resistance Scheme*

Because the value of insulator resistance cannot be tested directly, it is usually calculated by some formulas using the testing values of tiny-current or tiny-voltage. The methods can be classified as series process, such as multiple processes [1], and voltage-ratio method, and electric bridge method, and so on. In this design, series process is selected. This method is viewed as Fig. 1.

In Fig. 1, it defined U as input voltage, R_X as the insulator resistance to be tested, R_1, R_2 as measuring resistance (the action of R_2 is given as R_1 a suitable voltage), U_{IN} as the tested voltage.

Because of the values of U, R_1, R_2 are known and the value of U_{IN} can be tested, the value of R_x can be calculated as:

$$R_x = \frac{U - U_{IN}}{U_{IN}} R_1 - R_2 \quad (1)$$

(B.) *Analysis of the Inherent Resistance of Power*

The value of inherent resistance of power may affect the test results of insulator resistance, so an experiment is design to test this value. It takes the value of R_X as a known value, and selects the testing voltage as 100 V, and the value of R_X in ten groups among 0–50 MΩ. The value of R_S can be calculated by the tested voltage of R_X. From groups of calculated values, the relationship between R_X and R_S is as shown in Fig. 2.

The printing area is 122 × 193 mm. The text should be justified to occupy the full line width, so that the right margin is not ragged, with words hyphenated as appropriate. Please fill pages so that the length of the text is no less than 180 mm, if possible.

In Fig. 2, it shows that the relationship between R_X and R_S is approximate linear. So it can use $\hat{y} = \hat{b} + \hat{a}x$ to describe the relationship. Using the method of least square, it gets

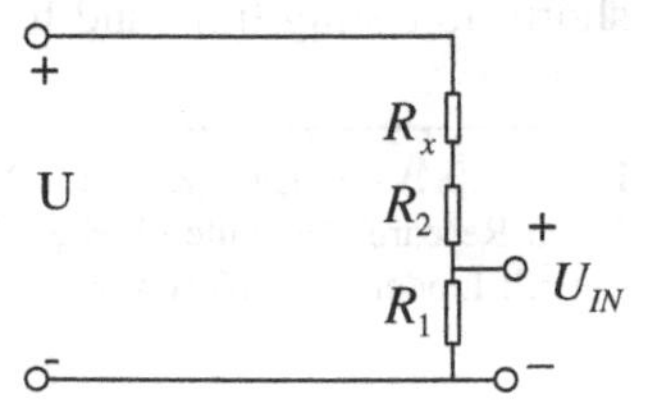

Fig. 1 The principle of series process

Fig. 2 The relationship between R_X and R_S

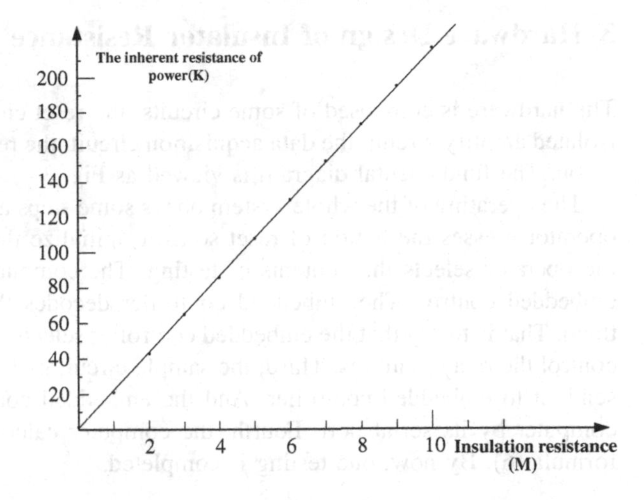

$$\begin{cases} \bar{x} = \frac{1}{n}\sum_{i=1}^{n} x_i \\ \bar{y} = \frac{1}{n}\sum_{i=1}^{n} y_i \\ \hat{a} = \frac{\sum_{i=1}^{n}(x_i - \bar{x})(y_i - \bar{y})}{\sum_{i=1}^{n}(x_i - \bar{x})^2} \\ \hat{b} = \hat{y} - \hat{a}\bar{x} \end{cases} \tag{2}$$

Taking the tested value in (2), it gets

$$\hat{y} = 3.622 + 21.90x \tag{3}$$

Changing the unit of R_S to MΩ, it gets

$$R_S = 0.003622 + 0.021900\ R_X \tag{4}$$

So the formula (1) can be refreshed as

$$R_X = 1.0224\left(\frac{U}{U_1}R_1 - R_1 - R_2\right) - 0.003703 \tag{5}$$

And it was the formula to calculate R_X.

3 Hardware Design of Insulator Resistance

The hardware is composed of some circuits: the reset circuit, the filter circuit and isolated amplify circuit, the data acquisition circuit, the relay matrixes circuits, and so on. The fundamental diagram is viewed as Fig. 3.

The operating of the whole system obeys some steps as follows. First of all, the operator presses the button of reset so as to initialize the whole system. Second, the operator selects the contents of testing. The computer sends its codes to the embedded control. The embedded controller decodes the codes and carries out them. That is to say that the embedded controller selects the right testing circuit by control the relay matrixes. Third, the sample circuit gathers the signal voltage and sends it to embedded controller. And the embedded controller sends the data to computer by its serial port. Fourth, the computer calculates the value of R_X by formula (6). By now, one testing is completed.

(C.) *Design of Reset Circuit*

The main unit of reset control circuit is MAX813L [2]. If we press the button of "reset" or the program is run in the wrong way, the reset control circuit will reset the whole system. The reset circuit is viewed as in Fig. 4.

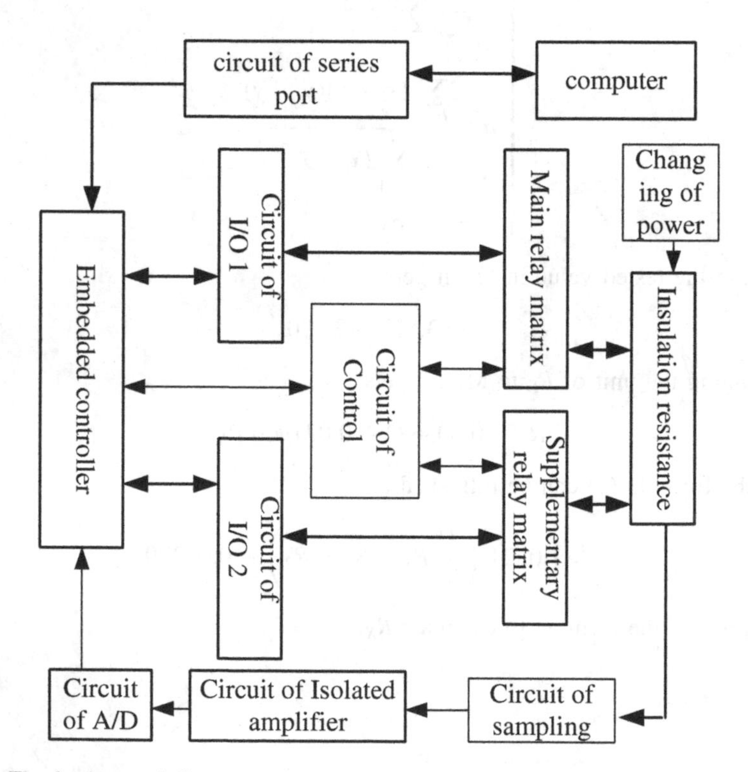

Fig. 3 The fundamental diagram

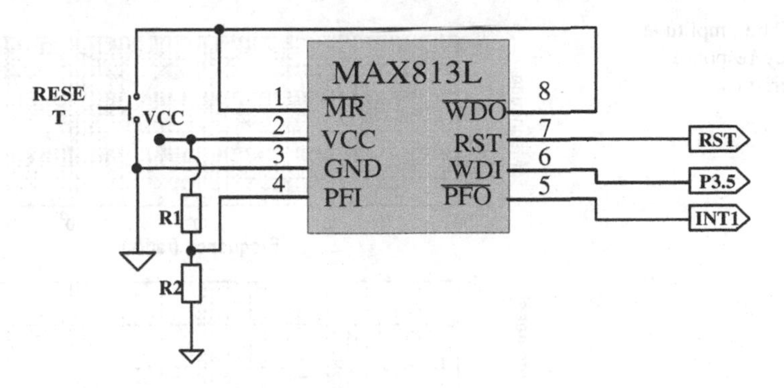

Fig. 4 The reset circuit

(D.) *Design of Filter and Insulating Transformer Circuit*

Because the value of U_{IN} is very small, the noise of power and the value of inherent resistance may influence the test result. So some filter circuit and insulating transformer circuit are designed. The principles of these circuits are viewed in Fig. 5.

The filter circuits are designed by two parts. The first part is low band filter as "RC" and the other part is band elimination filter as double "T" 。[3, 4]

The first part of no-power low band filter is made up by R_3 and C_1, its response is $F_1(s) = \frac{1}{1+sR_3C_1}$. And the other part is made up by R_4, R_5, R_6, C_2, C_3, C_4. Accord to the character of double "T" circuit, it gets: $R_4 = R_6 = 2R_5 = R$, $C_3 = C_4 = \frac{1}{2}C_2 = C$, and its response is

$$F_2(s) = \frac{\frac{1}{2}\left(R + \frac{1}{sC}\right)}{\frac{2R(1+sRC)}{1+(sRC)^2} + \frac{1}{2}\left(R + \frac{1}{sC}\right)} \tag{6}$$

So the response of whole filter circuit is

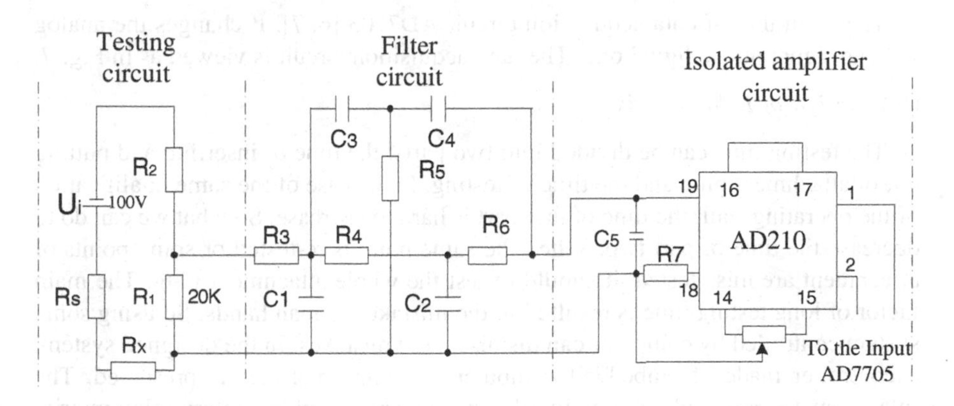

Fig. 5 The principle of filter and insulating transformer circuit

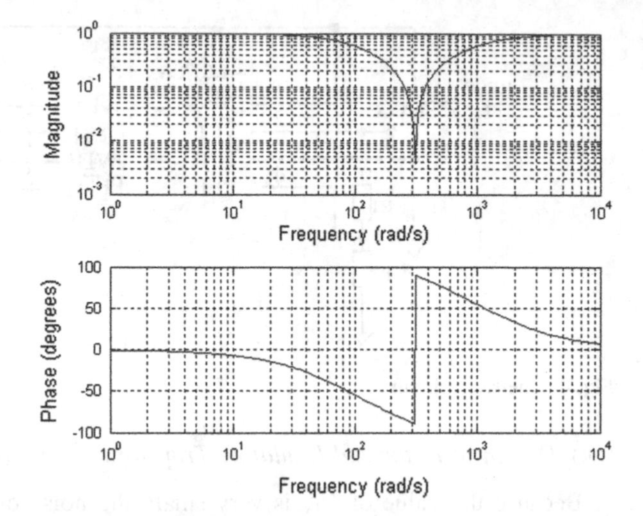

Fig. 6 The amplitude frequency response characteristics

$$F(s) = \frac{1}{1 + R_3C_1} \cdot \frac{R^3C^3 \cdot s^3 + R^2C^2 \cdot s^2 + RC \cdot s + 1}{R^3C^3 \cdot s^3 + 5R^2C^2 \cdot s^2 + 5RC \cdot s + 1} \quad (7)$$

Assume the frequency of the noise of power is 50 Hz, the ideal ω_n can be calculated as $\omega_n = \frac{1}{RC} = 50 \times 2\pi = 314.15$ rad/s. So the value of devices can be selected as follows: $R_3 = 51\,K\Omega$, $C_1 = 0.22\,\mu F$, $R = 14\,K\Omega$, $C = 0.22\,\mu F$. The amplitude frequency response characteristics are viewed as in Fig. 6:

In Fig. 6, it indicates that this filter circuit can decrease the error caused by power noise.

The main device of insulating transformer circuit is isolated amplifier AD210 [5]. AD210 and its circuit can output a high precision value of input signal as 1:1, so the circuit behind AD210 would not affect the input signal. The inherence resistance of this circuit is about 2 KΩ, and it can be a good source of A/D circuit.

(E.) *Design of Data Acquisition Circuit*

The main unit of data acquisition circuit AD7705 [6, 7]. It changes the analog voltage signal into a digital one. The data acquisition circuit is viewed as in Fig. 7.

(F.) *Design of Relay matrix*

The testing time can be divided into two parts: the time of inserting and putting out of attachment plug and the time of testing. In the case of the same qualification of the operating staff, the time of first part is hard to decrease. So what we can do to decrease the time of part two. When the same point is re-tested or some points of attachment are miss-tested, it should re-test the whole attachment plug. The main factor of long testing time is resulted in the mistakes of man hands. So using some system controlled by computer can dismiss these mistakes. In the designed system, a controller made of embedded computer and relay matrixes is presented. The relay matrixes are made of a main relay matrix and a supplementary relay matrix.

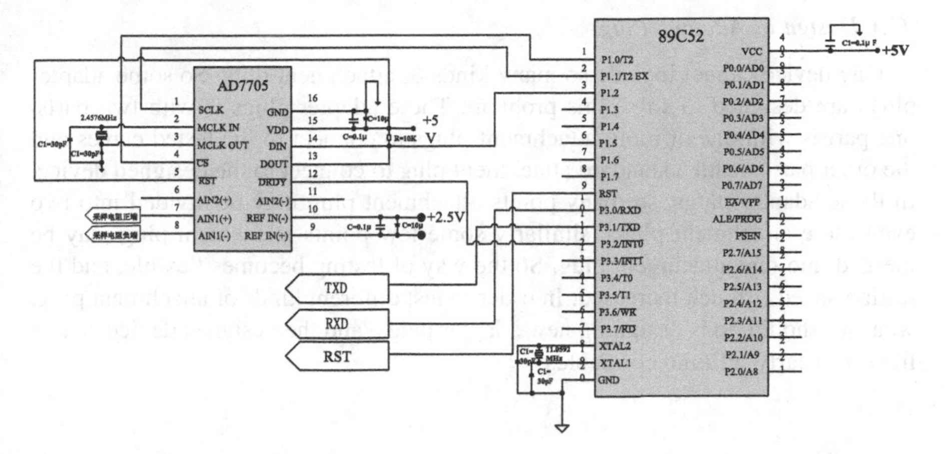

Fig. 7 The amplitude frequency response characteristics

It connects the similar point of these two matrixes to the same point of attachment plug. So the testing points can be chosen by relay matrixes which are controlled by embedded computer. The value of R_X can be calculated by formula (5).

In the designed system, the relay matrixes are both 12 × 12. So it can test more than 144 points at one time. And the testing time is much shorted. The logic chart of relay matrixes is viewed as in Fig. 8.

The main unit of relay matrixes control circuit is embedded controller AT89C52. The points of relay matrixes are controlled by its line voltage and row voltage. So the on–off operation of points of relay matrix can be controlled by the controlling of its line and row voltage by AT89C52 [8–10]. The voltage changing circuit can supply different values of the testing such as the testing voltage U and the units work voltage +5, +12 V, and so on.

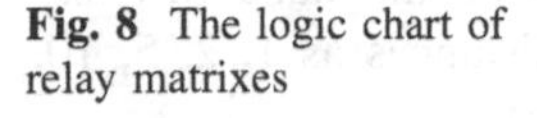
Fig. 8 The logic chart of relay matrixes

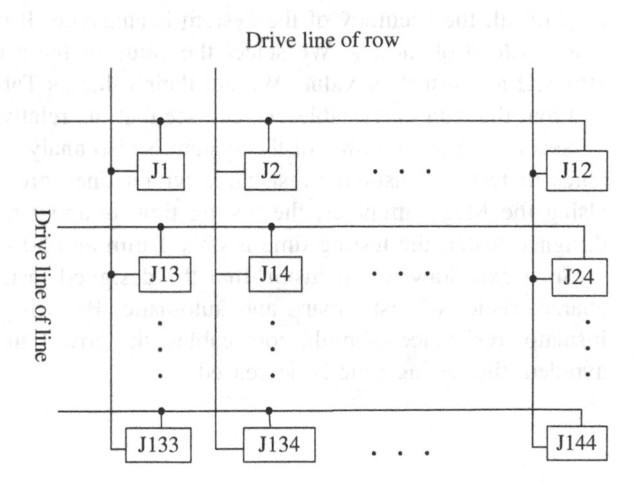

(G.) *Design of Adapter Plugs*

One device cannot include so many kinds of attachment plug. So some adapter plugs are designed to solve this problem. These adapter plugs is with two parts: one part is with one or more attachment plugs to connect to the tested cables and the other part is with a standard attachment plug to connect to the designed device. In those adapter plugs, so many points attachment plug may be divided into two even more attachment plugs, similarly some few points attachment plug may be merged into one attachment plug. So the way of testing becomes flexible, and the testing speed is much increased. In order to test different kinds of attachment plug, what we should do is designing new adapter plugs, and the designed device can be fitted to nearly all kinds of cables.

4 Software Design of Insulator Resistance

Because of the code compiled by computer language C running fast and compact, it takes computer language C to design the software. The software is made up with self-check procedures, data acquisition procedures, and control procedures, data base, and so on. The self-check procedures initialize the system and check the condition of system before testing. The data acquisition procedures gather the voltage signal so as to calculate the value of R_X. The control procedures control the relay matrixes so as to select the right circuit of testing. And the data base save the results of testing in storage.

The flow process of the testing is viewed as Fig. 9.

5 Test Result Analysis of Design Taster

First of all, the accuracy of the system is analyzed. It takes some standard resistance instead of the R_X. We select the value of these resistances as 1, 2, 5, 10, 100 MΩ and test their value. We get their value as Table 1.

From the data in the table; we can see that the relative error is less than 1.5 %.

Second the testing time of the system is also analyzed. It takes a cable with 50 cores to test the insulator resistance within one core and another for example. Using the Meg-ohmmeter, the testing time is about 6 min and 15 s. Using the designed tester, the testing time is only 1 min and 30 s.

So it can draw a conclusion that the designed insulator resistance tester has characteristics of fast, smart, and automatic. By using of this tester to test the insulator resistance of multi-core cables, the error cause by man-made factor is avoided, the testing time is decreased.

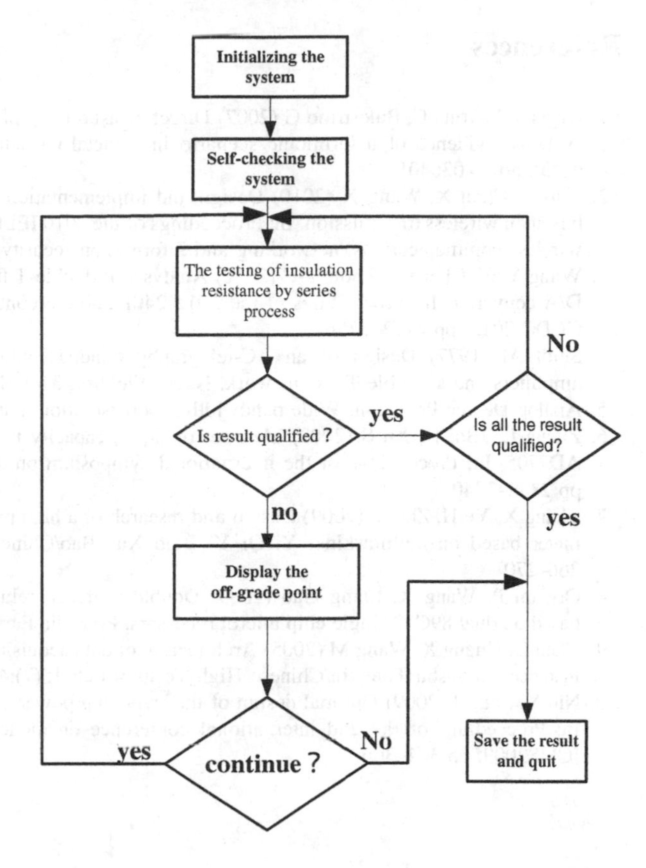

Fig. 9 The flow process of the testing

Table 1 The results of testing

Standard resistance (MΩ)	1	2	5	10	100
Test 1	1.00	2.03	5.01	10.12	101.42
Test 2	1.01	2.01	5.03	10.04	100.09
Test 3	1.00	2.00	5.13	10.22	101.03
Test 4	1.02	2.03	5.02	10.03	101.91
Test 5	1.00	2.02	5.12	10.12	101.52
Test 6	1.01	2.01	5.00	10.06	102.02
Test 7	1.01	2.03	5.09	10.00	100.05
Test 8	1.00	2.03	5.17	10.03	102.04
Test 9	1.03	2.00	5.08	10.11	101.51
Test 10	1.00	2.05	5.01	10.00	100.60
The value of average	1.008	2.015	5.066	10.073	101.219
Relative error	0.8 %	0.75 %	1.32 %	0.73 %	1.22 %

References

1. Orgiani P, Aruta C, Balestrino G (2007) Direct measurement of sheet resistance R in cuprate systems: evidence of a fermionic scenario in a metal-insulator transition. Phys Rev Lett 98(3), art.no.036401
2. Yao C, Chen X, Wang X (2010) Design and implementation of signal acquisition system based on wireless transmission. In: Proceedings of the 2010 IEEE international conference on wireless communications, networking and information security. WCNIS 2010, pp 121–125
3. Wang Y-W, Cheng F, Zhou Y-C (2012) Analysis of double-T filter used for PWM circuit to D/A converter. In: Proceedings of the 2012 24th Chinese control and decision conference. CCDC 2012, pp 2752–2756
4. Seidl Al (1977) Design of an AC-telegraphy bandpass filter incorporating operational amplifiers and a double-T RC network. Nachr Elektron 31(1):1–4
5. Analog Device Precision. Wide bandwidth 3-port isolation amplifier
6. Zheng L, JBai H, Xu S (2003) Measure of liquid capacity in pressure container based on AD7705. In: Proceedings of the international symposium on test and measurement, vol 4. pp 2838–2840
7. Zhang X, Ye H, Zhu Y (2009) Design and research of a high precision digital optical power meter based on multitasking. Yi Qi Yi Biao Xue Bao/Chinese J Sci Instrum 30(Suppl): 266–270
8. Orgiani P, Wang X, Liang Guo (2009) Double switched reluctance motors parallel drive based on dual 89C52 single chip microprocessors. Procedia Earth Planet Sci 1(1):1435–1439
9. Chen G, Huang X, Wang M (2005) Architecture of data acquisition and fusion system and its instance. Gaojishu Tongxin/Chinese High Technol Lett 15(7):40–44 (in Chinese)
10. Niu Y, Song J (2009) Optimal design of the switching power supply based on 89C52 SCM. In: Proceedings of the 2nd international conference on modelling and simulation, vol 8. ICMS2009, pp 304–306

A Flight-Test Information Reasoning System Combining Textual CBR and RBR Approaches

Bo Yang, Guodong Wang and Wentao Yao

Abstract This paper presents a reasoning method based on the combination of Textual CBR and RBR to deal with the case described by the natural language with the intention of resolving the information retrieval in the flight-test procedure. Then experiments are made by using the reasoning system based on the method. The results show that not only could the method retrieve the cases from the system correctly, but also the Textual CBR procedure could complement the RBR procedure when the latter one cannot retrieve the case correctly. Thus the method could efficiently increase availability of the reasoning system.

Keywords Flight-test information reasoning system · Textual case-based reasoning · Rule-based reasoning

This work is supported by the National Natural Science Foundation of China (No. 90820305, 60775040, 60621062, 61005085). State Key Laboratory Program of Pattern Recognition, Institute of Automation, Chinese Academy of Sciences. Aeronautical Science Foundation of China (No. 2010ZD01003).

B. Yang (✉) · W. Yao
State Key Laboratory of Intelligent Technology and Systems,
Tsinghua National Laboratory for Information Science and Technology,
Department of Computer Science and Technology, Tsinghua University,
100084 Beijing, China
e-mail: bo-yang07@mails.tsinghua.edu.cn

W. Yao
e-mail: ywt@mails.tsinghua.edu.cn

G. Wang
Shenyang Aircraft Design and Research Institute, Shenyang 110031, China
e-mail: wgddoctor@163.com

F. Sun et al. (eds.), *Knowledge Engineering and Management*,
Advances in Intelligent Systems and Computing 214,
DOI: 10.1007/978-3-642-37832-4_59,

1 Introduction

Performing all kinds of flight-test experiments in the real environment conditions is very important in design and appraisal process of advanced military aircraft, which is not only a comprehensive discipline of exploring unknown fields in the aerospace technology domain but also an important means of evaluating and finalizing the design. Flight test is indispensable, especially in the process of research and design of new aircraft. Generally speaking, flight experiment will inevitably involve a lot, even confused field and the domain knowledge including numerous and even complex dynamics knowledge. As a result, it is pivotal whether the knowledge and experience of test pilot and engineer that is relevant to the test flight task can be refined into automatic reasoning information system, which could provide optimal solutions for the problem during the flight test in addition to help pilots and flight-test engineer quickly understand and grasp and specific test flight task. The significance of the system is very profound.

So far there are many reasoning methods in the automatic reasoning area such as rule-based reasoning (RBR), case-based reasoning (CBR), and so on. Unfortunately, the classical reasoning methods cannot handle the rules or cases in the flight-test area as most of the expert experience is presented in human natural language, which cannot be used directly in the RBR system. Besides, the rules refined from the flight-test situation is limited by the knowledge and understanding of the pilot and engineers. So they cannot cover all kinds of situation that could happen in the flight-test experiment. Based on the condition above, the need of automatic refining of useful knowledge from the natural language described cases has been motivated.

Textural case-based reasoning (TCBR) is a subset of CBR method domain and TCBR is focused on the feature refining that is from the cases represented in the natural language. The main concept of TCBR method is to refine features and keywords with natural language processing (NLP) method, which is relatively mature in the computer science [1]. Since 1996 when Rissland implemented SPIRE system [2], TCBR has become one of the hot issues in the automatic reasoning domain. In 1997, Lenz et al. designed FAIIQ system [3], which referred to WordNet that used hierarchical structure representation to process the cases. Subsequently in 2004, Wiratunga et al. introduced a fully automated method for extracting predictive features to represent textual cases. The approach included feature selection with boosting and association rule induction to discover semantic relations between words [4]. Then Patterson et al. presented SOPHIA, a text clustering approach that does not require labeled data in 2005 by using term distributions to build themes, which are groups of words that appear in similar documents [5]. Recently more and more TCBR methods have been proposed on similarity calculation between textually represented cases, building structured case representations from texts [6] and adapting textual cases [7], etc.

Thus based on the above, in this work we use TCBR method to extract textual features from the cases represented in natural language combining rule-based reasoning procedure. The rest of this paper is organized as follows. First, we design the structure of the flight-test information reasoning system based on TCBR and RBR combination and implement the basic reasoning procedure. Then we focus on the concrete reasoning method based on the TCBR principle, especially on the similarity calculation procedure in mathematics. How to combine TCBR and RBR result is another key point of the work. Finally, a real flight-test information reasoning system is implemented on the computer to prove the methods we have been proposed.

2 Structure of Flight-Test Information Reasoning System

According to the actual need in the flight-test experiments, the system has three components: input component, which accepts user inputs in natural language and subsequently transfers them to the reasoning module after encoding; reasoning module, the most important component in the system, which analyzes the user input with some NLP approach and then transfers the result to the output component after reasoning; output component, which decodes the result from the reasoning module and then presents the reasoning result to the users in some way. Moreover, the system includes cases and rules maintaining functions, which can be concluded in Fig. 1.

Figure 2 shows typical procedure in the reasoning module. First, the user input is preprocessed to remove some special characters that do not meet the demand of the system. Then the module parallels the procedure of reasoning with TCBR and RBR method respectively. Finally, the results from TCBR and RBR are evaluated separately, then combined together to give the final reasoning representation by evaluation results. At last, the user can evaluate the reasoning result of the system, refine it, and then store it back to the cases database for further usage. In this way,

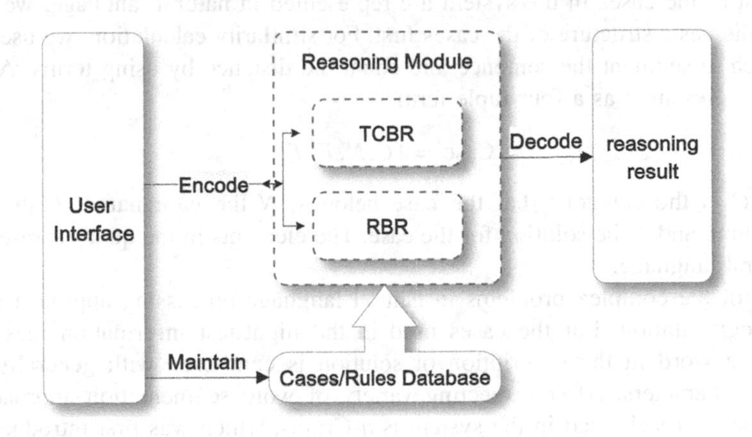

Fig. 1 The structure of the flight-test information reasoning system

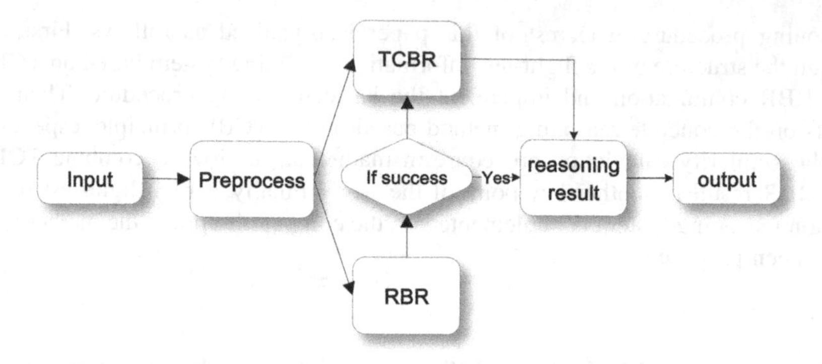

Fig. 2 The basic procedure of the reasoning module

the system can do its own optimizations as more and more cases refined by the users are saved into the cases database.

3 Approach Based on Combining TCBR and RBR Procedure

In this section we will take a deep look at the reasoning approach in the reasoning module of the system, especially the NLP approach for Chinese word segmentation and the combination method for TCBR and RBR results.

3.1 Case-Based Reasoning Upon TCBR Method

As most of the cases in the system are represented in natural language, we must define the basic structure of the cases first. For similarity calculation, we use NLP approach to segment the sentence and count the distance by using terms. A case can be represented as a four-tuple term

$$\text{Case} = \{C, N, F, T\} \tag{1}$$

where C is the category that the case belongs, N the case name, F the case description, and T the solution for the case. The elements in the quad represent all in natural language.

One of the complex problems in natural language processing approach is the word segmentation. For the cases used in the flight-test information reasoning system, a word in the description or solution is composed with generally 2–7 Chinese characters. After inspecting variety of word segmentation approach in NLP, the approach used in the system is n-Grams, which was first introduced in

1980 as a method of word prediction in natural language. In n-Grams procedure, it is first assumed that a word composed with N Chinese characters is only contextually relevant with the word composed with $N-1$ Chinese characters, that is

$$P(w_N|w_1, w_2, \ldots) = P(w_N|w_1, w_2, \ldots, w_{N-1}) \tag{2}$$

where w_i is the word after segmentation. From Eq. 2 we can conclude that n-Grams approach is a "regular grammar language with finite state". In the case of word segmentation in the flight-test information reasoning system, Kaze backtrace [8] is also used in order to avoid data sparsity. For n-order n-Grams, we calculate the probability as

$$P_n(w_n|\mathbf{w}_1^{n-1}) = P_{GT}(w_n|\mathbf{w}_1^{n-1}) + \alpha_n P_{n-1}(w_n|\mathbf{w}_2^{n-1}) \tag{3}$$

where P_{GT} is obtained by Good-Turing procedure and α_n is given by

$$\alpha_n = \begin{cases} 0 & c(\mathbf{w}_1^n) > 0 \\ \dfrac{\sum_{w_1^n} P_{GT}(w_n|\mathbf{w}_1^{n-1})}{1-\sum_{w_1^n} P_{GT}(w_n|\mathbf{w}_2^{n-1})} & c(\mathbf{w}_1^n) = 0 \end{cases} \tag{4}$$

Therefore, the words can be segmented with n-Grams approach according to Eqs. 2 and 3.

Similarly, this approach is also applied to word segmentation of the user input. As a result, all the cases in the system and the user input are composed with words. Then we define the similarity between two sentences based on n-Grams distance as the count of the words that are appeared in both sentences. That is, if there are two word strings A and B. We note G_A contains all unique words in A and G_B in B. c_i^A and c_i^B are the count of word i that appears in A and B separately. Then the similarity of word string A and B is given by

$$\gamma = \frac{\sum_{g_i \in G_A \cap G_B} c_i^A + c_i^B}{\sum_{g_i \in G_A \cup G_B} c_i^A + c_i^B} \tag{5}$$

where $G_A \cap G_B$ is intersection of G_A and G_B while $G_A \cup G_B$ is union of G_A and G_B. As a result, we can obtain a sequence of similarity between the user input and each case in the system.

At last, the case with maximal value of similarity is regarded as the TCBR result.

3.2 Rule-Based Reasoning Procedure

Different from the classical RBR approach, rules in the system is also presented in natural language. That is, a rule in the system can be described with a single production

$$R_i := \{A_1 \cap A_1 \cap \ldots \cap A_n \Rightarrow \text{Case}_i\} \tag{6}$$

where A_i is the property of rule and Case_i is the corresponding case. Because the properties of the rules are presented in natural language, the NLP approach for Chinese word segmentation can be used here. After that, for each rules we can calculate the similarity between the user input and the current rule. Then we obtain a sequence containing similarities for all the rules. At last the rule with maximal value of similarity is used for reasoning, and the corresponding case is regarded as the RBR result.

3.3 Combining TCBR and RBR Result

Because the rules in the system are summarized by the experts from the actual situation of the flight test, the information in the rules is more accurate than the one in the cases. So when the results from TCBR and RBR approach are combined, the RBR result is given higher priority than the TCBR result. That is,

If the results from both TCBR and RBR were the same, then any one of the results would be used for presenting system output. In another word, the system outputs the case from the TCBR or RBR procedure;
If the results were different, then the system output should be considered with their similarities. If there were only little gap between two similarities from TCBR and RBR procedure respectively, then the RBR result would be regarded as the system output. Otherwise, the result with maximal similarity would be used for generating system output.

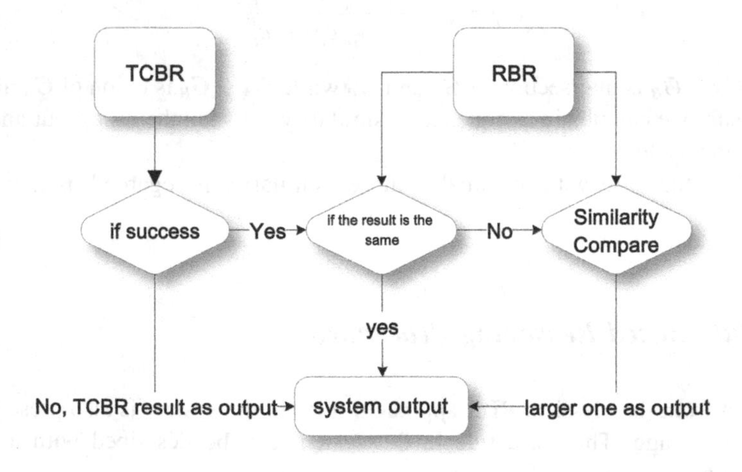

Fig. 3 The procedure used in combing the results of TCBR and RBR

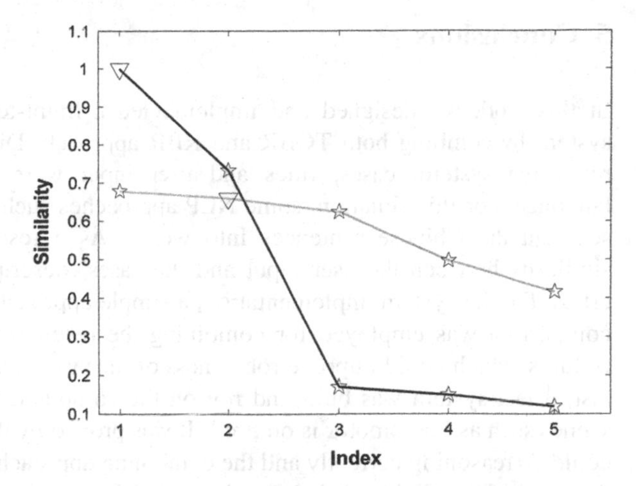

Fig. 4 The similarities of top 5 results from RBR approach (*triangles*) and TCBR approach (*stars*) in the first scene (*solid line*) and the second scene (*dash line*)

This approach is simple but more efficient and meets the need for the information reasoning system. The approach can be concluded in Fig. 3.

4 Experiments

Based on the above mentioned, we implemented the flight-test information reasoning system on the computer. In this system, we employed MySQL as the cases and rules database. For testing purposes, 280 cases and 30 rules out of six categories were encoded and put in the database.

Let us consider the scene: the user wants to acquire the solution for the situation when the aeromotor is on fire. Then he types down the description "red emergency light is on, voice say motor on fire, tail smoke" in Chinese on the input interface of the system. On our system, the result for this input was the case "aeromotor is on fire", which was just the pilot wanted. As the user input was reduced to "red emergency light is on" in Chinese, out system could still reason the same result. What is the difference between these two trails? For details, Fig. 4. shows the top 5 cases descended by similarity from both the TCBR and RBR approach. It is obvious that the system result for the first scene was the result from TCBR procedure while the one for the second scene was from RBR approach. In the first scene, the similarity of the TCBR result was lower cause the case description contains more words than those in the user input, but the properties of the "aeromotor is on fire" included exactly the same words of the user input. In the contrary, the similarity of the TCBR result was higher in the second scene, but much lower than that in the first scene, which we can conclude that the system robustness is improved by combing TCBR and RBR approach.

5 Conclusions

In this work we designed and implemented a flight-test information reasoning system by combing both TCBR and RBR approach. Different from the classical reasoning system, cases, rules and user input were all presented in natural language. For this situation, some NLP approaches such as n-Grams were used to segment the Chinese sentences into words. As a result, we can calculate the similarity between the user input and the cases' descriptions or the rules' properties. For the system implementation, a simple approach based on the similarities comparison was employed for combining the result from TCBR and RBR procedures, which could improve robustness of the system under some conditions. At last, a real system was built and run on the computer, which we tested in some scenes such as "aeromotor is on fire". It was proved by the results that our method could do reasoning correctly and the combining approach we used was efficient for the need of the flight-test information acquisition.

References

1. Rosina O, Weber KA, Stefanie B (2006) Textual case-based reasoning. J Knowl Eng Rev 20:255–260
2. Rissland EL, Daniels JJ (1996) The synergistic application of CBR to IR. J Artif Intell Rev 10:441–475
3. Lenz M, Burkhard HD (1997) CBR for document retrieval the FAllQ project. J Artif Intell 1266:84–93
4. Wiratunga N, Koychev I, Massie S (2004) Feature selection and generalisation for retrieval of textual cases. J Artif Intell 3155:806–820
5. Patterson D, Dobrynin V, Galushka M, Rooney N (2005) SOPHIA: a novel approach for textual case based reasoning. In: Kaelbling LP (ed) Proceedings of the eighteenth international joint conference on artificial intelligence. Morgan Kaufmann Press, San Mateo, pp 15–21
6. Brüninghaus S, Ashley KD (2005) Reasoning with textual cases. J Artif Intell 3620:137–151
7. Lamontagne L, Lapalme G (2004) Textual reuse for email response. J Artif Intell 3155:242–256
8. Katz SM (1987) Estimation of probabilities from sparse data for the language model component of speech recognize. J IEEE Trans Acoust Speech Signal Process 35:400–401

Vehicle Trajectory Collection Using On-Board Multi-Lidars for Driving Behavior Analysis

Huijing Zhao, Chao Wang, Wen Yao, Jinshi Cui and Hongbin Zha

Abstract In order to study the driving behaviors, such as lane change and overtaking, which concerns the relationship between ego and all-surrounding vehicles, it is of great demand in developing an automated system to collect the synchronized motion trajectories that characterize the full course of driving maneuvers in real-world traffic scene. This research proposes a measurement and data processing system, where multiple 2D-Lidars are mounted on a vehicle platform to generate an omni-directional horizontal coverage to the ego-vehicle's surrounding; focusing the driving scenario on motorway, two processing approaches in online and offline procedures are studied for vehicle trajectory extraction by fusing the multi-Lidar data that are acquired during on-road driving. A case study is conducted using a data set collected during 10 min' driving and lasted for 4.1 km long. The performance of trajectory extraction in online and offline procedures is comparatively examined. In addition, a reference vehicle participated in data collection too. The trajectory is analyzed to study its potential in characterizing the situations during vehicle maneuvers.

This work is partially supported by the Hi-Tech Research and Development Program of China (2012AA011801) and the NSFC Grants (61161130528, 91120010).

H. Zhao (✉) · C. Wang · W. Yao · J. Cui · H. Zha
Key Lab of Machine Perception, Peking University, 100871 Beijing, People's Republic of China
e-mail: zhaohj@cis.pku.edu.com

C. Wang
e-mail: wangchao@cis.pku.edu.com

W. Yao
e-mail: yaowen@cis.pku.edu.com

J. Cui
e-mail: cjs@cis.pku.edu.com

H. Zha
e-mail: zha@cis.pku.edu.com

F. Sun et al. (eds.), *Knowledge Engineering and Management*, Advances in Intelligent Systems and Computing 214,
DOI: 10.1007/978-3-642-37832-4_60, © Springer-Verlag Berlin Heidelberg 2014

Keywords Lidar sensing · Moving object detection and tracking · Trajectory collection · Driving behavior · Intelligent vehicle

1 Introduction

Driving is a daily behavior yet a highly complex task. For decades, large amount of research efforts have been devoted to study the mechanism how people perform driving tasks. The procedure is commonly acknowledged as an interactive result among three components, i.e., driver, vehicle, and environment. As the development of Advanced Driver Assistance Systems (ADAS), increasing needs are observed in awareness of a driver/vehicle's current situation, assessing potential risks, predicting future maneuvers etc., where a comprehensive understanding of driving behaviors in complex environments is indispensable.

Many researches have been studied on modeling different driving maneuvers at a tactical level using the data of either real traffic scene or from a driving simulator, where the vehicle's proprioceptive data such as brake, steering, gear, acceleration etc., the driver's gaze/head motion through in-vehicle sensing, as well as the parameters such as lane/geographical position, headway distance to the front vehicle through exteroceptive sensing are normally used [1–3]. However, many models are generated based on current or short period of history data, yielding the subsequent inference and decision, a purely reactive response to the instantaneous situation. While a real driving behavior is much complex, which conceives an action planning based on anticipation to future condition during a certain period, and the situation in which the ego-vehicle evolves and its negotiation with surrounding ones plays a crucial role. In order to study the effect of surrounding traffic characteristics in acceleration and lane change maneuvers, [4] explored the data set by FHWA 1983, which was collected through an aerial photography that covering a four-lane highway section of 997 m long. The data set lasts for one hour with a frame rate of 1 Hz, and each frame contains the observations of the position, lane, and dimensions of every vehicle within the section; [5] explored the trajectory data set, which were extracted from the videos that monitoring two highway sections of 640 and 503 m long in California. The data of each section lasts for 45 min with a frame rate of 10 Hz, and the synchronized trajectories comprising the lane changing maneuvers of 28 heavy vehicles and 28 passenger cars were analyzed. Due to restricted data and sensing techniques in literature, the driving behaviors concerning surrounding vehicles are far from being rigorously studied. It is of great demand in developing an automated sensing system, which acquire large amount of synchronized trajectories that characterize the interactions between subjects and surround vehicles in a variety of scenarios. An on-board system is thus studied in this research.

A key issue in acquiring trajectory data is moving object detection and tracking, which possesses a large body of research efforts. Most systems monitor the ego-vehicle's front zone using visual-approaches [6], Lidar [7, 8], or through multi-modal sensor fusion [9, 10]. In order to study the complex driving behaviors, such as lane change and overtaking, sensing to the ego-vehicle's omni-directional surrounding is important to capture the interactions/situations during the full course of a driving maneuver. An omni-video-based approach is proposed in [11] to generate a panoramic dynamic surrounding map; and a visualization tool is demonstrated in [12] with an iconic representation to the ego-vehicle's vicinity. However, the demonstrated trajectories are fragmentary, providing only short period of history data. The powerful 3D LiDAR, Velodyne HDL-64E [13, 14] is able to generate a high resolution 3D view of the surrounding with a frame rate up to 15 Hz, with which the motions of nearby moving objects can be clearly extracted. However, the issue of sensor cost can not be ignored.

Focusing a driving scenario on motorway, this research proposes an on-board multi-Lidar approach, aiming at collecting the motion trajectories of surrounding vehicles during on-road driving, which characterize the situations during full course of a vehicle maneuver, such as lane changing and overtaking. A measurement system is developed by mounting multiple low-cost 2D Lidars on a vehicle platform to compose an omni-directional Lidar sensing to its local surrounding. An example of the sensor layout can be found in Fig. 1. Two data processing approaches are studied for motion trajectory extraction in online or offline procedures. Experiments were conducted in the ring roads in Beijing. A comparative validation of the two approaches, as well as trajectory examination concerning its accuracy and performance in characterizing the course of driving maneuver, is studied using a data set during 10 min' driving and lasted for 4.1 km long. In the following, we first outline the measurement platform and sensing data in Sect. 2, address two data processing approaches in online and offline procedures for trajectory extraction in Sect. 3, present experimental results and discussions in Sect. 4, followed by conclusions and future work in Sect. 5.

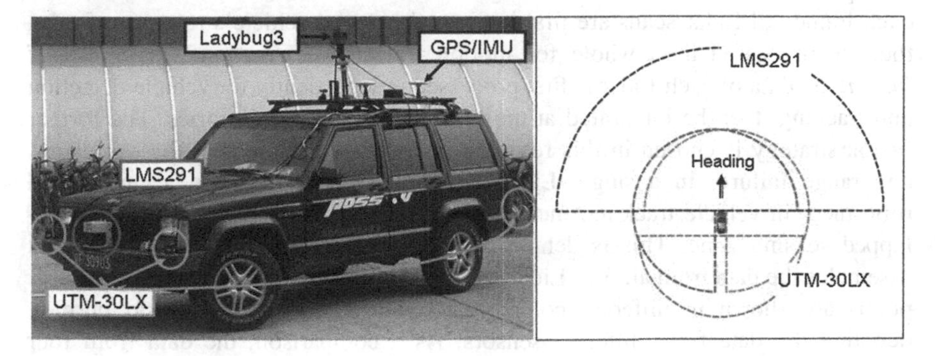

Fig. 1 Vehicle platform and sensor setting

2 Outline of the System

2.1 Measurement Platform

A measurement platform, as shown in Fig. 1, is developed to collect the motion trajectories of surrounding vehicles during its driving using on-board multi-Lidar sensors. Motion of the measurement platform is estimated by a GPS/IMU, which is used to geo-reference the Lidar data to a global reference frame, and output as a trajectory of the subject vehicle. Multiple 2D Lidars are used in the platform to achieve low-cost measurement to all-surrounding vehicles. As shown in Fig. 1, four Lidar sensors are mounted on the platform, where three Hokuyo UTM-30LX are set at the front-left, front-right, and middle-rear, a SICK LMS291 is supplemented at the middle-front bumper to cover a larger semi-circular zone in front of the vehicle. All Lidar sensors scan at a horizontal plane, so as to compose an omni-directional horizontal coverage to the ego-vehicle's vicinity. In such a setting, the Lidars have different viewpoints, and overlapped zone exist between the measurements of neighboring Lidars. After sensor setting, an external calibration is conducted by manually matching the data of different sensors to the same static objects within a horizontal plane, the parameters of which are used to align the laser points of all Lidar sensors to the vehicle's reference frame. An omni-directional vision system, Ladybug 3 (Point Grey Research Inc.), is mounted on the platform too, which is used in this research only for validation of the trajectory extraction algorithm. It will be used in developing a visualization tool for driving behavior analysis through future works.

2.2 Multi-Lidar Sensing Data

Motion trajectories of surrounding vehicles are extracted using multi-Lidar data, with the ego-vehicle's motion compensated by GPS/IMU input. In processing the multiple Lidar data, two fusion strategies are concerned: (1) *low-level fusion*: at each frame, all Lidar scans are first integrated into the vehicle's reference frame, then be processed as a whole for motion trajectory extraction; (2) *high-level fusion*: the data of each Lidar is first processed independently for vehicle detection and tracking, then be integrated at the level of object or trajectories. The former fusion strategy is chosen in this research in order to complement the occlusions and range failures in a single Lidar sensing, so as to improve accuracy and robustness in-vehicle tracking, meanwhile, reduce duplicated detections in over-lapped sensing zone. This is demonstrated in Fig. 2. An integrated frame that assembles the data from all four Lidar sensors is shown in the left sub-figure. Laser points are shown in different colors (e.g., yellow, pink, orange, and purple), denoting the data from different sensors. As a comparison, the data from four individual Lidars are also given in the middle sub-figure. Dotted lines are drawn

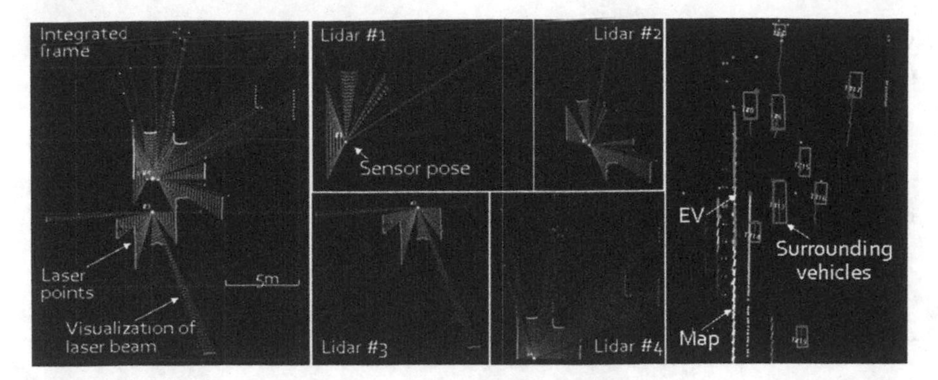

Fig. 2 Multi-Lidar sensing data. *Left* an integration of multiple horizontal Lidar data. *Middle* individual Lidar scans. *Right* A result after vehicle detection and tracking

between the sensor and the laser points, visualizing the laser beams of valid returns. By using multiple Lidars, sensing coverage is greatly enlarged to omni-directions around the ego-vehicle. Moreover, the data from different Lidars compensate each other. By integrating the measurements of multiple Lidars, a more complete sensing to the surrounding vehicles is achieved, enabling a more accurate estimation to the object state. Thus in this research, a low-level fusion of multi-Lidar sensing data is exploited, and approaches of extracting the motion trajectories is developed.

3 Multi-Lidar Processing for Trajectory Extraction

A Lidar scan is a mixture of the data of static and moving objects nearby. In order to extract the motion trajectories of moving objects, discriminating the data of static and moving ones are crucial. If a map of the static environment is known, the data of moving objects can be extracted by subtracting an observation with the map. However, in an unknown environment, an online extraction of moving object trajectories requires that a map be estimated simultaneously, which is formulated as below.

$$p(y_t, m_t | z^t, x^t). \tag{1}$$

where, at frame t, given sequences of observation $z^t = \{z_1, \ldots, z_t\}$ and ego-vehicle's pose $x^t = \{x_1, \ldots, x_t\}$, the problem is to find a simultaneous estimation to the state of surround vehicles y_t and a map of the static obstacles m_t. Based on the Bayes' rule and with the assumption that the observations to moving and static objects are independent, it can be further extended to

$$p(y_t | z^t, x^t, m_t) p(m_t | z^t, x^t). \tag{2}$$

where, $z^t = \{z^{mo}, z^{so}\}^t$ is a composition of the observations to moving and static objects. If they can be discriminated, (2) can further be converted to

$$p(y_t|z^{mo^t}, x^t)p(m_t|z^{so^t}, x^t). \tag{3}$$

In discriminating z^{mo^t} and z^{so^t}, below we study two approaches in online and offline procedures.

3.1 An Offline Approach: MMODT

MMODT—mapping and moving object detection tracking is developed as an offline approach for trajectory extraction, which separates map generation with moving object detect and tracking (MODT) in two subsequent procedures. As described in Fig. 3, a map estimation is conducted in the first step using all measurement data, e.g., *N* frames, where a robotic mapping approach [15] could be applied.

In the second step, after multi-Lidar integration and ego-motion compensation using GPS/IMU input, each Lidar frame is subtracted with the map to extract the data of moving objects, which is forwarded to the module of MODT. In an environment of motorway, they are the data of surrounding vehicles. The algorithm developed in the authors previous work [16] is applied for MODT, where a vehicle model is defined by characterizing the properties of 2D Lidar sensing,

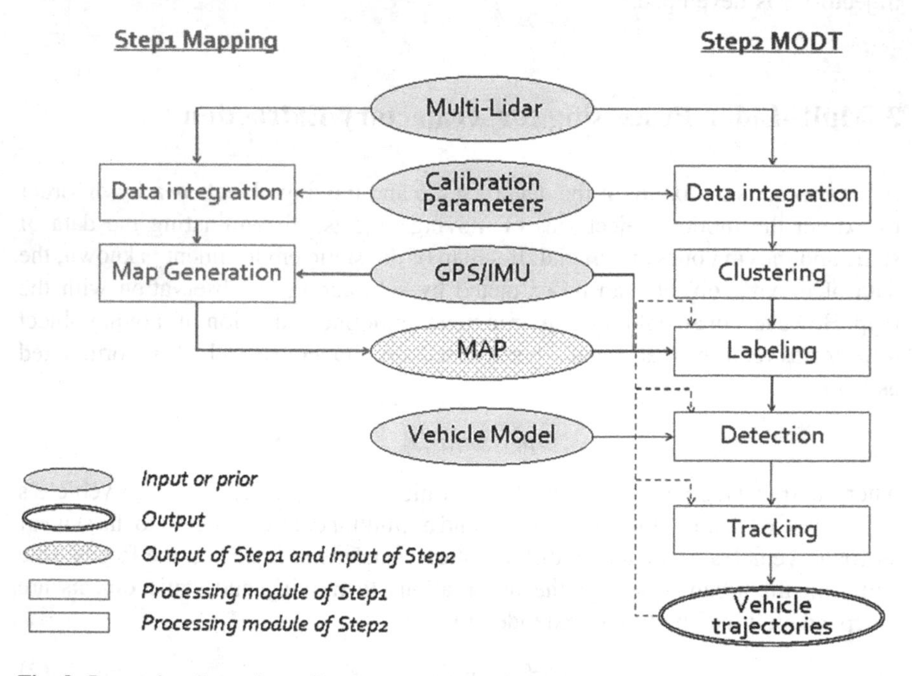

Fig. 3 Processing flow of an offline approach—MMODT

which approximates the horizontal contour of a surrounding vehicle using a rectangle with flexibilities in its width, length, corner, center points etc. The model is used to fit on the extracted Lidar data to detect and track all vehicle candidates.

This approach can also be applied to online processing if a map is given. However, no matter in which processing mode, the performance of trajectory extraction relies greatly on map accuracy. As demonstrated in experimental results (see Fig. 6), in a slow or jammed traffic environment, the data of surrounding vehicles might be misrecognized as those of static objects, degrading map accuracy dramatically. An advantage of the offline processing is that manual correction of map errors is allowed before forwarding it to the second step for MODT.

3.2 An Online Approach: SMMODT

SMMODT—simultaneous mapping and moving object detection tracking is developed as an online approach for trajectory extraction, where in an unknown environment, map and moving objects are estimated simultaneously as described in Fig. 4. As the static and dynamic environments are sensed in an integrated procedure, a practical solution is developed to discriminate the data, so as to divide the tightly correlated estimations in subsequent procedures.

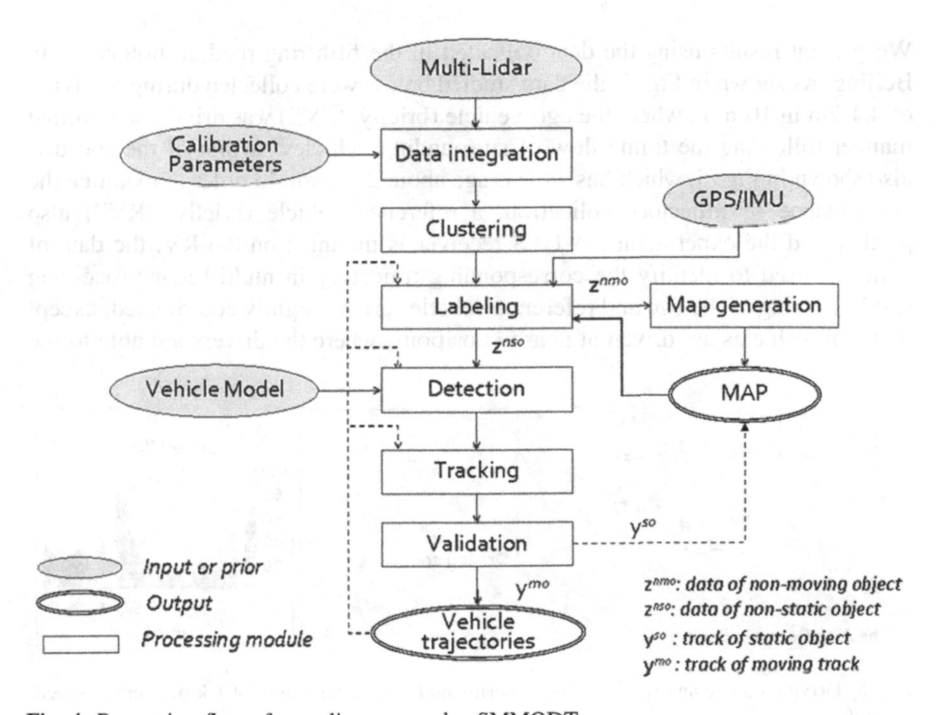

Fig. 4 Processing flow of an online approach—SMMODT

In this research, all candidates of surrounding vehicles are first detected and tracked. Validation is conducted before closing the processing at each frame, where each track is examined on its sequence of motion and shape parameters. If no motion is found on a track during a certain number of frames, the track is considered as a static one, subsequently its data is added into the map. If irregular changes on its motion and shape parameters are detected, the track is disposed as an erroneous estimation. The tracks with obvious motions, regular and reasonable changes in their parameters during a certain number of frames are validated as moving objects, while others are treated as seeds to be validated during future frames. So that the set of tracked objects y_t is divided in two groups, i.e., the tracks that have been validated as moving objects y_t^{mo}, and the seeds waiting for validation in future frame y_t^{so}. Given an observation z_t, the ego-vehicle pose x_t, and state of the validated moving objects y_{t-1}^{mo} at previous frame, z_t^{nso} is predicted of the data of non-static objects and subsequently, z_t^{nmo}, the data of non-moving objects, is estimated by subtracting z_t^{nso} from z_t. Be noticed that z_t^{nso} is a subset of z_t^{mo} and z_t^{nmo} is more than z_t^{so}. The difference comes from the uncertainties in seed objects. The data of z_t^{nmo} and z_t^{nso} are forwarded to modules of map generation and MODT, respectively, which are the same with those in MMODT.

4 Experimental Results and Discussions

We present results using the data collected in the fifth ring road, a motorway, in Beijing. As shown in Fig. 5, the data studied below were collected during a driving of 4.1 km in 10 min, where the ego-vehicle (briefly "EV") was driven in a normal manner following the traffic flow of surrounding vehicles. A plot of the speed is also shown in Fig. 5, which has an average about 27 km/h. In order to examine the performance of trajectory collection, a reference vehicle (briefly "RV") also participated the experiment . A GPS receiver is mounted on the RV, the data of which is used to identify the corresponding trajectory in multi-Lidar processing result. Driving of the ego and reference vehicles are not tightly coordinated, except that both vehicles are driven at nearby positions, where the drivers are able to see

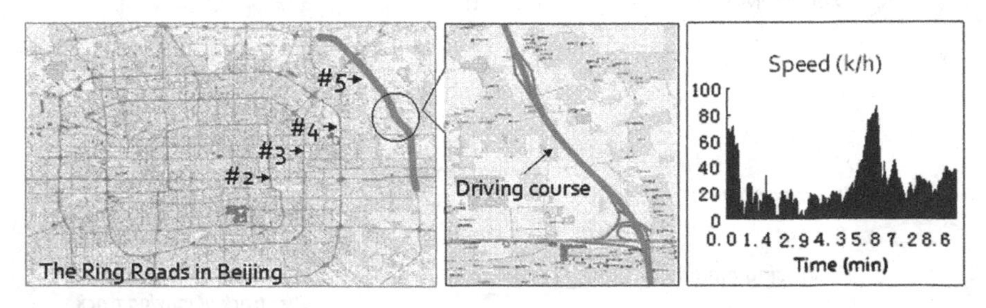

Fig. 5 Driving course and speed of the experimental data, total length: 4.1 km, average speed: 27 km/h

the vehicles of each other. However, it was not a mandatory requirement, and revealed from the experiment that coordinated driving in a normal traffic environment is quite difficult, where the direct insight between two vehicles were broken a number of times due to intrusions of other environmental vehicles. Below we compare the performance in trajectory collection using online and offline approaches, and analyze trajectory concerning its capability in representing full courses of vehicle maneuver.

4.1 Multi-Lidar Processing in Online and Offline Approaches

Two pairs of experimental results are shown in Fig. 6, which compares the results using SMMODT and MMODT in online and offline approaches, respectively. In online approach, a map is generated simultaneously with moving object detection and tracking, while it has the major challenge in discriminating the data of moving and static objects, which were measured in an integrated procedure. Such challenges are described in Fig. 6a and c. Generally speaking, the vehicles at nearby lanes, which are clearly measured by the Lidars, are detected and tracked correctly. While for those occluded or sparsely measured objects, map error and wrong trajectories happened. For example, laser beams may catch the objects such as buildings on road side, which are sparsely measured and failed in generating map data in case of no prior knowledge of the environment. Their data might be fit on the vehicle model, yielding wrong trajectories (see Fig. 6a for an example). On the other hand, when traffic speed is slow, if the motion of a vehicle track is not obvious enough concerning sensing and estimation errors, it is misrecognized as a static object, yielding erroneous map. See Fig. 6c for an example, where vehicle data in the past frames are remained on map, degrading the processing in subsequent frames. On the other hand, the processing results using MMODT in offline approaches are also demonstrated in Fig. 6b and d. As in this approach, map generation is an independent procedure with MODT, optimization of the map

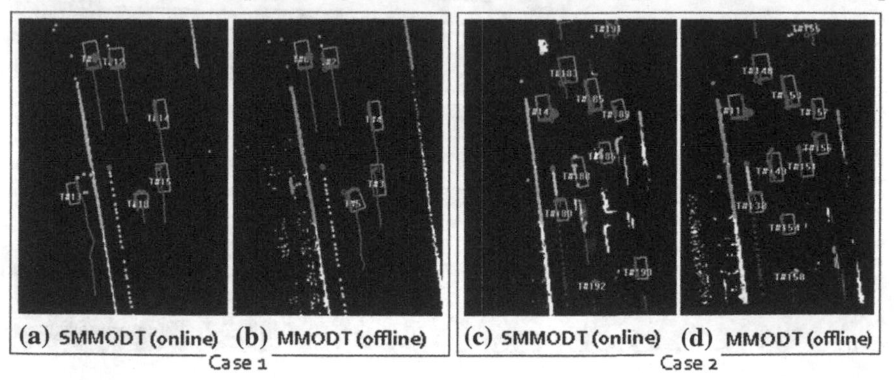

Fig. 6 Multi-Lidar processing results in online and offline procedures

Fig. 7 Visualization and validation the multi-Lidar processing results

using all measurement data, as well as manual editing to correct the erroneous map data, is allowed. Thus the data of surrounding vehicles can be extracted accurately, which is crucial in trajectory estimation.

In order for visualization and validation, multi-Lidar data and processing results are back projected onto the corresponding image frames of an omni-directional vision system, the Ladybug3 as shown in Fig. 1. In Fig. 7, points denote for Lidar data, different colors correspond to different sensors; blue cubes are the validated

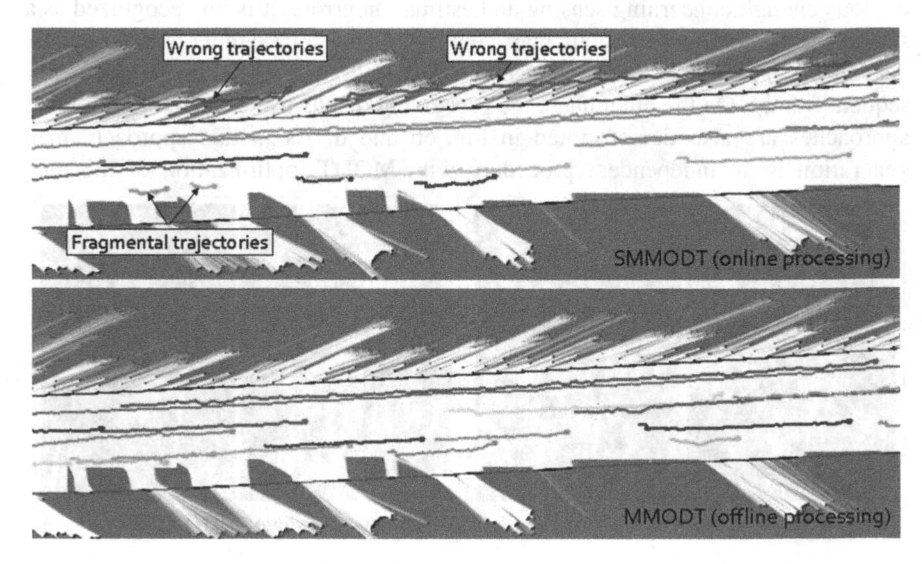

Fig. 8 Trajectory collection results

Table 1 Statistics of the trajectory collection results

		SMMODT (online)	MMODT (offline)
Number of trajectory	Total	482	395
	Invalid	135 (28 %)	
Number of trajectory pnts	Valid	54061	49458
	Invalid	8676 (16 %)	

vehicle tracks, where an universal height value (1.6 m in this result) is associated to the 2D state of each estimated vehicle.

4.2 Trajectory Collection Results

Trajectory collection results on two approaches are shown in Fig. 8 with its statistics listed in Table 1. The trajectories in online approach are manually labeled into valid and invalid, where a valid one must be that of a true vehicle, which is judged based on an assumption that it should not beyond road boundary. In addition, a valid trajectory must be long enough to characterize full courses of vehicle maneuver. Among the 482 trajectories, 135 (28 %) were labeled as invalid. In addition, trajectory length is depicted by its number of trajectory points. 28 % of Invalid trajectories hold 16 % of the total trajectory points, reflecting that most of the invalid trajectories are fragmental ones (Table 1).

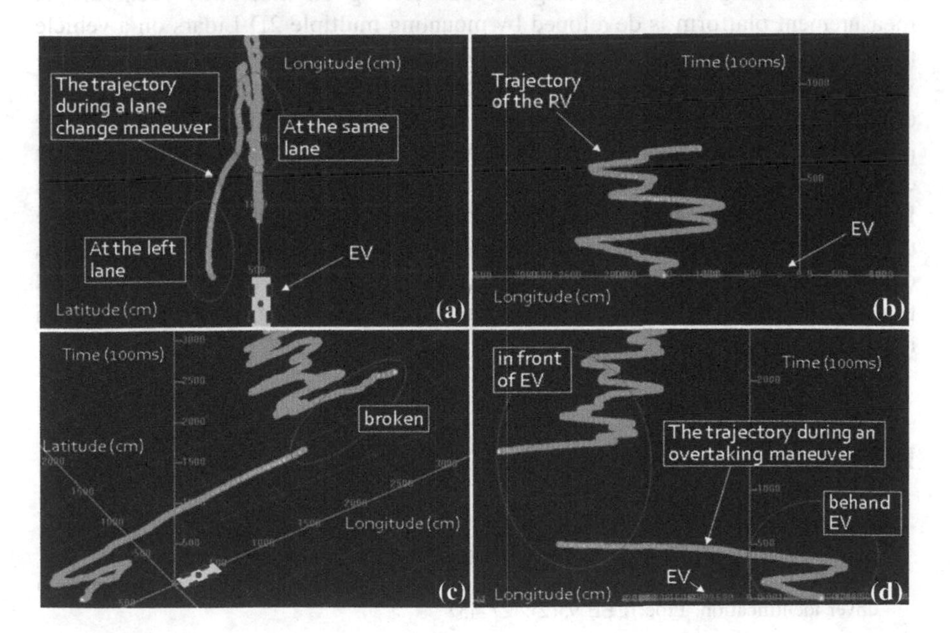

Fig. 9 Trajectory of the reference vehicle

4.3 Performance in Characterizing Complex Driving Behavior

As shown in Fig. 6, the pink dots are the coordinates measured by the GPS receiver on the reference vehicle. Trajectories of the reference vehicle are picked out, consisting of four pieces in the 10 min' data. The first broken is due to irregular jump in GPS positioning when the ego-vehicle went through an overpass, while two others are caused by the intrusion of other environmental vehicles. The trajectories are visualized in Fig. 9 in the ego-vehicle's frame in 3D, i.e., longitude, latitude, and time, which depicts the temporal variation of RV in its horizontal position with respecting to EV. There were many interactions between the ego and reference vehicles during on-road driving experiment. As shown in Fig. 9a, the reference vehicle was in the left-side lane at the beginning, then changed to the same lane with the ego-vehicle, the trajectory of which depicted a course of lane change maneuver. On the other hand, in Fig. 9d, the reference vehicle was in front at the beginning, then changed to the rear-side of the ego-vehicle, the trajectory of which depicted a course of overtaking maneuver.

5 Conclusion and Future Works

In this research, a system is proposed in order for collecting the motion trajectories of all-surrounding vehicles during on-road driving on motorway scenarios. A measurement platform is developed by mounting multiple 2D Lidars on a vehicle platform, which generates an omni-directional horizontal coverage to the ego-vehicle's surroundings. Focusing on motorway scenarios, two vehicle trajectory extraction approaches using multi-Lidar data are developed in online and offline procedures. A comparative study of the two approaches in trajectory extraction is conducted using a data set collected during a 10 min' driving on a ring road in Beijing, which lasted for 4.1 km long. Experimental results are presented, and the trajectory performance in characterizing the situations during vehicle maneuvers is demonstrated. Future studies will focus on developing a visualization tool and trajectory analysis algorithms in understanding and reasoning driving behaviors in real-world scenes.

References

1. Oliver N, Pentland AP (2000) Graphical models for driver behavior recognition in a smart car. In: Proceedings of IEEE intelligent vehicles symposium, pp 7–12
2. Miyajima C et al (2007) Driver modeling based on driving behavior and its evaluation in driver identification. Proc IEEE 95(2):427–437

3. McCall JC, Wipf DP, Trivedi MM, Rao BD (2007) Lane change intent analysis using robust operators and sparse bayesian learning. IEEE Trans Intell Transp Syst 8(3):431–440
4. Toledo T (2003) Integrated driving behavior modeling. Ph.D. thesis. Massachusetts Institute of Technology, Cambridge
5. Moridpour S, Rose G, Sarvi M (2010) Effect of surrounding traffic characteristics on lane changing behavior. J Transp Eng 973–985
6. Sun Z, Bebis G, Miller R (2006) On-road vehicle detection: a review. IEEE Trans Pattern Anal Mach Intell 28(5):694–711
7. Streller D, Dietmayer K, Sparbert J (2001) Vehicle and object models for robust tracking in traffic scenes using laser range images. In: Proceedings of IEEE international conference on intelligent transportation system, pp 118–123
8. Mendes A, Nunes U (2004) Situation-based multi-target detection and tracking with laserscanner in outdoor semi-structured environment. In: Proceedings IEEE/RSJ Conference on intelligent robots and systems, vol 1. pp 88–93
9. Kaempchen N, Dietmayer K (2004) Fusion of laserscanner and video for advanced driver assistance systems. In: Proceedings of IEEE international conference on intelligent transportation system
10. Floudas N, Polychronopoulos A, Aycard O, Burlet J, Ahrholdt M (2007) High level sensor data fusion approaches for object recognition in road environment. In: Proceedings of IEEE intelligent vehicle symposium, pp 136–141
11. Gandhi T, Trivedi M (2006) Vehicle surround capture: survey of techniques and a novel omni-video-based approach for dynamic panoramic surround maps. Proc IEEE Trans Pattern Anal Mach Intell 7(3):293–308
12. Morris, B., Trivedi, M.: Vehicle iconic surround observer: visualization platform for intelligent driver support applications. Proc. IEEE Intelligent Vehicle Symposium, 168–173 (2010)
13. Urmson C et al (2008) Autonomous driving in urban environments: boss and the urban challenge. J Field Robot 25(8):425–466
14. Montemerlo M et al (2008) Junior: the stanford entry in the urban challenge. J Field Robot 25(9):569–597
15. Thrun S (2002) Robotic mapping: a survey. CMU-CS-02-111
16. Zhao H et al (2012) Omni-directional detection and tracking of on-road vehicles using multiple horizontal laser scanners. In: Proceedings of IEEE intelligent vehicles symposium, pp 57–62

A Further Discussion on Cores for Interval-Valued Cooperative Game

Xuan Zhao and Qiang Zhang

Abstract In this paper, we give a further discussion on the core solution for cooperative games with fuzzy payoffs. Some notions and results from classical games are extended to fuzzy cooperative games. Using an example, we point out that the theorem about the nonempty of *I*-core proposed in 2008 was not sufficient. Furthermore, the equivalence relation between balanced game and nonempty core, which plays an important role in classic games, does not exist in interval-valued cooperative games. After all, the nonempty of *I*-core is proved under the convex situation. It perfects the theory of fuzzy core for interval-valued cooperative game.

Keywords Cooperative game · Interval-valued · Core · Balanced game

1 Introduction

Since von Neumann and Morgenstern's pioneering work (Theory of Game Economic Behavior [1]), game theory has been widely used to analyze conflict and cooperative situations in economics, sociology, politics, etc. In classical cooperative game theory, we mainly focus on the forming of coalitions and how they distribute their profits which are precise values. Thus several conceptions of solutions are proposed in distributing the payoffs such as imputation, core, stable set, Shapley value, and so on [2].

While in practice, the incompleteness and uncertainty information in cooperative games is unavoidable. Therefore, the behaviors of players as well as the solutions for the games become an important topic which is called the fuzzy cooperative game theory. Generally speaking, there are three kinds of fuzzy

X. Zhao (✉) · Q. Zhang
School of Management and Economics, Beijing Institute of Technology,
Beijing 100081, China
e-mail: zhaoxuan@bit.edu.cn

F. Sun et al. (eds.), *Knowledge Engineering and Management*,
Advances in Intelligent Systems and Computing 214,
DOI: 10.1007/978-3-642-37832-4_61,

cooperative games. The first kind is the case that players do not fully take part in a coalition, but do to a certain extent with a participation rate from 0 to 1 [3]. The second kind is the case that there is no method to predict the precise payoffs or the payoffs are disputable. In this case the characteristic functions given are fuzzy numbers, thus the values allocated to players are also fuzzy numbers [4]. The third kind is games with fuzzy coalitions and fuzzy payoff. In this paper, our concern is on the second kind of fuzzy cooperative game.

In the literature, many researchers have discussed cooperative games with fuzzy payoffs. As regards the core solution, Mareš [4] put up a core based on the order relation of possibility theory proposed by Dubios and Prade [5], whose calculation however is complex. Gök et al. [6] introduced the core-like solution using the selections of classic games, which is used in the two-person's interval-valued cooperative games. Later, they proposed another core-like solution called I-core and a condition for the nonempty of I-core called I-balanced was introduced [7]. Recently, Mallozzi et al. [8] proposed another core-like solution called F-core for games in which the value of any coalition is given by means of a fuzzy number. It generalizes the situation of classic cooperative games and interval-valued cooperative games.

This paper is organized as follows. In Sect. 2, we recall some basic definitions and results of interval numbers and cooperative games used in this paper. In Sect. 3, the crisp cooperative game is generalized to the situation with interval-valued payoffs. Whether the core solution is nonempty under different conditions is discussed. In Sect. 4, we point out the problems for further discussion.

2 Cores for Classic Cooperative Games

The definitions and conclusions used are from [9–11]. Let R be the set of all real numbers throughout the paper.

Definition 1 A classic cooperative game is a pair $<N, v>$ which contains:

(1) a finite set of players $N = \{1, 2, \ldots, n\}$;
(2) $v : 2^N \rightarrow R$ is a function assigning to each coalition $S \in 2^N$ a real number $v(S)$, such that $v(\emptyset) = 0$. v is called a characteristic function or the payoff of players.

We denote by G^N the family of all classical cooperative games with a player set N. The games referred in this paper are all TU (transferable utility) games without special explain.

Definition 2 A game $v \in G^N$ is superadditive, if

$$v(S \cup T) \geq v(S) + v(T), \quad \forall S, T \subseteq N, S \cap T = \emptyset.$$

A game $v \in G^N$ is convex, if

$$v(S \cup T) + v(S \cap T) \geq v(S) + v(T), \quad \forall S, T \subseteq N.$$

Definition 3 An imputation for the game $<N, v>$ is a vector $x = (x_1, x_2, \ldots, x_n) \in R^n$ satisfying

$$x_i \geq v(\{i\}) \qquad \text{for all } i \in N,$$
$$\sum_{i \in N} x_i = v(N).$$

Definition 4 The core of a game $<N, v>$ is the set of vectors satisfying

$$\sum_{i \in S} x_i \geq v(S) \qquad \text{for all } S \subseteq N,$$
$$\sum_{i \in N} x_i = v(N).$$

We write $v(i)$ instead of $v(\{i\})$ and $x(S)$ instead of $\sum_{i \in S} x_i$ in abbreviation.

Definition 5 Let $N = \{1, 2, \cdots, n\}$, $C_k = \{S_1, S_2, \ldots, S_k\}$ be the collection of nonempty subsets of N. If there exists a real number vector $(\lambda_{S_1}, \lambda_{S_2}, \ldots, \lambda_{S_k})$ satisfies

$$\sum_{S_j \in C_k} \lambda_{S_j} 1_{S_j} = 1_N,$$

where

$$1_{S_j}(i) = \begin{cases} 1 & i \in S_j, \\ 0 & \text{otherwise}, \end{cases}$$

then $(\lambda_{S_1}, \lambda_{S_2}, \ldots, \lambda_{S_k})$ is called a balanced vector of $<N, v>$.

Definition 6 A cooperative game $<N, v>$ is balanced if for any balanced vector $(\lambda_{S_1}, \lambda_{S_2}, \ldots, \lambda_{S_k})$,

$$\sum_{S_j \in C_k} \lambda_{S_j} v(S_j) \leq v(N).$$

Theorem 1 *Let $v \in G^N$ be a superadditive cooperative game, then the core $C(v)$ is nonempty if and only if $<N, v>$ is balanced.*

Remark 1 The proof of this theorem can be seen in [2]. Notice that $<N, v>$ is balanced can be derived from the nonempty of $C(v)$ without the condition that v is superadditive, but the converse is not true. An example is shown to illustrate it.

Example 1 Let $<N, v>$ be a three-persons cooperative game with

$$v(1) = 5, \quad v(2) = 5, \quad v(3) = 5,$$
$$v(\{1,2\}) = 10, \quad v(\{1,3\}) = 6, \quad v(\{2,3\}) = 6,$$
$$v(\{1,2,3\}) = 14.$$

Obviously, v is not superadditive. It is a balanced game as

$$\frac{1}{2}[v(\{1,2\}) + v(\{2,3\}) + v(\{1,3\})] \leq v(\{1,2,3\}).$$

But $C(v)$ is empty since the solution of

$$\begin{cases} x_1 \geq 5 \\ x_2 \geq 5 \\ x_3 \geq 5 \\ x_1 + x_2 \geq 10 \\ x_1 + x_3 \geq 6 \\ x_2 + x_3 \geq 6 \\ x_1 + x_2 + x_3 = 14 \end{cases}$$

doesn't exist.

Theorem 2 *If* $v \in G^N$ *is a convex cooperative game, then the core* $C(v)$ *is nonempty.*

3 Cores for Interval-Valued Cooperative Games

In this section, we try to find out if the theorem in crisp cooperative game theory still holds for cooperative games with interval-valued payoffs.

Definition 7 I is called an interval number if $I = [a, b]$ where $a \leq b$ and $a, b \in R$. Let IR be the set of interval numbers. For any $I = [a, b]$, $J = [c, d] \in IR$, we define

(1) $I + J = [a + c, b + d]$;
(2) $kI = [ka, kb], k > 0$.

Definition 8 The order relations between any two interval numbers are defined as follows, for any $I = [a, b]$, $J = [c, d] \in IR$:

(1) $I \succcurlyeq J$ if $a \geq c$ and $b \geq d$;
(2) $I \preccurlyeq J$ if $a \leq c$ and $b \leq d$;
(3) $I = J$ if $I \succcurlyeq J$ and $I \preccurlyeq J$.

Definition 9 A subset D of R^n is said to be convex if for any $x, y \in D$ and $0 < \lambda < 1$,

$$(1 - \lambda)x + \lambda y \in D.$$

All affine sets (include $\emptyset$ and R^n itself) are convex.

A subset K of R^n is called a cone if it is closed under positive scalar multiplication, i.e., $\lambda x \in K$ when $x \in K$ and $\lambda > 0$. A convex cone is a cone which is a convex set.

A basic result of convex sets is the separating hyperplane theorem.

Theorem 3 *Suppose A and B are two convex sets which do not intersect, i.e., $A \cap B = \emptyset$. Then there exist $\alpha \neq 0$ and b such that*

$$\alpha^T x \leq b$$

for all $x \in A$ and

$$\alpha^T x \geq b$$

for all $x \in B$.

In other words, the affine function $\alpha^T x - b$ is nonpositive on A and nonnegative on B. The hyperplane $\{x \mid \alpha^T x = b\}$ is called a separating hyperplane for the set A and B.

Definition 10 An interval-valued cooperative game is a pair $<N, \omega>$, where

(1) $N = \{1, 2, \ldots, n\}$ is the set of players;
(2) $\omega : 2^N \rightarrow IR$ is a map which assign to each coalition $S \in 2^N$ an interval number, which $\omega(\emptyset) = [0, 0]$. $\omega(s)$ is called an interval-valued character function and is expressed as $[\underline{\omega}(s), \overline{\omega}(s)]$.

We denote by IG^N the family of all interval-valued cooperative games with player set N.

Definition 11 Let $\omega \in IG^N$ be an interval-valued cooperative game, $<N, \omega>$ is superadditive if

$$\omega(S \cup T) \succcurlyeq \omega(S) + \omega(T), \quad \forall S, T \subseteq N, S \cap T = \emptyset.$$

Let $\omega \in IG^N$ be an interval-valued cooperative game, $<N, \omega>$ is convex if

$$\omega(S \cup T) + \omega(S \cap T) \succcurlyeq \omega(S) + \omega(T), \quad \forall S, T \subseteq N.$$

Definition 12 An imputation for the interval-valued game $<N, \omega>$ is an interval number vector $I = (I_1, I_2, \ldots, I_n) \in IR^n$ satisfying

$$I_i \succcurlyeq \omega(\{i\}) \qquad \text{for all } i \in N,$$
$$\sum_{i \in N} I_i = \omega(N).$$

The first condition means I is individual rational since no player will accept less than the minimum which he can obtain only by himself from the point of view of interval numbers. The second condition means I is efficient since the total utility $\omega(N)$ should be divided among all the players.

Definition 13 Let $\omega \in IG^N$ be an interval-valued cooperative game. Then the interval-valued core (I-core in short) is defined as $IC(\omega) = \{(I_1, I_2, \ldots, I_n) \in IR^n \mid \sum_{i=1}^{n} I_i = \omega(N) \text{ and } \sum_{i \in S} I_i \succcurlyeq \omega(S), \forall S \subseteq N\}$.

This implies that the I-core is a set of interval-valued distributions among players satisfying collective rationality and coalition rationality. Therefore, it is reasonable and has practical meaning.

Definition 14 We say that $<N, \omega>$ is I-balanced if for any balanced vector $(\lambda_{S_1}, \lambda_{S_2}, \ldots, \lambda_{S_k})$,

$$\sum_{S_j \in C_k} \lambda_{S_j} \omega(S_j) \preccurlyeq \omega(N).$$

Paper [7] comes to the conclusion that I-balance is equal to the nonempty of I-core (see Theorem 3.1). The example below shows that, unlike the crisp cooperative game, even under the condition ω is superadditive, I-balanced is still not sufficient for the existence of I-core elements.

Example 2 Let $<N, \omega>$ be a three-persons game with

$$\omega(i) = 0, \ (i = 1, 2, 3) \qquad \omega(\{1, 2\}) = [8, 9],$$
$$\omega(\{2, 3\}) = [8, 9], \qquad \omega(\{1, 3\}) = [4, 14],$$
$$\omega(\{1, 2, 3\}) = [10, 16].$$

It is easy to verify that ω is superadditive.

Suppose $([x_1, x_2], [y_1, y_2], [z_1, z_2]) \in IR^3$ is an element in $IC(v)$, then we have

$$\begin{cases} x_i, y_i, z_i \geq 0 \ (i = 1, 2) \\ x_1 + y_1 \geq 8 \\ x_2 + y_2 \geq 9 \\ y_1 + z_1 \geq 8 \\ y_2 + z_2 \geq 9 \\ x_1 + z_1 \geq 4 \\ x_2 + z_2 \geq 14 \\ x_1 + y_1 + z_1 = 10 \\ x_2 + y_2 + z_2 = 16 \end{cases}$$

There is no interval-valued solution of it, thus the I-core is empty.

Since

$$\omega(\{1,2\}) + \omega(\{2,3\}) \not\succeq \omega(\{1,2,3\}) + \omega(2),$$

it is not a convex interval-valued cooperative game.

Theorem 4 *Let* $<N, \omega>$ *be a convex interval-valued cooperative game, then its I-core is nonempty.*

Proof Since ω is convex, we have

$$\underline{\omega}(S \cup T) + \underline{\omega}(S \cap T) \geq \underline{\omega}(S) + \underline{\omega}(T)$$

and

$$\overline{\omega}(S \cup T) + \overline{\omega}(S \cap T) \geq \overline{\omega}(S) + \overline{\omega}(T),$$

for any $S, T \subseteq N$.

According to the crisp game theory, we get $<N, \underline{\omega}>$ and $<N, \overline{\omega}>$ have nonempty cores separately and both are balanced games. Using apagoge, we suppose there is no balanced vector $(\lambda_{S_1}, \lambda_{S_2}, \ldots, \lambda_{S_k})$ satisfying

$$\sum_{S_j \in C_k} \lambda_{S_j} \underline{\omega}(S_j) > \underline{\omega}(N) \tag{1}$$

or

$$\sum_{S_j \in C_k} \lambda_{S_j} \overline{\omega}(S_j) > \overline{\omega}(N). \tag{2}$$

Thus the sets

$$V_1 = \{y_L \in R^{n+1} \mid y_L = (1_n, \ \underline{\omega}(N) + \varepsilon), \ \varepsilon > 0\}$$

and

$$V_2 = \{z_L \in R^{n+1} \mid z_L = \sum_{S_j \in C_k} \lambda_{S_j}(1_{S_j}, \underline{\omega}(S_j)), \ \forall S_j \in C_k \ \text{ and } \ \lambda_{S_j} > 0\}$$

are both convex and disjoint.

It is easy to verify that V_1 and V_2 are convex sets, besides V_2 is a convex cone. In the following, we prove $V_1 \cap V_2 = \emptyset$.

Suppose their intersection set is not empty, we can get

$$1_n = \sum_{S_j \in C_k} \lambda_{S_j} 1_{S_j} \tag{3}$$

and

$$\underline{\omega}(N) + \varepsilon_0 = \sum_{S_j \in C_k} \lambda_{S_j} \underline{\omega}(S_j), \tag{4}$$

where $\varepsilon_0 > 0$.

Equation (3) indicates $(\lambda_{S_1}, \lambda_{S_2}, \ldots, \lambda_{S_k})$ is a balanced vector. Equation (4) shows that

$$\sum_{S_j \in C_k} \lambda_{S_j} \underline{\omega}(S_j) > \underline{\omega}(N).$$

It is a contradiction with Eq. (1). Therefore, V_1 and V_2 are disjoint.

In the same way, we conclude

$$V_3 = \{y_R \in R^{n+1} \mid y_R = (1_n,\ \overline{\omega}(N) + \varepsilon),\ \varepsilon > 0\}$$

and

$$V_4 = \{z_R \in R^{n+1} \mid z_R = \sum_{S_j \in C_k} \lambda_{S_j}(1_{S_j}, \overline{\omega}(S_j)),\ \forall S_j \in C_k \quad \text{and} \quad \lambda_{S_j} > 0\}$$

are both convex and disjoint.

Next, we prove that there is an element in I-core of $<N, \omega>$. By the separating hyperplane theorem, there is a nonzero vector $(\alpha_1, \alpha_2, \ldots, \alpha_n, \alpha) \in R^{n+1}$, for all $z_L \in V_2$, such that

$$(\alpha_1, \alpha_2, \ldots, \alpha_n, \alpha) \cdot z_L^T \geq 0 \tag{5}$$

and

$$(\alpha_1, \alpha_2, \ldots, \alpha_n, \alpha) \begin{pmatrix} 1_n^T \\ \underline{\omega}(N) + \varepsilon \end{pmatrix} < 0. \tag{6}$$

Since $<N, v_L>$ is balanced, $(1_n,\ \underline{\omega}(N)) \in V_2$, then,

$$(\alpha_1, \alpha_2, \ldots, \alpha_n, \alpha) \begin{pmatrix} 1_n^T \\ \underline{\omega}(N) \end{pmatrix} > (\alpha_1, \alpha_2, \ldots, \alpha_n, \alpha) \begin{pmatrix} 1_n^T \\ \underline{\omega}(N) + \varepsilon \end{pmatrix}.$$

The inequality above equals

$$\alpha \underline{\omega}(N) > \alpha \underline{\omega}(N) + \alpha \varepsilon.$$

For $\varepsilon > 0$, we see $\alpha < 0$.

Let

$$x'_L = -\frac{1}{\alpha}(\alpha_1, \alpha_2, \ldots, \alpha_n).$$

For $(1_{S_j}, \underline{\omega}(S_j)) \in V_2$, $\forall S \subseteq N$, from (6) we know

$$x'_L(S) \geq \underline{\omega}(S).$$

From the definition of x'_L and (5) we get

$$\underline{\omega}(N) \geq -\sum_{i=1}^{n} \alpha_i/\alpha = x'_L(N).$$

By modifying the vector x'_L, it is easy to get a vector $x_L \doteq (x_{1L}, x_{2L}, \ldots, x_{nL})$ satisfying

$$x_L(S) \geq \underline{\omega}(S), \quad \forall S \subseteq N$$

and

$$\underline{\omega}(N) = x_L(N).$$

As V_3 and V_4 are disjoint, similarly, there is a nonzero vector $x_R \doteq (x_{1R}, x_{2R}, \ldots, x_{nR})$ which is larger than x_L satisfying

$$x_R(S) \geq \overline{\omega}(S), \quad \forall S \subseteq N$$

and

$$\overline{\omega}(N) = x_R(N).$$

Thus $x = ([x_{1L}, x_{1R}], [x_{2L}, x_{2R}], \ldots, [x_{nL}, x_{nR}]) \in IR^n$ is in the $C(\omega)$.

Remark 2 Notice that the convexity of ω we defined is different from that in [11]. In that paper, the subtractions between interval numbers are defined although it does not always exist. As a result, some conclusions can be deduced only when the difference between the upper and lower boundaries is convex. We improve the results here.

Example 3 Let $<N, \omega>$ be a three-persons game with

$$\omega(1) = [0,1], \qquad \omega(2) = [0,2], \qquad \omega(3) = [0,3],$$
$$\omega(\{1,2\}) = [7,8], \qquad \omega(\{2,3\}) = [1,6], \qquad \omega(\{1,3\}) = [2,4],$$
$$\omega(\{1,2,3\}) = [9,13].$$

Since

$$\frac{1}{2}[\omega(\{1,2\}) + \omega(\{2,3\}) + \omega(\{1,3\})] \leq \omega(\{1,2,3\}),$$

it is a convex interval-valued cooperative game.

It is easy to verify that $([3,4],[4,5],[2,4])$ is an element in the $C(\omega)$.

4 Conclusions and Problems for Further Discussion

Under the condition of convex, we get the existence the interval-valued imputation of players satisfying collective rationality and coalition rationality. It is worthwhile to mention that we focus on the uncertainty of payoffs when it is given by means of only the lower and upper bonds without the intermediate values. In paper [8], it extends to the fuzzy numbers of payoff functions. Whether the condition mentioned in our paper is sufficient for the nonempty of F-core proposed in [8] is unknown.

Acknowledgments This work was supported by the National Natural Science Foundation of China and Specialized Research Fund for the Doctoral Program of Higher Education (No. 70771010, 71071018, 70801064, 20111101110036).

References

1. von Neumann J, Morgenstern O (1944) Game theory and economic behavior. Princeton University Press, Princeton
2. Owen G (1995) Game theory, 3rd edn. Academic Press, New York
3. Tijs S, Branzei R, Ishihara S, Muto S (2004) On cores and stable sets for fuzzy games. Fuzzy Sets Syst 146:285–296
4. Mareš M (2001) Fuzzy cooperative games: cooperation with vague expectations. vol 72, Springer, Heidelberg
5. Dubois D, Prade H (1983) Ranking fuzzy numbers in the setting of possibility theory. Inf Sci 30:183–224
6. Alparslan-Gok S, Miquel S, Tijs S (2009) Cooperation under interval uncertainty. Math Methods Oper Res 69:99–109
7. Alparslan-Gok S, Branzei R, Tijs S (2009) Cores and stable sets for intervalvalued games. Tilburg University
8. Mallozzi L, Scalzo V, Tijs S (2011) Fuzzy interval cooperative games. Fuzzy Sets Syst 165:98–105
9. Branzei R, Dimitrov D, Tijs S (2008) Models in cooperative game theory, vol 556. Springer, Berlin
10. Klir G, Yuan B (1995) Fuzzy sets and fuzzy logic. Prentice Hall, Upper Saddle River
11. Gok S, Branzei R, Tijs S (2008) Convex interval games

Current Status, Challenges, and Outlook of E-Health Record Systems in Australia

Jun Xu, Xiangzhu Gao, Golam Sorwar and Peter Croll

Abstract Australia is among a few countries that have developed national electronic health (e-health) record systems. From July 1, 2012, Australians can register with the Personally Controlled Electronic Health Records (PCEHR) system. Development of e-health systems is supported by stakeholders in Australia, however, the PCEHR system has been widely criticized. For the establishment of a matured e-health record system, this paper examines current status of Australia's development of e-health record systems, identifies the challenges encountered by the development, and analyses the outlook of the PCEHR system.

Keywords E-health · E-health record systems · PCEHR · Australia

1 Introduction

Systematic e-health records provide great opportunities to improve the quality and safety of healthcare, reduce costs, improve continuity and health outcomes for patients, save lives, time and money, make Australian health system more efficient, and provide every Australian with equitable access to healthcare [1, 2]. Technologically, Australia possesses the necessary foundation for implementing a national e-health system. On the one hand, almost all GPs and pharmacies are computer-assisted. Most public hospitals are in various stages of computerization/ digitization [3]. The majority of allied health professionals and medical specialists have regularly used computers for accessing online clinical reference tools, online training and education, billing and patient rebates, patient booking and scheduling as well as viewing diagnostic imaging results [4, 5]. More than 96 % GPs have

J. Xu (✉) · X. Gao · G. Sorwar · P. Croll
Southern Cross University, Lismore, Australia
e-mail: jun.xu@scu.edu.au

F. Sun et al. (eds.), *Knowledge Engineering and Management*,
Advances in Intelligent Systems and Computing 214,
DOI: 10.1007/978-3-642-37832-4_62, © Springer-Verlag Berlin Heidelberg 2014

access to the Internet in offices. 95 % of GPs use electronic patient medical records [2]. On the other hand, more than 80 % of Australians have access to the Internet [6]. The widely use of the Internet in Australian public contributes to the foundation for e-health system.

Consideration of e-health record systems in Australia started in 2000 [7]. MediConnect completed its field test in Launceston and Ballarat in 2004. This system-integrated medication data of healthcare providers for sharing data among doctors, pharmacists, and consumers. Key components of MediConnect would be incorporated into HealthConnect, which would be Australia's e-health system, according to [8]. The HealthConnect program, which completed in 2009, eventually consisted of a set of different projects undertaken individually by all the states/territories except Queensland. The program provided significant gains in the establishment of e-health infrastructure across Australia [9]. Since 2010, the Australian government has invested $467 million in the first release of the national Personally Controlled Electronic Health Records (PCEHR) system, which delivers core functionality and will grow over time [7]. The provider porter for health service providers is still being developed. More functions, elements, and links will be developed and made available gradually. Currently, many medical practices are updating or waiting for updated IT infrastructure for the system.

According to an estimate based on the economic modeling undertaken by Deloitte Consulting in 2010–2011, the PCEHR system could generate, approximately, $11.5 billion in net direct benefits over the period of 2010–2025, with $9.5 billion to Australian governments and $2 billion to the private sector including households, GPs, specialists, allied health clinics, private hospitals, and private health insurance providers [10]. Through innovative e-health initiatives like the PCEHR system, Australia will have developed legislations and infrastructure that will benefit future e-health projects [11]. A close example is the benefits of network infrastructure projects invested around 2000. According to Peter Fleming, CEO of the National E-health Transition Authority (NEHTA), Australia is expected to spend 20 % of gross domestic product on healthcare by 2020, and the estimated savings arising from e-health initiatives are $7 billion.

To achieve the benefits, there is a long way to go. The current release of the PCEHR system is far from mature and suffers criticisms from major stakeholders. The system is also facing the social challenges. Individuals are not enthusiastic in registration with the system.

2 Current Status

2.1 Adoption of E-health Record

On average, every year an Australian has 22 interactions with healthcare providers, including 4 visits to general practitioner (GP), 12 prescriptions, 3 visits to specialists, and most of the information of these interactions has been held in paper-based files or non-shared databases. The medical information may be inconsistent between files, inaccurate because of the lack of standards, incorrect because of manual operation, and is not available for emergent situations. According to [1, 2], 18 % of medical errors in Australia occur from inadequate information; nearly 30 % unplanned hospital admissions are associated with prescribing errors; approximately 13 % of healthcare provider consultation suffers missing information. One patient may have to undergo the same test for healthcare service from different providers. It is not usual that same questions are asked every time when people deal with a different healthcare provider. E-health records include patients' prescribed medication, test results, care plans, immunisations, and health alerts. Real-time and convenient access to such information will no doubt improve the quality of healthcare, especially in emergency situations and special conditions [12].

Reference [13] studied the use of health information technology in seven countries including US, Canada, UK, Germany, Netherlands, Australia, and New Zealand. According to their report, Australia, along with UK, Netherlands, and New Zealand, has nearly universal use of e-health records among GPs (more than 90 %), but the adoption rate of e-health records in hospitals is less than 10 %. Reference [13] argued that only those nationals willing to put in significant investments and take up the challenge of developing standards and interchanges will be able to succeed in their e-health record systems and provide the benefits of e-health records to their people. Australians are among these nationals. According to [14], Australian e-health spending in 2012 is over $2 billion. For the PCEHR system, on top of the $467 million invested for the period of July 2010 to June 2012, Australian government has allocated further $233.7 million for the period of July 2012 to June 2014 [15].

2.2 National Approach

E-health record initiatives have been widely viewed as an opportunity to have fundamental improvement in the public health sector. Most industrial countries have adopted the approach of national e-health record system [16]. Australia has taken a national approach in implementing an individual e-health record system. The Australian public has demonstrated strong support to the establishment of e-health record system and the national approach. An NEHTA study [17] showed that more than 80 % of nationwide 2,700 participants strongly supported the idea of establishing an individual e-health record system, and 90 % believed that the federal government should manage the implementation and operation of the system. In the same study, major advantages of such a system are identified, which include: (1) immediate access to important medical information to save lives and improve health services, (2) ease for patients, especially chronic illness sufferers, to know that their records are updated, accurate and readily available, and (3) ubiquitous health records (health records can be accessed everywhere). Major disadvantages include: (1) security breach of personal health information by hackers, (2) inappropriate use of personal health information (e.g., by insurance firms or employers against their clients or employees), and (3) high costs.

2.3 Progress of PCEHR

Main events leading to releasing the PCEHR system are as follows.

- In 2005, the NEHTA was established to coordinate the national e-health record project.
- In 2008, the National E-health Strategy was developed to present a blueprint for the national e-health record project.
- In 2010, development of the PCEHR system started based on the MediConnect and HealthConnect.
- In 2010, the Healthcare Identifiers Act 2010 was issued, which ensures that healthcare is matched to the health information that is created when the healthcare is provided. The correct match is achieved by assigning an identifier to each healthcare provider or healthcare recipient. Without obtaining the healthcare identifier, a healthcare provider or healthcare recipient cannot use the PCEHR system.
- In 2010, pilot-testing of the PCEHR system was conducted in 12 nationally selected sites. The participants are mothers and newborns, people with chronic disease, and those in aged and palliative care settings [1].
- In 2011, Privacy Impact Assessment Report regarding PCEHR was prepared for the Department of Health and Aging. The report analyses possible impacts on the privacy of personal information in the PCEHR system and recommends options for managing, minimizing, or eradicating negative impacts.
- In 2011, the Australian government released Concept of Operations: Relating to the Introduction of a Personally Controlled Electronic Health Record System. This document provides an overview of the PCEHR system, identifies stakeholders, and describes how the system would work.
- In 2012, the Practice Incentives Program (PIP) e-Health Incentive was announced by the Department of Health and Aging. General practices will be required to participate in the PCEHR program or become ineligible for the eHealth PIP [18].
- In 2012, the Personally Controlled Electronic Health Records Regulation 2012 was released to support the effective operation of the PCEHR system.
- The PCEHR system was released on 1 July 2012. Australians can register with the system, create and manage their personal health record online.

3 Challenges and Issues

3.1 Impact of Policy and Economic Uncertainty

Government policies affect the development of e-health record systems. Although both the current government and the opposition support e-health, the opposition has been debating against the PCEHR system, which was implemented by the current government. The federal opposition spokesman told eHealthspace.org that if coalition had won government in 2007, it would have done things differently for an e-health system [19]. If the opposition wins the next election in 2013, Australia's e-health record system may become obviously different from what it is today. The change in state/territory governments also impacts the delivery, management, and success of the PCEHR system. In addition, the uncertainty of Australia's economy, especially the unstable commodity market, will affect governments' financial ability and determination to see the project through, which will be a very long, expensive, and daunting journey. On the other hand, the full benefits of the PCEHR system will be realized by all the stakeholders over the next 10 years, according to [20].

3.2 Issue of Unified Approach

Federal government, state/territory governments, and local governments legislate or determine laws, regulations, and/or code of practices. Differences exist in privacy policies, community services, and other factors, which affect the PCEHR system. A unified approach should be adopted to ensure the successful implementation of the national PCEHR [21].

While the PCEHR approach provides the required leadership and the focus on delivering an integrated system nationally, it could very likely ignore certain important local elements. According to Deutsch [16], a review report by Boston Consulting Group had revealed that inadequate stakeholder management and regional cooperation is a problem with the PCEHR project. Looking after local needs is very critical for the success of the national PCEHR system since state/territory governments are managers and operators of actual public health services and are regulating local health professionals. A good balance has to be achieved, and a good combined solution of localization and centralization should be adopted, but without losing the focus on integration, interoperability, shared learning, and aligned implementation efforts [22].

3.3 Effectiveness and Usability Issues

Effectiveness and usability are the two most important characteristics for the success of any information systems. A good e-health record system should be effective to achieve its objectives and must be usable (ease of use). In a survey of 790 staff from 65 hospitals in Australia [23], usability was identified as a significant factor associated with use and frequency of use of e-health applications. The PCEHR system went live when it was far from mature [19]. The current release of the PCEHR system is not properly functioning and not easy to use. The recent unstable performance of the system has caused frustration among potential users. Many people simply quit during the registration process, which is not simple and streamlined, needs to go through a few steps with different websites, and requires latest medical information of the registrant. Confronting with users' complaints, NEHTA told that some PCEHR bugs would be fixed after go-live [24].

3.4 Opt-in Versus Opt-out Models

The PCEHR is an opt-in system, which allows individuals to make an explicit decision to participate. According to an analysis [25] of Australian government, personal privacy is the top concern for the PCEHR system. The privacy concern is a main reason for many Australians not to participate. Therefore, the opt-in system has made it a tough challenge to achieve a critical mass of adopters. According to [24, 26], only 803 people registered in the first week, and 4500 people in the first month all over Australia after the commencement of the PCEHR system. Medical professionals have reckoned that an e-health record system should be an opt-out one, which includes all Australians by default and allows individuals to quit if they decide not to participate [27, 28]. If this is the case, 'personally controlled' would have to be removed from current name of the system.

Moreover, there is a large private health sector, where some health practitioners view patient records as competitive information [29]. They do not share the information with others, and do not encourage the registration with the system.

According to one recent study on Australian public attitude and views toward privacy in healthcare, King et al. [30] pointed out that the most sensitive items of medical record have to be managed with extra caution, including sexually transmitted disease, abortion and infertility, family medical history, genetic disorder, mental illness, drug/alcohol incidents, and lists of previous operations/procedures. NEHTA [7] especially explained the security measures of the PCEHR system to protect privacy so as to remove the privacy concern, but the explanation is obviously not effective. Further PCEHR education for specific concerns, including privacy and usability, is necessary to encourage mass registration.

3.5 Issue of Transparent NEHTA Operation

There is a long way to go for the PCEHR system to be in its full operation, and it remains unclear when the project will be completed [24]. In the remaining process, the NEHTA should improve its operation. The PCEHR project has been blamed for lack of transparency. In the 2007 report by Boston Consulting Group, poor transparency and lack of communication regarding the operations of NEHTA was among its four major problems [16]. During the past 5 years, NEHTA has acted on this issue, however, it can be argued that there is still a lot of room for NEHTA to improve. In a recent Aus Health IT Poll, for the question 'Is government/NEHTA being transparent enough regarding the progress with the NEHRS/PCEHR?' 95 % of the participants choose 'They are excessively secretive' and the rest 5 % choose 'I have no idea'.

4 Outlook and Conclusion

4.1 Mobility and Social Computing

Accessing and managing personal health information via mobile devices should be taken into consideration. This consideration is essential for rural areas and remote communities, where physical Internet connection is not available, and is significant for the growing smartphone population. Currently, 52 % of Australians own smartphones [31]. Some healthcare providers have tapped into mobile health (m-health) records, such as the reported successful case of m-health record system implemented in Kimberly, Western Australia [32]. Like smartphones, social network sites and online communities, such as Twitter, Linked-in, Facebook, You-tube, and Wikis, have been widely used for sharing information, communication, co-creating content, cooperation, and collaboration. Nowadays, health social networks and online communities currently are mainly used for sharing ideas, discussing symptoms, and debating treatment options [33]. In the future, creative solutions for accessing health information via social networks and online communities should be explored for better efficiency and effectiveness of healthcare. There is a need for the development of regulations on and protection measures for accessing and managing health information via mobile devices and social networks.

4.2 Wider Adoption and Continued Use

By looking at international experiences of implementing national e-health record system (e.g., US, UK) it can be predicted that the uptake of PCEHR system will be slow because of the challenges and issues described in the previous section.

However, the NEHTA has been engaged in public consultations and refining the PCEHR system. The Concept of Operations [7] will be periodically updated with the further development of the PCEHR system. It is expected that the usability problem will be solved in the near future, and the privacy concern will be removed gradually.

From February 2013, an e-health PIP will only be available for medical practices that implement the PCEHR [34]. This indicates that PCEHR is mandatory for e-health PIP. In addition, the government will fund general practices to develop their e-health capability [18]. These incentive measures will motivate healthcare providers in using the system.

More importantly, when the users realize the benefits from the system, they will adopt and use it.

Acknowledgments This research is supported by Australian Government under the Australia–China Science and Research Fund.

References

1. Australian Nursing Journal (2011) News: national e-health rollout on. Aust Nurs J 18(10):5
2. Burmester S (2012) Review the progress of eHealth in Australia, presentation by national e-health transition authority, 20/03/2012, http://www.nehta.gov.au/publications/whats-new. Accessed 12 Mar 2012
3. Coiera EW, Kidd MR, Haikerwal MC (2012) Editorials: a call for national e-health clinical safety governance. Med J Aust 196(7):430–431
4. Department of Health and Aging (2011a) The eHealth readiness of Australia's allied health sector, department of health and aging, May 2011, http://www.health.gov.au/internet/publications/publishing.nsf/Content/CA2578620005D57ACA2579090014230A/$File/Allied%20Health%20ehealth%20readiness%20survey%20report.pdf. Accessed 20 Sept 2012
5. Department of Health and Aging (2011b) The eHealth readiness of Australia's medical specialists, department of health and aging, May 2011, http://www.health.gov.au/internet/publications/publishing.nsf/Content/CA2578620005D57ACA25790900158A0A/$File/Medical%20Specialist%20ehealth%20readiness%20survey%20report.pdf. Accessed 20 Sept 2012
6. Internetworldstats.com (2012) Australia: Internet usage stats and telecommunications market report, http://www.internetworldstats.com/sp/au.htm. Accessed 20 Sept 2012
7. NEHTA (2011) Concept of operations: relating to the introduction of a personally controlled electronic health record system. http://www.yourhealth.gov.au/internet/yourhealth/publishing.nsf/Content/PCEHRS-Intro-toc/$File/Concept%20of%20Operations%20-%20Final.pdf. Accessed 26 Sept 2012
8. Medicare (2012) MediConnect. 24/08/2012, http://www.medicareaustralia.gov.au/provider/patients/mediconnect.jsp. Accessed 27 Sept 2012
9. Department of Health and Aging (2009) Healthconnect evaluation, http://www.health.gov.au/internet/main/publishing.nsf/Content/B466CED6B6B1D799CA2577F30017668A/$File/HealthConnect.pdf. Accessed 27 Sept 2012
10. Department of Health and Aging (2012) Expected benefits of the national PCEHR system, Department of Health and Aging, May 2012, http://www.yourhealth.gov.au/internet/yourhealth/publishing.nsf/content/pcehr-benefits#.UGARmVElp8F. Accessed 20 Sept 2012

11. Glance D (2012) Everything you need to know about Australia's e-health record, The conversation, 01/03/2012, http://theconversation.edu.au/everything-you-need-to-know-about-australias-e-health-records-5516. Accessed 13 Aug 2012
12. Townsend R (2012) Doctors and patients uneasy about new e-health record system, The conversation, 05/07/2012, http://theconversation.edu.au/doctors-and-patients-uneasy-about-new-e-health-records-system-7706. Accessed 13 Aug 2012)
13. Jha AK, Doolan D, Grandt D, Scott T, Bates DW (2008) The use of health information technology in seven nations. Int J Med Inf 77:848–854
14. Lohman T (2010) Australian e-health spending to top $2 billion in 2010. 15/04/2010, http://www.computerworld.com.au/article/343220/australian_e-health_spending_top_2_billion_2010/. Accessed 27 Sept 2012
15. Dearne K (2012) Sceptics warn of risks and inadequacies in shared e-health records system, The Australian, 30/06/2012, http://www.theaustralian.com.au/australian-it/government/sceptics-warn-of-risks-and-inadequacies-in-shared-electronic-health-records/story-fn4htb9o-1226412887806. Accessed 11 Aug 2012
16. Deutsch E, Duftschmid G, Dorda W (2010) Critical areas of national electronic health record programs: Is our focus correct. Int J Med Inf 79:211–222
17. NEHTA (2008) Report for quantitative survey in July 2008. August 2008, http://www.nehta.gov.au/index.php?option=com_docman&task=doc_details&gid=585&Itemid=139&catid=130. Accessed 16 Aug 2012
18. McDonald K (2012) Use it or lose it—PCEHR for PIP', *PULSE + IT*, http://www.pulseitmagazine.com.au/index.php?option=com_content&view=article&id=983:use-it-or-lose-it-pcehr-for-pip&catid=16:australian-ehealth&Itemid=328. Accessed 25 Sept 2012
19. Gliddon J (2012) Opposition votes for ehealth, slams PCEHR, eHealthSpace.org, 15/02/2012, http://ehealthspace.org/news/opposition-votes-ehealth-slams-pcehr. Accessed 25 Sept 2012
20. NEHTA (2012) About the PCEHR system, http://www.ehealthinfo.gov.au/personally-controlled-electronic-health-records/about-the-pcehr-system. Accessed 25 Sept 2012
21. Dearne K (2011) Personally controlled electronic health record system coming, The Australian, 23/11/2011, http://www.theaustralian.com.au/australian-it/government/personally-controlled-electronic-health-record-system-coming/story-fn4htb9o-1226203867730. Accessed 14 Aug 2012
22. Sheikh A, Cornford T, Barber N, Avery A, Takian A, Lichtner V, Petrakaki D, Crowe S, Marsden K, Robertson A, Morrison Z, Klecun E, Prescott R, Quinn C, Jani Y, Ficociello M, Voutsina K, Paton J, Fernando B, Jacklin A, Cresswell K (2011) Implementation and adoption of nationwide electronic health records in secondary care in England: Final qualitative results from perspective national evaluation in early adopter hospitals. BMJ Group 2011, http://www.bmj.com/content/341/bmj.c4564. Accessed 16 Aug 2012
23. Gosling AS, Westbrook JI (2004) Allied health professionals' use of online evidence: a survey of 790 staff working in the Australian public hospital system. Int J Med Inf 73:391–401
24. Foo F (2012) Some PCEHR bugs will be fixed after go-live, says NEHTA, The Australian, 01/08/2012, http://www.theaustralian.com.au/australian-it/government/some-pcehr-bugs-will-be-fixed-after-go-live-says-nehta/story-fn4htb9o-1226440424847. Accessed 26 Sept 2012
25. eHealth Division (2011) Draft concept of operations: relating to the introduction of a PCEHR system', http://www.yourhealth.gov.au/internet/yourhealth/publishing.nsf/Content/CA2578620005CE1DCA2579040005A91C/$File/Draft%20Concept%20of%20Operations%20Feedback%20Report.pdf. Accessed 26 Sept 2012
26. McDonald K (2012) 803 Sign up in PCEHR's first week, 09/07/2012, http://www.pulseitmagazine.com.au/index.php?option=com_content&view=article&id=1068:803-sign-up-in-pcehrs-first-week&catid=16:australian-ehealth&Itemid=328. Accessed 27 Sept 2012

27. AMA (2012) Draft guide to using the PCEHR'. 04/04/2012, http://ama.com.au/media/draft-guide-using-pcehr. Accessed 26 Sept 2012
28. Taylor J (2011) Slow uptake better for e-health: Roxon, ZDNet.com.au, 13/09/2011, http://www.zdnet.com/slow-uptake-better-for-e-health-roxon-1339322154/. Accessed 02 Mar 2012
29. Winterford B (2009) Report urges electronic health records by 2012, itnews.com.au, 27/07/2009, http://www.itnews.com.au/News/151290,report-urges-electronic-health-records-by-2012.aspx. Accessed 02 Mar 2012
30. King T, Brankovic L, Gillard P (2012) Perspectives of Australian adults about protecting the privacy of their health information in statistical database. Int J Med Inf 81:279–289
31. Ross M (2012) Smartphone penetration booms down under, 16/05/2012, http://www.bandt.com.au/news/digital/smartphone-penetration-booms-down-under. Accessed 26 Sept 2012
32. Gliddon J (2012) Mobile health records connect outback communities, eHealthSpace.org, 24/02/2012, http://ehealthspace.org/casestudy/mobile-health-records-connect-outback-communities. Accessed 13 Aug 2012
33. Gaganayake R, Iannella R, Sahama T (2011) Sharing with care: an information accountability perspective. IEEE Internet Comput, pp 31–38
34. Jackson C (2012) The personally controlled electronic health record (PCEHR)—decision time approaching for general practitioners and practices, 22/05/2012, http://www.racgp.org.au/news/46973. Accessed 26 Sept 2012

Information Fusion: Popular Approaches and Applications

Min Wei

Abstract Information fusion is a hot topic in computer and related fields. It is widely used in military and civilian areas. In this paper, we first describe information fusion architectures of it to give a blueprint of information fusion approaches. Second, we review the most popular information fusion methods and analyze their advantages and disadvantages. Third, we outline their significant applications. Finally, conclusion remarks are drawn and some future prospects are given.

Keywords Information fusion · Architectures · Approaches · Applications

1 Introduction

Information fusion, also called data fusion or multi-sensor fusion, "is the process of combining information from a number of different sources to provide a robust and complete description of an environment or process of interest" [1]. It is a widespread phenomenon even for humankind and other animals. One fuses the information acquired by eyes, ears, nose, mouth, and other sensory distributed in our bodies for making decisions subconsciously every minute. In order to enlarge our perception, depth, and scope, a wide range of sensors, such as camera, radar, and satellite are utilized. To get more reliable and accurate information, fusion of information from multi-sensors is necessary.

M. Wei (✉)
State Key Laboratory of Intelligent Technology and Systems, Tsinghua National Laboratory for Information Science and Technology, Department of Computer Science, Tsinghua University, Beijing 100084, China
e-mail: weim10@mails.tsinghua.edu.cn

M. Wei
Department of Basic, Dalian Naval Academy, Dalian 116018, China

F. Sun et al. (eds.), *Knowledge Engineering and Management*,
Advances in Intelligent Systems and Computing 214,
DOI: 10.1007/978-3-642-37832-4_63,

Usually in systems-related computer science, the term information fusion includes multi-sensor modeling, classification, fusion, pattern recognition, and estimation. It was first proposed by the American Navy, for extension of sonar in the 1970s. With the emergence of new sensors, sophisticated processing techniques, and improved processing hardware, real-time fusion of data has become increasingly possible [2–4]. Just as the advent of symbolic processing computers in the early 1970s provided motivation to artificial intelligence [5], recent advances in computing and sensing have provided the ability to emulate, in hardware and software, the natural data fusion capabilities of humans and animals. Currently, information fusion systems are used extensively for target tracking, automated identification of targets, and limited automated reasoning applications. Information fusion approach has rapidly advanced from a loose collection of related techniques [6], to an emerging true engineering discipline with standardized terminology, collections of robust mathematical techniques, and established system design principles.

In the regaining period, to manage complicated fusion procedure efficiently, models of how to carry out tasks were created. One complex problem was divided into several parts. Relations of those reacted parts are hierarchy. Many kinds of architectures of information fusion have been put forward, such as the Intelligence Cycle [7], JDL [8], OODA [9], Omnibus [10], and Extended OODA [11]. However, the earliest model, e.g., the JDL fusion model, is the classical one [15].

Information fusion is found to have a wide range of applications such as military system, civilian surveillance and monitoring task, intelligent traffic system [12], process control systems, information system, medicine system [13], and environment perception [14]. Almost in all the fields worked with or assisted by electronic equipments, signs of information fusion can be found.

As stated above, one can state that information fusion is an important technology or branch of computer science and other related fields. Although there are many approaches to solve different problems in different applications, challenges such as accurate, robust, and real-time problems need to be addressed. In Sect. 2, we introduce information fusion architecture. Section 3, review the development and state of the art of a variety of information fusion methods, and compare the advantages and shortcomings of widely used approaches. In Sect. 4, we describe several application fields of information fusion. In Sect. 5, we draw the conclusions and put forward some interesting points to be investigated.

2 Information Fusion Architecture

The way in which sensors connect and share information is called information fusion architecture [15]. Information fusion architecture can be divided into distributed, centered, and mixed structures broadly. Distributed structure is designed to process local data which have been reprocessed. Centered structure deals with the original data from sensors. Mixed structure is a combination of the above two, and deals with original and reprocessed data.

In information fusion procedure, data from multi-sensors are handled in several layers. In each layer, original data is abstracted in different levels, including data detection, association, estimate, and combination, where each of them may have a different structure. Information fusion procedure can be divided into three layers, data (pixel) fusion, feature fusion, and decision-making fusion.

Data fusion is the lowest layer of the information fusion architecture, which is mainly applied to homogeneous sensors. It fusions all sensors' data first, then abstracts features from the fusion information, and identification. No information lost is its prominent advantage; but computation is its big problem.

Feature fusion is the middle layer. It abstracts features of sensors' original data first, then analyzes and processes the feature information. In this layer, the information has been compressed considerably, which advents to real-time process. The abstracted feature relates to decision analysis directly. Because of its high abstraction, accuracy becomes a problem.

Decision-making fusion is the highest layer. It observes one object with different type of sensors. Each sensor carries through the basic process, including reprocess and feature abstraction and identification, to get the preliminary conclusion of the object.

Most often, we consider the main functions and databases and relations of reacted parts of fusion procedure; the information fusion hierarchy which is prescribed by these factors is called functional model. More than 30 fusion architectures have been proposed [16], but the most widely cited model for data fusion is the JDL fusion model created by the American Joint Directors of Laboratories Data Fusion Subpanel [17]. In JDL (Fig .1) [18],data fusion process is divided into four levels, which make up a hierarchy of processing. Although this is by no means

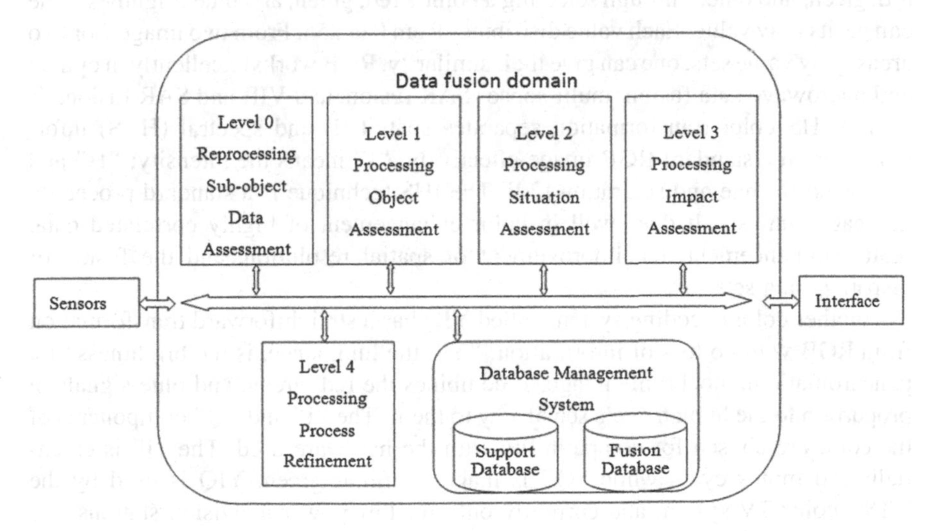

Fig. 1 JDL fusion model

the only hierarchy for data fusion and is primarily focused on military applications, it does provide a useful structure with which to classify fusion algorithms.

3 Information Fusion Approach

Compared with the appearance of new sensors, new technologies, and new applications, approaches of information fusion have mushroomed. From simple weighted average and conditional probability to complicated genetic algorithms and dynamic Bayesian networks, there are thousands of different approaches and editions of information fusion approaches. In this paper, the main clues of several important approaches are stated.

3.1 Fusion Methods for Images

In most fusion cases one needs to combine multi-source images, the use of advanced image processing techniques to integrate images emerges. Image fusion aims at integrating disparate and complementary data to enhance the information underlying images as well as to increase reliability of interpretation. For this, to fusion homogeneous or heterogeneous images acquired from cameras, radars, and infrared sensors, tasks such as sharpening, creation of stereo data sets, feature enhancement, classification, and overcoming gaps are demanded [19].

RGB (red–green–blue), IHS (intensity-hue-saturation) and YIQ are members of such methods. RGB allows one color to be represented by the three primary colors of red, green, and blue. Through selecting a point's red, green, and blue brightness, one can get its gray value. Each value distributes from 0 to 255. From two images' or two areas' gray value sets, one can give their similarity. RGB works excellently in optical and microwave data fusion, multi-sensor SAR fusion, and VIR and SAR fusion.

The IHS color transformation separates spatial (I) and spectral (H, S) information from a standard RGB image effectively. "I" means the intensity; "H" and "S" stand for hue and saturation [20]. The IHS technique is a standard procedure in image analysis. It does well in color enhancement of highly correlated data, feature enhancement, the improvement of spatial resolution, and the fusion of disparate data sets.

Another color encoding system called YIQ has a straightforward transformation from RGB with no loss of information. "Y", the luminance, is the brightness of a panchromatic monochrome image. It combines the red, green, and blue signals in proportion to the human eye's sensitivity to them. The "I" and "Q" components of the color are chosen for compatibility with the hardware used. The "I" is essentially red minus cyan, while "Q" is magenta minus green. YIQ is used by the NTSC color TV system and currently only for low-power television stations.

Image fusion algorithms are simple, effective, and popular, mostly used as pretreatment before fusion process. And in many cases, one algorithm is applied in combination with additional image fusion procedure, e.g., PCA, wavelets, adding, and multiplication [19].

3.2 Statistical Methods

For all descriptions of the sensing and information fusion process, uncertainty is an important problem. An unambiguous measure of this uncertainty must be provided to enable sensory information to be fused in an efficient and predictable manner. Although there are many methods to represent uncertainty, statistical methods based on probabilistic models are the core of almost all of them. The statistical approach has become a big family in information fusion.

3.2.1 Conditional Probability

Probabilistic models provide a powerful and consistent means to describe uncertainty in a wide range of situations, pulling naturally in ideas of information fusion and decision-making. Conditional probability is the soul of probabilistic models, which provides a method to measure sensor A's probability through sensor B's. If the two sensors are dependent, A's conditional probability equals A's probability. On the contrary to get A's probability, B's prior probability is needed except the conditional probability. Conditional probability enables uncertainties in the available data to be propagated through the calculations and to be represented in the output results. In practice, most of the time, the prior is given randomly and conditional probability is obtained from samples or calculated from another probability. Thus accuracy and complexity become big problems.

Harri Kiiveri et al. [21] proved CPNs (conditional probability networks) are useful tools for combining data for environmental monitoring problems, and analyzing sequences of full TM scenes with the method is computationally feasible. They integrate uncertain information from many different sources effectively and model parameters to be estimated from the data or supplied by experts in the area easily. Conditional probability models are also popular in most fusion processes, especially in target tracking [22–24], classification [25, 26], and remote sensing [27].

3.2.2 Bayesians

Bayesians is the advanced edition of conditional probability. Bayesian equation consists of likelihood, prior, posterior, and evidence, they are probabilities or conditional probabilities. Each probability can be calculated by the others. Bayesian equation is the base theory of conditional probability recursive models.

In the sensory network, one's posterior distribution can be another's prior. Hierarchical models and marginalization over the values of nuisance parameters are created. In theory, Bayesian is a perfect method to solve uncertainty in sensory information fusion system. However, in most cases, the computation is intractable.

Bayesian network (belief-network) and dynamic Bayesian networks are members of the Bayesian family. Bayesian network (BN) is a probabilistic graphical model that represents a set of random variables and their conditional dependencies via a directed acyclic graph. In the last decade, BNs have become extremely popular models, widely used in machine learning [28], speech recognition, text mining, medical diagnosis, weather forecasting, etc. The problem is that BNs can only deal with static problems. To those whose distributions are changed with time or other factors, they have no idea.

Dynamic Bayesian networks (DBN), are created to deal with sequences of variables. These sequences are often time-series or sequences of symbols. DBNs are more complicated than BNs, when nodes increase, the computation is impossible. To simplify DBNs is the prior problem. HMMs (Hidden Markova Models) are the simple formation of DBN. HMMs can be applied in many fields where the goal is to recover a data sequence that is not immediately observable. DBNs' application can be found in recognition, learning, and tracking areas.

Bayesian methods have two major limitations. One is the requirement of the assumption of mutual independencies among multiple classifiers. The other is that it cannot model imprecision about uncertain measurements.

3.2.3 D-S: Dempster-Shafer

The Dempster-Shafer (D-S) evidence theory [29], also known as the theory of belief functions, is a generalization of the Bayesian theory of subjective probability. It allows evidence from different sources to be combined and arrive at a degree of belief that takes into account all the available evidence. Belief in a hypothesis is a mass to each element of the power set assigned by evidence. It ranges from 0 to 1. Plausibility is 1 minus the sum of the masses of all sets whose intersection with the hypothesis is empty. It is an upper bound on the possibility that the hypothesis could be true. The D-S evidence theory is based on two steps: first obtaining degrees of belief for one question from subjective probabilities for a related question, second, combining such degrees of belief with Dempster's rule when they are based on independent items of evidence.

The D-S evidence theory's main advantage is its ability to deal with the probability masses from propositions that contradict each other [30], which can be used to obtain a measure of how much conflict there is in a system. This measure has been used as a criterion for clustering multiple pieces of seemingly conflicting evidence around competing hypotheses. Another advantage is computation; where there is no need to specify priors and conditionals, unlike Bayesian methods. Also, the D–S allows specifying a degree of ignorance in this situation instead of being forced to supply prior probabilities.

Zadeh [31] pointed out that counterintuitive results generated by Dempster's rule have a high degree of conflict. Wang [32] showed that the assertion "chances are special cases of belief functions" and the assertion "Dempster's rule can be used to combine belief functions based on distinct bodies of evidence" together lead to an inconsistency in the D-S.

To overcome its limitation, many tips are presented to improve its fusion ability. The D-S is used in prediction, recognition, and classification.

3.2.4 Kalman Filter

The Kalman filter (KF), also known as linear quadratic estimation (LQE), is an algorithm which uses a series of measurements observed over time, containing noise and other inaccuracies, to produce estimates of unknown variables. The KF is a recursive estimator. The state of the filter is represented by two variables: a posteriori state estimate at time k given observations up to and including at time k; a posteriori error covariance matrix. KF algorithms are mostly divided into two steps: predict and update. In predict step, the state estimate from the previous time is used to produce an estimate of the state of the current time. The estimate does not include observation information from the current time, so it is also known as a priori state estimate. In update step, the current observation is combined with a prior prediction to refine the state estimate that is termed as a posteriori state estimate. Then with the result of time k and observable information of time k, prediction of time $k + 1$ can be obtained. It tends to be more precise than those that would be based on a single measurement alone and efficient at estimating the internal state of a linear dynamic system from a series of noisy measurements.

The basic KF is limited to linear assumption. In fact, most complex systems are nonlinear, either with nonlinear process model or with nonlinear observation model or with both. Hence, nonlinear Kalman filters appear. EKF (Extended Kalman Filter) and UKF (Unscented Kalman Filter) are nonlinear KFs. In the EKF, the state transition and observation models need not be linear functions of the state but may instead be nonlinear functions.

When the state transition and observation models are highly nonlinear, the extended Kalman filter can give particularly poor performance [33]. This is because the covariance is propagated through linearization of the underlying nonlinear model. The UKF uses a deterministic sampling technique to pick a minimal set of sample points which are called sigma points around the mean. These sigma points are then propagated through the nonlinear functions, from which the mean and covariance of the estimate are then recovered. The result is a filter which captures the true mean and covariance more accurately. In addition, this technique removes the requirement to explicitly calculate Jacobians, which for complex functions can be a difficult task in itself, i.e., requiring complicated derivatives if done analytically.

With the development of applicable demanding, different editions of KF are generated. Kalman–Bucy filter, named after Richard Snowden Bucy, is a continuous-time version of the Kalman filter [34, 35]. Hybrid Kalman filter is a hybrid of nonlinear model and linear model [36], mostly used in physical systems which are represented as continuous-time models while discrete-time measurements are frequently taken for state estimation via a digital processor. The design steps for the hybrid Kalman filter are exactly the same as those for the general linear Kalman filter. First, a nonlinear target model is linearized at operating points. Then, Kalman gains are computed based on the linear representations of the plant model. When implemented, however, linear models and associated Kalman gains are integrated with the nonlinear target model. It is widely used in obstacle detection [37], sensor fault detection, positioning, etc.

In theory, KF is a good simulation for different systems. It has numerous applications in fields, commonly for guidance, navigation and control of vehicles, signal processing, and econometrics. But in fact, it takes a long time to calculate the iteration, and the result is not always accurate enough. Hence in practice, it is used in the lower layer of fusion architecture and as theory reasoning model.

3.2.5 Particle Filter

Particle Filter (PF), also known as Bootstrap or a sequential Monte Carlo method (SMC), is an optimal recursive Bayesian filter based on Monte Carlo simulation [38, 39]. Its main idea is to sample a set of points (particles) to similar probability distribution function, substitute the integration with sample mean, and then come to the state's minimum variance estimation.

In the early days, points of PF were stationary, which led to point's conflict and Filter divergence [39]. To solve this, Gorden brought up the Bootstrap filter. In Bootstrap, resample is added and to prevent divergence of filter, points with the lowest weights are deleted and new points added at the highest weights. Because of its excellent function, PF has been one of the hottest fusion approaches and applied to many fields, such as multi-target tracking [40, 41], maneuvering target identification [42–44], and lane detection [45, 46]. In the distributed sensor network, PF does excellent work [47].

PF can be applied to any nonlinear random system. In dynamic Bayesian network reasoning, it is the practical algorithm to replace the accuracy algorithm. And when there are large samples, one can find the global optimal solution too. However, the PF's runtime is long [48].

3.3 AI

In 1956, the term of artificial intelligence (AI) emerges. Not long after its appearance, AI was used for information fusion. It mainly includes methods based on neural network, expert system, fuzzy logic, and evolutionary computing.

3.3.1 Neural Network

In computational intelligence, neural network (NN), also called artificial neural network, is an interconnected group of artificial neurons that use a mathematical or computational model for information processing. Neural network algorithms attempt to abstract the complexity of biological neurons and model information processing in biological systems. NN is made of basically three types of layers: input, hidden, and output. In feed-forward networks, the signal flow is from one layer to another unidirectional; the data processing (hidden) can be extended over multiple layers of units, but no feedback. To improve NN's computability, feedback connections are added by John Hopfield [49]. Rumelhart et al. presented a type of NN with back-propagating errors, called BP. In BP "the procedure repeatedly adjusts the weights of the connections in the network so as to minimize a measure of the difference between the actual output vector of the net and designed output vector" [50]. Although BP has excellent ability to approximate the nonlinear mapping and solved many problems of machine learning, convergence speed and local minima and scale are its major problems.

NN is a method full of criticisms. For example, in 1997 A. K. Dewdney, a former Scientific American columnist said that, although the NN does solve a few toy problems, its powers of computation are so limited that he was surprised anyone took it seriously as a general problem-solving tool. But in practice, NNs have been successfully used to solve many complex and diverse tasks, ranging from autonomously flying aircraft to detecting credit card fraud. Because NN is nonlinear statistical data modeling or decision-making tool, it can be used to model complex relationships between inputs and outputs or to find patterns in data. Although it is true that analyzing what has been learned by an artificial neural network is difficult, it is much easier to do so than to analyze what has been learned by a biological neural network. Furthermore, researchers involved in exploring learning algorithms for neural networks are gradually uncovering generic principles which allow a learning machine to be successful [51].

Currently, artificial neural networks that are applied tend to fall within the following broad categories: function approximation, classification, sequence recognition, novelty detection, sequential decision-making, filtering, clustering, blind signal separation, and compression.

3.3.2 Genetic Algorithms

Genetic algorithm (GA) [52] is a search heuristic method which simulates the process of natural evolution, used to generate useful solutions for optimization and search problems. A genetic algorithm starts from a population of randomly generated individuals usually which is called the first generation. Then, after reproduction, crossover, and mutation the next population is generated. In each generation, the fitness of every individual in the population is evaluated, based on the value of their fitness multiple individuals they are stochastically selected from the current population, and modified to form a new population. The new population is then used in the next iteration of the algorithm. The algorithm terminates when either a maximum number of generations has been produced, or a satisfactory fitness level has been reached for the population.

Although GAs is a robust search method, most of them are simple and have strong function. The fitness is always problem dependent, when the algorithm terminates a satisfactory solution may not be reached. In many problems, GAs may have a tendency to converge towards local optima or even arbitrary points rather than the global optimum of the problem. Furthermore, it is difficult to operate GAs on dynamic data sets.

Currently, it is found in clustering [53], watermark detection [54], image segmentation [55], feature selection, prediction, decision-making, and parameter optimization.

3.3.3 Fuzzy Logic

According to Lotfi Askar Zadeh [56, 57], fuzzy logic is a problem-solving control system methodology that lends itself to implementation in systems ranging from simple, small, embedded micro-controllers to large, networked, multi-channel PC or workstation-based data acquisition and control systems. It was extended from fuzzy sets. In the fusion procedure, first represent multi-sensor's uncertainty with fuzzy sets, then deduce with multi value logic, and finally combine the related proposition by fuzzy sets theory rules [58]. When reasoning and modeling the uncertainty by some system methodologies, consistent results can be obtained.

Logic reasoning overcomes some problems of probability method, such as information presentation; it is closer to the habit of thinking of humankind. Its shortcomings are not mature enough, and there are too many objective factors in the logic reasoning procedure.

Fuzzy logic algorithms do well in pattern recognition [59], tracking [60], position, spatial road forecast [61], and decision-making [62].

3.4 Others

In DOD's JDL model, the five layers of reprocess, target assessment, state assessment, threaten assessment, and procedure assessment, are included. For different applications, different objects or problems exist. There is no approach that can solve all the problems efficiently. Hence, except for the approaches cited above, other information fusion approaches based on random theory, calculus, algebra, set theory, and graph theory are useful in some fields. In fact, with new sensors appearing, new methods will also be proposed.

4 Applications

Techniques which combine or fuse data are drawn from a diverse set of more traditional disciplines including: digital signal processing, statistical estimation, control theory, artificial intelligence, and classic numerical methods [63–65]. Historically, data fusion methods were developed primarily for military applications. However, in recent years these methods have been applied to civilian applications, and there has been bidirectional technology transfer [2]. Various annual conferences provide a forum for discussing data fusion applications and techniques.

The application is divided into military and non-military simply. More detailed, it can be classified as defense systems, geosciences, robotics and intelligent vehicles, medicine, and industrial engineering [66].

4.1 Defense Systems

Defense systems include automated target recognition (e.g., for smart weapons), guidance for autonomous vehicles, remote sensing, battlefield surveillance, and automated threat recognition systems, such as identification-friend- foe-neutral (IFFN) systems [67].

4.2 Geosciences

Geosciences concern the earth's surface with satellite images and remote sensing images. Fusion can be done on images from different sensors or on multi-data images. The main problems in this area are classification and interpretation of images [68]. The fusion of attributes allows detection of roads, buildings, farms, forest, mountains, rivers, or bridges. Also, sometimes one needs to aggregate the images of the same scene with different spatial resolutions.

4.3 Robotics and Intelligent Vehicles

Robotics and intelligent vehicles include two primary tasks: the perception of the environment that the robot or vehicle evolves and navigation. Environment perception, which is used to detect and recognize obstacles and around robot or vehicle—is a challenging problem. In different environments, obstacles range from the earth to the sky; they may be a kind of fish living in the sea or a stone in front. Hence, to identify the environment is never easy where several algorithms have to be solved. Information is usually obtained by means of encoder, camera, laser scanner, and sonar. In order to achieve robot or vehicle's location in real-time, grid has been used. Furthermore, the environment can be dynamic and it is necessary to include temporal coherence between data.

At the same time or after environment perception, navigation is held out. In addition to commanding the robot or vehicle in which direction to go or what to do, the collisions avoidance and the tracking of an object need to be investigated.

4.4 Medicine

Nowadays, medicine is information fusion's hottest application field. From diagnosis to the model of the human body, more and more doctors or medical scientists cure diseases that have been found and discover new diseases and explore the nature of the human body. New visualization and acquisition techniques allow data confrontation. Brain images are classified using multi-model images and tumor delineation is improved using radiographic images and ultrasonic images [69]. Liver, spleen, muscles, blood vessels and even protein, endocytosis, emotion, and intima–media thickness are considered by means of fusion of several images obtained by electrical impedance tomography, MR, PET, and CT [70–72].

4.5 Industrial Engineering

In the industry, applications such as non-destructive control use different measurements to validate soldering for example. Other applications such as industrial baking control of quality of rice and maize is predicted by fuzzy rule. Another application allows the detection of tool breakage [73].

5 Conclusion

Inspired from biological systems where multiple sensors with an overlapping scope are able to compensate the lack of information from other sources, the data fusion field has gained momentum since its appearance. The rapid spread from biometrics to security field, from robotics to everyday applications matches the rise of multi-sensor and heterogeneous-based approaches. In this regard, information fusion is viewed to be a popular and promising subject.

In this paper, we discussed several methods of information fusion; analyzed advantages and disadvantages of each method, described the relationship between the methods, and summarized their applications. Among these methods, conditional probability is regarded as the soul of KFs and Bayesians and PFs which are all popular methods in the fusion process. Specifically, PFs are more efficient and accurate than others in sense. Together with D-S, they are all statistical methods. Fuzzy logic and neural networks and genetic algorithm are called AI-based methods. They are applied in some fields where statistical ones cannot resolve effectively. Compared to statistical methods, AI ones need more time to compute and computational results' accuracy is low except in several applications such as robotics control and decision-making. Image-based methods are simple and time-consuming. To overcome the methods' shortcomings, some researchers combine two or several methods together and get better results.

The robust, correctness, accuracy, and timeliness are demanded and discussed in every fusion process. There are spaces for every information fusion method to improve performance.

References

1. Durrant-Whyte H (2001) Multi sensor data fusion, Springer handbook of robotics. Springer, Heidelberg
2. Hall DL, Llinas J (1994) A challenge for the data fusion community I: research imperatives for improved processing. In: Proceedings of 7th national symposium on sensor fusion, Albuquerque
3. Llinas J, Hall DL (1994) A challenge for the data fusion community II: infrastructure imperatives In Proceedings of the 7th national symposium on sensor fusion, Albuquerque
4. Jo K, Chu K, Sunwoo M (2012) Interacting multiple model filter-based sensor fusion of GPS with in-vehicle sensors for real-time vehicle positioning. IEEE Trans Intell Transp Syst 13(1):329–343
5. Gelfaud J (1992) Selective guide to literature on artificial intelligence and expert systems. In: American society for engineering education
6. Hall DL, Llinas J (1997) An introduction to multi-sensor data fusion. Proc IEE 85(1):6-23
7. Shulsky A (1993) Silent warfare: understanding the world of intelligence. Brassey's, Washington, D.C
8. White F (1998) A model for data fusion. In: Proceedings of the 1st international symposium on sensor fusion
9. Boyd J (1987) A discourse on winning and losing. Maxell AFB leture
10. Bedworth M, O'Brien J (2000) The omnibus model: a new model of data fusion? Proc IEEE AES Syst Magzine 15(4):30–36
11. Shahabzian E, Blodgett D, Labbe P (2001) The extended OODA model for data fusion systems. Fusion'2001, pp FrB1-19–25
12. El Faouzi N-E, Leung H, Kurian A (2011) Data fusion in intelligent transportation systems: progress and challenges—a survey. Inf Fusion 12:4–10
13. Plooij JM, Maal TJJ, Haers P, Borstlap WA, Kuijpers-Jagtman AM, Berge SJ (2011) Digital three-dimensional image fusion processes for planning and evaluating orthodontics and orthognathic surgery. A systematic review, Int J Oral maxillofac surg 40(4):341–352
14. Bertolazzi E, Biral F, Da Lio M, Saroldi A, Tango F (2010) Supporting drivers in keeping safe speed and safe distance: the SASPENCE subproject within the European framework programme 6 integrating project PReVENT. IEEE Trans Intell Transp Syst 11(3):525–538
15. Smith D, Singh S (2006) Approaches to multisensor data fusion in target tracking: a survey. IEEE Trans Knowl Data Eng 18(12):1696–1710

16. Salerno J, Hinman M, Boulware D (2004) Building a framework for situational awareness. Proceedings of seventh international conference information fusion, pp 219–226
17. US Department of Defense (1991) Data fusion subpanel of the joint directors of laboratories, technical panel for C3. Data fusion lexicon
18. Steinberg A, Bowman C, White F (1999) Revisions to the JDL data fusion model. SPIE 3719:430–441
19. Pohl C, Van Genderen JL (1998) Multisensor image fusion in remote sensing: concepts, methods and applications. Int J Remote sens 19(5):823–854
20. Harrison BA, Jupp DLB (1990) MicroBRIAN resource manual: introduction to image processing. Division of Water Resources, CSIRO, Australia
21. Kiiveri H, Caccetta P (1998) Image fusion with conditional probability networks for monitoring the salinization of farmland. Digit Signal Process 8:225–230
22. Mao L, Du J, Liu H, Guo D, Tang X, Wei N (2010) Two-stage target locating algorithm in three dimensional WSNs under typical deployment schemes. Wireless Algorithms Syst Appl 6221:172–181
23. Veitch D, Augustin B, Teixeira R, Friedman T (2009) Failure control in multipath route tracing. IEEE, pp 1395–1403
24. Jazayeri A, Cai H (2011) Vehicle detection and tracking in car video based on motion model. IEEE Trans Intell Transp Syst 12(2):583–595
25. Dembczy K, Cheng W, Hullermeier E (2010) Bayes optimal multilabel classification via probabilistic classifier chains. Proceedings of the 27th international conference on machine learning, Haifa
26. Yilmaz Isik (2010) Comparison of landslide susceptibility mapping methodologies for Koyulhisar, Turkey: conditional probability, logistic regression, artificial neural networks, and support vector machine. Environ Earth Sci 61:821–836
27. Zhang G, Jia X (2012) Simplified conditional random fields with class boundary constraint for spectral-spatial based remote sensing image classification. IEEE Geosci Remote Sens lett 9(5):856–860
28. Bishop CM (2006) Pattern recognition and machine learning. Springer, Heidelberg
29. Liu L, Yager RR (2008) Classic works of the Dempster-Shafer theory of belief functions: an introduction. Stud fuzziness soft comput 219:1–34
30. Sentz K, Ferson S (2002) Combination of evidence in Dempster-Shafer theory. SAND2002-0835 Technical report, Sandia National Laboratories
31. Zadeh L (1986) A simple view of the Dempster-Shafer theory of evidence and its implication for the rule of combination. Al Mag 7(2):85–90
32. Wang P (1994) A defect in Dempster-Shafer theory .In: Proceedings of the 10th conference on uncertainty in artificial intelligence. Morgan Kaufmann, San Mateo, pp 560–566
33. Julier SJ, Uhlmann JK (1997) A new extension of the Kalman filter to nonlinear systems. International symposium aerospace/defense sensing, simulation, and controls 3
34. Bucy RS, Joseph PD (2005) Filtering for stochastic processes with applications to guidance. Wiley, 1968; 2nd Edn. AMS Chelsea Publication, New York, ISBN 0-8218-3782-6
35. Jazwinski AH (1970) Stochastic processes and filtering theory. Academic Press, New York. ISBN 0-12-381550-9
36. Kobayashi T, Simon DL (2006) Hybrid Kalman filter approach for aircraft engine in-flight diagnostics: sensor fault detection case. NASA/TM—2006-214418
37. Einhorn E, Schr̈oter C, B̈ohme HJ, Gross HM (2007) A hybrid Kalman filter based algorithm for real-time visual obstacle detection. In: Proceedings of the 3rd ECMR, 2007, pp 156–161
38. Doucet A, De Freitas N, Gordon NJ (2001) Sequential Monte Carlo methods in practice. Springer, New York
39. Carpenter J, Clifford P, Fearnhead P (1999) Improved particle filter for non-linear problems. IEE Proc Radar Sonar Navig 146(1):2–7
40. Gordon NJ, Salmond DJ, Smith AFM (1993) Novel approach to non-linear/non-Gaussian Bayesian state estimation," IEE Proc F Radar Signal Process 140(2):107–113

41. Hue C, Le Cadre JP (2002) Sequential Monte Carlo methods for multiple target tracking and data fusion. IEEE Trans Signal Process 50(2):309–325
42. Veres GV, Norton JP (2001) Improved particle filter for multitarget-multisensor tracking with unresolved applications. IEE Target Tracking Algorithms Appl 1:12/1–12/5
43. McGinnity S, Irwin GW (2000) Multiple model bootstrap filter for manoeuvring target tracking. IEEE Trans Aerosp Electron Syst 36(3):1006–1012
44. Blom HAP, Bloem EA (2003) Tracking multiple manoeuvring targets by joint combinations of IMM and PDA. Proceedings of 42nd IEEE Conference Decision and Control
45. Liu G, Worgotter F, Markelic I (2010) Combining statistical Hough transform and particle filter for robust lane detection and tracking. In: 2010 IEEE intelligent vehicles symposium, WeF1.3, pp 993–997
46. Loose H, Stiller C (2009) Kalman Particle Filter for Lane Recognition on Rural Roads. Proceedings of IEEE intelligent vehicle symposium, pp 60–66
47. Blom HAP, Bloem EA (2003) Joint IMMPDA particle filter. In: Proceedings of sixth international conference information fusion, vol 2, pp 785–792
48. Liu J, Chu M , Liu J, Reich J, Zhao F (2004) Distributed state representation for tracking problems in sensor networks. In: Proceeding third international symposium information processing in sensor networks, pp 234–242
49. Hopfield J (1982) Neural networks and physical systems with emergent collective computational abilities. Proc Nat Acad Sci USA 79(8):2554–2558
50. Rumelhart DE, Hinton FE, Williams RJ (1986) Learning representations by back-propagating errors. Nature 323(9):533–536
51. Bengio Y, LeCun Y (2007) Scaling learning algorithms towards AI. MIT Press, Cambridge
52. Holland JH (1992) Adaptation in natural and artificial systems. MIT Press, Cambridge
53. Auffarth B (2010) Clustering by a genetic algorithm with biased mutation operator. WCCI CEC. IEEE, 18–23 July 2010
54. Davarynejad M, Ahn CW, Vrancken J, van den Berg J, Coello Coello CA (2010) Evolutionary hidden information detection by granulation-based fitness approximation. Appl Soft Comput 10(3):719–729
55. Maulik U (2009) Medical image segmentation using genetic algorithms. IEEE Trans Inf Technol Biomed 13(2):166–173
56. Zadeh LA (1965) Fuzzy sets. Inf Control 8(3):338–353
57. Zadeh LA et al (1996) Fuzzy sets, fuzzy logic, fuzzy systems. World Scientific Press, ISBN 981-02-2421-4
58. Dubois D, Prade H (2004) On the use of aggregation operations in information fusion processes. Fuzzy Sets Syst 142:143–161
59. Baraldi A, Blonda P (1999) A survey of fuzzy clustering algorithms for pattern recognition—part I. IEEE Trans Syst Man Cybern B Cybern 29(6):778–785
60. Garcia J, Patricio MA, Berlanga A, Molina JM (2011) Fuzzy region assignment for visual tracking. Soft Comput 15:1845–1864
61. Chow MY, Tram H (1996) Application of fuzzy logic technology for spatial load forecasting. In: IEEE, pp 608–614
62. Herrera F, Herrera-Viedma E, Mart L (2000) A fusion approach for managing multi-granularity linguistic term sets in decision making. Fuzzy Sets Syst 114:43–58
63. Hall DL, Linn RJ (1991) Survey of commercial software for multisensor data fusion. In: Proceedings of SPIE conference sensor fusion and aerospace applications, Orlando
64. Kessler H et al (1992) Functional description of the data fusion process, Technical report, Office of Naval Technology, Naval Air Development Center, Warminster
65. Wright F (1980) The fusion of multi-source data Signal, pp 39–43
66. Valet L, Mauris G (2000) A statistical overview of recent literature in information fusion. In: ISIF, MoC3, pp 22–29
67. Hall DL, Linn RJ Llinas J (1991) A survey of data fusion systems. In: Proceedings of SPIE conference on data structure and target classification, vol. 1470, pp 13–36

68. Solaiman B, Pierce LE, Ulaby FT (1999) Multisensor data fusion using fuzzy concepts: application to land-cover classification using ers-l/jers-1 sar composites. IEEE Trans Geosci Remote sens 37(3):1316–1325 (special issue on data fusion)
69. Nejatali A, Ciric IR (1998) Novel image fusion methodology using fuzzy set theory. Opt Eng 37(2):485–491
70. Collinet C, Stöter M, Bradshaw CR, Samusik N, Rink JC, Kenski D, Habermann B, Buchholz F, Henschel R, Mueller MS, Nag WE, Fava E, Kalaidzidis Y, Zerial M (2010) Systems survey of endocytosis by multiparametric image analysis. Nature 464(11):243–250
71. Molinaria F, Zengb G, Suric JS (2010) A state of the art review on intima–media thickness (IMT) measurement and wall segmentation techniques for carotid ultrasound. Comput methods programs biomed 100:201–221
72. Fisher LS, Ward A, Milligan RA, Unwin N, Potter CS, Carragher B (2011) A helical processing pipeline for EM structure determination of membrane proteins. Methods 55:350–362
73. Zaidi H, Montandon ML, Alavi A (2008) The clinical role of fusion imaging using PET, CT, and MR imaging. PET Clinics 2008, vol 3, no. 3, pp 275–291

SSE Composite Index Prediction and Simulation System

Multiple Regression and Time—Lag Cointegration Model

Dengbin Huang, Jingwei Jiang and Guiming Wang

Abstract Stock index forecasting is an important task in economic field; it is difficult to accurately predict index trends using the traditional prediction methods which are based on price and the quality. Studies on this field were described in this paper: historical data and the influence of macroeconomic factors to stock price, fits the mathematical models of the principal component of the multi-lag regression and Sequence Cointegration and stock index predicts; At the same time the author has done the system simulation, results show that the mathematical model is very efficient and practical.

Keywords Multi-lag regression · Timing cointegration · Stock index forecasting · Simulation system

1 Introduction

The stock market has 100 years history from appearance to now in developed country, its development has been quite mature, while Chinese stocks listed in Shanghai Stock Exchange on December 19, 1990 issue. It is difficult and important to do stock price forecasting. Now the stock price forecasting theory and its research methods are mainly based on price and quantity. Through the study of the history of the stock price to forecast the future price trend based on the price, mainly has the following three kinds of methods: technical analysis, fundamental analysis of stock method, and artificial intelligence analysis; through the analysis of the influence of the price trend of macroeconomic factors based on quantitative

An erratum to this chapter is available at 10.1007/978-3-642-37832-4_65

D. Huang (✉) · J. Jiang · G. Wang
College of Science, Naval University of Engineering, Wuhan 430033, China
e-mail: 15365286622@163.com

F. Sun et al. (eds.), *Knowledge Engineering and Management*,
Advances in Intelligent Systems and Computing 214,
DOI: 10.1007/978-3-642-37832-4_64, © Springer-Verlag Berlin Heidelberg 2014

research, use mathematical method to select the influential variables, determine the corresponding the regression equation to imitate and forecast, mainly has the following three kinds of methods: time series forecasting method, cointegration analysis, and variable analysis method.

Based on the analysis of price is mainly qualitative analysis, the analysis result is weak, lack of theoretical support, and with strong subjectivity. Based on the studies of volume, with limitations, sequential regression model assuming the time series pattern in the future and the past of the model is consistent with the hypothesis, in the long-term prediction is not practical. Multivariate analysis only considered and stock price index is related to several variables, as well as some of the information is not considered, along with the macroeconomic development and change, and stock index related variables are also changing. Cointegration analysis does not take consideration of time factor. Based on the above shortcomings, this paper considers the SSE Composite Index historical data and the influence of stock price volatility of macroeconomic factors were established based on the principal components regression and time series multivariate lag cointegration model, the prediction model of the rationality and accuracy of the verification process can obtain a good result.

2 Mathematical Model Based on the Principal Components Regression and Time Series Multivariate Lag Cointegration

2.1 Modeling and Simulation Flow Chart

Through the analysis of stock price index and historical stock data and the relationship among macroeconomic variables, and gradually establish based on the principal components regression and time series multivariate lag cointegration index prediction mathematical model, design of simulation system for stock index prediction The comprehensive index trend of the computer simulation experiment, verify the rationality and accuracy of prediction models, and finally the future stock index goes situation undertake forecasting, modeling, and Simulation of train of thought such as shown in Fig. 1.

2.2 Multivariate Regression and Time Series Model of Hysteresis

In order to establish multiple lag regression and time series model, we need to solve three major problems: (1) how to deal with the problem of many factors influence the target to establish multiple lag linear regression model; (2) how to identify the classification of time series data in which time sequence model; and (3) the model linear combination the cointegration test.

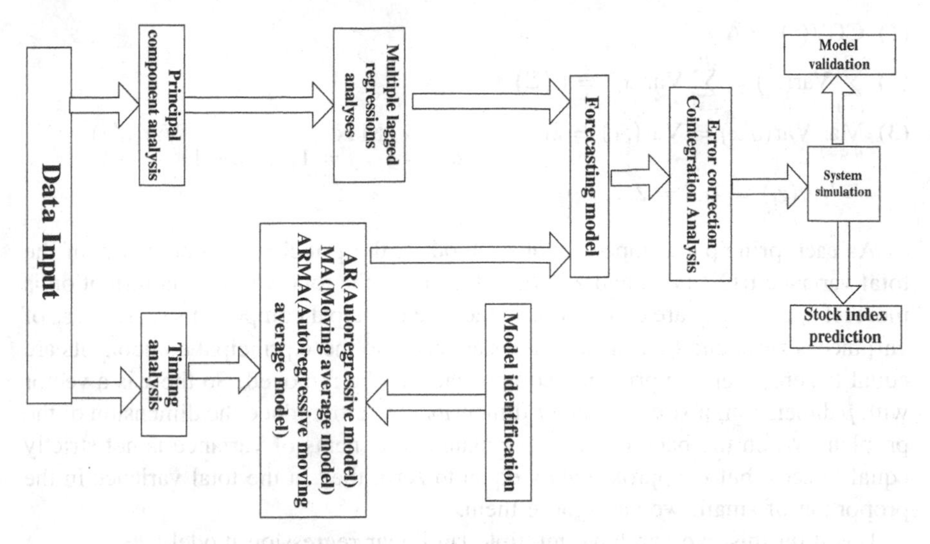

Fig. 1 Graph modeling and simulation of train of thought

2.2.1 Linear Regression Lag Model Based on PCA and Multivariate

The principal component analysis method is to use the idea of dimension reduction, by linear combination of the original number of indexes into a few unrelated index, and maintain the original index number information of a multivariate statistical analysis method. This paper selects the indexes of variance.

Assumed x to be $p \times 1$ random vector, $E(x) = \mu$, $\mathrm{COV}(x) = \sum > 0$, here μ for mathematical expectation, Σ covariance as. Recorded $\lambda_1 \geq \cdots \geq \lambda_p$ as the Σ characteristic root, $\varphi_1, \ldots, \varphi_p$ corresponding to the standard orthonormal eigenvectors, namely $\phi = (\varphi_1, \ldots, \varphi_p)$ the orthogonal array, and the

$$\phi' \sum \phi = \Lambda = \begin{pmatrix} \lambda_1 & & & \\ & \lambda_2 & 0 & \\ 0 & & \ddots & \\ & & & \lambda_p \end{pmatrix}$$

Call:

$$z = \begin{pmatrix} z_1 \\ \vdots \\ z_p \end{pmatrix} = \phi'(x - \varphi)$$

as principal component of random variables x, known $z_i = \varphi_i'(x - \varphi)$ as the i principal component, $i = 1, \ldots, p$. Principal component satisfies the following properties:

(1) $\mathrm{COV}(z) = \Lambda$

(2) $\sum_{i=1}^{p} \mathrm{Var}(z_i) = \sum_{i=1}^{p} \mathrm{Var}(x_i) = \mathrm{tr}(\Sigma)$

(3) $\underset{a'a=1}{\mathrm{Max}}\, \mathrm{Var}(a'x) = \mathrm{Var}(z_1) = \lambda_1 \quad \underset{\varphi_j' a=0,\, a'a=1 \quad j=1,\ldots,i-1}{\mathrm{Max}} \quad \mathrm{Var}(a'x)$

$= \mathrm{Var}(z_i) = \lambda_i, i = 2,\ldots,p$

As each principal component of each other, the principal component z_i of the total variance $\mathrm{tr}(\Sigma)$ is λ_i, and λ_i larger, therefore, the total variance contribution is more. If $\lambda_{r+1},\ldots,\lambda_p$ are equal to zero, then the principal components $z_{r+1},\ldots,z_p$ of variance is zero, and their mean values are zero, so these principal components are equal to zero, then the principal components can be removed. So the x is a vector with p dimension, just processing r dimensional vector reduce the dimension of the problem. When the back of the $p - r$ main components of variance is not strictly equal to zero, but is approximately equal to zero, they in the total variance in the proportion of small, we can ignore them.

Based on this, we can have multiple lag linear regression model

$$Y_t = \alpha_0 + \sum_{i=1}^{l} \sum_{j=0}^{k} \gamma_{ij} X_{i(t-j)} + \varepsilon_t, (k \in N)$$

In which k as model lag order, γ_{ij} as the lag factor, ε as random error, and is assumed to satisfy: $E(\varepsilon) = 0$, $\mathrm{var}(\varepsilon) = \sigma^2$.

2.2.2 Based on the Pattern Recognition Model for Time Series Analysis

Stationary time series models are of three main types: AR (regression) model, MA (moving average) model, ARMA (autoregressive moving average) model.

For real time series data, using the method of observation is difficult to determine the kind of model. By means of pattern recognition tools, we judge them using the autocorrelation coefficient and partial correlation coefficient.

Definition 1 $\{Y_t\}$ as time series, the t phase of the data Y_t, the lag d period of data for Y_{t-d}, $d \in N$

$$\rho_d = \frac{\mathrm{Cov}(Y_t, Y_{t-d})}{\sqrt{\mathrm{Var}(Y_t)}\sqrt{\mathrm{Var}(Y_{t-d})}}$$

Is defined as self correlation coefficient, the time lag d variable autocorrelation coefficient column is referred to self correlation function:

$$\{\rho_d\}(p = 0, 1, \ldots, D)$$

With the passage of time, if there is d greater than a certain constant, $\{\rho_d\}$ sequence value for all 0, it is called the censored, otherwise the tailing. In the lag d phase of autocorrelation coefficients, not get between Y_t and Y_{t-d} simple correlation, but also by the intermediate $d-1$ variables, purely for the sake of getting and relationship, following the introduction of partial autocorrelation function.

Definition 2 let partial autocorrelation coefficient

$$\phi_{dd} = \frac{D_d}{D} \quad \forall 0 < d < n$$

In which

$$D = \begin{vmatrix} 1 & \rho_1 & \cdots & \rho_{d-1} \\ \rho_1 & 1 & \cdots & \rho_{d-2} \\ \vdots & \vdots & \vdots & \vdots \\ \rho_{d-1} & \rho_{d-2} & \cdots & 1 \end{vmatrix}, \quad D = \begin{vmatrix} 1 & \rho_1 & \cdots & \rho_1 \\ \rho_2 & 1 & \cdots & \rho_2 \\ \vdots & \vdots & \vdots & \vdots \\ \rho_{d-1} & \rho_{d-2} & \cdots & \rho_d \end{vmatrix}$$

Then call $\{\phi_{dd}\}(d = 0, 1, \ldots, D)$ as the partial autocorrelation function, ϕ_{dd} and ρ_d has the same censored and tailing.

Autocorrelation function and partial correlation function for censored and tailing is a recognition model types of main basis, but in the actual process, truncation is not obvious, only with the aid of statistical means, the Table 1 shows the three kinds of stationary time series model of judgment principles.

2.2.3 Cointegration Analysis and Inspection

If the d time difference can be transformed into stationary sequence, it is called d order single make series, recorded as $Y_t \sim I(d)$; if $d = 0$, that Y_t is stationary process, namely $I(0)$ the process and smooth the process of equivalence.

If the two nonstationary time series linear combination is smooth, then called the two time series are cointegrated, i.e., nonstationary sequence X_t and Y_t a linear combination $X_t - \beta Y_t$ is the $I(0)$ process, so that X_t and Y_t are cointegrated, parameter β called the cointegration parameters.

We can use the DF test or ADF test to verify the cointegration. First to two order with the same number of non stationary time series least squares estimation, and then seek its residuals, finally by DF or ADF test residual series is stable can judge whether there exists cointegration relationship. The specific steps are as follows:

(1) For two time series show X_t and Y_t stationary test, and determine the order number. If both are the same number of nonstationary sequence order, then the second step;

Table 1 Three kinds of stationary time series model of judgment principles

Model type	Autocorrelation function ρ_d	Partial autocorrelation function ϕ_{dd}
AR(p)	Geometric attenuation (tailing)	$d > p$ order after closing to zero (censored)
MA(p)	$d > p$ Order after closing to zero (censored)	Is a geometric decay (trailer)
ARMA(p, q)	After $d > q - p$ geometric decay (tailing)	After $d > q - p$ the geometric decay (trailer)

(2) X_t and Y_t estimates with OLS, the model used here is: $Y_t = \beta_1 X_t + Z_t$;
(3) the residual Z_t of stationary test, you can use the following regression model:

$$\hat{Z}_t = \rho \hat{Z}_{t-1} + \beta_i \sum_{i=1}^{m} \Delta \hat{Z}_{t-i} + \varepsilon_t$$

$$\hat{Z}_t = \gamma_0 + \rho \hat{Z}_{t-1} + \beta_i \sum_{i=1}^{m} \Delta \hat{Z}_{t-i} + \varepsilon_t$$

$$\hat{Z}_t = \gamma_0 + \gamma_1 t + \rho \hat{Z}_{t-1} + \beta_i \sum_{i=1}^{m} \Delta \hat{Z}_{t-i} + \varepsilon_t$$

If the type does not include the lag, called the test method for EG test; if included lag, it is called AEG test.

2.2.4 Multiple Lag Regression and Time Series Model and Its Parameter Estimation

The integration of the multiple lag regression model and time series model is:

$$Y_t = \alpha_0 + \sum_{i=1}^{p} \beta_i Y_{(t-i)} + \sum_{l=1}^{m} \sum_{j=0}^{k} \gamma_{lj} X_{l(t-j)} + \varepsilon_t$$

in which Y_t as the index to t moment, X is the result of principal component analysis on stock index to produce the effect of macroeconomic variables. On the economic issues of the time sequence, the Y_t only historical information, the type $p \in N^+$, and the X effect can be lag, can also be advanced, so $k \in N$.

At the same time, we use correlation coefficient method to estimate the lag order number P, by Ad Hoc estimation method to estimate the lagging number k, using the least square method to estimate the parameters, α_0, $\beta_i (i = 1, 2, \ldots, p)$ and $\gamma_{lj} (l = 1, 2, \ldots, 6; j = 0, 1, \ldots, k)$.

3 Solution of the Multiple Regression and Time Series Model

3.1 Data Selection Instructions

This paper has the data from 05/2004 to 12/2010 SSE Composite index monthly, the monthly data to each month trading day closing index average, this can avoid the monthly in a random sample of one day to do a representative by the calendar effect, also taking into account the macroeconomic factors released is the whole reflection. Of course, also can choose to end at the end of the stock index, considering the average monthly such selection on the model solution is not influenced, therefore this article has taken months average data. Macroeconomic factors, data from the China Statistical Yearbook, website of the National Bureau of statistics, financial new web related statistical data.

3.2 Solution of Mathematical Model

The first step of suffrage is to sift macroeconomic factors; Through the analysis of stock price index, effect of macroscopical factors mainly include CPI, GDP, the amount of money in circulation M0, industrial added value, the RMB against the US dollar, total retail sales of consumer goods, new credit, economic boom index, the index of consumer confidence and PPI, respectively

$$X_1, X_2, X_3, X_4, X_5, X_6, X_7, X_8, X_9, X_{10}$$

then the principal component expressions for:

$$F_t = a_{i1}X_1 + a_{i2}X_2 + \cdots + a_{i10}X_{10} \quad i = 1, 2, \ldots, 10$$

(1) the standardization of original data;
(2) for the variables between the correlation coefficient matrix;
(3) correlation coefficient matrix R and the characteristic value and the characteristic vector.

Before the six eigenvalues of the cumulative contribution rate has reached 95.7 %, the description of the first six principal components basically contains all the indicators with the information, and the first six principal components for:

$$F_1 = 0.0333X_1 - 0.2549X_2 + \cdots + 0.2095X_{10}$$
$$F_2 = -0.5561X_1 - 0.2034X_2 + \cdots - 0.3748X_{10}$$
$$F_3 = 0.2229X_1 - 0.6877X_2 + \cdots + 0.4364X_{10}$$
$$F_4 = 0.2679X_1 + 0.3036X_2 + \cdots + 0.3652X_{10}$$
$$F_5 = -0.0786X_1 + 0.0463X_2 + \cdots + 0.4044X_{10}$$
$$F_6 = 0.1994X_1 - 0.2579X_2 + \cdots - 0.3455X_{10}$$

The second step, according to the first step the results of principal component analysis, time series analysis. The first unit root test, make a difference make it turn into a stationary sequence, then the lag analysis, according to the correlation coefficient, Ad Hoc estimation method to determine the lag order number p and k, using the least square method to estimate the parameters α_0, $\beta_i(i = 1, 2, \ldots, p)$ and $\gamma_{lj}(l = 1, 2, \ldots, 6; j = 0, 1, \ldots, k)$, finally, through cointegration analysis based on principal component regression and time series multivariate lag cointegration stock index forecasting model, as follows:

$$Y_t = \alpha_0 + \sum_{i=1}^{p} \beta_i Y_{(t-i)} + \sum_{l=1}^{m} \sum_{j=0}^{k} \gamma_{lj} X_{l(t-j)} + \varepsilon_t$$

Type of: $\alpha_0, \beta_i(i = 1, 2, \ldots, p), \gamma_{lj}(l = 1, \ldots, m; j = 0, 1, \ldots, k)$ for unknown parameters, ε_t random error, and satisfies $E(\varepsilon_t) = 0, \mathrm{Var}(\varepsilon_t) = \sigma^2$, and model coefficient will change according to different data selection.

To estimate the parameters, the final prediction model for:

$$\begin{aligned} Y_t = &-3535.2 - 0.1Y_{t-1} + 0.1Y_{t-2} - 0.3Y_{t-3} + 12X_{1(t-1)} + 2.3X_{1t} + 9.4X_{1(t+1)} - 0.1X_{2(t-1)} - 4.1X_{2t} \\ &- 10.4X_{2(t+1)} + 0.2X_{3(t-1)} + 0.2X_{3(t-2)} + 6.8X_{3t} - 6.9X_{3(t+1)} + 7.7X_{4(t-1)} - 12.3X_{4t} + 0.4X_{4(t+1)} \\ &+ 5.9X_{5(t-2)} - 15.8X_{5(t-1)} + 0.3X_{5t} - 0.2X_{5(t+1)} - 0.1X_{6(t-2)} - 7.1X_{6(t-1)} + 70.8 \end{aligned}$$

4 The Simulation Results Analysis and Accuracy Evaluation

Based on the prediction model, we design the simulation system software by using Matlab, Visual Basic 6.0 and Access. The software's interface is designed by VB. So that it is well user-friendly and easier to operate because of using some third-party controls. The simulation system software has the strong operation capability for using Matlab as computational tool. Software call Matlab program by using Active X part in VB. The database is established by Access. By hybrid programing, the simulation system software makes up the shortcomings of each software and achieves the desired function.

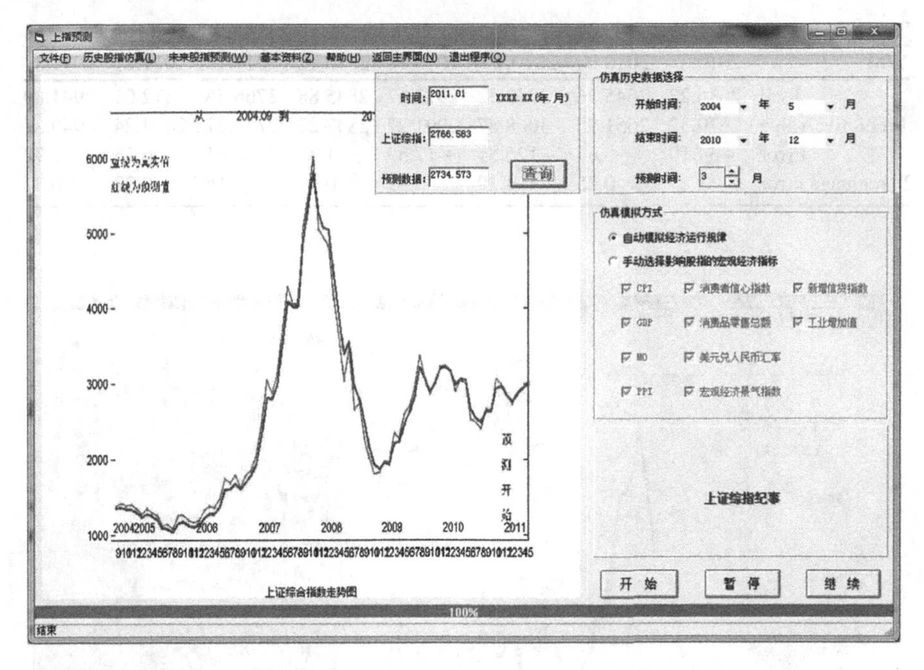

Fig. 2 Trend chart

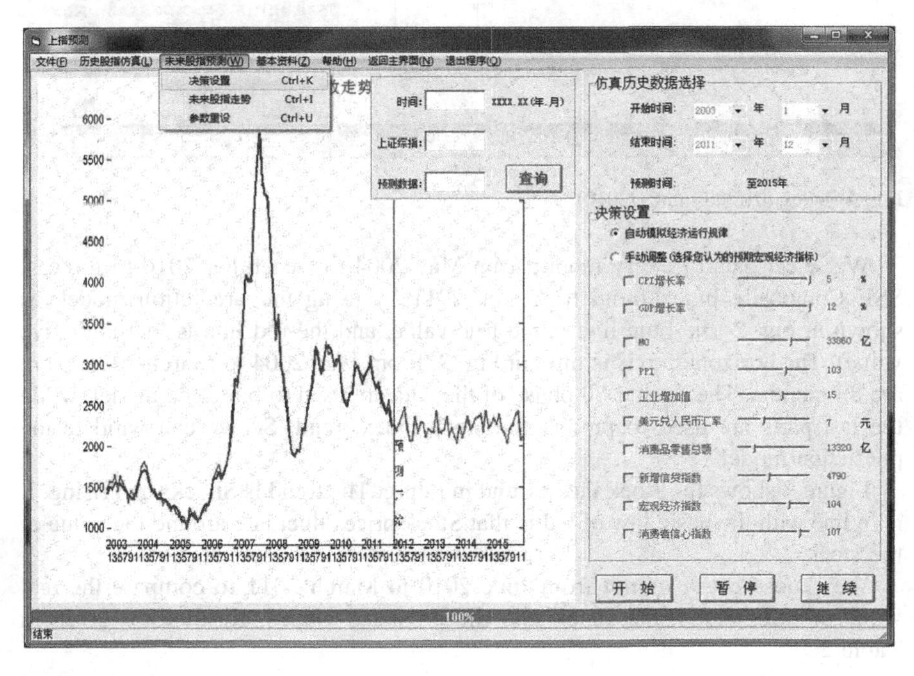

Fig. 3 The future stock index prediction

Table 2 Index comparison

Y/M	2010.8	2010.9	2010.10	2010.11	2010.12	2011.1	2011.2	2011.3
Real	2636.22	2645.26	2942.52	2973.67	2845.88	2766.58	2872.01	2941.89
Predictive value	2670.32	2651.95	3068.87	2991.27	2847.22	2734.57	2878.24	2949.81
Error	−34.10	−6.69	−126.35	−17.59	−1.33	32.01	−6.23	−7.92
Percentage error	1.29	0.25	4.29	0.59	0.05	1.16	0.22	0.27

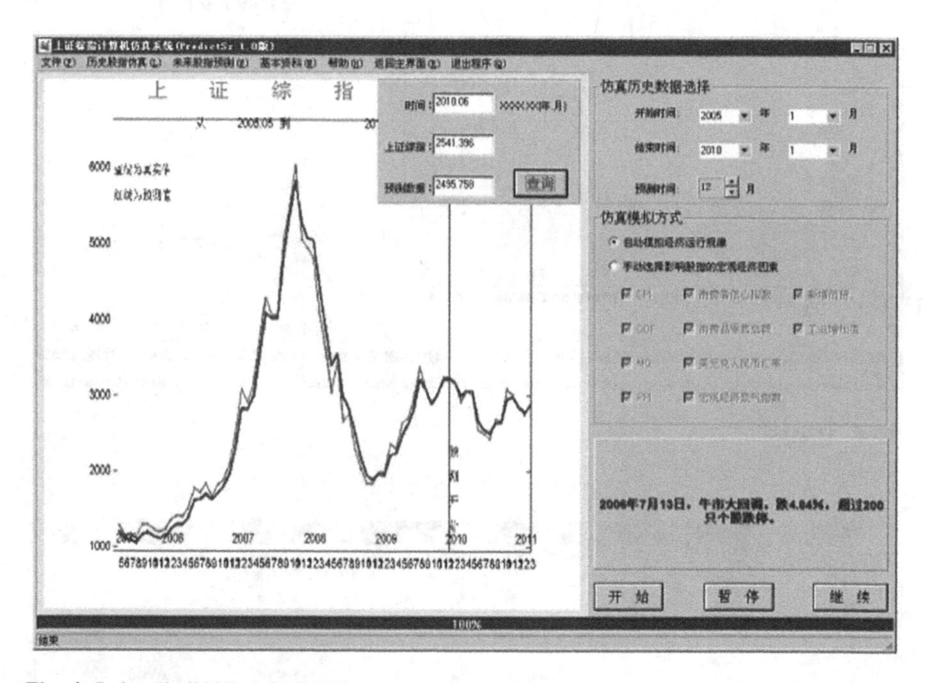

Fig. 4 Index (01/2005 to 01/2010)

We select data of every month from May 2004 to December 2010 to forecast SSE Composite Index trend to March 2011 by using the prediction model, as shown in Fig. 2 (the blue line is the true value, and the red line is the predictive value). The horizontal axis is time in Fig. 2. From May 2004 to March 2011, there are 83 months. The former 80 phase of the data are used to build the model, while the last parts are used to predict the stock index trend. So we can validate the prediction model.

Figure 3 shows the stock index trend in future. The trend is Shocks and rising. It is in line with the basic law of value that Stock prices fluctuate around the value of the stock.

We choose every month from June 2010 to March 2011 to compare the real value of SSE Composite Index and the predicted values. Results are given in Table 2.

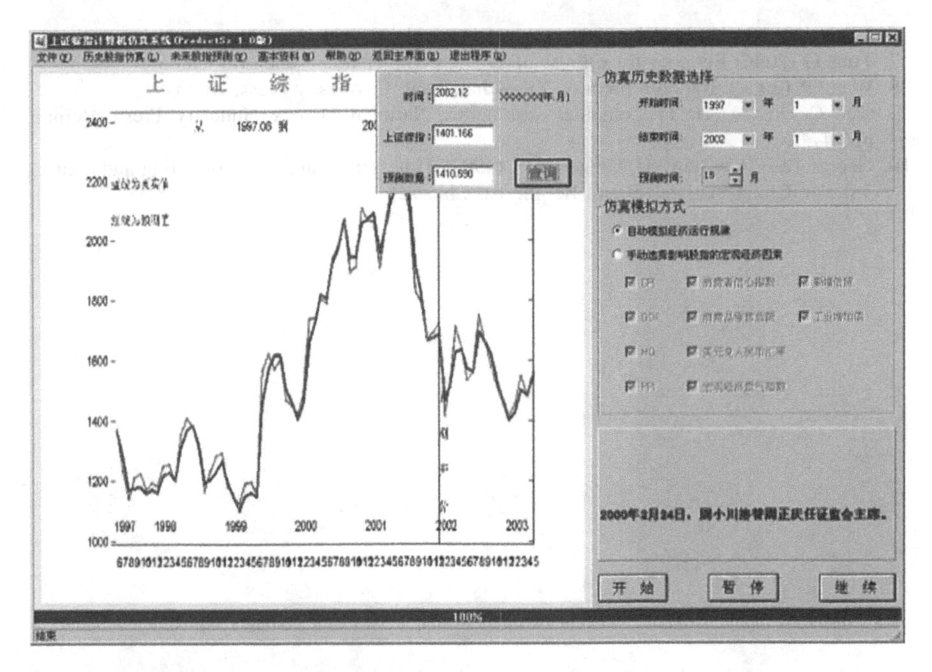

Fig. 5 Index (01/1997 to 01/2002)

Table 2 shows that the predictive value and the real value of the errors are within 5 %. There are higher accuracy of prediction.

We select different data to verify universal applicability of the model. We choose data of every month from January 2005 to January 2010 to predict the stock index trend of later 12 months (Fig. 4), and from January 1997 to January 2002 to predict the stock index trend of later 15 months (Fig. 5).

Acknowledgments The paper is supported by the National Natural Science Foundation of China, No. 61074191.

References

1. Trippi D (1992) Trading equit index futures with a neural network. J Portfolio Manag 19:25–35
2. Zhu K, Li J (2005) Empirical analysis on China stock index fluctuation characteristic. Appl Stat Manag 24(3):104–126
3. Poon WPH, Fung HG (2000) Red chips or H shares:which China-Backed securities process information the fastest. J Multinatl Financial Manag 10(3–4):315–343
4. Johnson R, Soenen L (2002) Asian economic integration and stock market comvement. J Financial Res 25(1):141–157
5. Chen X, Wang S (1987) Modern regression analysis. Anhui Education Press, Hefei

6. Clements MP, Hendry DF (2008)
7. Tong G (2006) Econometrics. Press of WuHan University, Wuhan (in Chinese)
8. Wang R (2006). Random process. Xi'an Jiao Tong University Press, Xi'an
9. Guo Q, Xu X (2011) Computer simulation. National Defense Industry Press, Beijing, pp 24–45
10. George EPB, Gwilym MJ, Gregory CR (1997) Time series analysis forecasting and control, 3rd edn. China Statistics Press, Beijing (in Chinese)

Erratum to: SSE Composite Index Prediction and Simulation System

Multiple Regression and Time—Lag Cointegration Model

Dengbin Huang, Jingwei Jiang and Guan Li

Erratum to: Chapter 'SSE Composite Index Prediction and Simulation System' in: F. Sun et al. (eds.), *Knowledge Engineering and Management*, DOI 10.1007/978-3-642-37832-4_64

The third author name 'Guiming Wang' in this chapter is incorrect. It should read as 'Guan Li'. The author is affiliated to: Accounting Center of Tsinghua University, Beijing 100084, China, Email: lguan@mail.tsinghua.edu.cn. Guan Li is the corresponding author of this chapter.

The online version of the original chapter can be found under DOI 10.1007/978-3-642-37832-4_64.

D. Huang · J. Jiang
College of Science, Naval University of Engineering, Wuhan 430033, China
e-mail: 15365286622@163.com

G. Li (✉)
Accounting Center of Tsinghua University, Beijing 100084, China
e-mail: lguan@mail.tsinghua.edu.cn

F. Sun et al. (eds.), *Knowledge Engineering and Management*,
Advances in Intelligent Systems and Computing 214,
DOI: 10.1007/978-3-642-37832-4_65,